BIODETERIORATION 7

Selected papers presented at the Seventh International Biodeterioration Symposium, Cambridge, UK, 6–11 September 1987

Topic Convenors

Biochemical Mechanisms—Dr B.F. Sagar
Biocides—Mr E.C. Hill
Biofilms—Dr L.H.G. Morton
Cosmetics and Pharmaceuticals—Dr M. Parker
Hydrocarbons—Dr R.J. Watkinson
Insects and Rodents—Mr P. Bateman
Leather and Textiles—Mr A.R.M. Barr
Marine Structures—Mr D.R. Houghton
Metals—Mrs C. Gaylarde
Museums and Archives—Mr D. Baynes-Cope
Paint and Concrete—Mr C. Grant
Paper and Packaging—Dr D.M. Holt
Physical Methods of Control—Dr H.O.W. Eggins
Plastics and Rubber—Dr K.J. Seal
Post-Harvest Decay—Dr B. Flannigan
Rapid Test Methods and Computers—Mr B.J. McCarthy
Solid Waste and Effluents—Dr J. Mills
Timber and Cellulosics—Dr R.A. Eaton

BIODETERIORATION 7

Edited by

D.R. HOUGHTON

Southsea, Hampshire, UK

R.N. SMITH

Hatfield Polytechnic, Hatfield, UK

and

H.O.W. EGGINS

Bioquest Ltd, St Peter's College, Birmingham, UK

ELSEVIER APPLIED SCIENCE
LONDON and NEW YORK

ELSEVIER SCIENCE PUBLISHERS LTD
Crown House, Linton Road, Barking, Essex IG11 8JU, England

Sole Distributor in the USA and Canada
ELSEVIER SCIENCE PUBLISHING CO., INC.
52 Vanderbilt Avenue, New York, NY 10017, USA

WITH 155 TABLES AND 140 ILLUSTRATIONS

Softcover reprint of the hardcover 1st edition 1988

British Library Cataloguing in Publication Data

International Biodeterioration Symposium
(7th: 1987: Cambridge, England)
Biodeterioration 7
1. Biodeterioration
I. Title II. Houghton, D.R.
III. Smith, R.N. IV. Eggins, H.O.W.
620.1'122

Library of Congress Cataloging in Publication Data

International Biodeterioration Symposium (7th: 1987:
Cambridge, Cambridgeshire)
Biodeterioration 7.

Bibliography: p.
Includes index.
1. Biodegradation—Congresses. I. Houghton, D.R.
II. Smith, R.N. (Robert Neil) III. Eggins, H.O.W.
IV. Title
QH530.5.157 1987 620.1'1223 88-11281

ISBN-13:978-1-94-010-7107-9 e-ISBN-13:978-94-009-1363-9
DOI: 10.1007/978-94-009-1363-9

Phototypesetting by Tech-Set, Gateshead, Tyne & Wear.

Preface

Because of the magnificent response to the call for papers for the 7th International Biodeterioration Symposium held at Cambridge, UK, some difficulties have been experienced in the editing of these proceedings. The numbers of papers submitted exceeded expectation and because of this it has been necessary to accommodate those not actually in the proceedings into the *International Biodeterioration* journal. A small number of papers were not suitable for publication and were therefore eliminated.

Many authors disregarded the guidelines laid down for the length of submitted papers. However, every attempt has been made to accommodate the maximum number of contributions in the proceedings. The original selection included those which most nearly conformed to the length requirement. Even so this has meant, in many cases, cutting down the text, eliminating tables and/or illustrations and pruning the reference list. When references have been trimmed a note has been included to the effect that an extended list may be obtained from the author/senior author. Where it was not possible to carry out these procedures without seriously altering the text and the import of the paper they have been included amongst those to be published in the journal *International Biodeterioration*.

The exceptions to the procedures outlined above are the invited review papers which have been presented in full. Happily, the authors have been conscientious in keeping to the guidelines laid down for these contributions.

In the attempt to utilise the maximum available space for the publication of submitted papers instead of listing convenors before each series of session papers, a list appears at the beginning of this volume. The order of presentation of papers at the Symposium has not been followed and the sessions have been arranged alphabetically.

It is considered that these proceedings reflect the excellent work of the Organising Committee of the Symposium as well as the general high quality of the papers and provides an invaluable volume for those seeking practical guidance as well as those engaged professionally in the field of biodeterioration.

D.R. HOUGHTON
R.N. SMITH
H.O.W. EGGINS

Contents

Computers

Concrete and Stone

Cosmetics and Pharmaceuticals

Ecology

Effluents

Hydrocarbons

Insects, Rodents and Birds

Marine

Mechanisms

Metals

Museums

Paint

Paper and Packaging

Physical Methods of Control

Plastics and Rubbers

Post-Harvest Decay

Rapid Methods

BIOCIDES

Developments in Microbicides for the Protection of Materials

W. Paulus

Bayer AG, Krefeld 1, Federal Republic of Germany

ABSTRACT

Today, the development of microbicides which satisfy the demands of our civilization with its heightened perception of environmental protection is a major task in maintaining functionality and quality of susceptible high-quality materials. For that purpose one utilizes the possibility to incorporate various toxophoric groups as well as toxophoric structural elements into a wide variety of molecules, which then make use of their specific properties to transport the toxophors to their site of activity, the microbial cell. The following toxophoric groups and structural elements and their mechanisms of action are examined: methylol groups, activated halogen groups, vinyl groups in α, β-position to an electronegative group and activated N–S bonds. It is shown that the knowledge of the mechanisms of activity of antimicrobials as well as of the structure and functioning of the various species of microbes allows the tailoring of microbicides which are adapted to the intended use and the species of microbes to be combatted.

INTRODUCTION

Around 3·5 thousand million years ago, the first living cells appeared on earth. These were microorganisms such as bacteria, algae, moulds and yeasts—the starting points for evolution and the variety of living species known to us today. However, the starting point, the microorganisms, still exist and are ubiquitous. But in the meantime, they too are made up of innumerable species. This allows them to exist under the most varied and extreme environmental conditions, and to utilize widely differing materials as their sources of food and energy. Microorganisms have learned to form spores, which permits them to survive not only periods of drought, but also of extremely high and low temperatures. We are familiar with acidophilic bacteria as well as thermophilic species which thrive in the boiling water of hot springs. We know of halobacteria, which possess powerful ionic pumps allowing them to survive in the saturated salt solutions of saline lakes.

We are familiar with the positive role of microorganisms. Not only are they the starting points for evolution and responsible for the maintenance of the metabolic cycle on earth, but they are also useful in such processes as fermentation which we utilize, for example, in the production of penicillin, in water purification at biological treatment plants, in production of 'biogas' from agricultural waste products, etc. Microorganisms are useful and vital in some respects, but destructive and harmful in others. The latter is the case when they appear as causes of infectious diseases, or as producers of poisonous and carcinogenic metabolites. This is also true when they decompose not only waste, but also high-quality materials, disrupt production processes, or impair the quality of products. It is estimated that microbial decomposition annually destroys materials valued at several thousand million dollars throughout the world.

Microbicides for control of microorganisms are thus useful and vital to man and beast. However, the primary purpose of the previous remarks was not to justify this trite statement, but rather to lead to the realization that location of a substance which truly earns the name microbicide—such as pentachlorophenol, which has fungicidal, algicidal, bactericidal, and insecticidal properties—is not the sole criterion in the essential development of microbicides. Today, microbicides can and must be developed which satisfy the demands of the heightened perception of environmental protection. Such materials include non-persistent products and products which display the desired—for instance algicidal—activity at very low concentrations. In other words, these are tailored microbicides which are adapted to the intended use and the species of microbes to be combatted. At present, the knowledge on structure and functioning of the various microbial species, as well as on the mechanisms of activity of antimicrobials is so voluminous that tailoring of microbicides is no longer utopian, if the principle of synergetics is followed. This principle states that appropriate interaction of individual entities or the combined action of suitable components leads to new structures and mechanisms of activity. This concept will be discussed using the development of microbicides for the protection of materials as an example.

METHODS

Minimum inhibition concentrations (MIC) are determined by preparing a concentration series in a nutrient solution or nutrient substrate. At the end of 1 or 2 weeks after infection of the nutrient media with the test organisms, the minimum inhibition concentration is recorded. The nutrient media for bacteria (pH 7·5) contain meat extract powder, neutral meat peptone, and D-glucose, whereas the nutrient media for fungi and yeasts (pH 5·6) contain malt liquor and peptone.

Biological oxygen consumption of a standard sewage in the presence of microbicides is determined using the Sapromat method.

RESULTS AND DISCUSSION

Most microbicides possess a toxophoric group (Fig. 1). This is a grouping in the molecule which is responsible for the antimicrobial activity.

However, the activity of a substance also depends on the overall structure, which can influence the selectivity, and on the lipid and water solubilities, the volatility, the reactivity, and the stability or instability of the molecule. The possibilities of influencing the activity became apparent to the synthetic chemist concerned with the development of microbicides. The degree of antimicrobial efficacy of substances which results from the presence of a given toxophor can be influenced by modification of that portion of the molecule which bears the toxophoric group. The above-mentioned chemical and physical properties, which contribute to the activity of the substances, are thus changed. This ability to modify the chemical and physical properties of substances while retaining the toxophor can be desirable not only with regard to increasing their activity, but also in tailoring to a specific application. Thus, for example, a fungicide for paints must not only possess a wide

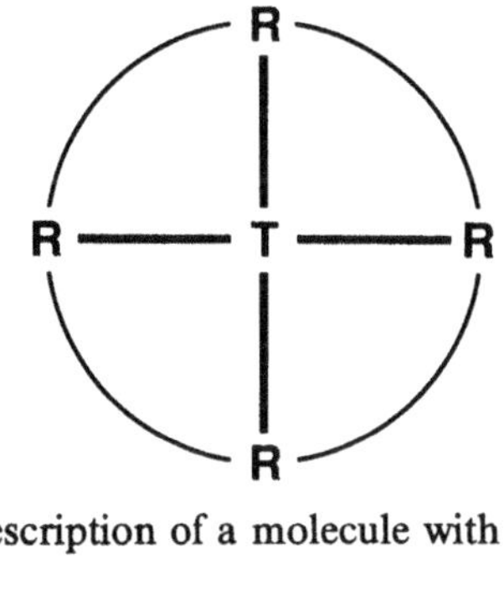

Fig. 1. Symbolized description of a molecule with a toxophoric group T.

spectrum of effectiveness but must also be colourless, stable to light, nonvolatile, insoluble in water, and stable in coating formulations.

Various toxophoric groups as well as toxophoric structural elements will be cited below. It will be demonstrated that these can be incorporated into a wide variety of molecules, which transport the toxophors to their site of activity, the microbial cell. There, the toxophoric group or structural element exerts its antimicrobial activity by reacting with specific entities of the cell.

Toxophoric groups—toxophoric structural elements

(A) Methylol groups capable of being cleaved off to give formaldehyde, or *Methylene groups capable of being hydrolysed to give formaldehyde*

$$R\text{—}CH_2OH \leftrightarrow R\text{—}H + CH_2O$$

$$R_1\text{—}CH_2\text{—}R_2 \underset{H_2O}{\leftrightarrow} R_1\text{—}CH_2\text{—}OH + R_2H \leftrightarrow R_1\text{—}H + CH_2O + R_2\text{—}H$$

TABLE 1
Examples of Formaldehyde-releasing Compounds

Hemiformals	R–O–CH$_2$OH
Hexahydro-s-triazines	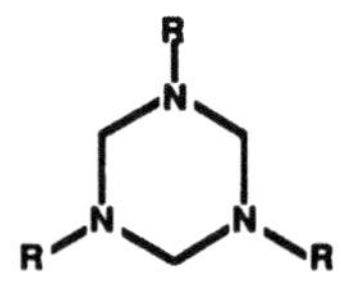
Oxazolidines	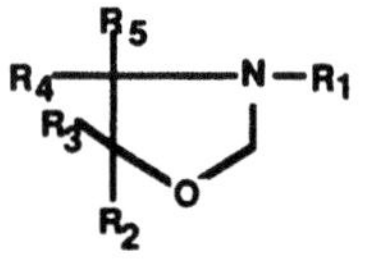
Dioxazolidinylmethanes	
N-Methylol amides	R–CO–NH–CH$_2$OH
Quaternary hexaminium salts	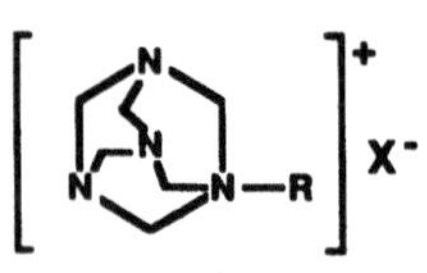

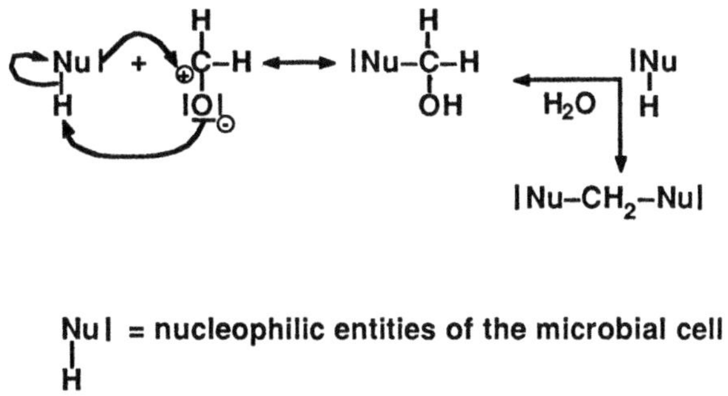

Fig. 2. Mechanism of activity of formaldehyde and formaldehyde-releasing compounds.

Substances of this type belong to the class of formaldehyde-releasing compounds (Table 1), of which numerous and widely differing members can easily be synthesized, since R can be varied almost at will (Paulus, 1980).

Formaldehyde-releasing compounds can be solids or liquids, water soluble or oil soluble, odourless, alkaline, neutral, or slightly acidic. They open up applications to the active agent—formaldehyde—which would otherwise be closed to it due to its unfavourable properties: its volatility, excessive reactivity (i.e. rapid inactivation), and its irritating odour.

Due to the electron deficiency at the carbonyl carbon atom, formaldehyde belongs to the group of electrophilic active agents, which can react with nucleophilic cell entities and thus exert antimicrobial activity (Fig. 2).

The methylol derivatives which arise can react with a second nucleophile of the same or a different type to form a methylene bridge.

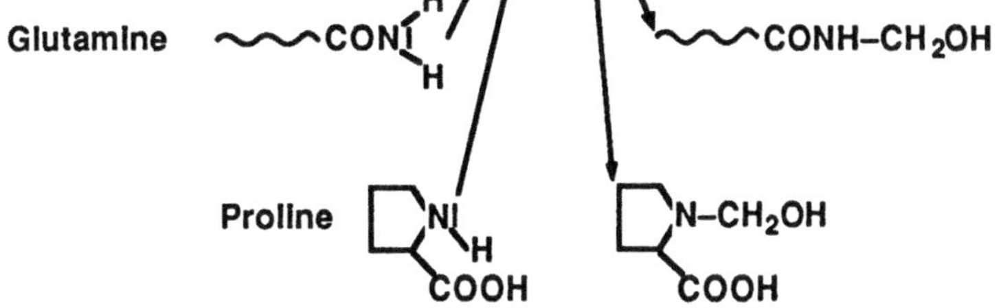

Fig. 3. Reaction of formaldehyde with amino acids.

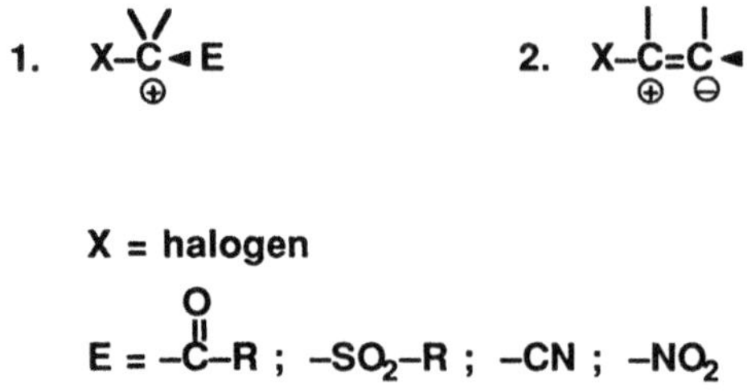

Fig. 4. Activated halogen groups.

Examples of nucleophilic reaction partners in the cell are amino and thiol groups, as well as the amide groups of amino acids or proteins (Fig. 3). These in turn are components of enzymes, which are inactivated by the reaction of their nucleophilic groups with formaldehyde (Ku & Billings, 1984).

(B) The activated halogen group in the α-position and in the vinyl position to an electronegative group E (Fig. 4)
It is apparent that a variety of modifications of both the electronegative group and the portion of the molecule bearing the halogen group are possible (Table 2). Thus, microbicides having a wide range of chemical and physical properties can also be synthesized within this class of substances.

The antimicrobial activity of these electrophilic substances similarly arises from the fact that nucleophilic entities of the microbial cell react with the positive carbon atom of the substances. This proceeds by way of an elimination reaction in the case of microbicides with activated

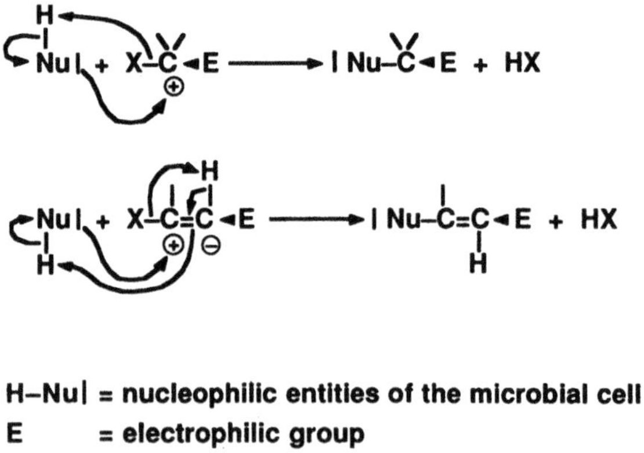

Fig. 5. Mechanism of activity of microbicides with activated halogen.

TABLE 2
Examples of Microbicides with Activated Halogen in the α-position to an Electronegative Group

$Br{-}CH_2{-}C(=O){-}O{-}CH_2{-}C_6H_5$	Liquid; of poor water solubility
$Cl{-}CH_2{-}C(=O){-}NH_2$	Solid; well soluble in water
$J{-}CH(J){-}SO_2{-}tolyl$ (J above, J–CH$_2$–SO$_2$–tolyl)	Solid; indissoluble in water
$N{\equiv}C{-}CH_2{-}C(Br)_2{-}C(=O){-}NH_2$	Liquid; moderately soluble in water
$Br{-}C(CH_2OH)_2{-}NO_2$	Solid; soluble in water

TABLE 3
Examples of Microbicides with Vinyl Activated Halogen

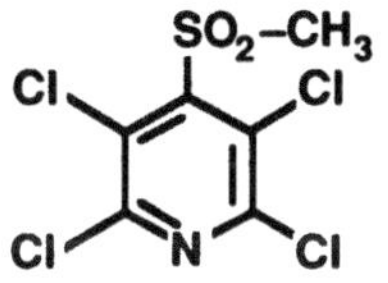	Solid; nearly indissoluble in H_2O
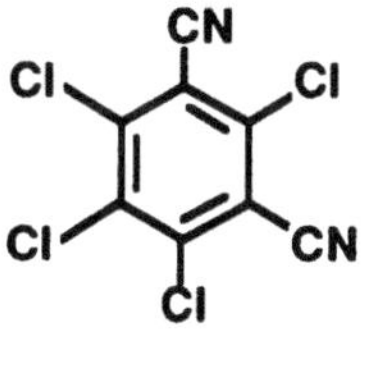	Solid; nearly indissoluble in H_2O
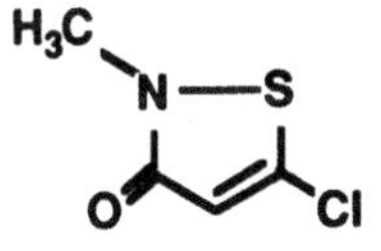	Solid; of poor water solubility

halogen in the α-position to an electronegative group (Fig. 5), whereas microbicides with vinyl activated halogen combine with nucleophiles by way of an addition–elimination reaction (Table 3).

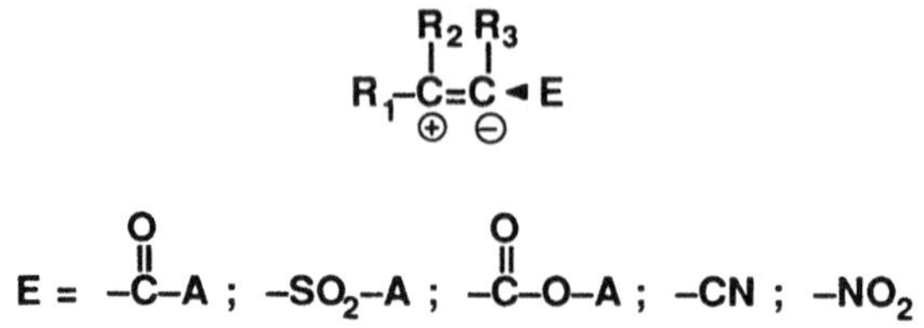

A = a resonance increasing group, e.g. aryl

Fig. 6. Reactive vinyl compound.

(C) The vinyl group in the α, β-position to an electronegative group E (Fig. 6) Depending on their structure, such vinyl compounds can be very reactive and easily polymerize, making them unsuitable as practical microbicides. However, if a so-called 'leaving group' (LG) is added to the unstable vinyl bonds, stable substances representing potential vinyl compounds are obtained (Fig. 7). These exhibit a microbicidal effect comparable to that of the vinyl compounds from which they originate (Paulus, 1976).

This effect is due to elimination of the leaving group to afford the free, reactive vinyl compound, which in turn reacts with the nucleophilic entities of the microbial cell to exert its antimicrobial action (addition reaction) (Fig. 8).

Cleavage of the potential vinyl compounds (elimination of the leaving group) to produce the active vinyl substances takes place even under physiological conditions (room temperature and approximately neutral pH). In general, the rate of cleavage is many times higher than those of the reactions leading to inactivation of the vinyl compound produced in the cleavage process. This represents an advantage in the practical use of

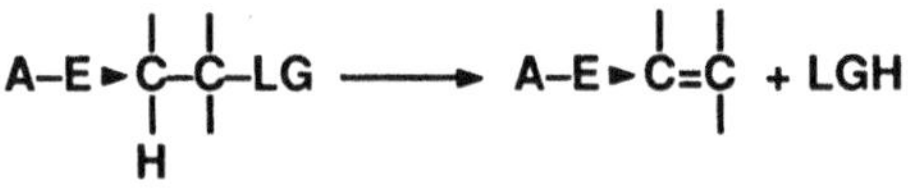

A = resonance increasing group
E = electrophilic group
LG = leaving group, e.g. halogen,
NR_2, $\overset{\oplus}{N}(R)_3$, S–R, $\overset{\oplus}{S}(R)_2$, $\overset{\oplus}{P}(R)_3$

Fig. 7. Cleavage of potential microbicidal vinyl compounds.

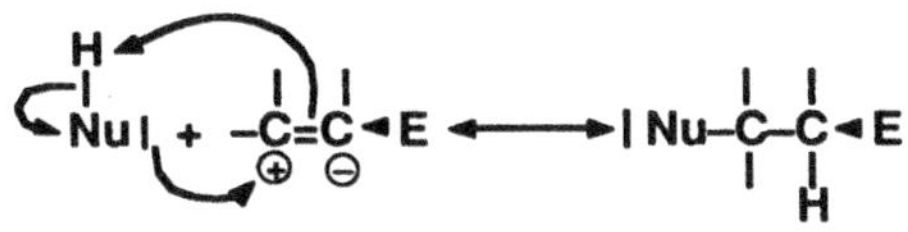

H–NuI = nucleophilic entities of the microbial cell
E = electrophilic group

Fig. 8. Mechanism of activity of microbicides with an activated vinyl group.

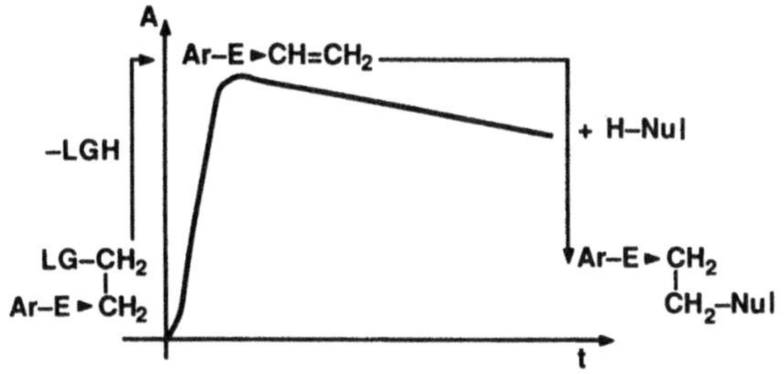

LG = leaving group
E = electrophilic group
H–NuI = nucleophilic entities of the microbial cell

Fig. 9. Antimicrobial activity (A) of an aqueous solution of a potential aryl vinyl ketone dependent on time (t) at pH 6·5.

TABLE 4
Examples of Microbicidal Vinyl or Potential Vinyl Compounds

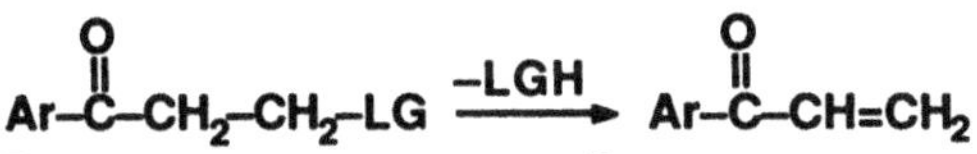	(Paulus, 1976)
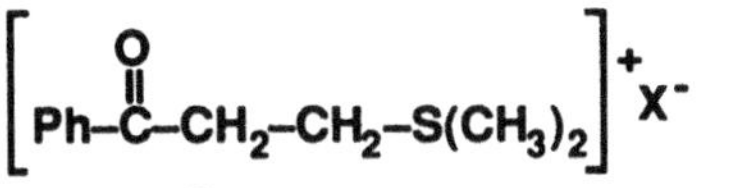	(Paulus et al, 1974)
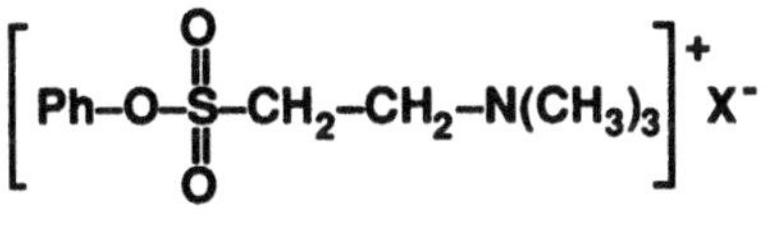	(Distler, 1965)
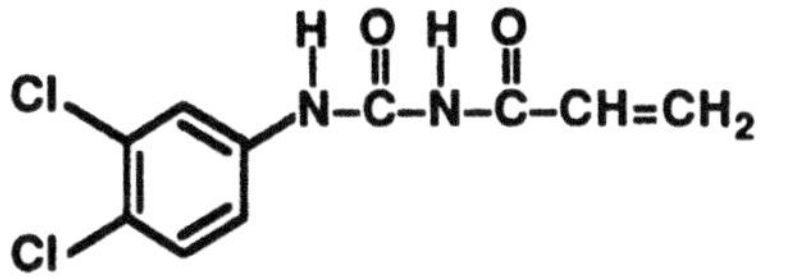	(Paulus, 1980)

potential vinyl compounds as microbicides. A biokinetic test produced a curve (Fig. 9) illustrating the above statements (Paulus, 1976).

If the possible variables in potential microbicidal vinyl compounds are considered—leaving groups, substituents at the leaving group, the electrophilic group and the resonance-increasing group—it is apparent that a wide scope exists for synthesis of antimicrobial substances having widely differing properties intended for use in a variety of applications (Table 4).

(D) The activated N-S bond (Fig. 10)
Activated N-S bonds are found in substances such as those containing the isothiazolinone ring and in N-trihalomethylthio compounds.

The isothiazolinone ring reacts with nucleophilic cell entities—especially thiol group-containing substances such as amino acids, proteins and enzymes—with opening of the ring (Crow & Leonard, 1965; Miller *et al.,* 1975), thus exerting antimicrobial activity (Fig. 11).

The antimicrobial activity of N-trihalomethylthio compounds is similarly based on the ability of the N-S bond to open and react with nucleophilic entities such as -SH groups (Owens & Black, 1959; Lucken, 1966). Such reactions apparently proceed by way of several steps and lead to disulphides. The course of the reaction can be summarized as follows:

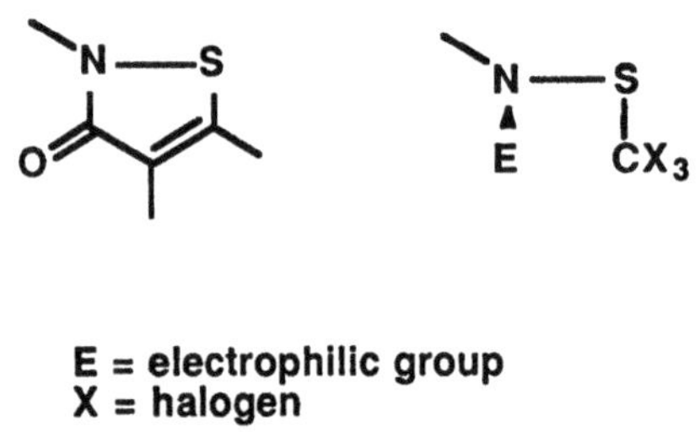

Fig. 10. Activated N-S bonds.

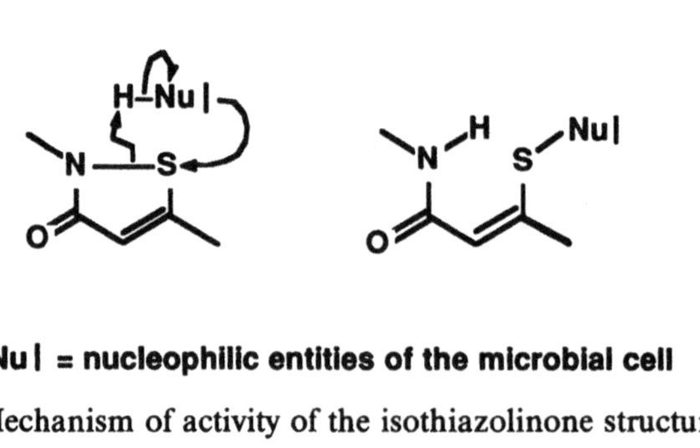

Fig. 11. Mechanism of activity of the isothiazolinone structure.

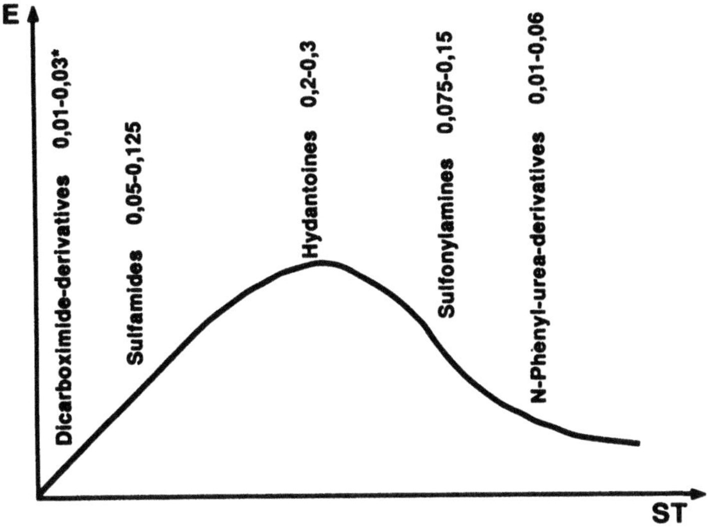

* reciprocal MIC's l. mg^{-1}• liter

Fig. 12.

$$E\blacktriangleright \underset{\substack{|\\ R}}{N}-S-CX_3 + \xrightarrow[H_2O]{H-S-R_2} R_2-S-S-R_2,\ HX,\ \overset{\substack{E\\ |}}{\underset{\substack{|\\ R}}{N}}-H,\ H_2S,\ CO_2$$

E = electrophilic group
X = halogen

A relationship exists between the antimicrobial activity (*E*) of N-trihalomethylthio compounds and the stability (ST) of their individual N-S bonds. This relationship is illustrated in Fig. 12.

According to this curve, N-trihalomethylthio compounds are especially active when their N-S bond displays a medium level of stability or reactivity. In this connection, it is important to note that the reaction of Cl-S-CX_3 with N-H compounds only leads to N-trihalomethylthio compounds with a significant microbicidal activity in cases where the N-H group has an acidic character.

$$\blacktriangleright \underset{|}{N}-H + Cl-S-CX_3 \xrightarrow{-HCl} \blacktriangleright \underset{|}{N}-S-CX_3$$

This is demonstrated in Table 5, which clearly shows that substitution of basic H atoms by the S-CX_3 radical produces inactive substances with

TABLE 5

Minimum Inhibition Concentrations (MIC's i. mg/litre) of R-S-CX_3 Compounds Produced from Substances with H-atoms of Differing Reactivity

R group in R—S—CCl_2F	*Test organism*							
	Alternaria tenuis	*Aspergillus niger*	*Aureobasidium pullulans*	*Chaetomium globosum*	*Coniophora puteana*	*Penicillium glaucum*	*Polyporus versicolor*	*Trichoderma viride*
$(n\text{-}C_4H_9)_2N$–	>1000	>1000	>1000	1000	100	>1000	>1000	>1000
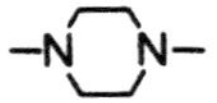		>1000		>1000		>1000		
$n\text{-}C_4H_9(H)N$–	1000	>1000	500	500	35	>1000	>1000	>1000
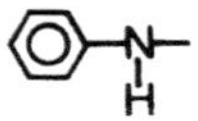	750	>1000	500	350	35	>1000	1000	>1000
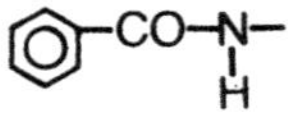		100		100		50		
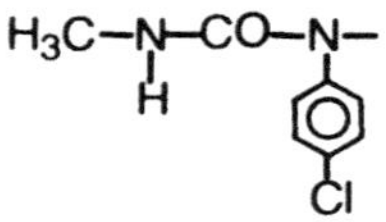	15	50	20	20	1	100	35	500

TABLE 6
Efficacy and Spectrum of Effectiveness of Different >N–S–CX_3 Compounds—MIC's (mg/litre) in Nutrient Agar

Test organism	*Dichlofluanide N,N-dimethyl-N'-phenyl-N'-fluoro dichloromethyl thio-sulfamide*	*N,N-dimethyl-N'-phenyl-N'-trichloro methyl thio-sulfamide*	*Fluorofolpet N-fluorodi-chloromethyl thio-phthal-imide*	*Folpet N-trichloro methyl thio-phthal-imide*
Alternaria tenuis	10	1000	10	750
Aspergillus niger	50	>4000	200	350
Aureobasidium pullulans	10	200	50	50
Chaetomium globosum	20	1000	100	50
Coniophora puteana	10	500	2	75
Lentinus tigrinus	2	200	2	35
Penicillium glaucum	20	4000	35	50
Polyporus versicolor	10	>1000	50	100
Sclerophoma pityophila	10	500	10	150
Trichophyton pedis	10		20	
Candida albicans	20	>5000	5	150
Candida crusel	20	>5000	20	500
Rhodotorula mucilaginosa	20	>5000	10	100
Saccharomyces cerevisiae	10	750	5	50
Torula rubra	20	>5000	15	50
Torula utilis	10	350	10	50

TABLE 7
Antimicrobial Effectiveness of Trihalomethylthio Compounds—Prototypes from Different Substance Classes—MIC's i. mg/litre

Substance class	*Prototype* $R = S\text{-}CCl_2F$	*Test organisms*								
		Alternaria tenuis	*Aspergillus niger*	*Aureobasidium pullulans*	*Chaetomium globosum*	*Coniophora puteana*	*Penicillium glaucum*	*Polyporus versicolor*	*Trichoderma viride*	*Algae*
Dicarbonic acid imides	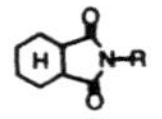	10	200	20	100	1	50		200	100
	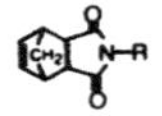	10	75	35	50	3	35	5	200	100
Ureas	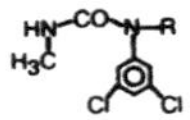	5	35	2	35	0.1	75	7.5	20	20
	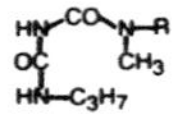	2	10	2	10	0.5	100	7.5	>1000	100
Hydantoins	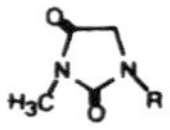	1	5	0.5	5	1	15	5	200	100

Sulfamides	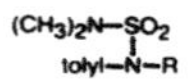	10	20	10	10	10	20	10	>5000	100
Sulfonylamines	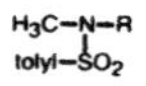	20	5	5	5	0.5	5	2	>1000	50
N-CF_3-amino benzoic acid derivatives	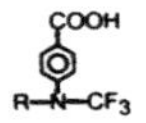	10	10	0.5	1	0.1	10		10	50
	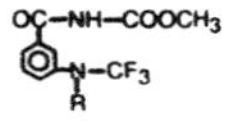	2	2	0.5	1	0.1	2		10	100
Benzpyrazoles	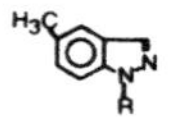		2		1		5			20
Benzimida-zinones	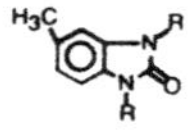	10	< 20	1	< 20	1.5	20	2	200	
Benzoxa-zinones	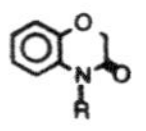	50		50		< 20				

N–S bonds of excessive stability. However, the MIC values also clearly demonstrate that active substances do result if the starting materials such as carboxylic acid amides contain an H atom with acidic character (Paulus & Kühle, 1986).

Another important finding is that >N–S–CCl_3 compounds are significantly less active than the corresponding >N–S–CCl_2F compounds (Kühle *et al.*, 1964) as shown in Table 6.

Since the N-trihalomethylthio compounds represent a large reservoir of active substances (cf. for example Table 7), it should not be excessively difficult to locate those which possess the chemical and physical properties required for any specific application.

At this point, the low level of environmental toxicity of many N-trihalomethylthio compounds, which is of great significance in cases such as use of the substances as algicides in antifouling paints, might be emphasized. Since N-trihalomethylthio compounds existing in a dissolved form in aqueous media hydrolyse at a rate which depends on the temperature and the pH (Paulus & Kühle, 1986), the environmental toxicity of the hydrolysis products is of interest, since only these have a lifetime which is of significance with respect to their possible environmental toxicity.

As the toxophoric S–CX_3 group is cleaved off during hydrolysis and decomposes to yield CO_2, H_2S and HX, it is necessary to study the hydrolysis product, the individual R_1R_2–N–H compound.

$$R_1R_2\text{–N–S–}CX_3 \xrightarrow{H_2O} R_1R_2\text{–N–H} + CO_2,\ H_2S,\ HX$$

The results of Ascomat tests are summarized in Table 8. They show that activated sludge organisms tolerate concentrations of the hydrolysis products which lie many orders of magnitude higher than the tolerated concentrations of the corresponding active agents. As shown in other studies, a similar situation applies with regard to the fish toxicity of the hydrolysis products.

In conclusion, let us return to a consideration of the toxophoric groups or structural elements responsible for the antimicrobial activity of a molecule. Here, we can add that the microbicidal activity or the spectrum of effectiveness of a molecule can in some cases be increased by incorporation of a further toxophoric group T. Such molecules (T_1–R–T_2) possess two reactive sites which can react with nucleophilic centres of the microbial cell. As a result, they are generally more active than the corresponding compounds with only one toxophoric group. Thus, for example, N-methylol chloroacetamide and dimethylol bromonitromethane possess

TABLE 8
Nontoxic Concentrations (NTC) of Trihalomethylthio Compounds and their Degradation Products for Activated Sludge Organisms in mg/litre

$R_1R_2N\text{-}S\text{-}CX_3$	*NTC*	$R_1R_2N\text{-}H$	*NTC*
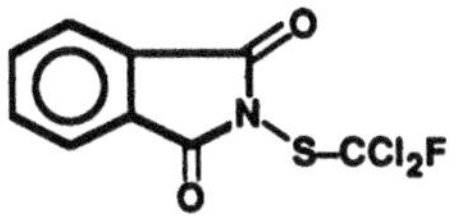	15	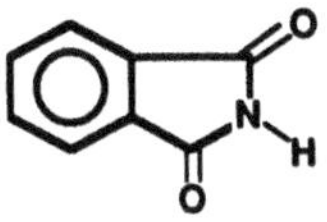	1000
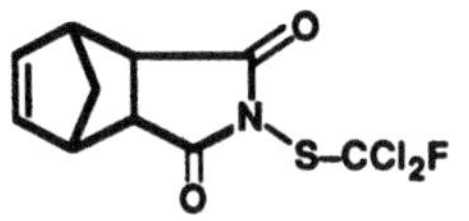	0·5	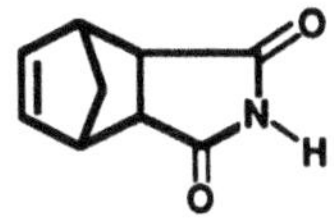	1000
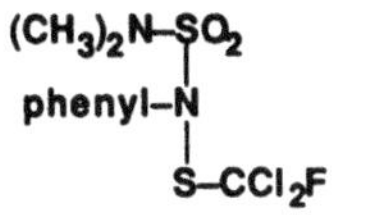	7	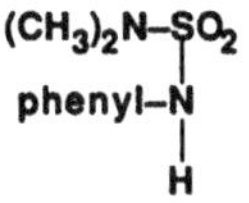	200
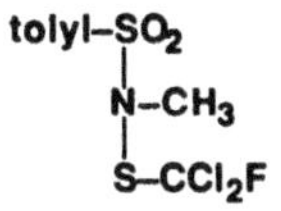	0·25–1	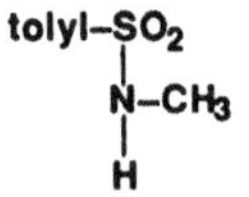	1000
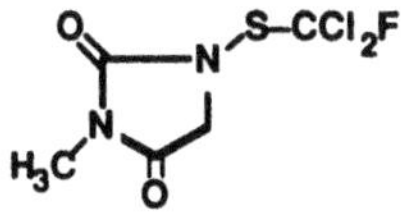	0·5	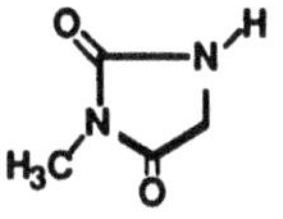	1000
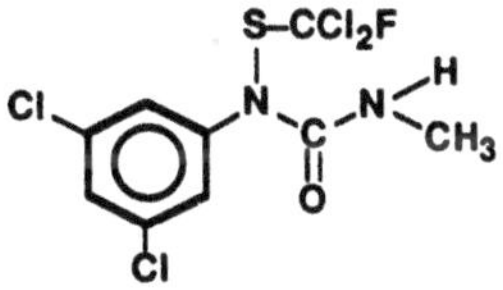	1	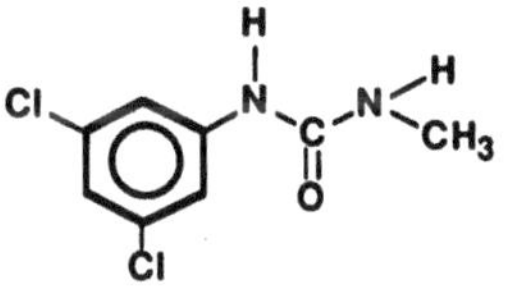	10

both an activated halogen atom in the 2-position to an electronegative group and reactive methylol groups (Fig. 13).

β-halovinyl aryl ketones display both a vinyl activated halogen atom and a reactive vinyl group in the α,β-position to an electronegative group (Fig. 14).

The 5-chloroisothiazolinones contain a vinyl activated chlorine atom as well as an activated N-S bond (Fig. 15).

The microbicides with toxophoric groups discussed here are substances with electrophilic character, a character which governs their reactivity and thus their wide spectrum of antimicrobial effectiveness. This in turn is a prerequisite for the use of microbicides for the protection of materials. All levels of stability, from labile to stable, can be found among the electrophilic agents. Anyone currently developing microbicides for the protection of materials must embark here upon a journey

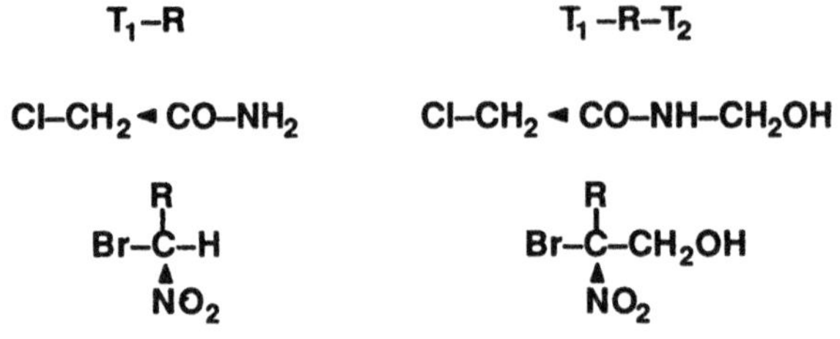

T = toxophoric group

Fig. 13. Molecules with two toxophoric groups (T).

T = toxophoric group

Fig. 14. Molecule with two toxophoric structural elements (T).

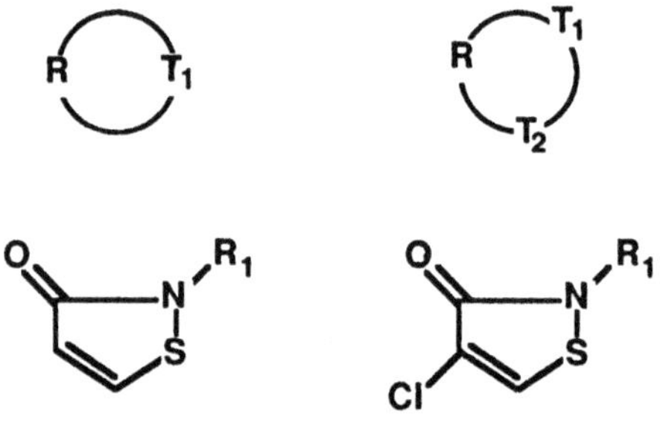

T = toxophoric group

Fig. 15. Molecule with two toxophoric structural elements (T).

of discovery, since the demand is not for persistence at any price, but for limited persistence graduated according to the application, for stability or instability as appropriate.

In consequence, it may be stated that development of material-protecting microbicides which can be handled without hazards to persons or the environment, and which can contribute to improving the quality of life, is by no means a utopian objective.

REFERENCES

Crow, D.W. & Leonard, N.J. (1965). *J. Org. Chem.*, **30**, 2660–5.

Distler, H. & Pommer, E.H. (1966). *Zentr. Bakteriol., Abt. II* (Parasitenkunde), **120**(1), 41–8.

Ku, R.H. & Billings, R.E. (1984). *Chem.-Biol. Interactions*, **51**, 25–36.

Kühle, E., Klauke, E. & Grewe, F. (1964). *Angew. Chemie*, **76**, 807–16.

Lucken, R.J. (1966). *J. Agric. Food Chem.*, **14**, 365–7.

Miller, G.A., Lewis, N.S. & Weiler, E.D. (1975). *U.S. Pat.*, **3**, 914 301.

Owens, R.G. & Blaak, G. (1959). *Boyce Thompson Inst.*, **20**, 475.

Paulus, W. (1976). *Proc. III. IBS*, Elsevier Applied Science Ltd, London, pp. 1075–82.

Paulus, W. (1980). *Biodet. Proc. IV. IBS*, Pitman, London, pp. 307–14.

Paulus, W. & Genth, H. (1980). *EP* 0 023 976 B1.

Paulus, W. & Kühle, E. (1986). *Biodet. 6, CAB Int.*, 79–88.

Paulus, W., Pauli, O. & Oeckl, S. (1974). *DE-05* 2 326 111.

Biodeterioration Problems in Formulated Products and Industrial Systems—Evaluation and Control

W.G. Guthrie, R. Elsmore & J.A. Parr

The Boots Company, plc, 1 Thane Road, Nottingham, NG2 3AA UK

ABSTRACT

This paper examines biodeterioration problems in a variety of systems and formulated products including cooling waters, industrial products, cosmetics and pharmaceuticals. The effect of factors influencing microbial activity such as chemical environment, pH and temperature is reported. Data are presented on the recovery of microorganisms from inherently hostile systems, e.g. alkaline cleaners, chiller systems. The effect of these environments on the efficacy of several preservatives and biocides is reported together with results on the varying susceptibilities of certain formulation ingredients to microbial insult.

INTRODUCTION

The problems caused by microbial contamination of pharmaceuticals, cosmetics and toiletries are well known. Many result in spoilage and/or growth of potential pathogens and are unacceptable for both economic and health reasons. (Happel, 1983; Sharpel *et al.*, 1983). Modern consumer products including liquid detergents, fabric conditioners, paints, etc., support microbial growth and are prone to biodeterioration problems. In industry, cooling waters, metalworking fluids and papermill process waters are also susceptible. This can result in biofouling, spoilage, growth of potential pathogens and corrosion (Hill, 1971; Purkiss, 1981; Parr, 1986; Elsmore *et al.*, 1987). The chemical nature, pH and temperature of many of these products and industrial processes may appear to represent a hostile environment for microorganisms. This paper reports on the effect of such factors on microbial activity. Data are presented on the susceptibility of formulations to microorganisms and on the effectiveness of various antimicrobial compounds.

METHODS

Challenge tests

The British Pharmacopoeia 1980 (Addendum 1986) was used for the challenge testing of the Standard BP formulations, aqueous cream, calamine cream, oily cream and magnesium trisilicate suspension.

A multistrain challenge test was used to test the alkaline cleaning product. This consisted of five mixed groups of test inocula each containing type species and contaminant isolates. The contaminant isolates were selected as strains showing the highest resistance to commonly used preservatives in minimum inhibitory concentration tests.

The five main groups consisted of:

1. Mixed Gram-negative bacteria. Routinely about 20 different species and strains are used including several *Pseudomonas* sp.; *Escherichia coli*; *Proteus* sp.; *Serratia marcescens*; *Enterobacter* sp.; *Klebsiella* sp.; *Salmonella typhimurium* and some factory isolates.
2. Mixed Gram-positive cocci. These routinely include *Staphylococcus aureus*; *S. epidermis*; *Micrococcus flavus* and *Streptococcus faecalis*.
3. Mixed yeasts. This group includes *Candida albicans*; *C. tropicalis*; *Saccharomyces cerevisiae* and some isolated spoilage yeasts.
4. Mixed moulds. These include *Aspergillus* sp.; *Stachybotrys atra*; *Amorphotheca resinae*; *Penicillium* sp.; *Trichoderma viride*; *Myrothecium resinae*; and several fungal contaminants isolated from infected products.
5. Alkaline product group. These include *Alcaligenes denitrificans* and a CDC group IV C-2 isolated from alkaline products.

The bacterial inoculum was prepared by washing off cultures from Tryptone Soya agar slopes. Growth was washed off with 0·1% peptone water and the resulting suspensions combined appropriately to form the two inoculum groups.

The yeast inoculum was a combined suspension, prepared by washing the spores from Sabouraud Dextrose agar slopes (grown 48 h at 25°C) with 0·1% peptone water containing 0·05% Tween 80.

The mould inoculum was a combined suspension prepared by washing the spores from Sabouraud Dextrose agar slopes (grown 7 days at 25°C) with 0·1% peptone water containing 0·05% Tween 80.

Aliquots (10 ml) of the diluted sample were challenged separately with 0·1 ml of inoculum from the four main groups of microorganisms. Challenge levels of 1×10^5–1×10^7 microorganisms per ml of product were achieved. At 0 h, 1 ml aliquots were withdrawn and the number of viable recoverable microorganisms/ml were determined by serial dilution and plate counts. The inoculated samples were held at 25°C and 1 ml aliquots were examined and assessed at 2 days, 1 week, 2 weeks and 4 weeks.

The alkaline product inoculum group was prepared by inoculating a 100 ml flask containing 20 ml of a mineral salts medium and 10 ml of alkaline cleaning fluid with 0·5 ml of a known contaminated product (containing *Alcaligenes denitrificans* and a CDC group IV C-2 isolate). The flask was incubated at 30°C for 48 h in an orbital shaker and 0·5 ml of this inoculum was used to challenge 20 g of test product. The sample was stored at 25°C and sampled at 0, 1 day, 2 days, 5 days and 7 days and plated into Todd Hewitt agar (Oxoid).

Cooling system study

During the cooling water survey, samples were removed and assessed for total viable count using a pour plate technique. Serial dilutions were made in ¼ strength Ringers solution. 1 ml volumes of this dilution series were plated in Tryptone Soya agar (TSA) to enumerate aerobic bacteria and Sabouraud Dextrose agar (SAB) to enumerate yeasts and moulds. The TSA plates were incubated at 32°C for 3 days and at 8°C for 14 days, and the SAB plates at 25° for 5 days. The number of sulphate reducing bacteria (SRB) were estimated using a modified iron sulphite agar tube technique. The tubes were incubated at 32°C for up to 21 days. Coliform bacterial numbers were estimated by plating dilutions of the sample in MacConkey Agar No. 3. The plates were incubated at 37°C for 18 h.

RESULTS AND DISCUSSION

Effect of chemical environment

The results of the BP Challenge tests on several standard BP formulations show a wide variation in the susceptibility of the products to the challenge organisms.

The unpreserved calamine cream resisted bacterial growth but supported yeast and mould growth while the unpreserved oily cream supported three of the four challenge organisms but did show some resistance to the Gram-positive insult. The aqueous cream BP resisted yeast, mould and Gram-positive bacterial growth but took 4 weeks to clear the Gram-negative insult despite the presence of the recommended preservative, chlorocresol.

The magnesium trisilicate mixture supported bacterial and mould growth but resisted the growth of yeasts. When chloroform water was included in the formulation the bacterial challenge was controlled but the product remained susceptible to moulds.

It is evident that even in relatively simple formulations, small changes in the chemical environment provided by the constituents can have a significant effect on the type of spoilage organism supported.

The results of a BP challenge test on three cream formulations immediately following manufacture and after 9 months storage at ambient temperature were devised to evaluate a mixture of BNPD (Bronopol-Boots) and dichlorobenzyl alcohol (Myacide SP) as an alternative preservative. In the aqueous and oily creams the combined preservative system was effective throughout the 9 month test. In the calamine cream, however, the efficacy of the mixture against yeasts and moulds was reduced over the test period.

The results of evaluating two preservative systems in the magnesium trisilicate mixture initially, and after 12 months storage at ambient temperature showed that there was some reduction in efficacy against yeast and moulds but a combination of BNPD plus Myacide SP was satisfactory against all four challenge organisms.

These results show the need to be aware of potential incompatibilities between preservatives and formulation constituents. The chemical environment in a formulation may alter on storage to allow growth of unwanted organisms.

Effect of pH

The preliminary screening of two alkaline hard surface cleansers (pH 10·6) indicate that neither product required preservation against the typical strains used in the challenge. However, the products were susceptible to microbial contamination when challenged with specially adapted spoilage organisms isolated from contaminated alkaline cleansers. They were not recoverable on normal recovery media.

It is clear that specialist microorganisms may be present in high pH products but may escape detection if inappropriate recovery media are used. It also emphasises the need to use appropriate organisms in challenge testing.

Effects of temperature

A recirculatory chilled water system serving a large dairy products manufacturing plant was studied (volume 200 m^3, temperature 6–8°C, pH 9–9·5). Cross contamination could sometimes occur between the chilled water and the dairy products and concern was expressed regarding overall levels of microbial contamination and the presence of potential pathogenic strains and coliform bacteria.

On initial examination, the system was contaminated with high numbers ($<10^7$ cfu/ml) of Gram-negative bacteria, anaerobic sulphate reducing bacteria (*c.* 10^5 cfu/ml) and lower numbers of coliform bacteria (*c.* 10^2 cfu/ml). Gram-positive bacteria and fungi were present in low numbers. The results of treating the system with a BNPD based biocide on a once weekly basis show that the coliform organisms were no longer isolated from the system and sulphate reducing bacterial numbers were reduced to low levels. Aerobic bacterial numbers were reduced by 3 log cycles over the month long study. The population of aerobic bacteria changed during the study from predominantly Gram-negative rod-shaped bacteria to mainly Gram-positive rod-shaped spore bearing bacteria, demonstrating that despite the hostile environment of pH 9 and a temperature of 6–8°C the chilled water system supported a large microbial population which presented a risk to the manufacturing process.

CONCLUSIONS

In nature, microorganisms are capable of surviving in a wide range of environmental niches. Many of these environments may initially be thought of as unfavourable but specialised microorganisms have developed which can exploit these extreme conditions.

In man-made products and industrial systems microorganisms have demonstrated the ability to proliferate in an equally wide range of physical conditions and chemical environments. The type of contamination can vary considerably and be difficult to predict. Even in relatively simple systems, minor changes in composition can have a significant effect on the types of organism which will infect a product. Preservatives can provide a hostile environment but care must be taken to ensure long-term efficacy and compatibility. Some formulation ingredients themselves possess the ability to resist certain types of microbial insult (e.g. anionic surfactants). It could be useful, therefore, to consider introducing a microbial susceptibility index for raw materials to aid formulators in choosing complementary ingredients to give maximum resistance to spoilage.

Contamination can also occur at extremes of temperature and pH. The formulator or operator must guard against overlooking the presence of organisms adapted to survive in these conditions.

REFERENCES

Elsmore, R., Guthrie, W.G. & Parr, J.A. (1987). *Speciality Chemicals*, **7** (3), 166–76.

Happel, J.A. (1983). *Antimicrobial Preservatives in Pharmaceuticals in Disinfection, Sterilization and Preservation*, 3rd edn., ed. S.B. Block. Lea and Febiger, Philadelphia, pp. 579–88.

Hill, E.C. (1971). *Process Biochemistry*, Nov., 17–34.

Parr, J.A. (1986). *Water and Waste Treatment*, **29** (1), 40–2.

Purkiss, B.E. (1981). *Speciality Chemicals*, **1** (4), 4–11.

Sharpel, M.S., Manowitz, M.S. & Manowitz, M. (1983). *Preservation of Cosmetics in Disinfection, Sterilization and Preservation*, 3rd edn., ed. S.B. Block. Lea and Febiger, Philadelphia, pp. 589–607.

BIOFILMS

Biofilm Development and Emulsifier Levels in Metal Working Fluid Systems

E.L. Prince & L.H.G. Morton

School of Applied Biology, Lancashire Polytechnic, Preston, UK

ABSTRACT

A technique for assessing the effective emulsifier concentration of industrial metal working fluid samples is described. It was used to study the effect of fungal biofilm formation in a soluble oil emulsion metal working fluid on emulsifier concentration. Results are presented which suggest microbial biodepletion of the emulsifier package of the fluid.

INTRODUCTION

Although the adhesion of bacteria to solid surfaces has received a great deal of attention (Costerton & Geesey, 1979.; McCoy *et al.*, 1981; Whittaker *et al.*, 1984) the role of fungi in biofilms has largely been overlooked.

The aqueous metal working fluids used in industry are highly susceptible to microbial infection and the resultant biofilm formation often leads to the production of microbial slimes, which may block pipes and filters (Hill, 1969).

The problems associated with microbial infection of oil emulsions are well documented, and have been reviewed by Hill (1978). One particularly undesirable consequence of such infections is the onset of emulsion instability which often results from biodeterioration of the emulsifiers and coupling agents present in the formulation (Hill, 1976).

Various methods of monitoring the effects of emulsifier degradation are available (Katinger *et al.*, 1970; Hill, 1976; Almen *et al.*, 1982) but each has limitations in the industrial environment.

Solutions of surface-active materials exhibit distinctive physical properties. At very low concentrations the emulsifier molecules, or their ions, are randomly arranged and act as normal solutes. As the concentration of emulsifier is increased, however, the emulsifier molecules begin to form spherical aggregates, termed micelles, in which the hydrophobic portions of the molecules are orientated towards the interior of the micelles, leaving the hydrophilic portions in contact with the aqueous medium. The concentration of emulsifier above which micelle formation becomes appreciable is termed the 'critical micelle concentration' (CMC) of the emulsifier, and is relatively constant for that emulsifier at any given temperature provided the concentration of simple salts within the system remains unaltered (Shaw, 1980). Several physical properties of the solution, such as osmotic pressure, turbidity, electrical conductance and surface tension, change abruptly at CMC and this phenomenon can be utilised in CMC determination.

Since the CMC of a given emulsifier, or combination of emulsifiers, occurs at a fairly well-defined concentration, the degree to which an emulsifier solution must be diluted in order to reach that concentration provides a measure of the amount of emulsifier present in the solution.

This paper describes a method of monitoring the effective emulsifier concentration of a soluble oil metal working fluid formulation using a modification of a system described by Cooper *et al.* (1979). The effect of fungal proliferation within the fluid on effective emulsifier concentration was studied.

MATERIALS AND METHODS

A sample of a biocide-free soluble oil emulsion concentrate was available and was added to tap water in the proportion of 33:1 in order to provide a typical end-use concentration. Aliquots of 150 ml of the diluted fluid were dispensed into four 250 ml flasks which were then plugged with non-absorbent cotton. Two of the flasks were inoculated with a spore suspension containing *c.* 1×10^6 propagules/ml of the following metal working fluid isolates;

Geotrichum candidum Link
Fusarium sp.

Cephalosporium sp.
Verticillium sp.
Scopulariopsis brevicaulis Bainier
Acremonium strictum W. Gams.

The remaining flasks were left uninoculated as controls. All four flasks were incubated at 30°C with continuous shaking until the inoculated flasks showed appreciable fungal growth. Both inoculated (test) and uninoculated (control) fluids were filtered through Whatman No. 1 qualitative filter paper to remove the majority of the biomass from the test fluids. Test and control samples were then centrifuged at 5000 G for 15 min using an MSE Europa 24 M centrifuge. The cell-free supernatant fluids were collected and used to prepare dilution series in distilled water between 1% v/v and 100% v/v. The absorbance of each dilution, at a constant 25°C, was measured at a wavelength of 540 nm using an LKB Ultrospec 4050 spectrophotometer.

RESULTS

The inoculated emulsion samples showed appreciable adherent fungal growth after 14 days incubation. The mean absorbance at each dilution

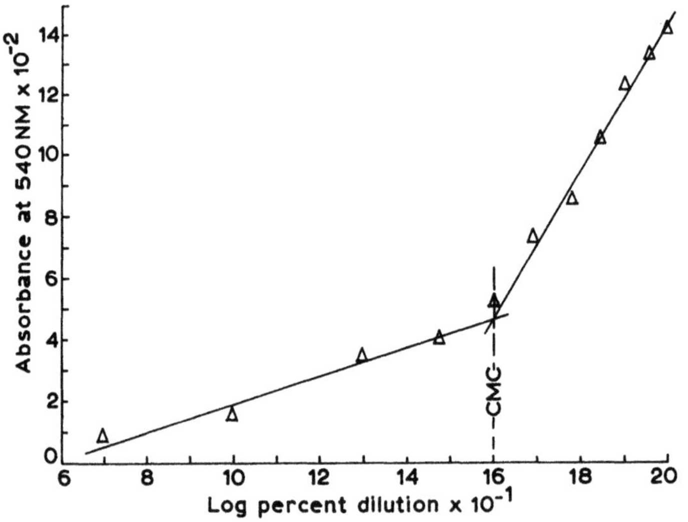

Fig. 1. The relationship between log % dilution and mean absorbance at 540 nm of dilutions of a fungally degraded coolant sample. Using this plot, CMC was estimated to occur at a log % dilution of 1·6.

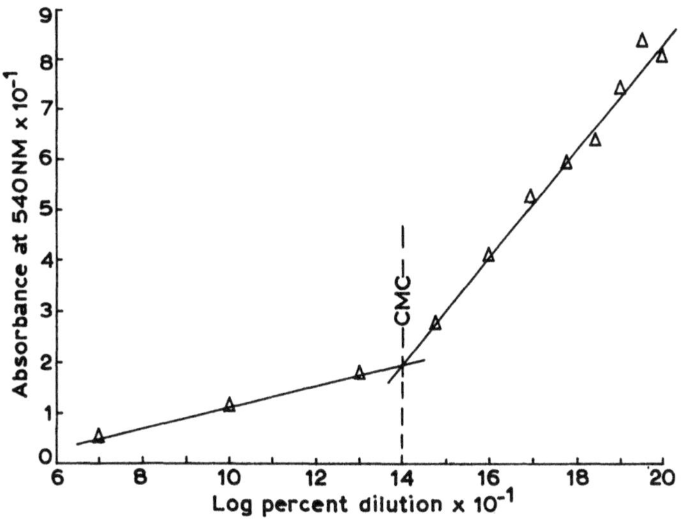

Fig. 2. The relationship between log % dilution and mean absorbance at 540 nm of dilutions of an uninoculated control coolant sample. Using this plot, CMC was estimated to occur at a log % dilution of 1·4.

for both the inoculated and control replicate samples was calculated by averaging the absorbance values obtained from each pair of replicate dilutions. Figures 1 and 2 show the relationships between log % dilution and mean absorbance for inoculated and control samples respectively. CMC of the inoculated fluids was estimated to occur at a log % dilution of 1·6, and that of the control fluids at a log % dilution of 1·4. It was therefore necessary to dilute the control samples to a greater degree than the inoculated samples in order to reach CMC. This result suggests a higher effective emulsifier concentration in the control samples than in the inoculated samples.

DISCUSSION

Results indicate a lower effective emulsifier concentration in fungally degraded emulsions than in control samples and it is suggested that this may be due to microbial biodepletion of the emulsifier package of the coolant. The fungally degraded samples all showed evidence of extensive biofilm development and it is considered that this technique might have potential application in measuring the effects of such biofilms in industrial fluids.

Work currently in progress seeks to correlate this phenomenon with other parameters and to assess the precise relationship between dilution to CMC and emulsifier concentration in various coolant samples and types of fluid.

REFERENCES

Almen, R., Mantelli, G., McTeer, P. & Nakayama, S. (1982). *Journal of the American Society of Lubrication Engineering*, **38**, 99–103.

Cooper, D.G., Zajic, J.E. & Gerson, D.F. (1979). *Applied and Environmental Microbiology*, **37**, 4–10.

Costerton, J.W. & Geesey, G.G. (1979). In *Surface Contamination, Vol. 1*, ed. K.L. Mittal. Plenum Publishing Corporation, New York.

Hill, E.C. (1969). *Transactions of the Society of Occupational Medicine*, **19**, 93–6.

Hill, E.C. (1976). In *Proceedings of the Third International Biodegradation Symposium*, ed. J.M. Sharpley & A.M. Kaplan. Elsevier Applied Science, London.

Hill, E.C. (1978). In *Developments in Biodegradation of Hydrocarbons*, ed. R.J. Watkinson. Elsevier Applied Science, London.

Katinger, H., Nobis, A & Meyrath, J. (1970). *Experentia*, **26**, 565–6.

McCoy, W.F., Bryers, J.D., Robbins, J. & Costerton, J.W. (1981). *Canadian Journal of Microbiology*, **29**, 910–17.

Shaw, D.J. (1980). *Introduction to Colloid and Surface Chemistry.* Butterworths, London, pp. 60–94.

Whittaker, C., Ridgway, H. & Olson, B.H. (1984). *Applied and Environmental Microbiology*, **48**, 395–403.

The Application of Interference Reflection Microscopy to the Study of Bacterial Adhesion to Solid Surfaces

Madilyn Fletcher

Center of Marine Biotechnology of the Maryland Biotechnology Institute, 600 E. Lombard Street, Baltimore, MD 21202, USA

ABSTRACT

Interference reflection microscopy was used to determine the influence of electrolyte concentration on the bacterium–substratum separation distance between attached Pseudomonas fluorescens *and glass. The addition of sodium chloride caused a decrease in separation distance, which again increased when electrolyte was replaced by water. Calcium chloride also caused a decrease in separation distance. However, with* P. fluorescens *strain H2S, a particularly adhesive strain, this effect was only partially reversible when calcium chloride was replaced by water.*

INTRODUCTION

A major obstacle to our understanding of the adhesive interaction at the bacterium/substratum interface has been our inability to observe living cells during adhesion and at the same time measure the progressive formation of the adhesive bond. However, this can be done applying an interference reflection microscopy (IRM) technique (Curtis, 1964). The technique allows the adhesive sites on the bacterial surface to be seen as attachment to the solid substratum is established. It is also possible to evaluate the change in distance and the time required as the cell proceeds from initial loose association with the surface to close and firm adhesion. IRM has been used previously to locate the adhesive sites on animal cells (Abercrombie & Dunn, 1975; Izzard & Lochner, 1976; Gingell & Vince,

1982). Although it is more difficult to apply to bacteria this and a previous study (Fletcher & McEldowney, 1984) demonstrate that the technique can be used to detect changes in bacterium/substratum separation distance that result from changes in environmental conditions.

With IRM, the darkness of the image of the attaching bacterium varies with its distance from the surface. Maximum brightness is obtained at a distance approximate to $\lambda/4n$ (i.e. the wavelength of light used for illumination/4 × the refractive index of the separating medium). In this study, $\lambda = 546$, and n is estimated to be 1·3, so that the distance of maximum image brightness is *c.* 105 nm. As the bacterium moves closer to the surface, the image becomes darker, until it reaches maximum darkness at its closest approach.

For this study, bacteria were prepared for observation by IRM by being placed in a 'flow chamber' in which the fluid could be replaced by test solutions. The influence of sodium and calcium electrolytes on adhesion was observed and recorded.

METHODS

The bacteria used were *Pseudomonas fluorescens* H2, the wild type strain which has moderate adhesion ability, and *P. fluorescens* H2 and H2S, a spontaneous mutant which has increased adhesion ability (Pringle *et al.*, 1983). These were grown to the early stationary phase in 50 ml of medium (0·1% peptone, 0·07% yeast extract, pH 7·4). The cells were collected by centrifugation and resuspended in an equal volume of distilled water.

The 'flow chamber' was constructed from three 18 mm, square, glass coverslips and a glass slide. One coverslip was supported on the other two, which lay flat on the glass slide about 8 mm apart. The coverslips were fixed in place with a cyanoacrylate adhesive. Fluid could run by capillary action beneath the supported coverslip between the two supports. After cleaning the chamber with detergent, a portion of the suspension was introduced and inspected immediately for attaching cells by alternating observations with IRM and phase contrast. An attaching cell would be located and observed as electrolyte solutions were introduced by adding a few drops of solution at one opening of the chamber and drawing it through by touching filter paper (Whatman No. 1) to the other opening. Progressively higher concentrations of electrolyte were added and any resulting changes noted and photographed. Electrolytes and their concentration ranges were: sodium chloride, 500 μM–10 mM; calcium chloride, 200 μM–50 mM. The influence of replacement of electrolyte with distilled water was also tested. New slides

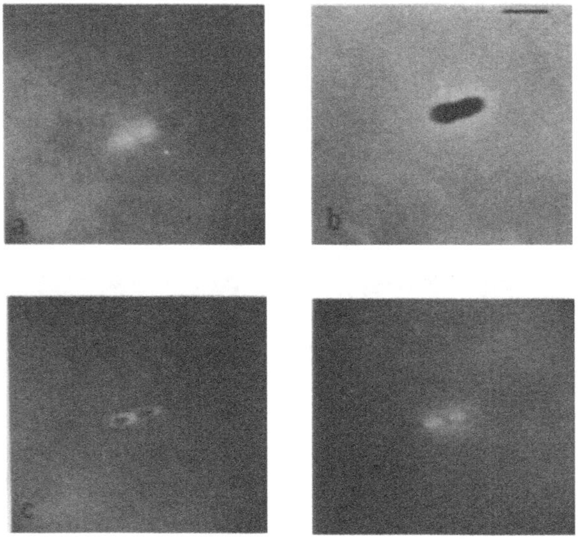

Fig. 1. IRM and phase contrast (Ph) micrographs showing the influence of sodium chloride on cell-surface separation distance with *P. fluorescens* H2. IRM (a) and Ph (b) of bacteria suspended in distilled water; IRM of bacteria suspended in 10 mM sodium chloride (c); after sodium chloride was replaced with water (d). Bar marker represents 2 μm.

were set up for each electrolyte, and observations were repeated numerous times with fresh batch cultures.

A Zeiss Standard 18 microscope was adapted for IRM as previously described (Fletcher & McEldowney, 1984). Data were recorded by photography with Technical Pan film, and film was developed with D-19 developer (both Kodak) for maximum contrast.

RESULTS

In the flow cell, some bacteria appeared to be firmly attached when observed by phase contrast, i.e. they did not exhibit Brownian motion and were not removed from the surface by replacing the liquid in the chamber with fresh distilled water. These were visible as bright images with IRM (Fig. 1(a)). This means that they were some distance from the surface (maximum brightness *c.* 105 nm), and were not within the primary minimum described by DLVO theory. However, because of the apparent firmness of adhesion, the gap between the cell and surface is probably bridged by an adhesive polymer (Fletcher & Floodgate, 1973),

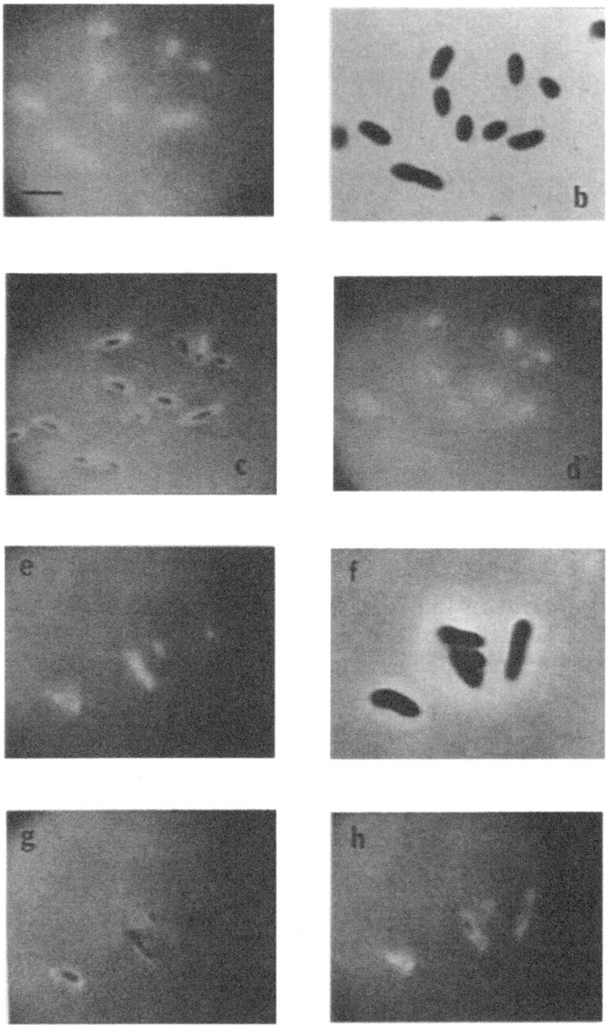

Fig. 2. IRM and Ph micrographs showing the influence of calcium chloride on cell-surface separation distance with *P. fluorescens* H2 (a–d) and H2S (e–h). IRM (a,e) and Ph (b,f) of bacteria suspended in distilled water; IRM of bacteria in 50 mM calcium chloride (c,g) and after calcium chloride was replaced with water (d,h). Bar marker represents 2 μm.

and this indicates that polymer bridging is important even at the very earliest stages of adhesion.

The introduction of either NaCl (Fig. 1) or $CaCl_2$ (Fig. 2) into the chamber caused the bacteria to move closer to the surface, demonstrated by the appearance of dark areas where the cell surface approached the glass. With both *P. fluorescens* H2 and H2S, the dark areas were conspicuous at cation concentrations of 5 mM or greater. With NaCl and with $CaCl_2$ and *P. fluorescens* H2, this effect was reversible, because replacing the NaCl solution with distilled water caused the disappearance of the dark areas and the return to a bright image. However, with $CaCl_2$ and H2S, the strain with greater attachment ability, the effect was only partially reversible (Fig. 2).

It is not surprising that the addition of electrolyte solutions resulted in movement of the bacteria towards the surface (Gingell & Vince, 1982), and this is predicted by DLVO theory. The cations cause a reduction in the repulsive force (i) between the negatively charged bacterial and glass surfaces and (ii) between negative groups on the adhesive polymer, resulting in its condensation and drawing the cell in towards the surface. This effect would also be expected to be reversible, as was found with sodium. However, the effect of calcium was particularly interesting, as it was reversible with strain H2, but only partially reversible with the more adhesive strain H2S. Such cation stabilization of adhesion might help to explain the greater attachment ability of H2S.

REFERENCES

Abercrombie, M. & Dunn, G.A. (1975). *Exptl. Cell Res.*, **92**, 57–62.
Curtis, A.S.G. (1964). *J. Cell Biol.* **20**, 199–215.
Fletcher, M. & Floodgate, G.D. (1973). *J. Gen. Microbiol.*, **74**, 325–34.
Fletcher, M. & McEldowney, S. (1984). In *Current Perspectives in Microbial Ecology*, ed. M.J. Klug and C.A. Reddy. American Society for Microbiology, Washington, D.C., pp. 124–9.
Gingell, D. & Vince, S. (1982). *J. Cell Sci.*, **54**, 299–310.
Izzard, C.S. & Lochner, L.R. (1976). *J. Cell Sci.*, **21**, 129–59.
Pringle, J.H., Fletcher, M. & Ellwood, D.C. (1983). *Appl. Environ, Microbiol*, **129**, 2557–69.

Legionella: a Biofilm Organism in Engineered Water Systems?

Jennifer S. Colbourne

Thames Water Authority, New River Head Laboratories, 177, Rosebery Avenue, London EC1R 4TP, UK

&

P. Julian Dennis

Quality Control & Safety Laboratory, PHLS Centre for Applied Microbiology & Research, Porton Down, Salisbury, Wiltshire, SP4 0PG, UK

ABSTRACT

Investigations of outbreaks of legionnaires disease have demonstrated that infection, in susceptible people, can occur by inhalation of water droplets containing viable legionellae. The appearance of the strain associated with disease in water in a building linked epidemiologically to the outbreak has been attributed in several instances to disturbance of the water system. As a result it has been postulated that legionellae may be associated with particulate matter and biofilms. New techniques for detecting legionellae in water samples by immunofluorescent microscopy (IFA) and systematic sampling were applied to two water systems linked with cases of disease. IFA results identified the location of the organism in each system—the biofilm adhering to the walls of a freshwater storage tank in a passenger ship and slime in the pond and pipework of two cooling towers. Legionella pneumophila *was recoverable by culture only from water. IFA detected the organism in water and biofilms even when control measures such as chlorine and heat were used. These results confirm that* L. pneumophila *can colonise biofilms, hence long-term prevention of legionellosis will come only with knowledge of the factors governing formation and inhibition of biofilms in water systems.*

INTRODUCTION

A decade has passed since *Legionella pneumophila* was first isolated from clinical material (McDade *et al.*, 1977). Since then legionellae have been recognised as aquatic bacteria which sometimes infect man. Infection occurs after release of water droplets containing viable legionellae and their inhalation by susceptible people (Backerville *et al.*, 1981). Current techniques for controlling legionellosis and investigating outbreaks of legionnaires disease rely on culturing the bacterium from water. Advances in methodology now permit the examination of large volumes of water and legionellae can be detected with relative ease when numbers in the water are as low as 10^3/litre. This concentration has become the standard for evaluating the efficacy of water treatment and, by default, is regarded also as a benchmark for risk assessment.

Historically water microbiologists have gathered knowledge of the flora of piped water systems by culturing water drawn from external access points. Unlike other aquatic environments sediments and surfaces are sampled rarely. Direct methods of looking for legionellae are not widely used. If *L. pneumophila* is isolated from a water sample or a building is implicated in an outbreak of disease then remedial measures, heat or chlorine, are instigated often before sampling. These measures result in cessation of disease but failure in vigilance in their application has been followed by a recurrence of cases.

This situation was identified during an investigation at the Wadsworth Memorial Hospital in Los Angeles between 1977 and 1980 and was apparent during investigations in the early 1980s at Kingston District Hospital and a hotel in Corby. The bacterium was cultured from water after disturbance of deposits and surfaces of water fitting components. These early outbreaks established two features of the ecology of legionella; first they can establish themselves in piped water systems and survive even when water conditions are hostile and secondly, the organism is associated with surfaces or particulate matter.

Considerable efforts have been made to establish effective preventative measures by the use of disinfectants, such as chlorine, or biocides. Generally these act on organisms in the water. The effectiveness of such measures against legionella in biofilms or particulate and organic matter has not been well studied but is likely to be much reduced. New microbiological techniques are required to establish the relationship between legionellae and biofilms, to aid outbreak investigations and to design effective long-term control measures. The study reported here, of a pair of cooling towers in a retail store and a passenger ship associated with 15 and 3 cases of legionnaires disease respectively, is a step towards this goal.

METHODS

64 samples from freshwater systems on the ship and from both cooling towers were collected and concentrated using the techniques described by Ashworth & Colbourne (1987) and Dennis *et al.* (1984). Water samples of 10 litres were concentrated by membrane filtration using nylon membranes (Pall) of 0·22 μm pore size. Slimes and deposits were collected with sterile swabs and spatulas and kept moist in screw capped containers. Water concentrates and specimens were examined for legionellae by direct culture on BCYE GVPC agar and by culture after treatment with acid and heat (Dennis, in press). Isolates were identified serologically.

In parallel, concentrates and specimens were examined microscopically using indirect immunofluorescent assay (IFA). Concentrates were fixed on multiwell slides by acetone and stained with monoclonal antibodies to *L. pneumophila* serogroup 1. The slides were examined for fluorescent cells using a Leitz SMLux Incident microscope. Samples yielding more than 10 fluorescent cells per well (10 microlitres) were scored as positive. IFA was repeated in duplicate using monoclonal antibodies to the Pontiac and Olda subtypes of *L. pneumophila* serogroup 1 (Watkins et al., 1985).

RESULTS AND DISCUSSION

During sampling (4 days) the water on the ship contained 1–2 mg/l of free residual chlorine and the hot water was held at 60 °C. *L. pneumophila* was not cultured from 51 samples despite treatment to neutralise the effects of chlorine and heat. These procedures were validated by culture of other *Legionella* spp. from 2 shower samples (Table 1). All water samples were also IFA negative; however, the Pontiac strain of *L. pneumophila* was observed in slime on the inner wall of a freshwater tank and deposits in the base of a calorifier. The cells in the calorifier deposit were small and rounded whereas those in the slime were typical rods. Survey and epidemiological information coupled with the appearance of the cells and culture negative results demonstrated that continuous chlorination (1–2 mg/litre prevents multiplication of *L. pneumophila* in

TABLE 1

Detection of legionellae by culture and IFA in samples collected from water systems on board a passenger ship

Sample	*Number of samples*	*Positive by culture*	*Positive by IFA*
Waters	38	2[a]	0
Slime deposits	13	0	2

[a] Isolates identified as *Legionella feeleii* and *Legionella bozemanii*.
Positive IFA = immunofluorescence with monoclonal antibodies specific for the Pontiac subtype of *Legionella pneumophila* serogroup 1 (Watkins *et al.*, 1985).

TABLE 2

Detection of *L. pneumophila* serogroup 1 by culture and IFA in samples from a pair of evaporative cooling towers in a retail store

Treatment	*Sample*	*Tower 1*		*Tower 2*	
		Culture	*IFA*	*Culture*	*IFA*
Cleaning 12 weekly	Pond water	+	+	—	+
	Pipe water	—[a]	+	—[a]	+
	Pond foam	+	+	—	+
Cleaning/chlorination	Pond water	—	+	—	—
After 3 months of cleaning 6 weekly	Pond water	+	+/−	—	+
	Foam	+	+	—	+
Cleaning/chlorination	Pond water	—	+	—	—
	Pipe water	—	+	—	+
	Pipe gaskets	—	—	—	+
Removal of pipes and slime	Pipe biofilm	—	—	—	+
	Pipe biofilm			—	+
	Pond slime	—	+		
After disturbance removal pipe/slime	Pond water	+	+	+	+/−
7 days of operation after chlorination	Pond water	—	—	—	—

Culture positive = *L. pneumophila* serogroup 1 subtype Pontiac.
[a]Culture positive for other legionellae.
IFA positive = fluorescent cells with monoclonal antibodies specific to *L. pneumophila* serogroup 1 subtype Pontiac.
+/− = number of fluorescent cells < 10 per well or detected with only one reagent.

cold water. Similarly holding water at > 50°C causes death or dormancy of the organism when it collects in calorifiers. Just prior to the onset of symptoms amongst passengers the free chlorine concentration was 0·2 mg/litre or less. A power failure, which shut down the treatment plant and pumps for 30 min, occurred.

The cooling towers were not dosed continuously or regularly with chlorine or biocides but periodically cleaned to remove scale. After isolation of the Pontiac strain of *L. pneumophila* from the pond water of one tower, both towers were investigated intensively for 6 months. The pond waters were regularly covered with foam and a black 'ringmark' slime existed in each pond. The rough surface of the concrete made the ringmark difficult to remove. Several dead-end sections of pipe connected to the condenser circuit conveying water back to the top of the towers were found. These were removed causing major disruption to the plant operation.

Initially the Pontiac strain was cultured from the pond water of one tower but was detected in both ponds by IFA (Table 2). Foam on the pond, which concentrated the organism 1000-fold was culture positive in one pond and IFA positive in both ponds. After cleaning and chlorination both became culture negative but remained IFA positive. The organism was not cultured from the ringmark, foam, materials or pipes but IFA revealed its presence in the ringmark in one tower and dead-end pipes and a rubber gasket in the condenser circuit of the other tower. Removal of dead-ends and slime initially caused both pond waters to become culture positive. After chlorination and normal operation for 7 days both ponds were IFA negative for the first time and foaming was reduced. The pair of cooling towers were situated in a city where a cluster of 15 cases of legionnaires disease occurred. Survey information revealed substantial differences in the mode of operation and construction of each system. In the tower where the organism occurred in the ringmark it was noted that cleaning, which focused on the pond, was followed by appearance of culturable cells in the water. By contrast in the tower where the organism had colonised the condenser pipework, culture from the water was possible only after a major disturbance. IFA, not culture, allowed effective decontamination measures to be devised.

The Pontiac strain of *L. pneumophila* is the strain most often associated with disease. This strain also exhibits a greater ability to survive in aerosols. In studies of outbreaks disturbance or lack of vigilance in control measures resulted in further cases of disease suggesting absence of the organism from the water by culture should not be regarded as a measure of safety. The study of two further outbreaks has confirmed the ability of the organism to grow or survive in surface biofilms in both domestic water services and cooling towers. When these biofilms were disturbed by routine physical cleaning or chemical treatment or by more dramatic engineering manoeuvres, legionella was released into the water.

Differences in virulence, survival and hydrophobicity of the Pontiac strain offer an explanation for the increased risk of infection which occurs when people are exposed to aerosolised water from systems colonised with this organism. It has been shown that viable but non-culturable forms of legionella can occur in water systems and this finding supports other data indicating that numbers of culturable bacteria in water cannot assure safety. If, when applied to other systems, IFA and systematic sampling yields similar data then the key to preventing legionellosis will be control of biofilm formation. More sophisticated mixed and continuous culture techniques are required to identify the factors which lead to colonisation of water systems by legionellae and these are now in use (Dennis & Colbourne, 1987). Our results confirm the value of IFA not only in tracking *L. pneumophila* in water systems, but in studying the behaviour of the organism in laboratory models.

ACKNOWLEDGEMENTS

Our thanks to Miss C. Berry, Mr R. Trew, Miss D. James (Thames Water) and Mr G. Vessey (PHLS, CAMR) for laboratory analysis. We gratefully acknowledge Dr Michael McIntyre PHLS Oxford for provision of reagents and Dr C.L.R. Bartlett (PHLS CDSC) and Mr. M. Bailey (London School of Hygiene) for initiating the investigations and their invaluable advice. As ever the studies could not have been completed without the unstinting efforts of the engineers responsible for the water systems.

REFERENCES

Ashworth, J.A. & Colbourne, J.S. (1987). In *Industrial Microbiological Testing, Society of Applied Bacteriology Technical Series No 23*, ed. J. W. Hopton & E.C. Hill. Blackwell, Oxford.

Baskerville, A., Fitzgeorge, R.B., Broster, M., Hambleton, P. & Dennis, P.J. (1981). *Lancet*, **ii**, 1389.

Dennis, P.J. (In press). In *A Laboratory Manual for Legionnaires Disease*, ed. T.G. Harrison & A.G. Taylor. John Wiley, Chichester.

Dennis, P.J. & Colbourne, J.S. (1987). In *Biodeterioration of Construction Materials, Proceedings of the Summer Meeting of the Biodeterioration Society*, ed. L.H.G. Morton. Publications Service, Lancashire Polytechnic.

Dennis, P.J., Bartlett, C.L.R. & Wright, A.E. (1984). In *Legionella, Proceedings of the 2nd International Symposium*, ed. C. Thornsberry, A. Balows, J.C. Feeley & W. Jakuboski. American Society for Microbiology, Washington, D.C.

McDade, J.E., Sheppard, C.C., Fraser, D.W., Tsai, T.R., Redus, M.A., Dowdle, W.R. and the laboratory investigation team (1977). *New England J. Med.*, **297**, 1197–203.

Watkins, I.D., Tobin, J.O.H., Dennis, P.J., Brown, W., Newnham, R. & Kurtz, J.B. (1985). *J Hyg.* (Camb.), **95**, 211–16.

An extended bibliography for this paper may be obtained from the senior author.

Non-biocidal Anti-fouling Coatings

Maureen E. Callow, R.A. Pitchers & Regina Santos*

Departments of Plant Biology and *Chemical Engineering, University of Birmingham, P.O. Box 363, Birmingham B15 2TT, UK

ABSTRACT

The dry weight of fouling on raft panels coated with silicone elastomers (low energy surfaces) has been measured at two-monthly intervals for one year. A flow cell is described for testing the adhesion of slimes to raft panels. Results on removal of slimes from silicone elastomers and formica are discussed.

INTRODUCTION

Testing the potential anti-fouling performance of non-biocidal surfaces necessitates the use of methods which measure the adhesion of organisms. Various methods have been employed in the evaluation of cell adhesion. Sophisticated devices for measuring adhesive strength of individual animals (barnacles and mussels) have been described by Bultman *et al* (1984) and Crisp and co-workers (e.g. Crisp *et al.*, 1985). The strength of adhesion of algal cells has been studied in the laboratory using water jets, a water broom (Charters *et al.*, 1973), a water tunnel (Jones *et al.*, 1982) and a radial flow chamber (Milne & Callow, 1985). Assessment of adhesion of fouling communities on raft panels or boats has been confined to arbitrary methods such as scrub brushes and water or water delivered using a conventional firehose (Bultman *et al.*, 1984). We have attempted to measure adhesion of organisms to non-biocidal 'non-stick' coatings on raft panels in a more controlled and reproducible manner using a flow cell.

MATERIALS AND METHODS

Formica panels 10 × 10 cm were coated with four experimental room temperature vulcanising (RTV) silicone elastomers (see Milne & Callow, 1985). Controls were untreated formica. The five types of surface were attached in groups of 25, using a Latin square arrangement, to both sides of wooden plates which were suspended vertically to a depth of 1·5 m from a raft in the Yealm Estuary, Devon. For each assessment 10 replicates were used. Panels were immersed in January 1986 and assessments made after 2, 4, 6, 8, 10 and 12 months. One hundred panels were recovered for each assessment. The fouling was removed from one group (5 surfaces × 10 replicates) with cotton wool buds and dry weight measured as described by French & Evans (1986). The remaining 50 panels were subjected to the flow cell described below before removal and dry weight determination of remaining fouling.

The flow cell consisted of two parallel perspex rectangular plates (0·1 m × 1·5 m). A slot in the rear perspex plate held an aluminium plate containing the five test panels. The front and rear plates were clamped together leaving a gap of 5 mm between them through which seawater was pumped. Metal bars were used to reinforce the structure of the flow cell because of the back pressure caused at high flow rates. The flow was adjusted using a rotameter up to a maximum flow of 79·45 litres/min.

$$Q = vA, \text{ therefore, } v = Q/A$$

Q = volumetric flow, A = cross-sectional area, v = velocity

Thus, A = 10 cm × 0·5 cm or 5×10^{-4} m^2.

For the experiments described here the flow cell was run for 10 min at maximum volumetric flow corresponding to a velocity of 2·65 m/s.

RESULTS

The total dry weight of fouling on panels immersed up to 16 weeks is shown in Fig. 1. The weight of fouling on the formica control shows an approximately 8-fold increase between 8 and 16 weeks due to rapid growth of macroalgae, chiefly *Laminaria saccharina*. The formica control was not assessed quantitatively after 16 weeks as the *Laminaria* plants were cut back to prevent the fronds sweeping the surface of the experimental panels (see Fig. 2). Although plants of *L. saccharina* up to 2 mm in length were present on the silicone elastomers at 8 weeks, these subsequently fell off due to poor adhesion leaving only a diatom slime.

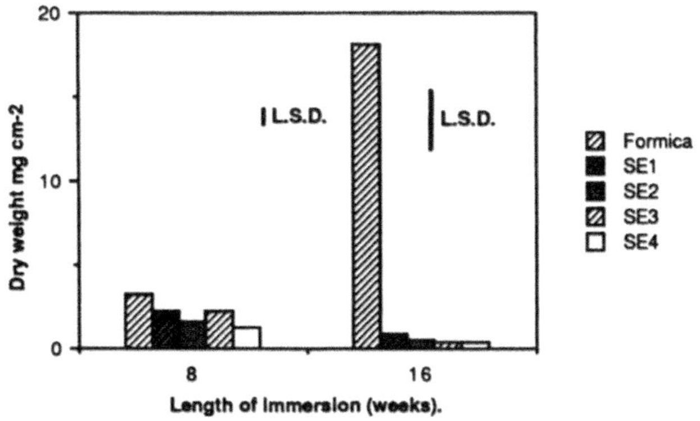

Fig. 1. Dry weight of fouling on raft panels. SE = silicone elastomer; LSD = Least Significant Difference. Each point is the mean of 10 replicates.

Diatom slimes were the only type of fouling present on the silicone elastomers between 16 and 48 weeks, in contrast to the formica panels which bore a typical climax community. At 48 weeks the dry weight of total fouling on the silicone elastomers was in every case less than 0·2 mg/cm². The dry weight of fouling on 8-week panels compared to

Fig. 2. Wooden holder with panels arranged in Latin Square, immersed 24 weeks. Top row, left to right: SE1; formica; SE2; SE3; SE4. Note heavy fouling growth on formica panels compared to the silicone elastomers which bear only a thin diatom slime.

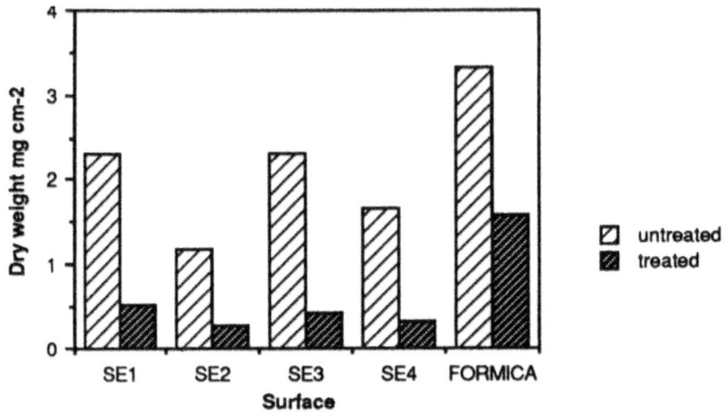

Fig. 3. Comparison between the dry weight of fouling on panels after treatment in the flow cell and untreated ones. SE = Silicone elastomer. All panels were immersed 8 weeks. Each value is the mean of 10 replicates.

similar panels subjected to the flow apparatus is shown in Fig. 3. Approximately 80% of the fouling is removed by the flow cell from all of the silicone elastomers compared to approximately 50% from the formica control. The formica control could not be used in the flow apparatus after 8 weeks due to the enormous quantity of fouling present. Diatom slimes which were present on the silicone elastomers after 6–12 months immersion were strongly adherent and less than 50% of the dry weight of fouling was removed by the flow cell. The major species present were *Cocconeis, Amphora* and *Achnanthes.*

DISCUSSION

Silicone elastomers are low energy surfaces which have been shown previously to have good antifouling properties by preventing the adhesion of most organisms and allowing only the weak adhesion of others (Callow *et al.*, 1986). In the present study we have attempted to compare the strength of attachment of slimes to silicone elastomers (low energy surfaces) and to formica (high energy surface). The wall shear stress for clean, i.e. smooth panels, at the operating velocity of 2·65 m/s is 20 N/m^2 which is of the same order of magnitude as the wall shear stress calculated (56 N/m^2) for an unfouled 300 m long ship operating at 15 knots (Schlichting, 1968). More detailed results could have been obtained if higher velocities could have been generated within the flow cell.

A number of low energy surfaces have been shown to reduce adhesion of organisms. The bond strength of mussels and barnacles is fifty times greater to slate (high energy surface) than to PTFE (low energy surface). Antifouling coatings have been made from PTFE and related fluoropolymers (Bultman *et al.*, 1984) and although these do accumulate fouling, attachment is weak. Silicone elastomers have also been shown to prevent mainly attachment of both animal and weed fouling (Callow *et al.*, 1986) and the present investigation indicates that the adhesion of slime to the silicone elastomers is considerably reduced compared to formica (high energy surface). Fletcher *et al.* (1984) showed that attachment and growth of macroalgal sporelings to low energy silane-based coatings resulted in altered morphology of the attaching rhizoids. The disappearance of *Laminaria saccharina* from the silicone elastomers after 8 weeks in this study, reflects the inability of macroalgae to adhere past the sporeling stage (1–2 mm length). Microscopic examination of rhizoids of young *Laminaria* sporophytes removed from the 8-week panels revealed that plants from the silicone elastomers bore a number of filamentous, branching, colourless rhizoids in contrast to the compact, disc-like rhizoids of plants from the formica panels. Similar changes in response to surface energy have been shown for *Enteromorpha* (Fletcher *et al.*, 1984). The recognition and response of organisms to low energy surfaces poses some interesting cell biological questions worthy of further investigation.

ACKNOWLEDGEMENTS

We are grateful to International Paint plc (MEC and RAP) and SERC (RS) for financial support and to Dr T.R. Bott for discussions on design of the flow cell.

REFERENCES

Bultman, J.D., Griffith, J.R. & Field, D.E. (1984). In *Marine Biodeterioration: An Interdisciplinary Study*, ed. J.D. Costlow & R.C. Tipper. U.S. Naval Institute, Annapolis, pp. 237–43.

Callow, M.E., Pitchers, R.A. & Milne, A. (1986). In *Algal Biofouling*, ed. L.V. Evans & K.D. Hoagland. Elsevier, Amsterdam, pp. 145–58.

Charters, A.C., Neushul, M. & Coon, D. (1973). *Limnology and Oceanography*, **18**, 884–96.

Crisp, D.J., Walker, G., Young, G.A. & Yule, A.B. (1985). *Journal of Colloid and Interface Science*, **104**, 40–50.

Fletcher, R.L., Baier, R.E. & Fornalik, M.S. (1984). In *6th Inernational Congress on Marine Corrosion and Fouling*, 5–8 August 1984, C.O.I.P.M., Athens.

Fletcher, R.L., Jones, A.M. & Jones, E.B.G. (1984). In *Marine Biodeterioration: An Interdisciplinary Study*, ed. J.D. Costlow & R.C. Tipper. Naval Institute Press, Annapolis, MD, pp. 172–82.
French, M.S. & Evans, L.V. (1986). In *Algal Biofouling*, ed. L.V. Evans & K.D. Hoagland. Elsevier, Amsterdam, pp. 79–100.
Jones, A.M., Fletcher, R.L., Daniel, G.F. & Jones, E.B.G. (1982). In *Fouling and Corrosion of Metals in Seawater*, ed. J. Mauchline. Oban, S.M.B.A., pp. 31–77.
Milne, A. & Callow, M.E. (1985). In *Polymers in a Marine Environment*, ed. R. Smith. The Institute of Marine Engineers, London, pp. 229-33.
Schlichting, H. (1968). *Boundary Layer Theory*, 6th edn.

An extended bibliography for this paper may be obtained from the senior author.

Microfouling Film Composition, Thickness and Surface Roughness on Ship Trial Antifouling Paints

D.C. Woods, R.L. Fletcher & E.B.G. Jones

The Marine Laboratory, Ferry Road, Hayling Island, Hampshire, PO11 0DG, UK

ABSTRACT

A range of antifouling panels were exposed under operational conditions on the bilge keel of an ocean going ship. The physical characteristics and the diatom composition of microfouling on these surfaces were shown to be influenced by the paint formulation.

INTRODUCTION

The introduction of increasingly efficient antifouling coatings has resulted in the effective control of most macrofouling organisms on in-service shipping; as a consequence microfouling has now become the more persistent problem (Woods *et al.*, 1986). Microfouling films develop rapidly on ships' hulls and can increase surface roughness and therefore skin friction resistance (Clitheroe, 1983). This results in reduced operational speed and/or greater fuel consumption (Gunn *et al.*, 1987; Woods *et al.*, 1986). The continued development and testing of antifouling formulations is therefore necessary to increase their efficacy against microfouling organisms. An established test method is the exposure of coated panels on the bilge keels of in-service vessels (Furtado & Fletcher, 1987). It permits a range of coatings to be tested and compared under identical operational conditions and subjected to the type of stresses generally encountered on in-service vessels but which are not simulated in static trials (Fischer *et al.*, 1984).

With the cooperation of W & J Leigh & Co. Ltd a programme of bilge keel panel trials was established to provide information on the influence of a range of antifouling paints on the development of microfouling communities. The physical and biological characteristics of the microfouling films, in particular film thickness, roughness and diatom composition were studied.

MATERIALS AND METHODS

A range of antifouling formulations were spray coated onto 15 × 15 cm plastic panels; Table 1 lists and gives details of 11 of these coatings which are considered in the present paper. These comprised both insoluble (1-5) and ablative (6–11) paints containing various copper and organotin biocides. They were exposed under oceanic in-service conditions for 35 months prior to examination.

Microfouling film thickness was measured by light section microscopy at 0·25 mm intervals for 1 cm transects on each surface. The individual measurements were used to quantify mean biofilm thickness and to produce a Roughness Increase Increment (RII) based on the difference in standard deviation (SD) values for the paint and biofilm surfaces. Subtraction of the paint surface SD from the biofilm SD removed the effect of both paint surface roughness and transect slope; the difference was the increase in roughness caused by microfouling.

Comprehensive lists of the diatom species present on each surface were compiled and the most prevalent species were weighed according to an arbitrary scale; Common (C) = 10–25% of diatoms per sample, Frequent (F) = 25–50% and Abundant (A) = over 50%. Also the relative percentage occurrence of the six main fouling species and seven genera was determined by recording their presence or absence in sub-samples containing 20–100 diatom frustules.

RESULTS AND DISCUSSION

Film thickness data

The following general observations were obtained from the film thickness measurements (Table 2).

(1) Mean biofilm thickness ranged from 47 to 562 μm.
(2) Generally thicker films (266–562 μm) occurred on ablative paints compared to insoluble matrix paints (47–261 μm).

TABLE 1

Paint code	*Matrix*	*Pigment/biocide*	*Ratio*
1	Vinyl	Cu_2O TBTF	1:1
2	Vinyl	Cu_2O TPTF	3:2
3	Chlorinated rubber	Cu_2O TPTF	7:3
4	Vinyl	Cu_2O TBTF	1:1
5	Vinylisobutylether	Cu_2O TBTF	1:1
6	Ablative (slow)	ZnO:TBTF	2:1
7	Ablative (medium)	ZnO:TBTF	1:1
8	Ablative (fast)	ZnO:TBTF	3·75:2
9	Ablative (medium)	CuSCN:ZnO:TBTF	2:1·5:2
10	Ablative (medium)	CuSCN:ZnO:TPTF	2:1·5:2
11	Ablative	Cu_2O TBTF	33·5:1

(3) The thinnest biofilm was recorded for the chlorinated rubber based paint which contained the highest Cu_2O to organotin ratio.
(4) The greatest mean thickness was recorded for the ablative paints containing CuSCN and TBTF or TPTF (562–540 μm respectively).
(5) The erosion rate of ablative paints influenced effectiveness, for example, results obtained for panels 6, 7 and 8 containing TBTF only showed that biofouling was least developed on those with a lower toxin loading/fast erosion rate compared to the medium eroding/higher toxin loading paint.
(6) Vinyl matrix paints provide better fouling control for the same toxin content than vinyl isobutylether (panels 4 and 5).

TABLE 2
Calculated Results for Microfouling Films

Panel code	*Mean biofilm thickness (μm)*	*RII (μm)*
1	100	21
2	104	57
3	47	6
4	119	44
5	261	71
6	433	116
7	390	92
8	266	55
9	562	197
10	540	75
11	319	86

(7) Insoluble vinyl matrix paints containing approximately equal proportions of Cu_2O and TBTF or TPTF (panels 1, 2, 4) all produced biofilms approximately 100 μm mean thickness, but RII values varied from 21 to 71 μm.
(8) RII values for thinner microfouling film generally reflected the mean thickness values. The erect and/or tufted nature of some of the constituents of the thicker biofilms, resulted in greater variation in RII values (for example, panels 9 and 10 exhibited similar mean thickness values but RII values of 197 and 75μm respectively).

Microfouling composition

(1) Microfouling films on the panels were dominated by diatoms; bacteria were also present on all surfaces along with reduced/ prostrate brown algae such as *Ectocarpus siliculosus* and *Streblonema* sp. respectively.
(2) The diatom composition studies revealed four prevalent species on these panels; the procumbent species *Amphora coffeaeformis* var. *perpusilla* and *Navicula corymbosa*, the erect, stalked species *Achnanthes parvula* and the polar attached *Synedra* sp.
(3) *N. corymbosa* was prevalent on all paints to varying degrees, whilst *Amphora coffeaeformis* was most widespread on the insoluble copper containing paints and less in evidence on the ablative paints, especially the TBTF only formulations (Table 3). *Achnanthes parvula* and *Synedra* sp. were 'common' or 'frequent' on the TBTF only panels (Table 4). Relative occurrence data showed that the former species was largely restricted to these coatings whilst an increase in the presence of *Synedra* sp. was recorded in the thicker biofilms on the ablative paints and insoluble coatings 4 and 5.
(4) A total of 74 diatom species or varieties were identified from these panels, distributed in 26 genera; sub-totals did not exceed 24 taxa. Most occurred in very low numbers.

Very little comparative work on microfouling film thickness on ships has been carried out although Gunn *et al.* (1987) published profiles of microfouling on paint samples from ships showing films ranging from 100 to 600 μm thick. In this respect it has been found that film thickness of only 90 μm can significantly affect drag (Loeb, 1981). Mean film thickness of the order recorded on these panels would, therefore, significantly influence skin friction resistance under operational conditions. RII values were necessarily low for thin biofilms but

TABLE 3
Relative percentage occurrence of the Main Fouling Species Identified from Bilge Keel Panels

Species	*Panel number and occurrence (%)*										
	1	*2*	*3*	*4*	*5*	*6*	*7*	*8*	*9*	*10*	*11*
Achnanthes parvula	0	4	0	64	4	100	96	98	8	2	2
Amphora coffeaeformis	100	100	100	100	92	72	94	90	94	70	100
Cocconeis pellucida	2	4	8	2	2	0	0	4	8	2	2
Navicula corymbosa	100	100	98	100	96	94	96	88	98	100	100
Synedra sp. (No1)	0	0	0	48	60	72	78	74	70	38	80
S. fasciculata	8	14	6	0	0	4	2	2	0	0	0
Genera											
Achnanthes	0	4	0	64	4	100	96	98	8	2	2
Amphora	100	100	100	100	92	72	94	90	94	70	100
Cocconeis	2	10	12	8	6	0	0	4	8	2	6
Licmophora	0	2	0	0	0	0	0	2	0	0	0
Navicula	100	100	98	100	96	94	96	88	98	100	100
Nitzschia	2	2	0	0	2	0	0	2	0	4	0
Synedra	8	14	6	48	60	72	78	74	70	38	80
Others	8	16	2	6	12	6	8	6	10	2	8

increased rapidly such that some biofilms of approximately 100 μm mean thickness (e.g. panel 2) exhibited RIIs similar to those for films of much greater mean thickness (e.g. panel 8). There was, however, a general increase and greater variation in the RII values recorded for thicker biofilms. The thinner microfouling films observed on ablative paints of high erosion rate despite a reduced toxin loading occurred as a result of better ablation of the softer coating ensuring more effective release of the biocide content and a less stable substratum for the attachment of fouling organisms.

TABLE 4
Diatom Composition of Microfouling Films

Species/panel	*1*	*2*	*3*	*4*	*5*	*6*	*7*	*8*	*9*	*10*	*11*
Achnanthes parvula						F	C	F			
Amphora coffeaeformis var. *perpusilla*	F	F	F	C	F	C	C				
Navicula corymbosa	F	F	F	F	A	C	C	C	A	A	F
Synedra sp.						C		C			
Taxa sub-totals	10	18	18	17	24	23	21	18	24	18	15
Totals: Genera = 26; Species = 70; Varieties = 10; Total taxa = 74											

Diatom films have been shown to act as a barrier to toxin release (Hendey, 1951) with members of the genus *Amphora* also shown to immobilise some toxins. It is likely that the presently described microfouling films exerted a similar effect on the action of these paints. Microfouling probably also acted as a barrier to effective ablation on panels 6–11 reducing toxin release rates to lower levels than for the insoluble paints which only required the maintenance of a toxin gradient to ensure leaching of soluble biocides.

The observed dominance of diatoms in these biofilms agrees with the work of Caron and Sieburth (1981) who found diatoms to constitute up to 99·9% of adhered organic material of microfouling films. Large numbers of diatom species have previously been identified from panels (Hendey, 1951) and in-service ships (Woods *et al.*, 1986). It is evident that a wide range of diatoms also colonised these surfaces. The majority were sporadic colonisers, including some pelagic forms, e.g. *Actinoptychus*, *Coscinodiscus* and *Thalassiosira* which were only present through chance entrapment in the biofilm. Similar pelagic diatoms have been recorded as constituents of microfouling films on ships (Woods *et al.*, 1986). Only the four main species appeared to extensively colonise these paints by cellular multiplication and colony formation.

The two major diatom species on these surfaces, *Amphora coffeaeformis* var. *perpusilla* and *Navicula corymbosa* were small, procumbent forms which produced copious quantities of unstructured extracellular mucilage. The former has frequently been observed on toxic surfaces (Woods *et al.*, 1986). The *N. corymbosa* has only recently been identified as an important fouling organism (Gunn *et al.*, 1987; Woods *et al.*, 1986); it constituted over 10% of the diatoms on all panels and over 50% on three. The prevalence of *Achnanthes parvula* on TBTF only paints and *A. coffeaeformis* on cuprous oxide containing paints agrees with previous work showing members of the genus *Achnanthes* to be more resistant to organotins than to cuprous oxide, whilst *Amphora* spp. were more resistant to Cu_2O (Callow, 1986). The *Synedra* sp. identified from these films exceeded 600 μm in length. The effect of fluid shear forces probably accounts for its reduced numbers in the thinner microfouling films.

CONCLUSIONS

(1) Stabilisation of ablative paints by microfouling films resulted in greater biofouling of these surfaces compared to insoluble paints.

(2) The use of CuSCN with TBTF in ablative formulations was less successful than with a higher TBTF content.
(3) Microfouling was reduced on the ablative paint with the faster erosion rate compared to the slower eroding paints with the same or higher toxin loading.
(4) Microfouling films of only 100 μm can cause marked increases in surface roughness along with a general increase in RII with greater mean film thickness.
(5) A large number of diatom species settle on antifouling paint surfaces although only a few actively and extensively colonise them.
(6) *A. coffeaeformis* var. *perpusilla* and *N. corymbosa* are the most successful colonisers of antifouling paints.
(7) Diatoms of different genera exhibit differential resistance to the major antifouling biocides. (i) *Amphora coffeaformis* is relatively resistant to cuprous oxide. (ii) *Achnanthes parvula* is more resistant to TBTF than copper based biocides.

ACKNOWLEDGEMENTS

We gratefully acknowledge the financial assistance of the Research Organisations of Ships Compositions Manufacturers (ROSCM) and the further practical assistance of W. and J. Leigh & Co. Ltd, particularly Dr O.P. Morris.

REFERENCES

Callow, M.E. (1986). In *Algal Biofouling*, Studies in Environmental Science 28, ed. L.V. Evans, & K.D. Hoagland. Elsevier, Amsterdam, pp. 1–20.
Caron, D.A. & Sieburth, J.M. (1981). *Appl. Environ. Microbiol.*, **41**, 268–73.
Clitheroe, S.B. (1983). *Australian O.C.C.A. Proc. & News*, **20** (1/2), 6–10.
Fischer, E.C., Castelli, V.J., Rodgers, S.D. & Bleke, H.R. (1984). In *Marine Biodeterioration: An Interdisciplinary Study*, ed. J.C. Costlow, & R.C. Tipper. Naval Institute Press, Maryland, pp. 265–299.
Furtado, S. & Fletcher, R.L. (1987). In *Preservatives in the Food, Pharmaceutical and Environmental Industries*, ed. R.G. Board, M.C. Allwood & J.G. Banks. Technical Series No. 22, Society for Applied Bacteriology.
Gillmer, T.C. (1970). *Modern Ship Design*. Naval Institute Press, Annapolis.
Gunn, N., Woods, D.C., Blunn, G., Fletcher, R.L. & Jones, E.B.G. (1987). In *Proc. Symp. on Problems in the Offshore Oil Industry*. Institute of Petroleum, London.

Hendey, N.P. (1951). *J.R. Microsc. Soc.*, **71** 1–86.
Loeb, G.I. (1981). Drag enhancement of microbial slime films on rotating discs. N.R.L. Rep. 4412.
Woods, D.C., Fletcher, R.L. & Jones, E.B.G. (1986). In *Proc. Ninth Int. Symp. on Living and Fossil Diatoms* (in press).

An extended bibliography for this paper may be obtained from the senior author.

Biofilms on Meat Processing Surfaces

A.H.L. Chamberlain & Surrinderjit Johal

Department of Microbiology, University of Surrey, Guildford, GU2 5XH, Surrey, UK

ABSTRACT

Physico-chemical surface parameters, wettability and net surface charge, were measured for Pseudomonas *cells and a range of surface materials used in the food industry. Data were obtained for surfaces and organisms before and after treatment with a solution of pork extract produced from fresh meat. Stainless steel (AISI 304) was hydrophilic and remained so after treatment, all the plastics surfaces were converted from hydrophobic to hydrophilic. Net surface charge on the stainless steel was markedly decreased by the meat juice but the plastics showed only minor changes.*

Examination of detachment of precolonised surfaces using a Fowler radial flow shear apparatus showed that all three surface materials performed in different ways both when untreated and treated. Possible reasons and the practical applications are briefly considered.

INTRODUCTION

Microbial biofilms are a natural result of contamination of any non-toxic, moist surface, particularly if there are large sources of nutrients and continued inocula. Thus almost all surfaces involved in food processing are likely to have very high biofilm-forming potentials unless the food materials have very low pH values. This will apply to pipework, tanks, machinery and flat exposed working surfaces.

The effects of colonisation of processing surfaces may result in contamination of food by organisms causing either food poisoning or food spoilage. In the latter case the spoilage may take the form of major product degeneration if well advanced, or at earlier stages may only result in organoleptically unacceptable products.

Many food products are handled at reduced temperatures in modern processing plants and this may also select which micro-organisms are most active; the psychrotrophs and psychrophiles being favoured. This is particularly true in the meat processing industries where carcasses are frequently spray-washed and rapidly chilled before transport or storage prior to butchering into joints.

Adhesion studies on bacteria associated with meat handling have normally concentrated upon attachment to the meat surfaces (Butler *et al.*, 1979; Notermans & Leemans, 1980). However, adhesion to the processing surfaces is also likely to be of considerable importance in cross-contamination with the freshly exposed meat surfaces being slid mechanically over the butchering surface.

This surface will inevitably carry some liquid derived from the chilled meats and it was the object of this paper to examine the effect of such a meat liquor on the attachment of one of the major meat spoilage organisms *Pseudomonas*. In this example a pork extract has been used.

MATERIALS AND METHODS

Organisms were initially isolated from commercial meat processing surfaces and maintained on tryptone soy agar. All experiments involving attachment were carried out with washed bacterial suspensions in 0·01 M phosphate buffer 6·4, at 20°C to preclude additional organic adsorption. Contact angle determinations were carried out essentially as described in Chamberlain & Van Woerkom (1986) but using a captive octane droplet advancing on surfaces immersed in 0·01 M phosphate buffer at pH 6·4. Microelectrophoretic mobilities were determined using a Rank Brothers (Bottisham) microelectrophoresis apparatus by kind permission of Dr D. Graham (BP Research, Sunbury-upon-Thames). Phosphate buffer was again the suspending medium.

Determinations of the radial flow shear forces required to prevent attachment were achieved by recirculating a phosphate buffer suspension of *Pseudomonas* through the apparatus at 2·5 litre/min for 4 h at 20°C to allow attachment and the uncolonised zones then detected by staining with crystal violet. Detachment experiments using the same apparatus were carried out for 2 h at 2·5 litre/min flow rate after precolonisation of test surfaces for 18 h at 20°C.

The pork liquor was produced by blending 20 g of lean meat in 100 ml distilled water for 1 min then filtering through a Whatman No. 1 paper. The liquid was applied as a 30 min pretreatment to sterile, chemically clean surfaces in all cases, then rinsed with a regulated stream of distilled water.

The test surfaces used were AISI 304 stainless steel and polyethylene (PE) (High Density Plastics Ltd) and polypropylene (PP) (Courtaulds Acetate Ltd) as supplied commercially to the food industry. These surfaces were cleaned in an ultrasonic bath in distilled water.

RESULTS AND DISCUSSION

Mean surface contact angles for clean and pork-liquor treated surfaces are shown in Table 1. The stainless steel appears hydrophilic due to the presence of a passivating film of mixed iron and chromium oxides. All the plastics were strongly hydrophobic but on treatment with the pork liquor all surfaces, plastics and stainless steel assumed a very similar hydrophilic surface value. A deposited film of *Pseudomonas* on glass gave a value of 30 ± 3·8°.

Table 2 shows the microelectrophoretic mobilities of finely divided substratum materials before and after coating with pork liquor. The effect on the values for stainless steel is very considerable but is barely significant with the plastics, although it is interesting to note that polyethylene values rise while those for polypropylene fall slightly bringing both surfaces to an almost common value. Values for untreated and treated *Pseudomonas* cell suspensions were −1·2 ± 0·02 and −0·9 ± 0·02 respectively.

The detachment of cells from a previously colonised surface using the

TABLE 1
Mean Surface Contact Angles Before and After Surface Treatment with Pork Liquor

Treatment	*Stainless steel*	*Polypropylene*		*Polyethylene*	
		A	*B*	*A*	*B*
Clean	35±3·0	127±4·1	139±3·9	96±3·0	137±3·3
Adsorbed pork liquor	30±3·0	39±3·0	36±3·1	36±4·0	36±3·1

All values in degrees (°), mean values from 60 determinations.

TABLE 2
Microelectrophoretic Mobilities of Clean and Pork-liquor-treated Processing Surfaces

Treatment	*Stainless steel*	*Polypropylene*	*Polyethylene*
Clean	−4·17±0·03	−1·12±0·08	−0·78±0·09
Adsorbed pork liquor	−0·48±0·03	−1·09±0·03	−0·98±0·01

All values × 10^{-8} m/s/v.

TABLE 3
Detachment from Precolonised Test Surfaces Using the Fowler Radial Flow Cell

Treatment	*Stainless steel*	*Polypropylene*	*Polyethylene*
Clean	25·8	9·6	> max
Pork liquor—bacteria	33·5	25·8	> max
Pork liquor—substratum	29·1	28·0	25·8
Pork liquor—both surfaces	7·8	25·8	> max

Values are surface shear stress N/m^2 required for removal.

radial shear flow apparatus gave results which varied depending upon which contacting surfaces were pork liquor treated (Table 3). Thus, in the cases with stainless steel, pretreatment of the bacteria alone caused a marked increase in the shear force necessary for detachment and a similar but slightly smaller effect was produced when the substratum surface alone was treated. However, if both surfaces were pretreated the strength of attachment was very much reduced. This is of practical importance because in carcass-breaking processes it will be inevitable that both organisms and surfaces are bathed in the meat juices and hence adhesion of *Pseudomonas* to a stainless steel substratum will be relatively poor. In the case of the plastics tested no clearance zone could be generated on the polyethylene except with 'substratum-only' pretreatment. The polypropylene surface initially required a low removal force but this was considerably increased to a similar set of values after any surface combination was pretreated.

CONCLUSION

The presence of pork liquor had a minimal effect on the hydrophilicity of the stainless steel but radically altered the microelectrophoretic mobility and detachment experiments showed a significant decrease in shear stress required for organism detachment. In contrast, meat juice pretreatment of both the plastics caused a major shift from hydrophobic to hydrophilic but had very little effect on the net surface charges and showed that, where both polypropylene and the bacteria were coated, removal of organisms from the surface required higher shear forces than for the cases where only one or neither surface was pretreated.

The results with polyethylene are more difficult to reconcile but the answer may be found in the differences in surface roughness of the two test materials. The average surface roughness value for polypropylene was 0·17 (±0·04) μm against the polyethylene which showed some striations, i.e. 0·25 (±0·11) μm along and 0·50 (±20) μm across the 'grain'. These values are too small to interfere with the contact angle determinations.

The results of the detachment experiments confirm the common observation that some bacteria attach more strongly to hydrophilic surfaces, although usually in lesser numbers, but that care must be exercised with predicting how organisms will interact if there are organic-containing fluids in contact with the organisms and/or substrata. In the food industry such surface preconditioning is probably universally inevitable yet is rarely considered. Recent work by Anderson *et al.* (1986) suggests that surfaces subjected to normal cleaning and sanitising procedures may still retain an adsorbed protein layer after cleaning. These aspects of microbial adhesion in the food industry are worthy of greater consideration.

REFERENCES

Anderson, M.E., Huff, H.E., Marshall, R.T. & Naumann, H.D. (1986). *Journal of Food Protection*, **49**, 342–6.

Butler, J.L., Stewart, J.C., Vanderzant, C., Carpenter, Z.L. & Smith, G.C. (1979). *Journal of Food Protection*, **42**, 401–6.

Chamberlain, A.H.L. & van Woerkom, R. (1986). In *Biodeterioration 6*, ed. S. Barry, D.R. Houghton, G.C. Llewellyn & C.E. O'Rear. CAB International, Slough, pp. 556–61.

Notermans, S. & Leemans, T. (1980). In *Microbial Adhesion to Surfaces*, ed. R.C.W. Berkeley, J.M. Lynch, J. Melling, P.R. Rutter & B. Vincent. Ellis Horwood, Chichester, pp. 524–6.

COMPUTERS

Biodeterioration: How Can Microcomputers Help?

T.N. Bryant

Medical Statistics and Computing, University of Southampton, South Academic Block, Southampton General Hospital, Southampton SO9 4XY, UK

ABSTRACT

The uses made of microcomputers in biodeterioration is presented, general and specific applications of microcomputers are outlined and possible sources of help are identified.

INTRODUCTION

The microcomputer is rapidly becoming a piece of standard equipment in the laboratories and offices of many scientists. Advances in computer technology are so rapid that the power of a modern desk top computer is equal to that of the large mainframe computers produced 10 years ago. The development of computer software and the attention to the interaction between the user and the computer is one of the reasons why microcomputers have been accepted so readily. Large, mainframe computers are perceived as being difficult to use, and in some cases they are, however the configuration of a microcomputer makes it far more suitable for certain applications.

Many people start to use a microcomputer in a general way, for example they may use it for word processing or as a database for the storage and retrieval of information. Others may purchase a piece of laboratory equipment in which a microcomputer is an integral part and is used to capture and subsequently analyse experimental data, or the microcomputer may be used as a control device which samples data and changes the experimental conditions according to specified limits.

DISCUSSION

What is a microcomputer?

The development of microcomputing dates from 1948 when Bell Telephone Laboratories developed the transistor; this device occupied 1 cm^2. By 1985 the same area could hold 1 million transistors and predictions are that by the year 2000 this will rise to 1 billion (Meindl *et al.*, 1981). The price, form and performance of integrated circuits has changed dramatically since their discovery by Texas Instruments in 1959. It has been said that 'if the automobile industry had achieved what the computer industry has achieved in the last 30 years then one could buy a Rolls-Royce motor car for three pounds and it would do 2 500 000 miles per gallon. Unfortunately it would be less than the size of a postage stamp. Some of the historical landmarks in microcomputing are listed in Table 1.

The essential element of any computer is the Central Processing Unit. This carries out two basic functions, it acts as a calculator and as a store of information. In a microcomputer the CPU is comprised of a microprocessor and memory. The intelligence and memory of the CPU alone is rather like the human brain without the rest of the human body; to be a useful machine input and output devices are required. The memory in most computer systems is volatile since once the power is switched off the contents of the memory are lost. Permanent storage of information is typically provided by some form of magnetic media although current developments could see optical devices taking over in the long term because optical storage media are inherently more reliable. A schematic diagram of the elements of a microcomputer is presented in

TABLE 1
Historical Landmarks in Microcomputing

1948	First transistor — Bell Telephone Laboratories
1959	First integrated circuit — Texas Instruments
1971	First microprocessor 4004 — Intel Corporation
1975	First microcomputer kit for $400 — Altair
1976	CP/M operating system and Microsoft Basic
1977	Apple II microcomputer
1979	WordStar word processing package
1981	IBM PC, MS DOS and Lotus 1-2-3
1982	BBC microcomputer — Acorn Computers
1987	IBM PS/2
198?	An operating system for Intel 80286 based systems

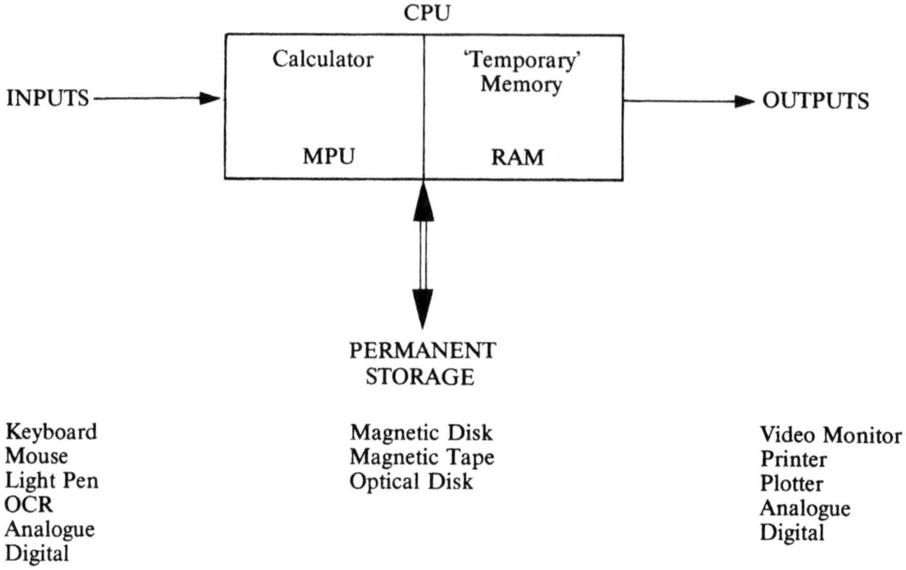

Fig. 1. A simple representation of a microcomputer.

Fig. 1, and a full discussion on the components of a computer system is given by London (1986). Two components go to make a usable computer system. Hardware, which consists of the physical elements such as integrated circuits and a power supply, and software, which is the term for the programmed instructions that tell the computer what and when to do things.

Why use a microcomputer?

Some people see microcomputers as the solution to all of their problems. However one needs to know what the problem is and how to solve it before a computer solution is introduced. Essentially it just speeds things up. However modern computers do this to such an extent that a human would never consider tackling some problems without one. A human has been said to be an error prone genius whilst a computer is a fast pedantic moron which, assuming correct programming and no mechanical breakdown, never makes a mistake.

Mainframe and minicomputers have been available since the 1960s so why the sudden interest in microcomputers? One major reason is cost. A system, with the power of a mainframe computer of 10 years ago, can be purchased for the same price as a television set and video recorder. Other reasons include ease of access; one can position a microcomputer where you want it. When using a larger computer one has to go to a terminal and although it is possible to use telephone dial up facilities these usually work at slower communication speeds. Ease of use is an important consideration in the design of microcomputers whereas larger systems are designed for use by trained personnel who would customise the environment for their colleagues. In theory large computer systems can be made easy to use; however, in practice this may not be done. The reliability factor is often introduced as a reason for the success of microcomputers. This is a 'red herring', since large modern computers are as reliable as microcomputers. It is the older systems that are less reliable and such systems often require routine maintenance which means they may be unavailable for a set period each week. If a large system breaks down many users are inconvenienced, but with microcomputers if one machine breaks down only one individual cannot work and even he may be able to borrow a machine from a colleague.

Microcomputers and larger computers are tools for different applications. For example, large computers are not technically suited for word processing since the way in which a Visual Display Unit (VDU) is usually connected to one cannot provide the speed at which information needs to be presented on the display screen for this task. Large computers deal with many users apparently simultaneously although in fact each user has access to the CPU for only a fraction of a second, before the next user is given priority. To a human working much more slowly than the computer it appears that they are the only user because it responds very quickly to each command. It would be impossible to use such a system to collect data from and control time dependent experiments because there is no guarantee that a variable would be recorded from a specific instrument at a precise time even if the interfacing problems could be sorted out. Low cost microcomputers dedicated to monitoring single pieces of laboratory equipment provide a realistic solution. On the other hand large computers provide a good solution to the storage of large volumes of information that many people may want to access simultaneously such as a database describing microbial cultures held by several organisations.

General applications of microcomputers in biodeterioration

There is a common core of applications that everyone who has access to a microcomputer can use. These are not restricted to scientists alone.

(i) Word processing

Many authors submit papers prepared using word processing software and the publishers of scientific journals and books are now accepting manuscripts on computer floppy discs. A development from this is Desk Top Publishing where graphical features are presented together with text. Some commercial publishers are now venturing into this area, investigating whether they can reduce their publication costs. An important feature to look for in these packages is how well the computer screen can display what the printed output will look like. Many packages include a spelling checker but when first used on a scientific manuscript these can be tedious because the dictionary is unlikely to contain technical words. However words can be added and so with usage this problem disappears.

(ii) Databases

A database is a large organised store of information which is available when you need it (Sharman, 1987). Database software ranges from list processing packages which mimic the 3 by 5 card index to true relational databases which seek to keep the amount of redundant information to a minimum. Applications are varied, one could keep track of a culture collection of bacterial isolates (Bryant, 1983) for example and specialised scientific databases are now being published. One example is in the area of genetic engineering where nucleic acid sequence information is now available. The amount of information held currently requires a large mainframe computer, but with the development of higher capacity storage devices for microcomputers this situation will change. Bibliographic database software is now available to store literature references. The quality of the packages available and the facilities they offer can vary. Most packages treat the book and journal entries differently, some limit the user in the number of authors or editors, keywords and the maximum number of characters for each of these as well as for the book or journal name. An example of the information one can enter is shown in Fig. 2. The more powerful packages will inspect word processing files and generate an automatic bibliography according to the style of the journal which is specified by the user. Another useful feature present in most bibliographic software is the ability to download citations from international databases such as MEDLINE or BIOSIS. Some reviews of some bibliographic databases appear in *Software Science Quarterly*, Volume I, Number 4, 1984 and Binary Issue 8, 1986.

(iii) Spreadsheets

One of the reasons for the success of microcomputers in the business community are spreadsheet programs. A spreadsheet is an electronic worksheet which provides a way of storing numbers or information in

Editing Journal Reference

RefID# : 7
Authors : Bryant, T. N.
Title : A microcomputer-based information storage and retrieval system for the maintenance of records from a culture collection
Keywords: Microcomputer, database, bacteriology, culture collection, DataStar
Journal : Journal of Applied Bacteriology
Pub Year : 1983
Volume : 54 : 191–107
Reprint : In File
Notes : A microcomputer-based system for the storage and retrieval of information on strains in a culture collection is described. The system was designed around commercially available software written for microcomputers.

Journal Edit Menu

Author Edit Title Edit Keyword Edit Misc. Edit Reprint Edit
Notes Edit Save Exit Quit Not save Go Main Menu Help
<Left> <Right>

[A] to edit this reference's authors

Fig. 2. A screen display of a reference entry in the bibliographic database 'Reference Manager'.

rows and columns. The intersection of a row and column specifies a unique location called a 'cell'. Data may be entered into a cell or a cell may offer information, some text for annotating the worksheet or a formula for calculating what value should be placed in that cell. Originally spreadsheet packages were promoted as enabling 'What – If' investigations to be performed on data, but they have been developed to include data management and graphical display facilities. They can also be used for statistical calculations (Kibby, 1986). Laboratory equipment manufacturers are now producing software that will capture data from chromatographic equipment, liquid scintillation counters or spectrophotometers and store it in a form that can be analysed using a spreadsheet package.

(iv) Graphics

This type of software can be classified into two categories. First, there is software available for producing graphs and charts from data files created by word processors, databases, spreadsheets and other programs. Secondly there are painting programs where the computer screen is the canvas and a pointing device, typically a mouse, is the brush. Subroutine libraries are also available for those who write their own computer programs, these routines relieve the programmer of the task of developing procedures for carrying out common activities such as drawing an axis with tick marks and a suitable scale. Output from graphical programs may be sent to a variety of devices, the computer screen, the printer, or plotter. There are also devices available for producing 35 mm colour transparency slides from screen images.

(v) Workstations

Microcomputers can be linked together to form a Local Area Network where they can share peripherals for producing printed and plotted output. Software and data files can also be shared and this can reduce the management problem of a microcomputer installation. If one introduces a revision to a program on a networked system only one copy of the software needs changing whereas on several independent systems each copy would need modifying. Users on a networked system can pass messages to each other using electronic mail. Microcomputers can also be connected to larger computers using Wide Area Networks so that they can act as intelligent terminals and provide electronic mail and transfer of data from the desk top to another computer located on the other side of the world.

Scientific applications of microcomputers in biodeterioration

There are numerous applications of microcomputers described in the papers of this symposium. They can be classified as follows.

(i) Data logging

The computer may be used to replace the X–Y plotter in the capture data from equipment, or it may be used to digitise an image recorded using a video camera.

(ii) Experimental control

The computer is programmed to collect data from an experiment, compare it with a set of predefined values and change the conditions of the experiment accordingly. This is achieved by sending signals to selected pumps, valves and switches.

(a) To enter results use the Function keys press:
F1 for NEGATIVE, F2 for POSITIVE & F3 for test NOT DONE

[Press F10 to Re-enter results from test 1 onwards]

1	Motility	+
2	Swarming	+
3	Luminescence	−
4	Growth at 4C	−
5	Growth in 0%NaCl	−
6	Moller's Arginin	−
7	Moller's Lysinin	+
8	Moller's Ornithi	+
9	Nitrate Reductio	+
10	Oxidase	+
11	Gluconate oxidat	−
12	Glucose Gas	−
13	Indole	+
14	ONPG	−
15	Voges Proskauer	−

Have these results been entered correctly ? Y

(b) Identification Level Not Reached

Strain	I.D. Score	Sum I.D. Score	Estimated Frequency of Occurrence
1 *V. parahaemolytic*	0·8934	0·8934	1 / 2.
2 *V. alginolyticus*	0·0756	0·9691	1 / 20.
3 *V. harveyii*	0·0306	0·9997	1 / 48.

Differences in test results between
the unknown and the more probable strains

Strain	Test	Your Result	Matrix Entry
V. alginolyticus	Voges Proskauer	—	89

Fig. 3(a) A screen display of the entry of test results from an unknown isolate into a computer-assisted bacterial identification program. (b) A screen display of the suggested identities of an unknown isolate.

(iii) Data analysis packages

Some packages have general application, such as statistical software, whereas other software, for the identification of unknown microbial isolates, is more specific to the subject area. Figures 3(a) and (b) show the output from a computer-assisted identification program developed for the IBM PC by the author.

Where to get help on microcomputers

One problem that many new microcomputer users find is that they need help. General applications are less of a problem as these are covered in the popular computing magazines, even so these publications seldom cover scientific word processing or review the bibliographic database software. Help may be found from various sources, institutional computer centres, user groups may exist within scientific societies, public user groups sometimes based on a range of microcomputers and scientific journals specialising in computer application are being published. Table 2 lists useful sources of information. Computer user groups are a source of useful software, which has been placed in the public domain. They often keep a library of such software. This software is of two types, 'Freeware' and 'Shareware' the latter may also be called 'User Supported' software. Freeware is completely free to use or copy, all that the author asks is that you acknowledge his programming efforts. Authors producing shareware maintain the copyright and encourage copying and sharing of their program. However, if the user finds the software useful then he is expected to send a contribution to the author, for which he then receives more extensive documentation, updates and in some cases telephone support. One shareware word processing package, PC-Write, was ranked in the top ten of all word processing packages. The author of PC-Write sends a commission to the original registered user each time a new user registers.

What does the future hold?

At the time of writing (1987) a major landmark in microcomputing has occurred. IBM produced the PS/2 models, which signals a change of direction. The original IBM PC and its compatibles are limited by design considerations to 640 kbytes of memory. This in recent years has been a problem for some applications. This memory barrier will be removed with the announcement of the OS/2 operating system. Microcomputers with the Intel 80286 microprocessor will be able to use 16 Megabytes of memory; programs which until now required the memory capacity of mainframe computers will begin to appear on desk top machines. The main alternative to the IBM PC–MS/DOS approach is offered by the Apple Macintosh. Its acceptance was hindered by an initial lack of suitable software but provides the basis for a good desk top publishing system. Image analysis requires fast computations which can be carried out in parallel, the development of the Transputer by Inmos may well take this subject into a new era.

TABLE 2
Sources of Help for Microcomputers

Scientific Societies		
Society for General Microbiology computer club Contact: Dr. B. J. Evans Department of Microbiology University College Newport Road, Cardiff CF2 1TA, UK Publication 'BINARY'	American Society of Microbiology microbiology computer users group Contact: c/o George Buck Clinical Laboratory NKC Hospitals P.O. BOX 35070 Louisville, KY 40232 USA Publication 'MCUG Newsletter'	
Both groups maintain a membership list and a program catalogue of microbiological programs		
Public microcomputer user groups		
IBM PC User Group, PO Box 830, London SE1 0DB, UK	COMPULINK User Group, 67 Woodbridge Road, Guildford, Surrey, GU1 4RD, UK	PD Software Library, Winscombe House, Beacon Road, Crowborough, Sussex TN16 1UL, UK
PC-Sig, 1030D East Duane, Avenue, Sunnyvale, California, USA	PC USER GROUPS see US phone book	
Journals covering computer applications		
Science Software Quarterly Dr Diana J. Gabaldon (Editor) Arizona State University, Center for Environmental Studies, Tempe, Arizona AZ85287, USA	Computer Applications in the Biosciences Dr R. J. Beynon (Editor) Department of Biochemistry, University of Liverpool, PO Box 147, Liverpool L69 3BX, UK	
BYTE — Europe McGraw-Hill Information Systems, McGraw-Hill House, Maidenhead, Berkshire UK	BYTE — USA One Phoenix Mill Lane, Peterborough, New Hampshire NH 03458-9990, USA	

It is clear that microcomputers are starting to make an impact in the study of biodeterioration, we can expect that in the future their impact will be even greater.

REFERENCES

Bryant, T.N. (1983). *Journal of Applied Bacteriology*, **54**, 101–7.

Kibby, M.R. (1986). *Computer Applications in the Biosciences*, **2**, 151–7.

London, K. (1986). *Introduction to Computers*, 5th edn. Faber and Faber, London.

Meindl, J.D., Ratnakumar, K.N., Gerzberg, L. & Saraswat, K.C. (1981). Digest of Technical Papers, *IEEE Solid-State Circuits Conference*, IEEE, New York, pp. 36–7.

Sharman, G. (1987). *Introduction to Database on Microcomputers*. Addison-Wesley, Wokingham.

Computer-assisted Taxonomy of Filamentous Microfungi

P.D. Bridge

CAB International Mycological Institute, Ferry Lane, Kew, Surrey TW9 3AF, UK

ABSTRACT

Numerical taxonomy has been widely used in microbiology to aid in the classification and characterisation of bacteria and yeasts, and computer-assisted identification of such organisms has become routine. Numerical taxonomic procedures may also be applicable to the study of filamentous microfungi, particularly in areas where the existing taxonomy is not clear. Information gained from taxonomic studies can enhance culture collection databases and can be useful in strain typing. Some numerical taxonomy methods are briefly reviewed and illustrated with data from the genus Penicillium.

INTRODUCTION

The taxonomy of filamentous microfungi encountered in applied microbiology is often difficult, however use of the correct name is very important as in depositing the organism in a culture collection, citing it as a source of a biochemical product, or in patent applications. The name of an isolate may imply further characteristics of importance, examples being mycotoxin production or resistance to particular biocides. The taxonomy of some of the commonly encountered microfungi is not always clear, as in *Penicillium* and *Fusarium* (Onions *et al.*, 1984; Samson & Gams, 1984), making identification difficult; computer-assisted numerical taxonomic studies can help clarify the situation.

GROUPING AND CLASSIFICATION

In order to produce a numerical classification, the first step is to arrive at an estimate of resemblance between organisms. Resemblance is most commonly represented as either a similarity or dissimilarity. There are many coefficients that may be used for this, but the simplest and probably most used, is the Simple Matching Coefficient. The calculation of this coefficient is simply the fraction of common results for two organisms (Sneath & Sokal, 1973).

When considering a group of organisms one must calculate the resemblance between each possible pair of organisms. The set of resemblance values are usually represented as a triangular matrix. Once the resemblances between isolates have been calculated they need to be represented in a form that shows any areas of high resemblance that may correspond with natural groups. One way of doing this is to order the isolates in the matrix so that areas of high similarity appear together. This poses problems with size and subsequent reproduction, and methods that show the best or average resemblances, such as clustering, are usually used. Clustering involves sequentially considering resemblance levels until all the organisms or all the values have been used. This is normally represented as a tree diagram, often referred to as a dendrogram or a phenogram, where the horizontal scale is resemblance and the vertical links mark the average resemblance between organisms (Sneath & Sokal, 1973).

A further aspect is the choice of coefficient for calculating resemblance. There are many different coefficients that may be used and these often consider different aspects of the data. As a result, one coefficient may be better suited to a particular set of data than another. Coefficients used include Jacquard's coefficient that discounts matching negative results (Sneath & Sokal, 1973) and the Pattern coefficient, which is able to minimise differences due to the vigour and growth rates of isolates (Sackin, 1981). It is desirable to obtain more than one representation of the data in order to increase confidence in any interpretation. Figure 1(a) shows the dendrograms obtained with 4 different coefficients from 81 test results for 12 strains of *Penicillium*. The salient point here is that with one coefficient (S_G) *P. aurantiogriseum* and *P. viridicatum* seem to form a group on their own, whereas with the others *P. aurantiogriseum* and *P. echinulatum* seem to form a group on their own. In reality these three species probably constitute one group.

The final picture, which may show distinct groups, is a convenient representation of the data. It is a clustering technique trying to show resemblances that will give the best groups using all of the organisms. As

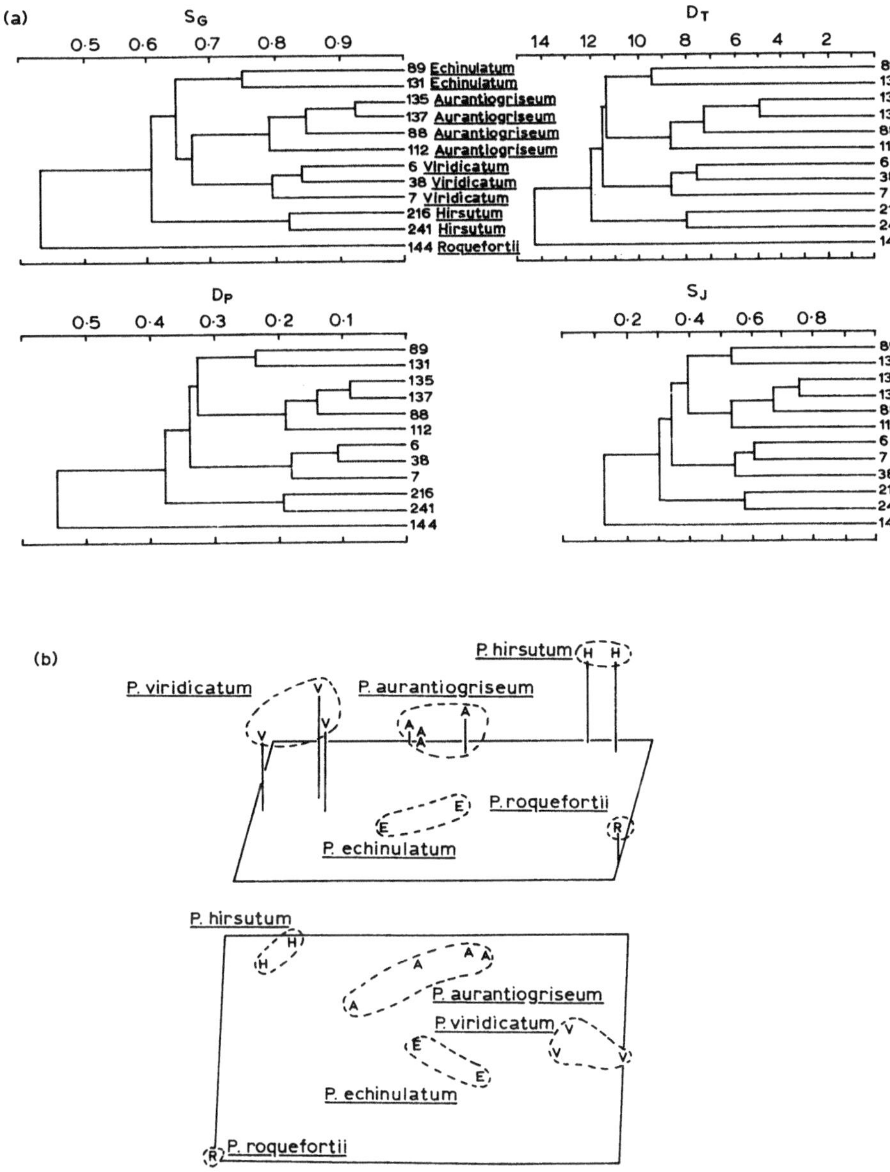

Fig. 1. (a) UPGMA clustering of 12 strains of *Penicillium* using four coefficients and 81 characters; (b) principal components plots showing two- and three-dimensional ordinations for the same data.

a result it is important to know to what extent the overall data has been distorted in drawing the dendrogram. One way to do this is to compare the final 'average' resemblances with the original ones by calculating a correlation coefficient to get an idea of whether there may be significant differences. This correlation gives a measure of confidence of any groups seen in the dendrogram (Jones & Sackin, 1980). Another way of assessing clustering is to measure the spread of each group and determine the likelihood of groups overlapping with each other (Sneath, 1977).

A slightly different approach involves ordination methods. These are non-hierarchic and based on the ordinations of the organisms in multi-dimensional A-space (see Sneath & Sokal, 1973). One can extract from the multidimensional system two- and three-dimensional ordinations that give the best practical representation. This is performed by procedures such as principal component or principal coordinate analysis. Each axis in this case is made up of linear combinations of the original characters that summarise the major variation between the isolates. Figure 1(b) shows two- and three-dimensional ordinations for the 12 penicillia illustrated earlier and the same 5 groups are visible.

IDENTIFICATION

A further aspect of numerical methods is computer-assisted identification, where data held in a computer may be used to construct identification keys and tables. The general procedure is to extract from the data the results that best characterise the taxa, which will be the statistically or practically best tests (Willcox *et al.*, 1980). These can then be used in a number of ways, as in polyclaves, where there may be added features such as the selection of the most useful tests, or identification matrices where resuilts are given as probabilities or likelihoods of identification (e.g. Feltham & Sneath, 1982; Willcox *et al.*, 1980). One advantage of this approach is that spurious identifications should be minimised.

CONCLUSIONS

Numerical procedures have become more widely available with the increase in sophistication of microcomputers. Commercial software such as CLUSTAN is now available for PC compatible computers as is sophisticated pattern recognition software such as SIMCA.

The use of numerical taxonomy procedures in future studies of microfungi may aid accurate taxonomic delineation of newly isolated strains. The data generated in such studies will enhance existing microfungi collections and databases and so lead to better characterised reference cultures, and the possibility of accurate strain typing.

ACKNOWLEDGEMENTS

This work was carried out in conjunction with Science and Engineering Research Council contract SO/17/84 'Systematics of microfungi of industrial and biotechnological importance'.

REFERENCES

Feltham, R.K.A. & Sneath, P.H.A. (1982). *Journal of General Microbiology*, **128**, 1815–30.

Jones, D. & Sackin, M.J. (1980). In *Microbiological Classification and Identification*, ed. M. Goodfellow & R.G. Board. Academic Press, London.

Onions, A.H.S., Bridge, P.D. & Paterson, R.R.M. (1984). *Microbiological Sciences*, **1**, 185–9.

Sackin, M.J. (1981). *Journal of General Microbiology*, **122**, 247–54.

Samson, R.A. & Gams, W. (1984). *Antonie van Leeuwenhoek*, **50**, 815–24.

Sneath, P.H.A. (1977). *Classification Society Bulletin*, **4**, 2–14.

Sneath, P.H.A. & Sokal, R.R. (1973). *Numerical Taxonomy*. W.H. Freeman, San Francisco.

Willcox, W.R., Lapage, S.P., Bascomb, S. & Curtis, M.A. (1973). *Journal of General Microbiology*, **77**, 317–30.

Willcox, W.R., Lapage, S.P. & Holmes, B. (1980). *Antonie van Leeuwenhoek*, **46**, 233–99.

CONCRETE AND STONE

Microbial Interactions with Mineral Materials

Wolfgang E. Krumbein

Geomicrobiology Division, University of Oldenburg, P.O. Box 2503, D-2900 Oldenburg, FRG

ABSTRACT

Microbes in the form of epilithic (on the surface), chasmolithic (in cleavages and fissures) and endolithic (within the material or interstitial space of the material) complex microbial communities interact in many different ways with mineral material and its external environment. Microbial films and microbial mats have been found on and within rock and concrete with dominantly phototrophic, chemoorganotrophic, chemolithoautotrophic as well as aerobic and anaerobic communities. Microbial systems may dominantly be embedded in mucous extracellular substances, may be attached to surfaces via fimbriae and fibrils, may be motile or sessile, and may interact in a biodeteriorating and in some cases also protective way with the material and its environment. Some examples are given of the catalytic effects of microbial systems on physico-chemical reactions between rock, concrete and the environment including detrimental chemical radicals in the environment. The latter may be biogenic or anthropogenic. The fractal physics of the mesoscopic system are discussed in relation to the microbial system and the interior surfaces of rock and concrete. In this respect diffusion-limited aggregation and pitting processes are discussed. The 'weathering' or 'transformation' processes and morphologies resulting are discussed in the light of mainly two questions: (1) Is it possible to define quantitative ratios of 'biodeterioration' versus physico-chemical deterioration? and (2) is it possible to define typical biogenic deterioration morphologies? The conclusion is that on increasingly larger (macroscopic) scales the answer is negative since the whole planet is a biologically controlled system and can only be regarded and analysed as a 'bioid' or 'bioplanet'. The individual physical, chemical and biological compartments already at mesoscopic levels are too interactive in many cases to give percentages or balanced calculations of mere physical or chemical reactions, especially because life itself is little more than accelerated or retarded physical and chemical reactions.

INTRODUCTION

Microbial interactions with materials is becoming an increasingly important subject of research and technology. One could even speak of the biotechnology or bioengineering of materials. Bioengineering of materials can also be defined as biophysics of solids. Surfaces and interior surfaces will eventually be colonized by microorganisms as long as a connection to the outside (permeability combined with porosity) and a minimum of humidity are given. The colonization will never be without influences on the physical and chemical properties of the material in question. Usually the colonization will be of utmost relevance. Figures 1–4 give an impression of the dimensions of microbial mats on and in rocks. The systems under magnification are similar to tropical rain-forests or jungles. The communities are as layered and as ecologically zonated as any gradient community in stagnant lakes or in microbial mat systems of sedimentary environments (Krumbein, 1983; Gerdes *et al.*, 1985). The rock community, however, is different and according to our knowledge, typical for its own environment (Giacobini, 1985; Jaton *et al.*, 1985; Lewis *et al.*, 1985).

In the interaction between rocks and their environment we are used to talking about weathering or erosion, terms with a physical-chemical connotation. In the recent past, however, we have also become accustomed to looking at mechanical, physical and chemical processes of material in biological terms, including the surface and the interior of our planet.

In several recent communications (Krumbein & Dyer, 1985; Krumbein 1986*a*, *b*, 1987*a*, *b*; Krumbein & Schönborn-Krumbein, 1987*a*, *b*) it is hinted that the biotransfer of materials on or near the surface of the earth is regulated directly or indirectly through biological (and mainly microbiological) pathways. Increasing evidence has been accumulated to show that the temperature and temperature changes of the surface are modified and reduced from theoretical extremes by biological activity, creating a 'greenhouse' effect in the atmosphere. It was also found that

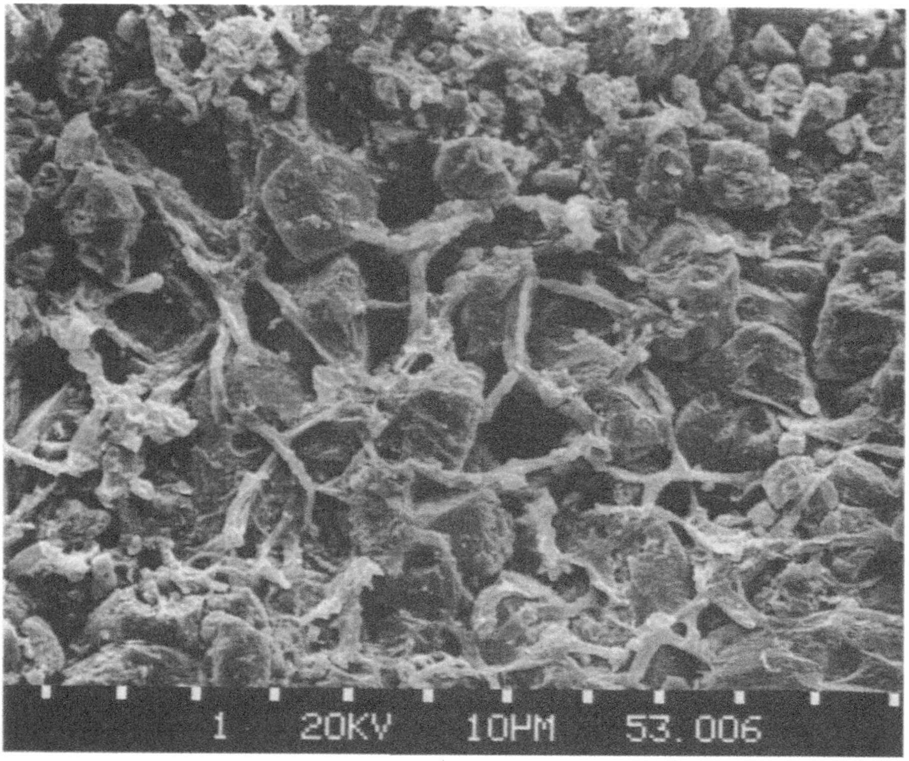

Fig. 1. Fungal 'rootwork', partially bioeroding and partially biostabilizing a porous upper Triassic (Keuper) sandstone.

the main reason for plant and animal populations of certain densities lies in the biotransfer of minerals into soluble nutrients in the pedosphere. Further, huge amounts of energy have been and are continually being stored and released by the biotransfer of sunlight and associated biological mineral cycles. Thus we regard the earth today as a bioplanet or a self-regulating, living entity of gigantic dimensions acting not unlike a laser. Life produces and maintains its own physical and chemical environment, favouring especially the maintenance and expansion of life. It is a self-enhancing process. Coherent self-enhancement at a sufficient pumping rate (sun light) and appropriate construction (geochemical reservoirs such as petroleum, gas, limestone, iron sulphide, phosphates, etc.) brings about a buffered biodynamic system in which the speed of transfer of materials (including minerals, rocks and their transformation into particles, solutes or gases) is biologically controlled. Weathering or erosion thus make part of an ingenious life support system and must be regarded as indirect or direct biotransfer. In other words the beneficial effect of the destruction of minerals or rocks is obvious when we look at the enormous amounts of fertilizer applied to soils in which the biotransfer of minerals (biomineralization and biomineralysis) seems to be insufficient to the exploiter (farmer). The principle, however, is a natural one and is also needed under natural conditions for the maintenance of the bioplanet's living activities.

In some natural environments therefore, the destruction of minerals and rocks is a necessary and positive process. Yet the very moment that a rock or mineral is used as a building stone, an ingredient of concrete, plaster, mortar, cement, a mural painting or any other cultural or technical object, it is considered that it should last forever, even if it is not exactly 'a thing of beauty'. In the eyes of our modern civilization its 'loveliness' even 'increases' with age and the patina of age. On the other hand patina is nothing else but the first steps of destruction and disappearance into nothingness.

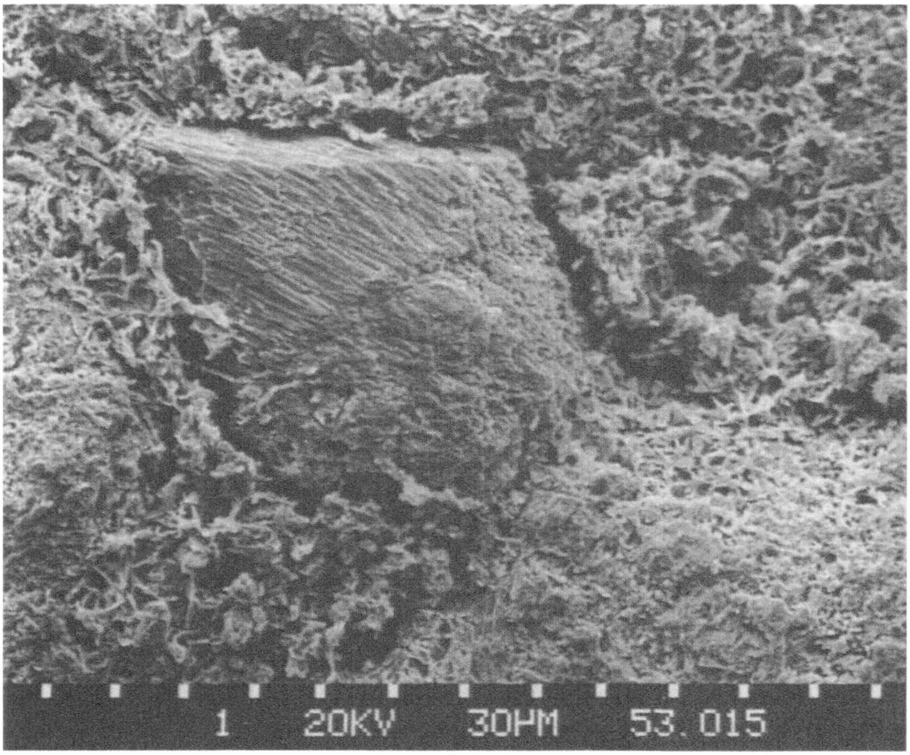

Fig. 2. 'Jungle' of fungal mycelia underneath an epilithic crustose lichen, exhibiting partially biocorroded sparry limestone.

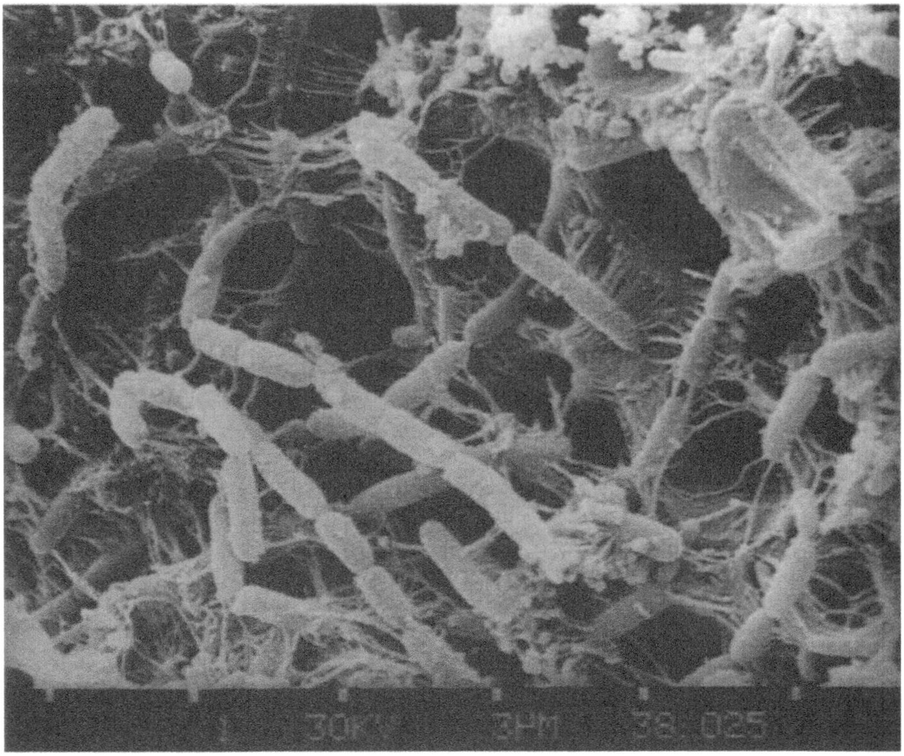

Fig. 3. Microbial 'jungle' on Keuper sandstone as in Fig. 1. In this section filamentous bacteria with strongholds and networks predominate.

Therefore the lesson to be learnt is to understand the biological processes which rule geophysiology. The term geophysiology is a new one and shall be appropriately defined: Physiology is the science of the normal functions and phenomena of living things (Greek: the study and description of natural bodies). Geophysiology is the science of the functions and phenomena of earth as a bioplanet (or 'bioid' or living body). In geophysiology, weathering is a necessary and a positive process. In the socio-cultural context of conservation and restoration it is a negative or destructive process. Therefore the term biotransfer or 'bioenhancement' of mineral cycles is often replaced by biodeterioration (i.e. a biological process that renders a thing worse than it was before).

Thus most contributions on the biodeterioration of concrete and stone focus on the following problems:

(1) What is rock and mineral deterioration?
(2) What is rock and mineral biodeterioration?

(3) Who or what produces biodeterioration?
(4) How can we measure biodeterioration?
(5) Does microbial growth on buildings always lead to biodeterioration?
(6) How can one differentiate between physical-chemical deterioration and biodeterioration?
(7) How can biodeterioration be stopped?
(8) How to improve materials through biodeterioration to enable bioenhancement
(9) Are air pollution and biodeterioration interconnected?
(10) Isn't air pollution itself in part a result of microbial biotransfer (and thus in part bioenhancement of the atmospheric buffer system)?

Answers to such questions have been sought for a long time. In recent communications the authors of the bible and Titus Lucretius Carus have been quoted. Both sources unanimously state that the biological attack

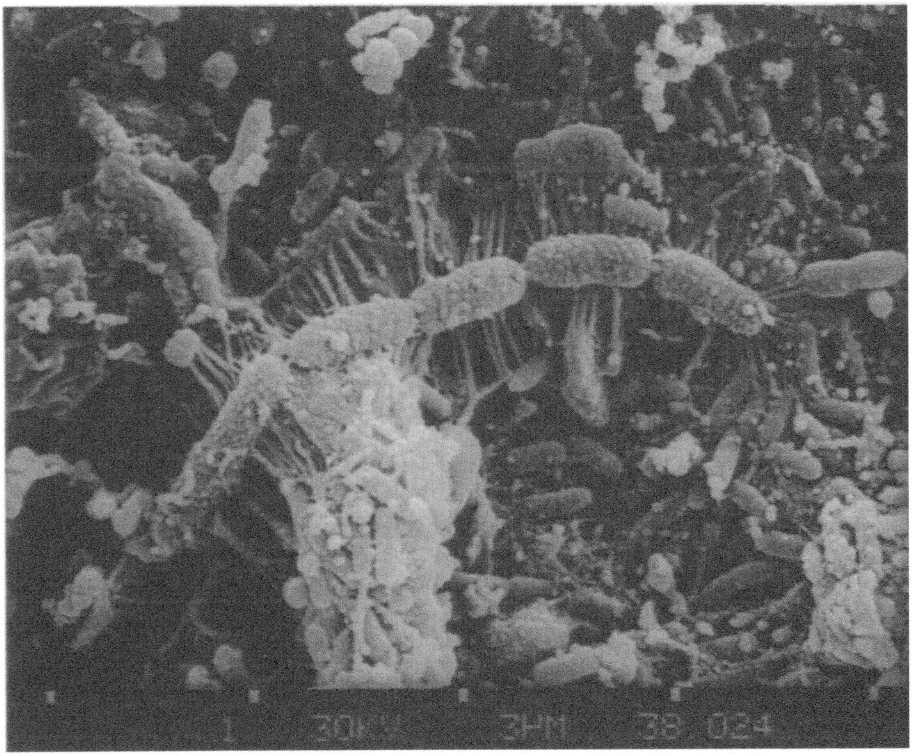

Fig. 4. Microbial 'jungle' on Keuper sandstone exhibiting partially motile species with unidirectional strongholds.

of rock is comparable to an infectious disease, often of an epidemic character. Although the first more or less detailed reviews of the influence of microorganisms on biotransfer and biodeterioration of mineral materials used in buildings and in art objects stem from almost 100 years ago (see reviews of Krumbein, 1972, 1973; Strzelczyk, 1981; Krumbein, 1987*c*) the scientists actively studying biodeterioration of mineral materials are still a minority compared to geologists, mineralogists, chemists dealing with the same problem of the decay of building stones and materials. This is clearly visible in the percentage of papers contributed to, for example, the last International Congress on the destruction of building stones in Lausanne.

It is expected that in the near future this situation may change, due to several very important findings which have now shown that biology may not only be a major independent reason for accelerated decay of building materials but also that interaction between air pollution and microorganisms tremendously enhances microbial populations and biodeterioration. Krumbein (1966, 1968), Lefevre *et al.* (1964) and Pochon *et al.* (1964) have already indicated the effect of air pollutants on the microflora active in biodeterioration. Recently, other workers have started to elaborate other pathways of interactive damage between air pollution and rock destruction (Bock, 1987; Krumbein & Petersen, 1987; Krumbein *et al.*, 1987).

Microorganisms are the most numerous organisms on Earth, reaching numbers of 1×10^{25}. They, however, are usually invisible and therefore seem often to escape attention (see Figs 1–4). To make things worse systems are dealt with that are badly defined physically and chemically. In a modern branch of physics—the physics of dynamic systems or fractal physics—it is increasingly the custom to separate the scales of view and of treatment of sections of the Earth in a way different from biologists and geologists. This 'modern' scale can be described as follows.

> 'In contrast to microscopic systems which can be described exactly by quantum mechanics on atomic scales of length and to macroscopic systems which can be described by everyday dimensions (foot, fathom, stone, etc.) we have no distinguished length for the mesoscopic system in between.'

Mesoscopic systems, however, are the most important ones in petrography, microbiology, fouling, biocorrosion waste water treatment, petroleum recovery and many other important fields in which geomicrobiologists are actively working. The mesoscopic system dealt with in this paper is microbial biodeterioration of rocks which sometimes reach more than 3 m^2 of internal surface and more than 1 million organisms/cm^3.

DEFINITIONS

The destruction of rocks and minerals in the above context and with reference to the problem of patina as a stabilizing change in the surface areas of some materials should no longer be defined as 'weathering' or 'erosion' or 'corrosion' alone. The connotation bioweathering, bio-corrosion, biodegradation, bioerosion, etc., is necessary in most cases. A general definition of rock decay was given by Krumbein (1966) and recently modernized and modified (Krumbein & Schönborn-Krumbein, 1987*a*, *b*; Krumbein, 1987*a*):

Weathering

(Bio-) corrosion (bio-) deterioration, (bio-) erosion, (bio-) weathering is an exchange of material and energy between two heterogeneous open systems: the solid substrate (rock, mineral, concrete, etc.) and the atmosphere. Both systems are defined by their physical properties (mass, volume, humidity, pressure), chemical composition, biological components (physiological, biochemical, biophysical) and through their innate and permanently added or subtracted energy. The mutual interaction of all components leads to a more or less complete turnover of the initial materials at the border between the two systems. The natural limit of turnover activities is given by the penetration depth of physical gradients, gases, solutions and organisms into the mineral material. This (bio-) deterioration or (bio-) decomposition process may come to a standstill for some time when conditions approach equilibrium (e.g. in the case of stabilizing patina or crusts). It may, however be revived if only one of the components involved changes or is submitted to change.

Following this general definition several additional modifications may be useful in order to clarify some points in the interaction of biological and inorganic processes. Several different subsections of biological and physical-chemical influences on rocks and mineral materials which further elaborate the above definition are listed alphabetically below (Krumbein & Dyer, 1985).

Abrasion and bioabrasion

Abrasion is a special case of erosion. It is defined as the friction or mechanical effect of any kind of solid or hard surface on another solid or hard surface leading to erosion. Thus abrasion is the destruction of a river bed or coastal rocky shore by stones, pebbles and sand; it also is wind polishing of desert rocks. Bioabrasion would be the action of animals on rocks with their hoofs, beaks and tongues. The chiton, for example, produces bioabrasion with its magnetite rasp or *Patella* drills a biogenic abrasive hole on the rocky shore with its shell. There are myriads of examples of quite effective bioabrasion.

Alteration and bioalteration

Alteration is the change of the mineralogical composition of any rock or mineral by chemical means. The term is mainly used for the attack of hydrothermal solution and alterations during metamorphism. In this case biological processes play a minor role. Bioalteration thus is not used in geological and geomorphological literature. Unfortunately alteration with accent aigue and bioalteration are common terms in the French language and may produce some confusion in language transfer. In French they have the meaning of weathering and bioweathering.

Decomposition and biodecomposition

In geological nomenclature decomposition is used as a synonym for chemical weathering by, for example, humic substances. The absurdity of the term weathering in this case calls for attention. Also it is highly questionable to talk about a chemical process initiated by, for example, humic acids without mentioning at the same time that this is a case of at least indirect biodecomposition. In biology the term decomposition is actually used in order to describe the process of formation of minerals from organic compounds via respiratory and fermentative pathways.

Degradation and biodegradation

Degradation is usually used as a term for the combined effects of erosion and weathering, i.e. the wearing down of the Earth's surface by physical, chemical and biological action in the case of biodegradation. Degradation of a soil describes the processes leading to the structuring of new lattices via base exchange and bioleaching in order to reduce (degrade) the exchangeability of cations.

Deterioration and biodeterioration

The terms deterioration and biodeterioration do not exist in the Glossary of Geology nor in the Oxford English Dictionary, although International Biodeterioration Societies and Journals exist. Deterioration means to make things worse, a status or act of decline to the negative. In this respect deterioration can barely be assigned to natural phenomena, since Man does not really understand what is better and what is good and what is bad for nature. Thus deterioration should be restricted to man-made substances and objects. Biodeterioration can only be related to weathering and erosion when the objects are exposed in the natural environment as, for example, the chemical deterioration of the Pyramids, the chemical and biological biodeterioration of the temples of Angkor Vat and the biodeterioration of stained glass windows or the bronze Quadriga at Venice.

Leaching and bioleaching (or biological leaching)

Leaching is a technical term derived from mining and metal processing. It can be effected by acids and bases or by electrochemical methods and when bioleaching or biological leaching is in question it will be effected by biogenic acids as, for example, sulphuric acid or by the direct leaching activity of chelating agents and direct surface contacts of microorganisms such as fungi, cyanobacteria and chemoorganotrophs. Leaching has been frequently observed under natural conditions and, mostly, biological leaching.

Solution and biosolution

Solution and biosolution are placed by some authors into the frame of erosion and by others into the effects of weathering alone. The action of water alone under the conditions prevailing at the surface of this planet is not very considerable. The point of discussion is, whether the solution effects of, for example, carbonic acid, oxygen, sulphuric acid, nitric acid, organic acids and chelators (solution/immobilization) should be classified as physical-chemical or biological factors. On second thought it turns out that all compounds dissolved in water or acting by diffusion into water from the atmosphere are ultimately biogenic compounds and that their respective concentrations in the system are as well more or less biologically produced. Therefore it is concluded that the major factors of solubilization of minerals and rock compounds occurring during weathering are biological and that the solubilisation of minerals should basically be termed biosolution. Some of the very soluble minerals such as halite may be an exception to this rule.

Processes

The most important processes involved in the biotransfer and biotransport of mineral materials and thus often in the biodeterioration of constructions, monuments, or objects of art have so far been identified as:

(1) Microbial production of protons.
(2) Microbial production of acids.
(3) Microbial production of complexing compounds.
(4) Microbial production of adhesives and surfactants.
(5) Microbial production of minerals (oxides, phosphates, carbonates, oxalates, etc.).
(6) Microbial production of ecological niches (e.g. pitting, crust formation, enlargement of crevices, production of water reservoirs).

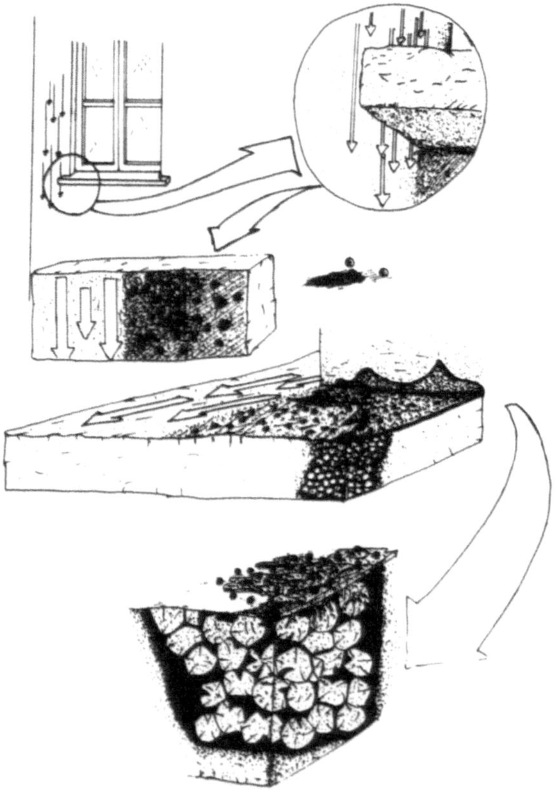

Fig. 5. Sequence of settlement of algae and lichen on a frequently humidified rock surface of a building leading to the attack of cement by rock penetrating fungi.

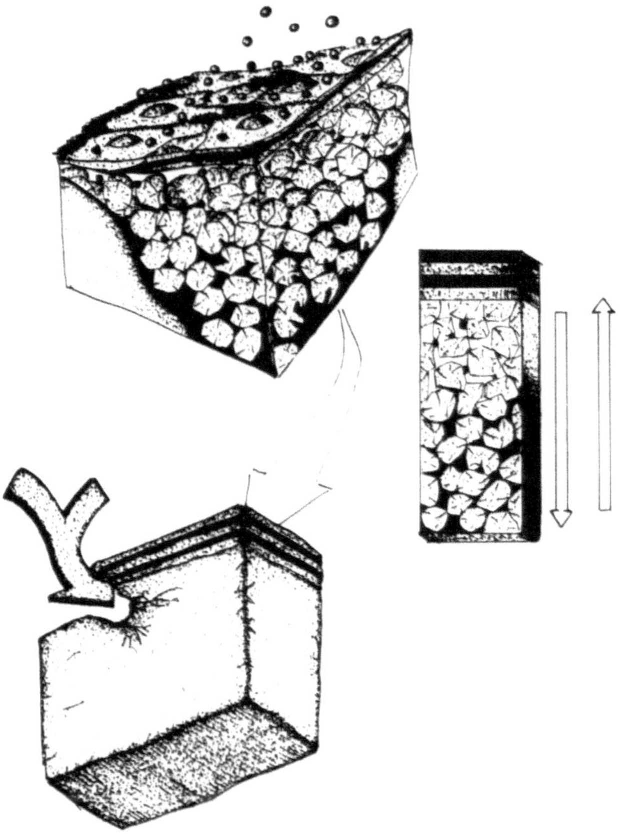

Fig. 6. Blow-up from Fig. 5. The flora may form several layers of epilithic crusts in several rainy seasons or through several more or less dry years. Thus a rock 'stromatolite' forms. Under certain conditions undercutting of the epilithic crust or desquamation of the crust with parts of the rocks will follow.

Some of the more specific actions of microbes in these fields are more explicitly documented in this volume. The question of proton production, for example, is described by Sand & Bock. The acid production problem is dealt with by Petersen *et al.*, and Lewis *et al.*, complex compounds are described and studied by Curri *et al.*, Jones *et al.* The production of adhesives and surfactants is discussed by Warscheid *et al.* The treatment of rock surfaces that are infected has been analysed so far by several groups among which are Emmel *et al.*, Lewis *et al.* and Warscheid *et al.* More or less recent contributions have been compiled by Krumbein (1986*a*).

Among the effects of microbial processes and actions on minerals, rocks and buildings or monuments made of mineral and rock which can

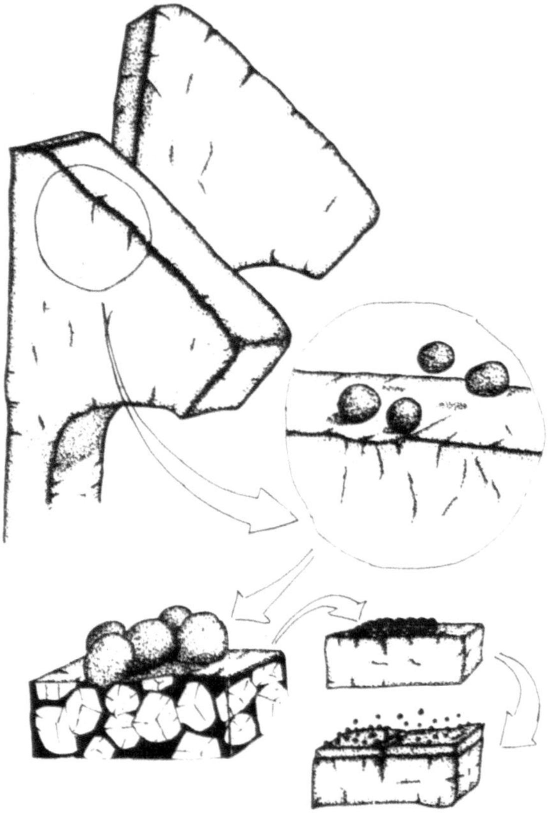

Fig. 7. Settlement of flora on drier and less humidified parts of the rock as compared to Figs 5 and 6. In this case the development of an endolithic microbial mat is more frequently observed.

be summarized as biocorrosion, bioerosion or biodeterioration are some that are also usually described as physical-chemical or pollution derived processes. These in fact are either completely or partially caused by microorganisms. The most prominent ones are:

(1) Epilithic crust formation and subsequent exfoliation of parts of the rock together with the epilithic crust (Figs 5 and 6).
(2) Endolithic crust formation and subsequent leaching and/or exfoliation of parts of the rock usually with parts of the endolithic crust (Figs 7 and 8).
(3) Pitting or cratering or corrosive pitting of glass, minerals or rocks via algae, chemoorganic fungi and bacteria or often lichens. This type of biotransfer or biocorrosion usually produces more or less

circular depressions in the rock surface as a consequence of former growth of epilithic crustose lichens or endolithic lichens (Figs 9 and 10). Some of the individual biological steps involved are described in the figure legends. Biogenic glass corrosion and pitting has also been documented recently by Callot *et al.* (1987).

The biocorrosive or pitting action of the lichen and fungal flora has also been described by Krumbein & Petersen (1987), Krumbein & Schönborn-Krumbein (1987*a*, *b*) and Petersen *et al.* (this volume).

In Figs 9 and 10 a general picture of the pitting action is given, while Figs 11–14 demonstrate some of the associated microorganisms. In Figs 14 and 15 the similarity of biopitting of glass and stone is shown.

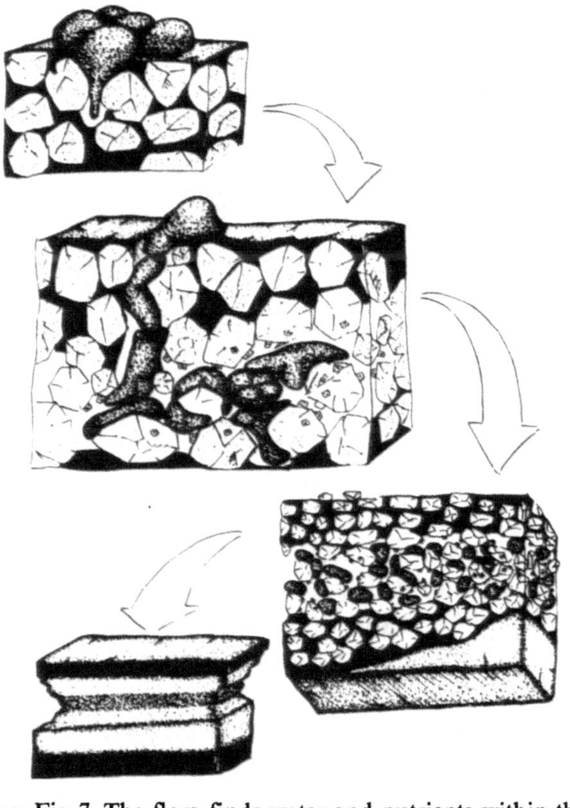

Fig. 8. Blow-up from Fig. 7. The flora finds water and nutrients within the rock at a certain depth. Detrimental and other metabolic products are transported to the surface layers of the rock (see Figs 16 and 17) and redeposited as a crust. The crust may close the capillaries and lead to biogenic exfoliation of large layers of the rock.

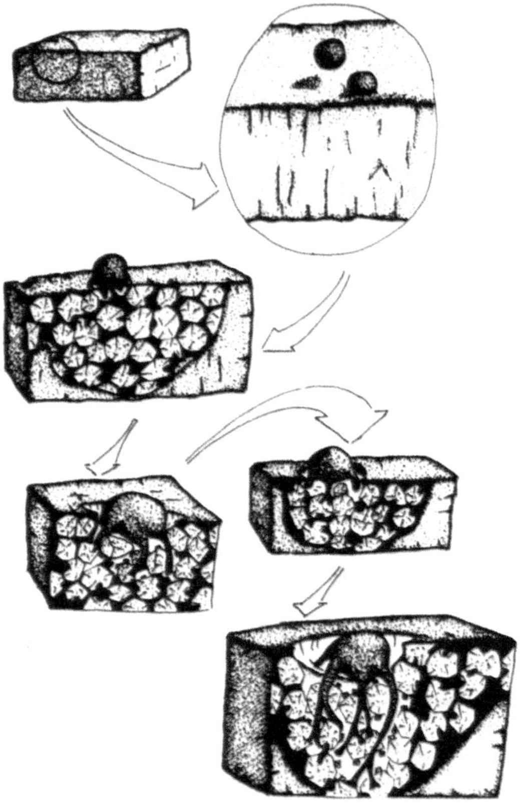

Fig. 9. Biocorrosion or biopitting. Crater shaped biocorrosion is very frequent in Roman and Greek marble monuments. The pitting is clearly associated with epilithic and endolithic lichens of various species.

Recently (Krumbein, in preparation) the pitting of steel, metal, and plastic surfaces in buildings and ships has been shown to be due to microbial communities. Similar crater formed depressions have also formed on old leather wall papers in Frisia (Krumbein, in preparation) and other places. Sometimes these phenomena are misinterpreted as sulphuric acid derived red rot phenomena or selective environmental pollutant attack.

Also crust formation and destruction (Figs 5–8) and subsequent exfoliation are often misinterpreted as abiogenic processes. For this reason a model of biosolution and biodeposition is presented in Figs 16 and 17.

A microbial community consisting of phototrophic algae at the surface of the rock material and of phototrophic cyanobacteria or lichen at (epilithic) or near (endolithic) the surface will have the tendency to be anchored deep within the rock (Fig. 16). Organisms will penetrate either by biosolution or by growth along capillaries to depths exceeding 10 cm in extreme cases. Usually between 2 and 8 mm depth will be reached. In the deeper parts minerals and particles will be dissolved. The materials will be transferred to the surface either by capillary water movement or by movement along the mycelia of fungi and actinomycetes typical for the rock environment (see Petersen *et al.* (this volume); Krumbein & Petersen, 1987). Hereby the inner parts of the rock are weakened, while the outer parts may get condensed and less permeable. This in turn will lead to exfoliation or desquamation of large chips of the rock.

The biotransfer of iron and manganese as some of the biologically metabolized elements which are readily redeposited is very important in

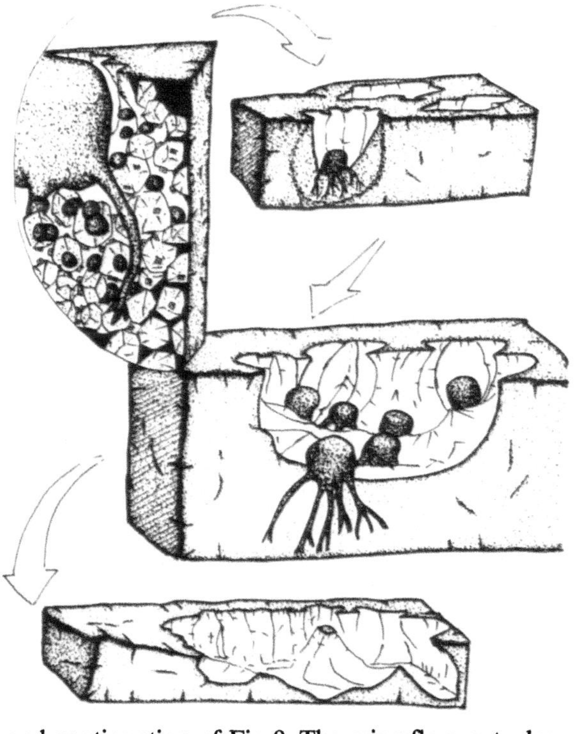

Fig. 10. Blow-up and continuation of Fig. 9. The microflora cuts deep holes into the marble in order to find sun-shield, humidity and nutrients. Eventually the substrate dries out completely or is poisoned by, for example, air pollution. The lichen or fungal colonies will subsequently die. In this case (lowermost part) it is difficult to attribute a biological origin to the corrosion pattern.

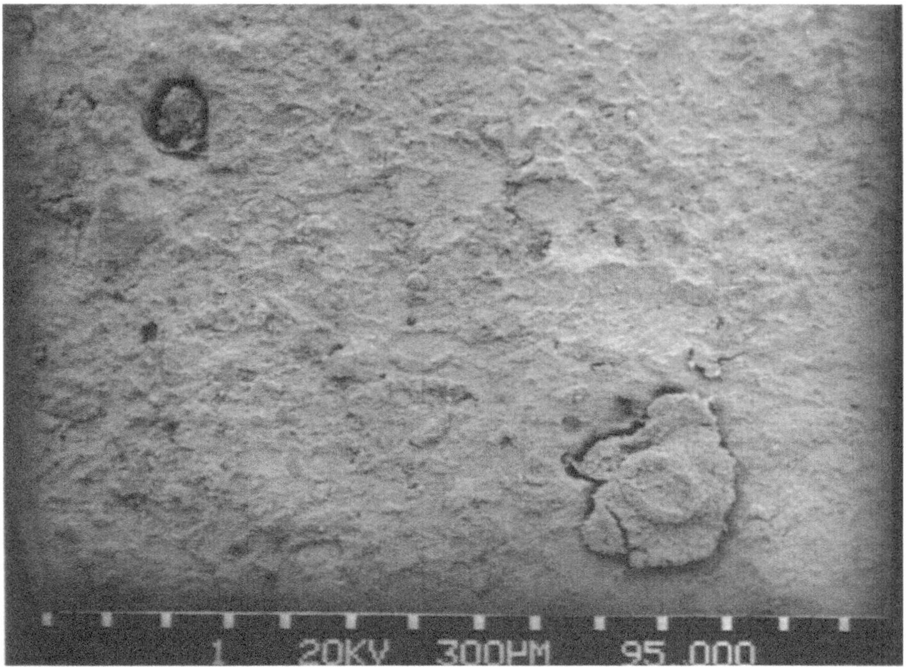

Fig. 11. Marble attacked by a lichen.

Fig. 12. Blow-up of Fig. 11 showing the depth of biosolution and biopitting as well as two micropits from which the lichen apothecium has been scraped off.

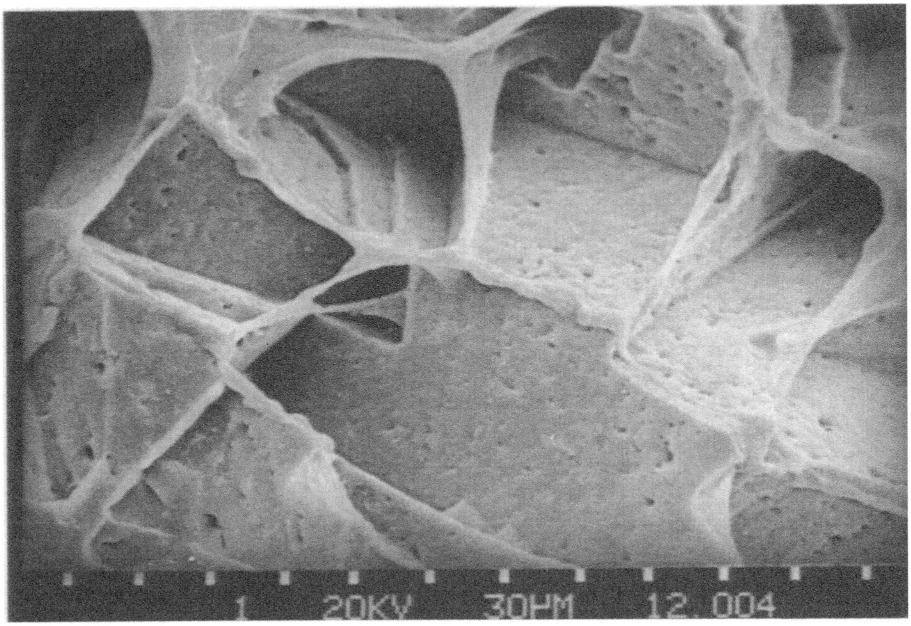

Fig. 13. By casting-embedding-etching techniques rock particles can be dissolved after impregnating the pores and the organisms with epoxy. After selective dissolution (e.g. carbonates), the organisms and the original or biocorroded pores remain. In this figure remains of fungi can be seen around the marble calcite.

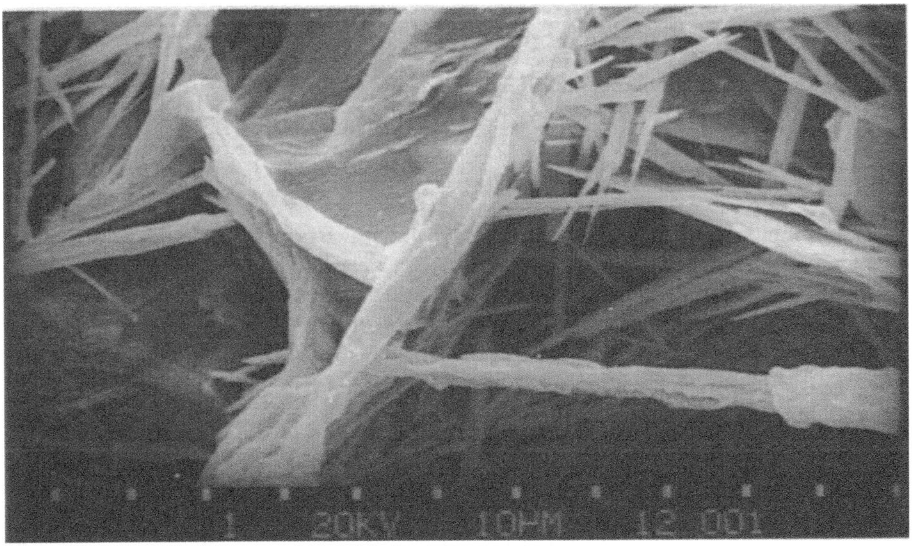

Fig. 14. This micrograph demonstrates (i) that fungal mycelia not only grow around rock particles but also penetrate them directly (hyphae in lower right). The needles are insoluble to weak acid and may be whewellite or a specific secondary silicate sometimes occurring with lichen growth and biocorrosion. The emplacement of three former calcite crystals is clearly discernible as holes.

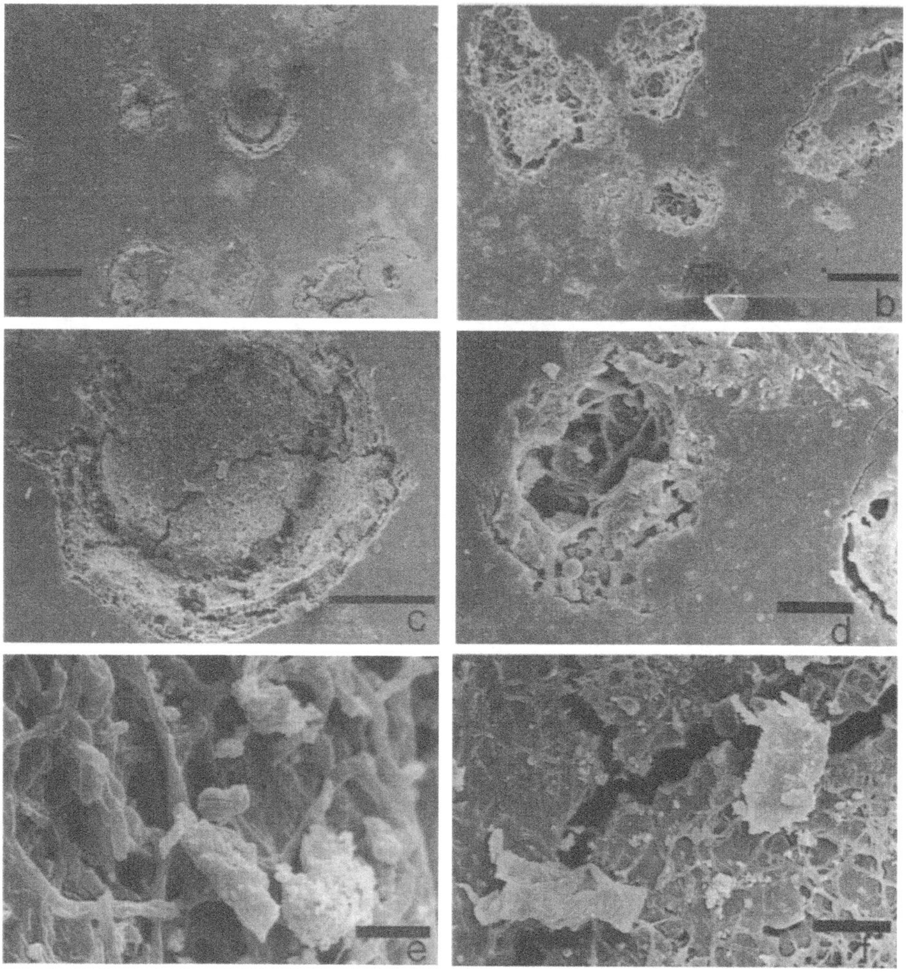

Fig. 15. Biopitting or biocratering by lichen growth. Examples of mediaeval church windows from Augsburg, Delmenhorst and Chartres is well documented. (a) clearly shows the subrosion through biopitting underneath an apparently intact surface (lower part). (b), (d), (e) and (f) give details of the fungal mycelium, while (c) shows how the undercutting in (a) lower part turns into the crater of (a) central part and (c). Bars: a = 4 mm, b = 1 mm; c = 1 mm; d = 30 µm; e = 20 µm; f = 100 µm.

this context and has recently been described as an important factor in crust formation and rock destruction (Krumbein *et al.*, 1987; Kraus & Krumbein, 1987; Krumbein, 1987*b*; Krumbein & Jens, 1981). The formation of sulphates and nitrates through the activity of chemolithoautotrophic microorganisms as well as associated crust development and crust morphologies have been described earlier (Pochon & Tchan, 1946; Kaufmann & Toussaint, 1954; Krumbein, 1968; Bock, 1987).

CONCLUSION

The study of biocorrosion, biotransfer, biotransport and biodeterioration of minerals and rocks is an expanding field of science. As long as engineers, constructors and conservators are not yet aware of the tremendous catalytic potential of microorganisms they will be misled and misguided

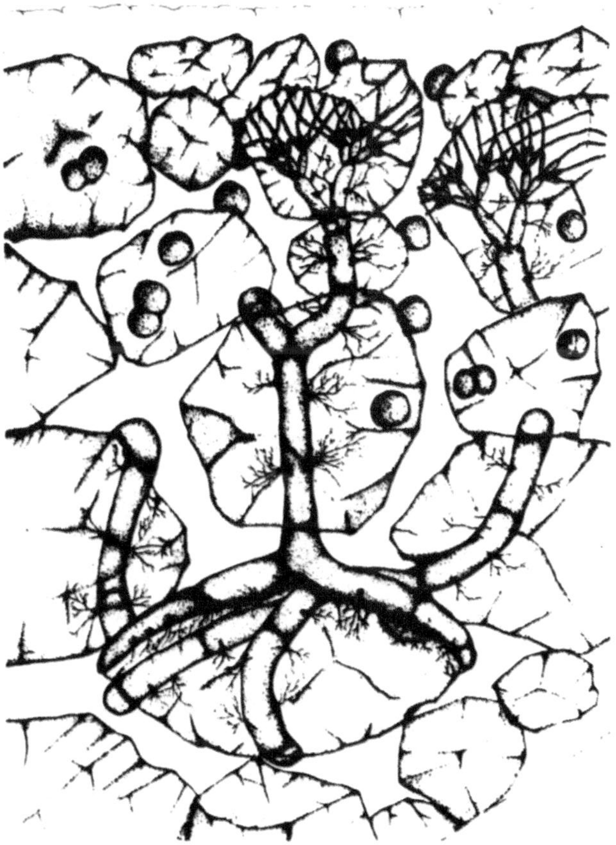

Fig. 16. Schematized presentation of a microbial community infecting a sandstone. Algae and/or cyanobacteria grow in the uppermost region where light still penetrates. The fungi are rooted deeper in the rock itself and absorb humidity, nutrients and detrimental cations which are detoxified by several different means. For simplicity bacteria are not represented (see, for example, Figs 3 and 4).

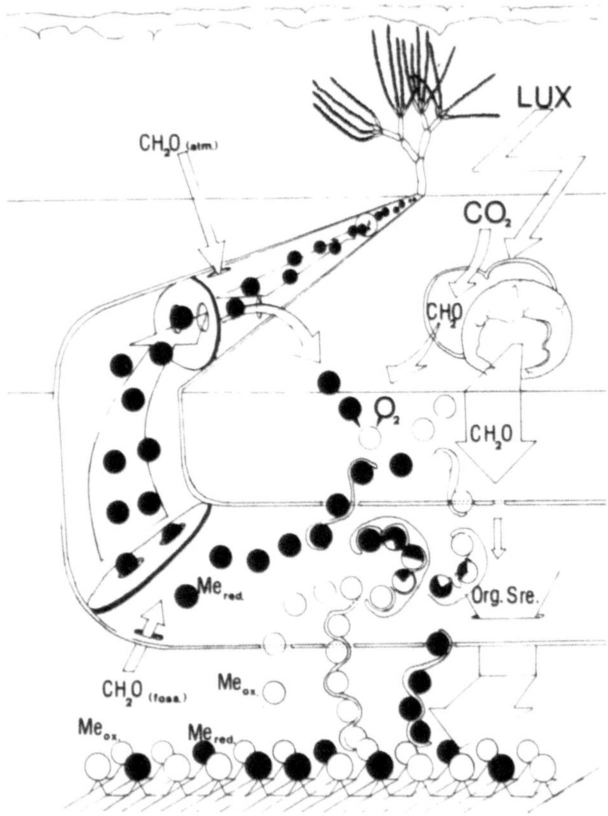

Fig. 17. Schematic pathway of events derived from Fig. 16. Light brings energy to algae and cyanobacteria. Through decay and lysis or by traustoria of lichen the energy is transferred to the fungus, which in turn delivers water and nutrients to the alga. Organic nutrients for fungi and bacteria may, however, settle from the atmosphere (especially as dust) or come from the original organic content of sandstones and limestones in particular. The fungus metabolizes organic matter under production of organic acids or siderophores which take up oxidized and/or reduced metals (white and black). Biotransformation (redox reactions), biotransfer (solution and precipitation) and biotransport of the elements in question has been observed, for example, in manganese and iron transferring, transforming and transporting fungi. The transport and oxidative precipitation of cations is also effected through capillaries and photosynthetically derived oxygen supersaturation in some pores of the rock.

into construction and treatment techniques which are not appropriate. The best examples of the importance of microbial biodeterioration are the biblical quotation of rocks with leprous growths, the bacterial attack on teeth, the well documented attack of sulphuric acid producing bacteria on concrete in anaerobic environments, and the 'green' and 'white' disease of the mural paintings of Lascaux. Fortunately we know practically everything of the terrible diseases of bone and teeth and we know the malign and the associated benign microflora very well. Unfortunately some of the microorganisms that are typical for rocks or happen to occur on and in rocks are only now being recognized. Bacteria and fungi are more familiar and have been isolated and analysed. Many of them are new species and genera which have never been seen or described in any other environment. The rock microflora is a fascinating one and is just starting to be understood with regard to the individual malign or benign processes. But the complex community relationships that occur in rocks undergoing biotransfer and biocorrosion are not well understood. Some of the figures in this paper and of Blaschke (1987) and Krumbein (1987*a*) have demonstrated the jungle like appearance of the rock microflora. The jungle of microbes, microbe-rock interactions, to say nothing of the remedies by which to stop the detrimental interactions without disturbing the positive ones are not yet understood.

ACKNOWLEDGEMENTS

The Federal Minister of Science and Technology, the EUREKA program EUROCARE and the Minister of Science and Arts of Lower-Saxony have largely contributed to our work through generous grants. V. Schostak has kindly prepared some of the figures. E. Johnston and A. Lovelock were so kind as to smooth out some of the germanisms in the text.

REFERENCES

Blaschke, R. (1987). *Bautenschutz Bausanierung*, **1**, 28–34.
Bock, E. (1987). *Bautenschutz Bausanierung*, **1**, 24–7.
Brimblecombe, P. (1987). *The Big Smoke*. Methuen, London, p. 185.
Callot, G., Maurette, M., Pottier, L. & Dubois, A. (1987). *Nature*, **328**, 147–9.
Curri, S.B. (in press) the diagnosis of biological attack of stone.
Emmel, T., Brill, H., Sand, W. & Bock, E. This volume, pp. 118–22
Gerdes, G., Krumbein, W.E. & Holtkamp, E.M. (1985). In *Hypersaline Ecosystems—The Gavish Sabkha*, Vol. 53, ed. G.M. Friedman & W.E. Krumbein. Berlin, Springer, pp. 238–66.
Giacobini, C. (1985). In *Enciclopedia delle Scienze*, ed. Istituto Geografico de Agostini, Novara, pp. 323–34.
Jaton, C., Orial, G. & Brunet, A. (1985). In *5th International Congress on Deterioration and Conservation of Stone*, Lausanne, 25–27 Sept. 1985, Vol. 2, ed. C. Felix. Presses Polytechniques Romandes, Lausanne, pp. 577–86.
Jones, D., Wilson, M.J. & McHardy, W.J. This volume, pp. 129–34.

Kaufmann, J. & Toussaint, J. (1954). *Corrosion Anticorrosion*, **2**, 240–4.
Kraus, K. & Krumbein, W.E. (1987). In *Naturwissenschaftliche Forschung an Kunstgütern aus Stein*, ed. H.J. Oel. Erlangen, pp. 124–34.
Krumbein, W.E. (1966). Zur Frage der Gesteinsverwitterung (über geochemische und mikrobiologische Bereiche der exogenen Dynamik). Würzburg, Dissertation.
Krumbein, W.E. (1968). *Z. allg. Mikrobiol.*, **8**, 107–17.
Krumbein, W.E. (1972). *Rev. Ecol. Biol.*, **9**, 283–319.
Krumbein, W.E. (1973). *Deut. Kunst Denkmalpflege*, **31**, 54–71.
Krumbein, W.E. (1983). *Precamb. Res.*, **20**, 493–531.
Krumbein, W.E. (1986*a*). *ICOMOS-Stone-Committee—Group Biology*—Literature List. Oldenburg, Universität., 121 pp.
Krumbein, W.E. (1986*b*). In *2nd Int. Kolloquium: Werkstoffwissenschaften und Bausanierung*, 2–4 Sept. 86, ed. F.H. Wittmann. Technische Akademie, Esslingen, pp. 45–62.
Krumbein, W.E. (1987*a*). Biotransfer in monuments—a sociobiological study. Durability of Building Materials.
Krumbein, W.E. (1987*b*). *Einblicke Forsch. Univ. Old.*, **5**, 4–9.
Krumbein, W.E. (1987*c*). *European Cultural Heritage*, **1**, 12–15.
Krumbein, W.E. & Dyer, B.D. (1985). In *The Chemistry of Weathering, Vol. 149*, ed. J.I. Drever. D. Reidel Publishing Company, Dordrecht, pp. 143–60.
Krumbein, W.E. & Jens, K. (1981). *Oecologia*, **50**, 25–38.
Krumbein, W.E. & Petersen, K. (1987). *Bautenschutz Bautensanierung*, 61–4.
Krumbein, W.E. & Pochon, J. (1964). *Ann. Inst. Pasteur*, **107**, 724–32.
Krumbein, W.E. & Schönborn-Krumbein, C.E. (1987*a*). *Bautenschutz Bausanierung*, **10**, 14–23.
Krumbein, W.E. & Schönborn-Krumbein, C.E. (1987*b*). *Bautenschutz Bausanierung*, **10**, 110–17.
Krumbein, W.E., Grote, G. & Petersen, K. (1987). In *The Biodeterioration of Constructional Materials*, Biodeterior. Soc. Occ. Publ., *Vol 3*, ed. L.H.G. Morton. Publications Service, Lancashire Polytechnic, pp. 15–28.
Lefevre, M., Pochon, J., Laporte, G. & Chalvignac, M.A. (1964). *Comp. Rend. Soc. Biol. Filiales*, 6576–8.
Lewis, F., May, E. & Bravery, A.F. (1985). In *5th International Congress on Deterioration and Conservation of Stone*, Lausanne, 25–27 Sept. 1985, Vol. 2, ed. C. Felix. Presses Polytechniques Romandes, Lausanne, pp. 633–42.
Lewis, F.J., May, E. & Bravery, A.F. This volume, pp. 107–12.
Lübbert, E. (1987). *Deutsche Kunst und Denkmalpflege*, **1**, 12–25.
Petersen, K., Kuroczkin, J., Strzelczyk, B.A. & Krumbein, W.E. This volume, pp. 123–8.
Pochon, M.J., Chalvignac, M.A. & Krumbein, W.E. (1964). *C.r. Acad. Sci., Paris*, **258**, 5113–15.
Pochon, J. & Tchan, Y.T. (1946). *Comptes Rendus des Seances de l'Academie des Sciences, Paris*, 695.
Strzelczyk, A.B. (1981). *Microbial Biodeterioration, vol. 6*, ed. A.H. Rose. Academic Press, London, pp. 62–80.
Warscheid, T., Petersen, K. & Krumbein, W.E. This volume, pp. 455–60.

Control of Microbial Growths on Stone and Concrete

B.A. Richardson

Penarth Research International Ltd, P.O. Box 142, Guernsey, Channel Islands, UK

ABSTRACT

The development, laboratory testing and service testing over the last 25 years of biocides for the control of microbial growths on stone and concrete are reviewed. It is concluded that there have been no significant developments over the last 20 years, and that certain borates, quaternary ammonium compounds and tri-n-*butyltin compounds remain the most effective materials available.*

INTRODUCTION

In 1967 The Commonwealth War Graves Commission was encountering increasing problems with the maintenance of the stonework in its care, particularly the control of algal and lichen growth on headstones. The headstones were cleaned by wet scrubbing followed by rinsing, with major cleaning by visiting teams every 4 years and some limited intermediate cleaning by local gardeners. The Commission instructed Penarth Research Centre (subsequently part of Penarth Research International Ltd) to investigate alternative cleaning processes and particularly the use of biocides to control algal, fungal, lichen and moss growth.

Available biocides were considered in relation to their apparent activity, mode of action, probable persistence, risk to operatives and risk to stone substrates. Only 3 groups of biocides, the borates, quaternary ammonium compounds and organotins were selected for in service trials on headstones in Europe and later in the tropics. Subsequently other amine and organotin compounds were proposed but in further service trials none of them proved as effective as the original compounds.

The Department of the Environment, Directorate of Ancient Monuments and Historic Buildings, was encountering similar problems and investigations were extended to ancient masonry with a requirement for eradication of existing as well as inhibiting subsequent growth. In trials at Tintern Abbey the same groups of compounds were most effective. Trials on hydraulic cement-based substrates such as concrete roadworks and asbestos cement roof tiles again confirmed their effectiveness.

BIOCIDES

Monuments of porous limestone and sandstone are particularly susceptible to biological growth but problems are also frequently encountered on buildings. Growth on masonry occurs particularly in areas of high rainfall, such as the west coast of Scotland, Ireland and Wales, although it occurs elsewhere on sheltered north elevations which are protected from the drying action of the sun and the prevailing south-westerly winds. Growth is not confined to the external elevations but is often found on the interiors as luxuriant growth of algae, bacterial slimes and even ferns, particularly on the inside of walls exposed to south-west winds. It can be avoided by keeping masonry dry, particularly by the use of external water repellent treatments, but some of the interior growth may not entirely disappear. In fact, many occur in situations completely isolated from the external walls, such as on granite arcade pillars, due to condensation resulting from periodic heating in buildings such as churches. However algicidal treatment can achieve dramatic results.

Silicone resin water repellents provide a realistic method for the inhibition of algal, lichen and moss growth, but a toxic biocidal treatment is necessary to eradicate established attack, and persistent toxic treatments can avoid the necessity to use water repellents. A wide variety of biocides is available but choice is limited by a number of factors.

Whilst the treatments must be toxic to the growth, they must also be completely harmless to the operatives and they must not damage the stone, either by direct action or by deposits which may result in damage; for example, by increasing water retention so that frost damage becomes more likely or by chemical reactions to give unsightly discoloration. The activity of many oxidising biocides, e.g. chlorinated lime paste, is only transient. Reliable control of biological growth consists of sterilisation at the time of application and inhibition. In the case of lichens, which grow very slowly, and mosses which prefer to colonise stone surfaces only after contamination with organic matter, sterilisation alone may result in freedom from apparent growth for several years. However, with algae, which rapidly colonise unprotected stone whenever the conditions of temperature and moisture content are appropriate, the advantages of such treatments may be lost after only a few weeks. In these cases treatment might be followed by a water repellent in order to maintain the stone in a dry condition and inhibit further growth.

The sodium salts of pentachlorophenol or *o*-phenylphenol may leave a deposit within the stone but it is slightly volatile and leachable. Whilst this type of system usually possesses activity against a wider variety of organisms, the activity decreases with time. These treatments should therefore be considered as eradicants prior to the application of a water repellent or perhaps a persistent toxic treatment.

One of the most efficient eradicant treatments consists of a mixture of sodium methyl siliconate and the sodium salt of a phenol, e.g. pentachlorophenol or *o*-phenylphenol. The siliconate aids the penetration of the phenol into the growth and the siliconate treatment of the growth tends to prevent subsequent water absorption, causing it to become dry and brittle so that it is readily removed by light brushing or becomes detached during periods of heavy weather. Unfortunately these treatments have serious disadvantages in that they are likely to cause severe staining on, for example, Portland limestone. This occurs through reaction of the alkali with iron in the substrate and the development of brown or purple colouration due to the reaction of the phenols with iron.

Staining is not the only danger. The introduction of alkali metal salts into stone may produce soluble salts which may cause crystallisation damage to the stone. In addition, caustic solutions present a hazard to operatives, as well as causing stripping of paint from joinery adjacent to the masonry. Chlorophenol systems are not widely used as they may contain dioxin impurities, which are very toxic.

Although phenols were most widely used as masonry biocides at the time of the original studies in 1967–69, inorganic metal salts had earlier been considered and were still extensively used. Copper, zinc and magnesium salts and also fluorides and silicofluorides were used, most of which provided a relatively transient eradicant action. Mercurial compounds were certainly most effective, although their action was again relatively transient. Their use was discontinued because of their high mammalian toxicity.

Many other compounds had been considered and tried at the time, but none of them combined the necessary biocidal activity and persistence with freedom from undesirable side effects. However, there were three groups of compounds which were proving very effective in similar applications.

Borates were widely used for wood preservation and fuel spoilage control. Sodium borates, for instance, borax, can be applied to stone, but reaction with atmospheric carbon dioxide produces sodium carbonate and boric acid which is almost insoluble in cold water. In polluted urban atmospheres the sodium carbonate reacts with atmospheric sulphur dioxide to form sulphites and, under the influence of sulphating bacteria, eventually sulphates which tend to crystallise and cause damage to the stone. Sodium salts are therefore best avoided. A useful compound is Polybor, a proprietary mixture of borates and boric acid. This simplifies formulation and treatment as it is much more soluble than either of its components, yet the sodium ion concentration is much lower than for borax, and reduces the risk of stone damage. In fact, stone damage has never been observed in field trials using Polybor. On reasonably porous stone, generous application of a 4% solution of Polybor in water will give freedom from algae for at least 2·5 years, and freedom from moss and lichen for a much longer period. Concentrations must be varied according to the porosity of the stone, reducing to *c.* 2% on a very porous stone but increasing to *c.* 10% on marbles and other stones of limited porosity. On building structures where moss and lichen are more important than algae, the effective life of Polybor treatment will be much greater, perhaps 10–20 years.

Although mercurials had been available for many years, other organometallic compounds were only introduced as industrial biocides a few years before the 1967–69 trials. The most important were the organotin and organolead compounds which were being used to an increasing extent at that time in wood preservatives and marine antifouling compositions. The activity of these compounds in relation to structure, and of other organic compounds of group IV metals, have been discussed in detail elsewhere by Richardson (1968*a*), the tin compounds being generally preferred because of their stability and the lower toxicity of their decomposition products. Optimum activity is associated with certain triorganotin compounds, particularly tri-*n*-butyltin and triphenyltin compounds, although only the former have been widely used commercially because of their greater availability. Preferred compounds, such as tri-*n*-butyltin oxide, are insoluble in water. They can be applied in kerosene and similar solvents but generally water-based formulations are more effective for the treatment of stone. Various solubilising systems can be used but quaternary ammonium compounds have proved most effective, partly because of their efficiency in solubilising tri-*n*-butyltin oxide but also because of their own biocidal activity.

Amines are an extraordinarily diverse group of compounds, but biocidal activity is best developed in the quaternary ammonium compounds. The mode of action seems to involve rupturing of the cell wall of fungi and algae, as well as by toxic action, and they are therefore useful alone or in combination with toxicants such as organotin compounds. The compounds have an affinity for a variety of substrates and are capable of providing a persistent inhibitory action. Desirable water solubility and powerful biocidal activity appeared to be most strongly developed in certain alkylbenzyl-dimethyl and benzalkyl-trimethyl ammonium compounds such as Benzalkonium chloride and Gloquat C. These can be used at concentrations as low as 1% to give an effective life against algae of about 2·5 years.

The most effective eradicant and persistent inhibitory action is associated with organotin biocides emulsified with quaternary ammonium compounds, particularly tri-*n*-butyltin oxide emulsified with Gloquat C. Formulations of this type are now widely used, although sometimes in combination with simple quaternary ammonium treatments. The latter are used both as pre-treatments to assist in the eradication of dense growth before application of the main treatment, and as algicides for simple spray maintenance treatments at intervals of several years following the main treatment.

RECENT DEVELOPMENTS

Following the original investigations in 1967–69, a series of further trials were carried out to assess new and apparently effective alternative treatments. The main developments were in amines where structures other than quaternary ammonium had been shown to be active against various fungi and bacteria. Parallel trials on stain control in wood preservation by Richardson (1975) suggested that the same quaternary ammonium compounds were still most effective, and this finding was confirmed by Richardson (1976) on stone by extending the original service trials to include a series of the most promising candidate compounds. Since then Grant and Bravery (1981*a*, *b*, 1985) have reached similar conclusions following their laboratory evaluation of algicidal biocides using the vermiculite bed technique.

CONCLUSIONS

Although it is now 20 years since the original review of biocides by Penarth Research Centre which resulted in the establishment of borates, quaternary ammonium compounds and organotin/quaternary ammonium formulations, none of the alternative biocides appear to be more effective or safer. Use of these systems in service conditions has generally confirmed the conclusions of Richardson (1973).

REFERENCES

Grant, C. & Bravery, A.F. (1981*a*). *Int. Biodet. Bull.*, **17**, 113–23.
Grant, C. & Bravery, A.F. (1981*b*). *Int. Biodet. Bull.*, **17**, 125-31.
Grant, C. & Bravery, A.F. (1985). *Int. Biodet. Bull*, **21**(4), 285–93.
Richardson, B.A. (1968*a*). *1st Int. Biodet. Symp, Biodeterioration of Materials*, Elsevier, London, pp. 498–505.
Richardson, B.A. (1968*b*). Cleaning Stone Buildings *Stone Industries*.
Richardson, B.A. (1973). *Stone Industries*, **8**(2) 22–6.
Richardson, B.A. (1975). Safer stain control using quaternery ammonium compounds and borates. IUFRO Conf. Abidjan.
Richardson, B.A. (1976). Control of moss, lichen and algae on stone. *Proc. Int. Symp. on Conservation of Stone, Bologna*.
Richardson, B.A. (1981). *Biological Aspects of Decay and Disfigurement of Stone.* UK Institute of Conservation.
Richardson, B.A. (1985). Conservation of masonry. Symp. on Building Appraisal Maintenance & Preservation, University of Bath.
Richardson, B.A. & Cox, T.R.G. (1974). Aqueous organotin wood preservatives. *Tin & Its Uses*, 102.

Metabolic Activities of Bacteria Isolated from Building Stone and their Relationship to Stone Decay

F.J. Lewis, E. May

School of Biological Sciences, Portsmouth Polytechnic, Portsmouth PO1 2DY, UK

&

A.F. Bravery

Building Research Establishment, Princes Risborough Laboratory, Aylesbury HP17 9PX, UK

ABSTRACT

Bacteria have been isolated from decaying stonework on ancient monuments in southern Britain and have been shown to possess different metabolic activities which can be related to stone deterioration in laboratory studies. Stone decay in bacterial cultures was assessed under different nutrient conditions by means of weight loss as well as changes in pH and soluble calcium. Bacteria were identified which could cause decay of both sandstone and limestone using either organic (peptone/glucose) or inorganic (thiosulphate) nutrients. Certain isolates produced a weight loss with both types of nutrients. Although large weight losses in the region of 5% were observed with most isolates, some produced a significant weight gain, greater than 1%, due to the formation of $CaSO_4$ on the stone. In most cases, however, the decay activity could be correlated to solubilisation of calcium from the stone. The calcium content of sandstone appeared to be a major factor determining the resistance of the stone to bacterial attack. These data demonstrate that whilst the metabolism of bacteria causing decay may be varied the mechanism of stone deterioration is likely to be similar.

INTRODUCTION

The extent to which bacteria have been involved in the process of stone decay has been the subject of some debate (Hueck van der Plas, 1968; Krumbein, 1972; Murr, 1980). However, there is general agreement that their decay activity will be largely dependent on the production of corrosive metabolites which can solubilise minerals in a manner similar to chemical agencies.

Bacteria can readily be isolated from decaying stonework (Lewis *et al.*, 1985, 1987). They include autotrophic types which are capable of generating strong acids, such as HNO_3 and H_2SO_4, from inorganic sulphur (Jaton, 1973) and nitrogen compounds (Kaltwasser, 1976) and heterotrophic types which use organic matter to produce organic acids (Wood & Macrae, 1972).

This study outlines laboratory experiments to assess the decay potential of bacteria isolated from stonework at two ancient monuments, Portchester Castle and Tintern Abbey. The aim was to relate the nutritional requirements of the isolates to the nature and extent of decay on sandstone and limestone from the monuments.

METHODS

Stone used

Three types of stone were used in the study. A green sandstone and Purbeck limestone from Portchester Castle, near Portsmouth, and a grey sandstone from Tintern Abbey, near Chepstow, quarried locally and used in conservation work.

Bacteria studied

Bacteria were isolated during bacteriological surveys of the study sites as detailed elsewhere (Lewis *et al.*, 1985). The isolation media were selective for sulphur-oxidising, nitrifying and heterotrophic bacteria.

Media

Decay was assessed under (a) heterotrophic conditions in a peptone (0·5%)/glucose (1·0%) medium and (b) autotrophic conditions using media incorporating thiosulphate for aciduric (Taylor & Hutchinson, 1947) and non-aciduric thiobacilli (Parker & Prisk, 1953).

Assay for decay

All bacterial isolates were screened for decay potential in a liquid, static culture system using 1 cm stone discs over 20 (heterotrophic) or 60 (autotrophic) days. Weight change was expressed relative to sterile controls. Changes in pH, soluble calcium (ion-selective probe) and thiosulphate (iodine titration) were monitored during incubation at 25°C.

RESULTS AND DISCUSSION

About 200 bacteria were isolated into pure culture during the surveys of the monuments and all were screened for decay potential in a liquid culture system developed to enhance growth and accelerate the decay process. The use of intact stone as a means of assessing decay is rarely recorded in the literature although Wood & Macrae (1972) reported weight losses in sandstone pebbles in culture with *Pseudomonas fluorescens*. Thirty isolates were able to cause a weight loss greater than 2% and often as high as 7% under heterotrophic conditions. Table 1 presents data for 6 isolates ranked according to weight loss in green sandstone. Similar weight losses were found with Purbeck limestone although in all cases the Tintern sandstone was highly resistant to degradation.

Isolates causing decay under autotrophic conditions are shown in Table 2. Some bacteria produced a weight loss in stone as high as 3·3%,

TABLE 1
Percentage Weight Losses by Three Types of Building Stone Incubated in Bacterial Cultures under Heterotrophic Conditions

Isolate	*% Weight loss*[a]		
	Green Sandstone	*Purbeck Limestone*	*Tintern Sandstone*
H30	7·0	3·8	+0·4
NB55	5·5	4·8	+0·2
H27	5·2	5·5	0·1
H24	5·2	3·0	+0·3
NB61	5·0	2·7	0·3
AS20	4·8	3·5	+0·2

[a]+ indicates weight gain.

TABLE 2
Percentage Weight Losses by Three Types of Building Stone Incubated in Bacterial Cultures under Autotrophic Conditions

Isolate	*% Weight loss*[a]		
	Green Sandstone	*Purbeck Limestone*	*Tintern Sandstone*
AS49	3·3	4·5	0·3
AS46	2·8	3·1	+0·1
AS39	2·1	1·3	0·4
AS37	1·7	1·7	0·1
AS25	1·6	1·6	0·0
S5	+1·3	0·4	+0·2
S59	+1·3	0·5	+0·6
S55	+1·3	0·5	+0·4
S52	+1·4	0·4	+0·4
S16	+1·5	0·3	+0·3

[a]+ indicates weight gain.

while others caused a significant weight gain resulting from the formation of a white precipitate, possibly $CaSO_4$, on the stone. This did not occur with either the limestone or Tintern sandstone, the latter being again resistant to damage.

Some of the bacterial types displaying heterotrophic activity had been isolated using only inorganic nutrients, suggesting a lack of selectivity of the medium and/or a capacity to tolerate low levels of organic nutrients. Another possibility was that some bacteria were able to grow mixotrophically using organic or inorganic nutrients according to their

TABLE 3
Culture Characteristics of Isolate AS46 Incubated under Autotrophic and Heterotrophic Conditions

Culture parameter	*Cultural conditions*	
	Autotrophic (60 days)	*Heterotrophic (20 days)*
Weight loss by stone	3·1%	1·8%
pH (initial in parentheses)	4·9 (6·8)	5·2 (7·4)
Soluble Ca^{2+} (initial 0 mg/ml)	0·7 mg/ml	0·3 mg/ml
Thiosulphate (initial 5 mg/ml)	0 mg/ml	NT

NT = not tested.

availability. Such an isolate was AS46, one of several bacteria which caused weight losses under both heterotrophic and autotrophic conditions. As Table 3 shows, these losses were accompanied by a large drop in pH even though the bacteria were supplied with different substrates, in these cases, peptone/glucose (heterotrophic) and thiosulphate (autotrophic). In each culture there was a rise in soluble calcium, undetectable at the start of the experiment. The implications are that certain bacteria on stone may be metabolically versatile, capable of growth, and perhaps damage, using a wide range of inorganic and organic nutrients, altering their activity to accommodate changes in the environment.

The stability of the grey Tintern sandstone contrasted sharply with the susceptibility of the green sandstone to weight loss in bacterial culture. Table 4 compares the performance of the two sandstones in culture with heterotrophic and autotrophic bacteria. As expected, the green sandstone exhibited a substantial weight loss not evident in the Tintern stone, in spite of a low pH in all cultures. This could be related directly to the calcite content of the stone and solubilisation of this calcium during decay. In these calcareous sandstones, calcite often forms the binding matrix for the silicate material. Acid attack of these cementing minerals may be occurring in both autotrophic and heterotrophic culture resulting in the observed weight loss. Therefore, although the nature of the nutrients available to the bacteria varied, the effect on the susceptible stone was the same. It is the structure and composition of the stone which will determine whether or not decay will occur.

These results suggest that bacteria on stone can be extremely versatile and could maintain their activity during nutrient perturbations, operating at low nutrient levels and utilising what the environment has to

TABLE 4

Culture Characteristics of Decay Isolates Incubated with Sandstones of Differing Calcite Contents

Isolate	*Sandstone type*	*Calcite (%)*	*Cultural parameter*		
			pH	*Ca^{2+} (mg/ml)*	*%wt. loss*
NB55[a]	Tintern	0	4·8	0	+0·2
	Green	20	5·8	0.85	5·5
AS49[b]	Tintern	0	3·1	0	0·3
	Green	20	4·6	0·55	3·3

[a] Grown heterotrophically for 20 days.
[b] Grown autotrophically for 60 days.

offer. As a consequence, bacterial populations may be able to maintain their involvement in the process of stone deterioration during periods of nutrient flux.

REFERENCES

Hueck van der Plas, E.H. (1968). *International Biodeterioration Bulletin*, **4**, 11–28.

Jaton, C. (1973). *Proceedings of the 1st International Symposium on Deterioration of Building Stone*, La Rochelle, Sept. 1972, pp. 149–54.

Kaltwasser, J. (1976). *European Journal of Applied Microbiology*, **3**, 185–92.

Krumbein, W.E. (1972). *Revue d'Ecologie et de Biologie du Sol*, **9**, 283–319.

Lewis, F.F., May, E. & Bravery, A.F. (1985). *Proceedings of the 5th International Congress on Deterioration and Conservation of Stone*, Lausanne, Sept. 1985. Presses Polytechniques Romandes, pp. 633–42.

Lewis, F.J., May, E. & Bravery, A.F. (1987). In *Biodeterioration of Constructional Materials*. The Biodeterioration Society, Occasional Publication No. 3, pp. 45–53.

Murr, L.E. (1980). *Material Sciences and Engineering*, **12**, 121–89.

Parker, C.D. & Prisk, J. (1953). *Journal of General Microbiology*, **8**, 344–64.

Taylor, C.B. & Hutchinson, G.H. (1947). *Journal of the Society for Chemical Industry*, **6**, 54–7.

Wood, P.A. & Macrae, I.C. (1972). *International Biodeterioration Bulletin*, **8**, 25–7.

Biogenic Sulphuric Acid Attack in Sewage Systems

Wolfgang Sand & Eberhard Bock

Institut für Allgemeine Botanik, Abt. Mikrobiologie, Ohnhorststr. 18, D-2000 Hamburg 52, FRG

ABSTRACT

A previous study indicated that thiobacilli were the most abundant microorganisms at corroded sites in concrete sewage systems, and Thiobacillus thiooxidans *the most abundant organism. The cell counts correlated with the extent of corrosion. As a result of the study the particular ecosystem designated the 'Biogenic Sulphuric Acid Corrosion of Concrete' was reproduced in a simulation apparatus. Samples were inoculated with thiobacilli and incubated for 9 months in a test chamber with a humidity > 95% and a temperature of 30°C. Three sulphur sources, hydrogen sulphide, thiosulphate, and methyl mercaptan were tested. Hydrogen sulphide at a concentration of 15 mg/m^3 resulted in severe corrosion and the dominant in the flora was* T. thiooxidans. *Hydrogen sulphide at a concentration of 2 mg/m^3 produced moderate corrosion and the dominants were* T. intermedius, T. novellus *and* T. neapolitanus. *Similar results were obtained with thiosulphate but methyl mercaptan at concentrations of 22 and 2 mg/m^3 caused negligible corrosion and only heterotrophic bacteria and fungi thrived on the concrete blocks, indicating that methyl mercaptan cannot be metabolized by thiobacilli. Field experiments showed that the natural corrosion process could be accelerated at least eight times by the simulation apparatus.*

INTRODUCTION

The involvement of thiobacilli in concrete corrosion has been known since the pioneering work of Parker (1945). Nevertheless scientists and engineers have restricted themselves to describing the phenomena and concrete quality was improved empirically. Thus an increased number of reports have been published, which deal with sulphuric acid corrosion of concrete (Milde *et al.*, 1983). Concrete is now the main construction material for sewage pipelines. In underground construction of new waste water trunks, pipes are pressed into the ground and concrete pipes are most suitable for this purpose.

Concrete is a mixture of calcium oxides, hydroxides and carbonates with quartz minerals and water. The calcium compounds may react with acids:

$$CaCO_3 + H_2SO_4 \longrightarrow CaSO_4 + H_2O + CO_2(\uparrow)$$

The calcium loss results in a reduction in the stability of the structure. Specialized microorganisms produce and excrete strong mineral acids. Thiobacilli oxidize inorganic sulphur compounds, e.g. sulphur or thiosulphate to sulphuric acid, while nitrifiers oxidize inorganic nitrogen compounds like ammonia to nitric acid. Both groups of microorganisms have been shown to cause biodeterioration of stone and concrete (Kirstein *et al.*, 1986; Sand, 1987). In sewage systems thiobacilli are responsible for the sulphuric acid corrosion of concrete. The source of sulphur compounds is the sewage, where aerobic and anaerobic processes yield methyl mercaptan, hydrogen sulphide, and other compounds (Pohl *et al.*, 1984). Knittel *et al.* (1985) have detected thiosulphate in corroded concrete sewers. In this study these three main sulphur compounds were evaluated with regard to their importance for the biogenic corrosion of concrete in simulation experiments, the results of which are reported here.

MATERIALS AND METHODS

Microorganisms: Strains of *T. thiooxidans, T. neapolitanus, T. intermedius* and *T. novellus* from corroded sites in sewage systems. Pure cultures of *T. thiooxidans* were grown on thiosulphate (10 g/litre) or sulphur (10 g/litre) (Milde *et al.*, 1983). Active cells were inoculated into 10 litre carboys and cultivated at 28°C for 10 days. The cells were harvested and sprayed on the concrete samples (Sand *et al.*, 1983; Sand, 1987). A total of 10^{13} cells as a mixture of the 4 species was used as inoculum.

Simulation apparatus: A test chamber of stainless steel (Weiss-Technik, Giessen) was used for the experiments. Test blocks, were stood on end with 10 cm immersed in water and incubated at 30°C and a

RH > 95%. The sulphur compounds were either added as gas (hydrogen sulphide and methyl mercaptan) or sprayed at intervals (5 min/h) on to the blocks (thiosulphate) (see Sand *et al.*, 1983). One experiment lasted for about 9 months.

Corrosion monitoring: The counts of thiobacilli, heterotrophic bacteria and fungi as well as the pH of the surface water and loss of substances were measured as described by Sand (1987). pH-indicators were used for pH-monitoring. Loss of substance was measured gravimetrically. The cell counts were determined either by Koch's plating technique (heterotrophs, fungi) or by a modified three-tube-MPN-techique with 3 selective media (Milde *et al.*, 1983).

RESULTS AND DISCUSSION

Hydrogen sulphide: Two experiments were run with a hydrogen sulphide concentration of 15 ± 3 mg/m^3 in the test chamber. This concentration was not considered to be limiting for growth of thiobacilli as a surplus of sulphur was noted on the test samples throughout the experiment. Hydrogen sulphide is decomposed by both chemical and biological reactions. The first step is a chemical autoxidation of hydrogen sulphide to sulphur. Sulphur is then oxidized biologically by thiobacilli to sulphuric acid. The acid reacts with the calcium compounds of the concrete to form gypsum, which is subsequently lost from the walls because of its solubility (0·5 g/litre) and crystallization effects (ettringite). With 15 g/m^3 H_2S severe corrosion was noted. Within 9 months the test blocks suffered an average weight loss of 3·3% (range 1–6%). The pH values of the surface water decreased to 1. The most abundant organism with 10^7 cells/cm^2 was *T. thiooxidans*. Nine out of ten organisms belonged to this most aggressive species. The cell counts of *T. thiooxidans* correlated with the degree of corrosion. In the second part of the investigation hydrogen sulphide was used at a concentration of 1·5 mg/m^3. In contrast to the previous experiment no surplus of sulphur could be detected. The cell counts of *T. thiooxidans* decreased to 10^6 cells/cm^2 while those for *T. intermedius*, *T. neapolitanus* and *T. novellus* increased to 10^7 cells/cm^2. The extent of corrosion also decreased to a moderate value of 2·1% for loss of weight. The pH-value of the surface water fell to 2–3.

Thiosulphate: The third part consisted of an experiment with thiosulphate as a source of sulphur. Again a surplus of substrate was noted on the test blocks (thiosulphate crystals). The results were comparable with those of the 1·5 mg/m^3 hydrogen sulphide experiment. *T. intermedius*, *T. novellus* and *T. neapolitanus* were dominant on the concrete blocks (10^7 cells/cm^2). The pH-value decreased to a range of 2–3, and the weight loss amounted to 1·8%.

Methyl mercaptan: The last sulphur compound tested was methyl mercaptan (22 mg/m^3). In contrast to the above results no corrosion could be demonstrated with this compound. Neither was there a measurable decrease in pH nor a loss of weight. The numbers of thiobacilli decreased below the detection limit of 10 cells/cm^2. Even though the inoculation was repeated the results were the same. Even when the concentration of methyl mercaptan was reduced to 2 mg/m^3, to exclude possibly poisonous effects of the gas, thiobacilli and hence corrosion were not detectable on the concrete blocks. Only heterotrophic bacteria and fungi grew on the test blocks but their cell counts were likewise reduced by 10^1–10^2.

All the data are summarized in Fig. 1. It is evident that a correlation exists between the cell counts of *T. thiooxidans* and the loss of weight due to biogenic sulphuric acid corrosion.

These laboratory experiments were accompanied by field experiments which were intended to determine the extent to which the simulation system accelerated corrosion compared to the natural sewer system. For this reason identical concrete blocks were incubated under a corrosive sewer atmosphere with an annual mean temperature of 15·5°C, a RH of 85–100%, and a supply of sulphur compounds from 1 to 18 mg/m^3. It took

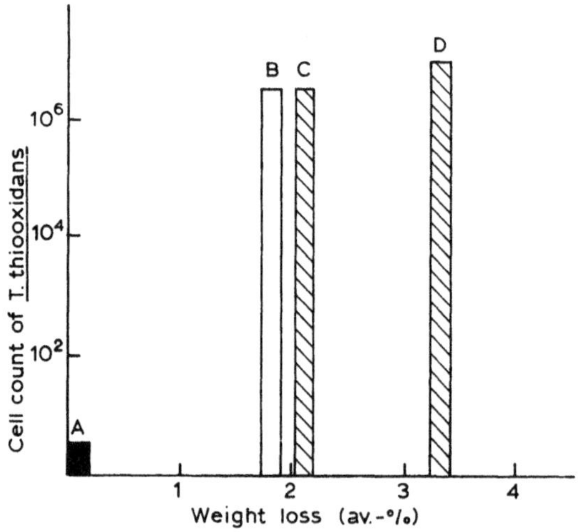

Fig. 1. The importance of hydrogen sulphide, thiosulphate, and methyl mercaptan in the sulphuric acid corrosion of concrete. A = methyl mercaptan (2 or 22 mg/m^3); B = thiosulphate (50 g/week); C = hydrogen sulphide (1·5 mg/m^3); D = hydrogen sulphide (15 mg/m^3).

6 years for the first signs of corrosion to appear compared with 9 months in the simulation experiment. Thus an acceleration factor of at least 8 was calculated in favour of the simulation apparatus. To summarize the results, it became evident that by means of the simulation chamber the biogenic sulphuric acid corrosion of concrete in sewage systems could be reproduced. The cell counts of *T. thiooxidans* can be used as an indicator for the extent of the corrosion. Hydrogen sulphide proved to be the most important sulphur compound for sulphuric acid corrosion. Severe corrosion was noted when a substrate was available in surplus. Thiosulphate and a limiting concentration of hydrogen sulphide were shown to be of some importance, although to a somewhat lesser extent and an intermediate amount of corrosion resulted from the use of this compound at this concentration. Methyl mercaptan has no importance in the biologically induced corrosion process. No corrosion could be demonstrated. By the use of the simulation apparatus the corrosion of concrete can be shown to be eight times as fast as with tests under natural conditions in a sewage system.

ACKNOWLEDGEMENTS

This work was supported by the Freie und Hansestadt Hamburg (Baubehörde—Stadtentwässerung) and by the Bundesministerium für Forschung und Technologie (FRG). The support of R. Bielecki and T. Emmel is gratefully acknowledged.

REFERENCES

Kirstein, K.O., Stiller, W. & Bock, E. (1986) *Beton- und Stahlbetonbau*, **81**, 202–5.

Knittel, D., Valenta, P., Aydin, M. & König, W.A. (1985). *Fresenius Zeitschrift für Analytische Chemie*, **322**, 581–2.

Milde, K., Sand, W., Wolff, W. & Bock, E. (1983) *Journal of General Microbiology*, **129**, 1327–33.

Parker, C.D. (1945). *Australian Journal of Experimental Biology and Medical Sciences*, **23**, 81–90.

Pohl, M., Bock, E., Rinken, M., Aydin, M. & König, W.A. (1984). *Zeitschrift für Naturforschung*, **39c**, 240–3.

Sand, W. (1987) *Applied and Environmental Microbiology*, **53**, 1645–8.

Sand, W., Milde, K. & Bock, E. (1983). In *Recent Progress in Biohydrometallurgy*, ed. G. Rossi & A.E. Torma. Iglesias, Associazione Mineraria Sarda, Italy, pp. 667–77.

Screening for Biocides to Inhibit Biogenic Sulphuric Acid Corrosion in Sewage Pipelines

Thomas Emmel,[a] Holger Brill,[b] Wolfgang Sand[a] &
Eberhard Bock[a]

[a]Institut für Allgemeine Botanik, Abt. Mikrobiologie, Ohnhorststr. 18, D-2000 Hamburg 52, FRG
[b]Schülke & Mayr GmbH, Robert-Koch-Str. 1, D-2000 Norderstedt, FRG

ABSTRACT

Sixteen biocides were examined for their ability to inhibit the growth of thiobacilli Biocide A inhibited the growth of T. thiooxidans, *and biocide B the growth of* T. neapolitanus, T. intermedius, *and* T. thiooxidans. *Biocides A and B were effective at concentrations of less than 0·001%. Tests with heterotrophic microorganisms resulted in minimal inhibitory concentrations, which were at least ten times higher. Concrete blocks were treated with two concentrations of A and B ($10\,g/cm^2$, $30\,g/cm^2$). The blocks were incubated in a H_2S test chamber for the simulation of biogenic sulphuric acid corrosion. During the 10 month test period the numbers of thiobacilli on the surface of the treated concrete blocks, with the higher concentration of biocide, remained 10^1–10^2 below the cell counts of the untreated control block and those treated with the lower concentration. The weight loss amounted to 2%, whereas the control lost 6%.*

INTRODUCTION

For economical and ecological reasons the interest in corrosion caused by microorganisms has increased in recent years. One well studied example is the biogenic sulphuric acid corrosion of concrete in sewage pipelines (Parker, 1945; Sand, 1987). This corrosion is caused by the action of the acidophilic thiobacilli. Several strains of the genus *Thiobacillus* have been isolated from corroded concrete walls of the Hamburg sewer system (Milde *et al.*, 1983). Simulation experiments in a H_2S test chamber (Sand *et al.*, 1983) showed that they produced sulphuric acid in the presence of sulphur compounds. Sulphuric acid weakens the concrete and endangers the structure. Reduced forms of sulphur, such as hydrogen sulphide, methyl mercaptan or dimethyldisulphide (Sand & Bock, 1984) occur in sewage pipelines as a product of the desulphurisation or decomposition of amino acids by aerobic and anaerobic microorganisms (Pohl *et al.*, 1984). One possibility to reduce the extent of the corrosion might be to inhibit the growth of thiobacilli by an application of biocides. This procedure would have two advantages. Old sewage pipelines could be kept at the status quo until sufficient money is available to repair them. New pipelines might be protected from the start. Such a biocide must only be effective against thiobacilli in order to avoid any adverse effect on the sewage and the treatment plant. Laboratory screening of 16 biocides and biocide-combinations was undertaken to find one which met the requirements. As a result two biocides were selected and subjected to a simulation experiment.

MATERIALS AND METHODS

Microorganisms: The experiments were performed with different strains of *T. thiooxidans*, *T. intermedius* and *T. neapolitanus*, isolated from the Hamburg sewer system. They were grown according to the method of Milde *et al.* (1983).

Inhibition experiments: Screening tests were performed with pure cultures in 100-ml Erlenmeyer flasks. Stock solutions of the biocides were prepared, sterilized by filtration, and added before inoculation. Assays were incubated for 3 weeks on a rotary shaker at 220 rpm and 28°C.

Biocide inhibition was determined to be positive:

1. if less than 30% of the thiosulphate had been oxidized or
2. if sulphate production amounted to less than 1 g/litre.

Assays without biocides served as controls. For the two selected biocides the minimal inhibitory concentration (MIC) was determined with six heterotrophic control organisms: *Staphylococcus aureus*, *Escherichia coli*, *Pseudomonas aeruginosa*, *Streptococcus faecium*, *Candida albicans* and *Aspergillus niger*.

Analytical methods: Thiosulphate was determined according to the

'Deutsche Einheitsverfahren' and sulphate was determined turbidimetrically as described by Dugan & Apel (1978).

Simulation experiments: Biogenic sulphuric acid corrosion was produced in a H_2S test chamber. A detailed description of the chamber and the simulation procedures has been given by Sand *et al.* (1983). Concrete blocks were treated by spraying with biocides A and B at 10 and 30 g/m^2. The blocks were incubated for 10 months with an untreated control in the H_2S test chamber at a H_2S concentration of 1·5 mg/m^3 ± 0·5 mg/m^3 at 30°C and a RH > 95%. Corrosion was determined according to Sand (1987) using a three tube most probable number (MPN) technique, with selective nutrient solutions, and weight loss determinations.

RESULTS AND DISCUSSION

Biocide screening: The 16 biocides were selected and tested for their inhibitory and selective action against thiobacilli, especially *T. thiooxidans*. The results are presented in Table 1. Most of the biocides were active against thiobacilli or at least *T. thiooxidans*, especially biocide A. This led to the selection of biocide A for the subsequent simulation experiment. Due to the limited space in the test chamber only one other biocide could be included, biocide B. It displayed inhibition of growth of all the thiobacilli, although the inhibitory concentration was somewhat increased. The MIC-values for the heterotrophic control organisms show that the two biocides are at least ten times more active against

TABLE 1
Screening of 16 Biocides for Inhibition of 3 *Thiobacillus* spp.

Inhibition of	*Biocides*															
	1	*2*	*3*	*4*	*5*	*6*	*7*	*8*	*9*	*10*	*11* *A*	*12*	*13* *B*	*14*	*15*	*16*
T. intermedius/ novellus	−	−	−	−	−	−	+	−	+	−	+	+	+	+	+	−
T. neapolitanus	−	−	−	−	−	−	+	−	+	−	+	+	+	+	+	+
T. thiooxidans	nd	+	−	−	+	+	−	+	+	+	+	−	+	+	−	−

\+ = growth inhibition at biocide concentrations of 0·001% and below.
− = growth inhibition not measurable at biocide concentrations of 0·001%.
nd = not determined.

thiobacilli than against the heterotrophic organisms (data not included).

Simulation experiment: Test blocks prepared from a concrete with low resistance against the biogenic sulphuric acid corrosion (blast-furnace-cement), were used for the biocide tests. They had been preconditioned for 28 days by exposure to standard conditions of temperature, CO_2 and humidity. The biocides were sprayed in two applications and the blocks stored for 14 days to allow for evaporation of the solvents, before being placed in the test chamber. The experiment started with the inoculation of the test blocks and monitoring began after 3 months. At 4 weekly intervals samples were taken from the blocks and tested for pH, the cell numbers of thiobacilli, heterotrophic bacteria, fungi, and loss of weight. In Figs 1 and 2 the results of the untreated block and treated block (30 g/m^2 of biocide A) are presented. The addition of biocide A resulted in a marked depression of the growth of thiobacilli on the concrete test blocks. On average the cell counts lie in the region of 10^5 cells/cm^2, whereas on the untreated block the cell counts amount up to 10^7 and higher. On the control, weight loss increased to 6% by the end of the experiment, whereas the biocide-treated block remained at *c.* 2% throughout. The pH of the surface water remained at 4, whereas on the biocide free block it decreased to 2.

The results obtained with biocide B are similar and are not therefore presented.

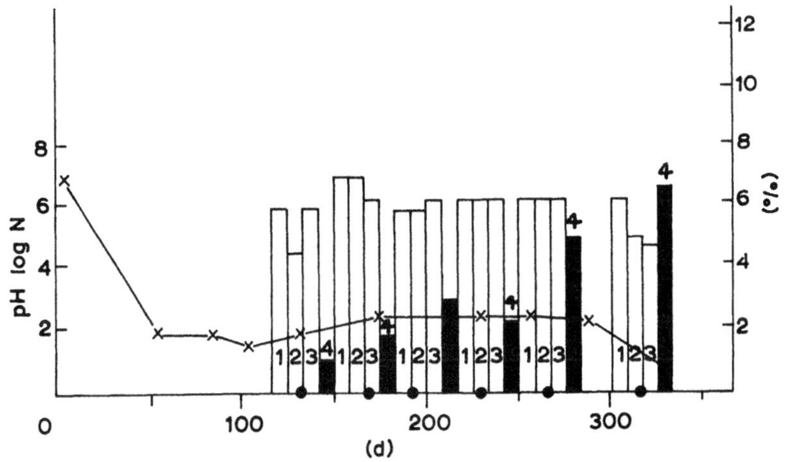

Fig. 1. pH-profile (x), weight loss (4, %-values), and log cell counts of *Thiobacillus (T.) intermedius/novellus* (1), *T. neapolitanus* (2) and *T. thiooxidans* (3) for the biocide-free test block during 10 months' simulation of biogenic sulphuric acid corrosion in the H_2S test chamber at a H_2S concentration of 1·5 mg/m^3.

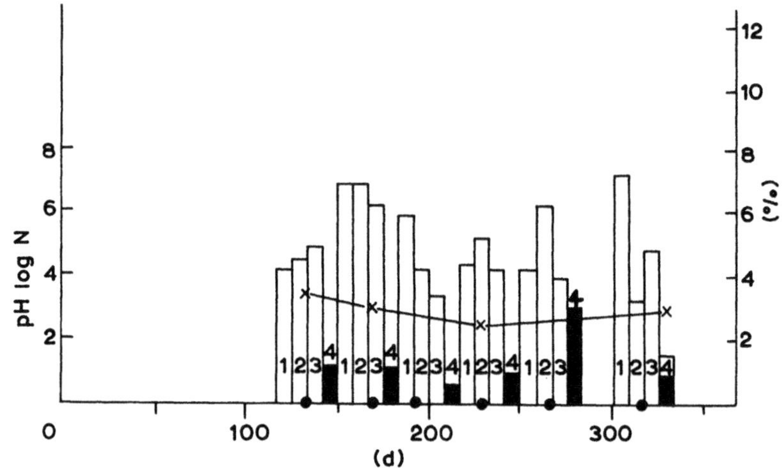

Fig. 2. pH-profile (x), weight loss (4, %-values), and log cell counts of *Thiobacillus (T.) intermedius/novellus* (1), *T. neapolitanus* (2), and *T. thiooxidans* (3) for the biocide treated test block (biocide A, 30 g/m^2) during 10 months simulation of biogenic sulphuric acid corrosion in the H_2S test chamber at a H_2S concentration of 1·5 mg/m^3.

As has been shown by Sand (1987), one simulation experiment is equivalent to 8 years of in-service testing. This is due to the fact that in the test chamber the conditions are optimized for the corrosion causing thiobacilli. The successful performance of biocide A (and B) suggests that they could be used successfully, in sewage works, to prevent or reduce biogenic sulphuric acid corrosion.

REFERENCES

Dugan, P.R. & Apel, W.A. (1978). In *Metallurgical Applications of Bacterial Leaching and Related Microbiological Phenomena*, ed. L.A. Murr, A.E. Torma & J.A. Brierley. Academic Press, New York, pp. 223–50.

Milde, K., Sand, W., Wolff, W. & Bock, E. (1983). *Journal of General Microbiology*, **129**, 1327–33.

Parker, C.D. (1945). *Australian Journal of Experimental Biology and Medical Sciences*, **23**, 81–90.

Pohl, M., Bock, E., Rinken, M., Aydin, M. & König, W.A. (1984). *Zeitschrift für Naturforschung*, **39c**, 240–3.

Sand, W. (1987). *Applied and Environmental Microbiology*, **53**, 1645–8.

Sand, W. & Bock, E. (1984). *Environmental Technology Letters*, **5**, 517–28.

Sand, W., Milde, K. & Bock, E. (1983). In *Recent Progress in Biohydrometallurgy*, ed. G. Rossi & A.E. Torma. Iglesias, Associazione Mineraria Sarda, Italy, pp. 667–77.

Distribution and Effects of Fungi on and in Sandstones

Karin Petersen,[a] Joanna Kuroczkin[b]* Alicja B. Strzelczyk[b] & Wolfgang E. Krumbein[a]

[a]Geomicrobiology Division, University of Oldenburg, P.O. Box 25 03, 2900 Oldenburg, FRG
[b]Institute of Conservation, University of Torun, Poland

ABSTRACT

Monuments built from different sandstones of the Triassic were analysed for microorganisms. Fungal colonies of the individual inoculates were isolated and identified. The species composition as well as the abundance of individual species are discussed.

Light and Scanning Electron Microscopy (SEM) revealed the mode and pattern of colonization of the uppermost stone layer by fungi.

The strains were analysed for their capacity for acid production on various carbohydrate sources. Their capacity for manganese oxidation and deposition of manganese oxides were also examined. This pattern, which in some cases is indicative of iron transfer and precipitation, is discussed in relation to crust formation and exfoliation.

INTRODUCTION

Involvement of fungi in the microbial decomposition of rocks has often been demonstrated (Henderson & Duff, 1963; Williams & Rudolph, 1974; Strzelczyk, 1981; Krumbein & Petersen, 1987).

Biodegradation is induced by acidolytic activity and complexing as well as by reduction and oxidation of mineral cations. Analyses of the distribution of fungi on weathered samples of sandstone monuments compared with their capacity for acid production and iron or manganese oxidation may throw light on the degradation process.

*Presently at the University of Oldenburg, Geomicrobiology Division, FRG.

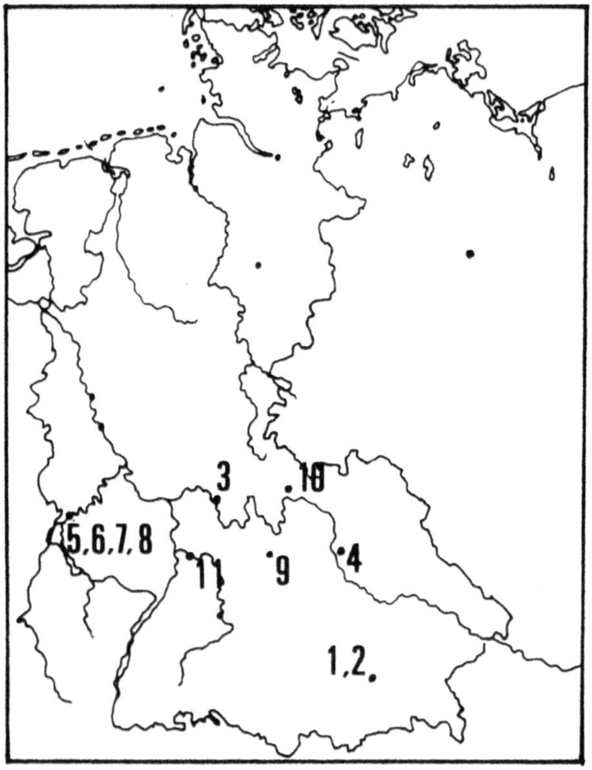

Fig. 1. Locations of monuments. 1, 'Alte Pinakothek', Munich—Cretaceous 'Green' Sandstone; 2, 'Neue Pinakothek', Munich—Cretaceous 'Green' Sandstone; 3. Castle, Aschaffenburg—Quartzitic 'Main' Lower Triassic Sandstone; 4, St Lorenz Church, Nuernberg—Quartzitic 'Burg' Sandstone; 5, St Gangolf Church, Trier—Quartzitic Lower Triassic Sandstone; 6, St Matthias Church, Trier—Quartzitic Lower Triassic Sandstone; 7, St Martin Church, Trier—Quartzitic Lower Triassic Sandstone; 8, 'Porta Nigra', Trier—Quartzitic Lower Triassic Sandstone; 9, Castle Schillingsfuerst, clay-cemented 'Schilf'-Sandstone; 10, Monastery, Birkenfeld, clay-cemented 'Schilf'-Sandstone; 11, Castle, Heidelberg—Quartzitic 'Main' Lower Triassic Sandstone.

Future investigations will therefore concentrate on the activities of those strains that are thought to be important and involved in the process of destruction. This may lead to a greater insight into the activity of microorganisms in the decomposition of rocks and minerals. For this reason the fungal population has to be analysed with respect to species composition, authochtony and aggressivity of the species.

METHODS

Samples which showed typical deterioration characteristics were analysed from the monuments indicated (Fig. 1).

Sandstone monuments from different sites were analysed (Warscheid *et al.*, this volume). Fungal colonies were isolated on Czapek-Dox-Agar (Merck, Nr. 5460, Darmstadt, FRG), identification was carried out at the Centraalbureau voor Schimmelcultures, Baarn, The Netherlands.

Acid production was examined on Czapek-Dox-Agar containing various carbohydrate sources: glucose, galactose fructose, mannose, maltose, saccharose, starch, glycogen and cellulose. The incubation temperature was 28°C. The plates were examined after 14 days for growth and acid production. Bromothymol-blue and Bromocresol-green were used as pH indicators.

Manganese oxidation was investigated according to Zavarzin (1982) and Bromfield (1956) and determined with Leuco Berbelin blue (Krumbein & Altmann, 1973).

RESULTS AND DISCUSSION

From the distribution patterns it is concluded that in further investigations, selected genera or species can be analysed which are most abundant on the monuments under investigation. This, however, does not exclude specific infections and biodeteriorating attacks at certain sites with special environmental conditions.

Some species (*Penicillium* spp., *Phoma* spp. and especially *Exophiala jeanselmei*) infect most of the objects examined, while others (e.g. a *Phaeotheca*-like fungus and *Cladosporium* spp.) are restricted to certain monuments.

Acidolysis by inorganic and organic acids is reported as a major source of biodeterioration of rocks (Strzelczyk, 1981; Bock, 1987). Excretion of organic acids by fungi, leading to acidolysis of stone was demonstrated by several authors (Williams & Rudolph, 1974). Di- and trivalent cations are reported to be solubilized by chelate forming organic acids (Krumbein & Petersen, 1987).

The authors cited stress the biodeteriorating activities of the following genera: *Cladosporium*, *Penicillium* and *Trichoderma*. Acid production by *Phoma* was also reported (Eckhardt, Personal communication). These genera have been demonstrated to be most abundant on the monuments investigated in our studies. They are therefore assumed to be destructive by acid production and/or complex formation. All species were screened for their capacity to decrease pH upon addition of different energy sources. About 50% of the strains decreased in pH value, at least on one carbohydrate source tested. The carbohydrates, if not anthropogenic in origin, may result:

from microbial intermediates, slimes or other extracellular polysaccharides (e.g. glucose galactose, fructose, mannose, maltose, saccharose);
from storage materials (like starch and glycogen);
or from algal cell walls (cellulose);

thereby presenting part of the interactions of microorganisms suggested earlier (Krumbein & Petersen, 1987). A more detailed report of the results will be published elsewhere.

Recently, mechanisms involving 'siederophores' (ferric-ion-specific ligands) of fungi in etching of microfractures were discussed (Krumbein & Schönborn-Krumbein, 1987) and demonstrated in *Penicillium notatum*.

Biotransfer of cations involved in crust formation inducing exfoliation has been demonstrated for famous sandstone monuments like Cologne Cathedral and manganese incrustations by fungal isolates from sandstone (Aschaffenburg Castle) have been reported (Krumbein & Petersen, 1987) (Figs 2 and 3).

Mn^{4+} formation from Mn^{2+} was proved for 21 of the 61 species

Fig. 2. SEM photomicrograph of manganese incrustations on fungal hyphae isolated from a sandstone monument (Aschaffenburg Castle).

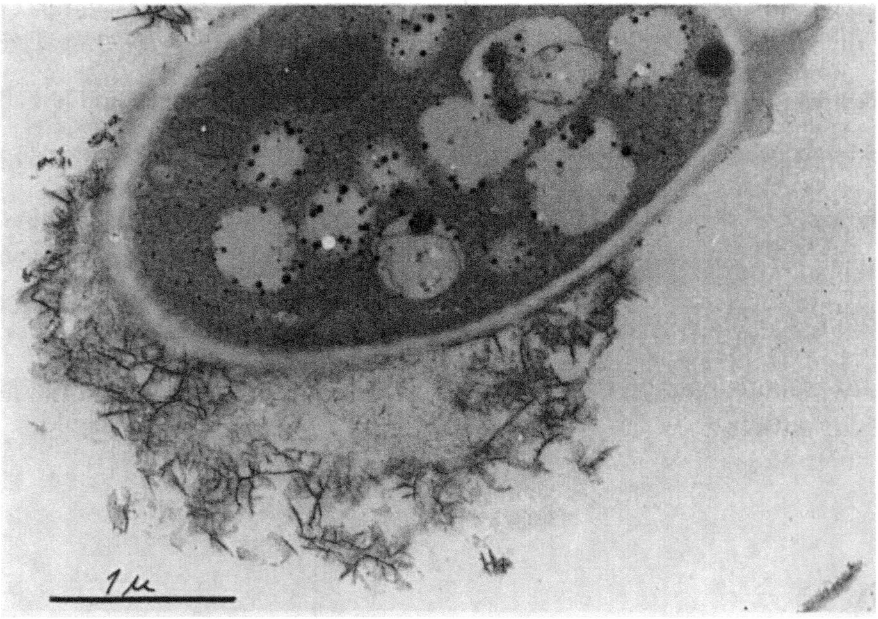

Fig. 3. TEM photomicrograph of manganese incrustations on fungal hyphae isolated from a sandstone monument (Aschaffenburg Castle).

examined, including those most abundant, like *Exophiala jeanselmei*, *Penicillium* spp. and *Phoma* spp. Manganese and iron biotransfer may thereby be involved in many biodeterioration and crust exfoliation processes.

The investigations reported have to be pursued and may outline interaction patterns when combined with bacterial and algal distribution data. This integrated research approach may also depict the influence of different environmental conditions and types of stones with their specific infections and biodeterioration patterns.

ACKNOWLEDGEMENTS

This work was supported by National Grant BAU 5015, of the Minister for Science and Technology, FRG. We are also grateful to S. Ammermann, G. Grote and E. Gründken for their skilful assistance.

REFERENCES

Bock, E. (1987). *Bautenschutz/Bausanierung*, Sonderheft, 42–5.
Bromfield, S.M. (1956). *Australian Journal of Biological Science*, **9**, 238–54.

Henderson, M.E.K. & Duff, R.B. (1963). *Journal of Soil Science*, **14**, 236–46.
Krumbein, W.E. & Altmann, H.J. (1973). *Helgoländer Wissenschaftliche Meeresuntersuchungen*, **25**, 347–56.
Krumbein, W.E. & Petersen, K. (1987). *Bautenschutz/Bausanierung*, Sonderheft, 61–4.
Krumbein, W.E. & Schönborn-Krumbein, C.E. (1987). *Bautenschutz/Bausanierung*, **1**, 14–23.
Strzelczyk, A.B. (1981). In *Economic Microbiology* ed. A.H. Rose. Vol. 6, pp. 61–80.
Williams, M.E. & Rudolph, E.D. (1974). *Mycologia*, **66**, 648–60.
Zavarzin, G.A. (1982). In *The Prokaryotes*, ed. M.P. Starr, H. Stolp, H.G. Trüper, A. Balows & H.G. Schlegel. Springer, Heidelberg, pp. 524–8.

An extended bibliography for this paper may be obtained from the senior author.

Effects of Lichens on Mineral Surfaces

D. Jones,[a] M.J. Wilson[b] & W.J. McHardy[b]

[a]Departments of Microbiology and [b]Mineral Soils, The Macaulay Land Use Research Institute, Craigiebuckler, Aberdeen AB9 2QJ, UK

ABSTRACT

The paper examines the effects and significance of lichens, in particular the crustose species, on natural rock and on building and monumental stones. Weathering and secondary products, as well as the effects of lichen acids, are documented and reference to control of lichen growth on mineral substrates is reviewed. A particular aspect of the diagnostic techniques referred to in the degradation effects of lichen activity is scanning electron microscopy with the associated electron probe micro-analysis facility.

INTRODUCTION

This paper considers the physical and chemical changes occurring in the mineral constituents of rock surfaces colonised mainly by crustose species of saxicolous lichens, as well as the secondary products formed as a result of lichen growth. Earlier studies did not have the benefit of scanning electron microscopes, therefore, research where such instruments have been used will be emphasized. Lichen growth on natural rocks, monumental and building stones are included because of their economic implications.

LICHEN WEATHERING OF ROCK-FORMING MINERALS

The effect of lichens on these minerals varies according to their chemical composition and crystallographic structure. In general, carbonate and ferromagnesian silicate minerals are highly weatherable and show characteristic solution features or are totally decomposed. Pomar *et al* (1975) demonstrated by scanning electron microscopy (SEM) and optical microscopy the solubilising effect of both endolithic (*Protoblastenia immersa*, *Verrucaria parmigera* and *V. sphinctrinella*) and epilithic (*Caloplaca callopisma*, *Aspicilia calcarea*) species on exposed carbonate surfaces in north-east Spain. The ferromagnesian minerals often show extensive surface corrosion even in the early stages of weathering. Olivine grains beneath crustose lichens are marked by deeply penetrating etch pits, whilst the surfaces of augite grains may reveal deep, cross-cutting, trench-like features. The latter probably represent the weathering out of exsolved lamellar intergrowths of a slightly different chemical composition (Jones *et al.*, 1981). With biotite, alteration may be so extensive that the major structural cations, including Mg, Fe, K and Al, are extracted from the structure leaving behind a coherent siliceous relict (Wilson & Jones, 1983). Similarly, the fibrous magnesium silicate mineral chrysotile may be altered to an X-ray amorphous silica gel that still retains the fibrous structure of the parent mineral (Wilson *et al.*, 1981). Feldspars are usually more resistant to lichen weathering although Ca-rich anorthite can be readily reduced to an isotropic, siliceous pseudomorph. The other plagioclase feldspars, such as labradorite and albite, are characterised by the extensive etching out of lamellar intergrowths and by the appearance of diamond-shaped etch pits. K-feldspar weathers in a similar way but is not so intensively corroded. Quartz is resistant to lichen weathering but it is not completely invulnerable. Thus, corrosion features were observed on the constituent grains of a lichen-encrusted quartzite (Jones *et al.*, 1981) substantiating earlier observations by Hallbauer & Jahns (1977) that the lichen *Dimelaena oreina* etched a highly quartzitic substrate.

WEATHERING AND SECONDARY PRODUCTS

Biomineralisation products have been characterised by SEM, transmission electron microscopy (TEM) and electron probe microanalysis. Thus, geothite (containing Al) was found in an ochreous deposit on the surface of the lichen *Tremolecia atrata* colonising a biotite-chlorite schist, identified by TEM with electron diffraction (Jones *et al.*, 1981). Another hydrous iron oxide, ferrihydrite, was identified in the lichen *Pertusaria corallina* colonising basalt (Jones *et al.*, 1980).

SEM and microprobe observations of the thallus of *Pertusaria corallina* colonising manganese ore revealed poorly formed, sub-equant manganiferous crystals with significant amounts of zinc. Clusters of such crystals gave an X-ray diffraction (XRD) pattern identifying them as manganese oxalate dihydrate and this was confirmed by infra-red spectroscopy. SEM examination of the rock-lichen interface of serpentinite and *Lecanora atra* revealed numerous distorted pyramidal crystals, 2–5 μm in diameter, rich in magnesium but containing iron, nickel and manganese as minor components (Wilson *et al.*, 1981). Such crystals were identified as glushinskite, the dihydrated form of magnesium oxalate. In a related study of the weathering of basalt, Jones *et al.* (1980). found masses of plate-like, calcium-rich crystals (from 2–6 μm diam. and 1–2 μm thick) associated with the hyphae of the fungal partner (mycobiont) of the encrusting lichen coloniser (*Pertusaria corallina*). These were identified as whewellite, the monohydrate form of calcium oxalate, by XRD. This form of oxalate was also identified at the rock-lichen interface of limestones by Pomar *et al.* (1975) and also by Ascaso *et al.* (1982) who, in addition, identified bipyramidal crystals of the dihydrate calcium oxalate, weddellite, at the limestone and dolomite-lichen interfaces in *Caloplaca callopisma* and *Diploschistes ocellatus*. It has been proved that the oxalates accumulated are the direct result of the oxalic acid released by the lichen mycobiont (Jones *et al.*, 1980; Wilson *et al.*, 1981).

EFFECT OF LICHEN ACIDS ON MINERAL WEATHERING

Iskandar & Syers (1972) proved that compounds produced by lichens (e.g. stictic and lecanoric acid) bring about chemical change in rocks and minerals, including biotite, granite and basalt. The chelating properties of lichen acids were later investigated by Ascaso *et al.* (1976) who reacted feldspars and mica with suspensions of the lichens *Rhizocarpon geographicum* and *Parmelia conspersa*.

LICHEN WEATHERING OF ROCKS IN BUILDINGS AND MONUMENTS

One of the earliest studies on the deleterious effects of lichens on man-made mineral material was by Mellor (1923) who concluded from her studies that alteration of stained glass in church windows was due to both chemical and mechanical agencies.

Seshadri & Subramanian (1949) described the occurrence of the foliose lichen, *Parmelia tinctorum*, defacing the sculptured panels in the walls of Barabudur, the world-famous Buddhist monument in central Java. Lloyd (1973) showed that porous sedimentary rocks of a calcareous nature are particularly susceptible to physical penetration by lichens but comments that lichen growth can sometimes provide protection against the scouring action of wind and weather and outweighs the damaging effects of the organism. The same author found that the white encrustation on a marble column of the porch of the Church of Santa Fosca on the island of Torcello, Venice, was associated with *Caloplaca aurantia* and *Physcia* sp. (probably *caesia*). Del Monte & Sabbioni (1983) also studied these encrustations, as well as those on the nearby Cathedral of Santa Maria dell' Assunta, and showed by XRD that they consisted almost entirely of weddellite. The stone columns were virtually covered with this material which SEM revealed as masses of rounded, or more frequently, idiomorphic crystals with monoclinic symmetry. The principal organisms associated with the crusts were blue-green algae, not lichens, and it was concluded that the attack on the limestone was the result of the reaction between the oxalic acid secreted by the algae and the calcite in the rock, leading to the precipitation of the insoluble weddellite. Recently, Del Monte & Sabbioni (1986) studied the alteration of marble and limestone surfaces on the facade of Reggio Emilia Cathedral in the central valley of the Po, Italy, and partly attributed the weathered features to algae and lichens, although air pollution now seems to prevail over biological agencies. Algae and lichens colonised internal microfractures (thought to have been enlarged through biochemical activity) of the limestone surfaces of the Cathedral facade and intergranular positions in the marble statue. Both weddellite and whewellite were again associated with the algae and lichens.

Brightman & Seaward (1977) have compiled a useful account of lichens on inorganic building materials derived from mineral sources. They refer to the fact that the lichen flora of abandoned quarries is often sparser and 'different in composition from that of nearby natural outcrops and boulders', suggesting that the man-made faces, caused by quarrying, are generally not aligned to the 'physical and biotic weathering processes that encourage lichen colonisation'. The authors also discuss the colonisation by lichens of bricks, plaster, cement and asbestos-cement sheeting, and refer to the buffering power of materials, particularly the last-named substrate, in influencing the colonisation by lichens in urban areas. Aspects of lichen damage to monuments, sculptures and old buildings are also referred to by Brightman & Seaward (1977) but they state that structural damage is most unlikely. In some cases lichen growth has an aesthetic appeal.

CONTROL OF LICHEN GROWTH

Perhaps one of the more spectacular examples of the damaging effects of lichen growth on sculptured surfaces of stonework is given by Hale (1983) who illustrates abundant lichen crusts on one of the Mayan ruins in Guatemala, Central America which, after treatment with mild fungicides, were largely removed revealing the elaborate carvings and erosive effect of the lichens (Fig. 1). Lloyd (1972) has discussed various approaches to the testing of lichen inhibitors to protect stonework and concrete finishes from lichens as spoilage organisms. One such testing on an asbestos-cement wall colonised by *Leconora dispersa*, using panels to which different chemicals (including ammonium sulphate, copper-8 quinolinolate, pentachlorophenyl laurate and trichlorphenoxy isopropanol) were applied, proved successful in that it was possible to determine 'differences in killing power between treatments'.

Richardson (1973) states that mosses, as well as lichens and algae, may be removed mechanically from monuments or structural masonry either by scraping or with a wire brush, the remaining debris being removed with clean water. The author further suggests that the use of silicone resin water repellents provides a realistic method for the inhibition of lichen and algae. Treatment of stone monuments with a proprietary borate, Polybor, seems to provide freedom from algal and lichen growth for a few years (between 10 and 20 years for lichens).

Fig. 1. Part of ancient stone monument (Stela K at Quirigia, Guatemala, Central America) after treatment with fungicide to kill lichen encrustations. Actual width of area illustrated is 40 cm.

CONCLUSIONS

The application of the scanning electron microscope, with its associated micro-analysis facility, is indispensable in the study of the colonisation of rocks by lichens and the resultant deterioration in the substrate.

ACKNOWLEDGEMENT

We are grateful to Dr Mason E. Hale, National Museum of Natural History, Smithsonian Institution, Washington D.C., for the photograph illustrated as Fig. 1.

REFERENCES

Ascaso, C., Galvan, J. & Ortega, C. (1976). *Lichenologist*, **8**, 151–71.

Ascaso, C., Galvan, J. & Rodriguez-Pascual, C. (1982). *Pedobiologia*, **24**, 219–29.

Brightman, F.H. & Seaward, M.R.D. (1977). In *Lichen Ecology*, ed, M.R.D. Seaward. Academic Press, London, pp. 253–93.

Del Monte, M. & Sabbioni, C. (1983). *Environmental Science & Technology*, **17**, 518–22.

Del Monte, M. & Sabbioni, C. (1986). *The Science of the Total Environment*, **50**, 165–82.

Hale, M.E. (1983). *The Biology of Lichens*. Edward Arnold, London.

Hallbauer, D.K. & Jahns, H.M. (1977). *Lichenologist*, **9**, 119–22.

Iskandar, I.K. & Syers, J.K. (1972). *Journal of Soil Science*, **23**, 255–65.

Jones, R.J. (1965). *Proceedings of the Geological Association*, **76**, 421–33.

Jones, D., Wilson, M.J. & Tait, J.M. (1980). *Lichenologist*, **12**, 277–89.

Jones, D., Wilson, M.J. & McHardy, W.J. (1981). *Journal of Microscopy*, **124**, 95–104.

Lloyd, A.O. (1972). In *Biodeterioration of Materials*, ed. A.H. Walters & E.H. Hueck-van der Plas, Vol. 2. Elsevier Applied Science, London, pp. 185–91.

Lloyd, A.O. (1973). *Atti Petrolio & Ambiente Congresso Roma*, 221–4.

Mellor, E. (1923). *Nature*, **112**, 299–300.

Pomar, L., Esteban, M., Llimona, X. & Fontarnau, R. (1975). *Instituto de Investigaciones Geologicas, Universidad de Barcelona*, **30**, 83–117.

Richardson, B.A. (1973). *Stone Industries*, **8**, 22–6.

Seshadri, T.R. & Subramanian, S.S. (1949). *Journal of Scientific and Industrial Research*, **B8**, 170–1.

Wilson, M.J. & Jones, D. (1983). In *Residual Deposits: Surface Related Weathering Processes and Materials*, ed. R.C.L. Wilson, Special Publication of the Geological Society, London; Blackwell, Oxford, pp. 5–12.

Wilson, M.J. & Jones, D. (1984). *Pedobiologia*, **26**, 373–9.

Wilson, M.J., Jones, D. & Russell, J.D. (1980). *Mineralogical Magazine*, **43**, 837–40.

Wilson, M.J., Jones, D. & McHardy, W.J. (1981). *Lichenologist*, **13**, 167–76.

COSMETICS AND PHARMACEUTICALS

Biodeterioration and Disinfectants

Sally F. Bloomfield

Chelsea Department of Pharmacy, Manresa Road, London SW3 6LX, UK

INTRODUCTION

The role of environmental sites and surfaces and skin surfaces (mainly the hands) in transfer of infection has long been recognised. Under normal circumstances, the healthy adult in the community is relatively resistant to infection but in the hospital environment, where resistance may be compromised, or in various situations where processing of foods, pharmaceuticals and so on provide an 'enrichment environment' for potentially pathogenic bacteria, the survival and transfer of microbial contamination is of considerable concern.

USE AND EFFECTIVENESS OF DISINFECTANTS

A review of the current situation suggests that within these environments disinfectants are used for three main purposes:

(1) To eliminate or prevent establishment of reservoirs of microbial contamination. Investigations of microbial contamination suggest that in any environment (domestic, hospital, catering, etc.) a number of wet or damp sites can be identified as permanent free-living reservoirs or sources of bacteria. These include not only sink surfaces and U-tubes, toilets, baths and drains, but also items such as damp dishcloths and cleaning cloths which may act both as reservoirs and also disseminators of bacterial contamination.

The frequency with which contaminated reservoirs are located at these sites is indicated by a recent investigation of some 200 homes by Scott *et al.* (1982). The results of this survey indicate frequency occurrence of enterobacteria species of the order of 24% (1 in 5) for domestic mops, up to 64% (2 in 3) for kitchen wastetraps, with occurrence of high bacterial counts (> 100 per sample) ranging from 10% for toilets up to 75–80% for bath waste traps and face cloths. Unfortunately, comprehensive data for other food processing environments and the hospital environment are not available, although similar results are indicated by individual studies of specific areas such as hospital and public toilets (Mendes & Lynch, 1976; Scott *et al.*, 1985) and hospital and domestic cleaning utensils (Davies *et al.*, 1968; Baird *et al.*, 1976).

Attempts to eradicate the sources of contamination by routine intermittent application of chemical disinfectants have proved largely unsuccessful. Typical investigations of hospital and domestic environments (Ayliffe *et al.*, 1966; Scott *et al.*, 1984) indicate that, although chemical disinfectants can be used to eliminate contamination from these sites and surfaces, recontamination usually occurs within as little as 1–3 h so that the protection afforded is relatively brief and therefore of little consequence. In some situations, these 'disinfection failures' occur by re-exposure to environmental contamination, whilst in others it results from the regrowth of residual resistant survivors not destroyed by the disinfection process. In order to achieve a permanent reduction in contamination at these sites, the development of sustained/continuous release formulations or disinfectants with residual action is required. Although some success has been achieved in the use of self-disinfecting cloths (Babb *et al.*, 1981) and continuous chlorine release toilet block formulations (Scott *et al.*, 1985), the development of hard surface disinfectants represents a more difficult problem although investigations of sustained action disinfectants have been reported by a number of workers (Klarmann *et al.*, 1953; Lester & Dunklin, 1955; Walter & Foris, 1963; Kingston & Noble, 1964).

It is perhaps regrettable that current pressure from hospital microbiologists against the use of chemical disinfectants for routine 'cleaning', although well-founded, has encouraged the assumption that environmental sources/reservoirs of contamination do not represent a hazard. An examination of the literature indicates substantial numbers of infection outbreaks in the hospital and the general community where environmental sites and surfaces were implicated as the source or means of transfer of an infection outbreak (Kohn, 1967; Taylor *et al.*, 1971; Thomas *et al.*, 1972; Teres *et al.*, 1973; Brown & Baublis, 1977; Zimakoff *et al.*, 1983; Millership *et al.*, 1986).

At the present time, the main areas of usage of chemical disinfectants are:

(2) To block routes of transfer or infection via hand contact and transfer surfaces. In the home and in the various commercial processing environments, critical contact surfaces include food and other preparation surfaces together with hand contact surfaces, particularly in the toilet and bathroom. In hospitals, whereas food preparation and similar hand contact surfaces are of importance, other surfaces such as trolley tops, surfaces where preparative work for patients is carried out by the nurses, and particularly the surfaces of medical and patient care equipment which come into direct contact with the patient must also be considered. Although contamination which occurs on these generally dry surfaces is rapidly and substantially reduced during even relatively short periods of 1–2 h drying time, our own laboratory studies (Bloomfield & Scott, unpublished data) and various hospital-based investigations (Lowbury & Fox, 1953; Pettit & Lowbury, 1968; De Wit *et al.*, 1979) indicate that drying alone is insufficient to guarantee elimination of contamination and that both Gram-positive and Gram-negative organisms may survive for substantial periods of time and in sufficient numbers to allow transfer of infection either directly or indirectly via inanimate objects or the hands.

As far as disinfection of these surfaces is concerned, whereas suspension tests may indicate the effectiveness of a particular agent, the same agent tested against surface contamination may give variable results even under carefully controlled laboratory conditions. Surface contamination is often found to be more resistant to disinfection. Werner and co-workers have demonstrated that the sensitivity of surfaces to chemical disinfection depends critically on the nature of the surface, the nature and number of contaminating organisms, and degree and type of soiling, the method of application, contact time, etc. (Werner *et al.*, 1977; Rathmachers & Borneff, 1977). Incidents in the hospital, domestic and food/pharmaceutical processing environments in which hand contact and transfer surfaces are implicated as the source or route of transfer of contamination/infection are regularly reported, thus indicating that appropriate decontamination procedures should be rigorously enforced (Datta & Pridie, 1960; Linton *et al.*, 1977; Steere *et al.*, 1978; Van Schothurst *et al.*, 1979; De Wit *et al.*, 1979; Palmer *et al.*, 1981). Outbreaks of infection attributed to inadequate disinfection of surgical and patient care equipment are also quite frequently reported.

(3) To block routes of transfer of infection via skin surfaces. Current evidence suggests that one of the most important routes for transfer of contamination, both in the hospital environment and elsewhere, is via the skin surface—mainly the hands. The subject of skin disinfection and its role in prevention of cross infection has been reviewed recently by Reybrouk (1983, 1986).

Removal or destruction of microbial flora from skin surfaces, as opposed to hard surfaces, represents a particular problem. Experimental investigations indicate that, although reductions in skin flora can reduce the probability of infection transfer, total elimination of bacteria from skin surfaces is not achievable. In general it is found that a large proportion of transient hand contamination acquired by contact with contaminated surfaces, food, patients, etc., may be removed by soap and water washing, although this effect can be substantially increased by the use of suitable skin disinfectants. In contrast with hard surfaces, it is also found that a greater reduction can be achieved by repeated application of certain types of agents over a prolonged period. This effect can be demonstrated with agents such as hexachlorophane, Irgasan, iodine and chlorhexidine, which leave an active residue on the skin after rinsing, to produce sustained action between applications. Investigations of the effectiveness of skin disinfectants are reviewed by Reybrouk (1986) and Russell *et al.* (1982)

DISINFECTANT AGENTS AND THEIR PROPERTIES

The majority of formulations which are currently used for disinfection purposes as described above are based on one or a combination of seven chemical agents or types of agents, namely chlorine, iodine, the phenolics, aldehydes, alcohols, quaternaries and chlorhexidine. A general review of these various agents (Russell *et al.*, 1982) indicates that their properties and suitability in different 'in use' situations vary considerably and can depend on any number of 'environmental' factors.

In general, the various agents fall into two groups; whereas the chemically reactive agents such as chlorine, iodine and glutaraldehyde have the advantage of good activity at relatively low concentrations against a wide spectrum of organisms, including spores and viruses, as well as vegetative bacteria and fungi, set against this is the fact that they readily react with other materials, producing loss of activity, toxicity, skin irritancy and deterioration of inanimate surfaces. Although these latter effects may be avoided by the use of more selectively based agents, such as the phenolics, quaternaries and chlorhexidine, which act by disruption of the microbial membrane, these agents in turn may have a limited spectrum of activity, with somewhat variable activity against Gram-negative organisms, spores, fungi and viruses.

Although disinfectant failures 'in use' associated with the occurrence of resistant organisms may be difficult to identify, the importance of this

TABLE 1
Microbial Species Isolated from Contaminated Disinfectants

Date	*Total number of contamination incidents reported*	*Frequency occurrence of:*				
		Total Pseudomonas *spp.*	Pseudomonas aeruginosa	Pseudomonas cepacia	*Other Gram-negatives*	*Gram-positives*
1951–60	5	3	2	0	2	0
1961–70	12	9	4	0	7	0
1971–80	15	12	1	9	3	0
1980–87	10	8	3	4	2	0
Total	42	32	10	13	14	0

'limited activity' is illustrated by reports in the literature in which typically resistant Gram-negative species such as the pseudomonads are found as contaminants of the disinfectant themselves, and in some cases may be implicated as the source of an infection outbreak (Table 1). Investigations associated with the use of contaminated disinfectants are summarised in Table 2. Of the 42 incidents investigated during the period 1951–87, 24 were reported to be associated with or implicated in colonisation of patients or outbreaks of infection. The majority of reports related to phenolic, quaternary ammonium or chlorhexidine formulations, although in 1981/82, there were three incidents of infection arising from contaminated iodophors. The contaminating organisms were mainly pseudomonads and in particular *Pseudomonas aeruginosa* and *P. cepacia*. Other Gram-negative species were also isolated, including *Alcaligenes*

TABLE 2
Reported Incidents of Microbial Contamination in Disinfectant Formulations

Date	*Total number of contamination incidents reported*	*Number of contamination incidents associated with infection*	*Disinfectant formulation involved*			
			Phenolic	*Chlor-hexidine*	*Quater-naries*	*Other*
1951–60	5	3	1	0	5	0
1961–70	12	5	5	5	5	0
1971–80	15	7	3	4	6	3[a]
1980–87	10	9	2	3	1	4[a]
Total	42	24	11	12	17	7

[a]Other disinfectants include Iodophors, and Resiguard (picloxydine and propamidine).

faecalis, *Enterobacter cloacae*, *Escherichia coli*, *Serrata marcescens*, *Flavobacterium meningosepticum* and *Enterobacter agglomerans*. No isolates of Gram-positive species were recorded. Perhaps one of the most worrying aspects is the fact that, despite considerable improvements in QC procedures involving handling, dilution, packaging, shelf life control, etc., there is little evidence of any reduction in the frequency of occurrence of contaminated disinfectants in recent years.

THE RESISTANCE OF MICROORGANISMS TO DISINFECTION

Although antibiotic resistance and its clinical implications have been extensively investigated over the years, problems associated with disinfection failures have received less attention. A review of the literature indicates that disinfection failures probably result from one or a combination of the following:

- (i) There is little doubt that failure to achieve an adequate bacterial action is frequently associated with inappropriate or incorrect use of a disinfectant. Thus, for example, with hard surface and particularly skin disinfectants, disinfection failures may result from failure to ensure adequate contact for the prescribed contact period. It is suggested also that a proportion of incidents results from failure to appreciate the possibility of recontamination during storage, particularly of damp items, such as cloths, where regrowth of residual survivors readily occurs. Current practice in domestic and other catering environments is to rely on disinfection after completion of food processing activities.
- (ii) Alternatively, disinfection failures may result from the presence of organisms which show unusually high resistance to the disinfectant agent.

This resistance to bactericidal agents may be demonstrated experimentally in one of two ways. Most usually the disinfectant agent is added to the test suspension, and the number of surviving bacteria in the reaction mixture determined, at intervals, by viable counting procedures. The typical 'tailing effect' obtained for the majority of chemical agents by plotting log survivors against contact time can be taken to indicate the presence of cells of above average resistance within the population. Alternatively, the presence of resistant organisms may be demonstrated by observing regrowth of residual survivors in the reaction mixture itself as determined by viable counting. This regrowth of residual survivors or 'rebound' effect is frequently reported as the cause of preservative failures in pharmaceutical formulations (Leak, 1983; Spooner & Corbett, 1985). It is probable that a similar effect is responsible for some incidents described previously where large numbers of survivors were recovered from disinfectant formulations. Although the precise biochemical mechanisms of these forms of resistance to disinfectant agents have not been elucidated it is generally assumed that survival and/or regrowth of resistant bacterial cells in disinfectant systems may be attributed to one or a combination of the following:

(a) It may be due to the presence of residual organisms within the population which are intrinsically resistant or which acquire resistance as a response to the presence of the disinfectant agent.

On the basis of current evidence it is concluded that intrinsic resistance to disinfectant agents as observed with certain Gram-negative bacterial species, mycobacteria, and also fungi, is associated with the outer cell envelope which may either constitute a physical barrier or may absorb and retain the agent, in both cases preventing access to the underlying sensitive protoplast. Impermeability of the cell envelope is also thought to be responsible for the natural resistance of some Gram-negative species to antibiotics.

In a similar way, there is strong evidence to suggest that acquired resistance may also result from changes in the physical or chemical structure of the cell envelope, thereby reducing permeability to disinfectants. Investigations of the relationship between the cell envelope and the occurrence of inherent and acquired resistance is reviewed in detail by Russell *et al.* (1982), Russell *et al.* (1986) and Gilbert (in press).

Alternatively, various workers (Beveridge & Hugo, 1964; Fewson, 1967; Beveridge & Tall, 1969) have demonstrated that resistance to phenolic and other bactericides may be associated with the production of oxygenase enzymes causing degradation of the aromatic ring. Organisms most frequently associated with this phenomenon are the pseudomonads, although various other species such as *Acinetobacter* have also been implicated. However, whether the production of such enzymes is responsible for the development of resistance to disinfectant agents is a matter for discussion. It is more likely that it is merely a property exhibited by certain inherently resistant Gram-negative species.

To determine experimentally whether the emergence of a resistant strain from a sensitive population is related to inherent resistance or is the result of a phenotypic adaptation (either due to changes in cell envelope structure or the production of inducible enzymes) presents a difficult problem. Experimental investigations indicate clearly that the antimicrobial resistance of a particular species can be affected by even relatively minor changes in 'growth conditions' (e.g. alterations in nutrients, growth temperature, pH, growth rate, etc.) and on this basis it seems logical to suggest that local variations with a growing culture could produce a population with varying antimicrobial sensitivity. Whether such resistance represents 'inherent' or 'acquired' resistance is then difficult to define. The effect of growth conditions on antimicrobial resistance is also reviewed by Russell *et al.* (1982), Russell *et al.* (1986) and Gilbert (in press).

On the basis that the resistance of these cultures is lost on subculture in the absence of the disinfectant agent, it is assumed that it must be phenotypic in nature. The plasmids may play a role in bacterial resistance to disinfectants. Although there is evidence to suggest that plasmid-mediated resistance may occur with certain metal compounds (mercury, arsenic, etc.) there is little to suggest that, in the clinical situation, plasmid or chromosome transfer is associated with the acquisition of resistance to disinfectants.

(b) Less obviously there are clear indications that the emergence of a resistant from an apparently sensitive microbial population may result from a change either in distribution of the agent or of the microbial cells themselves within the system to allow growth of cells originally held in stasis by the disinfectant agent.

For membrane-active agents such as the phenols and chlorhexidine, antimicrobial activity is thought to result from structural disorganisation of the membrane rather than physical disintegration. Experimental investigations have indicated that re-suspension of cells in water or saline produces substantial desorption of these agents from the cells, thereby indicating the reversibility of the adsorption process. In this situation it is logical to assume that cells originally held in stasis may recommence growth. In bactericidal experiments involving phenols, quaternaries, chlorhexidine, etc., Tween/lecithin systems are deliberately used to achieve desorption of bactericide to encourage colony formation from residual survivors.

Recent investigations also suggest that resistance to disinfectants may occur by changes, not in the distribution of the disinfectant, but in the distribution of the bacteria themselves. Experiments by Hugo *et al.* (1986) indicate that adsorption of *Pseudomonas cepacia* onto the surface of glass containers enables cells to survive in the presence of user-strength chlorhexidine (0·05% w/v). The organisms were firmly attached to the glass, and colonised glass slides could be used to infect uncontaminated chlorhexidine solutions. A similar mechanism was proposed to account for the prolonged survival of *Serratia mercescens* in chlorhexidine solution (Marrie & Costerton, 1981). Gwynn *et al.* (1981) showed that regrowth of *Pseudomonas aeruginosa* in antibiotic solutions resulted from adhesion of organisms to glass surfaces and subsequent dispersal of some of these cells into the culture medium.

Investigations by various workers (Costerton *et al.*, 1978) have indicated that, under suitable conditions, pseudomonads may secrete an extracellular lipopolysaccharide capsule or glycocalyx which enables cells to grow on surfaces as biofilms or microcolonies. These polysaccharide layers may function as a barrier to the uptake of disinfectants, thereby producing resistance. In situations such as water storage tanks, cooling towers, pipework, etc., it has been shown that contaminant organisms may be present as biofilms and that they may show increased resistance to disinfectant agents such as benzalkonium chloride (Marrie & Costerton, 1981). It is suggested that the contamination of poloxamer Iodine solutions, as reported by Berkelman *et al.* (1984) probably originated from massive numbers of bacteria embedded in the PVC pipe used in the manufacturing plant.

CONCLUSION

During the course of this paper, an attempt has been made to review all aspects which may contribute to disinfection failures in the practical situation. In general these failures may result either from inappropriate choice or incorrect use of agents, or alternatively by the presence of organisms of unusually high resistance. This resistance may occur by changes in structure or function of the cells or by a change in the distribution of the agent or the cells within the system.

In recent years, laboratory investigations have done much to increase our understanding of the properties of biocides and to facilitate better formulation and usage of these agents.

Despite this, there is still an urgent need for more detailed investigations, particularly with regard to identifying the underlying causes of disinfection failures and defining their relative importance in the practical situation.

REFERENCES

Ayliffe, G.A.J., Collins, B.J. & Lowbury, E.J.L. (1966). *British Medical Journal*, **2**, 442–5.

Babb, J.R., Collins, B.J., Davis, J.G. & Ayliffe, G.A.J. (1981). *Journal of Hospital Infection*, **2**, 267–72.

Baird, R.M., Brown, W.R.L. & Shooter, R.S. (1976). *British Medical Journal*, **1**, 511–12.
Berkelman, R.L., Anderson, R.L., Davis, B.J., Highsmith, A.K., Petersen, N.J., Bono, W.W., Cook, E.H., Machel, D.C., Favero, M.S. & Martone, W.J. (1984). *Applied and Environmental Microbiology*, **47**, 752–6.
Beveridge, E. & Hugo, W.B. (1964). *Journal of Applied Bacteriology*, **27**, 304–11.
Beveridge, E. & Tall, D. (1969). *Journal of Applied Bacteriology*, **32**, 304–11.
Brown, D.G. & Baubliss, J. (1977). *Journal of Paediatrics*, **90**, 453–7.
Costerton, J.W., Ingram, J.M. & Cheng, K.Y. (1978). *Scientific American*, **238**, 86–95.
Datta, N. & Pridie, R.B. (1960). *Journal of Hygiene, Cambridge*, **58**, 229–41.
Davies, J.G., Blake, J.R. & Woodall, C.M. (1968). *Medical Officer*, **120**, 29–32.
De Wit, J.C., Brockhuizen, G. & Kampelmacher, E.H. (1979). *Journal of Hygiene, Cambridge*, **83**, 27–32.
Fewson, C.A. (1967). *Journal of General Microbiology*, **46**, 255–66.
Gilbert, P. (In press). In *Microbial Quality Assurance in Pharmaceuticals, Cosmetics and Toiletries*. Ellis Horwood.
Gwynn, M.N., Webb, L.T. & Rolinson, G.N. (1981). *Journal of Infectious Diseases*, **144**, 263–9.
Hugo, W.B., Pallent, L.J., Grant, D.J.W., Denyer, S.P. & Davies, A. (1986). *Letters in Applied Microbiology*, **2**, 37–42.
Kingston, D. & Noble, W.C. (1964). *Journal of Hygiene, Cambridge*, **62**, 519–31.
Klarmann, E.G., Wright, E.S. & Shternov, V.A. (1953). *Applied Microbiology*, **1**, 19–23.
Kohn, J. (1967) *British Medical Journal*, **4**, 548.
Leak, R.E. (1983). Ph.D. thesis, London.
Lester, W. & Dunklin, E.W. (1955). *Journal of Infectious Diseases*, **96**, 40–53.
Linton, A.H., Howe, K., Bennett, P.M. & Richmond, M.H. (1977). *Journal of Applied Bacteriology*, **43**, 465–9.
Lowbury, E.J.L. & Fox, J. (1953). *Journal of Hygiene, Cambridge*, **92**, 345–55.
Marrie, J.J. & Costerton, J.W. (1981). *Journal of Clinical Microbiology*, **19**, 687–93.
Mendes, M.F. & Lynch, D.J. (1976). *Journal of Hygiene, Cambridge*, **76**, 183–90.
Millership, S.E., Patel, N. & Chattopadhyay, B. (1986). *Journal of Hospital Infection*, 7, 226–35.
Palmer, S.R., Jephcott, A.E., Rowland, A.J. & Sylvester, D.G.H. (1981). *Lancet*, **1**, 881–4.
Pettit, F. & Lowbury, E.J.L. (1968). *Journal of Hygiene, London*, **66**, 393–406.
Rathmachers, B. & Borneff, M. (1977). *Zentralblatt fur Bakteriologie, Parasitenkunde, Infektionskrankheiten und Hygiene* abt.I, Orig.B, **165**, 43–59.
Reybrouk, G. (1983). *Journal of Hospital Infection*, **4**, 103–10.
Reybrouk, G. (1986). *Journal of Hospital Infection*, **8**, 5–23.
Russell, A.D., Hammond, S. & Morgan, J.R. (1986). *Journal of Hospital Infection*, 7, 213–25.
Russell, A.D., Hugo, W.B. & Ayliffe, G.A.J. (1982). *Principles and Practice of Disinfection, Preservation and Sterilisation*. Blackwell Scientific Publications, London.

Scott, E., Bloomfield, S.F. & Barlow, C. (1982). *Journal of Hygiene, Cambridge*, **89**, 279–93.

Scott, E., Bloomfield, S.F. & Barlow, C. (1984). *Journal of Hygiene, Cambridge*, **92**, 193–203.

Scott, E., Bloomfield, S.F. & Barlow, C. (1985). *Journal of Applied Bacteriology*, **59**, 291–7.

Spooner, D.F. & Corbett, R.J. (1985). *Symposium on 'Formulating Better Cosmetics'*, University of Nottingham, March 1985.

Steere, A.C., Craven P.J., Hall, W.J., Leotsakis, N., Wells, J.G., Farmer, J.J. & Gangarosa, E.J. (1978). *Lancet*, **i**, 319–21.

Taylor, M.R.H., Keane, C.T., Kerrison, I.M. & Strone, J.L. (1971). *Lancet*, **i**, 865–7.

Teres, D., Schweers, P., Bushnell, L.S., Hedley-White, J. & Feingold, D.S. (1973). *British Hospital Journal and Social Services Review (suppl).*

Thomas, M.E.M., Piper, E. & Maurer, I.M. (1972). *Journal of Hygiene, Cambridge*, **70**, 63–73.

Van Schothurst, M., Huisman, J. & Van Os, M. (1979). *Nederlands Tijdschrift voor Geneeskunde*, **122**, 1121–5.

Walter, G. & Foris, S. (1963). *Soap and Chemical Specialities*, **22**, 77–101.

Werner, H.P., Swinke, U. & Werner, G. (1977). *Zentralblatt fur Bakteriologie, Parasitenkunde, Infektionskrankheiten und Hygiene* abt.I, Orig.B, **165**, 20–42.

Werner, H.P., Rathmachers, B. & Borneff, J. (1977). *Zentralblatt fur Bakteriologie, Parasitenkunde, Infektionskrankheiten und Hygiene* abt.I, Orig.B, **165**, 60–77.

Zimakoff, J., Holby, N., Rosendal, K. & Guilbert, J.P. (1983). *Journal of Hospital Infection*, **4**, 31–40.

A bibliography for incidents of microbial contamination in disinfectant formulations, 1951–1987, may be obtained from the author.

Clinical Consequences of Microbial Action on Medicines

S.P. Denyer

Department of Pharmacy, University of Nottingham, University Park, Nottingham, NG7 2RD, UK

ABSTRACT

The most obvious hazard from the microbial contamination of medicines is infection, but more subtle degradative changes can occur which may also be clinically significant. These changes can include the biodegradation of formulation excipients leading to an alteration in the physicochemical characteristics of a product, or metabolism of the drug itself. Such drug transformations, while infrequently found in contaminated pharmaceutical products, are known to occur in the laboratory and after drug administration in vivo *through the action of commensal organisms.*

INTRODUCTION

A medicinal product may encounter microorganisms during manufacture, storage and use. The outcome both for the product and the patient will vary depending upon the type and concentration of organism(s) present, their pathogenicity and metabolic versatility, and the nature of the product. This paper considers the metabolic consequences and possible clinical significance of microorganisms-product contact, whether arising through chance contamination or following introduction of the product to the commensal flora of the patient during therapy. The infective hazard of microorganisms in medicinal products is considered by R. Baird (this volume).

ACTION OF CONTAMINATING ORGANISMS ON MEDICINES

A medicinal product, whether formulated as a simple drug solution or a complex pharmaceutical preparation, possesses characteristics essential for its acceptability and efficacy in use. Given suitable conditions contamination with microorganisms may, by virtue of their growth potential, metabolic versatility and ability to act in concert, lead to product spoilage and subsequent loss of these characteristics. This may, in turn, have important clinical implications in loss of patient compliance and reduced therapeutic efficacy. Product changes can be manifest in several ways: observable physicochemical changes, changes in biological activity, or product toxicity (Parker, 1984).

Observable physicochemical changes

Smart and Spooner (1972) considered gross changes to be those detectable by the senses. Such obvious changes may lead to a justified reluctance on behalf of the patient to comply with dosage regimens and a desire to discontinue medication. Some instances of this type of spoilage are given in Table 1, the deleterious changes being typically demonstrated by turbidity, inhomogeneity, discoloration and aroma. Such phenomena may be secondary effects consequent upon the chemical changes wrought in the formulation by microbial activity, e.g. 'cracking' of an oil in water emulsion by acid production from lipolysis. Changes in the physicochemical nature of a product could conceivably also lead to inefficient dosage delivery and reduced bioavailability through, for instance, rapid sedimentation of drug particles in spoiled suspensions, poor distribution of oil and water phases in 'cracked' emulsions or premature drug release from damaged enteric coated tablets with subsequent drug degradation by gastric acid.

Change in biological activity

The transformation and inactivation of therapeutic compounds by potential contaminants has been demonstrated in the laboratory. Susceptible compounds include antibiotics, analgesics (Smart & Spooner, 1972), thalidomide (Midtvedt & Linstedt, 1970), alkaloids (Grant, 1973) and steroids (Cox & Sewell, 1968). Although there is little evidence to suggest that patients have been seriously affected in this way, it still affords the most direct mechanism by which contaminants can influence drug activity and several examples have occurred in practice (Table 2).

TABLE 1
Examples of Products Suffering Observable Degradation Following Microbial Contamination

Type of product	*Example(s)*	*Observed change*	*Spoilage organism(s)*	*References*
Solutions and mixtures	Simple aqueous solutions	Turbidity	Algae, moulds, bacteria, yeasts	Smart & Spooner (1972)
	Concentrated dill water	Visible growth	*Penicillium roqueforti*	Parker & Barnes (1967)
	Indigestion mixtures	Colour change (white to black)	Sulphate reducers	Spooner (1987)
	Magnesium hydroxide antacids	Ammoniacal odours	Coliforms	Robinson (1971)
Suspending/ thickening agents	Carboxymethyl-cellulose gels/ lubricants	Loss of viscosity	*Trichoderma* sp.	Beveridge (1975)
	Starch, tragacanth, acacia mucilages	Loss of viscosity	—	Beveridge (1975)
	Tragacanth mucilage	Fermentation	Yeasts	Denyer (unpublished)
Emulsions	o/w olive oil emulsions	Separation of phases, discoloration, unpleasant taste and smell	*Trichoderma viride*, *Ps. aeruginosa*, *Asp. flavus*, *Asp. niger*.	Beveridge (1975)
Creams	Calamine	Visible growth	Mould	Smart & Spooner (1972)
	Cosmetic	Visible surface growth	*Penicillium* sp.	Parker & Barnes (1967)
Tablets	Aspirin and Codeine BP	Surface discoloration and acetic acid smell	—	Denyer (Unpublished)
	Nitrazepam	Discoloration, strange smell	*Aspergillus* sp. *Penicillium* sp.	Anon. (1983)
	Prednisolone	Surface discoloration	*Aspergillus* sp.	Beveridge (Unpublished)
	Sugar-coated tablets	Disintegration and 'pitting' of coating	*Penicillium* sp. *Aspergillus* sp.	Parker & Barnes (1967)

TABLE 2
Examples of Microbial Transformation of Biologically Active Compounds in Formulated Products

Product	*Transformation*	*Organism(s)*	*Reference*
Syrup of Tolu	Cinnamic acid ⟶ toluene-like product	*Penicillium* sp.	Wills (1958)
Atropine eye drops	Loss of atropine	*Corynebacterium* sp. *Pseudomonas* sp.	Kedzia *et al.* (1961), Beveridge (1975)
Belladonna and Ipecacuanha paediatric mixture BPC	Loss of atropine	Mixed flora	Beveridge (1975)
Prednisolone tablets	Localised steroid transformation	*Aspergillus* sp.	Beveridge (1983) and unpublished
Hydrocortisone cream	Hydrocortisone ⟶ androst-4-ene-11β-ol-3:17 dione	*Cladosporium herbarum*	Cox & Sewell (1968)
Rose Hip Preparation	Destruction of Vitamin C	Anaerobic bacteria	Beveridge (Unpublished)

Toxicity

Microbial toxins or the products of microbial metabolism, if present in a contaminated pharmaceutical preparation, could conceivably give rise to adverse effects. Potent carcinogens are known to be produced by fungi (Wogan, 1975) and have been reported in artificially contaminated pharmaceutical starches (Fernandez & Genis, 1979). Fragments of dead microorganisms may produce pyrogenic reactions if administered intravenously or cause allergic reactions by other routes (Smart & Spooner, 1972).

ACTION OF COMMENSAL ORGANISMS ON MEDICINES

Systemic drug administration may be achieved by any one of several routes (enteral, intravascular, dermal or pulmonary), the primary objective being to ensure an adequate drug concentration at its site of action. Very often the drug or its metabolites will become exposed to the metabolic activity of the gastrointestinal flora, either as a result of incomplete absorption following oral administration or through metabolite secretion into the gastrointestinal tract via the bile. The products of bacterial metabolism may then be absorbed into the portal circulation, thereby extending the duration of drug or metabolite action. When re-absorption takes place following initial biliary excretion the phenomenon is termed enterohepatic cycling. Typically, the majority of microbial transformations are hydrolysis and reducing reactions, thus frequently reversing the predominantly oxidative and conjugative changes produced by the liver, the principal organ for drug metabolism. The evidence for microbial transformation of biologically active compounds within the digestive tract of man and animals has been extensively reviewed and significant effects on drug action, pharmacokinetics and toxicity have been reported (Goldman, 1982; Renwick, 1986).

In addition to the digestive tract, the skin surface also supports a resident microbial population (Noble, 1981). Members of this cutaneous flora have been shown to be capable of metabolising drugs which may be applied topically. These include steroid esters such as betamethasone 17-valerate and glyceryl trinitrate; prolonged clinical contact with the therapeutic agent may encourage selection of active metabolisers. Although in practice no evidence exists to indicate that these transformations are clinically significant, staphyloccal metabolism of glyceryl trinitrate can be demonstrated to have a theoretical relevance in percutaneous therapy (Denyer *et al.*, 1985).

DISCUSSION

Microorganisms have the potential to cause significant changes in formulated medicines and active ingredients both *in vitro* and *in vivo*. These effects may be clinically relevant for three reasons: overt spoilage may result in loss of patient compliance, changes in formulation characteristics could lead to inadequate dosage delivery, and transformation of therapeutic agents may alter pharmacological activity. In the latter instance, responsible organisms may be accidental contaminants or members of the patient's normal commensal flora.

REFERENCES

Anon. (1983). *Pharmaceutical Journal*, **230**, 300.

Baird, R.M. This volume, pp. 152–6

Beveridge, E.G. (1975). In *Microbial Aspects of the Deterioration of Materials*, SAB Technical Series No. 9, ed. D.W. Lovelock & R.J. Gilbert. Academic Press, London, pp. 213–35.

Beveridge, E.G. (1983). In *Pharmaceutical Microbiology*, 3rd edition, ed. W.B. Hugo & A.D. Russell, Blackwell Scientific Publications, Oxford, pp. 334–53.
Cox, P.H. & Sewell, B.A. (1968). *Journal of the Society for Cosmetic Chemists*, **19**, 461–7.
Denyer, S.P., Guy, R.H., Hadgraft, J. & Hugo, W.B. (1985). *International Journal of Pharmaceutics*, **26**, 89–97.
Fernandez, G.S. & Genis, M.V. (1979). *Pharmaceutica Acta Helvetiae*, **54**, 78–81.
Goldman, P. (1982). In *Microbial Transformation of Bioactive Compounds, Volume II*, ed. J.P. Rosazza. CRC Press Inc. Florida, pp. 43–65.
Grant, D.J.W. (1973). *Journal of Applied Bacteriology*, **36**, 47–59.
Kedzia, W., Lewan, J. & Wismienski, T. (1961). *Journal of Pharmacy and Pharmacology*, **13**, 614–19.
Midvedt, T. & Linstedt, G. (1970). *Acta Pathologica et Microbiologica Scandinavica, Section B*, **78**, 488–94.
Noble, W.C. (1981). *Microbiology of Human Skin*, 2nd edn, Lloyd-Luke, London.
Parker, M.S. (1984). *International Biodeterioration*, **20**, 151–6.
Parker, M.S. & Barnes, M. (1967). *Soap, Perfumery and Cosmetics*, December Issue, 1–4.
Renwick, A.G. (1986). In *Microbial Metabolism in the Digestive Tract*, ed. M.J. Hill. CRC Press Inc., Florida, pp. 135–53.
Robinson, E.P. (1971). *Journal of Pharmaceutical Sciences*, **60**, 604–5.
Smart, R. & Spooner, D.F. (1972). *Journal of the Society for Cosmetic Chemists*, **23**, 721–37.
Spooner, D.F. (1987). In *Microbial Quality Assurance in Non-sterile Pharmaceuticals, Cosmetics and Toiletries*, ed. S.F. Bloomfield. Ellis Horwood Ltd, Chichester (in press).
Westwood, N. & Pin-Lim, B. (1971). *Pharmaceutical Journal*, **207**, 99–102.
Wills, B.A. (1958). *Journal of Pharmacy and Pharmacology*, **10**, 302–5.
Wogan, G.N. (1975). *Annual Review of Pharmacology*, **15**, 437–51.

An extended bibliography for this paper may be obtained from the author.

Incidence of Microbial Contamination in Medicines in Hospitals

Rosamund M. Baird

North East Thames Regional Health Authority, Pharmaceutical Microbiology Laboratory at St. Bartholomew's Hospital, West Smithfield, London EC1, UK

ABSTRACT

In the past decade considerable improvements in the manufacturing facilities of hospital pharmacies have occurred. This investigation assesses whether the microbiological quality of 9479 non-sterile medicines, made in nine hospital pharmacies, over a 5 year period, had been affected by these improvements. During the study, the incidence of contamination was found to decrease to a reasonably constant level (average 9·32%) although rates varied between individual hospitals. In total 90% of products showed low levels of contamination (less than 10^2 *colony forming units/ml or g) and aerobic spore bearers accounted for 72·2% of isolates. A range of Gram-negative isolates, mainly* Pseudomonas *spp., were found in a small number of products, in some cases in high numbers up to* 10^5 *cfu/ml or g. The significance and acceptability of such isolates in pharmaceutical products are discussed.*

INTRODUCTION

The problem of microbial contamination has been appreciated for several decades, but much of the information has been anecdotal. Two main considerations have emerged, first concerning the spoilage of products, and any associated financial loss, and secondly the recognition of various potential health risks which may be associated with the use of contaminated products.

Contamination is known to occur both in manufacture and in use and is generally better understood and more easily controlled during manufacture. The incidence of contamination in non-sterile medicines has been reported in recent years (Kalling *et al.*, 1966). Concern was expressed in 1971, at the isolation of *Pseudomonas aeruginosa* from 2·7% of medicines in use in hospitals in England and Wales (Report of Public Health Laboratory Service Working Party, 1971). Subsequently *P. aeruginosa* has been shown to be a notorious contaminant of pharmaceuticals, having been isolated from a wide range of used and unused products (Robinson, 1971; Beveridge & Hope, 1971). In one survey, in the mid-1970s, 9% of hospital prepared products were found to be contaminated with this organism; at the same time a strong correlation was shown between medicament and environmental strains (Baird *et al.*, 1976).

In the past decade, however, significant changes have taken place. Manufacturing facilities have been improved considerably, potential sources of contamination have been well documented and the general awareness of the problem of microbial contamination and its control has increased (Underwood, 1983). One recent survey showed a significant improvement in the microbiological quality of hospital-manufactured products over a 9 year period (Baird, 1985). The purpose of the present investigation was to examine the incidence of contamination in non-sterile pharmaceuticals manufactured in nine hospitals over a 5 year period.

METHODS

Representative samples were collected aseptically from finished products. A 1 g or 1 ml sample was homogenised in 20 ml of nutrient broth containing 3% Tween 80 and appropriate neutralisers (Baird *et al.*, 1976). A total viable count was carried out by plating out 0·1 ml aliquots onto blood agar (Tissue Culture Services Ltd) and incubating aerobically at 37°C for 24 h. Resulting colonies were counted and a total viable count calculated. At the same time the broth homogenate was also incubated aerobically at 37°C for 24 h, and then subcultured onto blood agar and cetrimide agar (0·03%). All isolates were purified and identified (Baird, 1985).

RESULTS AND DISCUSSION

Between 1982 and 1986, 9479 non-sterile pharmaceutical products from nine hospitals were examined for microbial contamination, as shown in

TABLE 1
Incidence of Contamination in 9479 Products During a 5 Year Period

	1982	*1983*	*1984*	*1985*	*1986*	*Total*
No. examined	316	2 370	1 683	2 782	2 328	9 479
% contaminated	30·9	10·2	8·0	8·8	9·6	10·0
% Asb	25·3	8·1	5·6	5·7	6·7	7·2
% Gram + cocci	6·0	1·3	2·0	2·4	2·6	2·2
% Gram − rods	0	0·8	0·4	0·7	0·3	0·5
% with < 200 cfu/g or ml (minimum level of detection)	69·3	89·7	92·0	91·2	90·4	90·0

Table 1. In 1982, there was a high incidence of contamination, but subsequently this declined markedly and remained at a reasonably constant level. Although contamination rates varied considerably between hospitals, there was a marked improvement in the microbiological quality of products during the 5 year study (see Table 2). The differences may have been due to differences in manufacturing facilities or to the range of products tested, since they may vary in their susceptibility to contamination (Report of Public Health Laboratory Service Working Party, 1971). Aerobic spore bearers accounted for the majority of isolates (72·2%) but coagulase negative, Gram-positive cocci and Gram-negative bacteria were also found (22·2% and 5·5% respectively). The species of free living Gram-negative bacteria are shown in Table 3. *P. aeruginosa*

TABLE 2
Incidence of Contamination in Products from Nine Hospitals (%)

Hospital	*1982*	*1983*	*1984*	*1985*	*1986*
1	66·6	13·8	13·4	9·9	14·4
2	13·8	3·5	1·7	7·3	6·6
3	NT	7·5	8·2	14·0	7·3
4	25·0	13·9	7·5	4·8	5·3
5	18·0	4·2	1·6	2·2	2·9
6	NT	22·7	5·8	12·8	4·0
7	NT	15·9	10·9	6·3	8·5
8	34·2	12·1	7·1	8·3	15·2
9	47·0	25·7	20·3	19·3	21·3
10	30·9	10·2	8·0	8·8	9·6

TABLE 3
Distribution of Gram-Negative Isolates in 53 Products

	Creams	*Natural powders*	*Aqueous[a] products*	*Suspensions*
Acinetobactor sp.			1	1
Citrobacter sp.			1	
CDC sp.			4	
Enterobacter sp.		2	3	2
Klebsiella sp.			1	1
Moraxella sp.			1	
Serratia sp.			2	
Pseudomonas sp.	2		27	3
P. aeruginosa			2	
Total	2	2	42	7

[a] Disinfectants, mixtures, shampoos, solutions.

was found on only two occasions (soapless shampoo and alkaline nasal douche) but *Pseudomonas* spp. accounted for 62% of all Gram-negative isolates. The majority of isolates were found in aqueous products, such as disinfectants, etc., emphasising the need for special care when preparing these products, particularly in the control of raw materials, process water, and suitable bottle washing techniques. Total viable counts ranged from $< 10^2$ colony forming units g or ml (the minimum detectable level) to high counts in excess of 10^5 cfu/ml in some shampoos.

Extensive improvements in hospital manufacturing facilities in the past decade and increased staff awareness of the problems of microbial contamination, and its control, have had a significant effect on the overall microbiological quality of products made in hospitals. The results show a reduction in the overall level of contamination, 90% of products containing < 200 cfu/g or ml (compared with 68% with < 500 cfu/g or ml in the PHLS report), and a significant reduction in the incidence of *P. aeruginosa* contamination to 0·02% compared with earlier reports (PHLS, 1971; Baird *et al.*, 1976). Likewise *Pseudomonas* spp. were isolated from 0·35% of samples, compared with previous findings of 8·7% (PHLS, 1971). Nevertheless, the continuing presence of Gram-negative isolates cannot be condoned. For although such free-living opportunist pathogens are ubiquitously distributed in the hospital environment, they have also been held responsible for a growing number of patient infections, particularly in compromised and debilitated patients (Bassett, 1971; Ramphal & Kluge, 1971). Such organisms are easily identified nowadays using biochemical test kits. A concerted effort should now be made to reduce the risk of product contamination with these organisms, by applying well known and tested principles of contamination control to the manufacturing process.

REFERENCES

Baird, R.M. (1985). *Pharmaceutical Journal*, **234**, 54–5.
Baird, R.M., Brown, R.W.L. & Shooter, R.A. (1976). *British Medical Journal*, **1**, 511–12.
Bassett, D.C.J. (1971). *Proceedings of the Royal Society of Medicine*, **64**, 980–6.
Beveridge, E.G. & Hope, I.A. (1971). *Pharmaceutical Journal*, **207**, 102–3.
Kallings, L.O., Ringertz, O., Silverstolpe, L. & Ernerfeldt, F. (1966). *Acta Pharmaceutica Suecica*, **3**, 219–28.
Ramphal, R. & Kluge, R.M. (1971). *American Journal of Medical Science*, **277**, 57–66.
Report of the Public Health Laboratory Service Working Party (1971). *Pharmaceutical Journal*, **207**, 96–9.
Robinson, E.P. (1971). *Journal of Pharmaceutical Sciences*, **60**. 604–5.
Underwood, E. (1983). In *Pharmaceutical Microbiology*, 3rd edn, ed. W.B. Hugo & A.D. Russell, Blackwell Scientific Publications, Oxford, pp. 317–33.

Drug Interactions with Plastics and their Biological Consequences

M.C. Allwood

Regional Research & Development Laboratory, East Anglian Regional Health Authority, Central Pharmacy, Addenbrooke's Hospital, Cambridge, UK

ABSTRACT

Many types of plastic are now used to manufacture equipment used in the administration of injections. A number of interactions between parenteral drugs and particular plastics have been identified. Some drugs sorb to plastics, leading to loss of pharmacologically-active compounds. Major groups of drugs exhibiting this phenomena include polypeptides, nitrates and benzodiazepines. Some parenteral drug vehicles extract components from the plastics leading to potential hazards to patients.

INTRODUCTION

The injection of drugs, commonly referred to as parenteral administration, forms an essential part of modern therapy in hospitals. The potent nature of many of these drugs demands that great care is required to administer them safely and effectively. This is achieved by ensuring that the injection is delivered to the appropriate point in the body, at a predetermined rate and in a suitable vehicle. An essential aspect of effective, safe administration is the equipment used for drug delivery. Modern therapy places great demands on parenteral administration systems and plastics technology has made the greatest contribution to developments in use of these drugs. This explains the reason why virtually all injections are delivered to the patient using equipment with plastics as the major components. However, because the nature of the administration may vary widely, depending on the drug, disease state and type of patient, many different types of plastics have been employed. Whilst any polymer must comply with rigorous safety standards, specific requirements for particular applications may demand a material that is rigid or very flexible, thick or thin-walled, fabricated into a variety of physical forms and compatible with a wide variety of liquids and drug entities. It is impossible to achieve all of these requirements using one polymer and an increasing range of plastics are used. Whilst fulfilling the functional demands required it is inevitable that other aspects of drug delivery may be compromised. One particular concern is interaction with the plastics employed. This paper summarises the current state of knowledge of interactions between the many plastics now used in parenteral drug administration and the drugs themselves. The clinical consequence of these interactions is also assessed.

DRUG–PLASTIC INTERACTION MECHANISMS

Knowledge of drug uptake into plastic materials suggests that the various physicochemical processes involved may be described by three different mechanisms: absorption, adsorption or permeation.

Absorption

Absorption of specific drugs into a plastic depends on the nature of the plastic and the lipophilicity of the drug moiety. Because most drugs are weak acids or bases, pH has an important influence since only non-ionised drugs will absorb. In practice, drug–plastic absorption appears to occur almost exclusively to polyvinyl chloride (PVC) material. The plasticiser incorporated with the polymer may play an important role by providing the appropriate hydrophobic environment for drug migration. Some studies suggest that any un-ionised non-polar drug moiety will absorb into PVC (Moorhatch & Chiou, 1974; Kowaluk *et al.*, 1981). However, since most drugs in parenterals are either present as salts, and therefore ionised in aqueous, neutral vehicles, or are relatively polar, only a handful of therapeutic agents absorb to any significant degree on to PVC. These include benzodiazepines (in an aqueous vehicle (Mason *et al.*, 1981)) and vitamin A acetate (but not the palmitate (Moorhatch & Chion, 1974; Allwood, 1982)). The absorption of lignocaine, thiopentone and chlorpromazine to PVC has also been documented but the process is relatively minor and is highly pH-dependent.

The absorption process between a solid and aqueous solute is relatively slow and an equilibrium is only established after hours of contact. Similarly, saturation of binding sites in the plastic matrix usually requires long contact times. The process can be described by a constant partition model with very high partition coefficients. Consequently the fraction absorbed is relatively independent of concentration. In practical terms, the proportion of drug delivered is a relatively small fraction of the concentration in the original infusate. Also the dose delivered remains relatively constant over a long period, until saturation of the plastic is achieved. The fraction of drug delivered will then rise.

Adsorption

The binding of drugs to surfaces depends on ionisation, molecular polarity and surface charge on the plastics, and may be described as adsorption. The surfaces of most plastics are poor conductors, relatively non-polar and carry little or no surface charge unless generated as a consequence of the manufacture or storage of the item. Surface adsorption is likely to be non-specific and is a relatively rapid process (Whalen *et al.*, 1979). A particularly important example is insulin. This process is believed to involve relatively strong and non-specific forces leading to the rapid formation of an insulin monolayer. In fact insulin binds to many different types of surfaces including PVC, polyethylene and glass (Whalen *et al.*, 1979; Mitrano & Newton, 1982; Schildt *et al.*, 1978). The phenomenon can be described as Langmuirian; the amount absorbed is proportional to concentration. Surface saturation can be achieved rapidly in the presence of high concentrations of insulin. The process of adsorption of insulin is reduced in the presence of proteins (e.g. albumin) that compete for the same binding sites. The adsorption of other small peptides such as secretin and interferon have also been noted (Miyata *et al.*, 1974).

The clinical effects of drug adsorption are a consequence of the rapidity of the process, resulting in a rapid reduction during initial therapy. But, as the adsorbing surfaces become saturated, the dose of drug received by the patient may rise rapidly. The proportion and total quantity of insulin removed depends on the concentration, total surface area of the delivery system and the rate of administration. Losses can be minimised by using a 'high concentration — small container' approach, for example a syringe driven by a syringe pump is preferable to dilution into an infusion and delivery through an administration set.

Permeation

Some drugs are absorbed into the plastics matrix of PVC tubing and a proportion of the compound permeates through the matrix to be released from the outer surface into the atmosphere. The kinetics of chlormethiazole loss from solutions in contact with PVC sets indicates that this is likely to take place (Tswei *et al.*, 1980). The process is characterised by slow but continuous loss of drug during infusion, and that the plastic shows no evidence of becoming drug-saturated, despite long-term contact with it. Losses are almost continuous and the dose delivered does not change substantially during continuous therapy. It has also been noted that the distinctive odour of chlormethiazole can be discerned in a ward environment when it is being administered by slow infusion. Recent studies suggest that glyceryl trinitrate and isosorbide dinitrate also permeate through PVC tubing (Lee, 1986). Permeation can also take place through nylon and silicone tubing (Allwood, in press).

CLINICAL ASPECTS OF DRUG DELIVERY THROUGH PLASTICS

In practice only a small number of drugs exhibit significant interaction with plastic or other polymer-containing equipment. The details of these drugs are summarised in Table 1. The quantity of any drug actually delivered at any time will depend on a number of factors. These include the type of plastics in contact with the solution, the nature of the vehicle, the rate of administration and concentration of the drug. It also depends on the total time of administration, how frequently the set is changed, whether it is diluted in an infusion container, the chamber of a burette set

TABLE 1
Drugs with Clinically Significant Sorption to Plastics Used in Parenteral Drug Administration

Drug	*Evidence of significant sorption to:*					
	Glass	*PVC*	*PE*	*CP*	*PBD*	*EVA*
Insulin	Yes	Yes	Yes	Yes	Yes	
Secretin		Yes	Yes			
Interferon		Yes				
Diazepam		Yes				
Glyceryl trinitrate		Yes				Yes
Isosorbide dinitrate		Yes				Yes
Chlormethiazole		Yes				
Vitamin A acetate		Yes				

PVC: polyvinyl chloride; PE: polyethylene; CP: cellulose proprionate; PBD: polybutadiene; EVA: ethyl vinyl acetate.

or a syringe and the nature of the drug–plastic interaction. Because so many variables are involved, it is not possible to make accurate predictions of drug delivery if drug–plastic interactions take place. Drug sorption to the container or delivery system may not, in fact, be clinically hazardous providing its occurrence is recognised and understood by all concerned. Losses can be compensated for by careful monitoring of physiological parameters at regular intervals with appropriate adjustment of the drug delivery rate. Alternatively it may be possible to reduce sorption losses by simply changing the delivery system. For example, insulin should not be delivered after dilution in a large volume infusion and administered through a set, but by slow administration of a concentrated solution using a syringe pump.

EXTRACTION OF PLASTIC COMPONENTS BY DRUGS

It has been recognised for many years that blood stored in PVC bags will contain significant quantities of diethylhexylphthalate (DEHP), a plasticiser. Such compounds are potentially toxic, especially after chronic exposure (Kaul *et al.*, 1982). Patients infused with 2–3 litres of stored blood may receive up to 20 mg DEHP. Plasticisers are insoluble in water but can be extracted from PVC into lipid emulsions. Therefore, PVC bags should be avoided where possible. Administration sets containing PVC components will also release DEHP into infused fat emulsion. It has been estimated that up to 3 mg DEHP/day can be administered as a consequence of this effect (Allwood, 1986). For patients on long-term TPN, this represents a low level chronic exposure to these toxins.

Other polymers used in infusion administration equipment can release extractables into injections. Various rubber formulations used in plastic syringes or injection pots will release zinc-organic vulcanising agents. Dyes and filler may also be extracted. Silicone oil used to lubricate syringe barrels can often be dispersed in the injection. Most plastics are relatively inert although organic extractables have been observed in injections passed through nylon manometer lines (unpublished observation). More information concerning the occurrence and consequences of these effects is required.

CONCLUSIONS

The interaction of certain injections with plastic materials used in administration of the drugs is now a well documented phenomenon. It is important that these reactions are properly understood and possible risks to patients minimised to ensure safe therapy. Developments of new non-reactive plastic sets is to be encouraged. The use of such specialised equipment when dictated by the type of drug administered or the nature of the patient is to be commended.

REFERENCES

Allwood, M.C. (1982). *Clinical Nutrition,* **1**, 63–70.
Allwood, M.C. (1986). *International Journal of Pharmaceutics,* **29**, 233–6.
Allwood, M.C. (in press). *International Journal of Pharmaceutics.*
Kaul, A.F., Souney, P.F. & Osathanondh, R. (1982). *Drug Intelligence and Clinical Pharmacy,* **16**, 869–92.
Kowaluk, E.A., Roberts, M.S., Blackburn, H.D. & Polack, A.E. (1982). *American Journal of Hospital Pharmacy,* **38**, 1308–14.
Lee, M.G. (1986). *American Journal of Hospital Pharmacy,* **43**, 1945–50.
Mason, N.A., Cline, S., Hyneck, M.L., Berudi, R.R., Ho, N.F.H. & Flynn, G.L. (1981). *American Journal of Hospital Pharmacy,* **38**, 1449–54.
Mitrano, F.P. & Newton, D.W. (1982). *American Journal of Hospital Pharmacy,* **39**, 1491–5.
Miyata, M., Rayford, P.L. & Thompson, J.C. (1974). *New England Journal of Medicine,* **300**, 95.
Moorhatch, P. & Chiou, W.L. (1974). *American Journal of Hospital Pharmacy,* **31**, 72–8.
Schildt, B., Ahlgren, T., Berghen, L. & Wendt, Y. (1978). *Acta Anaesthesia Scandinavica,* **22**, 556–62.
Tswei, S.E., Nation, R.L. & Thomas, J. (1980). *European Journal of Clinical Pharmacology,* **18**, 333–8.
Whalen, F.J., LeCain, W.K. & Latiolais, C.J. (1979). *American Journal of Hospital Pharmacy,* **36**, 330–7.

The Degradation of Preservatives of Microorganisms

W.B. Hugo

618 Wollaton Road, Nottingham NG8 2AA, UK*

ABSTRACT

The ability of microorganisms to decompose aromatic compounds, many of them with intrinsic antimicrobial power, have been known from the beginning of the century. The early work arose from the finding that soil biocides disappeared from the soil and were traced to their metabolism by soil microorganisms. This finding gave rise to extensive biochemical investigations into the mechanisms of decomposition.

This background has been extended to the field of preservatives and is now well documented as one of the possible reasons for preservative failure.

This paper highlights some examples of this phenomenon.

INTRODUCTION

It may at first seem strange that chemical substances known for their ability to inhibit the growth of microorganisms may, for some strains, serve as a substrate for their growth.

This phenomenon has been well known to soil microbiologists who have investigated the breakdown of such substances as toluene, naphthalene and phenol in soils where they had been added as fumigants. Stomer (1907–8) isolated bacteria from the soil capable of

*Formerly Department of Pharmacy, University of Nottingham, Nottingham, UK.

destroying phenols and aromatic hydrocarbons such as toluene.

Gray & Thornton (1928) also studied the breakdown of soil fumigants by microorganisms and isolated bacteria capable of decomposing naphthalene, xylene and phenanthrene.

Fowler *et al.* (1911) isolated a phenol decomposing organism from a sewage purification plant and Wagner (1914) described the decomposition of phenol, cresol and phloroglucinol by microorganisms. Happold & Key (1932) isolated an organism from gas works effluent, designated by them Vibrio O/1 (oxidase 1), also capable of decomposing phenol.

The breakdown of naphthalene and phenanthrene had also been reported by Tausson (1927).

The ability of microorganisms to cleave aromatic rings including structures traditionally regarded as possessing antibacterial properties was thus well documented in early scientific literature.

Since then many other instances have been described and, in addition, fundamental studies to elucidate the metabolic pathways whereby these compounds are degraded have been initiated, and an extensive literature generated. The whole field has recently been reviewed in Gibson (1984).

THE DEGRADATION OF PRESERVATIVES

A table compiled from data obtained by Beveridge & Hugo (1964*a*) and describing the ability or otherwise of nine microorganisms to degrade eighteen aromatic compounds is given in Table 1.

Phenol

Stormer (1907–8) isolated a phenol decomposing organism and Happold & Key (1932) isolated an organism called Vibrio O/1 from phenol contaminated sewage effluent. Later they showed how phenol was first oxidised to catechol and thereafter degraded to 3-oxo-adipic acid and then to acetic and succinic acids both utilisable in the Krebs cycle. Davey (1961) described three pseudomonads capable of growing on phenol and *o*-, *m*- and *p*-cresols as sole carbon source while seven out of the nine microorganisms shown in Table 1 were able to utilise phenol. Buswell (1975) demonstrated the metabolism of phenol and cresols by *Bacillus stearothermophilis*.

Chlorinated phenols

The decomposition of chlorinated phenols has also been reported. Thus

TABLE 1

The Utilisation[a] of Aromatic Compounds by Bacteria as Sole Carbon Sources for growth (reproduced by permission of the authors and editor of the *Journal of Applied Bacteriology*)

Test organism[b]	*Test method*[c]	*p-Cresol*	*o-Cresol*	*M-Cresol*	*Benzoic acid*	*Phenol*	*p-Hydroxybenzoic acid*	*Salicylic acid*	*m-Hydroxybenzoic acid*	*Hydroquinone*	*Resorcinol*	*Catechol*	*Phloroglucinol*	*3,4-dihydroxy-benzoic acid*	*3,5-dihydroxy-benzoic acid*	*Gallic acid*	*Methyl gallate*	*Ethyl gallate*	*Methyl p-hydroxy-benzoate*
Pseudomonas	a	+	−	+	+	+	+	−	−	−	−	+	(+)	+	+	+	−	−	−
convexa X1	b	+	−	+	+	+	+	−	−	−	−	+	(+)	+	+	+	−	−	−
Ps. fluorescens	a	+	+	(+)	+	+	+	(+)	−	−	(+)	+	+	+	−	−	−	−	−
(NCIB 8248)	b	+	+	+	+	+	+	−	−	−	(+)	+	+	+	*d*	−	−	−	−
Vibrio 01	a	(+)	−	−	+	+	+	−	−	−	−	+	−	+	−	?	?	?	−
(NCIB 8250)	b	+	−	−	+	+	+	−	−	−	−	+	−	+	*d*	?	?	?	−
Pseudomonas sp.	a	+	−	−	+	+	+	+	−	−	−	+	−	+	−	−	−	−	−
(NCIB 8858)	b	+	−	−	+	+	+	+	−	−	−	+	−	+	−	−	−	−	−
Ps. fluorescens	a	−	(+)	−	+	+	+	(+)	−	−	−	+	−	+	−	−	−	−	−
(NCIB 8249)	b	−	−	−	+	+	+	−	−	−	−	+	−	+	−	−	−	−	−
p-Cresol organism	tb	+	+	−	+	(+)	+	−	+	−	−	+	−	+	−	−	−	−	−
(NCIB 8251)	b	+	+	−	+	(+)	+	−	+	+	−	+	−	+	*d*	−	−	−	−
V. cyclosites	a	+	−	−	+	+	−	−	−	−	−	+	−	−	−	?	−	−	−
(NCIB 2581)	b	+	−	−	+	+	−	−	−	−	−	+	−	−	−	?	−	−	−
Ps. desmolyticum	a	+	−	+	+	−	+	−	−	−	−	+	−	+	−	−	−	−	−
(NCIB 8859)	b	+	−	+	+	−	+	−	−	+	−	+	−	+	−	−	−	−	−
Pseudomonas sp.	a	+	−	−	+	−	(+)	−	−	−	−	+	−	+	−	−	−	−	−
(NCIB 8226)	b	+	−	−	+	−	(+)	−	−	+	−	+	−	(+)	−	−	−	−	−

[a] Utilisation: +, growth; (+), poor growth; −, no growth; ?, result uncertain.
[b] NCIB, National Collection of Industrial Bacteria.
[c] Test method: a, results obtained from test tube experiment; b, results obtained from auxanographic experiment.
[d] Experiment not done.

Erikson (1941) described the decomposition of *p*-dichlorobenzene by *Micromonospora chalaceae* while Walker (1954) showed that *o*-chlorophenol was decomposed by soil bacteria and *p*-chloro- and 2,4-dichlorophenol, under the experimental conditions used, were much more resistant to microbial degradation.

Salicylic acid

Salicylic acid is a key intermediate in the degradation of aromatic compounds with more than one ring. Beveridge & Hugo (1964*a*) found it to be utilised by four out of the nine organisms tested (Table 1).

Carboxylic acids

Benzoic acid is readily decomposed by ring degrading microorganisms (Table 1). Ingram (1959) described a yeast which could degrade benzoic acid but only with the concomitant degradation of sugars.

Esters of *p*-hydroxybenzoic acid

None of the organisms described by Beveridge & Hugo (1964*a*) (Table 1) were able to utilise methyl *p*-hydroxybenzoate but Hugo & Foster (1964) showed that *Pseudomonas aeruginosa* NCTC 7244, would grow in a mineral salt medium in the presence of 0·02% methyl *p*-hydroxybenzoate and 0·01% propyl *p*-hydroxybenzoate, the concentrations at one time employed in solutions used as a vehicle for the preparation of eye drops.

Beveridge & Hart (1970) made a further investigation of the decomposition of the esters of *p*-hydroxybenzoic acid by microorganisms. They found that *Alcaligenes* spp., *Aeromonas* spp., *P. aeruginosa*, *P. alcaligenes*, *P. fluorescens* and *Bacillus* spp. isolated from a variety of environments would degrade the esters.

Sokolski *et al.* (1962) demonstrated the hydrolysis of methyl *p*-hydroxybenzoate by *Cladosporium resinae*. The parent acid was decomposed by all but one of the nine organisms tested by Beveridge & Hugo (1964*a*) but these workers found no attack upon methyl *p*-hydroxybenzoate.

Gallic acid and esters

The alkyl esters of 3,4,5-trihydroxybenzoic or gallic acid are used as antioxidants for the prevention of rancidity in oil and fats.

Although gallic acid possesses antioxidant properties it is less active than the alkyl esters. Their stability towards microbial attack appears to have received little attention.

Beveridge & Hugo (1964*a*, 1965) have made a study of the stability of methyl, ethyl and propyl gallate and of gallic acid itself. In their studies the stability of the esters to attack by bacterial esterases and the utilisability of gallic acid and the esters as sole carbon source by microorganisms were investigated.

Stability towards esterase activity

Seven bacterial species previously shown to hydrolyse triglycerides and simple aliphatic esters (Table 2) were tested by Hugo & Beveridge (1962) for their ability to hydrolyse (saponify) the alkyl gallates. No such activity could be detected.

Gallic acid

The stability of gallic acid itself to microbial attack was first reported by Arima *et al.* (1954) who examined 130 strains of bacteria capable of degrading aromatic compounds; none could use gallic acid.

Beveridge & Hugo (1964*a*) showed that apart from *P. convexa*, the remaining species described in Table 1 were unable to utilise gallic acid.

Gallic acid was degraded rapidly, however, by *P. convexa*. The acid could serve as the sole carbon source for this organism when supplied in a mineral medium and was oxidised slowly but without a lag by a washed suspension of the organism after growth on nutrient agar. The activity of the cells could be dramatically increased by growth on nutrient agar containing gallic acid (Beveridge & Hugo, 1964*b*).

Several fungal species have been shown to metabolise gallic acid and Coupin (1927) showed that *Penicillium glaucum* could utilise gallic acid as sole carbon source for growth and since that date other species of moulds with gallic acid utilising ability have been described.

Beveridge & Hugo (unpublished) isolated a strain of *Penicillium steckii* capable of utilising gallic acid as sole carbon source.

Degradation of the alkyl gallates

Preparations containing the enzyme tannase, probably a mixture of enzymes, will degrade methyl gallate and an acetone powder from

TABLE 2

The Effect of Substrate upon the Lipolytic Power of Seven Organisms (reproduced by permission of the authors and editor of the *Journal of Applied Bacteriology*)

Test	*Reactions*[a] *on*							
	Victoria blue agar containing					*Tributyrin agar*	*Bromothymol blue broth containing*	
	Olive oil	*Coconut oil*	*Ethyl laurate*	*Butter fat*[b]	*Butter fat*[c]		*Triacetin*	*Ethyl butyrate*
Staph. aureus	++	++	+	++	++	++	++	0
Achr. fischeri	+	+	−	+	+	+	++	0
Ps. hydrophila	++	++	+	++	++	++	++	0
B. cereus	−	(+)	−	−	−	+	+	0
Serr. marcescens	++	++	+	++	++	++	−	0
Mic. freudenreichii	++	++	−	++	++	++	+	0
B. subtilis	−	−	−	−	(+)	++	++	−

[a]Reactions: ++, strong lipolysis; +, lipolysis; (+), weak lipolysis; −, no lipolysis; 0, no growth.
[b]Incubated at 25°.
[c]Incubated at 37°.

Aspergillus niger was able to decompose methyl gallate (Dykerhoff & Armbruster, 1933) due to the presence of a 'tannase'.

Beveridge & Hugo (1965) isolated two strains of the yeast-like fungus *Aureobasidiomyces* (formerly *Pullularia*) *pullulans*. These organisms used gallic acid and methyl, ethyl and propyl gallates as sole carbon source for growth in a mineral medium and would oxidise gallic acid in the Warburg apparatus after a preliminary lag. As with *P. convexa*, preliminary growth on a medium containing gallic acid abolished the lag period and produced cells with a greatly enhanced gallic acid oxidising ability. Adaptation to gallic acid also caused adaption to the alkyl gallates.

The total oxygen uptake using the alkyl gallates as sole substrate in the Warburg apparatus increased with increasing chain length and no trace of ester could be detected at the end of the experiment. It was thought that the alkyl side chain was being utilised and separate experiments showed that *A. pullulans* could use methanol, ethanol and propanol (also butanol and pentanol) as sole carbon source for growth.

ANAEROBIC METABOLISM OF PRESERVATIVES

The pathways discussed so far have all dealt with the decomposition of preservatives under aerobic conditions. In fact many of the studies and ancillary biochemical investigations were carried out in air by Warburg respirometry.

The anaerobic degradation of aromatic compounds has been reviewed by Sleat & Robinson (1984). This pathway could be of importance in preserved preparations packed in airtight containers.

Of the compounds decomposed anaerobically, the following preservatives were identified: benzoate, benzyl alcohol, phenol, the cresols, cinnamic acid and vanillic acid. The *o*-, *m*- and *p*-hydroxybenzoic acids were also metabolised and it would be interesting to see if their alkyl esters suffered decomposition.

It is clear that the ability to decompose compounds traditionally regarded as preservatives is a property of many microorganisms ranging from bacteria to fungi and this facet of microbial activity cannot be ignored during formulation and as a possible causative factor in microbial spoilage.

REFERENCES

Arima, K., Komagata, K. & Minoda, S. (1954). *Journal of the Agricultural Society of Japan*, **28**, 629–36.

Beveridge, E.G. & Hugo, W.B. (1964*a*). *Journal of Applied Bacteriology*, **27**, 304–11.
Beveridge, E.G. & Hugo, W.B. (1964*b*). *Journal of Applied Bacteriology*, **27**, 448–60.
Beveridge, E.G. & Hugo, W.B. (1965). *Journal of Applied Bacteriology*, **28**, 294–308.
Beveridge, E.G & Hart, A. (1970). *International Biodeterioration Bulletin*, **6**, 9–12.
Buswell, J.A. (1975). *Journal of Bacteriology*, **124**, 1077–83.
Coupin, H. (1927). *Comptes Rendus Hebdomadaires des Seances de l'Acadamie des Sciences,* **185**, 145–6.
Davey, B.B. (1961). *Journal of Applied Bacteriology*, **24**, 78–82.
Dykerhoff, H. & Armbruster, R. (1933). *Hoppe-Seyler's Zeitschrift fur Physiologisch Chemie*, **219** 38–56.
Erikson, D. (1941). *Journal of Bacteriology*, **41**, 277–300.
Fowler, G.J., Arden, E. & Lockett, W.T. (1911). *Proceedings of the Royal Society*, **83B**, 149–56.
Gibson, T.D. (ed.) (1984). *Microbiology Series, Vol. 13.* New York, Marcel Dekker.
Gray, P.H.H. & Thornton, H.G. (1928). *Zentralblatt fur Bakteriologie, Parasitenkunde, Infectionskrankheiten und Hygiene*, Arbeit II, **73**, 74–96.
Happold, F.C. & Key, A. (1932). *Journal of Hygiene, Cambridge,* **32**, 573–80.
Hugo, W.B. & Beveridge, E.G. (1962). *Journal of Applied Bacteriology,* **25**, 75–82.
Hugo, W.B. & Foster, J.H.S. (1964). *Journal of Pharmacy and Pharmacology*, **16**, 209.
Ingram, M. (1959). *Journal of Applied Bacteriology*, **22**, vi.
Sleat, R. & Robinson, J.P. (1984). *Journal of Applied Bacteriology,* **57**, 381–94.
Sokoloski, W.T., Chichester, C.G. & Honeywell, G.E. (1962). *Developments in Industrial Microbiology*, **3**, 179–87.
Stormer, K. (1907–8). *Zentralblatt fur Bakteriologie, Parasitenkunde, Infection-krankheiten und Hygiene*, Arbeit II, **iii**, 282–6.
Tausson, W.C. (1927). *Planta*, **4**, 214–56.
Wagner, R. (1914). *Zeitschrift fur Garungsphysiologie*, **4**, 289–319.
Walker, H. (1954). *Plant and Soil*, **5**, 194–204.

ECOLOGY

Studies on the Colonization and Degradation of Human Hair by *Streptomyces fradiae*

T.P. Carter, D.J. Best & K.J. Seal

Cranfield Biotechnology Centre, Cranfield Institute of Technology, Cranfield, Bedfordshire MK43 0AL, UK

ABSTRACT

Experiments in which Streptomyces fradiae *was grown on ether-extracted, ethylene oxide-sterilised human hair as sole carbon and nitrogen source have demonstrated the ability of the organism to bring about the complete structural degradation of the substrate. Electron micrographs reveal the colonisation and subsequent deterioration of the hair fibres. Photographic evidence shows that the microbial attack is uniform along the hair fibre and not localised to specific areas of colonisation. There is little evidence to suggest that degradation is initiated by mechanical disruption of the hair by penetrating hyphae. Time course studies clearly show the systematic removal of distinct histological regions of the hair fibres. Initially the epicuticle is clearly visible. This becomes obscured as colonisation proceeds. Removal of microbial growth by gentle washing reveals a layer that lies between the epicuticle and the cortex, possibly the exocuticle, endocuticle or outer regions of the cortical matrix. Later micrographs reveal the cortex below. Finally the disintegration of the cortex into individual cells is observed.*

Culture supernatants have been found to have an optimum keratinolytic activity at pH 9·0. At this pH the specific activity of culture supernatants is approximately ten times greater than that of commercially available keratinolytic enzymes.

INTRODUCTION

Due to stricter legislation and the increased cost of waste water treatment the British leather industry is seeking alternatives to some of the chemical methods presently employed in the treatment of raw hides. One example is the removal of hair from hides prior to the tanning process, which involves the use of proteolytic enzymes.

Noval & Nickerson (1959) showed that *Streptomyces fradiae*, in liquid culture, could bring about the complete digestion of a native keratin source, e.g. hair (as determined by substrate weight loss). An extracellular enzyme with substantial keratinolytic activity was purified and crystallised (Nickerson & Durrand, 1963). However, the enzyme obtained could only solubilise up to 60% of the substrate whereas whole cell cultures caused complete hair digestion. This discrepancy therefore suggested the possibility of the action of some other mechanism(s) in concert with the keratinolytic enzyme elaborated. There are a number of possible alternatives for the associated mechanism. First, as with dermatophytic fungi (English, 1962), hyphal penetration of the hair shaft would result in mechanical disruption of the keratin structure. Secondly, localised chemical reduction of the substrate may take place prior to keratinolytic action (Mathison, 1964) as seen in a similar reaction that takes place in the gut of the clothes moth larva *Tineola bisselliella* (Powning & Irzykiewicz, 1965). Therefore it is important to clarify this position so that a complete model of the mechanism of keratin degradation by *Streptomyces fradiae* can be constructed.

This paper describes the use of SEM, and dye-linked protein substrates to examine the structural disintegration of hair fibres as a result of enzymatic activity.

MATERIALS AND METHODS

Organism and maintenance. Streptomyces fradiae (ATCC 14544) was maintained on potato agar slopes at 4°C.

Culture conditions. Erlenmeyer flasks (250 ml) containing basal salts solution (100 ml, pH 7·5) were sterilised at 121°C for 15 min. Diethyl ether extracted hair (85 mg) was added, which had been sterilised in an atmosphere of ethylene oxide/air (50% v/v). The flasks were then inoculated (10% v/v) with a washed suspension of *Streptomyces fradiae* spores (10^8/ml), produced on potato agar medium. Controls were not inoculated. Flasks were incubated at 30°C on a rotary shaker (180 rpm).

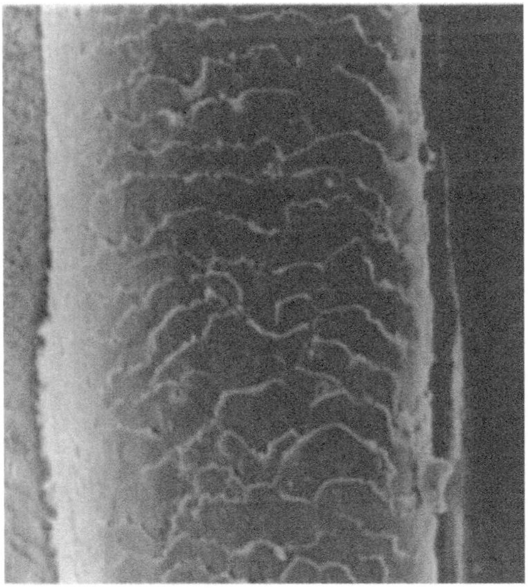

Fig. 1. Uninoculated control, magnification ×850.

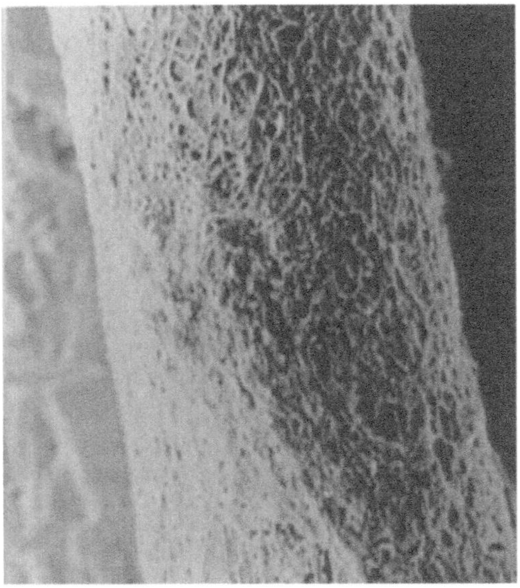

Fig. 2. Hair from 5 day culture, magnification ×850.

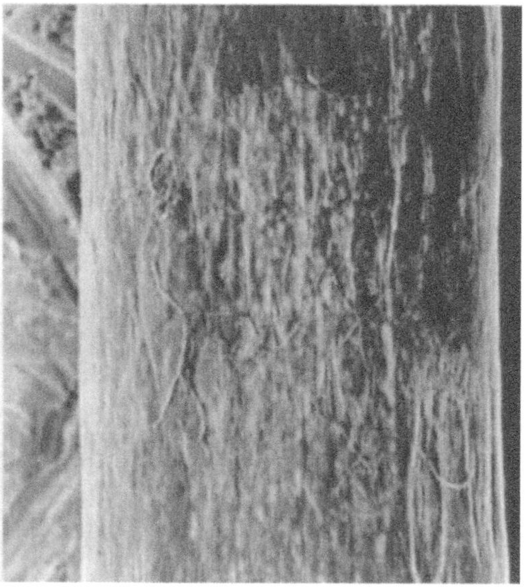

Fig. 3. Hair from 5 day culture, washed, magnification ×850.

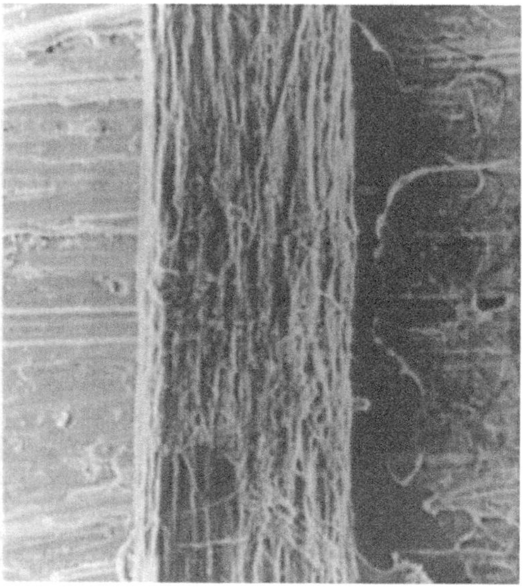

Fig. 4. Hair from 10 day sample, magnification ×400.

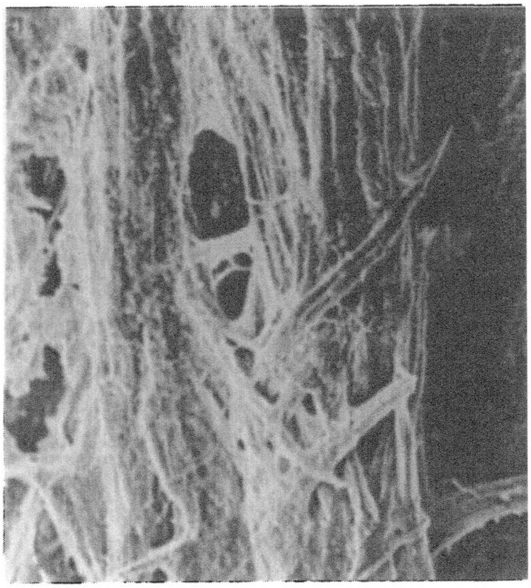

Fig. 5. Hair from 10–15 day culture, magnification ×850.

Fig. 6. Hair from 10–15 day culture, magnification ×850.

Preparation of samples for SEM. Samples of hair fibres were removed aseptically from control and inoculated flasks over a period of fifteen days and fixed in 10% (v/v) formaldehyde solution until required. Samples were mounted onto electron microscope stubs with conducting silver paint and allowed to dry thoroughly. They were then coated *in vacuo* with gold–palladium and observed using a Cambridge 6000 Stereoscan SEM.

Enzyme assays. Enzyme activity was measured using the dye-linked substrate keratin-azure. Buffer (1·4 ml, 50 mM) of appropriate pH (see Fig. 1) was added to microfuge tubes containing substrate (8 mg). The comparison of enzymes acting on the substrate was carried out at pH 9·0 using diethanolamine-HCl buffer (50 mM). These were then sealed and equilibrated at 37°C for 15 min. Culture supernatant (0·2 ml) was added to each tube. Controls involved the addition of culture supernatant denatured by boiling for 15 min. The tubes were then resealed and incubated in a water bath at 37°C for 1 h. After centrifugation to sediment the substrate, the absorbance (595 nm) of the supernatant was determined reading against a buffer only blank. Keratinase activity was calculated in Keratinase Units (KU). (1 KU = Change in absorbance (595 nm) between Test and Control samples/0·1 under standard conditions described above.)

Protein assay. Samples were assayed for protein using the method of Bradford (1976).

RESULTS AND DISCUSSION

From a series of samples taken throughout the incubation of *Streptomyces fradiae* with sterile hair, colonisation and degradation were observed. Figure 1 shows a hair fibre taken from an uninoculated control. No colonisation or degradation can be seen. Colonisation progressed rapidly, the whole of the hair fibres becoming completely covered with mycelium within 5 days (Fig. 2). Samples taken during the first four days displayed no evidence to suggest the occurrence of any mechanical disruption due to hyphal penetration of fibres. Gentle washing of 5 day samples revealed the extent of hair degradation (Fig. 3). The scale cells of the cuticle had been completely removed. Subsequent samples showed the exposed outer regions of the cortex, the cuticle having been completely removed (Fig. 4). Finally the cortical fibres were seen to disintegrate, resulting in the release of individual cells (Figs 5 and 6).

The scanning electron micrographs obtained demonstrate the ability of *Streptomyces fradiae* to cause the complete structural breakdown of native keratin in the form of human hair. Noval & Nickerson (1959) and Nickerson *et al.* (1963) showed that diethyl ether extraction and ethylene oxide sterilisation of hair causes no change in its chemical or physical structure. Any effect must therefore be due to the action, direct or indirect, of the microorganisms with which the culture was inoculated. The photographic evidence obtained corroborates the findings of these workers, in particular the suggestion (Noval & Nickerson, 1959) that hair digestion by this organism occurs in a sequence that can be divided into three stages. A lengthy lag period is followed in quick succession by enzyme production and hair digestion. Although in the early stages no mechanical disruption of the hair structure was seen, such effects cannot be discounted after the hair cortex has been exposed. Therefore it is assumed that the removal of the cuticular layers, demonstrably the most resistant regions of the hair, occurs in a manner similar to that proposed by Mathison (1964), i.e. by either the sole action of the enzyme, or enzymes, produced, or by chemical reduction of the substrate in a

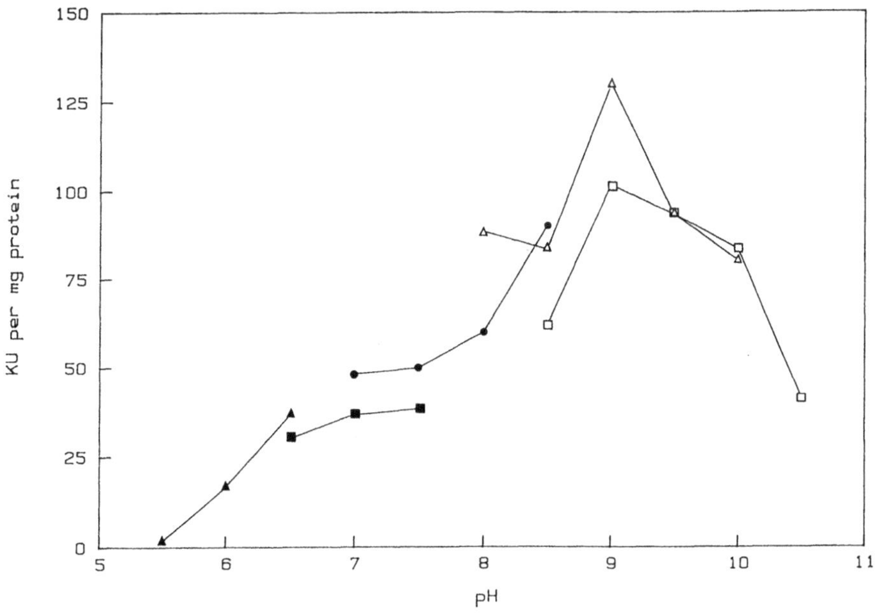

Fig. 7. pH profile of the keratinolytic activity of crude culture supernatant. All buffers were of strength 50 mM; ▲——▲ sodium hydrogen maleate-NaOH, ■——■ imidazole-HCl, ●——● tris-HCl, △——△ diethanolamine-HCl, □——□ glycine-NaOH.

TABLE 1
Comparison of Keratinolytic Activities, at pH 9·0, of Crude Culture Supernatant, Type XIV Protease from *Streptomyces Griseus* (Sigma) and Proteinase K from *Tritirachium Album* (Sigma)

Sample	*ABS (595 nm)*	*KU (total)*	*KU (ml^{-1})*	*KU (mg/protein)*
Proteinase K	0·152	1·52	7·6	19·49
Type XIV	0·146	1·46	7·3	14·04
Supernatant	0·078	0·78	3·9	130·00

localised area surrounding viable hyphae followed by enzymatic attack, as occurs in the gut of the clothes moth larvae *Tineola bisselliella* (Powning & Irzykiewicz, 1965).

Preliminary investigation of the enzymic activity of culture supernatants have shown a marked activity towards the dye-linked substrate keratin azure. Figure 7 shows a pH profile of the activity of crude culture supernatant towards keratin azure. Table 1 gives a comparison of keratinolytic activity at pH 9·0 between crude culture supernatant and two commercially available proteases. They show that the enzyme produced by *Streptomyces fradiae* has optimal activity at pH 9·0 and at this pH it has a specific activity up to ten times greater than other enzymes studied.

CONCLUSION

It has been shown that *Streptomyces fradiae* is capable of causing the complete structural breakdown of native fibrous keratin. This breakdown is facilitated by the production of an extracellular enzyme of relatively high specific activity towards the substrate.

ACKNOWLEDGEMENTS

We wish to thank Mr R. Webster and Mr M. Kirkland of the British Leather Confederation for their assistance.

REFERENCES

Bradford, M.M. (1976). *Analytical Biochemistry,* **72**, 248–54.
English, M.P. (1962). *Sabouraudia,* **2**, 115–30.

Mathison, G.E. (1964). *Annals of the Belgian Society for Tropical Medicine*, **44**, 767–91.
Nickerson, W.J. & Durrand, S.C. (1963). *Biochimica et Biophysica Acta*, **77**, 87–99.
Nickerson, W.J., Noval, J.J. & Robison, R.S. (1963). *Biochimica et Biophysica Acta*, **77**, 73–86.
Noval, J.J. & Nickerson, W.J. (1959). *Journal of Bacteriology*, **77**, 251–63.
Powning, R.F. & Irzykiewicz, H. (1965) *Australian Journal of Biological Sciences*, **13**, 59–68.

Microbiology of Automobile Electrocoat Paint Systems

L.A. Rossmoore, V.F. Riha & H.W. Rossmoore

Biosan Laboratories, Inc., Ferndale, MI 48220, USA and Department of Biological Sciences, Wayne State University, Detroit, MI 48202, USA

ABSTRACT

An experiment was conducted to evaluate paint application to a metal surface coated with bacteria. Automobile frames are exposed to a series of rinses prior to electrodeposition of undercoat. Bacteria isolated from pretreatment rinses were inoculated onto steel panels which, as in pretreatment, had been coated with zinc phosphate.

Inoculated and control panels were painted and assessed for adherence by a water soak and a corrosion cycling test. Scanning electron micrographs (SEM) were made on unpainted, inoculated panels to view the effects of bacteria on the zinc phosphate coating and to verify that they adhered to the metal. Results show that the bacteria did not interfere with paint adhesion. The SEM showed no adverse effects on the zinc phosphate.

INTRODUCTION

Electrodeposition of automobile undercoat is almost universally used in assembly plants. Prior to the application of paint, the car frame is carried through a series of pretreatment rinses (Table 1). These prepare the metal for the electrocoat process, both by cleaning and chemical treatment, that imparts corrosion protection and greater paint adherence.

There are nine pretreatment stages (Table 1). Many of the stages contain bacterial populations, the levels of which vary from around 10^2/ml to as high as 10^7/ml (Table 2). Thus, when a frame reaches the electrodeposition bath it may be coated with bacteria. This study assumed a worse case scenario in which the auto body would reach the paint bath with a significant bacterial coating. The effects of such a coating on paint adherence were examined.

METHODS

Samples were taken from each rinse system at two automobile assembly plants in the Detroit, Michigan area and serially diluted onto Standard Methods Agar (GIBCO Laboratories) for enumeration (Table 2). Once enumerated, subcultures were made of predominant colonial morpho-

TABLE 1
Typical Electrocoat Pretreatment

Rinse stage	*Process*	*pH*	*Time (s)*	*Temperature (°C)*
1	Spray clean	9·0– 9·5	30	49–52
2	Dip clean	9·0– 9·5	120	49–52
3	Spray rinse	8·0– 8·5	30	43–46
4	Dip rinse	8·5– 9·5	30	43 maximum
5	Dip phosphate	3·0– 3·5	120	43–46
6	Dip rinse	8·0– 8·5	30	ambient
7	Spray chrome, dip chrome	3·8– 5·0	30	ambient
8	Spray recirc. DI, dip recirc. DI	6·0– 7·0	30	ambient
9	Spray virgin DI	6·0– 7·0	10	ambient

TABLE 2
Bacterial Counts of Pretreatment Systems from Two Assembly Plants

	Bacteria/ml	
Rinse stage	*Plant A*	*Plant B*
1	<100	<100
2	<100	<100
3	<100	$1{\cdot}3 \times 10^7$
4	<100	$2{\cdot}1 \times 10^6$
5	<100	100
6	$1{\cdot}2 \times 10^7$	$1{\cdot}3 \times 10^4$
7	$1{\cdot}2 \times 10^3$	$4{\cdot}5 \times 10^7$
8	3×10^3	3×10^6
9	5×10^2	<100

TABLE 3
Selected Bacterial Identifications

Plant A		*Plant B*	
Organism	*Stage No.*	*Organism*	*Stage No.*
Pseudomonas picketii	6	*Moraxella* sp.	3
Pseudomonas aeruginosa	6	*Pseudomonas* sp.	3
Pseudomonas sp.	6	*Pseudomonas putida*	4
Pseudomonas picketii	7	*Acinetobacter lwoffi*	4
Pseudomonas picketii	8	*Flavobacterium* sp.	4
Pseudomonas sp.	8	*Pseudomonas cepacia*	6
		Flavobacterium multivorum	6
		Pseudomonas maltophila	7
		Pseudomonas paucimobilis	7
		Flavobacterium odoratum	8
		Pseudomonas aeruginosa	8

types. Identification of the organisms was made using the Micro-Scan Gram Negative ID Panel (American Hosp. Supply) (no Gram-positive organisms were found). Table 3 lists some of the dominant organisms from each stage.

Bacterial suspensions were made in sterile distilled water using pure culture isolates. Separate suspensions were made of organisms from each assembly plant. The initial inocula were made to contain 10^8

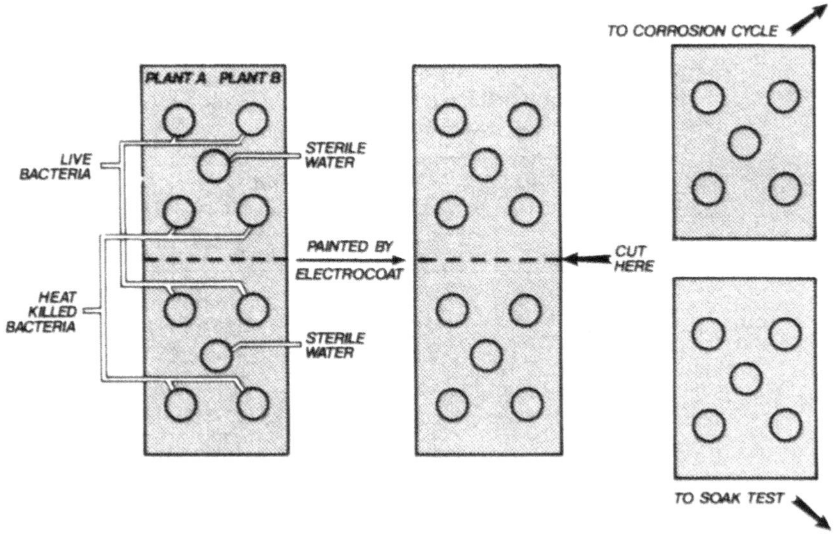

Fig. 1. Inoculation and treatment of panels.

bacteria/ml using a barium sulphate standard (Horwitz, 1975). The inocula were diluted to give suspensions of 10^4 bacteria/ml. A portion of each inoculum was removed and steam sterilized allowing each panel to be inoculated with both live bacteria and a heat-killed control.

Panels were inoculated as shown in Fig. 1. The samples were triplicated so that levels of 10^8, 10^4, and zero bacteria/ml could be studied. From each inoculum, 0·2 ml was removed and pipetted onto designated areas of the panel.

After inoculation and drying, the panels were painted and cut in half for testing. The top was subjected to a corrosion cycle test while the bottom portion was put into a water soak test. The latter involves immersion of the test panel in deionized water (52°C) for 72 h. At the conclusion of the test, the painted surface is visually examined and adverse conditions such as blisters, sags, rust spots, or scratches, are noted.

The corrosion cycle test combines the specifications of two American Society for Testing and Materials (ASTM) procedures (ASTM D1654-799, Painted or Coated Specimens Subjected to Corrosive Environments (ASTM, 1986*a*); and ASTM B117-73, Standard Method of Salt Spray [Fog] Testing (ASTM, 1986*b*)) with two General Motors specifications (GM 4298P: Salt Spray Testing; and GM 9102P: Corrosion Creepback Test Method). At time zero, a scribe is made in the test panel. The test, on a Monday to Friday cycle, involves immersion in a 5% sodium chloride solution each day for 15 min. Next, the samples are set out to dry at room temperature for 75 min and then placed in a humidity cabinet (60°C and 85% relative humidity) for 22 h and 30 min. On Monday only, the panels are also subjected to a twice repeated cycle of 1 h storage in a 60°C oven followed by 30 min in a −10°C cold cabinet. The test lasts 5 weeks.

After completion the panels are rinsed with warm water and visually examined for corrosion, blistering, peeling, or scribe line corrosion creepback (loss of adhesion between primer and steel).

One set of panels was inoculated according to Fig. 1 but were not painted. Instead, they were air dried and platinum coated (300 Å thick) using a Polaron SEM coating unit E-5000. They were then examined by a Phillips 505 SEM operated at 30 kV (Hyat, 1972).

RESULTS AND DISCUSSION

Visual observations of the panels following the 72 h soak test revealed no adverse changes in paint film adherence. Regardless of inoculation level,

(a)

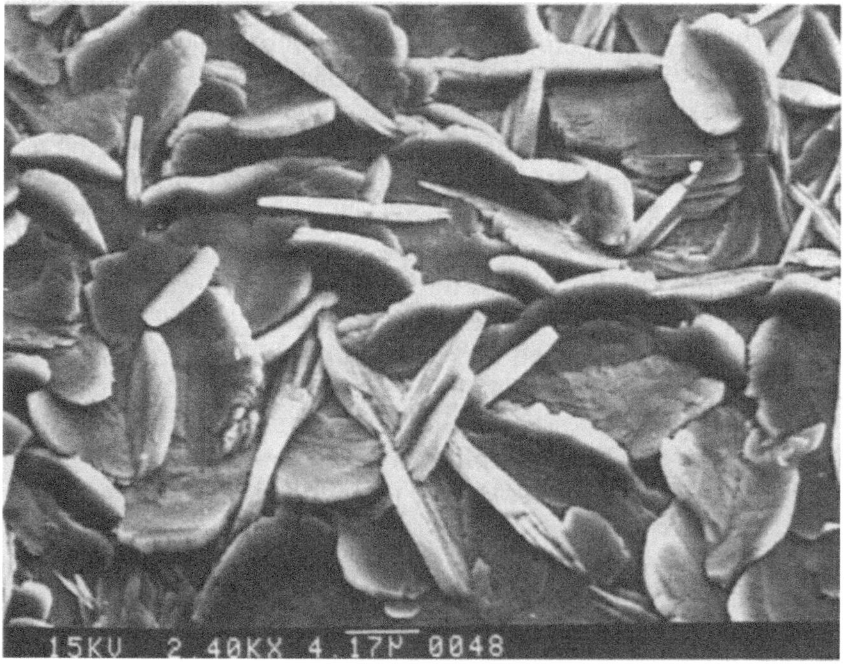

(b)

Fig. 2.

no increase in corrosion, blistering, or peeling was seen in the inoculated panels.

The corrosion cycling test yielded the same results. Increased loss of adhesion between primer and steel in those panels inoculated with bacteria was not demonstrated.

The SEM photos were able to show bacteria associated with the phosphate crystals (Figs 2(a) and 2(b)). However, based on previous data from SEM examination of phosphated surfaces as well as results from the present study, bacteria were judged to have no negative effects on the phosphated surface.

A worse case scenario was chosen to establish the protocol for this study. It is not likely that bacteria levels used in the experiments (10^8/ml) would ever be seen in the plants. It is also doubtful that the organisms would ever have a chance to dry and adhere to the metal, as they did in this test, simply because the auto frames are almost constantly moving from one rinse to the next.

Laboratory studies involved in studying biodeterioration and its prevention generally contrive a system of exaggerated conditions. If a substance can resist deterioration under harsh laboratory testing, successful resistance in the field can be predicted. It is on this basis we conclude that regardless of the degree to which pretreatment rinses are contaminated, adverse adherence of paint to metal is not expected.

REFERENCES

American Society for Testing and Materials (1986*a*). In *1986 Annual Book of ASTM Standards, Part 06.01*. ASTM, Philadelphia.

American Society for Testing and Materials (1986*b*). In *1986 Annual Book of ASTM Standards, Part 02.05, Part 03.02* and *Part 06.01*. ASTM, Philadelphia.

Horwitz, W. (ed.) (1975). *Official Methods of the Association of Analytical Chemists*, 12th edn. Association of Official Analytical Chemists, Washington, D.C.

Hyat, M.A. (1972). *Principles and Techniques of Electron Microscopy. Biological Applications*, Vol. II. Van Nostrand Reinhold Co., New York.

Incidence of *Aspergillus flavus* in Rice and Rice Products in Uttar Pradesh, India

Sadhana Srivastava, M.P. Tandon & Sudhir Chandra

Department of Botany, University of Allahabad-211002, U.P., India

ABSTRACT

One hundred and thirteen samples of rice and rice products collected from different places in thirty-three districts of Uttar Pradesh were screened for their associated mycoflora employing standard techniques. In addition to forms belonging to Penicillium, Alternaria, Curvularia, Cladosporium, Fusarium, Helminthosporium, Rhizopus *and* Trichoderma *species, a number of Aspergilli were found associated externally and internally with the samples, represented by* A. flavus, A. fumigatus, A. niger, A. flavipes, A. japonicus, A. terreus, A. parasiticus, A. tamarii *and* A. ochraceous. *Many of the samples especially those collected from damp areas of the state were heavily infested with* A. flavus.

INTRODUCTION

Mycotoxins have attracted considerable attention in recent years because of their potent toxic nature and common occurrence under natural conditions. Among all the mycotoxins known so far, aflatoxins occupy the most important position due to their strong hepatocarcinogenic property and frequent occurrence in higher concentrations in edible commodities (Bilgrami *et al.*, 1980). Aflatoxins arise naturally when a toxin producing strain of *Aspergillus flavus* or *A. parasiticus* grows on a substrate where environmental conditions are suitable for the development of moulds. The mycotoxin problem in relation to public health is alarming particularly in tropical countries like India. Climatic conditions coupled with socio-economic backwardness provide conditions conducive to mycotoxin production.

Uttar Pradesh is an important state where rice and rice products are important foodstuffs. Gangopadhyay and Chakrabarti (1981) reported mycotoxin production in stored rice due to *A. flavus*. In high humid areas of Orrisa state, rice and rice products have been infested with fungi which have had hazardous effects (Reddy *et al.*, 1983). Screening of fungal contaminants in samples of millets, pulses and oilseeds in Uttar Pradesh showed that the majority of the isolates of *A. flavus* were toxigenic. An extensive survey was conducted during 1986–87 and investigations were carried out to determine the incidence of various fungi, especially *A. flavus*, in samples of rice and rice products collected from different regions of the state. The findings are reported in the present paper.

MATERIALS AND METHODS

Samples of stored paddy, rice, crisp rice, flattened rice and puffed rice were collected from different regions of Uttar Pradesh (ISTA, 1966). In all, 33 districts were included in the survey. The samples collected were composited and reduced to a submitted sample, a subsample of which was taken as a working sample for that district. The samples were brought to the laboratory under aseptic conditions and stored in sterilized wide mouthed stoppered glass bottles at 4°C. They were examined within 15 days for the isolation of fungi using the agar plate (Muskett, 1948) and standard blotter (de Tempe, 1953) methods. In the agar method, 400 grains (4 replicates of 100 each) were plated in petri dishes (Corning, 100 mm diameter) containing potato dextrose agar (PDA) medium supplemented with aqueous streptomycin sulphate. In the blotter method, the same number of grains were plated in petridishes with three layers of sterilized moist blotters. In both methods, 10 grains were equispaced in each dish. They were incubated at 25-28°C in a chamber provided with daylight tubes (Phillips 40 W) and were examined after 8 days.

In order to isolate the internal mycoflora, seeds were surface disinfected in 2% sodium hypochlorite solution for 10 min before placing them on agar plates or blotters.

The incidence of a fungus in the samples was calculated as follows:

$$\%\ \text{Incidence} = \frac{\text{Number of samples showing association}}{\text{Total number of samples screened}} \times 100$$

On the other hand, the magnitude of infestation by a fungus was calculated as follows:

$$\% \text{ Infestation} = \frac{\text{Number of grains showing association}}{\text{Total number of grains screened}} \times 100$$

RESULTS AND DISCUSSION

In all, 17 genera of fungi were associated with rice and rice products. However, the frequency and magnitude of infestation varied with the substrate. The maximum incidence and infestation are shown in Fig. 1.

Mycoflora of paddy

Samples of paddy collected from 15 districts were analysed. The external mycoflora included forms belonging to *Aspergillus*, *Alternaria*, *Curvularia*, *Cladosporium*, *Fusarium*, *Helminthosporium*, *Penicillium*, *Rhizopus* and *Trichoderma*. In addition, there were two types (black and white) of sterile mycelium and six species of *Aspergillus*, viz., *A. flavus*, *A. fumigatus*, *A. niger*, *A. flavipes* and *A. terreus*. Forms belonging to *Helminthosporium*, *Alternaria* and *Trichoderma* were found only on the PDA plates.

Different forms showed a variation in their incidence. The highest was shown by *Rhizopus* sp., *Aspergillus niger* and *A. fumigatus*. The forms also differed in their magnitude of infestation. Highest infestation was shown by *Fusarium* sp., *Curvularia* sp., *Aspergillus niger* and *A. flavus*.

The internal mycoflora included species of *Alternaria*, *Curvularia*, *Cladosporium*, *Fusarium*, *Penicillium*, *Rhizopus* and *Aspergillus*. Black and white sterile mycelia were also present. All the forms were found on the PDA plates as well as the blotters. *Aspergillus* was represented by five species, i.e. *A. flavus*, *A. fumigatus*, *A. niger*, *A. japonicus* and *A. flavipes*.

Rhizopus sp., *Fusarium* sp., *Aspergillus niger* and *A. flavus* showed the maximum incidence while *Curvularia* sp., *Cladosporium* sp., *Fusarium* sp. and *Aspergillus niger* displayed the maximum infestation.

Mycoflora of rice

Externally all 23 samples were infested with a variety of forms including those belonging to *Alternaria*, *Botrytis*, *Bispora*, *Curvularia*, *Cladosporium*, *Fusarium*, *Mucor*, *Nigrospora*, *Penicillium*, *Rhizopus*, *Aspergillus* and black and white sterile mycelia. There were eight species of *Aspergillus*, viz. *A.*

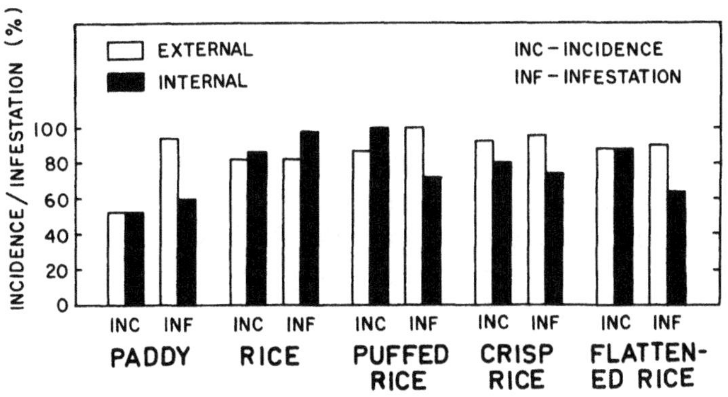

Fig. 1. Maximum incidence (%) and infestation (%) of *A. flavus* in rice and rice products.

niger, A. flavus, A. terreus, A. fumigator, A. parasiticus, A. japonicus, A tamarii and *A. ochraceous.* But for *Nigrospora* sp., all the forms could be recovered on PDA plates as well as blotters.

The percentage incidence varied with the forms. However, *Rhizopus* sp., *Aspergillus niger* and *A. flavus*, showed the highest incidence. On the other hand, maximum infestation was shown by *Rhizopus* sp., *Fusarium* sp. and *Aspergillus fumigatus.*

But for *Aspergillus parasiticus*, *A. tamarii*, *Botrytis* sp. and the white sterile mycelium, all the forms showed external and internal infestation. *Aspergillus clavatus* was only associated internally. Black sterile mycelium, *Nigrospora* sp., and *Bispora* sp. were recovered only on PDA plates and *Mucor* sp. on blotters. The rest of the forms were recovered by both the methods.

Maximum frequency was shown by *Rhizopus* sp., *Aspergillus niger* and *A. flavus*, and maximum infestation by *Aspergillus niger*, *A. flavus* and *Rhizopus* sp.

Mycoflora of crisp rice

Samples of crisp rice collected from 15 districts were analysed. Forms belonging to *Alternaria*, *Absidia*, *Curvularia*, *Cladosporium*, *Mucor*, *Penicillium*, *Rhizopus*, *Aspergillus* and black and white sterile mycelia constituted the external mycoflora.

Aspergillus was represented by six species, viz. *A. flavus*, *A. fumigatus*, *A. niger*, *A. flavipes*, *A. terreus* and *A. japonicus*. *Aspergillus flavipes*, *Curvularia* sp. and *Cladosporium* sp. were recovered only on PDA plates.

Maximum incidence was shown by *Penicillium*, *Aspergillus niger* and *A. fumigatus* and the highest infestation by *Aspergillus flavus*, *A. niger* and *Rhizopus* sp.

But for *Alternaria* sp. and *Absidia* sp. all the fungal forms were found externally and internally. In addition, *Helminthosporium* sp. was also found in internal association. *Aspergillus flavipes* was recovered only on blotters, while *Helminthosporium* sp. only on PDA plates.

Mycoflora of flattened rice

External mycoflora of the 18 samples included species of *Cladosporium*, *Curvularia*, *Fusarium*, *Monilia*, *Rhizopus* and *Aspergillus*. In addition, two types of sterile mycelium (black and white) were also recovered. *Aspergillus* was represented by eight species, viz. *A. flavus*, *A. fumigatus*, *A. flavipes*, *A. niger*, *A. terreus*, *A. japonicus*, *A. parasiticus* and *A. ochraceous*. *A. ochraceous* and *A. parasiticus* were recovered only on PDA plates. *A. niger*, *A. flavus* and *Rhizopus* sp. were the most frequent but maximum dominance was shown by *Rhizopus* sp., *Penicillium* sp. and *Aspergillus flavus*.

But for *Monilia* sp. and *Aspergillus ochraceous*, all the forms associated externally also showed internal association. *Aspergillus flavipes* and *Mucor* were recovered only on blotters. *Aspergillus flavipes*, *A. niger* and *Rhizopus* sp. showed maximum incidence, while *Rhizopus* sp., *Mucor* sp. and *Aspergillus flavipes* showed highest infestation internally.

Mycoflora of puffed rice

The samples from 8 districts were externally infested with a variety of fungi, viz. *Curvularia*, *Cladosporium*, *Fusarium*, *Monilia*, *Penicillium*, *Rhizopus*, *Aspergillus* and a sterile black mycelium. *Aspergillus* was represented by six species, viz. *A. flavus*, *A. fumigatus*, *A. japonicus*, *A. clavatus*, *A. terreus* and *A. niger*. *Aspergillus clavatus* and *Monilia* sp. appeared only on PDA plates while sterile black mycelium and *Cladosporium* sp. occurred only on blotters. One hundred per cent of samples were infested by *Aspergillus niger*, and *Rhizopus* sp. and *Aspergillus flavus* also showed high incidence.

But for *Aspergillus clavatus*, *A. terreus*, *Cladosporium* sp. and *Monilia* sp., all other forms associated externally were also present internally. *Penicillium* sp. was recovered only on PDA plates while *Fusarium* sp., *Aspergillus parasiticus* and *A. fumigatus* appeared only on blotters. Maximum incidence was shown by *Aspergillus flavus* and *A. niger*. These forms also displayed maximum dominance in the mycoflora.

An analysis of the data shows that, in general, the highest incidence and infestation was shown by *Aspergillus*, *Curvularia*, *Cladosporium*, *Fusarium*, *Penicillium*, *Rhizopus* and *Mucor*. *Aspergillus* was represented by 10 species but all the species showed variation in their frequency and extent of infestation on different substrates. Uniformly higher incidence and infestation was shown by *Aspergillus flavus* (Fig. 1). Its association with the majority of samples of rice and rice products collected from different regions of the state to an appreciable extent is alarming since some strains have been reported to be potent producers of aflatoxin (Acharya *et al.*, 1984). The occurrence of other species of *Aspergillus* (especially *A. parasiticus* and *A. ochraceous*) as well as those of *Fusarium* and *Penicillium* is also noteworthy since they are also known to produce mycotoxins (Bilgrami *et al.*, 1980).

CONCLUSION

The findings of the present study clearly show that the mycotoxins, particularly the aflatoxins, may pose a problem in the state of Uttar Pradesh because (1) rice and rice products are regularly consumed, (2) the consumption of mould contaminated commodities is inevitable due to food shortages and (3) there is no satisfactory measure to prevent the mould growth and to detoxify the contaminated commodity.

The mycotoxin producing potential of the moulds found associated with the rice and rice products is being evaluated. Attempts are being made to develop effective and economical methods for controlling mould growth and subsequent production of mycotoxins.

ACKNOWLEDGEMENTS

The authors are grateful to the Head of the Botany Department, University of Allahabad for laboratory facilities and to the State Council of Science and Technology for financial assistance.

REFERENCES

Acharya, P.B., Mohanty, C.R. & Gangopadhyay, S. (1984). *Indian Phytopathology*, **37**, 611–16.

Bilgrami, K.S., Misra, R.S. & Sinha, K.K. (1980). *Survey and study of mycotoxin*

producing fungi associated with grains in the standing maize crop. Final Report, ICAR Project, Allied Press, Bhagalpur.
de Tempe, J. (1953). *Proceedings of International Seed Testing Association*, **28**, 133–51.
Gangopadhyay, S. & Chakrabarti, N.K. (1981). *Current Science*, **50**, 272–5.
ISTA (1966). *Proceedings of International Seed Testing Association*, **31**, 1–152.
Muskett, A.E. (1948). *Transactions of British Mycological Society*, **30**, 74–83.
Reddy, B.N., Nusrath, M., Krishna Kumari C. & Saleba Nahdi (1983). *Indian Phytopathology*, **36**, 683–6.

EFFLUENTS

Effluent Treatment: An Introductory Review

A.D. Wheatley

Biotechnology Centre, Cranfield Institute of Technology, Cranfield, Bedford MK43 0AL, UK

ABSTRACT

The paper reviews the current methods of effluent treatment. The quantities, nature and costs of effluent disposal are briefly discussed followed by details of the criteria used to assess the most cost effective combinations of biological, chemical and physical unit processes.

Major changes have occurred in both the organisation and the technology of the industry over the last decade. The effects of these are noted together with the changes expected in the future.

The paper also discusses the impact of biotechnology on treatment and the opportunities this presents for improved efficiency, diagnostics, control and by-product recovery from effluents. The paper concludes that biological effluent treatment will remain the most common method of disposal for the foreseeable future and speculates on the likely future developments and research needs.

INTRODUCTION

Effluent treatment is the largest and one of the most important controlled uses of micro-organisms. Table 1 compares the quantity of sewage treated with other biological and non-biological products in the UK. There are about 9 million m^3 of domestic waste water produced a day, 7 million m^3 of industrial effluent and if rainwater and infiltration water are averaged and added this gives a total of 18 million m^3 for treatment.

About 5000 sewage works serve 80% of the population. A 1000 of which

TABLE 1
Quantities of Sewage Effluent Treated Annually Compared to other Common Industrial Products (UK Figures)

Product	*tonnes/annum*	*Price £/tonne*
Water as sewage	$6{\cdot}500 \times 10^6$	0·10
Milk	16×10^6	25
Steel	12×10^6	300
Beer	$6{\cdot}6 \times 10^6$	280
Inorganic fertiliser	$3{\cdot}3 \times 10^6$	200
Sugar	1×10^6	350
Cheese	$0{\cdot}2 \times 10^6$	1300
Baker's yeast	$0{\cdot}1 \times 10^6$	460
Citric acid	$0{\cdot}015 \times 10^6$	700
Penicillin	$0{\cdot}003 \times 10^6$	45 000

serve populations of over 10 000 and about 17 serve populations of over 300 000.

The cost of treatment of this waste in 1983/1984 was £1100 million (Water Authorities Association, 1985) (Table 2). This expenditure will continue to increase in real terms due to international agreement and popular demand for a better environment.

Most wastes are contaminated with organic matter and potential pathogens. The traditional aims of effluent treatment therefore, have been to remove the organic matter and reduce the number of faecal organisms. This allows the waste water to be discharged into the environment without adverse effect.

TABLE 2
Expenditure of UK Water Authority (1983/84) (£ million)

	Operating costs	*Total expenditure*	*Capital expenditure*
Water resources	105	153	44
Water treatment	231	348	53
Water distribution	516	792	246
Sewage systems	147	243	248
Sewage treatment	367	608	177
Drainage	53	86	78
Environmental services	37	59	33
Totals	1456	2289	879

WASTE WATER COMPOSITION

Waste waters are a complex mixture of suspended and dissolved materials. The polluting load is therefore defined by a range of tests designed to assess its physical, chemical and biological characteristics. These analyses are used to design suitable combinations of treatment equipment.

The biggest problem with domestic, agricultural and a wide range of industrial waste waters (e.g. from the food, drinks and fermentation industries) is caused by their organic content. The metabolism of this organic matter by the natural organisms contained in the receiving water causes rapid oxygen depletion and the consequential elimination of the normal aquatic flora and fauna. The organic content of the waste water is commonly determined by: Chemical Oxygen Demand (COD), Total Organic Carbon (TOC) and Biological Oxygen Demand (BOD).

Other important and common constituents of waste waters are suspended solids (SS) and nitrogen in its various forms. Suspended solids affect light penetration into receiving waters and also settle to disturb the benthic flora and fauna. They are measured by filtration or by light-scattering techniques. In domestic waste water, nitrogen is present in urea and protein. Residual ammonia is directly toxic to aquatic life and its oxidised products can cause nutrient enrichment and excessive algal growth (eutrophication). If effluents are recycled for water supply their nitrate also has to be controlled.

Very few waste waters have been characterised in detail but Painter (1971) and co-workers have managed to identify 75% of the organic matter in a domestic sewage. They found that 30–40% of the organic material was present in solution; 10–15% as colloidal solids (less than 0·1 μm); and 30–35% as settleable solids (more than 100 μm). The single largest organic constituent in both solution and suspension was the fatty acids, representing about 30% of the total organic carbon. The next largest constituent was the sugars: about 15% of the total organic carbon. It is likely that they are progressively converted into the fatty acids and that the amount of fatty acid varies according to the age of the waste water. The total fat and grease content of sewage was between 40 and 100 mg/litre and all the fatty acids up to C_{20} were identified. The most common were acetic, butyric and propionic acids derived from the simple sugars. Glucose, sucrose and lactose were the predominant sugars followed by galactose, fructose, xylose and arabinose. Carbohydrate in suspension was mainly cellulose and starch.

There are large changes in the flow and strength of waste waters according to patterns of behaviour and industrial production. Changes

TABLE 3
A Comparison of some Different Waste waters (in mg/litre except pH)

	Domestic	*Agricultural*	*Industrial*
pH	7·8	8·5	4·5
BOD	370	15 000	8000
COD	670	25 000	17 370
TOC	219	—	—
Total solids	1309	22 000	3200
Suspended solids	146	14 500	570
Ammoniacal nitrogen (as N)	46	2000	0·5

of up to 300% of average are common at domestic works when flows are affected by storm run-off. Table 3 gives a very general comparison of the three common types of waste water.

WASTE WATER TREATMENT

Municipal waste waters containing excreta, household wastes, industrial wastes and surface water run-off are often the most complex. Treatment of this type of waste water involves preliminary, primary and secondary stages and there may also be a tertiary stage when a high quality of final effluent is demanded (Fig. 1). A further stage may be added if an unusually high quality of effluent is required for re-use. Some of these unit processes can be omitted when treating more simple industrial waste waters.

Preliminary treatment, such as screening and grit separation, is intended to remove material which, although not exerting a major polluting load on subsequent treatment plant, can cause blockages, damage to equipment and other downstream problems. Screenings may be shredded and returned to the flow, or in-flow disintegration used. The screening from food processing effluents are often recovered as animal feed or other by-products.

Settlement of sewage can remove about 60% of the solids and about 35% of the polluting organic load. This reduces the load on subsequent biological treatment processes. The settlement of industrial waste waters often require chemical coagulation and flocculation because the particle size is too small to separate by gravity.

Biological treatment is an open mixed bacterial culture which also contains a diverse population of grazing organisms. There are two common types of reactor design (Fig. 2). Biological or percolating filters are fixed

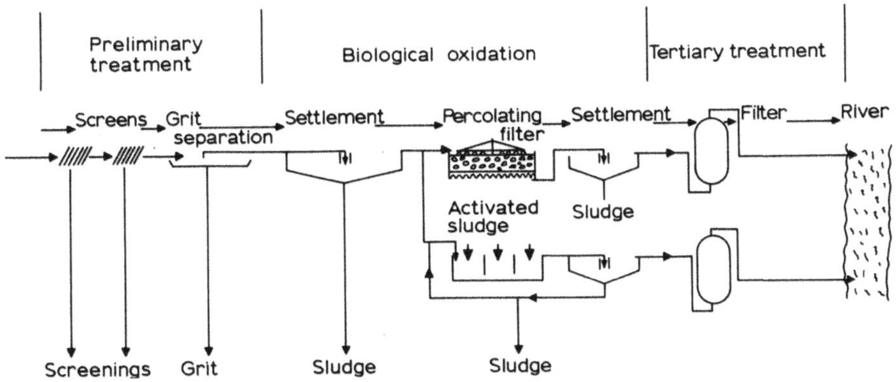

Fig. 1. Schematic diagram of sewage treatment.

film plug-flow reactors; activated sludge is a dispersed-growth system. Following primary settlement most effluents still contain both suspended and dissolved organic material and therefore the first stage of biological treatment is the adsorption of the substrate on to the biological surface. The artificial ecosystem can be almost completely balanced so that nearly all the available substrate and subsequent primary growth are consumed or leave the system. In this case, the products of waste water

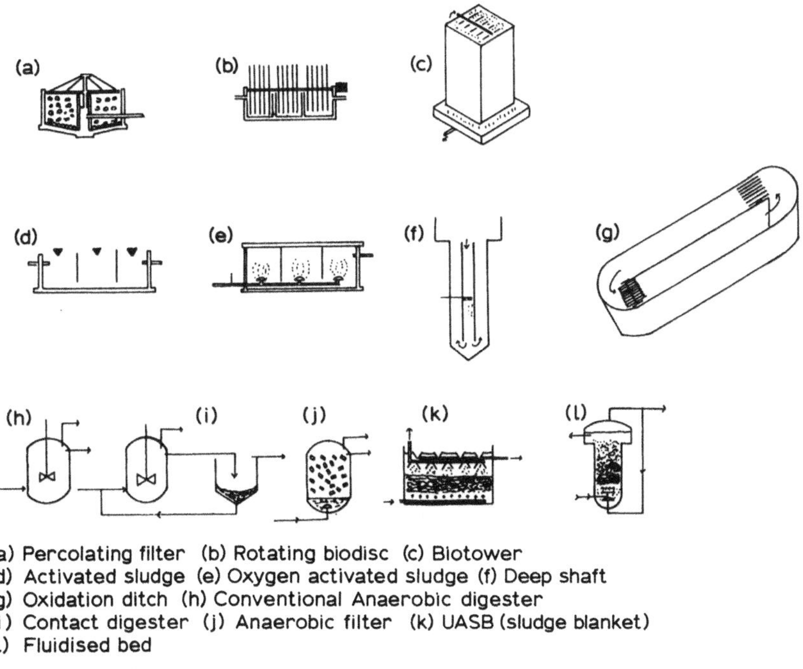

Fig. 2. Types of bioreactor used for effluent treatment.

treatment are simple inorganic salts and gases together with small quantities of residual inert solids.

The earliest type of treatment was the biological filter which developed directly from experiments carried out on soil percolation of sewage. The biomass is immobilised on an inert support medium that may be natural (e.g. stone) or synthetic. The waste water is sprinkled over the surface and allowed to percolate down through the tank and comes in intimate contact with the biomass. Biological filters are classified by the organic and hydraulic loads they receive and effluent quality deteriorates with increasing load. They are easy to maintain, incur low running costs and have a plant life of between 30 and 50 years. The process has been popular since its introduction in 1890.

The activated sludge process is a totally aquatic, dispersed biomass reactor. The bacteria are aerated and kept in suspension by bottom aerators or surface paddles. The process relies on the recycling of an active bios (the activated sludge) to mix with the incoming waste. A flocculated microbial culture is crucial to the operation since the treated effluent is separated from the active bios by settlement. Most of the settled sludge is then returned to mix with the incoming waste water. The process is more intense than biological filtration and is able to treat about ten times more effluent per volume of reactor; it is therefore cheaper in terms of capital cost. It is, however, more difficult to operate and maintain.

THE COST OF TREATMENT

A national system for the assessment of the cost of waste water treatment was introduced with the formation of the Water Authorities in 1974. Typically, costs to industry for water and effluent rose from a fractional 2% on product to about 8%. These costs have continued to rise at an average of 10% per annum in line with Government Policy, which suggests that charges still do not reflect the actual costs of environmental protection and water supply.

Food, drinks and fermentation is the largest industrial sector in the UK. It is also the largest water user. A recent MAFF review of waste (Gorsuch, 1987) showed that about 20% would benefit from introducing on site treatment (Table 4). A difficulty with this type of survey is that it requires constant review and can be effected drastically by non-cost factors such as legislation and public opinion. The Water Authorities have sought to overcome one of these difficulties by regularly publishing new unit cost factors for a standard computer model STOM (sewage

TABLE 4
Existing Treatments for Food Processing Effluents Costs (1982/83)

Treatment	*Typical load (kg BOD/m³/day)*	*Capital costs (£/m³) excluding settlement but inc. pipework*	*Running costs (p/m³) treated excluding amortization*	*Running costs (p/kg COD removed)*
Discharge to sewer	—	—	23	14·8
Settlement	—	200	0·5	0·9
Flotation (assisted)	—	250	30	20
High rate filters	2–3	300	20	10
Low rate filters	0·05–0·1	450	7	5
Activated sludge	0·5–1·5	200	45	40
Oxidation ditch	0·07–0·1	150	40	30
Lagooning	<0·05	100	0·5	0·9
High rate anaerobic treatment	5–7·5	250	20	6

treatment optimisation model) (Water Research Centre, 1977). STOM models combinations of various unit processes in various sizes to relate costs to design and operating conditions.

There is a strong public demand for improvements in and better protection of the environment. The European Commission is currently considering an Environmental Impact Directive which will make it mandatory to give a full inventory of raw materials, products and wastes. It will require a declaration of the disposal routes of wastes and their effects on the environment before planning consent will be granted. These factors have ensured a great deal of interest in environmental biotechnology and the possible improvements that new technology may offer.

ENVIRONMENTAL BIOTECHNOLOGY: NEW OPPORTUNITIES

The application of biotechnology depends on innovation and improved understanding in four key areas:

(1) Basic or fundamental biosciences leading to better control of industrial biochemistry.
(2) Novel bioreactors based on improved biochemical engineering.
(3) Genetically engineered, new, more potent micro-organisms.
(4) The development of biosensors and bioelectronics enabling better control of reactors.

Advances can offer new opportunities in waste treatment technology:

(a) Cheaper and more reliable/consistent treatment.

TABLE 5
Basic Research and its Potential Application to Waste Water Treatment

Area	*Applications*
1. Ecology and interdependence of microorganisms	Sludge bulking, less sludge, improved performance
2. Genetic improvement	Treatment of shock loads, biodegradation of toxins
3. Microbial physiology: structure of cell walls and extracellular polymers	Improved mass transfer adsorption characteristics
4. Nutrition and growth	Removal of nitrogen, phosphorus, better growth and reliability

(b) By-product recovery.
(c) Breakdown of persistent pollutants.

The remainder of this review will discuss examples of development in these areas and how they have been or could be applied to effluent treatment.

BASIC BIOSCIENCE

Table 5 divides current research and development into four areas with their possible impact on waste water treatment.

ECOLOGY

Most modern developments in effluent fermentation have increased process intensity by increasing biomass concentration but this has been at the expense of species diversity and balance between organisms. Although the activated sludge process (Fig. 1) is able to treat 10 times more effluent than biological filters it is more difficult to operate and generates large quantities of surplus biomass as sludge. Growths of fungi or filamentous bacteria, both of which occur in polluted waters and are components of biological filter films, seriously reduce the density of the activated sludge bios and interfere with the return of the active sludge to the aeration tank.

An understanding of waste characteristics that encourage filamentous organisms could bring about significant improvements in the operation of activated sludge plants.

Another consequence of the increase in process efficiency has been an increase in the amount of surplus biomass or sludge for disposal. Half the cost of effluent treatment in the UK is associated with sludge disposal. One interesting possibility is the uncoupling of anabolism from catabolism. This work is based on selective nutrient deficiencies and pulse feeding. More research and development is required.

GENETICS

The advances in molecular biology over the last 10 years allow in principle the enhancement of bacterial strains for waste treatment. There are two objectives:

(1) Improvement of overall performance, particularly the ability to deal with shock loads.
(2) The use of micro-organisms to degrade persistent (recalcitrant) or toxic materials.

To improve the rate of treatment or to decrease sludge production would need an effect on a wide range of micro-organisms acting on a large number of substrates. The problems associated with the production of new recalcitrant synthetic (xenobiotic) organic materials is likely to stimulate research into novel methods, such as biodegradation, for their disposal.

MICROBIAL PHYSIOLOGY AND NUTRITION

Basic research on microbial biochemistry has had an influence on process design and control. Research on the methane generating *Archaebacteria* for example has shown them to be very different from the common aerobic *Eubacteria* and *Pseudomonad* spp. Their cell walls contain no muramic acid and they exhibit resistance to antibiotics such as the penicillins and cephalosporins. Some species are extreme thermophils (60–97° C) and their lipids and enzymes differ from those of the classic bacteria. Several new coenzymes and factors have been discovered in association with their metabolism and the methanogens also require a wider range of trace elements.

Another example is the role of cell walls and the extracellular secretions of bacteria. Examination of treatment biofilms has shown that a large proportion of the bios is extracellular polymer. The biochemistry of the polymer is complex with a number of specific uptake systems for required nutrients. Field data from operating plants have indicated that the structural characteristics and quantity of polymer vary according to the type of substrate. Long-term research is directed towards altering the polymer to recover a wide range of useful or toxic materials.

Other projects include nitrogen stripping by utilising the facultative anaerobic metabolism and luxury phosphate uptake to avoid eutrophication.

NOVEL BIOREACTORS

In principle treatment efficiency is proportional to biomass concentration. Most developments in bioreactor design have been to increase the amount of biomass within the reactor. Plastic biosupport media was introduced in 1975 to increase the efficiency of biological filtration (Fig. 2). Plastic media could be fabricated into shapes with both a high surface area for biomass attachment and a high voidage for ventilation. Mineral media often clogged with excessive biomass when treating highly concentrated industrial wastes.

Another development in biological filtration was the Rotating Biological Contactor (RBC), in which a honeycomb of plastic sheets are slowly rotated on a shaft through a tank containing the waste water. About 40% of the surface is submerged in a typical unit. Rotation of the media ensures that uniform concentrations of dissolved oxygen and substrate are available to the total biomass. Eventually as biomass concentrations increase oxygen transfer becomes limiting and many of the changes to the basic activated sludge process were aimed at increasing the amount of air available. Thus one of the first possibilities for improving the activated sludge process was to use pure oxygen in completely enclosed tanks. The plant is very similar to conventional activated sludge but is able to operate at higher biomass concentrations and lower residence times without a loss in efficiency. Deep air lift fermenters have also been used in effluent treatment. The concentric tube reactor known as the 'deep shaft', can be 50–100 m deep. The shaft is partitioned into a downflow section, into which the incoming waste, returned active sludge and the process air are injected, and an upflow section. In the upflow section, bubbles of dissolved gases are released as the pressure decreases thus reducing the density of the circulating water. The difference in density between the downflow and upflow liquids produces the net driving force for the shaft. Dissolved oxygen and turbulence are high which permits elevated concentration of very active biomass to be maintained.

A combination of fixed film and activated sludge is the fluidised bed. These provide savings in capital because the plant does not require settlement. Sand, carbon or plastic are used for support particles and compressed air recycle or oxygen are utilised for fluidisation.

ANAEROBIC TREATMENT

Supplying the necessary air to guarantee aerobic conditions with a concentrated biomass and waste is very expensive. Anaerobic treatment is a possible alternative and there has been considerable effort recently to commercialise the process.

In a completely mixed system such as the traditional sludge digester

Hydraulic Retention Time (HRT) and Solids Retention Time (SRT) are the same and the minimum HRT is defined by microbial growth rate. To overcome the long generation time of anaerobic bacteria (several days) special reactors have been developed to retain most of the organisms inside the reactor or to recycle the bacteria after separation. Solids retention time is thus uncoupled from HRT and high bacterial concentrations can be obtained. There are four basic designs (Fig. 2): the biomass is held in the reactor by:

> Flocculation — Upflow Anaerobic Sludge Blanket (UASB), or by attachment to internal support media (fluidised beds, expanded beds or anaerobic filtration) or the biomass is physically separated from the effluent (by filtration or sedimentation) and then recycled to the reactor. Anaerobic filters can be either upflow or downflow.

BIOSENSORS

About 10% of the water research budget is currently allocated to Instrumentation Control and Automation (ICA). There are three potential advantages:

(1) It is cheaper than labour.
(2) Improved effluent quality.
(3) Better reliability.

The major block to further advances in control in the water industry are inadequate sensors. The crucial problems with typical electrochemical sensors are ageing and fouling. Biosensors and bioelectronic devices may be able to overcome this problem by continuous automatic replenishment. They are also the only techniques capable of responding to the very complex and changing variables typical of waste streams. A change in the activity of the biosensor would reflect changes in character and treatability of the waste. A variety of mediators are being researched:

(1) Respiration rate, e.g. O_2 consumption
CO_2 production
(2) Enzyme production
(3) ATP production
(4) Bioluminescence
(5) Co-enzyme fluorescence
(6) Photosynthetic activity

Respiration rate monitors for the 'on line' determination of sewage strength are already available in Japan and Germany.

BY-PRODUCT RECOVERY

The recovery of materials from waste is not as economically important as treatment. This is partly because of the complex heterogeneous nature of waste but also because of the low value of many of the recovered by-products. The large volumes, variability and chemical complexity of waste, makes the generation of single by-products with mono cultures very expensive. Difficulties also arise for recovery of very simple, low value by-products. The recovery of animal feed, biogas, water and fertiliser for example; involve the processing of very large volumes, incur very high capital costs, and can interfere with the efficiency of conventional waste treatment.

Simple new biotechnology such as genetically engineered ligands and bioaccumulation processes offer opportunities to recover more valuable by-products such as metals, special oils, peptides, vitamins and other materials which are difficult to synthesise. Research is being conducted on a laboratory scale and could be productive in the long term.

CONCLUSIONS

Aerobic effluent treatment is the largest practical use of micro-organisms. The diversity, volume and complexity of wastes means that there are few alternatives to this type of treatment for most waste waters. The size of the market will continue to stimulate development of more efficient processes. Improved reliability and reducing biomass yields are two areas to which biotechnology will make a contribution. Many difficulties have arisen in commercialising new reactor designs. There are important differences between conventional process engineering and the demands of effluent treatment. Traditional processes, although they need a lot of space, are often labour and trouble free.

The recycling of simple by-products such as fertiliser and water re-use will increase to conserve resources.

There is a good case for the anaerobic treatment of strong industrial waste and an expansion of the technology is likely. Current laboratory research on more valuable by-products and the removal of persistent chemicals should produce practical benefit in the long term.

REFERENCES

Gorsuch, T.T. (1987). Food Processing Research Consultative Committee: *Report to the Priorities Board, 1986.* MAFF HMSO, London.

Painter, H.A. (1971). In *Water and Water Pollution Handbook*, Vol. 1, ed. L.L. Ciaccio. Marcel Dekker, New York.

Water Authorities Association (1988). *Water Facts 1985*. Water Authorities Association, London, 47 pp.

Water Research Centre (1977). Cost information for water supply and sewage disposal. *Report TR61*, WRC 1977, 627 pp.

The Use of Chitin by Vibrios in Saline Waters

Rufus K. Guthrie & Daniel Q. Cofie

Disease Control, The University of Texas School of Public Health, UT Health Science Center at Houston, Houston, Texas, 77225, USA

ABSTRACT

Chitin is reported to be rather resistant to biodegradation, although in recent years there have been a number of reports which indicate the production of a chitinase by some microorganisms. Using naturally occurring chitin and practical grade granular chitin, we have studied the growth of V. cholerae *(both 01 and non-01) when utilizing chitin as a sole carbon source. In addition to growth characteristics, the effects of chitin presence on changes in agglutination patterns, and on the production of toxin have been studied. The implications of chitin biodegradation is considered both from the public health, and the environmental pollution standpoints.*

INTRODUCTION

The previous reports by Kaneko & Colwell (1978), Xu *et al.* (1982), Huq *et al.* (1983) and Guthrie *et al.* (1987) among others that populations of the genus *Vibrio* are associated with chitin bearing organisms in saline waters has led to the study of the growth of *Vibrio* species as it is related to biodeterioration of chitin. In previous studies the vibrios were most numerous and most easily cultured when chitinous organisms were present in saline water environment. Quite often, the only association noted, has been that the vibrios survived better in aquatic environments when sediment was present, either in estuaries, or in laboratory aquaria.

MATERIALS AND METHODS

Aquaria were set up containing filtered water from Galveston Bay, or artificial saline water prepared from 'Instant Ocean' salts. Differing concentrations of the following were added as nutrients; granular chitin, chitosan and N-acetyl-D-glucosamine (Sigma Chemicals, St Louis, Missouri); alkaline peptone (Difco); and crude shrimp shell chitin prepared by modification of the method of Hackman (1954). Each aquarium was inoculated with approximately 1×10^6 cells/ml of *V. cholerae*, 01, strain 569B, obtained from the Microbiology Department, University of Texas Medical Branch, Galveston, Texas; strains 7077, 27, and 83, obtained from the Louisiana Department of Health Laboratory in Lake Charles, Louisiana; and non-01 strains isolated from Galveston Bay at Kemah, Texas. The effect of each nutrient was studied at four different concentrations: 0·25 g/litre (designated low), 0·50 g/litre, 1·25 g/litre and 2·50 g/litre (designated high).

To determine whether the utilization of chitin by *V. cholerae* was dependent upon chitin concentration, microcosms were prepared using filtered bay water to which chitin was added at concentrations between 0 and 5 g/litre. The dose of *V. cholerae* inoculated into each was approximately 10^2 cells/ml.

At daily intervals, samples were aseptically drawn from each aquarium and plated for enumeration of vibrios. TCBS agar plates were used, and colonies were counted after incubation at room temperature for 24–48 h. The pH of the aquaria was adjusted initially, and remained approximately constant. Temperature of water in the aquaria was 22°C ±2.

RESULTS AND DISCUSSION

It was observed that the maximum solubility of chitin was 5 g/litre. The dose response is shown in Fig. 1. Concentrations of 5 g/litre, 1 g/litre, and 0·5 g/litre were tested with several strains of *V. cholerae* organisms. Growth of the organisms was dependent upon the amount of chitin available. In the absence of chitin, the organism count remained approximately stationary for 2 days, and then decreased to zero. In all the concentrations tested, the maximum count was reached on the third day, after which, the counts decreased, but even by the 8th day, they had not returned to the initial level within the aquaria. The highest count of organisms was reached with 5 g/litre, and the lowest when 0·5 g/litre was

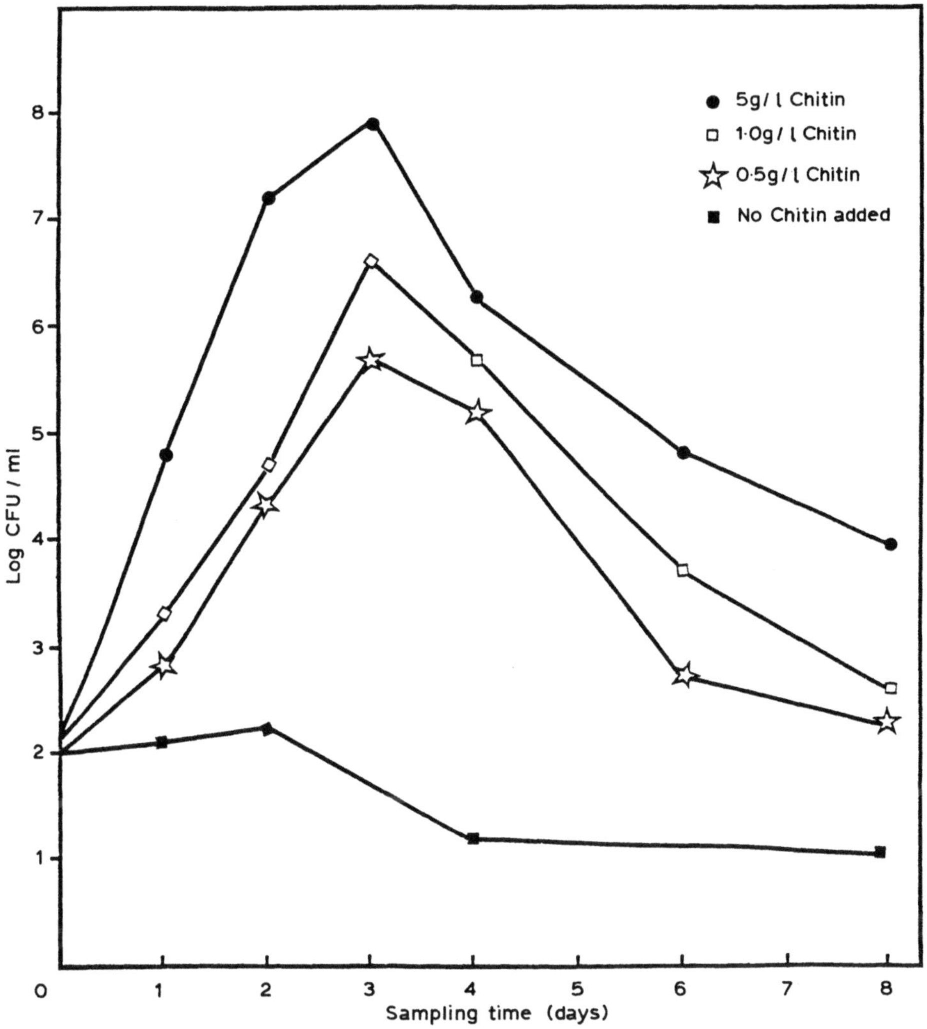

Fig. 1. Relationship of dose of chitin inoculated into aquarium to growth of vibrios.

added. Dose dependency was noted at each testing day from the first to the 8th.

In further tests (Figs 2 and 3) chitin degradation products, N-acetyl-D-glucosamine and chitosan were not as efficient in supporting growth of vibrios as the crude preparations. In low concentrations there was little difference in growth stimulation between crude shrimp chitin, and the practical grade granular chitin (crab chitin). However, in higher

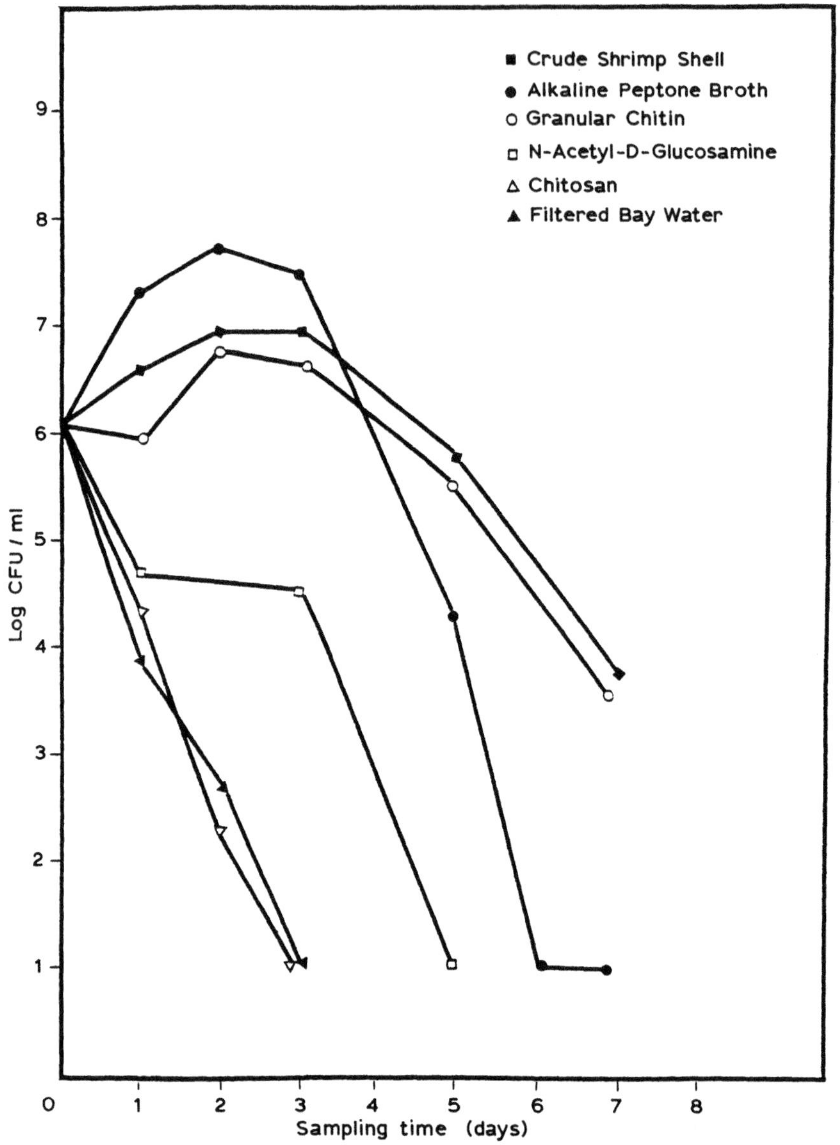

Fig. 2. The effects of low concentrations of chitin preparations on vibrio growth.

concentration, crude shrimp chitin was more stimulatory to vibrio growth than the partially purified granular chitin product. A reason for this was perhaps that the process of granularization of the chitin caused some biodegradation.

The solubility of chitin, and the various preparations used in this work

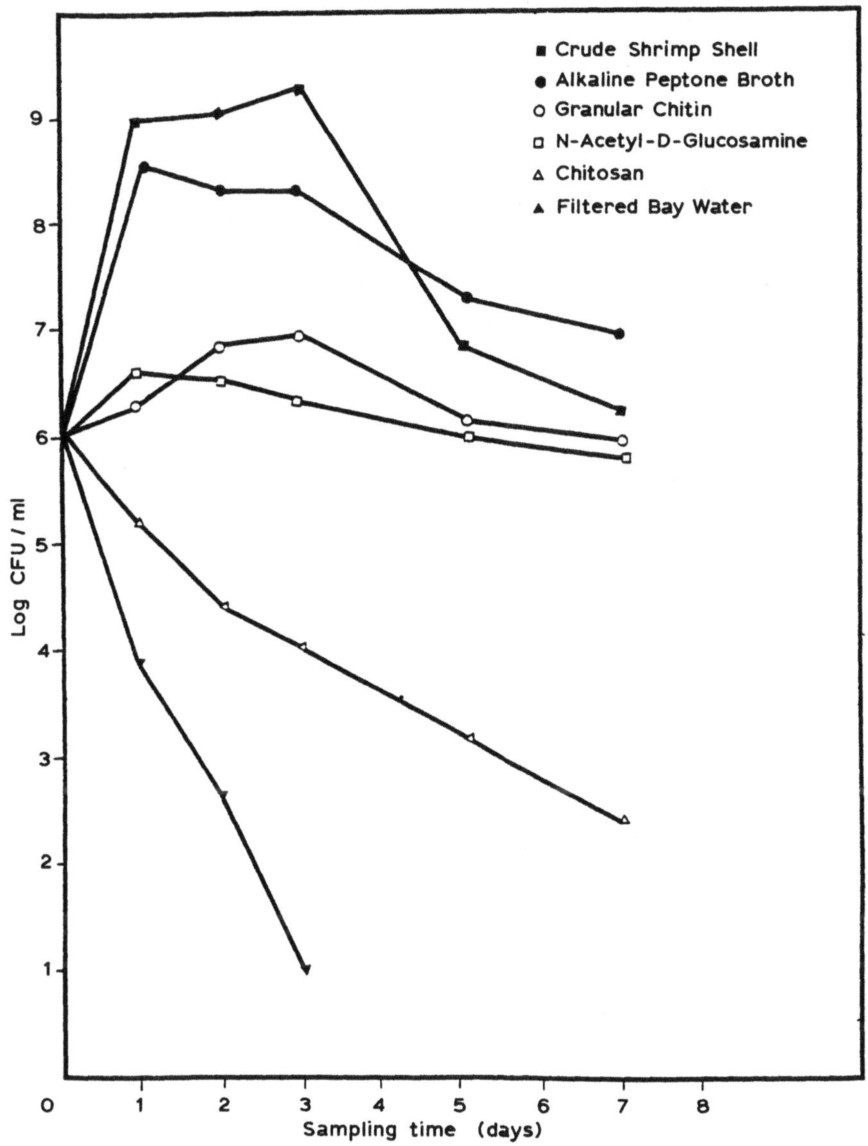

Fig. 3. The effects of high concentrations of chitin preparations on vibrio growth.

are critical to the availability of the chemicals to the organisms for growth. Further studies are continuing to detect the exact effects of these solubilities upon the growth of vibrios. It has been shown that solubility, and therefore availability of chitin, is more dependent upon the salinity of the waters, than upon its pH. Further studies are being carried out to confirm the relationships between these parameters.

Other studies are directed towards establishing more clearly the relationship between chitin utilization and the production of toxins and of the antigenic properties of the organisms. With the stimulation of growth, other characteristics such as toxigenicity and antigenicity are also affected as has been shown in preliminary tests.

From the standpoint of public health, it has been shown that the biodeterioration of chitin, is involved in the growth and presence of these pathogens in the estuarine environment. The results explain why the organisms have survived and have been characterized as natural inhabitants of these environments. More needs to be known about the potential usage of microorganisms in handling this natural waste product which is estimated to amount to millions of tons annually. Thought must be given to the usage of indicator organisms to detect waters polluted by sewage. If these pathogens are utilizing a natural product in the aquatic environment, then a way is needed to detect their presence, and to predict the potential for outbreak of infectious diseases.

REFERENCES

Amako, K., Shimodori, S., Imoto, T., Miake, S. & Umeda, A. (1987). *Appl. Environ. Microbiol.*, **53**, 603–5.

Guthrie, R.K., Cofie, D.Q. & Makututu, C. (1987). *Water Resour. Bull.* (in press).

Hackman, R.H. (1954). *Austr. J. Biol. Sci.*, **7**, 168–78.

Huq, A., Small, E.B., West, P.A., Huq, M.I., Rahman, R. & Colwell, R.R. (1983). *Appl. Environ. Microbiol.*, **45**, 275–83.

Keneko, T. & Colwell, R.R. (1978). *Microbiol. Ecol.*, **4**, 135–55.

Xu, H.S., Roberts, N., Singleton, F.L., Attwell, R.W., Grimes, D.J. & Colwell, R.R. (1982). *Microbial. Ecol.*, **8**, 313–23.

Microbial Degradation of Cyanide Containing Effluent from a Dye Industry

M.S. Andhale

Department of Microbiology, Institute of Science, Kile Ark, Aurangabad–431 001, India

ABSTRACT

Seventy samples of soils from the rhizosphere and rhizoplane of Sorghum vulgare *plants were examined to yield 80 cultures of sodium cyanide degraders by enrichment culture. After preliminary screening on the basis of their tolerance (100 mg/litre NaCN), a culture of* Arthrobacter *sp (A-3) and an* Alcaligenes fecalis *(AF-5) were selected for treating cyanide containing effluent (40 mg/litre) from a local dye industry. When these cultures were tried singly and in combination with activated sludge, it was found that they could bring down the cyanide concentration to 1 mg/litre level in 96 h. The effect was synergistic, and when used in combination a reduction to 0·05 mg/litre cyanide was achieved in 96 h. This combination showed a better degradation of cyanide (0·01 mg/litre) than activated sludge, indicating the beneficial role of the addition of such bacteria to the activated sludge inoculum. Based on these findings a bench scale activated sludge unit (17·5 litre tank) was set up in the laboratory to treat the cyanide waste from the dye industry. It had the following specifications; Feed volume 9 litres/day, mean detention time 13 h, temperature 28–30°C (room temperature). The results are being applied in the industry after further scale up studies. This study revealed the potential use of* Arthrobacter *sp (A-3) and* A. faecalis *(Af-5) in the effective disposal of cyanide in industrial wastes.*

INTRODUCTION

The presence of cyanide in industrial effluent waste presents a major environmental and ecological hazard. Although chemical methods of treating this compound are known, microbial degradation of cyanide is of interest both in understanding how cyanide may be dealt with in the environment and to evaluate the economic viability of microbial systems for cyanide degradation.

A number of microorganisms have been reported to possess the ability to produce, assimilate, or detoxify hydrogen cyanide (Andhale, 1986). Variations exist in the modes of assimilation of HCN by bacteria and fungi. *Chromobacterium violaceum, Bacillus megaterium* and *Escherichia coli* possess a β-cyanoalanine synthase activity which enables them to condense HCN with cysteine to form β-cyanoalanine. Another product of HCN assimilation of *C. violaceum* is D-cyano-α-aminobutyric acid. Some fungi incorporate HCN into the amino acids alanine, glutamine and asparagine (Akopyan *et al.*, 1975). *Pseudomonas fluorescens* converts HCN into ammonia and presumably further to carbon dioxide (Harris & Knowles, 1983). Rhodanese, which is widely distributed in bacteria and fungi, is also used for cyanide assimilation (Sorbo, 1975). Induction of HCN in sensitive respiration is an alternative device by which microorganisms can cope with HCN. Atkinson (1975) studied the capacity of *Bacillus stearothermophilus* possessing a rhodanese, to convert cyanide to thiocyanate by using a continuous reactor.

METHODS

Medium

A mineral salts medium (pH 8·5) was used for isolation of bacteria and the subsequent growth of cultures. This consisted of M-9 salts (Miller, 1972) with ammonium salts omitted and 1 ml trace metal/litre. Where required the medium was solidified with 1·5% (w/v) Oxoid purified agar. Glucose (10 g/litre) was routinely used as the carbon source. On plate growth the nitrogen source was either 5 g/litre NH_4Cl incorporated into the agar or 50 mg/litre NaCN supplied to medium

Isolation of cyanide utilising bacteria

The soil from sorghum plant rhizosphere was selected because it has a high content of cyanogenic glycosides such as linamarin, dhurrin, etc., which are important precursors of cyanide (Nartey, 1981). The rhizosphere contains a selection of microflora which attacks cyanide. Since cyanide degrading microorganisms were reported from soils used continuously for growing cyanogenic plants (Skowronski & Strobel, 1969), it was concluded that cyanide degrading microflora could be obtained if rhizosphere soils of sorghum plants were examined. The soil was air dried, sieved to remove large particles and stored in a plastic bag at 4°C.

An enrichment culture technique, as modified from the methods used by Schlegel & Jannasch (1967), Skowronski & Strobel (1969) and Agate (1977) was used to isolate the cyanide utilising microorganisms. To each Erlenmeyer flask (250 ml) containing 50 ml aliquot of glucose mineral salts medium (50 mg/litre) NaCN was added. Each flask was inoculated with 1 g of the sample under investigation. Shaking was carried out at 100 rpm on a rotary incubator shaker (Model E 25, EMNVEE, Pune) at 30°C. After 8 days, 2 ml of the culture was transferred to 50 ml of fresh glucose mineral salts medium with NaCN; the process was repeated for a total of four subcultures. The last subculture was checked for different types of microorganisms under the microscope by Gram staining. For the isolation of pure cultures the method of Harris & Knowles (1983) was followed. The isolates were identified according to standard methods.

Analytical methods

Growth in liquid culture was estimated from the absorbance 600 nm using 10 mm cuvettes. The Beer–Lambert Law was obeyed up to an absorbance of 1·0 (0·36 mg dry wt/ml). Samples were taken from the cultures and bacteria were removed by centrifugation for 5 min in a Remy High Speed refrigerated centrifuge and the cyanide content of the medium was determined colorimetrically.

RESULTS AND DISCUSSION

A total of seventy rhizosphere/rhizoplane soil samples were tested and 80 bacterial cultures were isolated having (NaCN) cyanide degrading capacity. The distribution of the isolates is given in Table 1. The 80 isolates belong to six different genera, viz. *Arthrobacter* (15%), *Alcaligenes* (16·25%), *Pseudomonas* (17·5%), *Micrococcus* (17·5%), *Bacillus* (26·75%) and *Klebsiella* (10%).

There were more cyanide degrading bacteria in rhizosphere soil than the rhizoplane soil. To differentiate the good cyanide degraders from the 80 isolates preliminary screening was carried out using growth in the presence of NaCN as a criterion. The isolates were allowed to grow at concentrations from 100 to 1000 mg/litre. It was observed that growth

TABLE 1
Genera of Cyanide Degrading Bacteria

Genus	*No. of strains*[a]
Arthrobacter sp.	12
Alcaligenes faecalis	13
Pseudomonas aureofaciens	14
Micrococcus sp.	10
Bacillus megaterium	16
Bacillus cereus	7
Klebsiella sp.	8

[a]Total number of isolates = 80

above the 500 mg/litre level hardly took place but some cultures grew poorly at 500 mg/litre. From each type of culture a strain was selected which showed such tolerance and was used in the experiments.

Optimum conditions for the growth of these cultures in the presence of NaCN with respect to tolerance to NaCN, temperature, pH and time of incubation period were provided. On the basis of cyanide degradation efficiency, out of the six type cultures, the two best cyanide degrading cultures were selected, viz. *Arthrobacter* sp. (A-3) and *Alcaligenes faecalis* (AF-5). They were able to degrade more than 90% of NaCN and were used for treatment of cyanide containing effluent.

Cyanide degradation of industrial effluent

The untreated industrial effluent containing cyanide was brought to the laboratory from a dye industry. The presence of cyanide was checked by a spot test with picric acid. The concentration was adjusted to 40 mg/litre by the picric acid method.

The experiments were performed using a single cyanide and mixed cyanide degrading cultures which were later mixed with the activated sludge. As a control activated sludge was tried alone. The pH of the effluent was adjusted to 8·5. An inoculum of 5% (v/v) was added and the flasks incubated on a rotary shaker at 30°C for 5 days. The residual cyanide was measured every 24 h by the picric acid and Lambert methods. The results are recorded in Table 2.

With the selected cultures, the residual cyanide after 120 h incubation was at the 0·02 mg/litre level, which is the permissible limit set by the Water Pollution Control Board. Whereas, when mixed cultures along with activated sludge were used the residual cyanide level was 2 mg/litre

TABLE 2
Effluent Treatment Using Selected Bacteria and Activated Sludge

Str. No.	*Culture No.*	*A-3*	*AF-5*	*A-3 + AF-5*	*(A-3 + AF-5 + Activated Sl[a]*	*Activated sludge*
	Incubation period (h)	*Residual cyanide (mg/litre)*	*Residual cyanide (mg/litre)*	*Residual cyanide (mg/litre)*	*Residual cyanide (mg/litre)*	*Residual cyanide (mg/litre)*
1	*2*	*3*	*4*	*5*	*6*	*7*
1	0	40·0	40·0	40·0	40·0	40·0
2	24	28·0	30·64	25·0	21·0	32·0
3	48	18·0	15·0	12·0	6·0	24·0
4	72	4·9	3·0	2·0	0·5	18·5
5	96	1·0	1·0	0·05	0·01	12·0
6	120	0·02	0·02	0·01	0·00	10·0

A-3: *Arthrobacter* sp. A-3.
AF-5: *Alcaligenes faecalis* AF-5.
[a]Sl: sludge.

after 72 h incubation and after 96 h it was 0·05 mg/litre. In the case of activated sludge alone, the cyanide degradation process was very slow and the residual cyanide level was 10 mg/litre after 120 h incubation time, which is too high for a safe disposal limit for such effluents. From this data a bench scale activated sludge unit was used based on a design by Ludzack & Bloomhuff (1959) for scale up studies on the degradation of cyanide waste from the dye industry. It has a 17·5 litre activated sludge stock tank which receives a continuous feed during 12–23 h per day. The usual feed volume was 9 litres/day and the mean detention time was 13 h. The unit was operated at room temperature (28–30°C) and worked successfully.

CONCLUSIONS

(1) A number of common soil bacteria were isolated which were capable of degrading cyanide.
(2) Strain selection was used to identify the best organisms and *Arthrobacter* sp. and *Alcaligenes faecalis* chosen as the best cyanide degraders.
(3) These organisms were added to a model activated sludge plant

treating dye waste. The degradation was better than with activated sludge alone.

(4) The experiments showed that the degradation of cyanide in industrial waste was possible in activated sludge and that the process benefited from bioaugmentation.

REFERENCES

Agate, A.D. (1977). *Proc. Sem. Environmental Impact on Developmental Activities and Adoption of Suitable Guidelines for Project Control in Selected Sectors such as Industries, Agriculture, Irrigation, Transport, Urbanisation, etc. Institute of Engineering,* Hyperabad, India, pp. 10.1–10.5.

Akopyan, T.N., Braustein, A.E. & Goryachenkova, E.V. (1975). *Proc. Nat. Acad. Sc. U.S.A.,* **72**, 1617–21.

Andhale, M.S. (1986). Ph.D. thesis, Marathwada University, Aurangabad, India.

Atkinson, A. (1975). *Biotechnol. Bioengng,* **17**, 457–60.

Harris, R. & Knowles, C.J. (1983). *J. Gen. Microbiol.,* **129**, 1005–11.

Howe, R.H.L. (1969). *Process Biochem.,* **25**, 25–8.

Ludzack, F.J. & Bloomhuff, R.N. (1959). In *Proc. 14th Ind. Waste Conf.,* Purdue University, pp. 547–65.

Miller, J.H. (1972). *Experiments in Molecular Genetics.* p. 431, Cold Spring Harbor Laboratory, New York, p. 431.

Nartey, F. (1981). In *Cyanide in Biology*, ed. B. Vennesland, E.E. Conn, C.J. Knowles, J. Westley & F. Wissing. Academic Press, New York.

Petter, E.E.J., & Ware, G.C. (1955). *Chem. and Ind.,* **40**, 1232–8.

Schlegel, H.G. & Jannasch, H.W. (1967). *Ann. Rev. Microbiol.,* **24**, 49–70.

Skowronski, B. & Strobel, G.A. (1969). *Can. J. Microbiol.,* **15**, 93–8.

Sorbo, B.H. (1975). In *Metabolic Pathways*, Vol. 7, ed. D.M. Greenberg. Academic Press, New York, pp. 433–56.

An extended bibliography for this paper may be obtained from the author.

HYDROCARBONS

Products of Hydrocarbon–Microorganism Interaction

Colin Ratledge

Department of Biochemistry, University of Hull, Hull HU6 7RX, UK

ABSTRACT

When microorganisms are grown on hydrocarbons, changes can be seen in cell morphology and composition, in the presence of specific surfactants which apparently act to solubilize the otherwise insoluble substrate, and finally in the presence of products arising directly from the oxidation of the hydrocarbons. This review discusses the occurrence of biosurfactants in microorganisms in general and offers a hypothesis on how they may function possibly in conjunction with amphipathic receptor molecules which may also act as channels for conveying the hydrocarbon into the cell membrane and thus into the cytosol. Dissociation of the hydrocarbon from the surfactant possibly occurs within such amphipathic molecules with the return of the surfactant to the outside of the cell. The hydrocarbon is considered to flow through the membrane, its return into the wall being prevented by lack of a solubilizing agent. Metabolism of the hydrocarbon is probably not necessary to achieve uptake but nevertheless does occur with alkanes and related molecules. Under certain conditions, usually achieved by using mutated organisms, products of alkane oxidation may accumulate. These are discussed in the latter half of this review.

INTRODUCTION

Hydrocarbons are still under active consideration as a potential feedstock for a variety of biotechnological processes (Drozd, 1987).

Although the original Single Cell Protein (SCP) production process developed by British Petroleum has long since been stopped, derivative processes are being used, and developed, for SCP in the USSR. Other biotechnological products which can be effectively produced from hydrocarbons include citric acid, glutamic acid, some carbohydrates, vitamins and even antibiotics. However, the difficulties which were experienced in gaining acceptance for SCP as being suitable for use as an animal feedstuff has made some manufacturers reluctant to use hydrocarbons as a feedstock in their processes. Furthermore the increased costs of alkanes since the early 1970s have now risen to the point where it is only slightly below the price of palm oil which can equally be used as a fermentation feedstock for the same processes.

In spite of these reservations, hydrocarbons, and especially alkanes, are still potentially valuable substrates for the biotechnological industry. In addition, problems of dealing with oil spillages and removal of oil-based pollutants (Gutnick & Rosenberg, 1977) as well as using microorganisms for enhanced oil recovery (Shennan & Levi, 1987) continues to focus attention on the mechanisms of hydrocarbon utilization. The routes of hydrocarbon metabolism are well described (Atlas, 1984; Bühler & Schindler, 1984; Boulton & Ratledge, 1984). The details of these routes will therefore not be reiterated in this short review. Instead, it is the intention to focus on the process of microbial interactions with hydrocarbons and to describe the various products which arise from this interaction. Such products may arise directly from the hydrocarbon itself or be associated with the physical process of allowing the microorganism to contact and solubilize an otherwise water-insoluble substrate. As the majority of work has been carried out with aliphatic hydrocarbons, this review concentrates upon alkanes as a typical hydrocarbon.

HYDROCARBON SOLUBILIZATION AND UPTAKE

Hydrocarbons by their very nature are insoluble in water. Their utilization by microorganisms that prefer water-soluble substrates immediately poses a problem of contact between two immiscible surfaces—that of the hydrocarbon droplet and that of the microbial cell. Three mechanisms for uptake have been described (Boulton & Ratledge, 1984, 1987):

(i) uptake of hydrocarbon dissolved in the aqueous phase—this is naturally a very low amount;

(ii) uptake via direct contact of the microbial cell with a hydrocarbon droplet; and
(iii) conversion of the hydrocarbon to an emulsified form which can then be taken into the cells as a quasi-soluble substrate.

Of these mechanisms, the first is of limited importance though it may be a mechanism which operates with short chain alkanes ($< C_{10}$) and with solid hydrocarbons including aromatic compounds. The second mechanism begs the question of how contact is achieved between a polar, hydrophilic cell surface and a non-polar, hydrophobic substrate. It is difficult not to assume that the contact is achieved by modification of the cell surface and that this therefore means that some type of microemulsion is created at the interface. This mechanism and the third one—that of overt solubilization—thus merge into each other and mechanisms (ii) and (iii) may be different aspects of the same phenomenon. However the occurrence of microbial flocs (cf. Glombitza, 1982) may argue in favour of two distinct mechanisms.

Solubilization of hydrocarbons: production of surfactants

Emulsification of hydrocarbons is a well-described process and a number of surfactants have now been isolated from microbial cultures following growth of bacteria, yeasts and moulds on a variety of (aliphatic) hydrocarbons. These emulsificants, which are mainly or entirely extracellular, may be relatively simple, low molecular weight glycolipids or complex high molecular weight materials of uncertain structures. The diversity of structures is illustrated in Table 1; for more complete information the recent monograph on this subject can be recommended (Kosaric *et al.*, 1987).

It should be emphasized that production of biosurfactants by growing microorganisms on alkanes can also be achieved by using either water insoluble substrates such as fatty acids or triacylglycerol oils. However, surfactant production can sometimes be achieved in the absence of such substrates (Mulligan *et al.*, 1984; Persson & Molin, 1987) and indeed it has been considered that it would be advantageous if commercial production of biosurfactants could be achieved in the absence of hydrocarbons (Guerra-Santos *et al.*, 1984). Thus Persson & Molin (1987) recently examined almost 2500 isolates of *Pseudomonas* and *Vibrionaceae* and found 30 organisms which were positive for production of a surfactant when grown on glucose. However it is likely that these isolates, which were obviously constitutive for surfactant biosynthesis, nevertheless relied upon the formation of such materials in order to achieve growth in their natural environment which was the skin of pigs or cattle. Here the nutrients available would be primarily the water-insoluble sebum secreted by the sebaceous glands of the skin and would require the presence of a surfactant in order for the organism to solubilize its carbon substrate. That about 1% of such organisms should prove to have a

TABLE 1

Examples of Biosurfactants Produced During Microbial Growth on Hydrocarbons or other Water-insoluble Substrates

Molecular species	*Organism*	*Surfactant*	*Reference*
Glycolipids			
	Arthrobacter sp.	Mono-, di- and tricorynomycolates of glucose, maltose, maltotriose, etc.	Li *et al.* (1984)
	Rhodococcus erythropolis	Trehalose-6, 6′-dicorynomycolates	Rapp *et al.* (1979)
	Pseudomonas aeruginosa	Mono- and di-rhamno-lipids	Hisatsuka *et al.* (1971)
	Torulopsis (Candida) bombicola and *T. apicola*	Sophorolipids	Cooper & Paddock (1984); Hommel *et al.* (1987)
	Ustilago maydis	Cellobiose glycolipid	Frantz *et al.* (1986)
Phospholipids			
	Acinetobacter HO1-N	Phosphatidylethanolamine	Käppeli & Finnerty (1979)
	Cladosporium resinae	Phosphatidylcholine and others	Siporin & Cooney (1978)
Amino acid-containing lipids			
	Bacillus subtilis	Surfactin (heptapeptido-3-hydroxy-14-methylpalmitate)	See Lang & Wagner (1987)
Complex molecules			
	Candida (Yarrowia) lipolytica	Liposan	Cirigliano & Carmen (1984, 1985) and Cameotra *et al.* (1984)
	Acinetobacter calcoaceticus (RAG-1)	Emulsan (see Table 2)	Zuckerberg *et al.* (1979)
	A. calcoaceticus (BD4)	Rhamnopolysaccharide	See Rosenberg (1986)

constitutive enzyme system for surfactant biosynthesis is not surprising.

Strategies for optimizing the production of biosurfactants in general have been recently reviewed by Kosaric *et al.* (1984) and Syldatk & Wagner (1987). A pilot-plant process for the production of the rhamnolipid surfactant of *Pseudomonas aeruginosa* has been described by Reiling *et al.* (1986) and Shabtai & Gutnick (1986) have described how enhanced production of Emulsan (see Table 2) may be achieved by using mutants of *Acinetobacter calcoaceticus* which had been selected for resistance to cetylammonium bromide. Hommel *et al.* (1987) have shown how important it is to design the appropriately balanced medium for good production of sophorolipid with *T. apicola* (see Table 3).

Two surfactants are now in commercial production, though for very different reasons. The sophorolipid from *Candida bombicola* is produced by Kao Soap Co. of Japan (see Falbe & Schmid, 1986) as an additive to lipsticks and other toiletries. Emulsan, the amphipathic polysaccharide emulsifier from *Acinetobacter colcoaceticus* RAG-1 is being used as an industrial oil cleaning agent (Gutnick & Shabtai, 1987). The properties of Emulsan are given in Table 2. Opportunities for further exploitation of this field evidently still exists as witnessed by the number of research groups engaged in these activities, many of whom are supported by industrial concerns. The applications of surfactants may therefore be seen as spanning the high-value low volume markets of cosmetics and toiletries to the lower-value, high volume market for industrial uses and even for the enhanced recovery of oil.

The concept of using microorganisms for enhanced recovery of oil from reservoir rock pools relies mainly on the innate ability of the microorganisms to produce biosurfactants *in situ* (Brown *et al.*, 1986;

TABLE 2

Properties of Emulsan (from Gutnick & Shabtai, 1987)

Producing organism: *Acinetobacter calcoaceticus* RAG-1
Mol wt ~980 00
Molecular form:
Polysaccharide complexed with Lipid
Sugars in polysaccharide backbone:
D-galactosamine
D-galactosaminuronic acid
+ unidentified sugar
Fatty acids present (total 5–15% w/w):
12:0, 14:0, 16:0, 18:0, HO-12:0
Function: binds tightly to oil interface and prevents coalescence of oil droplets
Used as industrial oil cleaning agent

TABLE 3
Effect of Yeast Extract on Formation of Glycolipids (GL) by *Tor. apicola* (from Hommel *et al.*, 1987)

	Growth substrate			
	Hexadecane[b]		*Glucose*[c]	
Yeast extract added (g/litre)	*Biomass yield (g/litre)*	*GL*[a] *yield*	*Biomass yield (g/litre)*	*GL*[a] *yield*
0·06	2	14	2	1
0·6	7	600	15	7
6	1	1	21	6

[a] Determined as critical micelle dilution factor.
[b] 10 days growth.
[c] 3 days growth.

Shennan & Levi, 1987). Thus surfactant production by microorganisms in response to their dealing with water-insoluble substrates is of economic importance. This is an area of current high interest with both oil companies and the oleochemical industries in general.

Uptake of hydrocarbons

The exact route of uptake of hydrocarbons into a microbial cell is still unresolved. Simple permeation of the hydrocarbon through the cell membrane may occur though this cannot be the entire explanation. How a 'solubilized' alkane in conjunction with a surfactant is taken up into a cell is still completely obscure and there appears to be no definitive evidence for dual uptake of hydrocarbon and surfactant. A hypothesis for this is, however, presented in Fig. 1. Käppeli & Fiechter (1977) have presented some evidence for the formation of a hydrophobic component of the cell wall of *Candida tropicalis* during growth on alkanes. They suggested that this complex of polysaccharide and fatty acids could act at the surface of the cells to give them an amphipathic layer. The hydrocarbon was then envisaged as becoming monodispersed in this layer. Uptake of the alkane, by an unspecified mechanism, was then presumed to follow. The polysaccharide–fatty acid complex which was identified was thus a surface biosurfactant though how the hydrocarbon would be released from this complex is not clear.

In the hypothesis advanced in Fig. 1, it is assumed that the surfactant and hydrocarbon are in a loose association and that together they would enter the lipophilic outer part of the cell envelope. Here the hydrocarbon would dissociate from the complex possibly within an amphipathic receptor/channel molecule as suggested by Käppeli & Fiechter (1977). The hydrocarbon would then partition into the membrane and the surfactant would be excluded from the channel and be returned to the exterior of the cell. As the concentration of hydrocarbon in the membrane builds up it would physically move out of the membrane in the only direction possible—to the cytosol. Being almost totally water-insoluble, the hydrocarbon would quickly coalesce into, first, microdroplets and then into larger agglomerations readily visible by electron microscopy (see Boulton & Ratledge, 1984).

It is clear from the available evidence (see Ito & Inoue, 1982) that surfactants have a specificity of action. They generally appear to promote growth of the producing species on alkanes and may be less effective, or even completely ineffectual, in stimulating growth of other microorganisms (see Table 4). Such a specificity of action could be easily accommodated into the proposed model (Fig. 1(B)) by requiring the surfactant/alkane complex to have the correct degree of solubility in the amphipathic receptor or when the complex had entered the channel for it then to

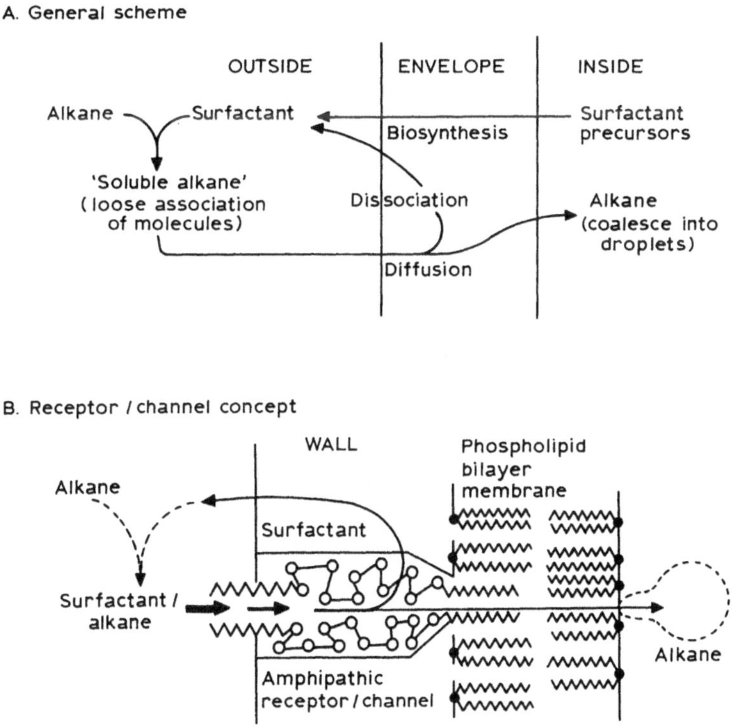

Fig. 1. Hypothesis for surfactant-linked uptake of alkanes into a microbial cell.

encounter the appropriate conditions for it to dissociate. Without dissociation, the hydrocarbon would not be released into the membrane.

Some indirect support for this hypothesis—which is proposed simply as a model awaiting testing—may be seen in the recent publications of Lindley & Heydeman (1983, 1986). Lindley & Heydeman (1986) have described alkane uptake by *Cladosporium resinae* as being initially by passive adsorption of the alkane onto the cell surface. (A surface receptor site?) The nature of the cell surface is then crucial to efficient uptake. This process is then superseded by an energy-requiring transfer of alkane into the cytosol of the cell—this could be an active transport process akin to uptake of sugars, amino acids, etc., or could be by pinocytosis for which energy would also be required. (The latter process would involve physical rearrangement of the phospholipid membrane.)

The results of Lindley & Heydeman (1983, 1986) clearly indicate that the surface of the microbial cell is important when considering the uptake of hydrocarbons. Changes in the membrane structures of bacteria, yeasts and moulds following growth on alkanes have often been described (see Boulton & Ratledge (1984) for review and photomicrographs). The invaginations of the cytoplasmic membrane, which increases the phospholipid content of the cells (Ratledge, 1987), presumably create additional points for hydrocarbon passage into the cell. Rosenberg *et al.* (1982) have described numerous surface fimbriae on the cell surface of *Acinetobacter* sp. and suggested that these played a part in achieving adhesion of cell to the hydrocarbon droplets. Outgrowths from yeasts cultured on alkanes have also been reported (Osumi *et al.*, 1975). These could be considered as alternative mechanisms for achieving initial attachment of the cell to the alkane droplet—perhaps without the need for biosurfactant synthesis. In addition, several authors have identified channels running through the cell walls of yeasts grown on hydrocarbons

TABLE 4

Effect of Sophorolipids from *T. apicola* on Growth of other Microorganisms on Hexadecane (from Hommel *et al.*, 1987)

	Lag phase	*Sp. growth rate (h^{-1})*	
		Alone	*+ Glycolipid*
Acinetobacter calcoaceticus	Shortened	0·67	0·75
Corynebacterium fascians	Lengthened	0·52	0·19
Torulopsis apicola	Shortened	0·01	0·01
Candida lipolytica	Lengthened	0·2	0·03

In addition, the glycolipid exhibits antibiotic effects against *B. subtilis*, *Rhodococcus* spp., etc., and antiphage activity with *Plectonema* and cyanophage LPP-1.

(e.g. Meisel *et al.*, 1976).

All modifications of the cell structure appear to reflect adaptations of the microorganism to deal with the problem of growing on a water-insoluble substrate and probably encompass several distinct features: solubilization or improved contact, uptake and, finally, metabolism. It is therefore not possible to say with any degree of certainty what are the respective functions of the observed changes in cell morphology. However the hypothesis advanced above and in Fig. 1 could bring together many of the above observations into a tentative unified scheme.

One change in cell morphology, which is now well described in eukaryotic cells, does however have a clear function ascribed to it. This is the formation of peroxisomes which are not evidenced in cells grown on glucose and other carbohydrates.

Peroxisomes are responsible for the oxidation of fatty acids which arise from the alkanes (or other related substrates). Their appearance is induced therefore by growing yeasts (and presumably fungi also) on alkanes, fatty acids, vegetable oils or animal fats. Catalase which is associated with the peroxisomes is thus elevated in activity when such substrates are being used and other enzymes associated with the peroxisomes (see Boulton & Ratledge, 1984) can be similarly enhanced in activity (Gmünder *et al.*, 1981; Holdsworth, 1987). A discussion of the role of the peroxisomes is however outside the scope of this review but the subject has been reviewed by Fukui & Tanaka (1979, 1981) and Numa (1981).

To summarize: although there is evidence favouring a more complex mechanism of alkane uptake than direct diffusion, most microorganisms when grown on alkanes will show the presence of accumulated droplets of alkane within the cells and indeed Lindley & Heydeman (1986) have been able to show that the amount of sequestered alkane exceeds that in the surrounding medium. These inclusions of hydrocarbons can be seen in most electron micrographs of alkane-grown cells. Such an acquisition could be taken as indicative of a facilitated diffusion process in which the hydrocarbon diffuses into the cell—perhaps by virtue of its association with the appropriate surfactant (see Fig. 1). The loose association of surfactant and hydrocarbon could dissociate (or disaggregate) within the cell envelope or membrane by means of entering a specific receptor/channel created by an amphipathic molecule (Käppeli & Fiechter, 1977). The surfactant would be returned to the outside and the alkane would enter the cytoplasm. The alkane would then be effectively trapped within the cell as the surfactant would not be present within the cell itself and so the process could not be reversed.

The ensuing metabolism of the hydrocarbon, which is probably not necessary to achieve uptake, may then give rise to a number of direct products which are discussed below. However, the rate of hydrocarbon metabolism is evidently less than the rate of uptake to account for the ready intracellular accumulation of hydrocarbons. The hydrocarbons themselves are seemingly without effect on the cell structures and organelles as they will readily coalesce into droplets keeping the concentration of free hydrocarbon extremely low.

DIRECT PRODUCTS FROM HYDROCARBON OXIDATION

General considerations

Figure 2 shows the principal routes of alkane oxidation and the various products which arise. Some products are found extracellularly and these have been indicated in the figure. However other products, such as phospholipids, occur both intra- and extracellularly. Extracellular lipids, including phospholipids, may function in the solubilization of the hydrocarbon substrate (see above and Table 1). Microbial attack on an alkane usually (but not always—see below) commences at a methyl terminal producing, in sequence, a fatty alcohol, fatty aldehyde and, finally, fatty acid. Attack may also occur at the second methyl terminus to give dicarboxylic acids though diols as well as the ω-hydroxyfatty acids may occur as intermediates. Unless β-oxidation is prevented—either by mutation to remove an appropriate gene product or by use of an inhibitor—both mono- and dicarboxylic acids will be readily degraded by the β-oxidation cycle and little, if any, product can be recovered directly. Fatty acids may be found in the free form outside the cell but if they remain intracellular they must be esterified first to give the thioester of coenzyme A and then to more permanent molecules such as the triacylglycerols and glycerophospholipids. Thus the fatty acyl groups of the various lipids within the cell are found to correspond in chain length to the alkane used as the original substrate (see Boulton & Ratledge, 1984, 1987). There is apparently no necessity for a microorganism to synthesize its own lipids when supplied with a gratuitous supply of fatty acids derivable from an alkane substrate and consequently if odd chain length alkanes are used (especially C_{15} or C_{17}) then the fatty acyl groups recovered from the cell lipids will likewise be of an odd chain length. There are limits, however, to the fatty acids which are able to be utilized by cells for biosynthesis of phospholipids, etc., and usually fatty acids must be between C_{14} and C_{18} in length to be accommodated. Small

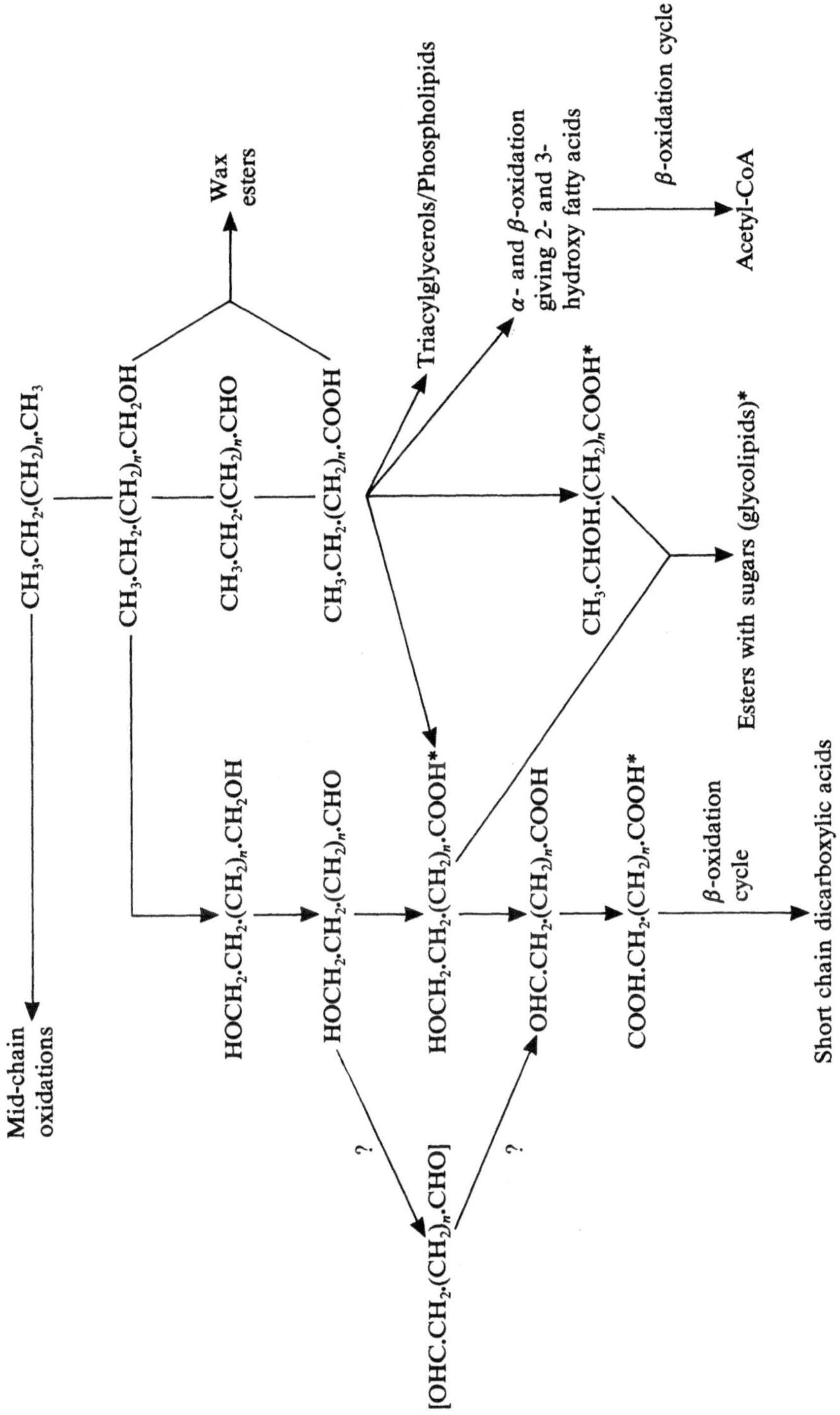

Fig. 2. Primary pathways of alkane oxidation in microorganisms '*' indicates mainly found as extracellular products, '?' indicates uncertain route.

amounts of both shorter and longer chain length fatty acids can be tolerated. If a short chain alkane, say dodecane, is used as substrate, elongation of the ensuing fatty acids by addition of C_2 units (acetyl-CoA) must occur. Longer chain length alkanes must be similarly decreased in length, again usually by removal of C_2 units.

Desaturation of the saturated fatty acids derived from an alkane must occur as cells require their phospholipids to be mobile and this would not occur if they were composed entirely of saturated acids. Consequently, growth of a microorganism on, say, hexadecane will result in the formation of palmitic acid (16:0) and also palmitoleic acid [16:1(9)], stearic acid (18:0) and oleic acid [18:1(9)]. With yeasts and fungi, polyunsaturated acids, linoleic [18:2(9,12)] and linolenic acids [18:3(6,9,12) or 18:3(9,12,15)] will also be formed. As these fatty acids can also be recovered following growth of a microorganism on penta- and heptadecane (although pentadecanoic and heptadecanoic acids will also be present in some abundance) this would indicate that α-oxidation, i.e. loss of a C_1 unit (as CO_2), probably occurs at some stage. The possible sequences for fatty acid chain length modification are thus:

$$C_{15}\rightarrow C_{17}\rightarrow C_{16}\rightarrow C_{18} \text{ or } C_{15}\rightarrow C_{14}\rightarrow C_{16}\rightarrow C_{18}\text{; and } C_{17}\rightarrow C_{19}\rightarrow C_{18}\rightarrow C_{16} \text{ or } C_{17}\rightarrow C_{16}\rightarrow C_{18}.$$

In general, growth of microorganisms results in an increase in the amount of lipid within the cell and importantly there are significant increases in the proportion of phospholipid in such cells which when carbohydrate substrates are used remains approximately constant (Ratledge, 1987).

Products of alkane oxidation

Numerous attempts have been made to use alkanes, and occasionally alkenes, as substrates to produce intermediates which are normally produced chemically. These may be described as the direct products of the interaction of a microorganism with a hydrocarbon.

Little attempt is made to produce fatty acids by microbial oxidation of alkanes as these materials can be readily obtained from animal or plant sources and are relatively cheap. However, a process to produce the dietarily essential fatty acid, γ-linolenic acid [18:3(6,9,12)], by growing *Mortierella isabellina* on dodecane has been reported (Yokochi & Suzuki, 1983) though it is likely that commercial production of this fatty acid will be achieved using carbohydrates as feedstocks (Meers & Prescott, 1987).

Conversions of alkanes to a wide variety of products have been frequently reviewed (Rehm & Reiff, 1981; Bühler & Schindler, 1984; Ratledge, 1984). Formation of fatty alcohols has been frequently reported by various workers though yields are usually low (~ 10 mg/litre). The fatty alcohols, being the primary product of alkane oxidation (see Fig. 2), are readily oxidized via the fatty aldehyde to the fatty acid. Attempts to delete the long chain alcohol dehydrogenase from bacteria or yeasts, and thus achieve accumulation of the fatty alcohol, normally fail because of the multiplicity of alcohol dehydrogenases with broad specificities in the cells (see Bühler & Schindler, 1984). Aldehydes being more unstable than alcohols tend either to autooxidise to fatty acids during isolation and work-up or be readily converted enzymatically *in situ* to fatty acid by aldehyde dehydrogenases which again may occur in multiple forms with broad specificities.

Diterminal oxidation of alkanes to yield dicarboxylic acids or ω- and sometimes ω-1 hydroxyfatty acids (see Fig. 2) have been reported on many occasions. The route to dicarboxylic acid formation may proceed by a variety of routes (see Fig. 2) in which the α,ω-diol route is probably a minor pathway (Yi & Rehm, 1982*a,b,c*). Transformations of long chain alkanes (C_{12}, C_{14} and C_{16}), fatty alcohols and fatty acids to the corresponding dicarboxylic acids was originally reported in high yields by Uchio & Shiio (1972*a*) using mutant cells of the yeast *Candida cloacae*. Other reports and patents have since indicated similarly high conversions of alkanes, alkyl halides, α-olefins and even various triacylglycerols to dicarboxylic acids (Patents, 1982*a,b*, 1986; Liu *et al.*, 1984; Furukawa *et al.*, 1986) by both yeasts as well as bacteria. Yields of over 60 g/litre have been reported on several occasions (Uchio & Shiio, 1972*b*; Shen *et al.*, 1977). Hill *et al.* (1986) have reported besides the formation of dicarboxylic acids, the formation of small amounts of 3-hydroxy derivatives of the dicarboxylic acids. Shorter chain dicarboxylic acids were also isolated in this work. The former products probably arise by the dicarboxylic acid undergoing partial oxidation via the β-oxidation cycle and presumably the 3-hydroxy dicarboxylic acid is not tightly bound to the enzyme complex and so dissociates to be excreted in the medium.

In all the above cases, the yeast or bacterium concerned with the production of dicarboxylic fatty acids have been mutated to prevent further degradation of the product. Presumably one of the enzymes of the β-oxidation cycle is deleted to achieve this objective. Thus the monocarboxylic fatty acid arising as the primary product of alkane oxidation could not be degraded in the conventional manner and thus became the new substrate for the terminal alkane oxidase system. (It is not known with any certainty however if the alkane hydroxylase or oxidase which is responsible for the initial attack on the alkane is also used to oxidize the ω-CH_3 group of the accumulating fatty acid.)

With some yeasts, the ω-oxidation stops at the formation of the ω-hydroxyfatty acid. These yeasts belong to a small group of related species: *Candida* (*Torulopsis*) *bombicola* (=*T. magnoliae*, *T. apicola* and *T. gropengiesseri*) and *Candida bogoriensis* which are characterized by the formation of sophorolipids arising by condensation of sophorose with the hydroxyfatty acids. (This surfactant was mentioned above, see also Table 1.) These yeasts also produce the ester involving ω-1 hydroxyfatty acids, which presumably arise by virtue of a lack of specificity in the hydroxylase accomplishing the second hydroxylation reaction.

Subterminal oxidation of alkanes has been found to occur with both bacteria and fungi (Rehm & Reiff, 1981). Fredricks (1967) was amongst the first workers to characterize 1-, 2-, 3-, and 4- and/or 5-decanols along with the corresponding ketones after growth of *Pseudomonas aeruginosa* or *Mycobacterium* (now *Rhodococcus*) *rhodochrous* on decane. Since this early report, numerous other examples have been published (see Bühler & Schindler, 1984; Boulton & Ratledge, 1984) and the list of organisms showing a propensity for attack at the 2-, 3-, 4-, 5- or 6-positions of an alkane is now relatively long and encompasses several bacteria (*Pseudomonas* spp, *Arthrobacter*, *Rhodococcus*, etc.), a number of yeasts mainly of the *Candida* genus and several moulds belonging both to the Phycomyces and Ascomycetes. Both secondary alcohols and ketones have been recognized though yields do not appear to be particularly high and no reports of commercialization of such a process have appeared.

Mechanisms for the subsequent degradation of mid-chain oxidation products have been covered in other reviews (Rehm & Reiff, 1981; Ratledge, 1984).

CONCLUSIONS AND CODICIL

The interactions of microorganisms with hydrocarbons creates a complex situation which becomes confused by the hydrocarbons being metabolized to a number of products. The products may occur extracellularly (and intracellularly) but may have nothing to do with achieving solubilization or uptake of the hydrocarbon. It may well be asked how does a microorganism achieve solubilization of a hydrocarbon, such as an alkane, if it is unable to produce the necessary components because the alkane cannot first gain entry into the cell and be oxidized to the desired product. The situation may be further confused by mistaking cause and effect. Thus, are the increased levels of membrane structures which are frequently seen in hydrocarbon-grown cells induced by a need to prevent undue membrane damage or do they represent a genuine increased surface area for achieving permeation of the alkane and sites for alkane oxidation? Occurrence of extracellular surfactants does not always depend upon the presence of hydrocarbons, so do the surfactants really play a role in solubilization, uptake and transport? Could the surfactants merely be an overspill of 'neutralized fatty acids' which the cell cannot deal with in any other way? The view presented in this review is an attempt to produce a little coherence into a confused area. These views at best may stimulate some fresh insight into the problem of hydrocarbon transport; at worst they can do no more than stir up an already muddy pool.

Our status of knowledge of hydrocarbon transport is at least 30 years behind our understanding of carbohydrate transport and at the current rates of progress is not likely to improve. This is primarily because hydrocarbon biochemistry is highly specialized, attracts few quality researchers but, perhaps, most importantly does not occur in either of the two favourite organisms of the geneticists: *Escherichia coli* and *Saccharomyces cerevisiae*. Many of the problems alluded to in this review could be overcome if good genetic mutants could be produced but it will obviously take time and effort to develop these with the organisms currently being used. But it could be done.

With hydrocarbons being the speciality product of just a relatively few large petrochemical companies, it should perhaps be these companies that take up the challenge of solving some of these basic problems even though they are unlikely to achieve a commercial process out of the work. If such companies are not interested in solving the fundamental problems associated with hydrocarbon utilization why then should anyone else? Without a serious commitment into this area of research, a future reviewer of this topic in 20 years' time is likely to comment, as has been done here: 'Not much is known'.

REFERENCES

Atlas, R.M. (ed.) (1984). *Petroleum Microbiology*, Macmillan, New York.

Boulton, C.A. & Ratledge, C. (1984). *Biotechnology*, **9**, 11–77.

Boulton, C.A. & Ratledge, C. (1987). In *Biosurfactants and Biotechnology*, ed. N. Kosaric, W.L. Cairns & N.C.C. Gray. Marcel Dekker, New York, pp. 47–87.

Brown, M.J., Moses, V., Robinson, J.P. & Springham, D.G. (1986). *CRC Critical Reviews in Biotechnology*, **3**, 159–97.

Bühler, M. & Schindler, J. (1984). In *Biotechnology, Vol. 6A*, ed H.J. Rehm & G. Reed. Verlag Chemie, Weinheim, pp. 329–85.

Cameotra, S.S., Singh, H.D. & Baruah, J.N. (1984). *Biotechnology and Bioengineering*, **16**, 554–6.

Cirigliano, M.C. & Carmen, G.M. (1985). *Applied and Environmental Microbiology*, **49**, 846–50.

Cooper, D.G & Paddock, D.A. (1984). *Applied and Environmental Microbiology*, **47**, 173–6.

Drozd, J.W. (1987). In *Carbon Substrates for Biotechnology* (Special Publication of the Society for General Microbiology), in press.

Falbe, J. & Schmid, R.D. (1986). *Fette Seifen Anstrichmittel*, **88**, 203–12.

Frantz, B., Lang, S. & Wagner, F. (1986). *Biotechnology Letters*, **8**, 757–62.

Fredricks, K.M. (1967). *Antonie van Leeuwenhoek*, **33**, 41–8.

Fukui, S. & Tanaka, A. (1979). *Trends in Biochemical Sciences*, **4**, 246–9.

Fukui, S. & Tanaka, A. (1981). *Advanced Biochemical Engineering*, **19**, 217–37.

Furukawa, T., Matsuyoshi, T. & Kise, S. (1986). *Journal of Fermentation Technology*, **64**, 97–101.

Glombitza, F. (1982). *Acta Biotechnologica*, **2**, 43–50.

Gmünder, F.K., Käppeli, O. & Fiechter, A. (1981). *European Journal of Applied Microbiology and Biotechnology*, **12**, 135–42.

Guerra-Santos, L., Käppeli, O. & Fiechter, A. (1984). *Applied and Environmental Microbiology*, **48**, 301–5.

Gutnick, D. & Rosenberg, E. (1977). *Annual Review of Microbiology*, **31**, 379–96.

Gutnick, D.L. & Shabtai, Y. (1987). In *Biosurfactants and Biotechnology*, ed. N. Kosaric, W.L. Cairns & N.C.C. Gray. Marcel Dekker, New York, pp. 211–46.

Hill, F.F., Venn, I. & Lukas, K.L. (1986). *Applied Microbiology and Biotechnology*, **24**, 168–74.

Hisatsuka, K., Nakahara, T., Sano, N. & Yamada, K. (1971). *Agricultural and Biological Chemistry*, **35**, 686–92.

Holdsworth, J.E. (1987) Ph.D. thesis, University of Hull, Hull, UK.

Hommel, R., Stüwer, O., Stüber, W., Haferburg, D. & Kleber, H.-P. (1987). *Applied Microbiology and Biotechnology*, **26**, 199–205.

Ito, S. & Inoue, S. (1982). *Applied and Environmental Microbiology*, **43**, 1278–83.

Käppeli, O. & Fiechter, A. (1977). *Journal of Bacteriology*, **131**, 917–21.

Käppeli, O. & Finnerty, W.R. (1979). *Journal of Bacteriology*, **140**, 707–11.

Kosaric, N., Cairns, W.L., Gray, N.C.C., Stechey, D. & Wood, J. (1984). *Journal of the American Oil Chemists' Society*, **61**, 1735–43.

Kosaric, N., Cairns, W.L. & Gray, N.C.C. (eds) (1987). *Biosurfactants and Biotechnology*. Marcel Dekker, New York.

Lang, S. & Wagner, F. (1987). In *Biosurfactants and Biotechnology*, ed. N. Kosaric, W.L. Cairns & N.C.C. Gray. Marcel Dekker, New York, pp. 21–45.

Li, Z.-Y., Lang, S., Wagner, F., Witte, L. & Wray, V. (1984). *Applied and Environmental Microbiology*, **48**, 610–17.

Liu, Z.T., Yi, Z.H., Li, K.Y., Han, H.Z., Hia, Y.M. & Fang, H.F. (1984). *Third European Congress of Biotechnology: Proceedings Vol. 2*, pp. 71–76.

Lindley, N.D. & Heydeman, M.T. (1983). *Journal of General Microbiology*, **129**, 2301-5.

Lindley, N.D. & Heydeman, M.T. (1986). *Journal of General Microbiology*, **132**, 751-6.

Meers, J.L. & Prescott, W. (1987). *Chemspec Europe 87 BACS Symposium*, pp. 1-3.

Meisel, M.N., Medvedova, G.A. & Kozlova, T.M. (1976). *Mikrobiologiya (USSR)*, **45**, 844-51.

Mulligan, C., Cooper, D.G. & Neufeld, R.J. (1984). *Journal of Fermentation Technology*, **62**, 311-14.

Numa, S. (1981). *Trends in Biochemical Sciences*, **6**, 113-15.

Osumi, M., Fukuzumi, F., Yamada, N., Nagatani, T., Teranishi, Y., Tanaka, A. & Fukui, S. (1975). *Journal of Fermentaion Technology*, **53**, 244-8.

Patent (1982*a*). Japanese Patent to Diacel Chemical Industries Ltd. JP 82 65194 (see *Chemical Abstracts*, **97** (1982) 108552).

Patent (1982*b*). Japanese Patent to Mitsubishi Petroleum Co. JP 8279889 (see *Chemical Abstracts*, **97** (1982) 108553).

Patent (1986). US Patent 4624920 to Research Association for Biotechnology, Tokyo.

Persson, A. & Molin, G. (1987). *Applied Microbiology and Biotechnology*, **26**, 439-42.

Rapp, P., Bock, H., Wray, V. & Wagner, F. (1979). *Journal of General Microbiology*, **115**, 491.

Ratledge, C. (1984). *Journal of the American Oil Chemists' Society*, **61**, 447-53.

Ratledge, C. (1987). In *Lecithin: Technological, Biological and Therapeutic Aspects*, ed. I. Harin & G.B. Ansell. Marcel Dekker, New York.

Rehm, H.J. & Reiff, I. (1981). *Advances in Biochemical Engineering*, **19**, 175-215.

Reiling, H.E., Thanei-Wyss, U., Guerra-Santos, L.H., Hirt, R., Käppeli, O. & Fiechter, A. (1986). *Applied and Environmental Microbiology*, **51**, 985-9.

Rosenberg, E. (1986). *CRC Critical Reviews in Biotechnology*, **3**, 109-32.

Rosenberg, M., Bayer, E.A., Delarea, J. & Rosenberg, E. (1982). *Applied and Environmental Microbiology*, **44**, 929-37.

Shabtai, Y. & Gutnick, D.L. (1986). *Applied and Environmental Microbiology*, **52**, 146-51.

Shen, Y.-Ch. *et al.* (1977). quoted from Bühler & Schlinder (1984) loc. cit.

Shennan, J.L. & Levi, J.D. (1987). In *Biosurfactants and Biotechnology*, ed. N. Kosaric, W.L. Cairns & N.C.C. Gray. Marcel Dekker, New York, pp. 163-81.

Siporin, C. & Cooney, J.J. (1978). *Applied Microbiology and Biotechnology*, **42**, 605-9.

Syldatk, C. & Wagner, F. (1987). In *Biosurfactants and Biotechnology*, ed. N. Kosaric, W.L. Cairns & N.C.C. Gray. Marcel Dekker, New York, pp. 89-120.

Uchio, R. & Shiio, I. (1972*a*). *Agricultural and Biological Chemistry*, **36**, 1169-77.

Uchio, R. & Shiio, I. (1972*b*). *Agricultural and Biological Chemistry*, **36**, 1169-75.

Yi, Z.H. & Rehm, H.J. (1982*a*). *European Journal of Applied Microbiology and Biotechnology*, **14**, 254–8.

Yi. Z.H. & Rehm, H.J. (1982*b*). *European Journal of Applied Microbiology and Biotechnology*, **15**, 244.

Yi, Z.H. & Rehm, H.J. (1982*c*). *European Journal of Applied Microbiology and Biotechnology*, **15**, 175.

Yokochi, T. & Suzuki, O. (1983). *Kagaku Gijutsu Kenkyusho Kokoka*, **78**, 137–40.

Zuckerberg, A., Diver, A., Peeri, Z., Gutnick, D.L. & Rosenberg, E. (1979). *Applied and Environmental Microbiology*, **37**, 414–20.

Microbiology and Distribution of Hydrocarbons in the Marine Surface Oil Film

G.D. Floodgate & M.G. Route

School of Ocean Sciences, University College of North Wales, Menai Bridge Anglesey, Gwynedd, UK

ABSTRACT

The amount of carbon tetrachloride soluble material varied in the surface film or at 1 m depth across the northern part of the Irish Sea, as did the concentration of the aromatics taken as a measure of oil pollution. Considerable differences in both concentration and composition were detected over short distances both vertically and horizontally. GLC signatures indicated that possibly both biogenic and anthropogenic hydrocarbons were present. There was no clear seasonal pattern, and frontal influences were not unequivocally detected when the western Irish Sea front was firmly established. The bacterial variation showed the expected greater concentration in the surface film on most occasions. The front as expected had no discernible effect on the number of bacteria. Inorganic nitrogen was markedly increased in the surface film as compared with the bulk water but appears to have had little effect on the rate of hydrocarbon degradation.

INTRODUCTION

Since polluting oil has a lower specific gravity than sea water, it might be expected that the surface film would be more heavily loaded than the water 1 m below. There is also the possibility that the converging currents associated with frontal systems might cause an accumulation of oil in the frontal region, though the wind velocity might well be the determining factor. In this paper we present work designed to explore these hypotheses.

METHODS

Five cruises were undertaken during 1985 on a line across the Irish sea which transverses the thermally induced front (Fogg *et al.*, 1985) which appears each summer. Surface water samples were collected by a Garrett sampler (Garrett, 1965), while another sample was collected from a depth of 1 m using a chemically clean NIO sampler. The samples were immediately extracted using redistilled carbon tetrachloride. Care was taken to ensure that all containers were chemically clean, and any losses were monitored using 'artificial' samples containing known amounts of crude oil. The carbon tetrachloride extracts were taken to dryness at 76°C under reduced pressure in a gentle stream of oxygen free nitrogen, and weighed. The samples were redissolved in carbon tetrachloride and fractionated on silica gel columns to separate hydrocarbons from other carbon tetrachloride soluble substances. The hydrocarbons were further separated into aliphatics and aromatics by eluting with *n*-pentane and dichloromethane respectively. Both fractions were then taken to dryness under reduced pressure and weighed. Samples for gas chromatography were taken up in hexane. Approximately 0·5μ ml (depending on concentration) was injected on to a 25 m fused silica (SE 52 stationary phase) capillary column of a Carlo Erba gas chromatograph fitted with a flame ionisation detector. The GC was programmed from 60 to 300°C with a ramp rate of 5 deg/min with a flame ionisation detector. Bacterial counts were measured using epifluorescent microscopy with acridine orange stain (Egan & Floodgate, 1985).

RESULTS

The concentration of carbon tetrachloride soluble substances in the Irish Sea was higher than anticipated, the range being 1·75–243·83 mg/litre. These values are thought to reflect the fact that the area is one of intense shipping activity. These analyses were also expected to reveal that the concentration in the surface film was higher than at 1 m depth and that the converging currents at the frontal zone would have increased the loading at the front as compared to either side. In fact neither of these expectations was realised. Table 1 shows data from Cruise 2 as an example. This cruise was undertaken in July in calm weather. In seven out of ten stations the concentration was higher at 1 m depth than in the surface film. The situation was however very variable. During a rough early April cruise, the surface loading was always higher at every station. Taken over a whole year 60% of the stations gave the expected result. The

TABLE 1
Weight of Carbon Tetrachloride Soluble Material (mg/litre) in Surface Water and at 1 m Depth

	Station									
Depth	*1*	*2*	*3*	*4*	*5*	*5·5*	*6*	*7*	*8*	*9*
0	3·79	3·22	3·55	5·01	6·20	4·53	2·75	5·24	2·55	12·65
1	12·61	13·46	5·94	6·49	3·56	1·56	5·37	9·22	8·38	1·05

frontal station, 5·5, did not show any accumulation even when the front was strongly established. Neither sea state nor season had any significant effect. Besides considerable variation within the 1 m depth difference, the concentration varied laterally from station to station. Again no pattern to the results was observed. The weight of aromatic hydrocarbons can be taken as a measure of the degree of oil pollution since the aromatics tend to be conservative as compared to the alkanes, and aromatic hydrocarbons are rarely formed biogenically. The amount ranged from undetectable to 4·24 mg/litre. Over the year, at 18% of the stations, the surface layer had less oil than the water 1 m below it. As with the carbon tetrachloride soluble fraction, the aromatics showed a distribution that was heterogeneous and patchy. The pattern of bacterial distribution followed that found by Egan & Floodgate (1985), that is the bacteria in the bulk water are fairly homogeneous. The earlier work had shown that the effect of the front on bacterial numbers was very modest with the surface stratified sea water having two or three times the number of bacteria as the mixed water. This was not borne out in these cruises since the number was not significantly different on either side of the front. It is also generally accepted that micro-organisms tend to be concentrated in the surface film (Harvey, 1966). The present study broadly confirmed these findings in that the ratio of the surface film bacterial numbers to those at 1 m depth was greater than one in 60% of the samples examined (Table 2). In the remaining 40% the surface had less organisms, the ratio dropping to as low as 0·24. Kassab *et al.* (1985) argued from statistical analysis that marine bacterial distributions are contagious, that is very patchy; a conclusion supported by these findings. The variations do not appear to be related to the sea state. Analyses of the carbon tetrachloride soluble material by GLC following preparative chromatography again illustrated the highly variable nature of the hydrocarbons in sea water. Full details will be reported elsewhere, but are illustrated by Fig. 1 which are the signatures from surface and depth at station 8 on cruise 5. It will be seen that there are considerable

TABLE 2
Ratio of Bacterial Numbers at Surface and 1 m Depth

Station	*Cruise* 1	2	3	4	5
1	0·99	1·22	0·97	0·94	2·44
2	3·04	0·46	0·24	1·04	0·52
3	0·91	0·73	0·45	8·78	1·35
4	2·01	6·47	0·42	6·00	1·91
5	1·48	0·26	0·61	4·97	0·62
5·5	1·20	4·82	0·63	1·15	1·42
6	0·85	5·04	0·88	9·33	0·64
7	1·18	6·44	0·71	2·66	1·14
8	1·33	4·22	0·33	4·06	1·43
9	1·63	3·30	1·43	—	1·73

differences in the concentration of each alkane within the 1 m change in depth. Similar variations were observed from station to station and with time. However a consistent pattern was never detected, and it must be assumed that chance is the major factor governing which suite of hydrocarbons is found at any one location. The most likely source of hydrocarbons is undoubtedly pollution since the sampling area is close to busy shipping lanes and the Mersey estuary which receives many industrial effluents. One of the interesting features of the GLC traces is the preponderance of even numbered carbon chain alkanes in the C_{16}–C_{22} range which appeared in many samples. As there is no evidence of preferential sparing by hydrocarbonoclastic bacteria of alkanes in this

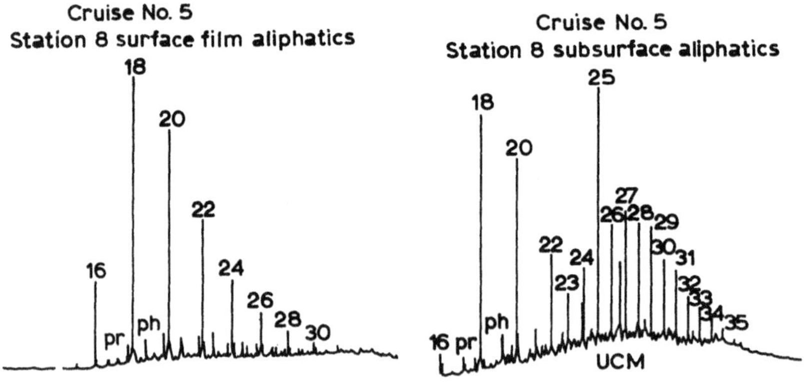

Fig. 1. GLC signatures of aliphatic fraction.

range, it seems unlikely that they are to be considered as conservative components of polluting oil. The possibility of in situ synthesis is intriguing. The alkanes produced by planktonic algae and bacteria usually have about this length of carbon chain but odd carbon chain lengths predominate, especially C_{17} and C_{19}. Nevertheless even numbered chains are formed by *Escherichia coli* (n-C_{18}) and *Desulphovibrio* spp. (n-C_{20}). The presence of the C_{16}–C_{20} alkanes in samples taken in November diminishes the possibility that they are algal in origin, and suggests that bacteria may be responsible. Various workers (for references see Floodgate (1984)) have reported the neo-synthesis of new alkanes from polluting oil. The surface film was generally enriched with inorganic nitrogen, particularly ammonia. In one instance the surface layer was enriched by a factor of 127. Again about 60% of the surface films had greater amounts of inorganic nitrogen than the 1 m depth water. However no correlation was found between those stations which had higher bacterial loadings and those with higher concentrations of inorganic nitrogen or organics. It can be concluded that frontal systems have little or no effect on the concentration of oil in the surface film, and although hydrocarbons tend to accumulate in the surface film the amount is not always greater than in the water immediately below. The composition of the oil varies very markedly within a short distance, so that bacteria are in fact presented with a heterogeneous substrate to degrade. There is some evidence that biological transformation or *de novo* synthesis should be taken into account even when considering a polluted area.

REFERENCES

Egan, B. & Floodgate, G.D. (1985). *Phil. Trans. R. Soc. Lond.*, **B310**, 435–44.

Floodgate, G.D. (1984). In *Petroleum Microbiology*, ed. R.M. Atlas. Macmillan, London, pp. 355–97.

Fogg, G.E., Egan, B., Floodgate, G.D., Jones, D.A., Kassab, J., Lochte, K., Scrope-Howe, S. & Turley, C.M. (1985). *Phil. Trans. R. Soc. Lond.*, **B310**, 555–71.

Garrett, W.D. (1965). *Deep Sea Res.*, **14**, 221–7.

Harvey, G.W. (1966). *Limnol. Oceanogr.*, **11**, 608–13.

Kassab, J.Y., Whittaker, C., & Floodgate, G.D. (1985). *Phil. Trans. R. Soc. Lond.*, **B310**, 521–54.

The Microbial Activity of Glutaraldehyde in Chain Conveyor Lubricant Formulations

K. Rossmoore

Diversey Wyandotte Corporation, Wyandotte, MI 48192, USA

ABSTRACT

Water-based chain conveyor lubricants are biosusceptible to microbial contamination, including environmental pathogens. Glutaraldehyde was evaluated as a control agent in use-diluted lubricant formulations against dairy and beverage field isolates, including Gram-positive and Gram-negative bacteria and fungi. Further efficacy was established against Listeria monocytogenes, *a ubiquitous environmental pathogen in dairy, meat and related processed food operations. Field tests were carried out in dairy applications comparing untreated and treated conveyor line contamination levels. The results of laboratory and field studies indicate that glutaraldehyde can be considered an effective biocide in this area of application.*

INTRODUCTION

The function of chain conveyor lubricants in the food, dairy and beverage industry is to lubricate chain surfaces, reduce friction, and ensure product and container transfer throughout the operation. Environmental contamination from air, water and product spillage can result in slime buildup, offensive odour, unpleasant appearance and corrosion (Rossmoore & Lichorat, 1986). In addition conveyor lubricants may harbour human pathogenic organisms. *Listeria monocytogenes* has been isolated from dairy chain conveyor lubricant. This organism is a recently recognized food-borne pathogen, the presence of which cannot be tolerated in the dairy and food industry environment.

The purpose of this study was to find an effective biocide in order to (1) increase bioresistance of the diluted conveyor lubricant and (2) control environmental contamination *in situ* along the conveyor line, thus minimizing the potential for the distribution of contaminants. The list of candidate biocides is very limited. In addition, to satisfy compatibility and EPA approval, the candidate should be effective within the 'resident' time (i.e. between 30 min and 1 h).

The chemical that seemed to fit these requirements was a commercial preparation of glutaraldehyde which, at alkaline pH, is a potent microbicidal agent (Russell & Hopwood, 1976). It has good resistance to neutralization by organic matter when compared to other biocides (Gelinas & Goulet, 1983).

Glutaraldehyde, besides having a broad spectrum, has a reasonably fast rate of kill and is classified as a chemosterilant. Factors influencing activity have been reviewed (Gorman & Scott, 1980). For the present application, the level is at least 10 times less than the requirement for medical purposes (Rubbo *et al.*, 1967).

METHODS

Microbiological media

Tryptic soy agar (TSA), sabouraud agar (SAB), brain heart infusion agar (BHI) and the neutralizing buffer were obtained from DIFCO Laboratories, Detroit, Michigan. Neutralizing buffer with 1% glycine was used for swab sample treatment.

Test organisms

All of the biodeteriogens were isolated from dairy conveyor lubricants, and identification was carried out by pure culture isolation technique on TSA, SAB and BHI prepared with 1% conveyor lubricant. Identification of pure cultures of bacteria and fungi was based on microscopic examination, Gram staining, and commercial biochemical test systems (API 20E, API 20C, Analytab Products; and Rapid NFT, DMS Laboratories, Inc.). All incubation was at 28°C for a minimum of 48 h and up to 4 days for fungi. Representative members of field isolates were selected for future efficacy studies; in addition, *L. monocytogenes* (ATCC 8766), *Zygosaccharomyces bailii* (ATCC 8766) (a resistant 'wild' yeast in the beverage industry), and *Staphylococcus aureus* (ATCC 6538) were added to the biodeteriogen list. The list of field isolates includes *Pseudomonas fluorescens*, *Klebsiella oxytoca*, *Serratia marcescens*, *Micrococcus* sp., *Geotrichum candidum*, and *Candida lambica*. *L. monocytogenes* was kept and plated on BHI agar.

Inoculum preparation

Bacterial and fungal inocula were prepared from TSA and SAB slants, respectively. After two transfers, the growth was washed off with 2 ml of phosphate buffer and diluted to 50 ml, 2 ml of which were inoculated into 250 ml French square bottles containing TSA or SAB with 3% agar. The cultures were incubated at 32°C for 24 h (bacteria) and at 25°C for 48 h (fungi). The surface growth was washed off with 3 ml of phosphate buffer, pH 7·2, and diluted so as to yield 10^7 cfu/ml (bacteria) and 10^5–10^6 cfu/ml (fungi). The bacterial inoculum was adjusted using a Spectronic 20 Photometer (Bausch & Lomb) at 500 nm.

Compatibility of lubricant formulations with glutaraldehyde

Eight lubricant formulations were tested at their use dilution levels, and 250 ppm of glutaraldehyde was added to 500 ml of each. At 0, 10, 30 min and at 1, 3, 5, 24 h 50-ml aliquots were taken and inoculated with 5×10^7 *Klebsiella oxytoca*. After 10 min contact time, the activity loss was tested by plating on TSA; inactive systems did not give microbial control. A good compatibility lubricant formula (L-4) was selected for laboratory efficacy studies (Table 1).

TABLE 1
Antibacterial Efficacy of Glutaraldehyde[a] in a Compatible Lubrication Formulation

Test organisms	*Inoculum (cfu/ml)*	*Log reduction in 30 min*[b]
Pseudomonas fluorescens	2×10^7	>5
Klebsiella oxytoca	8×10^7	>5
Serratia marcescens	5×10^8	>5
Staphylococcus aureus	4×10^7	>5
Listeria monocytogenes	4×10^7	>5

[a]Glutaraldehyde minimum effective concentration: 25 ppm.
[b]Based on $\log_{10}$ reduction.
Efficacy was tested in lubricant alone and in the presence of 1% beer and 1% milk, respectively. All permutations were equally effective.

Efficacy of glutaraldehyde

Diluted lubricant (100 ml) was inoculated and glutaraldehyde was added. At the appropriate intervals, 1 ml was taken into 9 ml of neutralizer solution and plated on TSA for bacterial and SAB for fungal counts. In addition, the lubricants were further stressed with 1% milk or 1% beer to simulate the dairy or brewery environment.

Field efficacy studies

Two dairy field studies were conducted with compatible lube systems, and one of these studies is presented here. Surfaces were sampled by swab using two cotton applicator sticks pre-wet with neutralizer prior to use. Five links (60 cm^2) were swabbed at each of the 21 sampling sites along the floor conveyor line. After swabbing, the applicator was placed into 10 ml neutralizer for transporting to the laboratory in ice and plated in 24 h to assess microbial population levels. The line was sampled before biocidal treatment ('base line' contamination) and repeated after 2 weeks of continual biocidal treatment ('treated' line contamination). The glutaraldehyde concentration was 85 ppm at the dosing pump at the time of the field test.

RESULTS AND DISCUSSION

The microflora of the dairy conveyor lubricant was very similar to the microflora of the brewery conveyor lubricant (Rossmoore & Lichorat, 1986). The results of fluid compatibility tests indicated good to poor stability of glutaraldehyde in different formulations.

Considering the relatively short time needed for this special application, the systems that showed longer than 24 h of biological activity were considered good compatibility systems; between 5 and 24 h activity was considered fair; and less than 30 min activity was classified as poor. The results of biological compatibility of fluids were in good correlation with the results on chemical compatibility of glutaraldehyde. Monoethanolamine is a major factor of chemical incompatibility in lube systems. Cross-link formation with glutaraldehyde results in subsequent activity loss.

The results of laboratory efficacy studies indicate that glutaraldehyde is effective at a dose level of 25 ppm against deteriogens and pathogenic bacteria in compatible fluid (Table 1). A higher level and longer contact time are needed for the control of fungi (Table 2).

TABLE 2
Antifungal Efficacy of Glutaraldehyde[a] in a Compatible Lubricant Formulation

Test organisms	*Inoculum (cfu/ml)*	*Log reduction in 30 min*[b]
Geotrichum candidum	1×10^5	>3
Zygosaccharomyces bailii	9×10^6	>3
Candida lambica	2×10^6	>3

[a]Glutaraldehyde minimum effective concentration: 250 ppm.
[b]Based on $\log_{10}$ reduction.
Efficacy was tested in lubricant alone and in the presence of 1% milk and 1% beer, respectively. All permutations were equally effective.

The results of field studies show that 85 ppm of glutaraldehyde reduced surface contamination along the dairy floor conveyor an average 4·4 $\log_{10}$ for bacteria and 3·4 $\log_{10}$ for fungi per 60 cm^2 sampled. Greater effectiveness under field conditions against fungi could be attributed to a much longer contact time.

Visual evidence of successful biocide treatment is shown in Fig. 1. Lubricant spray heads have a different appearance after glutaraldehyde treatment and is indicative of limitation of biofilm formation (Eagar *et al.*, 1986).

Fig. 1. Lubricant spray nozzles from treated (right) and untreated (left and center, respectively) lubricant.

REFERENCES

Eagar, R.G., Jr., Theis, A.B., Turakmia, M.H. & Characklis, W. (1986). In *Corrosion/86*, Paper No. 125. National Association of Corrosion Engineers, Houston.

Gelinas, P. & Goulet, J. (1983). *Journal of Applied Bacteriology*, **54**, 243–7.

Gorman, S.P. & Scott, E.M. (1980). *Journal of Applied Bacteriology*, **48**, 161–90.

Rossmoore, K. & Lichorat, J.J. (1986). In *Biodeterioration VI, Proceedings of the 6th International Biodeterioration Symposium*, ed. Sheila Barry & D.R. Houghton. C.A.B. International Mycological Institute, The Biodeterioration Society, pp. 448–53.

Rubbo, S.D., Gardner, Joan F. & Webb, R.L. (1967). *Journal of Applied Bacteriology*, **30**, 78–87.

Russell, A.D. & Hopwood, D. (1976). In *Progress in Medicinal Chemistry, Vol. 13*, ed. G.P. Ellis & G.B. West. 271-301. North Holland Publishing Co., Amsterdam, pp. 271–301.

Control of Microbial Contamination of Fuels in Storage

Jean L. Shennan

BP Research Centre, Sunbury on Thames, UK

ABSTRACT

Recently, problems of microbial contamination in stored fuels, especially middle distillate fuels, appear to be on the increase. While changes in fuel composition and in product distribution systems may be contributing causes, it is recognised that more attention must be given to the control of fuel storage systems to minimise free water accumulation and consequent microbial growth. Good housekeeping, improved maintenance and management are the most cost-effective measures. When problems persist or clean-up procedures are required, resort is often made to use of biocides. The caution with which engine manufacturers approve the use of fuel containing biocides is a severe restriction on choice. Although the existing market for fuel biocides is small, biocide manufacturers are actively developing and promoting products for this sector of the oil industry.

INTRODUCTION

Microbial growth in the water bottoms of storage tanks is a well known problem of fuel storage and can cause a variety of deleterious effects, including pump, filter and coalescer blockage, tank corrosion and 'off-specification' fuel product. Fuel which has been contaminated at some stage in distribution will infect all downstream systems. The management of this problem is familiar to the aviation industry, where good housekeeping to prevent water accumulation at all stages of aviation turbine kerosene storage, from refinery to wing tank, together with fine filtration and careful monitoring, ensures a safe fuel supply.

In former years, problems arising from microbial contamination of other fuels, those not subject to the stringent quality controls imposed on jet fuel, were infrequently reported. Remedial measures were taken when contamination could no longer be tolerated.

Recently, there appears to have been an increase in the extent of microbial problems in fuel storage. Also, petroleum microbiologists report that the principal contaminants can be yeast or even bacteria in contrast to the previously predominant moulds. It has been speculated that changes in crude oil sources, in refinery practices, in fuel additives and in fuel distribution patterns have affected the biological stability of fuels, especially middle distillate products (Smith, 1987). Such changes may lead to an increase of water content of fuels or to altered nutrient status for potential contaminants introduced during the more complex fuel transportation and storage patterns.

The increasing incidence of microbial problems in middle distillates has brought with it the need to use suitable remedial and control measures. The problems are varied and the question of whether or not to use biocides and, if so, which antimicrobial chemicals to choose, may depend upon a number of factors. These include the complexity and economic value of the system to be treated, the ease with which alternative methods of clean-up and control can be used and, not least, the acceptability of a biocide in the engine.

CAUSE AND CONSEQUENCES OF MICROBIAL CONTAMINATION OF FUELS

Microorganisms will always be present in fuel storage. If free water accumulates, by condensation, through rain water ingress or in water-displaced tankage, then growth is likely to develop. The major factor is the formation of a discrete water phase. Often moulds, able to use hydrocarbons as a carbon source, develop as a thin mat of fungal threads floating at the oil/water interface. As biomass accumulates, bacteria and yeasts will also be found growing in the mat and, eventually, dust and grit become entrapped. Microbial activity is also likely to produce polymeric products, increasing the bulk of the interfacial biomass, and surface-active metabolites which may create an emulsified phase (water-in-oil grading to oil-in-water) between the bulk fuel and the water bottom. Organic matter from the interface will eventually sink to the base of the storage tank to build up a sludge layer together with other debris. Anaerobic conditions under this sludge blanket and under any wall growth of microbial slime will provide a favourable environment for the anaerobic sulphate reducing bacteria (SRB). SRB are active in sulphide formation and are implicated in certain corrosion mechanisms.

It is the consequence of microbial growth, therefore, not just the presence of the organisms as biomass, that give rise to problems. The potential effect of this contamination has to be assessed in the light of the value of the fuel product and the sensitivity of the system in which it will be used.

Contaminated storage can have a number of effects on a fuel. If viable cells or spores are picked up by the fuel phase they will contaminate water bottoms in static storage tanks or vessels at each subsequent stage of the fuel distribution line. Any substantial interfacial material carried with the fuel off-take may block filters, coalescers, valves and control systems. The emulsification of water in the fuel phase may affect the combustion properties while surfactants or acid corrosivity in the water phase may affect distribution system instrumentation. Pitting corrosion of tank walls is often associated with the activities of SRB under bottom sludge or wall slimes. While microbial activity at a water bottom interface is unlikely to affect the chemistry of the bulk fuel in high turnover systems (providing mixing of the phases does not occur), chemical contamination of fuel may occur in the longer term, e.g. strategic, storage if significant amounts of H_2S are produced by SRB activity.

Fuel storage systems vary widely. They include: refined fuel storage tanks at refineries, terminals, retail sites and customer installations on land and in ships, aviation fuels and gas oil in long-term strategic storage, aviation fuel in service, marine gas turbine fuels transported and used from dedicated shipboard tanks and diesel fuels stored on off-shore platforms in sea-water ballasted tanks.

CONTROL MEASURES AGAINST MICROBIAL CONTAMINATION OF FUELS

Good housekeeping and system management

Good housekeeping, careful design and attention to system maintenance and management are the principal methods for controlling microbial problems, especially in short-term storage.

The first and most obvious strategy is to exclude, or drain, accumulated free water wherever possible. This may in practice be thwarted by poor design, old or poorly maintained tankage. Tanks, pipelines, etc., should be constructed to minimise the ingress of water and allow water removal by providing drainage points at all low points in a system.

Tank cleaning is important to remove scale and rust particulates which create micro-environments for microbial growth.

Management of a fuel installation involves avoidance of water and oil mixing, allowing fresh product to settle before withdrawal, careful use of slops pits to prevent recycling of contaminated fuel and the use of mixers, where installed, to prevent sludge build-up.

Filtration or centrifugation will decrease the microbial load of contaminated fuel as viable cells attached to particulate matter will be removed from the bulk fuel. In marine diesel installations, water settling and drainage can be improved by heating to about 50°C.

Chemical methods of contamination control

Good housekeeping practices together with intermittent use when necessary of an approved biocide have been effective in controlling microbial problems in aviation fuels. However, the producers, distributors and users of other fuels do not have to be so concerned by safety. Nonetheless there are attractions in using biocides to control fuel contamination.

Where tankage can be taken out of use, a combination of physical cleaning, with the use of biocides in the first rinse waters is advisable. Any effective antimicrobial chemical could be used, provided it is non-corrosive and acceptable for discharge into surface drainage systems. Suitable safety precautions need to be taken, especially when handling biocide concentrates, including controlled tank entry procedures and the use, as necessary, of appropriate protective clothing. Thorough rinsing with potable quality water is then required to remove all traces of biocide. All water introduced in cleaning should be of such a standard.

Biocides can only be added to a storage tank in service if suitable dosing points are available and their position with respect to the height of the water bottom is known. Decisions must be made as to whether the biocide has to be added to the bulk oil phase to partition into the water phase or whether the water phase can be dosed directly. Consideration must also be given to the safe disposal of water drained from treated tanks.

BIOCIDES FOR FUEL SYSTEMS

Selection criteria

Many factors bearing on safety, environmental hazard and compatibility, in addition to efficacy, have to be borne in mind when selecting biocides for use in any system.

Information on biocides which have been proposed or used in fuels can be found in the manufacturers' literature (for a selection see Table 1), in reviews on industrial biocides (e.g. Russell *et al.*, 1982) and in publications dealing specifically with laboratory or field tests on biocides in fuels (Rogers & Kaplan, 1968; Hill, 1970; Smith & Crook, 1983; Andrykovitch & Neihof, 1987; Herbert *et al.*, 1987; Rossmoore *et al.*, 1987).

An *ideal* biocide for addition to a fuel system would meet a number of criteria, including the following (in no particular order of rank). It:

—must not affect the combustion properties of the fuel;
—must not produce deposits or emissions which may damage the engine;
—must not emulsify the fuel or cause water retention;
—should be adequately soluble in fuel but partition into water to exert its biocidal effect;
—alternatively, if required as a fuel preservative, should remain in the oil phase;
—should have a wide antimicrobial spectrum;
—must be effective at low concentrations (say 200–300 ppm) in fresh or sea-water environments to cells in suspension;
—must be effective, with longer contact times or higher doses, to cells in surface slimes;
—must be cost-effective;
—must be stable in storage;
—must have a comprehensive data package provided including good toxicological information;
—should be backed up by technical service from the manufacturer/supplier.

No commercially available biocide meets all these criteria, but there are several products for use in various aspects of fuel system treatment (Table 1).

The search for a safe fuel biocide began with jet fuel. For 30 years the scene has been monopolised by the organoborianane fungicide, Biobor JF, hitherto the only biocide permitted for use in jet fuel by aircraft engine manufacturers. Its use is, however, restricted to intermittent addition, when contamination is expected or is known to be present. The anti-icing additive, ethyleneglycolmonomethylether (EGME) added to jet fuel for some end-users, is also slowly biocidal at the relatively high levels used (about 0·15% volume)(Neihof & Bailey, 1978).

When other fuel systems have required treatment Biobor JF has been the only biocide which could be recommended as safe, from the aspect of final fuel quality. However it is only effective under restricted water/oil ratio conditions, can take several days to be completely effective and is inactive against bacteria.

Several biocide manufacturers and oil industry service companies have begun to look for effective biocides for treating fuel systems among their current and development product ranges and are actively seeking approval from engine builders and fuel end-users.

Partition effects

When deciding on the suitability of biocide treatment, the main differences between fuel storage and distribution systems are associated with the amount of water present and whether it is static or continually renewed, as in water-displaced tanks.

TABLE 1

Biocides which Have Been Used or Proposed for Use in Various Fuel Storage Systems

Company	*Biocide*	*Chemical Type*
Angus Chemical Company	FUELSAVER	Morpholines
Applied Chemicals Ltd	PETROTEC B3-389	Isothiazolinones
British Drug Houses Ltd	PANACIDE	Bisphenol
Bode Laboratories (UK)	BAKZID 80: BODOXIN	Formaldehyde release
Boots Company plc	MYACIDE	Bromonitrogen
Dupont (UK) Ltd	BENOMYL	Imidazolcarbamate
IDF Production Chemicals	BIOTREAT 1488	Benzisothiazolinone
	BIOTREAT 1489	Isothiazolinones
ICI plc	PROXEL AS	Benzisothiazolinone
Nalfloc Ltd	NALFLOC 303-MC	Formaldehyde release and imidazoline
Olin Chemicals	OMADINE TBAO	Pyridinethiol
Petrolite Ltd	PETROLITE X520	Organoborinanes
Rohm & Hass (UK) Ltd	KATHON FP1.5	Isothiazolinones
Sterling Industrial Ltd	MAR 71	Oxyazolidines
US Borax & Chemical Corp.	BIOBOR JF	Organoborinanes
Vecom-Houseman Marine	HATACIDE 9 (Biofree OP)	Isothiazolinones
Water Management Chemicals Ltd	OIL AID 82	Isothiazolinones

This list is not claimed to be exhaustive and does not imply endorsement for use. Readers should apply to the individual biocide manufacturers or suppliers for further information on the products.

There are four possible categories of biocide with respect to their partitioning characteristics in fuel/water systems:

—Oil dispersible products which partition completely into the water phase, are active there at low concentrations and are, therefore, suitable for treatment of water bottoms of already contaminated tanks.
—Oil soluble products partitioning into the water phase. These are best suited for 'protective' use in clean systems where only low levels of water are expected.
—Oil soluble blends which are claimed to separate into oil phase-active and water phase-active components protecting the oil/water interface and the water bottom.
—Water soluble products for addition to tank water bottoms: this is seldom advised initially since the first line of control is to minimise water ingress and accumulation.

MONITORING OF FUELS FOR MICROBIOLOGICAL CONTAMINATION

The recognition of an incipient problem is preferable to system malfunction caused by microbial activity. This implies that simple and reliable sampling and monitoring methods should be available and that 'alarm' levels be known for a given storage system and fuel combination.

Methods of testing fuels are described in the literature (e.g. Rogers & Kaplan, 1968) and give reasonably reliable results for samples presented. However, there is still a need for the development of more accurate test kit methods for field use.

Obtaining representative samples and interpretation of test results remain fraught with difficulty. Judgements are often made on the basis of insufficient information, including a lack of base line data on the 'normal' microbial status of the systems.

CONCLUSIONS

Vast quantities of fuels are transported and stored without encountering microbiological problems. When necessary, however, the recommendation for biocide treatment of fuel storage systems is still a matter for individual consideration of each problem presented.

Progress towards finding alternative biocides for jet fuel and suitable biocides for other fuels is progressing but a number of pragmatic factors need to be recognised (in addition to cost considerations):

—Only an intermittent requirement exists as problems are often transitory and, in non-aviation fuels, not a hazard to safety.
—Caution among fuel suppliers regarding addition of biocides as a preservative to fuels has usually limited use to 'shock' dose treatments.
—Physical cleaning, good housekeeping and maintenance are usually the most cost-effective measures for coping with contamination. This implies a commitment to regular inspection and to taking tankage and distribution equipment out of commission for cleaning.
—It is virtually impossible to control microbial growth or activity in water-displaced storage systems or cavern storage.
—The crucial factor for approval of a biocide for use in turbine fuels is its acceptance by engine manufacturers.

ACKNOWLEDGEMENTS

The author wishes to thank the British Petroleum Company plc for permission to publish and to her colleagues in BP Oil International and BP Oil (UK) whose efforts have contributed to the paper.

REFERENCES

Andrykovitch, G. & Neihof, R.A. (1987). *Journal of Industrial Microbiology*, **2** (1). 35–40.

Herbert, B.N., Hill, E.C., Oates, P.D., Powell, D., Shennan, J.L. & Whittle, R. (1987). In *Industrial Microbiological Methods*, ed. E.C. Hill & J. Hopton. Academic Press, New York, in press.

Hill, E.C. (1970). *Journal of the Institute of Petroleum*, **56**, 138–46.

Neihof, R.A. & Bailey, C.A. (1978). *Applied & Environmental Microbiology*, **35** (4), 698–703.

Rogers, M.R. & Kaplan, A.M. (1968) *Developments in Industrial Microbiology*, **9**, 448–76.

Rossmoore, H.W., Wireman, J.W., Rossmoore, L.A. & Riha, V.F. (1987). Factors to consider in testing biocides for distillate fuels. Presented at ASTM Symposium of Distillate Fuel System Contamination, Cincinnati 24–25 June, 1987.

Russell, A.D., Hugo, W.B. & Ayliffe, G.A.J. (eds) (1982). *The Principles and Practice of Disinfection, Preservation and Sterilisation*, Blackwell, Oxford.

Smith, R.N. (ed) (1987). *Microbiology of Fuels*. Institute of Petroleum, London.

Smith, R.N. & Crook, B. (1983). In *Biodeterioration-5*. ed. T.A. Oxley & S. Barry. John Wiley, New York, pp. 486–93.

Bacterial Extracellular Polymers: a Major Cause of Spoilage in Middle Distillate Fuels

R.N. Smith

Biodeterioration Centre, The Hatfield Polytechnic, Hatfield, Herts AL10 9AB, UK

ABSTRACT

Spoilage of middle distillate fuels by certain bacteria has been associated with the formation of a membranous material believed to be a polysaccharide or lipopolysaccharide which acts as a solid surfactant. This material forms a skin at the oil/water interface to which the bacteria (scinnogens or skin producers) are attached. It may accumulate to produce a thick layer of slime at the oil/water interface extending up into the overlaying oil. The membrane material is strongly lipophilic and so occurs suspended in the fuel; it is also heavily hydrated and contains relatively large quantities of water. The consequences of spoilage by membrane polymer forming bacteria are: blockage of fuel filters by the sheets of suspended polymer; turbidity due to suspended polymer; an unacceptably high water content in the fuel and when held at low temperatures a severe increase in viscosity. These phenomena have been observed in Jet A1, Derv, marine gas turbine oil, and crude oil.

INTRODUCTION

One of the most frequent symptoms of microbial spoilage of middle distillate fuels oils such as Derv, jet fuel, gas oil and similar products is the formation of a heavy layer of slime at the oil/water interface, this slime is also found attached to the walls of fuel tanks and the inside of pipes where it entraps water and leads to microbial corrosion of the metal surface. Such slime formation is symptomatic of spoilage by *Hormoconis resinae*, *Paecilomyces* sp. and other filamentous fungi which occur as fuel deteriogens and has been extensively reported. However, spoilage can also be caused by bacteria (Williams & Lugg, 1980). The symptomatology, aetiology, deteriogenesis and ecology of these bacterial slimes are discussed below. Such oil degrading bacteria are known to produce a range of lipopolysaccharide biosurfactants (Zajic & Mahomedy, 1984).

VISUAL AND MICROSCOPIC OBSERVATIONS

This phenomenon was first recognised in jet fuel and gas oil from large storage tanks. A heavy slime layer was present at the oil/water interface which could be mistaken for fungal spoilage. When interface samples were examined under the microscope no mycelial growth could be detected but there appeared a translucent membraneous material which resembled pieces of 'clingfilm' (Fig. 1). This was recognisable by a slight increase in opacity and stretch lines such as occur when a sheet of clingfilm is held under uneven tension. After observing a number of samples it became apparent that this membrane material could collect resin, dust and other deposits from the fuel and become folded on itself with age to produce relatively thick opaque tan or dark brown pieces at the oil/water interface (Fig. 2). Some time later a sample of Derv was

Fig. 1. Initial stage in membrane formation (×400) with what is probably a single sheet of polymer showing the stress lines similar to those seen in a stretched plastic film.

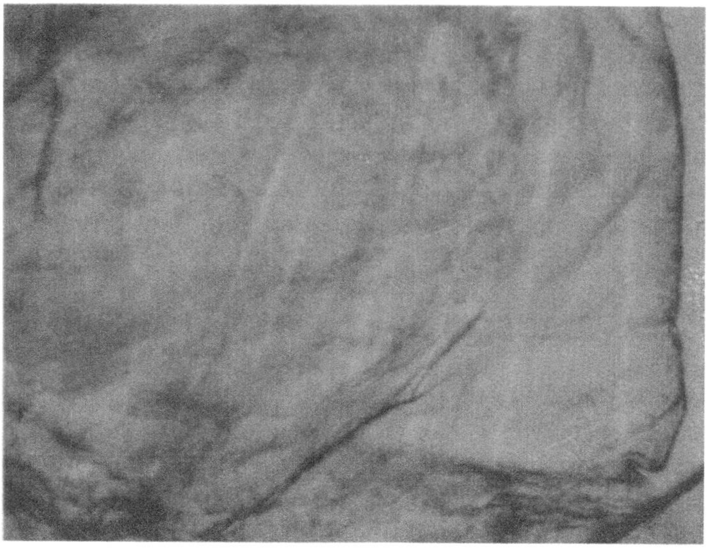

Fig. 2. An older piece of membrane polymer (×100) in which several layers have become folded on one another. The edge of the folded polymer sheets are visible on the left hand side of the photograph.

received for examination, following complaints from users of filter blockage resulting in diesel engine failure and comments from the suppliers that the water bottom of the storage tank was covered by what could be best described as a layer of 'frog's spawn'. Microscopic examination of this slime showed that fungal filaments were absent and the slime consisted of an amorphous mass of tan to dark brown material. A close examination showed that this slime was composed of a mass of the membrane material observed earlier. When the samples were stained, short rod-shaped bacteria were seen attached to the sheets. Samples of membrane material were analysed by X-ray diffraction and SEM probe analysis and found to contain appreciable quantities of iron.

Similar bacterial spoilage and severe membrane formation has also been recorded in samples of North Sea crude oil which had been held in tanks for some time. The oil layer just above the water bottom showed a considerable increase in viscosity and contained large quantities of water (10%) in a form which was quite stable with no tendency for the water to separate out. When this layer was centrifuged a water layer formed at the bottom under a layer of tan to brown coloured solid layer (compounded membrane material) with apparently normal crude oil above.

CULTURAL EXAMINATION

Samples of the water bottoms were diluted in 0·1% Corexit and spread on tryptone soya agar plates. A list of the major isolates is given in Table 1.

GROWTH OF BACTERIAL SCINNOGENS IN SHAKE FLASK CULTURE

Isolates were grown individually and in mixed culture in 250 ml flasks containing 35 ml of sterile Derv over 35 ml of sterile Bushnell and Haas solution (Bushnell & Haas, 1941) with and without iron added as iron filings. The flasks were incubated for 7 days at 25°C on an orbital shaker at 150 rpm. After 7 days the contents of each flask were transferred to boiling tubes and allowed to stand for a further 7 days for observation.

RESULTS

Bacterial numbers were generally greater in the Derv + iron (10^9 against 10^8) and the water bottoms in Derv + iron treatments were slightly less acid. *Pseudomonas fluorescens, P. maliphora* and *Alcaligenes* sp. were found to be the most prolific membrane producers. After transferring to boiling tubes the Derv appeared totally emulsified. However subsequent observations showed that the apparent emulsification was due to drops of oil (0·5–1·0 mm diameter) enclosed in a skin or membrane secreted by the bacteria. This phenomenon was termed ‘pseudoemulsification’. When these droplets were stirred with a metal rod they burst releasing the oil and the membrane was seen floating in the oil and gradually settled to the bottom of the oil phase. In the Derv + iron treatment there was a heavy deposit of membrane in the water bottom but less ‘pseudo-emulsification’ or membrane present in the oil. Thus ferrous or ferric

TABLE 1

Isolate	*Source*
1. *Serratia odorifera* (Ser)	Derv
2. *Pseudomonas* sp. (Ps)	Derv
3. *Pseudomonas maliphora* (B152)	Jet A1
4. *Pseudomonas fluorescens* (B170)	Derv
5. *Alcaligenes* sp. (B23)	Marine gas oil

ions present in the water bottom appear to have reacted with the membrane, perhaps increasing its density and causing it to sink and collect in the water bottom. The membrane material was seen to be sticky and often became attached to the wall of shake flasks and boiling tubes.

GROWTH OF SCINNOGENIC BACTERIA IN STIRRED TANK FERMENTERS

The three scinnogens which were deemed to be the most prolific membrane producers were cultured for 3 weeks in a 7 litre stirred tank fermenter with forced aeration and impeller agitation and containing 2·5 litres sterile Jet A1 and 2·5 litres sterile bacterial Bushnell and Haas solution.

After 3 weeks the oil phase appeared to be full of suspended membrane material. When oil containing this material was cooled to 0–5°C there was a marked increase in viscosity and it became virtually gelatinous. The oil phase was then separated and the membrane concentrated in the bottom of the oil phase by centrifugation, after which the oil/membrane suspension separated into three layers; a clear oil layer at the top, a layer of compacted tan coloured membrane material at the bottom of the oil and below this a layer of water. The oil and water layers were removed, the membrane was resuspended in petroleum ether and centrifuged. The compacted membrane was then collected, dehydrated by shaking in acetone, centrifuged and the acetone allowed to evaporate off and the dried residue weighed. The dry weight was only 0·12 g. When the dehydrated material was rehydrolysed by shaking in a water and oil mixture the membraneous mass reformed at the oil/water interface.

DISCUSSION

The results show that when certain oil degrading bacteria grow in the presence of oil a membrane polymer is produced which forms a solid skin at all oil/water interfaces. The nature of this material is probably polysaccharide or lipopolysaccharide. Oil deteriogens have been shown to produce biosurfactants which are lipopolysaccharide. The walls of such organisms have also been shown to be lipophilic causing the cells to adhere to the oil/water interface. It is suggested that the membrane polymer may be a polysaccharide slime which has been modified by the addition of lipophilic groups or it may result from a biosurfactant which

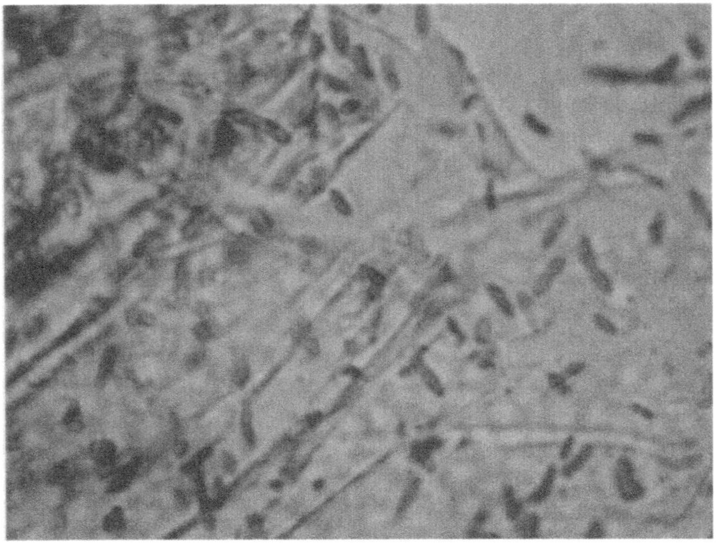

Fig. 3. A sheet of membrane polymer showing short rod-shaped bacteria (stained) attached to the surface.

is secreted as the monomer and subsequently polymerises on the oil/water interface to form a thin membrane. It is thus able to form a 'solid' surface at the oil interface to which bacteria may become attached. They have in fact been recorded in this situation. These membranes hold relatively large quantities of water, and also move up into the oil phase thereby increasing the effective oil/water interface for microbial growth. Their ability to support such growth was demonstrated by the fate of one sample of membrane, produced in Jet A1, which subsequently became infected by *H. resinae*. The fungus grew on membrane suspended in oil to a height 150 mm above the oil/water interface. These properties and the ability to hold large amounts of water in the oil phase must provide a considerable ecological advantage to oil deteriogens.

The presence of iron reduced the amount of membrane formation in the oil but there was a very heavy deposit of material in the water bottom extending from the oil/water interface. Thus iron as ferrous or ferric ions appears to react with the membrane polymer causing its density to increase so that it sinks into the water bottom.

The consequences of this form of spoilage for fuels are: blockage of fuel filters by the sheets of suspended polymer, turbidity due to suspended polymer; an unacceptably high water content in the fuel and when held at low temperatures a severe increase in viscosity sufficient to prevent the flow of fuel to the engine.

REFERENCES

Bushnell, L.D. & Haas, H.F. (1941). *Journal of Bacteriology,* **41**, 653–73.
Williams, G.R. & Lugg, M. (1980). *International Biodeterioration,* **16**, 103–5.
Zajic, J.E. & Mahomedy, A.J. (1984). In *Petroleum Microbiology*, ed. R.M. Atlas. Collier Macmillan, London.

Application of Formaldehyde-releasing Compound in Oil in Water Emulsions in a Deep Coal Mine

B.J. Zyska, Helena M. Wichary, Z.T. Cieplik & Danuta Kwiatkowska

Central Mining Institute, Plac Gwarków 1, 40–951 Katowice, Poland

ABSTRACT

A 140-day trial was carried out using microbiocide in the oil in water emulsion used for a powered support hydraulic system. Microbial spoilage of the emulsion was evaluated at five points of the system. For aerobic bacteria the viable number was $3{\cdot}5 \times 10^5$*–*$2{\cdot}5 \times 10^7$*, for moulds* $6{\cdot}5 \times 10^2$*–*$4{\cdot}0 \times 10^3$ *and for yeasts and yeast-like fungi* $1{\cdot}0 \times 10$*–*$1{\cdot}6 \times 10^4$*. The presence of sulphate-reducing bacteria (SRB) was confirmed. The concentration of formaldehyde-releasing compound TL-7, inhibiting the coal mine flora, was estimated to be 0·1% and the total effective inhibition time was found to be over 100 days. During the trial the concentration of microbiocide was controlled by a microbiological assay. The level of microbial infection of emulsion in the hydraulic system was almost zero. Using this microbiocide in the emulsion the corrosion hazard for the powered support installation was significantly reduced.*

INTRODUCTION

Microbial infection and deterioration of oil in water emulsions used in the coal mines in hydraulic props and the self-advancing support, has for many years been the subject of investigations at the Central Mining Institute, Poland (Zyska, 1968; Zyska *et al.*, 1975; Bohdanowicz-Murek *et al.*, 1980). The contamination of oil in water emulsion by microorganisms results in its irreversible decomposition, unpleasant odour, clogged filters, generation of acid metabolites, decreasing pH, corrosion of props and the self-advancing support, giving overall serious economic losses. Moreover, the presence of pathogenic microorganisms in the oil in water emulsion represents a hazard for the miners handling the support, resulting in severe infections (Pivnick *et al.*, 1954; Tant & Bennett, 1956; Weidle, 1973; Genner & Hill, 1981). Requirements for a preservative to ensure the resistance of the oil in water emulsion to microbial deterioration should include:

- —no adverse effect to the properties of the hydraulic fluid;
- —a broad spectrum of effectiveness against both aerobic and anaerobic microorganisms;
- —favourable solubility properties;
- —low toxicity to human beings; and
- —biodegradability.

The formaldehyde-releasing compounds have a less pronounced effect against fungi than against bacteria and promote the development of formaldehyde-resistant bacteria (Druskeit & Eggensperger, 1971).

The aim of this paper is to present the case history of a trial in a deep coal mine to examine the effectiveness of a microbiocide in the hydraulic system of the mine support.

MATERIALS AND METHODS

The trial was performed in the Szombierki coal mine, Bytom, Poland, on two longwalls with self-advancing support, at a depth of 730 m. In the first part of the trial, up to the 12th week the total volume of hydraulic fluid in the system was 19·2 m^3 and in the period from the 12th to the 20th weeks 40·0 m^3. From the 13th week onwards the volume of hydraulic fluid had to be increased to serve an additional longwall. The microbiocide used in the trials was a formaldehyde-releasing compound TL-7, produced on an industrial scale in Poland since 1986. Methods for estimation of level of microbial spoilage of emulsion and required concentration of microbiocide to eliminate the microflora colonising the mine emulsions, were presented earlier in detail (Bohdanowicz-Murek *et al.*, 1980).

In industrial trials it is important to ensure continuous estimation of the microbiocide concentration in the hydraulic system. Inevitably some leakage of emulsion from the system occurs, this has to be made up with fresh emulsion plus microbiocide. For continuous estimation of the microbiocidal concentration of the compound in the emulsion the filter paper diffusion procedure was used, similar to that for testing the resistance of antibiotics to microorganisms (Kosikowski & Ledford, 1960). This assay was based on the growth inhibition of a very sensitive test strain of bacteria isolated from 300 strains by replicate plating. The MIC for the compound TL-7 against the isolated strain 613 was 0·01% in agar. The zones of inhibition produced within the zones of exhibition are measured and compared with those produced by standard preparations to give a semi-quantitative estimate of the microbiocide concentration in the emulsion. Strain 613 gave a minimum zone of inhibition at a 0·05% concentration of TL-7. During the whole trial period the level of copper and steel corrosion and the changes in swelling of rubber seals was regularly checked, according to the Polish Standard BN-78/0535/ (1978).

RESULTS AND DISCUSSION

The viable number of aerobic bacteria and fungi in the oil in water emulsion was estimated at five points of the pipelines in the mine and the results are given in Table 1. Sulphate-reducing bacteria (SRB) were present at three of the five points. Laboratory tests on the inhibiting effect of the microbiocide on the emulsion flora of the mine were made in 2000 ml jars. The results are given in Table 2. Microbiocide TL-7 at a concentration of 0·1% was chosen for this trial. Table 3 presents data on the effectiveness of TL-7 in the emulsion of the hydraulic system in the

TABLE 1

Viable Number of Aerobic Bacteria and Fungi in the Oil in Water Emulsion in Szombierki Coal Mine

Sampling point	*Microorganisms in 1 ml emulsion*		
	Aerobic bacteria	*Moulds*	*Yeast and yeast-like fungi*
Surge chamber 3	$1{\cdot}5 \times 10^6$	$4{\cdot}0 \times 10^3$	$1{\cdot}0 \times 10^3$
Longwall support 214	$3{\cdot}5 \times 10^5$	$2{\cdot}5 \times 10^3$	$2{\cdot}5 \times 10^3$
Longwall support T	$1{\cdot}0 \times 10^6$	0	$1{\cdot}0 \times 10^1$
Surge chamber 730 m	$7{\cdot}5 \times 10^6$	$7{\cdot}0 \times 10^2$	$1{\cdot}6 \times 10^4$
Longwall support outflow	$2{\cdot}5 \times 10^7$	$6{\cdot}5 \times 10^2$	$1{\cdot}4 \times 10^4$

TABLE 2
Effect of the Formaldehyde-releasing Compound TL-7 on the Emulsion Microflora of the Szombierki Coal Mine

Concentration in emulsion	*Period of total inhibition of microorganisms (days)*	
	Aerobic bacteria	*Fungi*
0·025	0	0
0·050	7[a]	0
0·100	100[a]	100[a]
0·125	100[a]	100[a]
0·150	100[a]	100[a]

[a] Still inhibitory when taken off test.

TABLE 3
Effectiveness of the Formaldehyde-releasing Compound TL-7 in the Hydraulic System of the Self-advancing Support in Szombierki Coal Mine, Poland

Duration of the trial (weeks)	*Concentration of the microbiocide TL-7 (%)*	*Level of microorganisms in 1 ml emulsion*		
		Aerobic bacteria	*Moulds*	*Yeasts and yeast-like fungi*
1	0·10	0	0	$8{\cdot}0 \times 10^2$
2	0·11	0	0	0
3	0·12	0	0	0
4	0·11	0	0	0
5	0·09	0	$1{\cdot}8 \times 10^1$	0
6	0·11	0	$1{\cdot}5 \times 10^1$	0
7	0·08	0	$1{\cdot}5 \times 10^1$	0
8	0·07	0	$2{\cdot}5 \times 10^2$	0
9	0·07	0	$2{\cdot}5 \times 10^2$	$2{\cdot}0 \times 10^1$
10	0·06	0	$1{\cdot}5 \times 10^1$	0
11	0·06	0	$1{\cdot}0 \times 10^1$	0
12	0·06	0	$1{\cdot}2 \times 10^1$	0
13	0·05	$3{\cdot}5 \times 10^6$	$1{\cdot}5 \times 10^2$	$1{\cdot}5 \times 10^3$
14	0·11	$3{\cdot}5 \times 10^2$	$2{\cdot}0 \times 10^2$	$1{\cdot}4 \times 10^3$
15	0·11	0	$3{\cdot}5 \times 10^1$	0
16	0·10	$8{\cdot}0 \times 10^1$	$5{\cdot}0 \times 10^1$	$4{\cdot}5 \times 10^1$
17	0·07	$2{\cdot}0 \times 10^1$	$2{\cdot}0 \times 10^1$	0
18	0·06	$2{\cdot}0 \times 10^1$	$3{\cdot}0 \times 10^1$	0
19	0·05	$1{\cdot}8 \times 10^1$	$1{\cdot}5 \times 10^1$	0
20	0·05	$1{\cdot}2 \times 10^6$	$1{\cdot}5 \times 10^1$	0

self-advancing support. There were two trial periods. In the first, after the 5th week a formaldehyde-resistant strain was found to be established among the moulds. In the second period strains of formaldehyde-resistant bacteria were also established. The increase in viable number of microorganisms in the 13th week was due to connecting the hydraulic system to the second longwall, where the emulsion had not previously been treated with the microbiocide. During the whole period of the trial the emulsion was assayed for SRB but no evidence of their presence was detected. For technical reasons it is important to know the behaviour of the steel and copper support surfaces in contact with the emulsion, and of the rubber seals in the support. It was found that the corrosivity of the oil in water emulsion containing the microbiocide TL-7 was low, i.e. under the permissible values. The swelling of the rubber seals was also negligible. This trial indicates the satisfactory effect of the microbiocide TL-7, produced by the Polish chemical industry. The next step is to undertake industrial trials in the coal mine environment with a modified microbiocide including a carbamate compound, effective against fungi colonizing the emulsions in deep coal mines.

REFERENCES

BN-78/0535 *Fireresistant hydraulic fluids containing water. Requirements and tests.* In Polish.

Bohdanowicz-Murek, K., Zyska, B.J., Zankowicz, L.P., Wawrzyniak, E. & Terech, I. (1980). In *Biodeterioration*, ed. T.A. Oxley, G. Becker & D. Allsopp. Pitman, London, pp. 357–62.

Druskeit, W. & Eggensperger, H. (1971). *Erdöl-Kohle-Erdgas-Petrochem*, **24**, 103–5.

Genner, C. & Hill, E.C. (1981). In *Economic Microbiology, Vol. 6. Microbial Biodeterioration*. Academic Press, London, pp. 259–306.

Kosikowski, F.W. & Ledford, R.A. (1960). *Journal of American Veterinary Medicine Association*, **136**, 297–99.

Pivnik, H., Engelhard, W.E. & Thompson, T.L. (1954). *Applied Microbiology*, **2**, 140–3.

Tant, C.O. & Bennett, E.O. (1956). *Applied Microbiology*, **4**, 332–8.

Weidle, H. (1973). *Seifen-Oele-Fette-Wachse*, **99**, 725–30, 741.

Zyska, B.J. (1968). In *Biodeterioration of Materials*, ed. A.H. Watters & J.J. Elphick. Elsevier Applied Science, London, pp. 111–9.

Zyska, B.J., Bohdanowicz-Murek, K., Terech, I. & Zankowicz, L.P. (1975). In *Modelling of Environmental Effects on Electrical and General Engineering Equipment*, Vol. 1. Czechoslovak Academy of Sciences, Praha, pp. 264–9.

The Bacterial Metabolism of Monocyclic Aromatic Hydrocarbons

D.J. Hopper

Department of Biochemistry, University College of Wales, Aberystwyth, Dyfed SY23 3DD, UK

ABSTRACT

The general features of aerobic, bacterial catabolism of aromatic hydrocarbons are reviewed and some recent developments in the anaerobic catabolism of benzene and toluene are discussed.

AEROBIC METABOLISM

The pathways for the aerobic catabolism of several simple aromatic hydrocarbons are relatively well defined and are described in a recent review by Gibson & Subramanian (1984). One feature of these pathways is the use made of oxygenases both in the preparation of the compound for ring-fission and in the ring-cleavage reaction itself. This latter reaction gives rise to aliphatic compounds that can be converted into intermediates of central metabolic pathways. The ring-cleavage in these aerobic pathways is achieved by dioxygenases acting on a fairly restricted range of compounds which are common with pathways for catabolism of aromatic compounds other than pure hydrocarbons. With one or two exceptions they contain at least two hydroxyl groups situated either *ortho* or *para* to each other.

Thus the initial steps in catabolism involve modification of the aromatic hydrocarbon to give an appropriate ring-fission substrate. For many compounds this is achieved by hydroxylation of the ring usually by an NADH-requiring dioxygenase that inserts both atoms of O_2 giving a *cis*-dihydrodiol intermediate. An NAD^+-linked dehydrogenase then converts this into a catechol as ring-fission substrate (Fig. 1). The dioxygenases that hydroxylate benzene and toluene have been isolated and shown to be complex multicomponent systems (Axcell & Geary, 1975; Subramanian *et al*., 1980) similar in some respects to alkane monooxygenases.

Some bacteria, however, initiate catabolism by hydroxylation of the alkyl substituent apparently using monooxygenases again similar to alkane hydroxylase. In an alternative pathway for toluene it is oxidised to benzoic acid by hydroxylation of the methyl group to give benzyl alcohol and subsequent oxidation to the acid via the aldehyde. Similarly, *p*- and *m*-xylene can give rise to the corresponding toluic acids. The acids are then converted into catechols via diol intermediates which undergo simultaneous dehydrogenation and decarboxylation (Fig. 1) although for cumic acid derived from *p*-cymene the carboxyl group is retained (DeFrank & Ribbons, 1977).

ANAEROBIC METABOLISM

The importance of oxygenases in the aerobic catabolism of aromatic hydrocarbons has been emphasised but clearly such enzymes cannot operate for anaerobic catabolism. Less is known overall about this aspect of aromatic metabolism although pathways involving the reduction of the ring of benzoate and subsequent cleavage without oxygen, or hydration of the ring have been demonstrated (see review by Berry *et al*., 1987). Knowledge of anaerobic pathways has been based on studies of compounds that already contained partially oxidised groups but recently there have been reports of degradation of aromatic hydrocarbons under

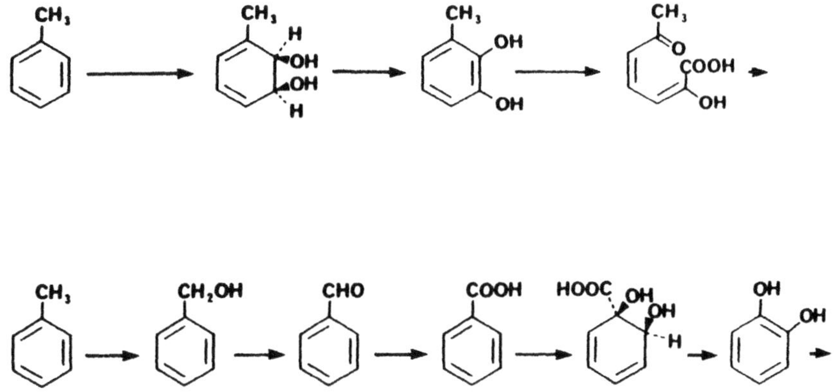

Fig. 1. Aerobic pathways for the metabolism of toluene.

anaerobic conditions. In one of these Grbić-Galić & Vogel (1987) demonstrated the transformation of benzene and toluene by a methanogenic consortium that was isolated originally from sewage sludge as a ferulic acid-degrading culture. These cultures converted more than 50% of added hydrocarbon into methane and CO_2 and in experiments where culture fluids were sampled during incubations a range of putative intermediates were identified by GC-MS. From benzene these included phenol and cyclohexanone whereas those from toluene included *o*- and *p*-cresols, benzyl alcohol, benzaldehyde, benzoic acid, 2-methylcyclohexanol, 4-methylcyclohexanol, methylcyclohexanol and 2-hydroxybenzoic acid. Of particular importance seems to be *p*-cresol which accumulated transiently early in the incubation before the appearance of benzoic acid and it was suggested that the one may be a precursor of the other. A plausible scheme can be put forward to accommodate this (Fig. 2). The increased percentage of $^{14}CO_2$ when [methyl-^{14}C] toluene was used as substrate rather than ring-labelled toluene also suggested oxidation of the methyl to carboxyl. This is supported by the results of another group from a comparison of the fates of the methyl carbon of *m*-cresol again by methanogenic consortia (Roberts *et al.*, 1987). For [methyl ^{14}C] *p*-cresol 85·8% of added label was recovered as $^{14}CO_2$ and 1·9% as methane, whereas for [methyl ^{14}C]*m*-cresol only 4·8% was recovered as $^{14}CO_2$ and 85·9% as $^{14}CH_4$. Furthermore the disappearance of the *p*-cresol was accompanied by a rapid production of $^{14}CO_2$ indicating possible early decarboxylation whereas the release of $^{14}CH_4$ from *m*-cresol occurred only after most of the substrate had disappeared from the medium.

Several other groups have reported the anaerobic catabolism of *p*-cresol under a variety of conditions. Bossert *et al.* (1986) described the isolation from polluted river sediment of a culture capable of catabolism of *p*-cresol under nitrate-reducing conditions. The transient appearance

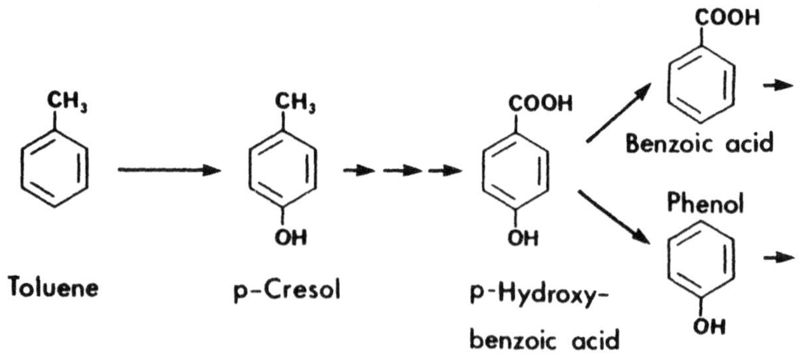

Fig. 2. Proposed routes for the anaerobic metabolism of toluene.

of *p*-hydroxybenzoate in the medium was demonstrated. The culture contained two bacterial species, only one of which, PC-07, was capable of modifying *p*-cresol. It converted *p*-cresol into *p*-hydroxybenzoate with concomitant reduction of nitrate to N_2 and intermediate formation of *p*-hydroxybenzylalcohol and *p*-hydroxybenzaldehyde (Bossert & Young, 1986). It was not, however, able to degrade the acid further and relied for this on the second organism, PB-04.

Biodegradation of *p*-cresol under sulphate-reducing conditions has also been described (Smolenski & Suflita, 1987) and although all three cresol isomers were eventually degraded, *p*-cresol was most rapidly metabolised. Again there was transient formation of *p*-hydroxybenzoic acid and, under appropriate conditions, *p*-hydroxybenzylalcohol and *p*-hydroxybenzaldehyde also accumulated, all indicative of oxidation of the methyl group. Both decarboxylation and dehydroxylation of substituted aromatics by anaerobes has been described. Thus phenol or benzoate could be formed from *p*-hydroxybenzoate (Fig. 2) and the catabolism of *p*-cresol via phenol has been suggested (Young & Rivera, 1985). However, in these sulphate-reducing systems acclimatised to *p*-cresol, benzoate was degraded without lag, whereas phenol was not metabolised even after prolonged incubation.

These reported transformations of aromatic hydrocarbons include reactions that in aerobic metabolism might be expected to have oxygenases as catalysts, the ring hydroxylation, for example, or the hydroxylation of the methyl group of toluene. This raises the question as to how such reactions are achieved anaerobically and what types of enzymes are involved.

Some light may be thrown on the methyl hydroxylase from studies of the *p*-cresol methylhydroxylase (EC 1.17.99.1) from aerobic organisms. Although many methylhydroxylases in aerobes appear to be monooxygenases, probably similar to alkane hydroxylases, those examined that convert *p*-cresol to *p*-hydroxybenzylalcohol are quite different. The important point in this context is that they are not oxygenases but can act anaerobically and it is, therefore, interesting to see *p*-cresol implicated in these anaerobic pathways. Several such enzymes have been purified, mainly from species of *Pseudomonas*, and all are very similar. The enzyme acts as a dehydrogenase, probably giving rise to quinone methide which is hydrated to yield the alcohol (Fig. 3). It has been shown, using $H_2{}^{18}O$ that the oxygen of the hydroxyl group is derived from water (Hopper, 1978). Toluene is not a substrate and the enzyme requires both a hydroxyl and methyl (or substituted methyl) group in the ring; these must be orientated *para* to each other to allow formation of the quinone methide. Thus *m*-cresol is not a substrate and, if a similar enzyme was

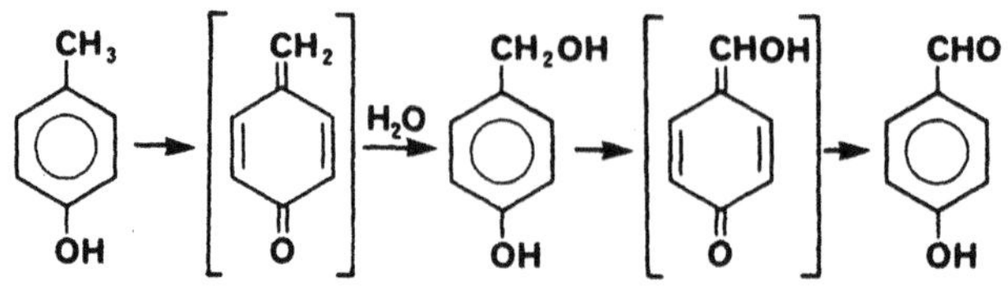

Fig. 3. Reaction catalysed by *p*-cresol methylhydroxylase.

present in the methanogenic cultures of Roberts *et al.* (1987), this might account for differences seen in the catabolism of *m*-cresol and *p*-cresol. The enzyme has quite a broad specificity, tolerating substituents in other ring positions and alkyl groups other than methyl in the *para* position (McIntire *et al.*, 1985). *p*-Hydroxybenzylalcohol is itself a substrate and the same enzyme will convert *p*-cresol through the alcohol to *p*-hydroxybenzaldehyde (Fig. 3). It is a flavocytochrome with an $\alpha_2\beta_2$ quaternary structure. As a dehydrogenase it requires an electron acceptor which in assays is provided by phenazine methosulphate. There is evidence that for hydroxylase A from *Pseudomonas putida* NCIB 9869 the natural acceptor is azurin, a small copper-containing protein. The electrons are passed to a membrane component and thence to cytochrome oxidase.

Such an enzyme would account for the ability of bacteria to oxidise the methyl group of *p*-cresol anaerobically and there is evidence that a similar type of enzyme is present in PC-07 of Bossert & Young (personal communication). It can also be seen how the introduction of hydroxyl into the ring of toluene, even though this may be removed later, would facilitate oxidation of the methyl group. The mechanism for anaerobic hydroxylation of the ring remains unclear although it has been shown that the oxygen in the phenol and *p*-cresol, produced from benzene and toluene respectively, is derived from water (Vogel & Grbić-Galić, 1986). Possibly hydration and dehydrogenation are involved but there is still much interesting enzymology to be done in this area.

REFERENCES

Axcell, B.C. & Geary, P.J. (1975). *Biochem. J.*, **146**, 173–83.
Berry, D.F., Francis, A.J. & Bollag, J.M. (1987). *Microbiol. Rev.*, **51**, 43–59.
Bossert, I.D. & Young, L.Y. (1986). *Appl. Environ. Microbiol.*, **52**, 1117–22.
Bossert, I.D., Rivera, M.D. & Young, L.Y. (1986). *FEMS Microbiol. Ecol.*, **38**, 313–19.
DeFrank, J.J. & Ribbons, D.W. (1977). *J. Bacteriol.*, **129**, 1356–64.
Gibson, D.T. & Subramanian, V. (1984). In *Microbial Degradation of Organic Compounds*, ed. D.T. Gibson. Marcel Dekker, New York, pp. 181–252.

Grbić-Galić, D. & Vogel, T.M. (1987). *Appl. Environ. Microbiol.,* **53**, 254–60.
Hopper, D.J. (1978). *Biochem. J.,* **175**, 345–7.
McIntire, W., Hopper, D.J. & Singer, T.P. (1985). *Biochem. J.,* **228**, 325–35.
Roberts, D.J., Fedorak, P.M. & Hrudey, S.E. (1987). *Canad. J. Microbiol.,* **33**, 335–8.
Smolenski, W.J. & Suflita, J.M. (1987). *Appl. Environ. Microbiol.,* **53**, 710–16.
Subramanian, V., Liu, T.N., Yeh, W-K. & Gibson, D.T. (1979). *Biochem. Biophys. Res. Commun.,* **91**, 1131-9.
Vogel, T.M. & Grbić-Galić, D. (1986). *Appl. Environ. Microbiol.,* **52**, 200–2.
Young, L.Y. & Rivera, M.D. (1985). *Water Res.,* **19**, 1325–32.

INSECTS, RODENTS AND BIRDS

Bird Damage: a Constant Threat in an Ever Changing Scene

Christopher J. Feare

MAFF Worplesdon Laboratory, Tangley Place, Worplesdon Surrey GU3 3LQ, UK

ABSTRACT

Human demographic, social and political pressures impose demands on land resources that are met by changing land use. The pace and timing of this change has varied in different parts of the world. For example, the transition from temperate forest to grassland (including cereal cultivation in large parts of Eurasia and North America) occurred some centuries ago, while the loss of tropical rain forest is a major current issue of environmental concern (World Wildlife Fund 1986). Today's land use changes can, due to increased mechanization, be much more rapid than in the past: this factor applies particularly to rain forest removal, where the timber has great commercial value which provides an added incentive to increase the rate of its removal during the opening up of the land.

ENVIRONMENTAL CHANGE AND BIRD DAMAGE

The major factors underlying gross changes in land use are urbanization and agricultural production and both processes lead to the creation of environments that are generally lower in ecological diversity than the habitats that they replaced. In both the tropics and in temperate regions the simpler agricultural and urban ecosystems support fewer bird species but some of those that do establish themselves become exceptionally numerous. They tend to be omnivorous or granivorous and their large populations may lead them into conflict with man for a number of reasons. In urban areas, Feral Pigeons *Columba livia* and House Sparrows *Passer domesticus* rely heavily on waste and spillage from human activity and create cleansing and potential health problems (Feare, 1986). In farmland, birds with specialized diets such as the granivorous Red-billed Quelea *Quelea quelea* and herbivorous Woodpigeon *Columba palumbus* and Eared Dove *Zenaida auriculata*, together with generalists like the European Starling *Sturnus vulgaris* and Red-winged Blackbird *Agelaius phoeniceus*, have attained huge populations that can cause considerable damage, though usually only locally (e.g. Murton, 1965).

In developed countries in particular, there is a further tendency to intensify production within agricultural habitats. This involves the creation of large areas of monoculture and the treatment of these with fertilizers, herbicides and insecticides. This exacerbates the paucity of bird species in such habitats but those that remain can again attain large populations. Breeding populations of the generalists, such as Rooks *Corvus frugilegus* and European Starlings, fare less well in the most intensive agricultural regimes (Brenchley, 1984). In Ohio, however, Stone & Danner (1980) found the reverse for Red-winged Blackbirds, where bird density and levels of damage were higher in intensive corn growing areas. Out of the breeding season birds are more able to move to areas where food is most readily available and such itinerant populations appear less sensitive to intensification.

Even within intensively farmed areas, large changes can be brought about by new developments due to the introduction of new cultivars or crops or new husbandry techniques. These may arise from scientific innovation or political priorities (Lane, 1983), but the birds already predominant in agricultural ecosystems can often adapt and take advantage of the new circumstances. In France, European Starlings shifted their wintering area to the north when maize silage became readily available following increased intensification of husbandry in Brittany and Normandy (Gramet & Dubaille, 1983). Now, however, this aspect of Starling damage is ameliorating following declines in breeding populations in northern Europe (Feare, 1987). In Britain, oilseed rape became popular as a break crop with the advent of an EEC subsidy and with the development of high yielding autumn sown cultivars (Lane, 1983). Woodpigeon populations responded to the exponential increase in the area of oilseed rape grown and this crop became a major winter food for the birds. It replaced clover, whose demise had been responsible for a decrease in the British Woodpigeon population in the 1960s (Inglis, 1986).

More subtle changes in agriculture can also affect bird behaviour and lead to damage. Skylarks *Alauda arvensis* are particularly common in cereal growing areas where in winter they eat seeds and leaves. In fields of sugar beet, the use of selective herbicides removes seeds and weed leaves that comprise the major part of the birds' diet and they may then devote more time to grazing sugar beet cotyledons (Green, 1980). The damage inflicted can be exacerbated by other changes in crop husbandry, particularly precision drilling (Dunning, 1982), which results in a lower density of sown plants, thereby reducing the possibility of compensation in the event of plant removal.

Alteration in the timing of events during the agricultural calendar can influence the occurrence and extent of bird damage. Rooks caused extensive damage to spring-sown cereals in north-east Scotland, where the late sowing of the crop coincided with the birds' breeding. In the milder climate of southern England, however, earlier sowing removed this coincidence and Rooks rarely fed on spring sown cereals (Feare, 1978). Since those studies were undertaken, there has been a considerable shift to autumn sown barley throughout the cereal growing areas of Britain, thereby depriving breeding Scottish birds of the food associated with spring cultivation. However, autumn sown barley ripens early and the milky seed is available in summer, at a time when Rooks experience food shortage. Rooks have now learned to exploit this food supply and summer now constitutes the period of most severe damage to barley (personal observation).

ECOLOGICAL SOLUTIONS TO BIRD DAMAGE

From the foregoing it is apparent that many bird damage situations have their origin in the ecological simplification of the environment. Those birds that become pests often display considerable adaptability within these less diverse habitats, enabling them to respond rapidly to changes within them by modification of behaviour and population size. Most bird pests are, in fact, r-selected species with a high population turnover and a potential for rapid population increase. This militates against the use of killing as a damage control tactic: such control operations are expensive, often ineffective in lowering population size and reducing damage, and can be environmentally damaging. Furthermore, serious bird damage is often localized, so that many of the birds killed may not have contributed to the crop losses.

The adaptability and mobility of pest bird populations implies that other traditional bird damage control strategies, such as scaring, will have limited success and may not be cost-effective. The knowledge that much bird damage is a consequence of adaptation to habitats of low ecological diversity clearly indicates that long-term solutions should be sought through ecological changes, rather than through *ad hoc* measures to counteract a particular damage problem. While it is difficult to determine the extent to which sufficiently large ecological alterations are practicable and cost-effective, there are indications that, under some circumstances, environmental modifications can provide realistic solutions to damage problems. This approach is advocated by scientists closely involved with bird damage research (Feare *et al.*, in press).

In view of the generally local nature of bird damage, it is unlikely that gross changes in farm environments will be made with the sole purpose of combatting bird depredations. However, food surpluses in Europe and North America, together with an increasing concern in these areas for environmental welfare, are leading to plans for diversification of farm activities. These will undoubtedly increase ecological diversity, which could help to reduce some bird pest populations. Greater crop diversity can offer other advantages on a more local scale through its provision of alternative feeding sites for birds that might otherwise concentrate on one crop. The world population of the Dark-bellied Brent Goose *Branta bernicla bernicla* has increased dramatically since the early 1970s, partly as a result of the cessation of hunting but also possibly as a result of a change in winter diet. The geese used to feed on saltmarsh plants but during the early years of population increase, feeding on winter cereals began. The recommended damage reduction strategy is to divert the geese from cereal fields on to grassland, but the areas around the east and south coasts of Britain, where the geese winter, are good cereal-growing land. Farmers are under economic pressure to maximize the area of cereals grown, which leads to a reduction in grassland. In this situation, any diversification that produced more grazing land would clearly assist with damage alleviation. Similarly, in the United States it has been found that the planting of decoy crops of sunflower and corn in strategically placed set-aside fields could reduce damage by Blackbirds (Icteridae) to commercial sunflower crops and was cost effective. The creation of attractive refuges through crop diversification could assist in the solution of many agricultural bird problems, although the potential for increasing pest populations must be appreciated.

Traditionally, diversionary tactics relied on scaring or chemicals, usually insecticides or fungicides, that have been found to have bird repellent properties when offered at sub-lethal doses. More realistic approaches, involving cultural practices that take account of the birds' perceptual abilities are now being investigated. For example, certain maize hybrids possess resistance to Red-winged Blackbird attack. Many bird pests show cultivar preferences among susceptible crops and these could be exploited in attempts to reduce damage. Furthermore, these preferences rely on naturally occurring physical and chemical differences between cultivars, an understanding of which will form the basis of non-toxic and environmentally safe chemical repellents, to which birds respond. Future development should involve the manufacture and testing of synthetic analogues, together with selective breeding or genetic manipulation to produce cultivars with naturally-occurring deterrents.

Other modifications of cultural practices can be used to interrupt birds' feeding routines and thereby reduce damage. It has been shown that Bullfinch damage to fruit buds can be spread more evenly over orchards by leaving space around them. The birds fly further from the cover, in which they spend much of their time, to the fruit trees and this results in a reduced tendency to concentrate their damage in the outer trees. Thus fewer trees suffer levels of bud loss high enough to reduce yields.

The change in Britain from spring to largely autumn cultivation has helped to resolve some bird problems (e.g. Rook damage to spring sown cereals) but has created others (e.g. Brent Goose damage to winter wheat). This illustrates the potential of altering the timing of farming events to reduce the coincidence between these and times of bird activity and suggests a further way of offering practical solutions to some bird problems. In many situations, however, the timing of farm activities is determined by a range of factors and it is difficult to time precisely events such as sowing and harvest. In some years the harvest of irrigated rice could be timed to coincide with absences of migratory Quelea, while in other years meteorological factors delayed crop growth so that maturation occurred after the arrival of the birds. In East Africa, damage to sorghum and sunflower by the doves *Streptopelia senegalensis* and *S. capicola* could be avoided by early planting and harvest so that ripening occurred when bird numbers were low. In New Zealand early planting of oilseed rape would permit harvest before Greenfinches *Carduelis chloris* formed post-breeding flocks capable of causing damage. Where farming activities are seasonally restricted, however, it is important that farmers should synchronize their operations as closely as possible, for crops sown out of phase can be particularly vulnerable to bird damage.

Clearly, there are limitations on the use of cultural practices in the resolution of bird problems, for bird damage is only one, often localised, factor that affects crop yield. However, there are equal, if not greater, limitations on the capabilities of the more traditional approaches of killing and scaring to reduce damage in a cost-effective way. In seeking solutions to bird problems it is essential to make maximum use of current knowledge of bird behaviour and ecology and to understand the effects of farming practices on them. This will allow us to use proposed changes in land use to help alleviate bird damage; the planned diversification of agriculture in Europe presents opportunities for the integration of changes that will help to combat damage with those of broader environmental benefit. The knowledge must be used to predict problems that might arise through proposed changes in farm management. In many areas our knowledge of bird ecology is insufficient to permit this. But broad analyses are invaluable in improving our understanding of the responses of birds to agricultural change. Similarly, experimental work that seeks least vulnerable options will make an important contribution to the alleviation of future potential problems.

REFERENCES

Brenchley, A. (1984). Regional variations in the breeding distribution of the Rook (*Corvus f. frugilegus L.*) in relation to differing agricultural regimes within Great Britain. *Ph.D. thesis*, University of Aberdeen.

Dunning, R.A. (1982). *Proceedings of the British Crop Protection Council Symposium*, BCPC Publications, Croydon, pp. 143–54.

Feare, C.J. (1978). *Annals of Applied Biology*, **88**, 329–34.

Feare, C.J. (1986). *Proceedings of the Seventh British Pest Control Association Conference*, Guernsey, 14 pp.

Feare, C.J. (1987). *B.T.O. News*, **149**, 6.

Feare, C.J., Greig-Smith, P.W. & Inglis, I.R. (in press). *Proceedings of the Nineteenth International Ornithological Congress*, Ottawa.

Gramet, P. & Dubaille, E. (1983). *Bulletin du Société des Sciences Naturelles de la Ouest de la France*, **5**, 163–72.

Green, R.E. (1980). *Journal of Applied Ecology*, **17**, 613–30.

Inglis, I.R. (1986). *The Winter Atlas of Birds in Britain and Ireland*, ed. P.C. Lack. Poyser, Berkhamstead.

Lane, A.B. (1983). *Agriculture, Ecosystems and Environment*, **10**, 299–309.

Murton, R.K. (1965). *The Woodpigeon*. Collins, London.

Stone, C.P. & Danner, C.R. (1980). *American Midland Naturalist*, **103**, 196–9.

World Wildlife Fund (1986). *Leaflet on Tropical Rain Forest*, 27 pp.

An extended bibliography for this paper is available from the author.

Damage Prevention through Improved Detection of Insect Pests in Stores

D.B. Pinniger

Ministry of Agriculture, Fisheries and Food, Slough Laboratory, London Road, Slough, Berks, UK

INTRODUCTION

There have been major changes in storage practices over the last 10 years which have influenced strategies for the prevention of damage by insect pests.

The objective of a store is to maintain goods in a stable and unchanged condition. A wide range of insect pests can attack or spoil stored materials and because many of them have very different dietary and environmental needs, successful control requires an informed approach. 'Blanket' treatments with persistent pesticides may kill many insects but they may be ineffective because some survive and pesticide residues in food, and the environment are increased. Similarly, fumigation can kill insects but the effect of the fumigant gas on stored materials and the resultant residues in food or artefacts may be unacceptable.

Because insect infestations may remain undetected until serious damage has occurred, many control programmes rely on high doses or on widespread application. Efficient insect detection techniques can be used as the basis of store management systems. Integrated use of insect detection and targeted pesticide application can prevent damage and reduce pesticide usage.

THE ROLE OF DETECTION

Large numbers of insects are unacceptable and require remedial action because if they are easy to find, damage may have already occurred. The pest status of insect species varies widely, at low levels of infestation the acceptance or lack of tolerance of insects depends upon the species, the commodity and the environment. For example, one booklouse may not cause serious concern in a flour mill warehouse whereas a single live grain weevil in a 20 tonne load of grain will result in it being rejected. The temperature and humidity will also determine the importance of pest presence.

Practical and effective pest detection relies upon knowledge of the ecology, behaviour and potential of the pest species for causing economic or physical damage together with information on the material at risk and the storage environment.

DETECTION TECHNIQUES

Visual inspection still remains the most widely used approach for finding insects. At best, an experienced observer can be extremely effective; at worst, high levels of infestation can remain undetected until serious damage has occurred.

Physical traps

Pest insects trapped in webs can be the first sign of a problem. Sticky traps have been successfully adapted for use in stores. Vertically suspended 'fly-papers' have been used in food stores for moths (*Ephestia* sp.) and in museum stores for furniture beetle (*Anobium punctatum*). They will trap insects which blunder into them and can be useful for pests which fly actively. However, they may be unacceptable in food processing areas. Horizontal sticky traps are very effective for trapping cockroaches (*Blattella germanica* and *Blatta orientalis*) and have been used in food stores and museums. The adhesive does not retain some of the smaller beetle species which may also be deterred from entering the trap. Traps with an improved adhesive particularly if combined with an effective attractant could be useful.

Pitfall traps have been used to estimate populations of ground living insects and Cogan & Wakefield (1987) evaluated 1 pint plastic beer mug pitfall traps and probe traps for detection of the major stored grain pests *Oryzaephilus surinamensis, Sitophilus granarius* and *Cryptolestes ferrugineus*. Probe traps are plastic cylinders perforated with holes which are inserted below the grain surface. They will trap insects moving through grain rather than those wandering on the surface. Cogan & Wakefield (1987) demonstrated that the use of both pitfall and probe traps increased the chances of detection of the major grain pests by at least tenfold compared to conventional spear and sampling techniques.

Electric grid traps which utilise a UV light source are marketed for the control of flying insect pests. Examination of the catch tray under one of these traps can reveal hitherto unsuspected pests.

Traps with attractants

Traps with food lures have been evaluated and use of a perforated plastic mesh bait bag containing wheat, groundnuts and carobs (locust beans) was successful. The food in these bags encompasses a range of food preferences of insect pests and 60 different species have been found in these bags in a variety of environments.

Corrugated card Storgard™ traps with a central pitfall containing cereal oil have been described. Although they are not as generally effective where bait bags cannot be used.

Studies on insect pheromones have resulted in many proposals for traps enhanced with pheromone components or their analogues.

Pheromones appear to offer a means of attracting and detecting insects at very low population levels but the results of successful laboratory assays may not be confirmed in practice because of the complexity of the pheromones, the habits of the pests and the nature of the environment. Traps baited with sex pheromone lures for the detection of the moth *Ephestia* sp. have been described by Cogan & Hartley (1984). However, Storgard™ traps with agregation pheromone lures have been much less successful in detecting the flour beetle *Tribolium castaneum* (Newton, 1987).

DETECTION STRATEGY IN PRACTICE

There are a number of clear examples where trapping programmes have led to improved control of insects.

Although full inspection for cockroaches in hospitals and prisons cannot be dispensed with for the initial determination of population spread and density, it can be supplemented by use of sticky traps. The numbers of insects caught subsequently after insecticide treatment is used to determine the rate of decline of the population. They have been adopted by most professional servicing companies.

Pitfall and probe traps have been adopted by commercial grain traders for monitoring large bulks of grain. *Oryzaephilus surinamensis, Sitophilus granarius* or *Cryptolestes ferrugineus* have been found in traps well before they have been detected by conventional methods. This early warning enables the storekeeper to implement cooling or fumigation strategies and prevent infestation developing into a hot spot. In a recent case, where only low doses of fumigant were achieved, conventional sampling pronounced treatment a success whereas pitfall traps showed that the insects had survived (Cogan, personal communication).

Moth traps designed around TDA (Z,E-9,12-tetradecadienyl acetate) pheromone lures have been used by the food industry for a number of years. The successful operation of detection and monitoring programmes depends upon regular and accurate trap recording and an understanding of the storage environment. Cogan & Hartley (1984) have shown that infestation in a number of situations can be dramatically reduced when trap catch information is used to implement cleaning and pesticide treatment. For example, Cogan (1983) describes a system operated by a large chocolate manufacturer. Machines adjacent to traps which contain *Ephestia* adults are intensively cleaned and treated. Table 1 shows the success of this strategy. The difficulties of inspecting some mill and food production areas have placed increased reliance on trap catches for assessing *E. kuehniella* infestation. Accurate monitoring coupled to cleaning and use of pesticides have eliminated the need for costly annual methyl bromide fumigation.

The identification and evaluation of dominicalure, the aggregation pheromone of the lesser grain borer *Rhizopertha dominica* has resulted in the production of commercial lures. But its use in traps for detection of the larger grain borer *Prostephanus truncatus* has proved to be of greater value. The spread of this pest across Africa has been monitored by dominicalure traps to give warning of the presence of the beetle so that control programmes can be implemented. Tests are currently in progress to evaluate Trunc-call 1 and 2, the pheromone components of *Prostephanus truncatus*.

TABLE 1

Effect of a Control Programme Based on Catch of *Ephestia* in a Chocolate Factory on Customer Complaints of Infested Chocolate (from Cogan, 1983)

Year	*Complaints*	*Complaints/100 tonnes chocolate*
1978–80[a]	440	10–15
1981	100	4
1982	10	0·3
1983	8	0·3

[a] Gradual introduction of monitoring system.

Storgard™ traps have been used in a number of museums and museum stores for early warning of pest presence. In limited trials in a museum store, 18 standard traps were compared with 18 with an additional lure of 0·5 g undecanoic acid. However, although very few adult *Anthrenus* were found in the traps, Table 2 shows the numbers of larvae of *Anthrenus verbasci* and *Attagenus pellio* captured in these traps over 12 months. They show that live pests were present in places thought to be free of infestation and that undecanoic acid clearly enhanced the catch of larvae.

FUTURE DEVELOPMENTS

Figure 1 shows the major components which must be considered in evaluating traps in the laboratory and in practice together with commercial priorities. Specificity and complexity of the lure or trap design will determine whether many cheap traps or a few complex traps need to be deployed. Clear guidelines are necessary for the interpretation of trap catch.

A number of techniques and discoveries showing considerable potential need further development if they are to be useful commercially. These include the isolation and identification of stegobinone, the pheromone of the drug-store beetle *Stegobium paniceum* which has also been isolated from *Anobium punctatum*. Suitable traps could be baited with a single pheromone lure and used for the detection of both these species. The series of macrolide lactone aggregation pheromones associated with *Oryzaephilus* sp. and *Cryptolestes* sp. are currently being evaluated but a major factor which influences the use of these pheromones is their chemical complexity. Seven different complex stereospecific macrolide lactones are produced by five different beetle species. Inter and intra-specific behavioural responses and the difficulties of synthesis may preclude the commercial use of these pheromones.

The sex and species specificity of the insect pheromones and disappointing performance in some field trials encourages the continuing

TABLE 2
Number of Insects Trapped in a Museum Store

	Standard Storgard traps	*Storgard traps with undecanoic acid*
Anthrenus verbasci larvae	42	64
Attagenus pellio larvae	40	84

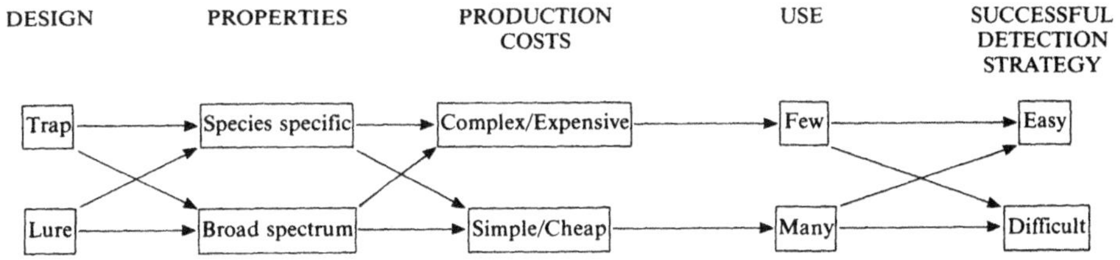

Fig. 1. Factors which influence trap development.

development of food lures, including those based on carob volatiles. Their identification offers the possibility of a broad spectrum attractant for a wide range of food store pest species.

Effective lures and traps which are commercially viable are needed. It is also essential to produce strategies for trap use in collaboration with the storage industry. Pressure for higher standards and lower levels of infestation is coupled with pressure for reduced levels of pesticide use and residues.

REFERENCES

Cogan, P.M. (1983). *Mitteilungen von der Deutsche Gessellschaft für Allgemeine und Angwandte Entomologie*, **4**, 108–10.

Cogan, P.M. & Hartley, D. (1984). *Proceedings 3rd International Conference of Stored Product Entomology, Kansas, 1983*, pp. 631–9.

Newton, J. (1987). Traps for detecting stored product insects—are they reliable? Symposium on Stored products Pests Control, Reading, 1987.

An extended bibliography for this paper can be obtained from the author.

Review of Rodent and Insect Damage to Stored Products and Non-pesticidal Methods of Control

J.A. Gibson

Pest Control Consultant, 72 Robins Way, Hatfield, Herts AL10 9QG, UK

ABSTRACT

The mounting public concern for environmental contamination by poisons and the demand for additive-free food, coupled with the resistance of pests to pesticides have led to increased interest in physical and biological control techniques: a summary is presented. Methods of damage assessment and detection of low level infestations of insects and mites in bulk commodities are compared and the characteristics of rodent damage behaviour considered. Attractants, repellents and the susceptibility of different types of packaging materials to rodent and insect attack and the effects of temperature, modified atmospheres and other physical techniques in reducing damage are reviewed. The role of chemosterilants for rodent control and insect growth regulators for control of stored product insects are evaluated. The consequences of elimination of the use of conventional pesticides in food processing, warehousing and retail premises and the prospects for pesticide-free control are appraised.

INTRODUCTION

Intensive monoculture agriculture and an increase in urban centres has simplified a formerly variable habitat along with its natural checks and balances, and substituted uniformity supported by the constant addition of outside resources. As a result pest species find nearly unlimited nutritious food and favoured habitats in these situations. Man has been forced to control these pests to protect food and property and such control programmes, in the recent past, have implied use of pesticides often as the only means of control. In the two decades after World War Two screening programmes for pesticides were expanded and 739 new ones reached commercial development by 1975 (Corbett, 1979) and many of these are still in use today. However two adverse factors have emerged: the spread of many strains of pesticide resistant pests and the widespread concern about the toxic hazards. Resistance to insecticides particularly by common hygiene and stored product insects has become difficult to overcome due to multiple resistance where the selection of resistance to certain pesticides has conferred a cross reference to compounds never used in control programmes for those species (Collins, 1973; Rust and Reierson, 1978). This phenomenon is even more acute in rodenticide resistance where control is largely dependent on one group, the anticoagulants, and the current status of these materials in the UK has been reviewed (MacNicoll, 1986).

Public concern about pesticides generated by Rachel Carson's Silent Spring in 1962 was sustained by later Americans like Ralph Nader and more recently by Van der Bosch whose book the *Pesticide Conspiracy* (1978) implies all commercial experts are dishonest, emphasised again by two American sociologists writing in *Nature* 1981. This concern has been compounded by the general disgust about additives in food. These public pressures combined with increased legal requirements for pesticide registration have increased the cost of development to a near prohibitive level.

Consequently the demands for alternative techniques which combine safety to man and efficacy in control have increased. This paper reviews damage assessment and detection techniques and considers possible methods of exclusion and control without resort to conventional pesticides.

DAMAGE ASSESSMENT AND PEST DETECTION

The first step in combating pests of store food is to detect them and assess the damage they cause. The damage ranges between nuisance value of contamination by small numbers of pests or their excreta to the total destruction of large tonnages of product.

Traditionally damage and loss caused by infestation of insects and mites are measured in terms of quantity of grain consumed or rendered unfit for uses in human or animal feed (Freeman, 1973). However rejection of grain which is not insect free has required the development of sensitive detection techniques. Until recently the grain spear and vacuum sampler have been used but Chambers (1987) suggested these are time consuming and inefficient. He advocates the use of pitfall traps and grain probes in combination with an insect attractant such as carob extract resulting in more effective detection. The use of near infra-red reflectance for detection and measurement of mite infestation (Wilkin *et al.*, 1968) overcomes many of the difficulties created by their small size and could be used to monitor grain on delivery. In food processing plants the use of stored product moth and beetle pheromone traps (Cogan & Wakefield, 1987; Pinniger & Wildey, 1983) enables successful monitoring and detection of infestation of specific insects.

Unlike insects which often are disturbed throughout food stores, rodents will be at the periphery of bulk stores and often non-randomly distributed through bagged or boxed products. This complicates any statistical approach to sampling and assessment. Again estimates of damage are varied although Dubock (1984) estimates that in the People's Republic of China 13 million tonnes of grain are lost to rodents which equals the quantity imported. In practice estimates of damage are often based on population censuses. An established technique (Jackson, 1979) relies on correlating the results of two types of estimates: one live trapping and the other using droppings, runways, burrows and gnawed food to estimate population. This is obviously time-consuming and expensive and more research is needed in this area.

THE EFFECT OF PACKAGING MATERIAL ON REDUCTION DAMAGE

Measurement methods have been devised for biting action of rodents using the Gnathodynameter (La Voie *et al.*, 1976) and for insects (Highland & Wilson, 1981). Batth (1970) compared the damage by three species of insect to aluminium packed soup mix and found *Tenebroides* to be more destructive than *Trogoderma* larvae and *Oryzaephilus* unable to penetrate the packs at all. Cline (1978) compared the resistance of six packaging materials to penetration by a range of insects and found polypropylene to be most resistant. The effectiveness of packs in resisting insects is dependent on the seal.

Little work has been done on the damaging behaviour of rodents. La Voie and Glahn (1975) concluded hardness and diameter of pack to be the most important physical parameters in rodent attack. Teichner *et al.* (1969) emphasised that all exploratory activities in the rat are part of a threat evaluation process and Glahn and La Voie (personal communication) suggested that gnawing is a learned behaviour and that prior to attack the container is identified by olfactory cues. Gibson (1980) found that for mice, olfactory cues alone are not the dominant ones for gnawing and it may be a random exploratory process. La Voie (personal communication) has suggested application of repellent chemicals to packaging materials. For example, lithium chloride produces a learned taste aversion and Nachman & Hartley (1975) report that a diet containing 4% lithium chloride was completely avoided by rats. However any application would have to ensure that the rodent tasted the additive before the packaging material was rejected from the mouth.

Reduction of insect damage in stored food is likely to rely on physical modification of the environment as addition of CO_2 to bulk grain or use of non-ionising radiation rather than additions to packaging materials. The techniques are reviewed (Fleurat-Lessard, 1987) but are fairly costly and labour instensive.

CHEMOSTERILANTS AND INSECT GROWTH REGULATORS

Resistance to insecticides and a requirement for insect free stored food has stimulated research into alternative methods of protection. Insect juvenile hormone analogues were first investigated by Strong & Dieckman (1973) but only recently have the problems of their instability and the insensitivity of some insect species been overcome. They produce lethal disruption of insect development and in some species reduce the reproductive capacity of adults. They have the advantage of negligible toxicity to vertebrates and exert their effects only on arthropods. The most effective compound to date is fenoxycarb which has been shown to be effective against a range of common stored product pests including *Sitophilus* spp., at levels between 5 and 10 mg/kg in grain (Edwards & Short, 1984) and larger scale tests (Edwards *et al*., 1987) show fenoxycarb to be comparable with or superior to chlorpyrifos-methyl admixture treatment.

Prospects for use of chemosterilants or pheromones in rodent control are more limited. The chemical which has received most attention is alphachlorohydrin (Bowerman & Brooks, 1971) which causes epididymal lesions at lower doses and death at higher doses. However bait acceptance has been a problem which may be overcome by micro-encapsulation. In any case chemosterilants can only be used as a supplement to conventional control programmes in any situation where one rodent is unacceptable as in a kitchen or food factory.

Application of pheromone signals has had some success with insects with the development of pheromone traps for the Mediterranean flour moth (Hartley, 1983). However it remains virtually unexplored for rodent pests (Shumake, 1977). Some possibilities have been identified for the European Rabbit (Bell, 1987) using male urine to discourage feeding and adult female urine in suppressing male development, but these are still at an early stage.

In both insects and rodents, in order to be effective, sterilisation would need to reach every individual of the target sex and therefore interference with development to sexual maturity would appear to be a more promising avenue for research.

DISCUSSION AND CONCLUSION

The reduction and ultimate elimination of conventional pesticides for pest control relies on exclusion, sophisticated monitoring techniques to detect pests at low levels and high standards of hygiene and plant design to remove potential harbourages.

Modern building materials and construction techniques for food premises permit exclusion of rodents and flying insects from surrounding areas. However incoming raw materials need to be inspected for signs of rodents and detection techniques such as infra-red reflectance used to detect insect infestation. Continuous monitoring of raw materials in store is then required for insects and rodents. High risk products such as grain may need to be treated with insect juvenile hormone analogues to prevent the growth of infestation.

Criticial control points which are particularly vulnerable to infestation need to be identified (in the production process) and these should be closely monitored for infestation and priority given to high hygiene standards and good plant design in such areas. Finally packaging materials should be selected which most effectively resist rodent and insect attack.

A major change is required in the outlook of food manufacturers and retailers to pest control if pesticide free pest control is to be achieved.

REFERENCES

Batth, S.S. (1970). *J. Econ. Entomol.*, **63** (2), 653–5.

Bell, D.J. (1987). In *Control of Mammal Pests*, ed. C.G.J. Richards & T.Y. Ku. Taylor & Francis, London, New York and Philadelphia.

Bowerman, A.M. & Brooks, J.E. (1971). *Wild. Mgmt,* **35** (4), 618–24.

Chambers, J. (1987). *B.C.P.C. Stored Product Insect Monograph, No. 37.*

Cline, L.D. (1978). *J. Econ. Entomol.*, **71** (5), 726–9.
Cogan, P. & Wakefield, M. (1987). *B.C.P.C. Stored Product Insect Monograph, No. 37.*
Collins, W.J. (1973). *J. Econ. Entomol.*, **66**, 44–7.
Corbett, J.R. (1979). *Chem. and Industry*, **22**, 772.
Dubock, A.C. (1984). *Proc. Conf. Organisation & Practice of Vertebrate Pest Control*, pp. 315–26.
Edwards, J.P. & Short, J.E. (1984). *J. Stor. Prod. Res.*, **20**, 11–15.
Edwards, J.P., Short, J.E. & Abraham, L. (1987). *B.C.P.C. Stored Product Insect Monograph, No. 37.*
Fleurat-Lessard, L. (1987). *B.C.P.C. Stored Product Insect Monograph, No. 37.*
Freeman, J.A. (1973). *Ann. Technol. Agric.*, **22**, 509.
Gibson, J.A. (1980). The behaviour of the house mouse *Mus musculus* L in relation to the development of a new control method using contact rodenticides. Ph.D. thesis, University of Aston.
Hartley, D. (1983). *Food & Flavour Processes*, **5**, 53–5.
Highland, H.A. & Wilson, R. (1981). *J. Econ. Entomol.*, **74** (1), 67–70.
Jackson, W.B. (1979). *E.P.P.O. Bull.*, **7** (2), 439.
Voie La, G.K. & Glahn, J.K. (1975). *Proc. 3rd Int. Biodegradation Symposium*, pp. 297–301.
Voie La, G.K., Glahn, J.K. & Cogelia, N.J. (1976) *Proc. 25th Int. Wire, Cable Symp.*, pp. 117–24.
MacNicoll, A.D. (1986). *Proc. 7th B.P.C.A. Conf.*
Nachman, M. & Hartley, P.L. (1975). *J. Comp. Physiol. Psych.*, **89** (9), 1010–18.
Pinniger, D.B. & Wildey, K.B. (1983). *Proc. 5th B.P.C.A. Conf.*
Rust, M.K. & Reierson, D.A. (1978). *J. Econ. Entomol*, **71**, 704–8.
Shumake, S.A. (1977). In *Chemical Signals in Vertebrates*, ed. D. Muller-Schwarze & M.M. Mozell. Plenum Press, New York.
Strong, R.G. & Dieckman, J. (1973). *J. Econ. Entomol.*, **66**, 1167–73.
Teichner, W.H., Warranch, R., Lo Piccols, M. & Campbell, C. (1969). Mode of action of rodent repellents and attractants. Tech. Report 70–69 P.R. Natick Mass Pioneering Res. Lab. US Army Natick Lab. pp. 1–161.
Wilkin, D.R., Cowe, I.A., Thind, B.B., McNicol, J.W. & Cuthbertson, D.C. (1968). *J. Agric Sci. Camb.*, **107**, 439–48.

MARINE

The State of the Art in the Protection of Marine Structures from Biodeterioration

Giulio Relini

Laboratori di Ecologia Animale e Biologia Marina—Istituto di Anatomia Comparata dell'Università—Via Balbi 5—16126 Genova, Italy

ABSTRACT

The biodeterioration of structures immersed in sea water or in which sea water is circulating is caused by marine borers and fouling organisms. Biofouling can be divided into two main categories, macro and micro, with prevention of the latter remaining an unsolved problem in many cases. The principal methods for control of fouling and borers are reviewed with special attention to ships, offshore platforms, intakes, condensers, fishing gear and piling. The materials involved are mainly timber, concrete, metals, plastics and paints. In the light of economic and performance requirements, only a few methods are applied successfully in general practice: impregnation of substrata by chemicals, antifouling coatings and elastomeric materials, treatment of circulating sea water, mechanical removal techniques, thermal-control and velocity-control techniques. Antifoulant substances are by design toxic and have been developed to be released into the environment at toxic concentrations, so they may become a source of pollution. A biological approach to fouling control has been explored seriously recently but at present biological control is not of wide industrial use.

INTRODUCTION

The biodeterioration of marine structures immersed in the sea or in which sea water is circulating is mainly due to biofouling on the surface and to borers, i.e. organisms able to bore, which penetrate deeply into the substrate and destroy material such as timber and calcereous stone. Furthermore, the decay of several materials in the marine environment is brought about by microorganisms such as bacteria and fungi. Consequently, marine structures need to be protected.

Marine biofouling is caused by the settlement or presence of marine organisms on surfaces in contact with sea water and can be divided into micro and macrofouling. The latter is composed of sea weeds (filamentous or branching) and macroscopic animals including barnacles, mussels, bryozoans, serpulids, hydroids, etc.

In the 1950s, 2000 species were identified in the 'fouling community' in the classic publication *Marine Fouling and its Prevention* by the Woods Hole Oceanographic Institution (1952) (see Table 1). Twenty years later over 4000 species were described as belonging to the fouling community (Crisp, 1972). This is not surprising because fouling depends on the locality and the biological environment so that almost all benthic organisms are potential foulers and the list of the species will undoubtedly increase in the future, in particular because microorganisms have been seriously underestimated. Fouling is a complex phenomenon resulting from several processes, the rate and extent of which are influenced by numerous physical, chemical and biological factors in the immediate environment of the surface (Figs 1 and 2). Consequently there may be as many fouling control strategies as there are fouling situations, though the only practical method to avoid such a community is to prevent any of its members from settling on the surface. The mechanism of marine fouling were discussed at the first Biodeterioration Symposium by Houghton (1968). The list of the main wood borers is referred in Table 2.

The brief overview presented here will not be a comprehensive review of the recent literature but will be basically a summary of the main biodeterioration control technologies based on the extant literature and of my experience mainly with the Mediterranean Sea (see list of papers in Relini (1983)).

The main information and references referred to in this paper, derive

TABLE 1

Number of Species in Each Group Reported in *Marine Fouling and its Prevention* (1952)

Bacteria	37	Flat worms	12	Arthropoda	292
Fungi	14	Nemertea	11	Mollusca	212
Algae	563	Rotifers	5	Echinodermata	19
Protozoa	99	Bryozoa	139	Tunicates	116
Sponges	33	Brachiopoda	1	Fish	11
Coelenterata	286	Annelida	108	Total	1958

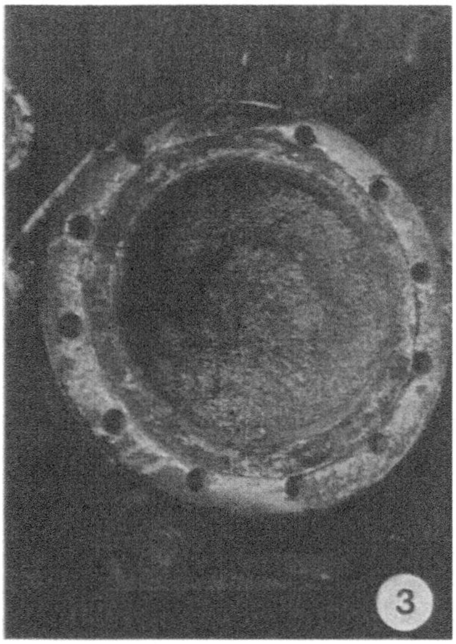

Fig. 1. Asbestos panel (20 × 30 cm) immersed for 6 months at 1 m depth, 6 km offshore of Ravenna (Adriatic Sea). Mussel settlement is about 58 kg/m^2.

Fig. 2. Fouling on a structure immersed during 1 year at 65 m depth, 7 km offshore of Crotone (Ionian sea). The main organisms are *Picnodonta cochlear* (Bivalve) and *Salmacina dysteri* (Serpulid).

Fig. 3. A pipe in a power station completely occluded by fouling, mainly Serpulids.

Fig. 4. Heavy settlement of Serpulids (35 cm thickness) on the walls of power station intakes in Genoa. The type of fouling is influenced by chlorination.

from, first the book *Marine Biodeterioration: An Interdisciplinary Study*, edited by J.D. Costlow & R.C. Tipper (Naval Institute Press, 1984). (It is a successful attempt to bring up to date the well-known *Marine Fouling and its Prevention*, 1952), secondly from the six *Proceedings of the International Congresses on Marine Corrosion and Fouling* and also from the six *Proceedings of the International Biodeterioration Symposia*.

MAIN PROTECTION TECHNOLOGIES

The principal protection technologies still in use are listed in Table 3 (for references see Fischer *et al*., 1984) and their application to marine structures is listed in Table 4 using a key derived from Table 3.

Chemical control

Primary toxic compounds

At the present time the most common and successful means of control remains the use of chemicals mainly based on toxic compounds that are added to organic-matrix paint systems (soluble or insoluble), as a part of the composition of materials, metal alloys, plastics, concrete, as injectable substances in sea water circulating systems and as impregnants of synthetic and natural materials like plastic and timber. By the early 1900s the most frequently used toxicants were compounds of copper, mercury and arsenic. The latter two have been banned in many countries

TABLE 2
A Check-list of the Most Common Wood-boring Animals (after Eaton, 1985)

Molluscs		*Crustaceans*
(a) *Teredinids* (shipworms)	(b) *Pholads* (piddocks)	(a) *Isopods*
Bactronophorus	Lignopholas	(i) *Limnoriidae* (gribble)
Bankia	Martesia	Limnoria
Licyathifer	Xylophaga	Paralimnoria
Lyrodus		Phycolimnoria
Nausitoria		(ii) Sphaerimatidae (pill-bug)
Neoteredo		Cymodoce
Nototeredo		Exosphaeroma
Psiloteredo		Sphaeroma
Teredo		
Teredora		(b) *Amphipods*
Teredothyra		(i) *Cheluridae*
Sphathoteredo		Chelura
Uperotus		

because of the danger to workers' health and their contribution to pollution. Since the publication of *Marine Fouling and its Prevention* (1952), many toxic agents and better means of utilizing them have been tested. A review of antifouling agents is reported by Fischer *et al.* (1984) and by Johnsen & Rendback (1976), and of impregnants and preservatives by Barnacle (1976), Bultman & Parrish (1979) and Eaton (1985).

TABLE 3
Main Protection Technologies (AF = Antifouling) (mainly from Fischer *et al.*, 1984)

(1) Chemical control technology	(A) Primary toxic compounds	(a) AF paint/coatings
		(b) AF elastomers
		(c) Injectable biocides
		(d) Impregnation
		(e) Gas diffusion
	(B) Inherently antifouling structural materials	
	(C) Advanced chemical-control concepts	(a) Strippable plastic films
		(b) Acrylic films
		(c) Flocculants
		(d) Deaeration
		(e) Repellents
		(f) Inhibitors of attachment
		(g) Inhibitors of shell formation
		(h) Hormonal control
(2) Scrubbing	(A) Exterior surface	(a) Ship hulls
		(b) Other ship devices
	(B) Interior surfaces	
(3) Thermal-control method		
(4) Osmotic-control methods		
(5) Velocity-control techniques		
(6) Air bubbles		
(7) Surface modification		
(8) Sonics		
(9) Electric currents and fields		
(10) Magnetic fields		
(11) Optical methods		
(12) Nuclear methods		
(13) Explosive removal method		

TABLE 4
Marine Structures and their Main Antifouling Control Systems (numbers refer to Table 3)

Ship and submarine hulls	1Aa	1Ab	2Aa		
Sea water piping system	1Ac	2B	3	4	5
Condenser	1Ac	1B	1Ccd		
Sonar domes	2A	2Ab			
Navigational buoys	1Aa	1Ab	2Ab		
Offshore platforms	2Ab	1Aa	1Ab		
OTEC	2Ab	3			
Moored oceanographic instruments	1B	2			
Telephonic and electric cables	1Ab	1B			
Wood structures	1Ad	1Ae	1Ab		
Fishing gear	1B	1Aa			
Aquaculture facilities	1Aa	1Ab	1B		
Concrete	1Aa	1Ad			

In the 1960s a very successful new group of toxicants was developed, the organotins (De la Court 1976, 1980). In particular, tributyl tin oxide and tributyl tin fluoride (TBTO and TBTF) were rapidly accepted by the marine paint industry, in combination with copper oxide or without. As is well known, the efficacy of antifouling coating depends not only on the toxicity of the active agent but also upon its leaching rate. In fact, if the leaching rate is too slow the paint will not prevent fouling settlement and if it is too fast the coating will only be effective for a short time. It is reported that the minimum critical rate for cuprous oxide must be 10 μg Cu/cm^2/day (Van Londen, 1963) and for organotin compounds from 1–2 to 5 μg Sn/cm^2/day (De la Court & Devries, 1976). However to prevent settlement of algal spores a leaching rate of 9 ug/cm^2/day is required (De la Court, 1980).

De la Court (1987) has pointed out that effective leaching rates are higher, 16 μg/cm^2/day of TBTO are required to prevent settlement of algae and barnacle, 25 μg/cm^2/day of CuO_2 μg for algae and 18 for barnacles.

Organotin coatings belong to two basic types:

— Coatings that incorporate organotin compounds into the composition as discrete materials.
— Coatings based on film forming resins that contain a chemically bound organotin moiety.

From the latter derive self-polishing copolymer antifouling paints in which the movement of the water slowly erodes and dissolves the coating, constantly presenting a new active surface. These coatings, marketed in the late 1970s are very effective because they keep the ship's hull clean and reduce the drag for long periods.

Elastomers
A further improvement in the protection of materials was achieved with the use of antifouling elastomers (nitrile rubber, neoprene, etc.) because of their ability to provide an increased reserve of toxicant. At the present time the elastomeric materials provide an acceptable surface film performance and the longest service lives of any antifouling coating.

Injectable biocides
In pipes, heat exchangers, and other flowing sea water systems, the prevention of fouling settlement is provided by chemicals added to the circulating water. For this purpose the earliest and most widely employed biocide was chlorine (in particular, hypochlorous acid) derived from a large variety of sources. An inappropriate schedule of chlorination can change the fouling community, allowing some species to grow in large amounts as has happened with Serpulids in the Mediterranean Sea. In some power station intakes Serpulid settlement can reach a thickness of 35 cm and up to 1105 g/dm^2 of net-weight in a year (Relini, 1977*b*) (Figs 3 and 4). Other injectable biocides have been used. The most commonly employed are acrolein, ozone, inorganic ions, bromine and recently organotins. These biocides have been compared by Fischer *et al.* (1984), who reported on their dosage, primary modes of action, advantages and disadvantages. Characklis *et al.* (1985) has discussed the control of microbial fouling in heat exchangers.

The research carried out on all the above systems has shown significant benefits when they are used with additional chemicals or other stress factors such as water velocity or heat. More recently, because of the environmental impact of almost all the above mentioned biocides, chlorine dioxide has been reintroduced. It is more costly and more effective than chlorine but seems less dangerous to the environment (Mayack *et al.*, 1984).

Impregnation

The capacity of porous materials, such as wood, concrete and anodized aluminium to absorb and retain liquid biocides is used for fouling prevention and against decay Good results have been achieved with creosote, 70/30 creosote/coal tar, copper salts, mercuric compounds, tributyl compounds and different compounds responsible for the natural bioresistance shown by some tropical hardwoods (Barnacle, 1976; Furtado *et al.*, 1976; Bultman & Parrish, 1979; Webb *et al.*, 1984; Eaton, 1985).

Gas diffusion

Vapam (Na-N-methyldithiocarbamate), Vorlex (methylisothiocyanate) and chloropicrin (trichloronitromethane) placed in piles to diffuse through the wood as gases are remedial treatments used against marine borer attack and the internal decay of wood.

Inherently antifouling structural materials
In some cases, such as the interiors of pipe systems, screens, nets, etc., materials must be developed and employed that are intrinsically antifouling because it is not possible to use coatings, impregnants, etc. The most widely used AF structural materials are copper and some of its alloys, especially those containing nickel.

According to Moreton & Glover (1980), 90–10 copper–nickel in the form of expanded metal mesh, for fish cages, shellfish trays and sea water intake screens offers both practical and economic advantages. The lining of sea water intake ducts, the sheathing of offshore structures and cladding of ships' hulls are examples of other new developments in the application of copper–nickel alloy suggested by the above-mentioned authors.

Advanced chemical and biological control concepts
The aim of these methods, which are based on different chemical and biological principles, is to limit or to avoid the use of toxic agents. Unfortunately, at the present time they are not of wide practical use and generally are less effective than those based on conventional toxicants; they are listed here because of their possible interest and future improvement.

Strippable plastic films. Plastic films of 'Saron' and polytetrafluoroethylene (PTFE) are applied to the surface. After some months they are removed by stripping off the coating together with the fouling.

Acrylic films. Conventional AF paints are covered by ablative plastic films whose function it is to reduce drag and to control release of toxicants under turbulent conditions.

Flocculants. The use of non-toxic additives such as cationic flocculants (acrylate additives) can improve the overall performance of chlorination without changing the discharge water quality. Only small quantities of these flocculants (2–4 mg/litre) added on a daily basis improve the efficacy of chlorination in deposit removal in cooling systems and eliminate the need for non-oxidizing biocides and chemical dispersants.

Deaeration. In closed systems the removal of air from cooling water for heat exchangers has been suggested. Deaeration, obtained by physical or chemical methods, has several advantages (it is environmentally safe, prevents fouling, reduces corrosion of some alloys and is independent of the design of the plant). It also has disadvantages the power requirement is high and the effect of the anaerobic organisms accelerates the corrosion of some alloys.

Repellents and others. New alternative approaches to fouling control are based on non-toxic substances able to prevent the settlement of micro- and macrofouling. Repellents such as tannic acid, benzoic acids and acrylamide have been used in non-toxic paints. Hormonal control was tested against barnacles to inhibit shell formation, and some chelating agents and substances extracted from borer-resistant woods were studied for their ability to repel barnacles and teredinids. The finding that settlement, attachment, and metamorphosis of the planktonic larvae of many species are subject to exogenous (as well as endogenous) biochemical factors and mechanisms suggest that the operative biochemical control sites may be especially useful targets for a new generation of effective and specific countermeasures to control marine fouling (Morse, 1984).

Scrubbing

Mechanical cleaning by scrubbing is probably the oldest practical method known to man to remove fouling and is still widely applied on external and internal surfaces. The basic justification for scrubbing is to restore smooth surfaces and rejuvenate partially spent AF paints. Because of the development of sophisticated technologies, scrubbing is possible during drydocking and afloat by divers and by special equipment capable of cleaning the underwater hull.

Scrubbing is also used for fixed structures like underwater platform legs. It is estimated that every year about 900 t of mussels are scraped away by divers from gas-platforms situated off Ravenna in the Adriatic Sea (50 t on one platform) (Relini, 1977*a*).

Heat-transfer surfaces of OTEC systems and optical surfaces for submerged instruments require special cleaning systems that are an integral part of the device itself.

The mechanical method for removing fouling from the interior of piping systems and heat exchangers is still one of major importance and in some cases it is the only one possible. An up-to-date review on the subject was presented by Bell *et al.* (1985).

In pipe systems the following methods are employed: (a) foam-rubber balls of various composition which are slightly larger than the internal diameter of the tube; (b) flow-driven brushes whose oscillating action scrubs the surfaces of the pipe, removing any accumulated material; (c) other physical cleaning methods use abrasives (sand, fly ash, diatoms).

Thermal control method

Thermal control methods, based on heating the water, are generally confined to water-cooling systems with reverse flow capability. They are still very successful and are sometimes combined with chemicals such as chlorine.

They have been employed at very high temperature to prevent fouling and also slime formation on optical surfaces. The main problem with these methods is their environmental impact because of the discharge of water at high temperature.

Osmotic control

This is based on the change in salinity of the water, especially decreasing it by mixing it with fresh water. Lower salinities can kill many fouling organisms, as happens when a ship sails in the fresh water of a river, but calcareous shell and other hard structures remain attached to the hull, influencing the performance of the ship.

Velocity

The relative velocity of the water film between sea water and substrate influences the settlement of fouling and especially of macrofouling. This approach is useful in systems in which sea water is circulating (such as pipes and power station intakes). For example at Vado Ligure (SV) power station (1320 MW), four culverts (1500 m long, 2·2 m diameter) in which the sea water flow was at 2·9 m/s (44 m^3/s) remained fouling free for 14 years.

Air bubbles

Air bubbles released from pipes laid alongside the keel kept the hull clean by passing over it at a fairly rapid rate (Houghton, 1970). However, where the bubbles are not closely adherent to the surface fouling will occur.

Surface modification

It is well known that surface properties (texture, chemical composition, electrical charge, etc.) are fundamental in the settlement both of macro- and microfouling. A surface immersed in the sea absorbs organo-compounds from the water and this macromolecular layer generally is the precursor for bacterial slime settlement and the accumulation of fouling. The antifouling technique is based on the maintenance of a critical surface energy which renders it unsuitable for the settlement of organisms.

Sonic

Sonic antifouling, mainly based on high-frequency sounds, in general requires a high expenditure of energy and is very localized. Ultrasonic cleaning is widely used in industry but only for small components. Recently however piezoelectric polymer hull vibrators have been described as being successful in fouling prevention (Murphy *et al.*, 1980).

Electric current, electric fields, magnetic fields

These have limited application, can only be used in very special circumstances and are not reliable.

Optical methods

These are based on the response of the organisms to the variation of the intensity and wave length of light. They have no practical application except in relation to ultra-violet radiation in enclosed systems.

Nuclear and explosive removal methods

These are based respectively on the incorporation of radioactive materials into fouling coating and the detonation of a network of explosives applied underwater to the ship's hull, but are not generally practical and effective methods of control.

CONCLUSION

Many failures in biofouling control are due to improper design, selection of materials, preparation of materials or construction techniques. Chlorination in cooling water systems and antifouling coatings on most marine structures remain the main and most economical control methods of biofouling. However, because of their toxicological and environmental impact (Laughlin & Linden, 1985; Frenet-Piron & Alliot, 1987; Short & Thrower, 1986), new non-toxic technologies are required. The prevention of microfouling, in many cases, is still an unsolved problem.

Often antifouling technologies have been developed without the necessary knowledge of the fundamental processes involved in fouling settlement and in particular as Gareth Jones (1987) has pointed out—of adhesion mechanisms. Probably in the future the prevention will be based on biochemical or physical (like non-stick surfaces) control of the organism settlement, but we have to await for several decades.

ACKNOWLEDGEMENTS

The author is grateful to Professor E.B. Gareth Jones, J. De Palma and D.R. Houghton for suggesting improvements to the manuscript.

REFERENCES

Because of restricted space, a small portion of the literature is reported as a guide to further reading.

Barnacle, J.E. (1976). *Proc. 4th International Congress on Marine Corrosion and Fouling.* C.R.E.O., France, pp. 57–66.

Bell, R., Lee, R. & Torbin, R. (1985). *Condenser Biofouling Control Symposium: The-State-of-the-Art*, June 18–20, EPRI, 43 pp.

Bultman, J.D. & Parrish, K.K. (1979). *Inter. Biodet. Bull.*, **15** (1), 19–27.

Characklis, W.G., Bukke, R. & Yeh, A.I. (1985). *Condenser Biofouling Control Symposium: The State-of-the-Art*, June 18–20, EPRI, 30 pp.

Chow, W. (1985). *Condenser Biofouling Control Symposium: The-State-of-the-Art*, June 18–20, EPRI, 30 pp.

Crisp, D.J. (1972). In *Proc. 3rd International Congress on Marine Corrosion and Fouling*, National Bureau of Standards Gaithersburg, Maryland, pp. 88–93.
De la Court, F.H. (1976). *Proc. 4th International Congress on Marine Corrosion and Fouling*, C.R.E.O. pp. 113–18.
De la Court, F.H. (1980). *J. Oil and Colour Chemists Association*, **63**, 465–73.
De la Court, F.H. (1987). *7th International Biodeterioration Symposium*, Cambridge.
De la Court, F.H. & De Vries, H.J. (1976). *Proc. 4th International Congress on Marine Corrosion and Fouling*, C.R.E.O., pp. 113–18.
Eaton, R.A. (1985). In *Preservation of Timber in the Tropics*, ed. W.P.K. Findlay. Martinus Nijhoff, Dr. W. Junk Publishers, pp. 157–91.
Fischer, E.C., Castelli, V.J., Rodgers, S.D. & Bleile, H.R. (1984). In *Marine Biodeterioration: An Interdisciplinary Study*, ed. I.D. Costlow & R.C. Tipper. Naval Institute Press, pp. 261–349.
Frenet-Piron, M. & Alliot, A. (1987). *Oceanologica Acta*, **10** (1), 73–82.
Furtado, S.E.J., Jones, E.B.G. & Bultman, J.D. (1976). *Proc. 4th International Congress on Marine Corrosion and Fouling*, C.R.E.O., pp. 187–92.
Houghton, D.R. (1968). *Proc. of the 1st International Biodeterioration Symposium*, Southampton, pp. 55–61.
Houghton, D.R. (1970). *Underwater Science and Technology Journal*, pp. 100–4.
Johnsen, S. & Rendback, V. (1976). *Proc. 4th International Congress on Marine Corrosion and Fouling*, France C.R.E.O., pp. 271–7.
Jones, E.B.G. (1987). *7th International Biodeterioration Symposium*, Cambridge.
Laughlin, R.B. & Linden, O. (1985). *Ambio*, **14** (2), 88–94.
Mayack, L.A., Soracco, R.J., Wilde, E.W. & Pope, D.H. (1984). *Water Research*, **18** (5), 593–9.
Moreton, B.B. & Glover, T. (1980). *Proc. 5th International Congress on Marine Corrosion and Fouling*, Biologia Marina, Ed. Garsi, pp. 267–78.
Morse, D. (1984). In *Marine Biodeterioration: An Interdisciplinary Study*, ed. J.D. Costlow & R.C. Tipper. Naval Institute Press, pp. 134–40.
Murphy, P.V., Michel, P., Guelorgat, O. & Latour, M. (1980). *Proc. 5th International Congress on Marine Corrosion and Fouling*, Ed. Garsi, pp. 293–8.
Relini, G. (1977*a*). *Rapports Commission internationale Mer Méditerranée*, **24** (4), 175–6.
Relini, G. (1977*b*). *Atti 7° Simposio sulla conservazione della Natura, Bari*, 20–23 Aprile, 1977, pp. 179–85.
Relini, G. (1983). *Bulletin de liaison du COIPM*, **15**, 1–31.
Short, J.W. & Thrower, F.P. (1986). *Marine Pollution Bulletin*, **17** (12), 542–5.
Van Londen, A.M. (1963). A study of ship bottom paints. Report 54C of Netherlands' Research Center TNO for Shipbuilding and Navigation, pp. 1–34.
Webb, A.D., Parrish, K.K., Graham, R.D. & Bultman, J.D. (1984). In *Marine Biodeterioration: An Interdisciplinary Study*, ed. J.D. Costlow & R.C. Tipper. Naval Institute Press, pp. 41–6.

The Minimum Leaching Rate of some Toxins from Antifouling Paints Required to Prevent Settlement of Fouling Organisms

F.H. de la Court

TNO Paint Research Institute, P.O. Box 203, 2600 AE Delft, The Netherlands

ABSTRACT

In this paper the determination of the critical leaching rate to prevent settlement of algae and barnacles has been described for cuprous oxide and triphenyltinfluoride. To prevent algal fouling a copper leaching rate of at least 22 μg Cu/cm²/day (2·6 μg/m²/s) is required and for TPTF, at least, 5 μg Sn/cm²/day (0·6 μg/m²/s). Barnacle settlement is prevented by a leaching rate of 16 μg Cu/cm²/day (1·9 μg/m²/s) and 2 μg Sn/cm²/day (0·25 μg/m²/s) respectively. These amounts are considerably higher than have been used up till now. The higher toxicity observed for TPTF does not imply that much lower concentrations can be used because it has a low solubility.

INTRODUCTION

Pollution of harbours and coastal waters and the measures to be taken to prevent it, is an important issue nowadays. Antifouling paints contribute to the problem because they contain toxins which leach from the paint film, in some instances affecting shellfish, e.g. oysters. Furthermore during application they may endanger the health of the applicators and other workers in the drydock. It is therefore understandable that a number of governments have already restricted the use of some types of antifouling paints under certain conditions. It is doubtful, however, whether the measures taken are always based on reliable information regarding the toxicity, the amount of toxin present in the paint, their leaching rate and the minimum rate necessary to prevent settlement. This information is not readily available (manufacturers do not normally disclose details of their paints) and it is difficult if not impossible to obtain it from the literature.

In this paper a study has been made of the leaching rates of cuprous oxide (Cu_2O) and triphenyltinfluoride (TPTF) and has been found to differ from those usually accepted. The experiments carried out to determine these rates are described and discussed.

METHODS

Three different kinds of Cu_2O were used to prepare paints of the contact leaching type to Am. Spec 121, containing 46·5% by volume of the toxin. The tricresylphosphate, however, was replaced by dicresylphosphate (de la Court & de Vries, 1973). A further two paints were prepared containing 35·7% and 31·4% TPTF by volume respectively (Kielhorn, 1983). All the paints were applied by spray gun to test panels at a thickness of *c.* 100 μm. They were attached to a rotor apparatus (de la Court & de Vries, 1973) and rotated in natural seawater at 25°C at a circumferential speed of 17 knots (8·7 m/s). After various intervals samples were taken to determine the amount of Cu and Sn remaining in the paint film. For Sn a neutron activation analysis was employed (de la Court, 1980), and Cu was estimated by macerating a small piece of paint film and determining the amount of Cu by an iodometric titration method. In addition after different periods of rotation panels were removed and exposed on a raft in den Helder harbour to assess algal and barnacle resistance (de la Court, 1980). The analyses of Cu and Sn were used to determine the leaching/time relationship from which the leaching rate was calculated for the period of fouling exposure.

RESULTS AND DISCUSSION

From the Cu_2O containing paints the leaching of Cu took place in the usual manner (de la Court & de Vries, 1973). Because the mechanism is of the diffusion type the leaching should decrease with time. For all the Cu_2O paints the best fit through the observed values seems to be a curve of the form

$$\mu g\ Cu/cm^2 = a \,.\, T^b \qquad (1)$$

in which $\mu g\ Cu/cm^2$ is the amount of Cu in micrograms per square centimetre that has been leached after T days; a and b are constants depending on the paint formulation and the conditions of ageing (rotor speed, seawater temperature and pH, etc.). In this study the conditions were always the same. The formulations differed only with regard to the kind of Cu_2O which produced only minor differences in leaching rate behaviour and were not significant. Taking the 17 results together a value of $1{\cdot}9 \times 10^3$ for a was obtained and 0·35 for b (using SI units μg/m^2 and T in seconds a becomes $3{\cdot}5 \times 10^5$).

The leaching of TPTF is linearly related to rotor time as shown earlier (de la Court & de Vries, 1976). The leaching rate of the paint with 35·7% of the toxin was 3·4 μg Sn/cm^2/day (0·39 μg/m^{2}s) and for that containing 31·4% the rate was 2·6 μg Sn/cm^2/day (0·30 μg/m^2/s). From the results presented above it is concluded that the leaching rate for Cu is time dependent whilst that for TPTF is not, confirming earlier findings. From the leaching curve for each paint the leaching rate was calculated for the period over which the fouling exposures were made. Algal fouling has been plotted against the corresponding Cu and Sn leaching rates and the same has been done in respect of barnacles. Since it was not quite clear, for Cu, what the leaching rate value is which will prevent algal settlement, a number of earlier results have been incorporated. The same procedure has been adopted with the results for Sn. The leaching rate of Cu which will prevent barnacle fouling is easily established but again data from earlier experiments have been used to establish the value for Sn. The critical leaching rate values have been indicated in both figures. It appears that the critical leaching rate figure for Cu is higher than that normally accepted, i.e. 10 μ/cm^2/cm^2/day (1·2 μg/m^2/s) (Woods Hole, 1952) and furthermore a different value has to be adopted for algal and barnacle fouling respectively. The same situation is apparent when TPTF is used as the toxicant (de la Court & de Vries, 1976).

Whilst the critical leaching rates were determined under dynamic conditions, in other words in flowing seawater at 17 knots (8·7 m/s), the fouling resistance was determined on a raft where the flow rate is much lower. This reflects the situation on a ship's hull where the leaching rate (and therefore ageing) is highest when the ship is under way and when no fouling occurs. When it is stationary the leaching rate is much lower and is more like that found on a raft. Under these conditions settlement may occur. Whether or not this happens depends on the concentration of toxin at the paint film/seawater interface. When it remains above the lethal level for algal and animal settlement the surface will remain free of macro-fouling. In a successful antifouling paint this concentration is maintained continuously through leaching of the toxin though it is decreased by the flow of water over the surface. When the ship is stationary, and under raft conditions, the flow of seawater is variable but is of little or no consequence if the leaching rate is above the critical level. Below the critical level conditions may be such that because of a low flow rate the leaching is still high enough to maintain a lethal concentration which it could not do under an increased flow rate.

Although the toxin concentration build-up at the interface is more complicated than described here and biological factors play a role, the foregoing may, in essence, explain why below the critical leaching rate settlement does not occur on every occasion. In any case the critical leaching rate determined under dynamic conditions can be considered as the level above which no fouling occurs.

It has been found that TPTF is more efficient as a toxin than Cu_2O though the difference is smaller than it appears. This is because the leaching rate shows the amount of metal leached and TPTF contains only 31·5% Sn while Cu_2O contains 89% Cu. The remaining advantage of TPTF over Cu_2O is probably not real because TPTF is only very slightly soluble and the amount included in a paint cannot be lowered if fouling is to be prevented.

REFERENCES

De la Court, F.H. (1980). *Journal of the Oil & Colour Chemists Association*, **63**, 465–73.

De la Court, F.H. & de Vries, H.J. (1973). *Journal of the Oil & Colour Chemists Association*, **56**, 388–95.

De la Court, F.H. & de Vries, H.J. (1976). *Proceedings of the 4th International Congress on Marine Corrosion and Fouling*, Juan les Pins France, pp. 113–118.

Kielhorn, F.W. (1983). Report V-83-111 of the Paint Research Institute TNO (confidential).

Woods Hole Oceanographic Institution (1952). *Marine Fouling and its Prevention*. US Naval Institute, Annapolis.

The Effect of Microbial Biofilms on Organotin Release by an Antifouling Paint

James W. Mihm & George I. Loeb

David Taylor Naval Ship Research and Development Center, Code 2841, Annapolis, Maryland 21402, USA

ABSTRACT

The effect of microbial films on the release rate and speciation of tributyltin toxin from an organotin paint was determined. Test panels were coated with organotin copolymer ablative antifouling paint. The release rates of the panels were measured before and after biofilms were formed, and after they were mechanically removed. Results show that the bacterial films decreased the release rate but did not degrade the tributyltin. Algal films showed an increase in the release rate with an accompanying degradation of the tributyltin to di- and monobutyltin. For all panels the release rates significantly increased after the biofilms were removed.

INTRODUCTION

Organotin containing antifouling paints are very effective against marine hard fouling (Anderson & Dalley, 1986). However, microbial biofilms have been found on such coatings during service. The environmental assessment of the implementation of organotin containing antifouling requires knowledge of the release rate of this toxin into the environment. Therefore, any differences in organotin release rates or chemical speciation due to the formation of biofilms on the coating should be taken into account when environmental impacts are assessed.

Measurements of release rates made on painted surfaces which have been exposed to natural waters, and so support a microbial biofilm, would have to be made first with the biofilm in place and then after the biofilm had been scraped from the surface. The physical removal of the biofilm would disturb the painted surface and not give a true measurement of the release rate of a biofilm free coating. The significant requirement for this work was to maintain a non-biofilmed surface long enough to establish a baseline release rate for the unfilmed paint. This work consisted of preparing panels coated with an organotin antifouling paint whose release rate would be measured before and after biofilms had been formed on the coated panels. Determining the release rate after the biofilm was removed would allow comparisons with data taken in field experiments.

METHODS

The experimental panels consisted of 3 by 5 in steel Q-panels. Panels were cleaned with xylene and then coated with 3 mils of F-150 epoxy primer Mil 24441, a 1 mil tack coat of the primer, followed by a 3 mil coating of organotin copolymer ablative antifouling paint. Panel immersion chambers consisted of 940 ml glass jars which contained 850 ml ASTM class 2 seawater at 34‰ salinity. Polycarbonate hangers were used to suspend the panels in the seawater. The glass jars were placed in a 22°C temperature control system containing a reciprocal shaker with a 4 mm stroke at 72 strokes/min to thoroughly mix the seawater around the panel.

Stock ASTM seawater was prepared by dissolving the salt mix in distilled water to a salinity of approximately 70‰ and stored in covered glass aquariums with constant aeration. Microoganism growth was prevented by passing the seawater through an ultra-violet aquarium sterilizer at a rate of 100 ml/min. Prior to use, working seawater solutions were diluted to 34 ppm with distilled water. Periodic checks of the microorganism population in the aquarium holding tanks were conducted.

After the panels were placed in the test jars the 850 ml of seawater was changed twice weekly to prevent buildup of organotin concentrations. Water from these test jars (200 ml) were used for analysis of the amount and speciation of the released organotin. The tin extraction method, taken from Olson *et al.* (1985) consisted of an initial extraction with 10 ml methylene chloride, 10 min of vigorous shaking, 10 min phase separation, followed by recovering the methylene chloride from the water sample. These steps were then repeated with a fresh portion of methylene chloride and the two 10 ml aliquots combined. A modification to Olson's method involved the omission of the sodium borohydrate during the initial extraction process (Matthias, personal communication), since our extracted samples were to be stored (0°C) for 2–3 months prior to analysis, and sodium borohydrate during storage could have caused degradation of the sample. The sodium borohydrate was added immediately before separation of the mono-, di- and tributyltin by gas chromatography and measurement of the tin concentrations with flame ionization detection.

Midway into the experiments temperature control in the laboratory failed, leading to widely varying release rates of organotin. The test jars and shaking device were then placed in a temperature controlled chamber at 22°C, and resulted in a stable release rate of organotin.

After a stable release rate was obtained panels were inoculated with bacterial and algal cultures consisting of organisms from the hull of a ship which is coated with an organotin antifouling paint. Bacteria were cultured in Difco marine broth (2261), while the algae were cultured in formula F2, both standard culture media. Biofilms were formed on the panels by inoculations, each followed by a change of seawater after a 24 h filming period.

These films were allowed to remain on panels A, B, C and F for 38 days, with samples for analysis being taken twice a week when the seawater was changed. After 38 days the thickness of the biofilm was measured on the panels using a light microscope, and the biofilms removed and samples taken for identification. Panels were then placed back into the test jars to determine the release rates after the films were removed. On other panels, D, E and F, biofilms were not removed after the 38 days, but remained on the panels for the 110 day duration of the experiments.

RESULTS

Stable release rates for the unfilmed panels were established at 0·11–0·16 μg/cm^2/day TBT (av. TBT 0·135 + 0·06 μg/cm^{2}day), with a release of dibutyltin of <0·001 μg/cm^2/day (Table 1). Panels A–E were inoculated with the bacteria culture and panels F and G with the algal culture. After the bacterial films formed on panels, the release rate decreased 60% to an average of 0·07 + 0·03 μg/cm^2/day TBT. Release rates remained at this level for the 38 days the bacterial films were on panels A, B and C. During this period the dibutyltin release rates remained at <0·001 μg/cm^2/day. On panels D and E, with the bacterial films not removed, the release rates remained lower than the initial rates for the remaining 110 days.

TABLE 1

Average Tributyl and Dibutyltin Release Rate for Panels without Biofilms, after Biofilms Developed, and after these Films Had Been Removed from the Panels. Also Included are Biofilm Thickness Measurements

Panel number	*Film type*	*Release rate (μg/cm²/day TBT)*						*Biofilm thickness (mm)*
		Without biofilm		*With biofilm*		*Biofilm removed*		
		Tri	*Di*	*Tri*	*Di*	*Tri*	*Di*	
A	Bacteria	0·16 ± 0·02	<0·001	0·04 ± 0·004	<0·001	0·36 ± 0·06	<0·001	$3·5 \times 10^{-3}$ cm
B	Bacteria	0·14 ± 0·02	<0·001	0·06 ± 0·11	<0·001	0·22 ± 0·11	0·004 ± 0·002	$2·2 \times 10^{-3}$
C	Bacteria	0·12 ± 0·02	<0·001	0·04 ± 0·01	<0·001	0·29 ± 0·14	0·005 ± 0·002	$3·8 \times 10^{-3}$
D	Bacteria	0·20 ± 0·02	<0·001	0·11 ± ID[a]	<0·001	ND[b]	ND[b]	ND[b]
E	Bacteria	0·18 ± 0·02	<0·001	0·09 ± ID[a]	<0·001	ND[b]	ND[b]	ND[b]
F	Algal	0·11 ± 0·02	<0·001	0·16 ± 0·13	0·02 ± 0·01	0·53 ± 0·26	0·03 ± 0·02	$2·7 \times 10^{-3}$
G	Algal	0·17 ± 0·04	<0·001	0·75 ± 0·24	0·06 ± ID[a]	ND[b]	ND[b]	ND[b]

[a] Insufficient data.
[b] No data.

In contrast algal films on panels F and G caused an increase in the release rate to an average of 0·46 + 0·42 μg/cm²/day TBT. Also, the dibutyltin concentrations significantly increased from <0·001 up to 0·02–0·06 μg/cm²/day DBT, with a few measurements showing low amounts of monobutyltin. In addition, there were many other chromatograph peaks which are unidentified, but demonstrate the breakdown of the organotin. These substances may be other degradation products, or reaction products of tributyltin metabolites, such as methyl-tin compounds. Identification of these substances will require additional effort. The dominant algal form on these samples was the blue-green alga *Anacystis montana* identified by Dr Phillip Sez, Georgetown University.

After the biofilms (bacteria and algal) were removed, release rates increased approximately 2–5 times the original stable rate.

During GC analysis a very large peak occurred after the tributyltin peak for all paint extracts analysed. This peak prevented detection of tetrabutyltin, or clear resolution of any peaks at longer retention times. Since the detector is biased toward tin detection, it is possible that this signal represents a significant amount of tin containing material. However, this has not yet been demonstrated and further studies will be required before any clear identification of this material can be achieved.

The observable thickness of the biofilms measured with a light microscope confirms that biofilms had developed on the panels (Table 1). Comparisons can be made with biofilm thickness measurements made in the natural environment.

DISCUSSION

These results indicate that biofilms have the ability to alter release rates from the paint coating. The bacterial and algal films developed on these samples had quite different effects. Bacterial films decreased the amounts being released. This reduction could be due to an interaction of released material with the extracellular materials produced by the bacteria. Alternatively, diffusion could be significantly reduced in the biofilm, or the bacteria could create an environment which inhibits the release of the organotin from the paint. These effects could be important in designing antifouling paints that would have very low release rates yet still be effective antifoulants because the toxin would be present in the biofilm.

The opposite effect is seen with the algal films, with increased release rates and concurrent breakdown of the tributyltin. This may also be due to the local environment created by the algal metabolites. The nature of the metabolites or their influence on the local water chemistry in the biofilm (such as pH) facilitates the increased release of the toxin from the pain and its passage through the biofilm. It is not clear at this point what the mechanisms are for the breakdown of the tributyltin, however, this type of biodegradation has been reported by other workers (Olson & Brinckman, 1986).

Effects of algal films are considered important to the environmental impacts of tributyltin. Algae can be highly efficient detoxifying agents in the aquatic environment, especially at installations designed to capitalize on this property. Algae can be grown in ponds receiving effluents from a number of processes. Both dispersed cell populations and attached forms are well known to have great versatility. Their use has been evaluated in European power stations for this purpose, with highly encouraging results of more than 90% conversion to inorganic tin in short times (Saroglia *et al.*, 1982). However, we may expect that the implications of biotransformation of even inorganic tin to methyl tin or similar materials in estuarine sediments need to be considered.

The availability of algal detoxification is of great significance for biotechnological processing of organotins. The metabolic processes of naturally occurring bacteria have not been found suitable for organotin detoxification, however the detoxifying capability of naturally occurring algae allows an immediate biotechnological approach without relying on the development of genetically engineered bacterial strains.

The increase in the release rate after the biofilms were removed is likely due to increasing the paint surface area from the scraping process. This is a very important consideration for other field and laboratory work which involved biofilm removal prior to release rate measurement.

REFERENCES

Anderson, C.D. & Dalley, R. (1986). *Proceedings, Oceans 86, Vol. 4*, pp. 1108–13.

Olson, G.J. & Brinckman, F.E. (1986). *Proceedings, Oceans 86, Vol. 4*, pp. 1196–201.

Olson, G.J., Brinckman, F.E., Matthias, C.L. & Bellama, J.M. (1985). A comprehensive method for the determination of aquatic butyltin species at ultratrace levels using simultaneous hybridization/extraction with GC-FPD (NBSIR Tech. Rep. 85-3295). National Bureau of Standards, Gaithersburg.

Saroglia, M.G., Ambrogi, R. & Scarano, G. (1982). *Can. Tech. Rep. Fish. Aquat, Sci.*, 1163–70.

MECHANISMS

Biodegradation of Biomass—Mechanisms

Douglas E. Eveleigh

Department of Biochemistry & Microbiology, Cook College, Rutgers University, New Brunswick, New Jersey 08903, USA

ABSTRACT

The efficiency of microorganisms in the recycling of organic materials both by individual cells and by consortia, is illustrated by their effective mineralization of resistant materials ranging from cellulose, to lignin and to novel recalcitrant xenobiotics. The mechanisms for such transformations that are gradually being unravelled, are discussed. Such understanding will permit rationale approaches for gaining greater catalytic efficiency and conversely, for development of more selective inhibition of degradative processes.

INTRODUCTION

Biodegradation is dependent upon 'enzymes'. This presentation considers aspects of enzymology in relation to mineralization of biomass (cellulose, hemicellulose, lignin) at mechanistic and evolutionary levels; the whys and hows. Why can enzymes attack such a broad spectrum of substrates? How did the diverse array of enzymes arise? How are newly evolved enzymes maintained? As this research area has been the focus of several recent reviews (e.g. Hartley *et al.*, 1987) focus is on the mechanism of enzyme action in the degradation of biomass polymers and the evolution of enzymes.

Estimates of the annual world production of biomass (cellulose, hemicellulose and lignin) are about $1{\cdot}8 \times 10^{12}$ t, of which only about 150×10^{6} t is used by the pulp and paper industry. This massive production attests indirectly to the efficacy of the microbe's degradative power, while the industrial usage figures lend perspective to man's activities and the global picture. Polymers of biomass can be of a repeating homo- (starch, cellulose, pectin) or heteropolymeric (hemicelluloses and lignin) nature. Lignin is of greater complexity, comprised of randomly linked oxygenated phenyl propane units. Even a cursory consideration of the degradative enzymes acting towards a homopolymer such as cellulose emphasizes the diversity of enzyme complexes. For cellulase there are several interacting exo- and endo-splitting glucanases which may be secreted or occur as large cell surface associated complexes termed the cellulose (for review see Eveleigh, 1987). Cellobiose alone can be degraded through the action of at least six enzymes; cellobiase, cellobiose, oxidase, cellobiose kinase, cellobiose phosphorylase, cellobiose quinone oxido-reductase (Eveleigh, 1983; Woodward & Wiseman, 1982). Analogously, the attack of pectin, a polymer of methylated alpha-1,4-galacturonic acid (up to 95% methylation) is also through the action of a battery of enzymes; four depolymerases (exo- and endo-splitting eliminases and hydrolases) attack the natural polymer, while following demethylation (pectin methyl esterase), a further set of four enzymes, exo- and endosplitting pectate lyases and polygalacturonidases attack the derived acidic polymer. It is also manifest that as the main chain has interspersed 1,2 linked rhamnose moieties, rhamnosidase is a further component of this degradative battery of pectinases (Cooper, 1987; Linhdart *et al.*, 1986). The message 'bounteous enzymes' is repeated yet again by consideration of hemicellulases. Arabino-xylans are attacked by a consortium of xylanases and arabinosidases (Dekker, 1985; Reilly, 1981). The important role of deacetylases in making hemicelluloses susceptible to poly-saccharases has also been emphasized (Morris & Bacon, 1977; Biely, 1985; Wood & McCrae, 1986). The interacting synergistic nature of these enzymes is yet another added level of complexity. It is necessary to travel this jungle of enzymes in order to unravel the bases of degradation and it is from this perspective that the mechanism of action and the specificity of enzymes that degrade biomass are discussed.

LYSOZYME

The singular model of polysaccharase mechanism is the bacteriolytic enzyme, lysozyme, which Fleming discovered in 1922. It was in the 1960s that dramatic enzymological advances were made through elucidation of its amino-acid sequence, its analysis by X-ray crystallography and from these studies the formulation of its detailed mechanism of action (for review see Osserman *et al.*, 1974; Fersht, 1977; Jolles & Jolles, 1984). Lysozymes (1,4-N-acetylmuramidases) cleave the C_1–O glycosidic bond of N-acetylmuramic acid (NAM) connecting to C-4 of N-acetylglucosamine (NAG) in bacterial mucopeptides. The enzyme cleft has six sites, A, B, C, D, E, F. NAG can bind to all sites, but due to steric effects, NAM can only bind to positions B, D and F. Lysozyme cleaves the oligomer D–E bond which is positioned between the carboxyl groups of Glu-35 (unionized) and Asp-52 (ionized). The proposed sequence of attack involves a carbonium ion intermediate stabilized through the ionized Asp-52 carboxylate ion, and general acid catalysed removal of the alcohol by the unionized carboxyl of Glu-35. There was also the prediction that during carbonium ion formation the sugar ring at D is distorted into the 'sofa' form. This mechanism has been generally agreed upon except for some doubt with regard to the original 'sugar ring distortion' mechanism. Levitt (1974) through analysis of detailed substrate (NAG_6)-enzyme coordinates suggested that the sugar at site D binds in the normal chair conformation. Thus the transition state (carbonium ion which includes the sofa conformation) is achieved but not through straining and distortion of the sugar ring. However, this idea resulted in little modification of the basic model of mechanism of attack. Post & Karplus (1986) now suggest a completely novel mechanism based on a molecular dynamic simulation of lysozyme action. Their analyses indicated that the Glu-35 carboxyl group could form hydrogen bonds with the endocyclic D ring oxygen O_5 and the hydroxymethyl oxygen O_6 but not with the exocyclic oxygen O_4. The NAG D site residue again is in the undistorted chair conformation. This led to a proposed mechanism of protonation of ring O_5 by Glu-35, followed by cleavage of the endo C_1–O_5 bond yielding an acyclic oxocarbonium ion intermediate stabilized by Asp-52. With attack by water, the C_1–O_4' bond is cleaved, plus ring closure. Distortion of the ring in site D is not required (Post & Karplus, 1986). This novel mechanism will stimulate reanalyses of the mechanism of glycosidic hydrolyses.

LYSOZYME AND GLYCOSIDASES

The basic roles of the catalytic amino acids Glu-35 and Asp-52 are analogous in both the original and the new Karplus reaction schemes. The role of these residues is taken up by, respectively, Glu-73 of goose lysozyme and Glu-11 of T-4 phage lysozyme, and of Asp-86 of goose and Asp-20 of phage (Weaver *et al.*, 1985). This leads to a consideration of the evolution of lysozymes, and also their relationship to glycosidases in general. Mackay *et al.* (1985) note that in general there is great diversity of glycosidase proteins but suggest that there are common and conserved regions with similarities between amylase, cellulase and xylanase. Thus some degree of homology has been shown between the active site sequences (Glu-33-Tyr-51) of *Schizophyllum commune* beta-(1,4)-endoglucanase and those of hen egg white lysozyme including the catalytic residues Glu-35 and Asp-52 (Yaguchi *et al.*, 1983). Earlier, Bause & Legler (1980) had noted a similarity of the *Aspergillus wentii* Ser-Asp-Thr beta-glucosidase sequence associated with the active site, to that occurring in hen egg white lysozyme (Thr-Asp-Thr). Further structural similarities have been shown between *Trichoderma* endoglucanases and cellobiohydrolases (Paice *et al.*, 1984; Knowles *et al.*, 1987). Our studies from sequencing the gene of endoglucanase of *Microbispora bispora* show three putative active sites analogous to those of endoglucanase of *Trichoderma reesei* (Yablonsky *et al.*, 1987). These molecular approaches are providing considerable impetus to the study of the evolution of glycosidases (cellulase and xylanase) and their mechanism of action (for review see Aubert *et al.*, in press).

ANOMERIC CONFIGURATION OF THE PRODUCT AND MECHANISMS OF ACTION

As noted above the mechanism of attack by lysozyme is well studied and it is agreed that the products retain the beta-configuration at the C_1 carbon. This is theorized to occur due to the formation of a carbonium ion intermediate coupled with a unidirectional nucleophilic attack. Retention of configuration is generally found with endo-acting glycosidases (alpha-amylase; endo-beta-(1,4)-glucanase) and it is assumed with mounting evidence that their mechanisms follow the lysozyme pathway. In contrast, enzymes acting in an exo-splitting manner, invert the anomeric configuration of the product. Beta-amylase yields beta-maltose, glucamylase yields beta-glucose while cellobiohydrolase produces alpha-cellobiose. In these instances, it is inferred that hydrolysis involves a single step mechanism. These concepts of anomeric inversion/retention actions of glycosidases have been illustrated by classical optical rotational (King, 1963; Baker & Himmel, 1986) and by NMR (Eveleigh & Perlin, 1969) studies.

The anomeric configuration of the glycosidase product in the examples noted above, appear rigidly defined. However, Hehre's recent studies question this rule. For instance, his group noted that *Bacillus pumilus* beta-xylosidase acts towards beta-D-xylopyranosides and lower xylooligosaccharides with absolute specificity for the beta-D-xylopyranosyl configuration, with configurational inversion and without glycosyl transferring ability (Kasumi *et al.*, 1987). In contrast, with glycosyl fluorides as substrates, the enzyme will attack both beta- and alpha-anomeric substrates. Further, though the beta-xylosidase hydrolyses beta-D-xylosides with configurational inversion, the reaction with beta-D-xylosyl fluoride yields a retention of configuration. From the relationships between the initial attack of alpha-D-xylosyl fluoride in relation to substrate concentration, the possible occurrence of two substrate molecules bound to the active site was suggested. They postulate a scheme in which alpha-D-xylopyranosyl fluoride in the presence of xylose yields the disaccharide 4-*o*-beta-D-xylopyranosyl-D-xylose. Fluorine is displaced by the 4-CH group of the acceptor in this formation of xylobiose. The beta-xyloside product is now hydrolysed to yield beta-D-xylose, i.e. retention of configuration. Hehre's group here reported similar findings with 'inverting glycoside hydrolases' (beta-amylase, glucoamylase, glucodextranase). Thus most, if not all, glycosylases may possess functionally flexible catalyic groups and thus have the ability to act on different substrates by different catalytic mechanisms. Glycosylases appear not to be alone in this respect' (Kasumi *et al.*, 1987). With such a broad mechanism of action, it is not surprising also to find glycosidases that have either restricted or broad specificity. For instance beta-glucosidase kinase has rigid requirements for a beta-glucoside and yet is lax with regard to the aglycone moiety (aliphatic, aromatic or hexitol) (Palmer & Anderson, 1972). Callobiohydrolases appear poorly named when their variety of cleavage actions towards methyl-umbelliferyl oligosaccharides is observed (van Tilbeurgh *et al.*, 1982). Glycosidases are also well known for their transfer reactions with a variety of substrates (Aerts *et al.*, 1985).

BINDING

A variety of approaches are being used to clarify the bases of glycosidase action, especially such essential concepts of binding to an insoluble substrate. For instance cellobiohydrolase I of *Trichoderma reesei* has been shown to possess two domains, one at the carboxy terminal acting as a binding site to the insoluble cellulose substrate, while the other in the core protein contains the hydrolytic site (van Tilbeurgh *et al.*, 1986). These two domains were shown through isolation of two distinct components following partial digestion with papain. The same enzyme has been shown by small angle X-ray scattering to be tadpole shaped (18 × 4·4 nm) (Schmuck *et al.*, 1986) and these two groups of workers are deciphering whether the tadpole 'head' or 'tail' contains the enzyme active site (H. Esterbauer, personal communication). The concept of distinct binding and cleavage domains for a *Trichoderma* cellobiohydrolase is quite clear from such studies. Support for such concepts is found at the gene level. Identical repetitive amino-acid sequences have been described in *Clostridium thermocellum* endoglucanases from the otherwise dissimilar celA and celB genes, and these have been proposed as binding sites (Béguin *et al.*, 1987). A similar notion has been ascribed to repetitive sites (proline-threonine) that occur on *Cellulomonas fimi* exo- and endoglucanase (O'Neill *et al.*, 1986; Wang *et al.*, 1986). Essentially similar PT sequences to those of *C. fimi* occur on the *M. bispora* endoglucanase (Yablonsky *et al.*, 1987).

PROTEIN MODIFICATION

Glycosidases, in attacking insoluble substrates, must be secreted and hence have to withstand the rigors of the external environment, and in general have evolved to be relatively stable. The rationales for stability are elusive. Matthew's group have recently shown through cloning and sequencing of 25 temperature sensitive mutants of T4 phage obtained by site directed mutagenesis, that changes of a single amino acid can yield marked change in temperature stability, yet without apparent major change in the three-dimensional structure of the enzyme (Alber *et al.*, 1987). Such studies are providing new approaches to gain an understanding of temperature stability, catalytic efficiency and substrate specificity. It should be possible to develop more efficient enzymes via site specific mutagenesis even though the necessary amino acid substitutions may occur in regions that can be in 'rigid' or buried sites. A recent example is modification of subtilisin. These serine proteases of *Bacillus licheniformis* and *B. amyloliquefaciens* are quite distinct, differing in k_{cat}/K_m to a range of substrates and in greater than a 60-fold difference in catalytic efficiency. However, by consideration of the active sites, three aminoacid replacements obtained via site directed mutagenesis within the active site of *B. amyloliquefaciens* (predicted from the *B. licheniformis* active site), yielded novel subtilisins with altered and enhanced activity with regard to substrate specificity (Wells *et al.*, 1987). Such approaches confirm the feasibility of producing novel enzymes with optimal pH, temperature stability and substrate specificity profiles.

LIGNINASES

The discovery of ligninases originally described as 'hydrogen peroxide dependent oxygenases' opened up a further conceptual avenue in biodegradation (Glenn *et al.*, 1983; Tien & Kirk, 1983). Subsequently, it was shown that ligninase did not require molecular oxygen, and that it was indeed a peroxidase but will transfer a single electron from substrate to enzyme using H_2O_2 as an oxidant (Kirk, 1987; Palmer *et al.*, 1987). Ligninase can also catalyse the removal of single electrons from non-phenolic aromatic moieties (e.g. electronegative methoxylated rings of lignin). This specific activity of lignin peroxidase is of great interest as such a loss of single electrons yields radical cations. These positively charged radicals can oxidize further substrate, to yield directly cationic and radical moieties. At the onset of ligninase production veratryl alcohol is produced which is rather surprising, for one would expect this non-phenolic aromatic to be oxidized by the enzyme. However, oxidation of veratryl alcohol produces a cation that can yield veratrylaldehyde. Should a second oxidizable compound be present (further non-phenolic moieties abound in lignin), the radical cation can act as a single electron oxidant and the veratryl alcohol is regenerated via its action in oxidizing the second compound. Veratryl alcohol could thus act as a mediator for ligninase (Harvey *et al.*, 1986). These concepts suggest that lignin degradation can now occur at distances well beyond the ligninase initiating site through continued free radical action. The broad specificity of such reactions has prompted study of ligninases (Hammel *et al.*, 1986; Bumpus & Aust, 1987; Preiss *et al.*, 1987) and of Mn^{2+}-dependent peroxidases (Paszczynski *et al.*, 1985) in the degradation of other recalcitrant compounds, including DDT, benzo-(a)pyrenes, polychlorinated biphenyls and even dioxins.

EVOLUTION

Evolutionary approaches are of major interest in biodegradation in both naturally occurring (Clarke, 1984; Mortlock, 1984) and recombinant DNA derived strains (Ramos & Timmis, 1987). At present the evolutionary relationships between cellulolytic and hemicellulolytic microbes and of diverse lysozymes are restricted to direct comparative analyses of genes or products (Mackay *et al.*, 1985; Weaver *et al.*, 1985). The study of newly evolved cellulases or xylanases is in its infancy, except perhaps for consideration of latent (cryptic) cellobiase genes.

These latter phenotypically silent DNA sequences can be activated and expressed as rare events by a few members of large populations through mutation, recombination, insertion elements or other genetic mechanisms (Hall *et al.*, 1983). Such cryptic systems can be complex, for example for cellobiose utilization in *E. coli*, cellobiose (Cel^+) mutants that use cellobiose, arbutin and salicin have a specific gene (phosphotransferase celT) and phosphoglucosidase (celH) in one operon besides other genes (arbT) that specify transport and use of arbutin (Kricker & Hall, 1987). Indeed four gene clusters appear evident in *E. coli* for beta-glucoside utilization (Hall & Betts, 1987). The persistence of such cryptic genes is suggested to occur through occasional selection (Hall *et al.*, 1983). Li (1984) calculates that such cryptic genes can be retained without having any advantage over a useless DNA sequence, provided reactivation occurs for substantial periods. These studies will spur exploration of further evolutionary research avenues.

In review, mechanistic studies of glycosidase action indicate the broad potential for their being able to attack diverse substrates, while the concept of ligninase acting through the generation of free radicals yields an understanding of how such peroxidases can transform refractory xenobiotics. For the industrial transformation of biomass into chemical feedstocks, and also for degradation of industrial pollutants, the major stumbling block appears as the need for more efficient enzymes. From the recently developing concepts of the mechanisms of action of glycosidases and of peroxidases, it should be possible to design more effective enzyme systems through rational site specific mutagenesis and protein modification, and also by careful selection of evolving strains.

ACKNOWLEDGEMENTS

This study was supported through research funds of the New Jersey Agriculture Experiment Station (K-01111-05-87) and through the US Department of Energy Contracts XX-4-04150-1 and DE-AS05-83ER13140.

REFERENCES

Aerts, M., van Opstal, O. & De Bruyne, C.K. (1985). *Carbohyd. Res.*, **138**, 127–34.

Alber, T., Dao-pin, S., Nye, J.A., Muchmore, D.C. & Matthews, B.W. (1987). *Biochemistry*, **26**, 3759–62.

Aubert, J.-P., Béguin, P. & Millet, J. *FEMS Symposium* (in press).

Baker, J.O. & Himmel, M.E. (1986). *J. Chromatog.*, **357**, 161–81.

Bause, E. & Legler, G. (1980). *Biochim. Biophys. Acta*, **626**, 459–65.
Béguin, P., Millet, J. & Aubert, J.-P. (1987). *Microbiol. Sci.*, **4**, 277–80.
Biely, P. (1985). *Trends Biotechnol.*, **3**, 286–90.
Bumpus, A. & Aust, D. (1987). *Bioessays*, **6**, 166–70.
Clarke, P.H. (1984). In *Microbial Degradation of Organic Compounds*, ed. D.T. Gibson. Marcel Dekker, New York, pp. 11–27.
Cooper, R.M. (1987). In *Genetics and Plant Pathogenesis*, eds. P.R. Day & G.J. Jellis, Blackwell, Oxford. pp. 261–81.
Dekker, R.H.F. (1985). In *Biosynthesis and Biodegradation of Wood Components*, ed. T. Higuchi. Academic Press, New York, pp. 505–33.
Eveleigh, D.E. (1983). In *Biomass Utilization*, ed. W.A. Cote. NATO Advanced Study Institute. Plenum Press, New York, pp. 365–91.
Eveleigh, D.E. (1987). *Phil. Trans. Roy. Soc.*, Lond. **A 321**, 435–47.
Eveleigh, D.E. & Perlin, A.S. (1969) *Carbohyd. Res.*, **10**, 87–95.
Fersht, A. (1977). *Enzyme Structure and Mechanism*. W.H. Freeman, Reading, UK, 371 pp.
Glenn, J.K., Morgan, M.A., Mayfield, M.B., Kuwahara, M. & Gold, M.H. (1983). *Biochem. Biophys. Res. Comm.*, **114**, 1077–83.
Hall, B.G. & Betts, P.W. (1987). *Genetics*, **115**, 431–9.
Hall, B.G., Yokoyama, S.Y. & Calhoun, D.H. (1983). *Mol. Biol. Evol.*, **1**. 109–24.
Hammel, K.E., Hammel, B.K. & Kirk, T.K. (1986). *J. Biol. Chem.*, **261**, 16948–952.
Hartley, B.S., Broda, P.M.A. & Senior, P.J. (1987). *Phil. Trans. Roy. Soc., Lond.*, **A 321**, 405–568.
Harvey, P.J., Schoemaker, H.E. & Palmer, J.M. (1986). *FEBS Lett.*, **195**.
Jolles, P. & Jolles, J. (1984). *Molec. Cell. Biochem.*, **63**, 165–89.
Kasumi, T., Tsumuraya, Y., Brewer, C.F., Kersters-Hilderson, H., Claeyssens, M. & Hehre, E.J. (1987). *Biochemistry*, **26**, 3010–16.
King, K.W. (1963). In *Advances in Enzymic Hydrolysis of Cellulose and Related Materials*, ed. E.T.Reese. Pergamon Press, New York, pp. 159–69.
Kirk, T.K. (1987). *Phil. Trans. Roy. Soc., Lond.*, **A 321**, 461–74.
Knowles, J., Lehtovaara, P., Teeri, T., Pentilla, M., Salovuori, I. & Andrè, L. (1987). *Phil. Trans. Roy. Soc., Lond.*, **A 321**, 449–54.
Kricker, M. & Hall, B.G. (1987). *Genetics*, **115**, 419–29.
Levitt, M. (1974). Peptides, Polypeptides & Proteins, ed. E.R. Blout, F.A. Bovey, M. Goodman & N. Lotan. John Wiley, New York, p. 99.
Li, W. (1984). *Molec. Biol. Evol.* **1**, 213–19.
Linhardt, R.J., Galliher, P.M. & Cooney, C.L. (1986). *Appl. Biochem. Biotechnol.*, **12**, 135–55.
Mackay, R.M., Baird, S., Dove, M.J., Errat, J.A., Gines, M., Moramelli, F., Nasim, A., Willick, G.E.. Yaguchi, M. & Seligy, V.L. (1985). *Biosystems*, **18**, 279–92.
Morris, E.J. & Bacon, J.S.D. (1977). *J. Agric. Sci. Camb.*, **85**, 327.
Mortlock, R.P. (ed.) (1984). *Microorganisms as Model Systems for Studying Evolution*. Plenum Press, New York, 326 pp.
O'Neill, G., Goh, S.H., Warren, R.A.J. Kilburn, D.G. & Miller, R.C. Jr (1986). *Gene*, **44**, 325–30.
Osserman, E.F., Canfield, R.E. & Baychok, S. (eds) (1974). *Lysozyme*. Academic Press, New York.

Paice, M.G., Desrochers, M., Rho, D., Jurasek, L., Roy, C., Rollin, C.F., deMiguel, E. & Yaguchi, M. (1984). *Bio/Technology*, **2**, 535-9.
Palmer, R.E. & Anderson, R.L. (1972). *J. Biol. Chem.*, **247**, 3415-19.
Palmer, J.M., Harvey, P.J. & Schoemaker, H.E. (1987). *Phil. Trans. Roy. Soc., Lond.* **A 321**, 495-505.
Paszczynski, A., Huynh, V.B. & Crawford, R.L. (1985). *Arch. Biochem. Biophys.*, **244**, 750-65.
Post, C.B. & Karplus, M. (1986). *J. Am. Chem. Soc.*, **108**, 1317-19.
Preiss, U., Engelhardt, G., Wallnofer, P.R. & Mucke, W. (1987). *Chemosphere*, **16**, 963-8.
Ramos, J.L. & Timmis, K.N. (1987). *Microbiol. Sci.*, **4**, 228-37.
Reilly, P. (1981). In *Trends in the Biology of Fermentations for Fuel and Chemicals*, ed. A. Hollaender. Plenum Press, New York. pp. 111-29.
Schmuck, M., Pilz, I., Hayn, M. & Esterbauer, H. (1986). *Biotechnol. Lett*, **8**, 397-402.
Tien, M. & Kirk, T.K. (1983). *Science*, **221**, 661-3.
van Tilbergh, H., Claeyssens, M. & De Bruyne, C.K. (1982). *FEBS Lett.*, **149**, 152-6.
van Tilbeurgh, H., Tomme, P., Claeyssens, M., Bhikhabhai, R. & Pettersson, G. (1986). *FEBS Lett.*, **204**, 223-9.
Wang, W.K.R., Gerhard, B., Gou, Z.M., Kilburn, D.G., Anthony, R., Warren, J. & Miller, R.C. Jr (1986). *Gene*, **44**, 315-24.
Weaver, L.H., Grütter, M.G., Remington, S.J., Gray, T.M., Isaac, N.W. & Matthews, B.W. (1985). *J. Molc. Evol.*, **21**, 97-111.
Wells, J.A., Cunningham, B.C., Graycar, T.P. & Estell, D.A. (1987). *Proc. Nat. Acad. Sci., USA*, **84**, 5167-71.
Wood, T. & McCrae, S. (1986). *Phytochemistry*, **25**, 1053-5.
Woodward, J. & Wiseman, A. (1982). *Enzyme Microbial Technology*, **4**, 73-9.
Yablonsky, M.D., Bartley, T., Elliston, K.O., Kahra, S.K., Shalita, Z.P. & Eveleigh, D.E. (1987). *Biochemistry & Genetics of Cellulose Degradation*, ed. J.-P. Aubert, J. Millet & P. Béguin. *FEMS Symposium*, Paris, France, Sept. 6-9 1987.
Yaguchi, M., Roy, C., Rollin, C.F., Paice, M.G. & Jurasek, L. (1983). *Biochem. Biophys. Res. Comm.*, **116**, 408-11.

Mechanisms of Bacterial Biodegradation of Alkyl Sulphate and Alkylpolyethoxy Sulphate Surfactants

Graham F. White & Nicholas J. Russell

Department of Biochemistry, University College, P.O. Box 78, Cardiff, CF1 1XL, UK

ABSTRACT

Sulphated surfactants, including long-chain alkyl sulphates and polyethoxy sulphates, are important components of modern detergent formulations. Depending on the alkyl chain these compounds are readily biodegradable by microorganisms in soil and natural waters. From a biochemical viewpoint, biodegradation is initiated by different mechanisms. Alkyl sulphates are attacked initially by a range of sulphatases of differing specificities, which liberate long-chain alcohols that can be assimilated via central metabolic pathways. In contrast the biodegradation of alkylpolyethoxy sulphates is less straightforward. Our studies using sodium dodecyl triethoxy sulphate (SDTES) have shown that primary biodegradation can be initiated by sulphatase or etherase attack. Some bacteria possess an etherase system which produces a mixture of mono-, di- and triethylene glycol monosulphates, whereas in others the etherase acts almost exclusively on the alkyl–ether bond, generating triethylene glycol monosulphate. In pure cultures of single isolates these sulphated glycols and their oxidised derivatives are not broken down further. However, in environmentally-relevant test systems of mixed cultures of glycol sulphates are metabolised to CO_2 *and inorganic sulphate. Hydrophobic metabolites containing the alkyl portion of the parent surfactant are readily biodegradable in mixed culture systems. Therefore, in environmental terms, the alkylpolyethoxy sulphates can be regarded as completely biodegradable.*

INTRODUCTION

Synthetic surfactants belong to one of four types depending on the nature of the hydrophilic head-group, viz. anionic, cationic, non-ionic or amphoteric (zwitterionic). Commercially, anionic surfactants are by far the most important. Their applications include light domestic products (washing-up liquids, solid-surface cleaners, etc.), heavy industrial cleaners and personal hygiene products such as shampoos and toothpastes.

Among anionic surfactants there is considerable diversity of chemical structure and, consequently, of biodegradability. For example, in the tetrapropylene-based alkyl benzene sulphonates, which were produced in quantity immediately post-World War II, the branched alkyl chains and aryl C-SO_3^- hydrophilic group make the molecule resistant to degradation. In the 1950s these compounds caused severe pollution of receiving waters with formation of blankets of foam, which led to a ban on their use in countries subscribing to OECD guidelines. In contrast, surfactants such as alkyl sulphates (e.g. sodium dodecyl sulphate, $C_{12}H_{25}OSO_3^-Na^+$, SDS) and alkylpolyethoxy sulphates (e.g. sodium dodecyltriethoxy sulphate, $C_{12}H_{25}(OCH_2CH_2)_3OSO_3^-Na^+$, SDTES), in which the chains are predominantly linear and the SO_3^- unit is linked via oxygen in a C–O–SO_3^- ester bond, undergo rapid primary biodegradation and generally pose no foaming pollution problems.

The loss of surfactant properties is regarded as primary biodegradability. Although this is an important consideration, there is growing concern about the persistence and potential toxicity of intermediates derived from primary biodegradation. This paper summarises some of our studies on the biodegradative pathways for alkyl and alkylpolyethoxy sulphate surfactants in pure and mixed bacterial cultures, and their relevance to the environmental situation.

METHODS

Two bacterial strains have been used extensively in the studies reported in this paper. *Pseudomonas* C12B was isolated from soil by enrichment culture on SDS. Its surfactant-degrading properties, especially the relevant enzymology, have been reviewed in detail elsewhere (e.g. Dodgson & White, 1983). *Pseudomonas* DES 1 was isolated from sewage cultures adapted to commercial alkylpolyethoxy sulphate surfactants.

The methods for determining metabolic pathways in pure and mixed (river water) cultures using ^{35}S- and ^{14}C-labelled surfactants are detailed in Hales *et al.* (1986) and Griffiths *et al.* (1987). Briefly, cells were grown in defined media containing radio-labelled surfactants, and radioactive fractions analysed routinely by TLC using silica gel or cellulose plates and a wide variety of solvent systems. Metabolites were identified by co-chromatography with synthetic standards, isotope dilution analysis, NMR and mass spectroscopy.

RESULTS AND DISCUSSION

Biodegradation pathways of alkyl sulphates

Cell extracts of *Pseudomonas* C12B grown on SDS as sole source of carbon hydrolyse [^{35}S]SDS to liberate $^{35}SO_4^{2-}$ and dodecanol (Payne *et al.*, 1965). The extracts also contain NAD-linked dehydrogenase activities that oxidise primary and secondary alcohols (Williams *et al.*, 1966). Collectively, these early studies together with parallel work by Hsu (1965) indicated that degradation occurred by initial release of SO_4^{2-} from SDS followed by oxidation of the liberated alcohol.

Recent work in Cardiff using [1-^{14}C]SDS has shown that during the early stages of degradation, the major radiolabelled metabolite (up to 20% of incorporated radioactivity) was [1-^{14}C]dodecanol with small amounts of [1-^{14}C]dodecanal (< 0·2%), which were converted to ^{14}C-labelled free fatty acids (mainly 12:0 and 14:0). Subsequently, the radiolabel appeared in phospholipids and in water-soluble products, probably intermediates of β-oxidation and the Krebs cycle. Ultimately, 70% of the radiolabel appeared as $^{14}CO_2$, a value consistent with substantial and possibly complete biodegradation (Payne & Wiebe, 1978). This and the earlier work of Payne and Hsu has led to formulation of the pathway shown in Fig. 1.

Secondary alkyl sulphates (i.e. with sulphate attached to non-terminal carbon atoms) are degraded by an analogous, though more complex, pathway. Extracts of bacteria grown on Oronite, a commercial mixture of secondary alkyl sulphates, hydrolysed pentan-3-yl sulphate and a mixture of secondary decyl sulphates (predominantly decan-2-yl sulphate plus small amounts of the 3-, 4- and 5-isomers). Induced bacteria could also oxidise linear secondary alcohols. Thus the initial stages of degradation (sulphate liberation, then alcohol oxidation) were analogous to those of the primary isomers. Using another *Pseudomonas* isolate, Lijmbach & Brinkhuis (1973) showed that the full sequence was sulphate ester → alcohol → ketone → hydroxy ketone → dione → aldehyde plus acid.

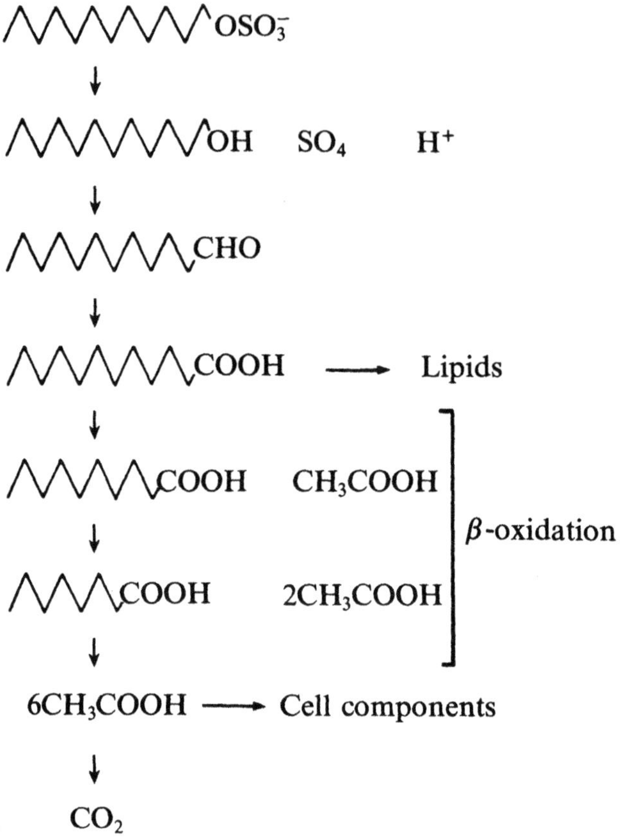

Fig. 1. Metabolic pathway for the biodegradation of SDS by *Pseudomonas* C12B.

Biodegradation pathways of alkylpolyethoxy sulphates

Studies with pure bacterial cultures

Alkylpolyethoxy sulphates differ from alkyl sulphates by the possession of ethylene glycol unit(s) between the alkyl chain and the $-SO_3^-$ moiety (e.g. SDTES, $C_{12}H_{25}(OCH_2CH_2)_3OSO_3^-Na^+$). This modification produces surfactants which combine good foaming properties with low skin irritancy, making them very suitable for domestic and personal product use. Compared with alkyl sulphates, this structural change, especially the presence of ether linkages, complicates the biodegradation pathway. Despite this fact, alkylethoxy sulphates undergo rapid primary biodegradation and it is not difficult to isolate bacteria that can use the compounds as sole carbon and energy sources.

One such isolate, *Pseudomonas* DES1, when grown on [^{35}S]SDTES, produced up to five ^{35}S-labelled organic metabolites in addition to $^{35}SO_4^{2-}$. NMR analysis of the pooled metabolites indicated the absence of alkyl chains. Moreover these metabolites did not accumulate during growth on [1-^{14}C]SDTES, indicating that primary cleavage occurred between the sulphur atom and C-1 of the alkyl chain. Obvious potential sites of attack were the ether links in the polyethylene glycol chain; by using a combination of co-chromatography with chemically-synthesised and authenticated standards, and isotope dilution analysis, the five organic metabolites were identified as mono-, di- and triethylene glycol monosulphates and the corresponding oxidised forms (-CH_2OH to -COOH) of the latter two esters, namely acetic acid 2-(ethoxysulphate) and acetic acid 2-(diethoxysulphate) respectively. The sequence of formation of ^{35}S-labelled metabolites during growth indicated that mono- and diethylene glycol monosulphates were derived directly from SDTES rather than from triethylene glycol monosulphate. Conversely, acetic acid 2-(ethoxysulphate) and acetic acid 2-(diethoxy sulphate) were produced by oxidation of the parent glycol monosulphates, not directly from SDTES. Inorganic sulphate only increased in amounts when

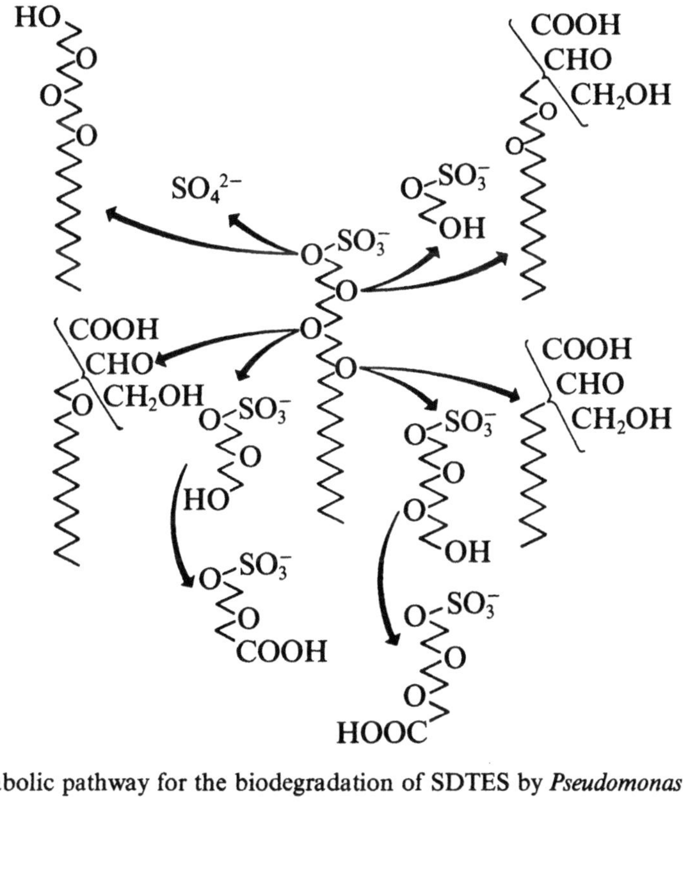

Fig. 2. Metabolic pathway for the biodegradation of SDTES by *Pseudomonas* DES1.

SDTES was still present in the medium; none was released from any of the five organic metabolites. This biodegradative sequence is summarised in Fig. 2.

The relative contributions of etherase(s) and sulphatase(s) to formation of organic and inorganic ^{35}S-labelled metabolites, respectively, in *Pseudomonas* DES1 were 2:1 when bacteria were grown on [^{35}S]SDTES in the presence of Empicol ESB/3S, a commercial detergent containing alkylethoxy sulphates. Because the organism cannot cleave ether or sulphate ester bonds in any of the glycol sulphate metabolites, the amounts of each of the three pools of glycol chain lengths reflected the extent to which SDTES was cleaved at each ether bond (Fig. 3). The relative proportions of cleavage at the ether bonds were remarkably constant under a variety of growth conditions, suggesting that a single etherase enzyme was responsible. *Pseudomonas* C12B also displayed a 2:1 ratio of its etherase/sulphatase activities, and had similar ether-cleaving specificity (Fig. 3). Two other bacterial isolates, SC25A and

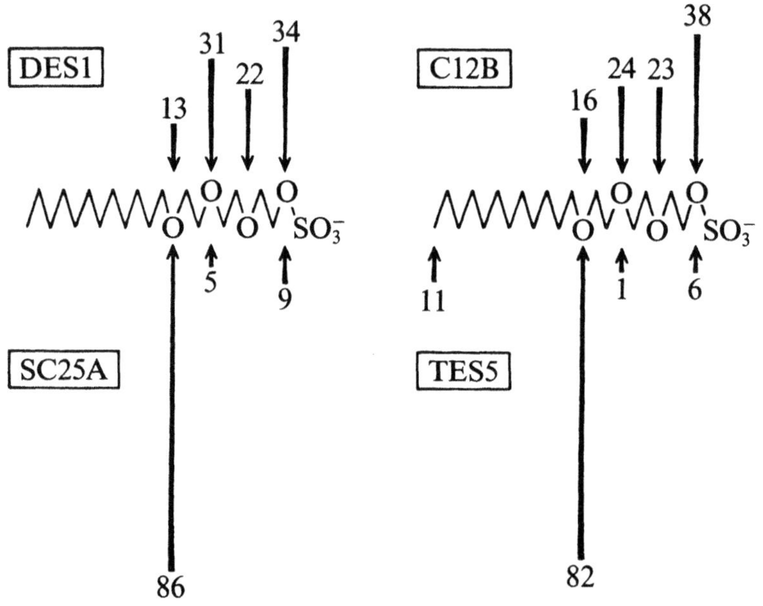

Fig. 3. Relative contributions of sulphate hydrolysis, ether cleavage and ω-/β-oxidation to the biodegradation of SDTES in four bacterial isolates. Lengths of the arrows indicate relative percentages of surfactant (quoted at base of arrow) which undergo primary biodegradation at the indicated points of attack. In DES1 and C12B (isolated on sulphated surfactants), etherase is relatively non-specific and complemented by a significant contribution from sulphatases. In SC25A and TES5 (isolated on polyethylene glycols), the etherase is predominant and highly specific for the alkyl ether bond. ω-/β-oxidation also occurs in TES5.

TES5, which both utilise polyethylene glycol (Hales *et al.*, 1986), produced a totally different pattern. Each employed predominantly the etherase route with only minor ($< 10\%$) contribution from sulphatase(s). Etherase specificity was very high, favouring almost exclusively ($> 95\%$) the alkyl–ether bond. Furthermore, TES5 also produced metabolites that were tentatively identified as products of ω-/β-oxidation. Thus this organism was notable for employing all three possible degradation routes simultaneously.

Etherase/sulphates action produces, in addition to the sulphated metabolites, a series of hydrophobic compounds containing the alkyl chain and (for SDTES) up to three ethoxylate residues. In *Pseudomonas* DES1 up to twenty hydrophobic metabilites are formed from [1-^{14}C]SDTES (Griffiths *et al.*, 1987). They comprise three groups—alcohols, carboxylic acids and (probably) aldehydes—representing the different oxidation states arising from the action of dehydrogenases on the initial etherase products. The main hydrophobic metabolites (see Fig. 2) were tri-, di- and monoethylene glycol dodecyl ethers ($C_{12}[EO]_3$; $C_{12}[EO]_2$; $C[EO]_1$) and dodecanol, together with their corresponding carboxylic acids, namely 3,6,9-trioxa-heneicosanoic acid ($C_{12}[EO]_3COOH$); 3,6-dioxa-octadecanoic acid ($C_{12}[EO]_2COOH$); 3-oxa-pentadecanoic acid ($C_{12}[EO]_1COOH$); and dodecanoic acid. In addition, there were metabolites which contained alkyl-glycol ether bonds and carbonyl groups but no hydroxyls.

Studies with mixed microbial populations

Studies on axenic cultures are essential for elucidating the biochemistry and enzymology of biodegradation pathways. Although the organisms used are isolated from natural sources, it remains to be established that metabolism observed in pure cultures also occurs in environmental situations. Moreover, in the present study it was important to show that metabolites accumulating in pure cultures did not persist in the environment. To assess the environmental acceptability of surfactants, mixed cultures are usually employed in biodegradability tests which fall into two groups (Gilbert, 1979): the first indicate susceptibility to microbial degradation (e.g. OECD screen tests), and the second assess biodegradation rates under laboratory simulation of environmental conditions (e.g. river-water die-away tests).

In both OECD screen and river-water die-away tests [^{35}S]SDTES, either pure or as a component of commercial surfactant mixtures, undergoes rapid primary biodegradation. As in pure cultures, ether cleavage predominated over sulphatase activity; the main bond broken is the alkyl–ether bond and the same series of ^{35}S-labelled glycol sulphates and their oxidation products (carboxylate derivatives) were formed as seen previously. Biodegradation via ω-/β-oxidation of the dodecyl chain was a minor route, which correlates with the observation that only one out of four bacterial isolates tested possessed this ability (Hales *et al.*, 1986). In contrast, Steber & Wierich (1985) found that biodegradation of the non-ionic surfactant octadecyl heptaethoxylate was principally by ω-/β-oxidation, possibly reflecting its longer alkyl chain length compared with SDTES.

A major distinction between the pure and mixed cultures was that the glycol sulphates and their oxidation products, which persisted in pure cultures, were completely mineralised to CO_2 and inorganic sulphate in 28–35 days in river-water die-away tests. Moreover, the overall biodegradation rates were faster in cultures from river-water and activated sewage which had been acclimatised to surfactant, especially mixtures, indicating that induction of enzymes and/or selection of specific strains occurs naturally.

ACKNOWLEDGEMENTS

This and related work was supported by grants from SERC and NERC which are gratefully acknowledged.

REFERENCES

Dodgson, K.S. & White, G.F. (1983). *Topics Enz. Ferment, Biotechnol.*, **7**, 90–155.

Gilbert, P.A. (1979). *Ecotoxicol. Environ. Safety*, **3**, 111–5.

Griffiths, E.T., Hales, S.G., Russell, N.J. & White, G.F. (1987). *Biotechnol. Appl. Biochem.*, **9**, 217–29.

Hales, S.G., Watson, G.K., Dodgson, K.S. & White, G.F. (1986). *J. Gen. Microbiol.*, **132**, 953–61.

Hsu, Y.-C. (1965). *Nature*, **207**, 385–8.

Lijmbach, G.W.M. & Brinkhuis, E. (1973). *Ant. van Leeuwen.*, **39**, 415–23.

Payne, W.J. & Wiebe, W.J. (1978). *Ann. Rev. Microbiol.*, **32**, 155–83.

Payne, W.J., Williams, J.P. & Mayberry, W.R. (1965). *Appl. Microbiol.*, **13**, 98–101.

Steber, J. & Wierich, P. (1985). *Appl. Environ. Microbiol.*, **49**, 530–7.

Williams, J.P., Mayberry, W.R. & Payne, W.J. (1966). *Appl. Microbiol.*, **14**, 156–60.

Cellulase Mechanisms

T.M. Wood

Rowett Research Institute, Bucksburn, Aberdeen AB2 9SB, UK

ABSTRACT

Although it is well established that a cellulase that can hydrolyse hydrogen-bond ordered cellulose (crystalline) comprises endo-1,4-β-glucanases, exo-1,4-β-glucanases (normally cellobiohydrolases) and β-glucosidases, the various enzymic interactions are still not well understood. The principal areas of disagreement are summarized. New data are presented showing that extensive hydrolysis of crystalline cellulose in the form of the cotton fibre requires the presence of two immunologically distinct cellobiohydrolases and at least one endoglucanase: an hypothesis to explain this new observation is presented.

INTRODUCTION

Microorganisms producing cellulase enzymes are found scattered amongst the actinomycetes, the true bacteria and the gliding bacteria, but amongst the fungi the capacity to effect the breakdown of cellulose is widespread. However, only a few of the fungi appear to excrete large amounts of extracellular cellulase, and it is from these cellulases that most of our understanding on the mechanism of cellulase action has been obtained. The availability of information on the mode of cellulase action operating in the bacteria is poor in comparison, because the enzymes, with one or two notable exceptions, are cell-wall bound.

FUNGAL CELLULASES

The extracellular cellulases from the aerobic fungi *Trichoderma viride, T. reesei, T. koningii, Fusarium solani, Penicillium funiculosum, P. pinophilium, Sporotrichum pulverulentum* and from the anaerobic fungus *Neocallimastix frontalis* are the best known examples of cellulases that can degrade crystalline hydrogen bond-ordered cellulose (Wood & McCrae, 1979; Wood *et al.*, 1986). Most other extracellular cell-free cellulases from fungi have little action on crystalline cellulose, but they can hydrolyse soluble derivatives of cellulose such as CM-cellulose (carboxymethylcellulose) or chemically 'swollen' celluloses which are model substrates for measuring attack on the amorphous or paracrystalline areas of the cellulose. Fractionation studies carried out on the cellulases of certain of the above mentioned fungi, have shown that the distinguishing feature of a cellulase that can degrade crystalline cellulose is that it contains a cellobiohydrolase enzyme in addition to the randomly-acting endoglucanases and β-glucosidases that are found in cellulases which can degrade only partially disordered amorphous cellulase. There is a consensus that these enzymes by their synergistic action can together effectively render crystalline cellulose soluble, but there is considerable debate on the extent of this co-operation, on the number of enzymes required, and on the role each enzyme plays in the process. These uncertainties have caused continuing debate and speculation regarding the mechanism of cellulase action, and have raised questions regarding the purity and substrate specificity of the various enzymes.

Table 1 summarizes widely held views on the substrate specificities of the various types of enzyme present (Wood, 1985). However, as 'anomalous' results would now appear to abound in the literature, particularly in relation to specificities of the cellobiohydrolase (Chanzy *et al.*, 1984; Chanzy & Henrissat, 1985), it is becoming clear that either these definitions are no longer acceptable or, as demonstrated by Sprey & Lambert (1983), the various 'purified' enzymes may exist in some circumstances as multicomponent enzyme complexes which are only separable electrophoretically after treatment with suitable dissociating agents.

Despite these problems, a significant feature of most studies is the synergism observed between the various enzymes in solubilizing crystalline cellulose. Early fractionation studies carried out in this laboratory on the cellulases of *T. koningii* and *F. solani* (Wood & McCrae, 1979) show that the activity to crystalline cellulose in the form of the cotton fibre is virtually completely dependent on the co-operative action

TABLE 1
Action of Cellulase Components on Different Substrates

Enzyme	*Crystalline cellulose*	*Amorphous 'swollen' cellulose*	*CM-cellulose*	*Cello-oligo-saccharide*	*Cellobiose*
Cello-biohydrolase	Slow	Very active	NIL	Active	NIL
Endo-glucanase	NIL	Very active	Very active	Active	NIL
β-glucosidase	NIL	NIL	NIL	Active	Active

of a fraction containing cellobiohydrolase and a fraction containing endoglucanase and β-glucosidase enzymes (Table 2).

Thus, the activity to cotton is lost when the fraction containing cellobiohydrolase, endoglucanase and β-glucosidase enzymes are separated, but it is recovered quantitatively when the various fractions are recombined in their original proportions. Reconstituted mixtures of fractions containing endoglucanase and cellobiohydrolase can account for most of the cellulase activity of the unfractionated cellulase system. However, maximum efficiency in rate and extent of hydrolysis is obtained only when β-glucosidase is also present in the mixture of

TABLE 2
Cellulase Activity of the Components of *Trichoderma koningii* and *Fusarium solani* Cellulase When Acting Alone and in Combination[a]

Enzyme	*Cellulase activity (solubilization of cotton, %)*	
	T. koningii	*F. solani*
Cellobiohydrolase (CBH)	1	2
Endo-1,4-β-glucanase (Endo)[b]	1	1
β-Glucosidase	Nil	Nil
Endo + CBH	55	58
Endo + β-glucosidase	6	4
CBH + β-glucosidase	20	18
Endo + CBH + β-glucosidase	72	71
Original unfractionated enzyme	71	71

[a] All enzymes were present in the same proportions in which they were present in the original unfractionated enzyme (from Wood & McCrae (1979*b*)).
[b] This was the enzyme fraction left after the removal of β-glucosidase and cellobiohydrolase. The fraction was rich in activity to CM-cellulose and contains many protein components identified by isolectric focusing in polyacrylamide gels.

enzymes (Wood & McCrae, 1979). The role of β-glucosidase is to remove the inhibitory effects of cellobiose which is one of the major products of the action of the other two enzymes (Wood, 1981). In some enzyme systems cellobiose oxidase (Eriksson, 1981) and cellobiose dehydrogenase (Dekker, 1980; Eriksson, 1981; Coudray *et al.* 1982) are also present, and these together with the β-glucosidase play important roles in regulating the rate and extent of hydrolysis by the cellobiohydrolase and endoglucanase enzymes.

MULTIPLE FORMS AND THEIR PROPERTIES

A characteristic property of each of the cellulases studied is that each type of enzyme exists in multiple forms (Wood, 1985). However, the nature of, the origin of, and the requirement for these multiple components are not yet known; and these considerations are still the subject of intense debate in the literature. In this debate the possibility has been addressed that the multiplicity could be genetically determined, or could result from partial proteolysis (Nakayama *et al.*, 1976; Gong *et al.*, 1979) or differential glycosylation of a common polypeptide chain (Gunn & Brown, 1979). In reality, all of these possibilities seem to operate in *T. reesei* and other cellulases.

The multiplicity of the cellulase components of *T. reesei* (Fagerstam & Pettersson, 1979; Gritzali & Brown, 1979), *T. koningii* (Wood & McCrae, 1972, 1978) and *P. pinophilium* (Wood *et al.*, 1980; Wood & McCrae, 1986) has been particularly well studied. *T. reesei* cellulase has been shown to contain between five and eight electrophoretically distinct endoglucanases, two immunologically distinct cellobiohydrolases (I and II) and two or three β-glucosidases. *T. koningii*, *T. viride* and *P. pinophilium* cellulases are equally heterogeneous (Wood & McCrae, 1979).

Heterogeneity in cellobiohydrolase activity has been intensively investigated, particularly the immunologically distinct cellobiohydrolases which have been shown recently to act synergistically in solubilizing Avicel, and, to a lesser extent, cotton fibre. Cellobiohydrolase I and II in both *T. reesei* (Enari & Niku-Paavola, 1987; Gritzali & Brown, 1979; Fagerstam & Pettersson, 1979) and *P. pinophilium* (Wood *et al.*, 1980; Wood & McCrae, 1986*a*) have been shown to be glycoproteins which differ both in the amount of carbohydrate associated with the protein and in the amino acid composition. In the case of *T. reesei* cellobiohydrolase I and II, no homology has been found to exist in the amino acid sequence (Enari & Niku-Paavola, 1987; Pettersson *et al.*, 1981), but in each case the carbohydrate moiety has been found to be required for binding of the enzyme to the substrate (Van Tilbeurgh *et al.*, 1986). Both cellobiohydrolases have been reported recently to have some properties consistent with the conclusion that attack on the cellulose chain is not exclusively from the end (Henrissat *et al.*, 1985; Enari & Niku-Paavola, 1987; Kyriacou *et al.*, 1987). However, as the contamination of cellobiohydrolase with only minute quantities of endoglucanase would produce the same effect, these claims must be considered with caution. Complex formation between cellobiohydrolase and endoglucanase has been shown in this laboratory to be a feature of the cellulases of *T. reesei* and *P. pinophilium*, and in another laboratory between endoglucanase, xylanase and β-glucosidase (Sprey & Lambert, 1983).

Multiplicity in terms of endoglucanase activity is also the norm in most cell-free cellulase preparations examined thus far (Wood & McCrae, 1978; Shoemaker & Brown, 1978; Gong *et al.*, 1979; Gong & Tsao, 1979). However, opinion as to the nature and origin of the heterogeneity in this activity is particularly diverse and controversial. The different endoglucanases found in any one cellulase system are reported to differ in the M_r values, in their isoelectric points, and in their association with carbohydrate (Coughlan, 1985). A characteristic property of each endoglucanase is the slope of the line obtained when the fluidity (1/viscosity) of a solution of carboxymethylcellulose incubated with enzyme is plotted against the reducing power of the solution (Wood, 1985). Because viscosity is a parameter related to chain length, the differences in the slopes of the lines have been interpreted to indicate the existence of enzymes which vary in their mode of attack. By this criterion, most cellulase studies contain several different endoglucanases, but a notable exception in this regard was found in the cellulase of *Talaromyces emersonii*, when Maloney *et al.* (1985) established that the four endoglucanases isolated appeared to effect changes in the solution of CM-cellulose that produced identical fluidity/reducing power relationships. It was concluded that differential glycosylation of a single enzyme form rather than genetically determined differences in primary structure accounted for the heterogeneity.

Evidence that heterogeneity is the result of proteolytic modification of one or more endoglucanases has been provided by Nakayama *et al.* (1976) and by Gong *et al.* (1979), but this is disputed by Labudova and Farkas (1983), Dunne (1971) and Kolar *et al.* (1985). Gritzali & Brown (1979) claim, on the basis of an examination of the properties of the cellulase system induced in *T. reesei* by the disaccharide sephorose, that only a single endoglucanase is produced by the fungus.

Thus, the nature and origin of the heterogeneity continues to be uncertain in the case of both endoglucanase and cellobiohydrolase activities. On balance it would seem, however, that there is much evidence for post-secretional modification of the enzymes synthesized by the microorganisms, and it seems certain that some heterogeneity at least is artefactual. Nevertheless, it is not known with any degree of certainty just how many enzymes are genetically determined, and it is unlikely that this problem, like the problem of the specificity of individual enzymes, will be resolved until a more detailed analysis has been done on a variety of enzymes produced by a number of genes cloned in heterologous hosts. Studies involving the isolation and cloning of cellulase genes are now being carried out in several laboratories (Shoemaker *et al.*, 1983; Teeri *et al.*, 1983; Schwarz et al., 1985; Joliff *et al.*, 1986).

Wood (1981) has argued from theoretical considerations of the stereochemistry of the cellulose chains in the cellulose crystallite, that a minimum of two stereospecific cellobiohydrolases and two stereospecific endoglucanases may be required for maximum efficiency in the solubilization of crystalline cellulose.

SYNERGISTIC ACTION

Several hypotheses have been formulated in the last 25 years in an attempt to describe the various enzymatic interactions involved in cellulolysis (Reese *et al.*, 1950; Wood & McCrae, 1972; Reese, 1975; Wood, 1981; Chanzy *et al.*, 1983; Ryu *et al.*, 1984; Wood, 1985; Enari & Niku-Paavola, 1987). Reese and his colleagues (Reese *et al.*, 1950) envisaged (Fig. 1) that attack was initiated by a non-hydrolytic chain-separating enzyme (C_1), producing a reactive 'swollen' cellulose which could then be attacked by the hydrolytic endoglucanases (C_x). This hypothesis, which dominated the thinking and guided the experimental approach for many years, was abandoned by most workers with the isolation of a cellobiohydrolase enzyme which acted in synergism with the endoglucanases to solubilize crystalline cellulose. Reese was not convinced that cellobiohydrolase was the C_1-factor, and he later modified his hypothesis to suggest that C_1 might be a special type of endoglucanase that was capable of cleaving a few covalent links resulting in some 'swelling', but having no detectable hydrolytic activity (Reese, 1975). In this new hypothesis, cellobiohydrolase was relegated to the role of hydrolysing the partially degraded cellulose along with the endoglucanases.

There is as yet no good evidence to support the 'C_1 hypothesis', although Chanzy *et al.* (1983) have recently found in their studies with the

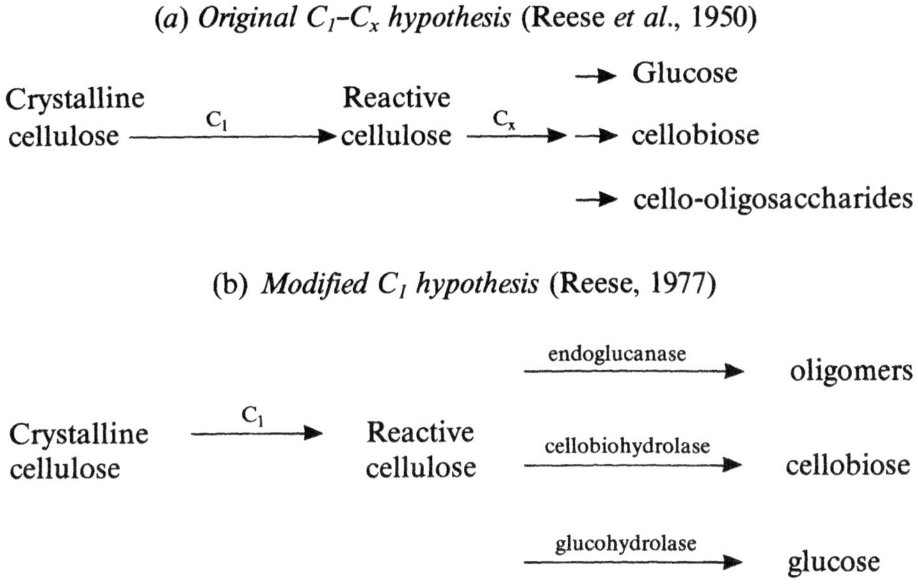

Fig. 1. Original and amended hypotheses for the mechanism of cellulase action involving a C_1 component.

electron microscope that crystalline cellulose from the alga, *Valonia macrophyta*, can be 'swollen' by the cellobiohydrolase I of *T. reesei*, and, in the presence of high concentrations of the enzyme, to be substantially hydrolysed. However, the presence of large amounts of glucose in the hydrolysate along with the cellobiose may indicate that the cellobiohydrolase was associated with a contaminating endoglucanase. Moreover, the results contrast with those of White & Brown (1981), who could find no evidence in the electron microscope for the independent action of cellobiohydrolase. Instead, purified endoglucanase produced some splaying of ribbons into microfibril bundles, supporting the observations of Wood & McCrae that endoglucanase increases the capacity for uptake of alkali and initiates the attack.

The production of short fibres from cotton has also been studied as a manifestation of early attack, but doubt has been expressed about the significance of these observations in the context of attack on the cellulose crystallite (Wood, 1988). Vaheri (1982) considers that oxidation of the cellulose and the formation of onic acids precedes the generation of short fibres. She suggests that it is the endoglucanases that produce the short fibres, but views on this are conflicting. Thus, the formation of short fibres has been reported to be a property of both cellobiohydrolases I (Nummi *et al.*, 1983) and II (Enari & Niku-Paavola, 1987) of *T. reesei*, but Wood &

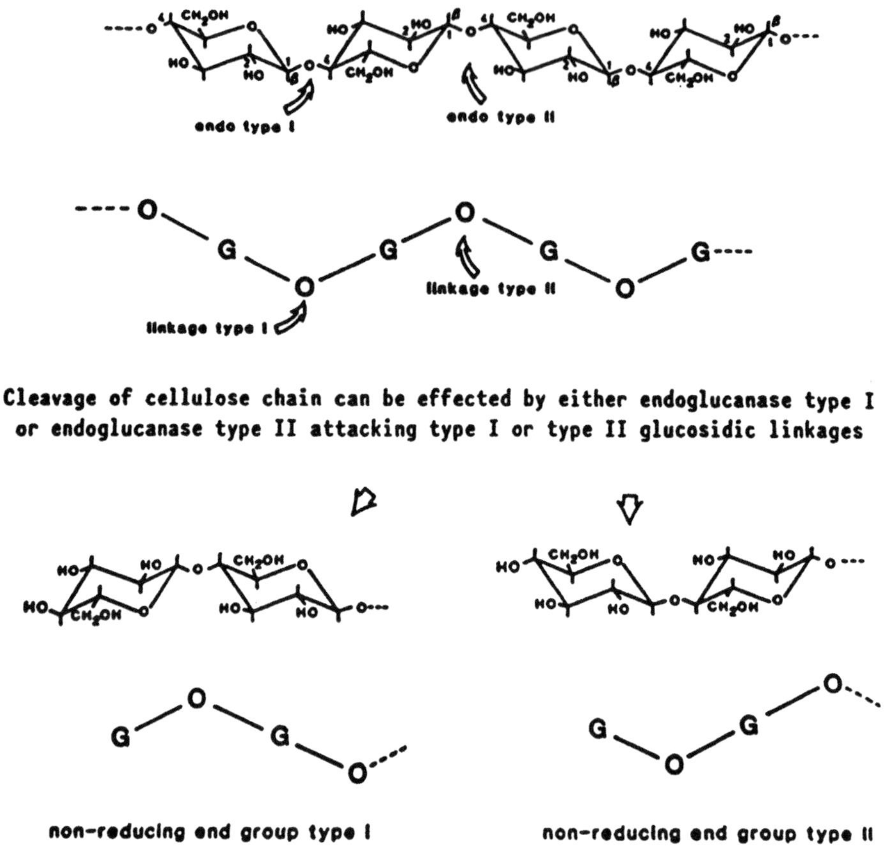
endo type I
endo type II
linkage type I
linkage type II
Cleavage of cellulose chain can be effected by either endoglucanase type I or endoglucanase type II attacking type I or type II glucosidic linkages
non-reducing end group type I
non-reducing end group type II

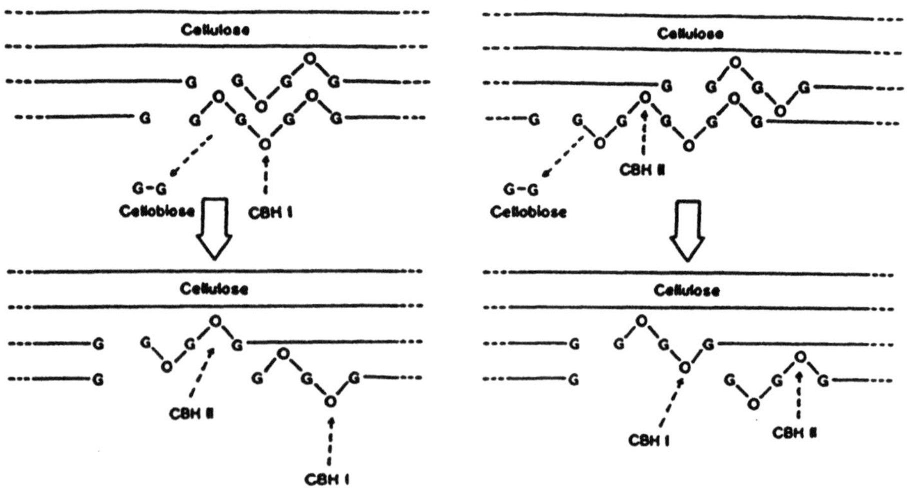
Cellulose
G-G
Cellobiose
CBH I
CBH II

McCrae report that neither cellobiohydrolase I (Wood *et al.*, 1980) nor cellobiohydrolase II (Wood & McCrae, 1986*a*) of *P. pinophilium* were found to produce these short fragments.

Thus, the question of which enzyme initiates the attack on the cellulose crystallite, and how attack is manifested, is the subject of some controversy.

The data of White & Brown (1981) obtained in the electron microscope are consistent with the model presented on the basis of biochemical data to explain synergism between endoglucanases and cellobiohydrolases (Wood & McCrae, 1972). Thus, it is considered that endoglucanase initiates the attack on amorphous regions of the cellulose and the new non-reducing chain ends formed are then hydrolysed by the end-wise acting cellobiohydrolase (Wood *et al.*, 1987). However, as there are several reports where little or no synergism was observed between endoglucanases and cellobiohydrolases (Wood & McCrae, 1978, 1979), it has now been suggested that this model is an oversimplification (Wood, 1981). Wood (1981) has attempted to explain these 'anomalies' in terms of the stereochemistry of the cellulose chains, and he has used a similar explanation for the synergism observed between the two immunologically unrelated cellobiohydrolases reported to exist in the cellulase systems of *T. reesei* and *P. pinophilium* (Wood, 1985). Ryu *et al.* (1984), on the other hand, have demonstrated the importance of differential adsorption of cellobiohydrolase and endoglucanase on the synergetic process.

New data (Wood *et al.*, 1987) obtained using the various purified components of the cellulase system of *P. pinophilium* now present a clearer picture of the mechanism of action. Wood *et al.* suggest that while the model involving sequential action of endoglucanase and cellobiohydrolase is essentially true, it would now appear that little or no

---

Fig. 2. Proposed mechanism of cellulase actions involving sequential action between endoglucanase and cellobiohydrolase. Two types of synergism are shown: that between endoglucanase and cellobiohydrolase and that between two immunologically unrelated cellobiohydrolases. It is assumed that all the enzymes are stereospecific and that each is capable of attacking only one of the two types of glycosidic linkage existing in the cellulose chain. Endoglucanase can attack linkages remote from the chain-end and the linkage hydrolysed will depend on the stereospecificity of the enzyme; cellobiohydrolase attacks the penultimate glycosidic linkage at the non-reducing chain-end and the linkage hydrolysed will depend on the stereospecificity of the enzyme.

Synergism between endoglucanase and cellobiohydrolase will depend on the endoglucanase generating the particular configuration of non-reducing end group which is a substrate for the cellobiohydrolase. Synergism between cellobiohydrolase I and II will depend on the action of one cellobiohydrolase (I) on one particular configuration of non-reducing end group (I) exposing a naturally-occurring chain-end with a configuration which is a substrate for the other cellobiohydrolase (II).

TABLE 3
Synergism between Endoglucanases and Cellobiohydrolases (CBH) of *P. pinophilium* in Solubilizing Cotton Fibre

Enzyme combined with CBH enzymes	*Solubilization (%)*		
	+CBH I	*+CBH II*	*+CBH I + II*
None	4	6	13
E I[a]	6	4	17
E II[a]	5	1	13
E III[a]	5	6	35
E V[a]	3	3	16
E VI[a]	5	11	38
I + II + III + V + VI[b]	10	4	56

[a] Endoglucanase (1 IU CM-cellulase).
[b] Endoglucanases (0·2 IU of each).

hydrolysis is effected unless *both* the immunologically distinct cellobiohydrolases are present along with specific endoglucanases (Table 3). The fact that only one stereospecific endoglucanase is required for extensive hydrolysis, is consistent with the explanation that the particular configuration of the non-reducing end-group generated by the endoglucanase (type I, Fig. 2) in the first stage of the hydrolysis would be attacked by the stereospecific cellobiohydrolase for which this chain-end was a substrate (CBH I, Fig. 2). The successive removal of cellobiose units from this chain-end would then eventually expose on another cellulose chain-end a naturally-occurring chain-end of different configuration which would be a substrate for the other stereospecific cellobiohydrolase (CBHII, Fig. 2). Hydrolysis of the two chains in tandem would explain the synergistic activity and the dependence on the two cellobiohydrolases.

Attack by the other stereospecific endoglucanase (type II, Fig. 2) would generate a non-reducing end of different configuration (type II, Fig. 2), and this would make the hydrolysis more efficient.

BACTERIAL CELLULASES

Most cell-free enzyme preparations from cellulolytic bacteria are found to be unable to solubilize highly ordered native celluloses to an extent that denotes that the crystalline component is being solubilized. It is even true of the *Sporocytophaga myxococcoides* which has been described as the most powerfully cellulolytic bacterium known. It has been suggested that the fact that some bacterial cellulases are highly active only when associated with the bacterial cell, may be explained by their localized high concentration and favourable alignment on the surface of the cell (Hofsten, 1975). This inability of the cell-free enzyme preparations to attack the cellulose may be because the necessary topological conformation is not attained (Hofsten, 1975). In these circumstances it is clear why there is no relationship between the ability of the bacterium to effect the breakdown of cellulose and the ability of the cell-free enzyme to do the same.

Despite these problems, some cell-free culture filtrates of cellulolytic bacteria have been shown recently to be able to solubilize crystalline cellulose. The most notable example is the anaerobe *Clostridium thermocellum* (Johnson *et al.*, 1982; Lamed *et al.*, 1985), but it seems that the cell-free enzymes of *Acetivibrio cellulolyticus* may also be of interest now in this respect (MacKenzie *et al.*, 1987). Remarkably, in both cases the activity to crystalline cellulose was dependent on the presence of Ca^{2+} and a thiol reducing agent such as dithiothreitol.

The extracellular cellulase of *C. thermocellum* has been studied in some detail, and it has been found that the component that can degrade crystalline cellulose is a multicomponent–enzyme complex comprising a cellulose binding factor in addition to multiple endoglucanase components (Bayer *et al.*, 1983; Lamed *et al.*, 1983*a*). Some of the properties of the complex are also compatible with the presence of an exoglucanase (Lamed *et al.*, 1983*b*), but so far no enzyme with this activity has been isolated. The multienzyme complex, which has a molecular weight of approx. 2 million, is termed a 'cellulosome' (Lamed *et al.*, 1983*b*). It has been demonstrated that this 'cellulosome' mediates attachment of the bacterium and the cellulose. Attempts to disrupt the complex have resulted normally in diminished activity on crystalline cellulose.

The cellulolytic bacterium *Ruminococcus albus* (Wood *et al.*, 1982) also forms high molecular weight complexes which contain endoglucanase, but it has not been shown conclusively that they contain exoglucanase activity. In another rumen bacterium, *Bacteroides succinogenes*, vesicular structures associated with cellulase activity have been reported (Groleau & Forsberg, 1981). Cellulose-grown cells of this bacterium released 90% of the endoglucanase into the supernatant, and 50–60% of this was associated with sedimentable membranous fragments.

In all of these bacterial cellulase systems then, it would appear that the existence of the various components in a multicellulase complex is essential for cellulolysis of crystalline cellulose. In the circumstances, it seems possible that this complex must be carefully constructed to present the various enzymes required for cellulolysis in the correct order and in the correct orientation for maximum efficiency.

REFERENCES

Bayer, E.A., Kenig, R. & Lamed, R. (1983). *Journal of Bacteriology*, **156**, 818–27.

Chanzy, H. & Henrissat, B. (1985). *FEBS Letters*, **184**, 285–7.

Chanzy, H., Henrissat, B., Vuong, R. & Schulein, M. (1983). *FEBS Letters*, **153**, 113–17.

Chanzy, H., Henrissat, B. & Vuong, R. (1984). *FEBS Letters*, **172**, 193–7.

Coudray, M.R., Canavascini, G. & Meier, H. (1982). *Biochemical Journal*, **203**, 277–84.

Coughlan, M.P. (1985). *Biotechnology and Genetic Engineering Reviews*, **3**, 39–109.

Dekker, R.F.H. (1980). *Journal of General Microbiology*, **120**, 309–16.

Dunne, C.P. (1982). *Enzyme Engineering*, **6**, 355–5.

Enari, T.-M. & Niku-Paavola, M.-L. (1987). *CRC Critical Reviews in Biotechnology*, **5** (1), 67–87.

Eriksson, K.-E. (1981). In *Trends in the Biology of Fermentations for Fuel and Chemicals*, ed. A. Hollaender & R. Robson. Plenum Press, New York, pp. 19–32.

Fagerstam, L.G. & Pettersson, L.G. (1979). *FEBS Letters*, **98**, 363–7.

Fagerstam, L.G. & Pettersson, L.G. (1980). *FEBS Letters*, **119**, 97–100.

Gong, C.S. & Tsao, G.T. (1979). *Annual Reports on Fermentation Processes*, **3**, 111–39.

Gong, C.S., Ladisch, M.R. & Tsao, G.T. (1979). In *Advances in Chemistry Series*, Vol. 181, ed. R.D. Brown & L. Jurasek. American Chemical Society, Washington, D.C., pp. 261–87.

Gritzali, M. & Brown, R.D. (1979). In *Advances in Chemistry Series*, Vol. 181, ed. R.D. Brown & L. Jurasek. American Chemical Society, Washington, D.C. pp. 237–60.

Groleau, D. & Forsberg, C.W. (1981). *Canadian Journal of Microbiology*, **27** (5), 517–30.

Gunn, E.K. & Brown, R.D. (1977). *Biochemica et Biophysica Acta*. **492**, 225–31.

Henrissat, B., Driguez, H., Viet, C. & Schulein, M. (1985). *Biotechnology*, **3**, 722–6.

Hofsten, B. (1975). In *Symposium on Enzymatic Hydrolysis of Cellulose*, ed. M. Bailey, T.-M. Enari, M. Linko. SITRA, Finland, pp. 281–95.

Johnson, E.A., Sakajah, M., Halliwell, G., Madia, A. & Demain, A.L. (1982). *Applied and Environmental Microbiology*, **43**, 1125–32.

Joliff, G., Beguin, P. Juy, M., Millet, J., Ryter, A., Poljak, R. & Aubert, J.P. (1986). *Biotechnology*, **4**, 896–900.

Kolar, H., Mischak, H., Kammel, W.P. & Kubicek, C.P. (1985). *Journal of General Microbiology*, **131**, 1339–47.

Kyriacou, A., Mackenzie, C.R. & Neufield, R.J. (1987). *Enzyme and Microbial Technology*, **9**, 25–32.

Labudova, I. & Farkas, V. (1983). *Biochemica et Biophysica Acta*, **774**, 135–40.

Lamed, R., Setter, E. & Bayer, E.A. (1983*a*). *Journal of Bacteriology*, **156**, 828–36.

Lamed, R., Setter, E., Kenig, R. & Bayer, E.A. (1983*b*). *Biotechnology and Bioengineering Symposium*, Vol. 13, pp. 163–81.

Lamed, R., Kenig, R. & Setter, E. (1985). *Enzyme and Microbial Technology*, **7**, 37–41.

MacKenzie, C.R., Patel, G.B. & Bilous, D. (1987). *Applied & Environmental Microbiology*, **53**, 304–8.

Moloney, A.P., McCrae, S.I., Wood, T.M. & Coughlan, M.P. (1985). *Biochemical Journal*, **225**, 365–74.

Nakayama, M.J., Tomita, Y., Suzuki, H. & Nisizawa, K. (1976). *Journal of Biochemistry*, **79**, 955–66.

Nummi, M., Niku-Paavola, M.-L. Lappalainen, A., Enari, T.-M. & Raunio, V. (1983). *Biochemical Journal*, **215**, 677–83.

Pettersson, L.G., Fagerstam, L.G., Bhikhabhai, R. & Leandoer, K. (1981). In *The Ekman Days International Symposium on Wood and Pulping Chemistry*, Vol. 3, SPCI, Stockholm, pp. 39–42.

Reese, E.T. (1975). In *Recent Advances in Phytochemistry*, Vol. 11, ed. F.A. Loewry & V.C. Runeckles. Plenum Press, New York, pp. 311–67.

Reese, E.T., Siu, R.G.H. & Levinson, H.S. (1950). *Journal of Bacteriology*, **59**, 485–97.

Ryu, D.D.Y., Kim, C. & Mandels, M. (1984). *Biotechnology and Bioengineering*, **26**, 488–96.

Schwarz, W., Bronnenmeier, K. & Staudenbauer, W.L. (1985). *Biotechnology Letters*, **7**, 859–64.

Shoemaker, S.P. & Brown, R.D. (1978). *Biochimica et Biophysica Acta*, **523**, 147–61.

Shoemaker, S., Schweickart, V., Ladner, M., Gelfand, D., Kwok, S., Myambo, K. & Innis, M. (1983). *Biotechnology*, 691–5.

Sprey, B. & Lambert, C. (1983). *FEMS Microbiology Letters*, **18**, 217–22.

Teeri, T. Salovuori, I. & Knowles, J. (1983). *Biotechnology*, 696–9.

Van Tilbeurgh, H., Tomme, P., Claeyssens, M., Bhikhabhai, R. & Pettersson, G. (1986). *FEBS Letters*, **204**, 223–7.

Vaheri, M.P. (1982). *Journal of Applied Biochemistry*, **4**, 153–8.

White, A.R. & Brown, R.M. (1981). *Proceedings of National Academy of Science (USA)*, **78**, 1047–51.

Wood, T.M. (1981). In *The Ekman Days International Symposium on Wood Pulping Chemistry, Vol. 3*, SPCI, Stockholm, pp. 31–8.

Wood, T.M. (1985). *Biochemical Society Transactions*, **13**, 407–10.

Wood, T.M. (1988). In *Biosynthesis and Biodegradation of Cellulose and Cellulolic Materials*, ed. C.A. Weimer & P.J. Weimer. Marcel Dekker, New York, in press.

Wood, T.M. & McCrae, S.I. (1972). *Biochemical Journal*, **128**, 1183–92.

Wood, T.M. & McCrae, S.I. (1978). *Biochemical Journal*, **171**, 61–72.

Wood, T.M. & McCrae, S.I. (1979*a*). In *Advances in Chemistry Series*, Vol. 181, ed. R.D. Brown & L. Jurasek. American Chemical Society, Washington, D.C., pp. 181–209.

Wood, T.M. & McCrae, S.I. (1979*b*). In *Developments in Food Science 2*, ed. H. Chiba, M. Fujimaki & K. Iwai. Elsevier, Holland, pp. 257–63.

Wood, T.M. & McCrae, S.I. (1986*a*). *Carbohydrate Research*, **148**, 331–44.
Wood, T.M. & McCrae, S.I. (1986*b*). *Biochemical Journal*, **24**, 93–9.
Wood, T.M., McCrae, S.I. & Macfarlane, C.M. (1980). *Biochemical Journal*, **189**, 51–65.
Wood, T.M., Wilson, C.A. & Stewart, C.S. (1982). *Biochemical Journal*, **205**, 129–37.
Wood, T.M., Wilson, C.A., McCrae, S.I. & Joblin, K.N. (1986). *FEMS Microbiology Letters*, **34**, 37–40.
Wood, T.M. McCrae, S.I., Wilson, C.A., Bhat, K.M. & Gow, L.A. (1987). *Proceedings Symposium, Biochemistry & Genetics of Cellulose Degradation*, Paris, Academic Press, New York (in press).

Modes of Microbial Attack on Lignin

Paul Broda

Department of Biochemistry & Applied Molecular Biology, University of Manchester Institute of Science and Technology, PO Box 88, Manchester, M60 1QD, UK

ABSTRACT

The best understood system for lignin biodegradation is that of the white rot fungi, e.g. Phanerochaete chrysosporium. *These organisms produce extracellular ligninases which are peroxidases that have a wide substrate specificity appropriate to the range of inter-monomeric bonds found in lignin. The ligninolytic activity is triggered by depletion of either carbon or nitrogen nutrients. Considerable effort is now being made to elucidate the molecular genetics of this system and its control More recently attention has been attracted to actinomycete bacteria which apparently solubilise lignin in lignocellulose. There are indications that these organisms degrade such material in a fundamentally different manner. Thus, lignin solubilisation occurs during normal growth and one of the products is a polymeric derivative of lignin. Some suggestions are made on imminent developments in the field, and their relevance to the problem of biodeterioration.*

Lignocellulose consists of three major polymers, cellulose, hemicellulose and lignin. In lignin the monomers are three cinnamyl alcohols, and it is assembled by an oxidative polymerisation of free-radical species of these monomers to yield a highly irregular polymer with a number of intermonomeric bond types. The predominant class is the β-aryl ether bond, but others include direct linkages and ethers between aromatic rings. Any system that frees lignin from lignocellulose must be able to attack at least a high proportion of the bonds. The resistant nature is due to this extreme heterogeneity, but there are organisms, e.g. white rot fungi, that can degrade it. By study of modes of lignin degradation by different microbes it is possible to devise more effective methods of protection of lignocellulosic materials from degradation. Hartley *et al.* (1987) is a recent source of information on the topic of lignocellulose degradation.

The degradation of wood takes place in a number of clearly defined stages, each due to a particular group of organisms (Levy, 1987). The initial colonisers are bacteria, some of which destroy the pit membranes, thereby increasing aeration of the wood. This makes possible access by subsequent scavengers, the primary moulds and sapstain fungi. After these, the true wood-rotting fungi, the soft, white and brown rots, are evident. White rots destroy both the holocellulose and lignin, bleaching the wood. In contrast the brown rots only utilise the holocellulose, leaving the lignin as a brown, brittle and friable residue. The white and brown rots are responsible for opening up the structure of the wood cell walls. This then allows the cellulolytic secondary moulds to gain access to and therefore compete for the carbohydrates that are present.

In nature the function of lignin solubilisation may be primarily to allow the degrading organism to obtain access to the carbohydrates since the energetic investment in lignin solubilisation is substantial; thus the white rot basidiomycete *Phanerochaete chrysosporium* will only degrade lignin if it is depleted for carbon and nitrogen sources. *P. chrysosporium* produces apparent ligninases extracellularly. The type enzyme is a peroxidase that has an associated haem group and requires H_2O_2 as a cosubstrate.

The key catalytic reaction of these glycoproteins is the one-electron oxidation of aromatic nuclei, generating short-lived aryl cation radicals (Kirk, 1987). These decompose by a number of reactions, which have been investigated using dimeric model compounds containing bonds represented in lignin. It has been shown that a classical peroxidase-type catalytic cycle is involved. However, degradation of lignin into smaller moieties has not been demonstrated using such isolated ligninases or by the crude extracellular mixture of enzymes that are secreted during the ligninolytic phase. There is therefore the need to establish the full complement of components needed for biological activity.

Recently molecular genetical methods have started to be applied to this problem. It has been shown that control of ligninolytic attack is likely to be at the level of gene expression, mediated through cyclic AMP. A number of ligninase-specifying enzymes have been cloned and sequenced (Tien & Tu, 1987; Broda *et al.*, in press). Furthermore, a cellulase gene has been cloned and sequenced (Broda *et al.*, in press).

This is important since it is evident that in *P. chrysosporium* degradation of lignin and of cellulose are coupled processes, and can now both be studied in this organism.

Recently the discovery that *P. chrysosporium* has extensive restriction fragment length polymorphisms (RFLPs) have been exploited to construct a genetic map (Broda *et al.*, in press). This has allowed mapping of the ligninase-related and cellulase genes. It also forms the basis for a possible breeding programme for strains with enhanced lignin-degrading capability but unable to degrade cellulose. Other work is directed towards developing a transformation system that would allow site directed mutagenesis. The kind of objective envisaged is the construction of strains able to degrade lignin during growth rather than only when severely depleted for a nitrogen source.

As with cellulose degradation, organisms other than fungi may be important in lignin breakdown. It has recently been shown that some actinomycete bacteria can solubilise lignin (Crawford, 1978; McCarthy & Broda, 1984). The mode of attack is different from that by *P. chrysosporium*, since here a high molecular weight soluble product is formed from lignin. In addition, the control of this attack is different, since it can occur even in a rich medium, in normal growth (McCarthy & Broda, 1984).

The solution of basic scientific and technological questions with respect to lignocellulose degradation would yield handsome dividends. The amount of work that has been done, particularly that using molecular genetical methods, is very small compared with the effort expended on health care topics. A number of growing points in this area can be identified:

(i) It has been shown that it is possible to establish rather quickly molecular genetics for *P. chrysosporium*, an organism with considerable technological potential. The next few years will see an elucidation of the components of the overall process of lignocellulose degradation and their control.

(ii) The nature of lignin solubilisation by actinomycete bacteria will be analysed. The recent demonstration of an extracellular protein, distinct from cellulases and xylanases (Mason *et al.*, in preparation), opens up the way to establishing molecular genetics here too.

(iii) Analytical techniques, e.g. NMR spectroscopy (including solid-state NMR) and liquid chromatography will be developed further to assess the nature of the chemical changes in the lignocellulose substrate and the nature of the degradation products.

(iv) As more experience is gathered, projects on the molecular biology of lignin degradation by the termite microflora might be embarked upon.

(v) Cell biology offers at least two tools for the study of biodeterioration of lignocellulose. One is monoclonal antibodies, and the other is DNA probes. The demonstration that at least one lignin-degrader is extensively polymorphic with respect to its DNA opens the way to detailed tracking of infestations deriving from particular sites, since the methods for isolating and handling such DNA are now relatively straightforward (Raeder & Broda, 1985).

(vi) As more is known about the modes of lignin degradation by different organisms, one can seek to design specific inhibitors in a rational manner, analogous to rational drug design.

ACKNOWLEDGEMENTS

Work in the author's laboratory is part of a programme supported jointly by the Agricultural and Food Research Council and British Petroleum's Venture Research Unit.

REFERENCES

Broda, P., Raeder, U., Brown, A., Thompson, W., Schrank, A. & Sims, P. (In press). In *FEMS Symposium on Biochemistry and Genetics of Cellulose Degradation*.

Crawford, D.L. (1978). *Applied Environmental Microbiology*, **35**, 1041–5.

Hartley, B.S., Broda, P.M.A. & Senior, P.J. (1987). *Phil. Trans. Roy. Soc.*, **Series A 321**, 403–568.

Kirk, T.K. (1987). In Hartley *et al.* *(ibid.)*.

Levy, J.F. (1987). In Hartley *et al.* *(ibid.)*.

McCarthy, A.J. & Broda, P. (1984). *Journal of General Microbiology*, **130**, 2905–13.

Raeder, U. & Broda, P. (1985). *Letters in Applied Microbiology*, **1**, 17–20.

Tien, M. & Tu, C-P.D. (1987). *Nature*, **326**, 520–3, and corrigendum **328**, 742.

Mechanistic Aspects of Polyurethane Biodeterioration

D.S. Wales & B.F. Sagar

Shirley Institute, Wilmslow Road, Didsbury, Manchester M20 8RX, UK

ABSTRACT

Polyester-based polyurethanes are particularly susceptible to biodegradation by a wide range of micro-organisms. Many of them produce extracellular esterases, which have been used to study the degradation of model polyurethanes. The polymer containing only urethane links, and the polymers containing ester and urethane links were degraded, indicating that both are susceptible to enzymatic hydrolysis. A polyester-based polyurethane containing an effective biocide was still degraded by the isolated enzyme preparations. Partial characterisation of the crude esterase preparations from four microfungi by gel permeation chromatography has indicated possible differences in the mechanism of polyurethane degradation by different micro-organisms. Analysis of the MW distributions of a polyester-based polyurethane before and after microbial degradation showed that bulk MW, although reduced, is not unduly affected by the action of micro-organisms, but polydispersity of the polymer increases.

INTRODUCTION

Although all organic polymers contain a potential carbon source for microbial growth, most synthetic polymers are resistant to biodegradation (Potts *et al.*, 1973). Polyester-based polyurethanes, however, are particularly susceptible to biodegradation of their polymer structure by a wide range of micro-organisms (Darby & Kaplan, 1968; Kaplan *et al.*, 1968); the mechanism is believed to be mediated through hydrolytic enzymes that attack the polyester component of the polyurethanes (Pathirana & Seal, 1985; Wales & Sagar, 1985). However, since polyether-based polyurethanes also exhibit limited susceptibility to microbial degradation (Kaplan *et al.*, 1968), it is likely that the urethane links of the polyurethane structure are also hydrolysed enzymatically.

MATERIALS AND METHODS

Production of crude esterase preparations

Aspergillus terreus, Chaetomium globosum, Cladosporium cladosporioides, Rhizopus stolonifer and *Trichoderma viride* were grown in batch culture in 5 litres stirred fermenters at 30°C in PL medium, and crude esterase preparations isolated by acetone precipitation (Wales & Sagar, 1985).

Commercial enzymes

The following commercial enzymes were obtained: Palatase (lipase) and Alcalase (protease)-(Novo Industri A/S); Lipase A (John & E. Sturge Ltd); Papain (protease) and Takamine fungal protease (Miles Kali-Chemie GmbH).

Estimation of esterase activity

Esterase activity was estimated by the spectrophotometric method described by Wales & Sagar (1985).

Synthesis of polyurethanes

Four polyurethanes have been studied. PU1 (derived from butane-1,4-diol and isophorone diisocyanate) and PU2 (from poly(ethylene adipate) and toluene diisocyanate) were synthesised as described previously (Wales & Sagar, 1985). PU3 was synthesised similarly from poly(ethylene adipate), butane-1,4-diol and toluene diisocyanate; a biocide, Actifresh PD1116 (British Sanitised), was added to a portion of this polymer at a concentration of 0·5% by weight of solids before the films were cast. PU4 was a commercial polyester-based polyurethane.

Degradation of polyurethanes on agar plates

Dumb-bell shaped specimens (length 12 cm, gauge width, 0·6 cm) of PU3 (with and without biocide) or PU4 were placed on two types of agar plate, one a mineral salts agar (MSA) as described in BS6085 (Anon., 1981), the other a minimal nutrient agar (MNA) containing the same salts as MSA but with added mycological peptone (1% w/v) and malt extract (1% w/v). The plates containing PU3 were inoculated with single spore preparations of *Chaetomium globosum, Cladosporium cladosporioides* or *Trichoderma viride*, prepared as described in BS6085 (Anon., 1981). The PU4 specimens were inoculated with a mixed spore preparation of *Aspergillus amstelodami, A. niger, Aureobasidium pullulans, Penicillium funiculosum, P. ochro-chloron, Rhizopus stolonifer, Stachybotris atra* and *Ulocladium oudomansi*. After incubation at 28 ± 1°C for 4 weeks, the dumb-bells were removed from the plates, washed, and their tensile strength determined (Anon., 1976).

Enzyme treatment of polyurethanes

Dumb-bell shaped specimens of polyurethanes were exposed to enzyme solutions (0·4 units/ml) as described by Wales & Sagar (1985).

Gel permeation chromatography of crude enzyme preparations

The crude enzyme preparations from *A. terreus, C. globosum, R. stolonifer* and *T. viride* were subject to gel permeation chromatography (Wales & Sagar, 1985).

Gel permeation chromatography of PU4

Molecular weight distribution measurements of PU4 before and after biodegradation were made by gel permeation chromatography of the polymer, dissolved in tetrahydrofuran (THF) on a 60 cm mixed pore size column of PL-GEC (styrene based) with a mobile phase (THF) flow rate of 1 ml/min; the eluate was passed through a UVIKON ultraviolet detector (280 mm) and the resulting signal analysed on a BBC 128K microcomputer. Molecular weights quoted from this analysis should not be taken as absolute values; they have been calculated using the Mark-Houwink constants for polystyrene (the calibration standard), since constants for PU4 were not available, but comparisons between undegraded and degraded samples are valid.

RESULTS AND DISCUSSION

Enzymatic degradation of polyurethanes

The esterase elaborated by *A. terreus* and the commercial protease Alcalase were most effective in degrading PU1; significant reductions in tensile strength were also effected by the esterase produced by *C. globosum* and Lipase A (Table 1). The breaking extension of this polymer was not significantly affected, however, by treatment with any of the enzyme preparations except for papain. These results demonstrate the susceptibility of the urethane groups to enzymolysis, since they are the only hydrolysable bonds in PU1. Whether both esterases and proteases are implicated remains uncertain since the protease Alcalase possesses 4-nitrophenyl acetate hydrolase activity, presumably due to a contaminating esterase, and the possibility that both the crude laboratory esterase preparations from *A. terreus* and *C. globosum* contain protease activity has not been eliminated.

Only the esterase from *R. stolonifer* and Takamine fungal protease failed to reduce significantly the tensile strength of PU2 (Table 1). Thus

TABLE 1
Tensile Properties of PU1 and PU2 after Enzyme Treatment

Enzyme	*PU1*			*PU2*		
	Tensile strength (MN/m^2)	*Loss over control (%)*	*Breaking extension (%)*	*Tensile strength (MN/m^2)*	*Loss over control (%)*	*Breaking extension (%)*
Buffer	14·1		5·5	35·8		513
A. terreus esterase	8·8	37·6[a]	8·2	28·8	19·6[b]	447
C. globosum esterase	11·3	19·8[a]	8·5	28·4	20·6[a]	495
R. stolonifer esterase	15·2	+7·8	2·5	29·9	16·5	519
T. viride esterase	13·1	7·1	4·8	32·4	9·5[a]	508
Lipase A	11·4	19·2[a]	2·2	31·8	11·2[c]	511
Palatase	11·9	15·6	6·7	24·3	32·2[c]	446
Alcalase	8·6	39·0[c]	1·7	29·5	17·6[c]	482
Papain	13·1	7·1	0·3	29·6	17·3[c]	489
Takamine fungal protease	14·1	0·0	2·3	35·0	2·2	512

[a]Significantly different from the buffer control at the 95% confidence level.
[b]Significantly different from the buffer control at the 98% confidence level.
[c]Significantly different from the buffer control at the 99% confidence level.

there is ample evidence from these results and the results of Pathirana & Seal (1985) that the ester links in polyester-based polyurethanes are susceptible to enzymatic hydrolysis.

Biodegradation of a polyester-based polyurethane containing a biocide

PU3, without the inclusion of a biocide, was degraded by *T. viride, C. globosum* and *C. cladosporioides* on MSA and MNA plates, and by aseptic treatment with their crude esterase preparations (Table 2). No significant degradation of PU3 containing Actifresh PD1116 was observed on the agar plates but aseptic treatment with the crude esterase preparations isolated from these three microfungi again caused substantial reductions in the tensile strength (Table 2). These results have important implications concerning the degradation of polyurethane coated fabrics in the field; growth of micro-organisms on surface contamination may produce extracellular esterase, which can migrate to the polyurethane surface, and cause degradation despite the polymer being protected with a biocide.

Gel permeation chromatography

Esterase enzymes

Separation of the crude laboratory esterase preparations from *C. globosum, R. stolonifer* and *T. viride* (Fig. 1) produced only single peaks of esterase activity, coinciding with major protein peaks (MWs 135 000, 150 000 and 160 000 daltons respectively). Separation of the crude esterase from *A. terreus*, however, resulted in two peaks of esterase activity with MWs of 104 000 and 160 000 daltons; protein peaks were associated with both of these esterase peaks, and a third protein peak (MW 40 000 daltons) was also observed. These results indicate that the complexity of enzymatic breakdown of polyurethane varies from one micro-organism to another. The two peaks of esterase activity from the separation of the crude enzyme preparation from *A. terreus* may indicate the presence of both endo- and exo-activity. Endo-acting esterase would have a marked effect on the degree of polymerisation by cleaving bonds randomly along the polymer chain, resulting in appreciable loss in tensile strength but relatively little concomitant weight loss, in contrast to the action of an exo-esterase, which removes successive monomer units from the chain ends, resulting in a disproportionate weight loss relative to the effect on tensile strength. Work carried out previously by the authors has indicated that biodeterioration of polyester-polyurethanes by *C. globosum* or *T. viride* produces significant losses in tensile strength with little or no weight loss, whereas *A. terreus* caused appreciable losses in both tensile strength and weight.

TABLE 2
Biodeterioration of PU3 with and without Biocide

Micro-organism	*Treatment*	*No additive*			*Biocide treated*		
		Tensile strength (MN/m^2)	*Loss over control (%)*	*Breaking extension (%)*	*Tensile strength (MN/m^2)*	*Loss over control (%)*	*Breaking extension (%)*
Control (0·5% sodium azide)	MSA	14·8		652	7·3		848
	MNA	12·8		660	7·2		764
	Buffer	12·1		668	6·5		923
T. viride	MSA	12·9	12·8[a]	550	7·4	+1·3	804
	MNA	7·5	41·4[b]	568	7·2	0·0	900
	Esterase	6·9	43·0[b]	800	2·1	67·7[b]	1120
C. globosum	MSA	15·1	+2·0	604	7·0	4·1	816
	MNA	6·5	49·2[b]	520	7·9	+9·7	763
	Esterase	7·5	38·0[b]	650	3·6	44·6[b]	900
C. cladosporioides	MSA	9·5	35·8[b]	532	7·8	+6·8	846
	MNA	11·0	14·1[a]	620	7·9	+9·7	867
	Esterase	7·8	35·5[b]	834	2·3	64·6[b]	1076

[a]Significantly different from the control at the 95% confidence level.
[b]Significantly different from the control at the 99% confidence level.

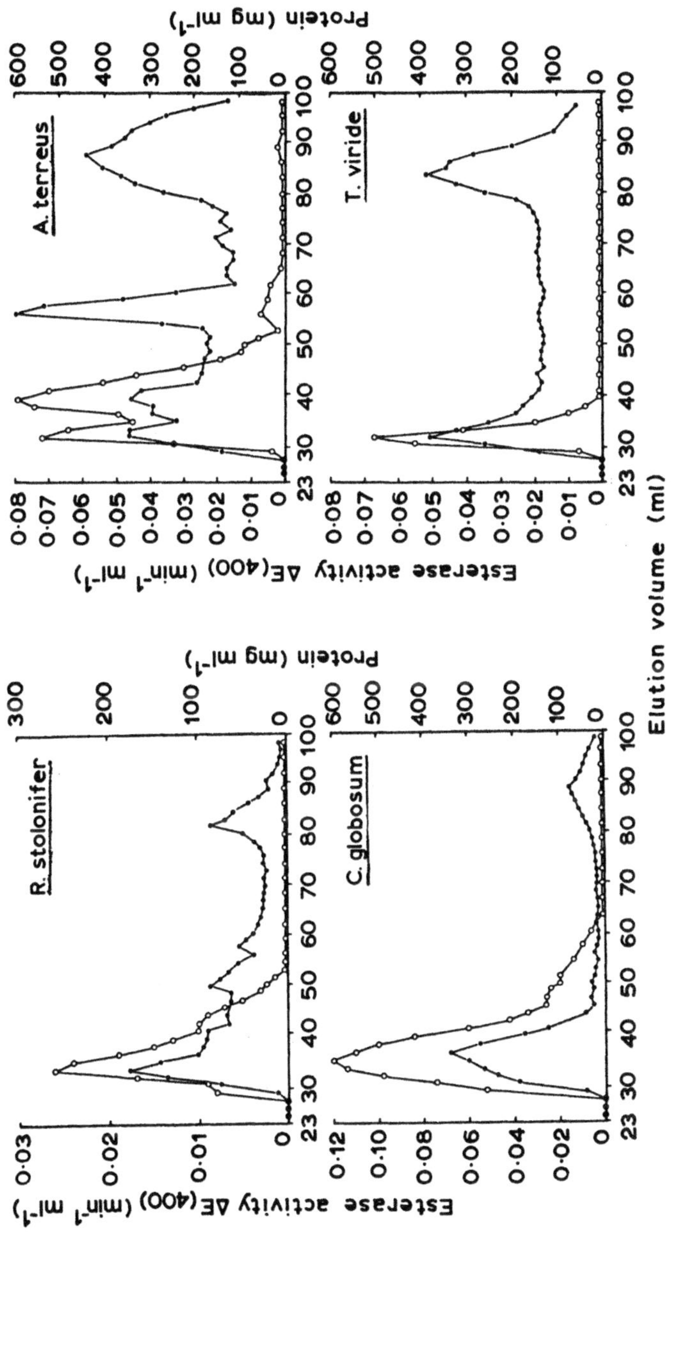

Fig. 1. Esterase activity (○–○) and protein (●–●) after separation of crude enzyme preparations from *Aspergillus terreus*, *Trichoderma viride*, *Rhizopus stolonifer* and *Chaetomium globosum* on a Sephadex G-100 column.

Molecular weight distribution of biodegraded PU4

GPC analysis of the MW distribution of PU4, after exposure to the mixed inocula on MSA and MNA plates, indicated (Table 3) a greater reduction in MW and increase in polydispersity on MSA than on MNA; the polymer on MNA was hardly affected in terms of bulk MW but showed an increase in polydispersity. These results, taken in conjunction with the changes in tensile strength (Table 3), show that microbiological degradation need not unduly affect the bulk of the polymer. In this experiment it seems likely that the extensive reduction in strength results from the initiation of superficial cracks, which propagate through the film under tension. The negligible losses in weight *vis-à-vis* the large reductions in tensile strength of the microbiologically-degraded polymer (Table 3) suggest that much of the degradation is mediated by endo-acting enzymes. The pattern of reduction in molecular weight could also be expected from attack by endo-acting enzymes.

TABLE 3
Molecular Weights of PU4 by GPC and Comparison with Tensile Strength and Weight Loss

Treatment	*Relative MW*	*MW MN*	*Polydispersity MW/MN*	*Relative tensile strength*	*Relative weight loss*
Control	1	1	1·58	1	1
Mixed spore MSA	0·86	0·72	1·89	0·30	1
Mixed spore MNA	0·98	0·90	1·73	0·34	1

REFERENCES

Anon. (1976). *Plastics-tensile strength, elongation and elastic modulus*, British Standard 2782: Part 3. British Standards Institution, London.

Anon. (1981). *Determination of the resistance of textiles to microbiological deterioration*, British Standard 6085. British Standard Institution, London.

Darby, R.T. & Kaplan, A.M. (1968). *Applied Microbiology*, **16**, 900–5.

Kaplan, A.M., Darby, R.T., Greenberger, M. & Rogers, M.R. (1968). *Developments in Industrial Microbiology*, **9**, 201–17.

Pathirana, R.A. & Seal, K.J. (1985). *International Biodeterioration*, **21**, 123–5.

Potts, J.E., Clendinning, R.A., Ackart, W.B. & Niegisch, W.D. (1973). In *Polymers and Ecological Problems*, ed. S. Cruillet. Plenum Press, New York, pp. 61–80.

Wales, D.S. & Sagar B.F. (1985). In *Biodeterioration and Biodegradation of Plastics and Polymers*, ed. K.G. Seal. Biodeterioration Society, C/o C.M.I., Kew, Surrey, UK, pp. 56–69.

METALS

Electrochemical Interpretation of the Role of Microorganisms in Corrosion

Hector A. Videla

Bioelectrochemistry Section, INIFTA, Faculty of Exact Sciences, University of La Plata, C.C.16.Suc.4.1900, La Plata, Argentina

ABSTRACT

The evolution of studies of biologically-induced corrosion over recent years shows an increased understanding of the complexity of the interaction between microorganisms and the metal/solution interface. The corrosion process is an electrochemical phenomenon even though the microorganisms have an important influence. The improvements of electrochemical and microbiological techniques have provided a better understanding of the corrosion process, but it is difficult to justify a single mechanism for biologically-induced corrosion. The anaerobic corrosion of ferrous metals by sulphate-reducing bacteria is analysed.

INTRODUCTION

Special attention has been paid during the last decade to the study of biologically induced corrosion (BIC) of metals and alloys used in industrial installations. There is an increasing recognition of the importance and frequency of the failures of metal structures, caused by the active participation of microorganisms usually present in many natural waters or soils (Kobrin, 1976; Tatnall, 1981; Pope, 1986; Stoecker, 1986).

Although the electrochemical nature of corrosion is unquestionable, microorganisms play an important rôle in the process and the electrochemical techniques and parameters used to interpret a case of biologically induced corrosion (BIC) have to be applied with care. Some concepts valid in the assessment of inorganic corrosion must be revised and redefined when applied to the study of BIC.

This paper analyses the basic aspects of BIC from an electrochemical point of view emphasizing the ways in which microorganisms can interact with the metal/solution interface as well as how biological activity can be evaluated and interpreted through the use of electrochemical techniques.

A RETROSPECTIVE OVERVIEW OF THE ELECTROCHEMICAL INTERPRETATION OF BIC

Several excellent reviews have been published recently on the anaerobic corrosion of iron by sulphate reducing bacteria (SRB) (Cragnolino & Tuovinen, 1984; Hamilton, 1985; Tiller, 1986). A brief retrospective overview is presented here of those studies in which electrochemical techniques were used. The research at the Bioelectrochemistry Section of INIFTA since the end of the last decade on the effects of sulphur compounds (inorganic or biogenic) on localized corrosion of iron and steels is also reviewed.

The cathodic depolarization theory (CDT) envisaged by von Wolzogen Kuhr & van der Vlugt (1934) can be considered as the first attempt to interpret microbial corrosion electrochemically. The three elements that participate in BIC are shown by the authors in a sequence of chemical reactions. The basic idea of the CDT was that the removal of hydrogen from a cathodic area on the iron surface by the hydrogenase enzyme of the bacteria coupled to the reduction of sulphate to sulphide could account for the corrosion of iron in water-logged clay soils. In such environments due to their neutral reaction and low content of oxygen, corrosion of iron would be negligible or absent because neither of the two common reactants which bring about depolarization of the cathodic reaction would be available. The bacterial effect described would cause iron depolarization favouring its dissolution.

The sequence of reactions proposed was:

Overall reaction: $4Fe + SO_4^{2-} + 4H_2O = 3Fe(OH)_2 + FeS + 2OH^-$

Metal

Iron dissolution: $4Fe = 4Fe^{2+} + 8e^-$ (Anode)

Electrochemical cell

Hydrogen reduction: $8H^+ + 8e^- = 8H_{ad.}$ (Cathode)

Electrolyte
Water dissociation: $8H_2O = 8H^+ + 8OH^-$

Microorganism — Cathodic depolarization
Microbial consumption of hydrogen: $SO_4^{2-} + 8H = S^{2-} + 4H_2$

Corrosion products: $Fe^{2+} + S^{2-} = FeS$ $3Fe^{2+} + 6OH^- = 3Fe(OH)_2$

The term depolarization was not used in a strictly electrochemical sense but merely to indicate that there was an undefined change in the electrochemical behaviour of the system (Duquette, 1986).

The rate at which hydrogen ions are reduced to hydrogen depends on several factors such as nature of the metal, hydrogen concentration and temperature.

In modern electrochemical terms the rate determining step for the cathodic reaction in deaerated aqueous solutions is the hydrogen reduction reaction, because of the large activation energy required by the recombination of adsorbed hydrogen atoms to produce molecular hydrogen (Bockris & Reddy, 1970). SRB, by using adsorbed hydrogen in the sulphate reducing reaction, would increase the cathodic reaction rate.

The indirect role of the bacteria in accelerating corrosion, as proposed by the CDT, is very hard to accept in the light of present electrochemical knowledge.

Several years later, electrochemical techniques such as polarization measurements and corrosion potential versus time records were used for the first time to assess the action of SRB on iron corrosion (Horvath & Solti, 1959; Booth & Tiller, 1960). Experimental results were interpreted as a confirmation of the cathodic activity although corrosion rates obtained in the laboratory were substantially lower than those measured in the field. Later Horvath (1960) reported the variation of corrosion potential with time for sterile and inoculated culture solutions. A sudden fall in potential values after SRB inoculation was attributed to the depolarizing effect of the bacteria. After reaching a minimum the potential changed to become stable at a positive value. This effect was assigned to a passivating action of the iron sulphide formed on the electrode surface. In Horvath's paper as well as in other publications produced later by the research group at Teddington (Booth *et al.*, 1967), electrochemical experiments were used to follow a time dependent polarization behaviour of the steel in the test solutions. This type of assessment was used later by several authors in BIC studies under different experimental conditions (Costello, 1974) and using different types of microorganisms (Westlake *et al.*, 1986; Edyvean & Terry, 1983). During the 1970s Costello (1974) assumed that bacterial action on corrosion was indirect but was due to the effect of hydrogen sulphide which would be used as cathodic reactant. According to the electrochemical knowledge on the effect of sulphur compounds on localized corrosion, hydrogen sulphide is able to accelerate the rate of both cathodic and anodic reactions (Iofa, 1980). The complexity of biological environments involved with SRB activity make it very difficult to assess a microbial effect by means of electrochemical methods when the chemical composition and pH of the medium is continuously varying due to microbial metabolism. One important fact generally not taken into account in the formulation of a mechanism to explain the anaerobic corrosion of iron is that in practical situations metal surfaces are rarely free from deposits of several types (oxides, sulphides, hydroxides or even biofilms). Under these circumstances, the explanation must include the breakdown of passivity by metabolic products of aggressive characteristics poured into the medium by microbial activity. In addition, other aggressive anions like the widely distributed chlorides are frequently already present in many natural and industrial environments.

With the aim of assessing the effect of several sulphur compounds produced either as final or intermediate products of SRB metabolism on the passivity breakdown of mild steel, a series of studies were made in the laboratory in well defined experimental conditions without the presence of bacteria. These results were later compared with those obtained in similar experimental conditions but in the presence of SRB or its metabolites. Localized corrosion measurements such as E_p values, repassivation potentials, induction times, kinetics of pit growth, pit density and morphology of pits were employed. The first advance in this direction was made using a marine strain of SRB to assess the electrochemical behaviour of 1020 SAE carbon steel in artificial sea water or 0·5 M sodium chloride solutions at different concentrations of sulphides (Salvarezza & Videla, 1980). E_p values and corrosion potentials changed through more active values in the presence of both types of sulphides incorporated in the medium in similar concentrations. In addition, the breakdown of passivity was accomplished more easily in deaerated solutions where lower levels of sulphides were needed. To give fundamental support to the research in the BIC field we studied the electrochemical behaviour of mild steel in alkaline and neutral buffered solutions of phosphate-borate by means of potentiostatic and potentiodynamic techniques. In this electrolyte the electrodissolution and passivation processes of mild steel have been well documented in the literature (Ogura & Takesue, 1980) in a pH range similar to that usually found for SRB growth.

There is a general agreement on the nature of the initial corrosion product formed either in weak acid, neutral or alkaline solutions containing sulphides. This product corresponds to mackinawite, which forms a poor protective layer on the metal surface. In relation to SRB action it has been reported that the corrosion rate of mild steel in the presence of iron sulphides increases with iron sulphide concentration (King & Wakerley, 1973). The iron sulphide interrelationships in aqueous solutions and their biogenic counterparts have also been described (Rickard, 1969). According to our results obtained in alkaline and neutral phosphate-borate solutions the anodic behaviour was interpreted in terms of a competitive adsorption of H_2O,OH^- and HS^-. The various processes like mackinawite formation, development of pitting or passivation of the metal surface by an oxide layer depends on the HS^-/OH^- concentration ratio in the solution (Salvarezza *et al.*, 1982, 1983). The metal surface at low potentials is covered by an iron hydroxide layer. The latter becomes a complex film at high potentials approaching the $Fe_3O_4/Fe_2O_3 \times H_2O$ structure (Salvarezza *et al.*, 1982).

The effects of sulphides can also explain pit morphology observed in the presence of SRB. In this case the distinctive corrosion pattern on carbon steel is the presence of pits as disc-shaped concentration rings which are similar to those obtained in the presence of sulphides due to the fast radial pit growth.

Intermediates of microbial sulphate reduction involve compounds which are metastable and susceptible to microbiological oxidation (Cragnolino & Tuovinen, 1984). Both thiosulphate and tetrathionate can be detected transiently in culture solutions. To assess the effect of several of the metabolic intermediates of sulphate reduction by SRB, a comparative pitting corrosion study of mild steel in different alkaline solutions containing salts with sulphur-containing anions was made (Vasquez Moll *et al.*, 1984).

In the presence of either thiocyanate or thiosulphate anions, iron pitting takes place through sulphide formation, which reacts at the electrode surface yielding a poorly protective mackinawite film. The general effect of these anions is to increase the dissolution current in the active direction hindering passive film formation. Confirming these effects of sulphur compounds other than sulphides on the corrosion behaviour of iron, a mechanism of microbial pitting in stainless steel has been recently published (Newman *et al.*, 1986). The authors found that oxidation of mixed sulphate–sulphide solutions readily produces an environment that is able to pit AISI 304 stainless steel, even in the absence of chloride ions. Thiosulphate seems to be a more effective pitting agent than sulphides at low concentrations.

A potentiating effect of chloride anions on the passivity breakdown of mild steel by sulphide anions was reported previously (Acosta *et al.*, 1984).

A typical experiment to assess the effect of sulphur anions on the passive behaviour of mild steel is depicted in Fig. 1. A passive layer in the absence of sulphur species is formed at a constant potential and then the sulphur containing salt is added to the electrolyte. When mild steel electrodes are anodized at −0·64 V (SCE) only a $Fe(OH)_2$ film was formed. The addition of several concentrations of Na_2S produces a marked current increase depending on sulphide concentration and leading to a total iron sulphide coverage of the metal surface. An increase in Na_2S concentration and decrease in the anodizing potential result in a decrease of both current and induction time for pitting. As during the stationary phase of growth SRB produce sulphide concentrations ranging from 0·001 M to 0·01 M (Videla, 1986*a*), to assess the effect of biogenic sulphides in comparison with that of inorganic sulphides, mild steel electrodes were first passivated at −0·64 V in artificial sea water. This potential corresponds to a passive behaviour of the metal in this medium. There was a continuous current decrease with time (Fig. 2). The

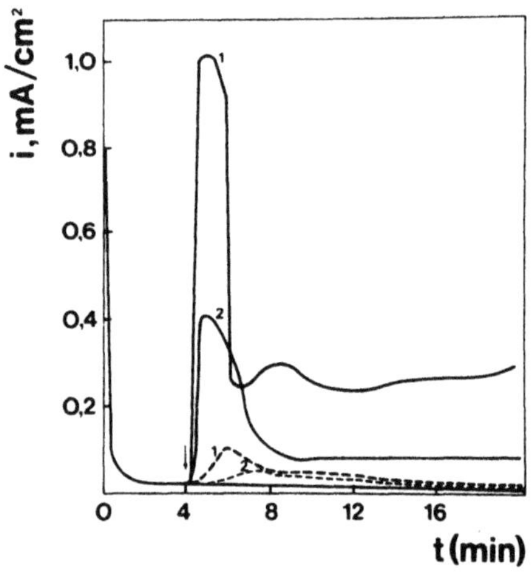

Fig. 1. Current-transients in phosphate–borate solutions at two different potentials. The arrow indicates the instant of Na_2S addition to produce different concentrations: (a) buffer solution without sulphide. Electrode potentiostatized at −0·64 V (full lines): (1) buffer + $1{\cdot}6 \times 10^{-2}$ M Na_2S; (2) buffer + 8×10^{-3} M Na_2S. (b) Electrode potentiostatized at 0·20 V (dashed lines): (1) buffer + $1{\cdot}6 \times 10^{-2}$ M Na_2S; (2) buffer + $0{\cdot}8 \times 10^{-3}$ M Na_2S (Videla, 1986*a*). (With permission of the National Association of Corrosion Engineers.)

addition of 10 ml of SRB culture to give a final sulphide concentration of $1{\cdot}2 \times 10^{-3}$ M produced a sudden current increase. Taking into account the present state of understanding of BIC of iron and steels by SRB, a bioelectrochemical interpretation of the passivity breakdown process can be summarized in this way.

(a) The metabolic activity of SRB brings to the metal/solution interface several sulphur containing anions of corrosive characteristics either as final metabolic products (sulphides, bisulphides or hydrogen sulphides) or intermediate metabolic compounds (thiosulphates, polythionates).

(b) These compounds are corrosive to mild steel mainly through the transformation to sulphide anions. The characteristics and the intensity of sulphide action on mild steel is closely related to the nature of the passive film already present on the metal. The formation of a poorly adherent film of mackinawite facilitates the anodic breakdown of passivity that could be considered as the first stage in the corrosion process. Cathodic depolarization, attributed to SRB or to iron sulphide films in the literature, develops later than passivity breakdown, while the corrosion process is in progress. In addition, the action of sulphides in the corrosion process can be enhanced by other aggressive anions such as chlorides, already present in the environment (Salvarezza & Videla, 1984; Videla, 1986*b*). The effect of SRB can be evaluated experimentally through the assessment of changes in the passive film, induction times for pitting, E_p variation and type of corrosion products.

(c) Physicochemical characteristics of the liquid environment (pH, ionic composition, oxygen levels) modify the SRB effects which could eventually change from corrosive to passivating.

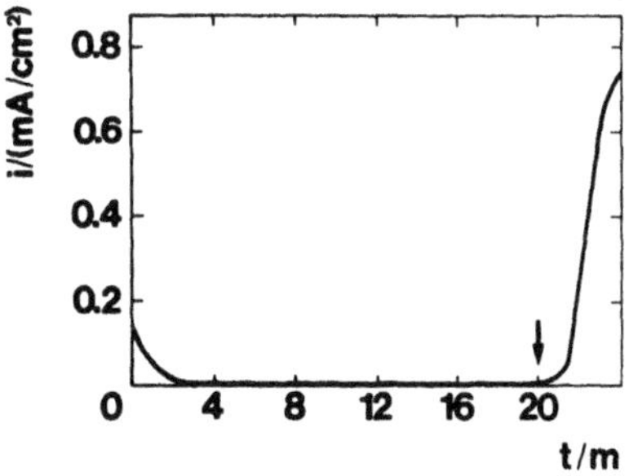

Fig. 2. Current transient at constant potential (−0·62 V) in artificial sea water. The arrow indicates the addition of 10 ml of SRB culture. Final sulphide concentration: $1{\cdot}2 \times 10^{-3}$ M (Videla, 1986*a*). (With permission of the National Association of Corrosion Engineers.)

(d) The action of biogenic sulphides produced by SRB can also be enhanced by the activity of other bacteria present simultaneously at the metal solution/interface.

Dramatic differences occurred in corrosion of mild steel coupons as measured by mean weight loss, within three weeks, when SRB colonized steel in association with *Vibrio anguillarum* despite the fact that this microorganism has no corrosive effects in pure culture (Gaylarde & Johnston, 1982). The latter effect was recently analysed using electrochemical polarization techniques to evaluate quantitatively the action of both bacteria on corrosion and documented graphically by SEM (Gaylarde & Videla, 1987). These results show that a marine *Vibrio* is able to corrode mild steel by itself in Postgate C medium (supplemented with 3% NaCl to provide marine chloride concentration) but the pitting attack is more intense in SRB cultures and even more in the presence of *Vibrio* plus SRB. Microbial colonies were detected on the metal surface after 24 h of incubating mild steel electrodes in a laboratory batch fermenter inoculated with *Vibrio algynoliticus* (Fig. 3). After removal of the colonies the metal underneath showed localized areas of attack where generalized corrosion and micropitting were visible (Fig. 4). These results suggest that joint microbial action on the metal/solution interface could be the depassivation of the surface by one microbial species facilitating the action of aggressive anions present in the medium or

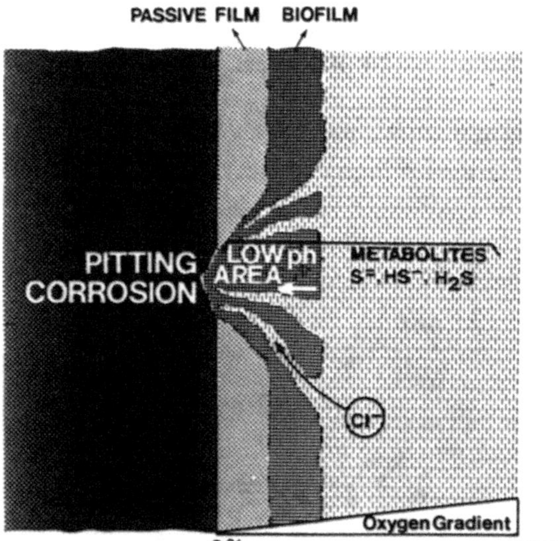

Fig. 3. Idealized scheme of the passivity breakdown of mild steel according to a bioelectrochemical interpretation of the process.

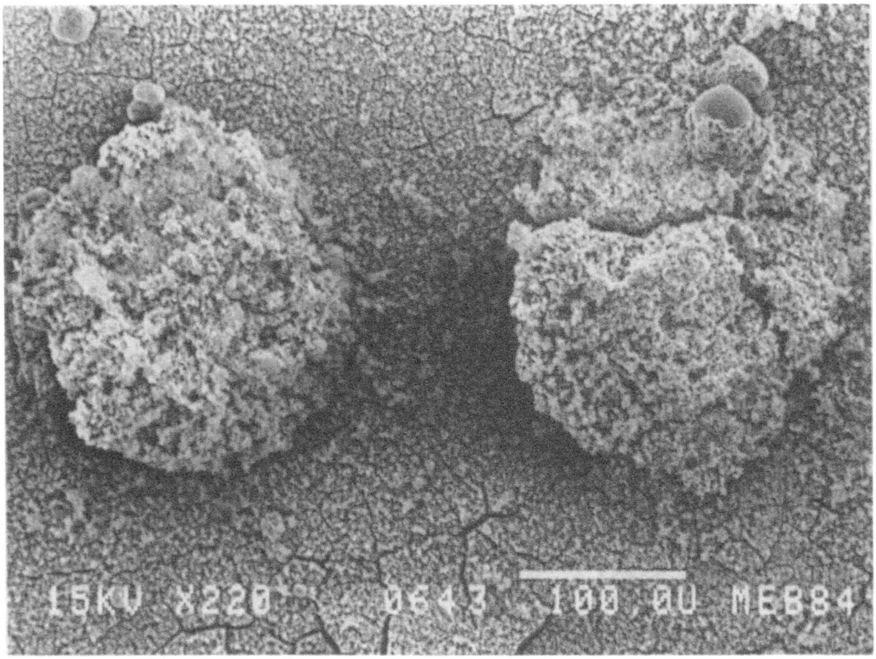

Fig. 4. Scanning electron micrograph showing colonies of cells of *V. algynoliticus* on the metal surface. (Magnification ×220.)

produced metabolically by other microorganisms (SRB). This suggestion has also been made by Westlake *et al.* (1986).

According to a simplified model of the metal/solution interface the bioelectrochemical interpretation of mild steel passivity breakdown by SRB follows the scheme shown in Fig. 5.

FORTHCOMING RESEARCH

The present state of the art in BIC interpretation demands recognition of the importance of fouling of metal surfaces in natural or industrial environments and the formation of biofilms. It is necessary to develop methods to characterize the biofilm/metal interface and its interaction with the liquid environment. This research on both sides of the interface (metal and solution) will need an adequate evaluation of the corrosion process, the biofilm population and its metabolic activity. The significance of biofilm modification of the transport processes at the interface, oxygen levels, biofilm patchiness, microbial consortia and pH gradients are some of the aspects that will need careful consideration. In spite of the literature available on BIC and biofouling, information on the

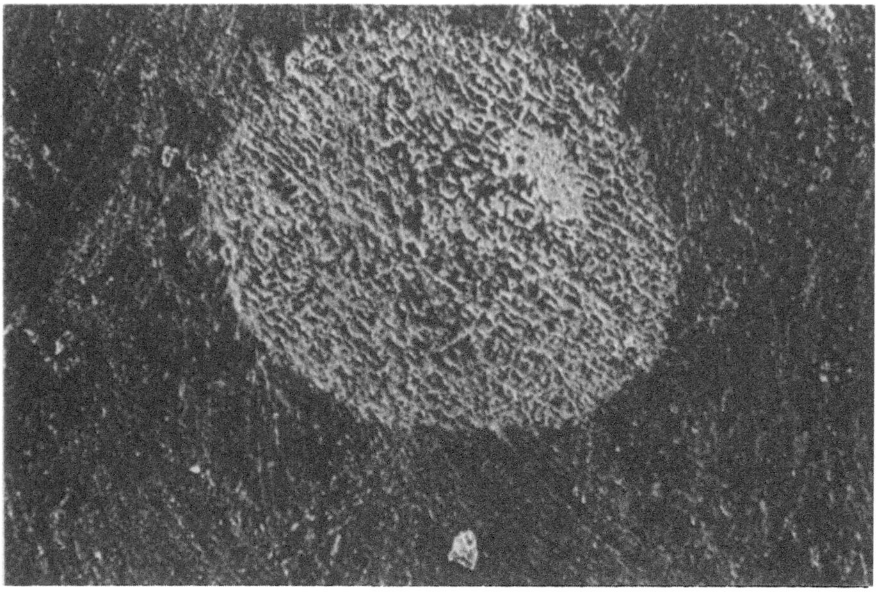

Fig. 5. Scanning electron micrograph showing mild steel surface beneath a microbial colony after cleaning. Generalized attack and micropitting are found under the *Vibrio* colony. Occasional larger pits seen in these areas are due to anodic polarization above the E_p. (Magnification ×240.)

relation between fouling and corrosion is by no means complete. Previous attempts to correlate fouling with corrosion of the metal surface have generally been made under conditions where a considerable biomass had already accumulated.

Some recent attempts to interpret electrochemically the effect of biofouling on corrosion have been made on stainless steels exposed to sea water using corrosion potential versus time measurements (Scotto *et al.*, 1985; Johnsen & Bardal, 1986). The corrosion potential of AISI type 316 steel and the rate of oxygen reduction, as measured by the cathodic polarization curve, have been determined with and without the formation of a natural marine bacterial film on the metal surface (Dexter & Yao, 1987). Data obtained show that the amount of scatter in the corrosion potential increases dramatically in the presence of the biofilm and that the spotty nature of the film encourages pit initiation. It is assumed that bacterial films on stainless steel can stimulate localized corrosion in natural sea water mainly in two ways: by initiating corrosion through the formation of oxygen concentration cells and by increasing the rate of the cathodic reduction reaction. These two effects combine to make the corrosion potential of type 316 stainless steel highly variable within 7 days of exposure to natural sea water (Fig. 6). When localized

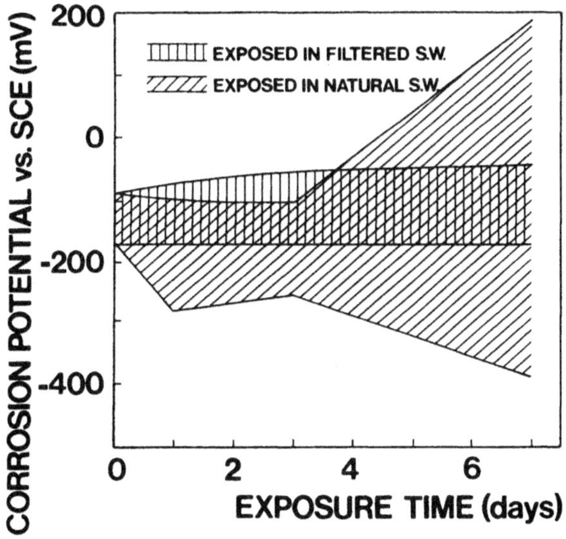

Fig. 6. Open circiut corrosion potentials of type 316 stainless steel during biofilm forming exposures in natural and filtered sea waters (Dexter & Yao, 1987). (With permission of the National Association of Corrosion Engineers.)

corrosion is initiated the potential becomes active, but on electrodes that remain passive the biofilm causes the potential to shift in the noble direction, probably through a change in the kinetics of the oxygen reaction.

A comparative study of the effects of microfouling on the corrosion of several metals in contact with polluted sea water (Videla *et al.*, 1987) revealed that the nature of the metal surface determines microbial attachment. Less easily corrodible metals present an ideal substrate for microbial colonization.

In addition, the increase in pollutant content of the water shortens the period of time needed for the formation of a biofilm on the metal surface. The results of electrochemical measurements obtained with artificial sea water, sodium chloride solutions or membrane filtered sea water were very different from those obtained with natural sea water due to the absence of biofouling activity.

CONCLUSION

The microbial colonization of metals and alloys is able to change the corrosion behaviour and resistance to breakdown drastically, by changing the type and concentration of ions, pH and oxygen levels as well as introducing diffusional barriers to the transport of chemical species at the metal/solution interface. These varied effects result in changes in passive film composition, type of corrosion products and electrochemical parameters used in the assessment of BIC. Electrochemical measurements can be made *in situ* under natural conditions or in the laboratory, but in this case, any alteration of the biological activity must be avoided. A new concept of the electrical double layer, at the biologically fouled metal/solution interface, is needed. The use of electrochemical methods, coupled with an adequate characterization of the type of microorganisms and their interrelationships as well as an evaluation of the effect of metabolic activity and role of biofilms on the corrosion behaviour of each metal, will be necessary to reach a better understanding of BIC.

ACKNOWLEDGEMENTS

The author wishes to thank the Organization of the American States (OAS) and Aquatec Quimica S.A., Sao Paulo, Brazil for providing travel funds.

REFERENCES

Acosta, C.A., Salvarezza, R.C., Videla, H.A. & Arvia, A.J. (1984). In *Passivity of Metals and Semiconductors*, ed. M. Froment. Elsevier, Amsterdam, pp. 387–92.

Bockris, J.O'M. & Reddy, A.K.N. (1970). *Modern Electrochemistry, Vol. 2.* Macdonalds, London, 1970.

Booth, G.H. & Tiller, A.K. (1960). *Transactions of the Faraday Society*, **56**, 1689–96.

Booth, G.H., Cooper, A.W. & Cooper, P.M. (1967). *Chemistry and Industry*, **49**, 2084–5.

Costello, J.A. (1974). *South African Journal of Science*, **70**, 202–4.

Cragnolino, G. & Tuovinen, O.H. (1984). *International Biodeterioration*, **20**, 9–26.

Dexter, S.C. & Yao, G.Y. (1987). *Corrosion 87*, paper no. 377. NACE, Houston.

Duquette, D.J. (1986). In *Proceedings of the Argentine-USA Workshop on Biodeterioration*, (CONICET-NSF), published by Aquatec Quimica S.A., Sao Paulo, Brazil, pp. 15–32.

Edyvean, R.G.J. & Terry, L.A. (1983). *International Biodeterioration Bulletin*, **19**, 1–11.

Gaylarde, C.C. & Johnston, J.M. (1982). *International Biodeterioration Bulletin*, **18**, 111–16.

Gaylarde, C.C. & Videla, H.A. (1987). *International Biodeterioration*, **23**, 91–104.
Hamilton, W.A. (1985). *Annual Reviews of Microbiology*, **39**, 195–217.
Horvath, J. (1960). *Acta Chimica Hungara*, **25**, 65–78.
Horvarth, J. & Solti, M. (1959). *Werkstoffe und Korrosion*, **10**, 624–30.
Iofa, Z.A. (1980). *Protection of Metals*, **16**, 220–5 (English translation).
Johnsen, R. & Bardal, E. (1986). *Corrosion 86*, paper no. 227. NACE, Houston.
King, R.A. & Wakerley, D.S. (1973). *British Corrosion Journal*, **8**, 41–5.
Kobrin, G. (1976). *Materials Performance*, **15**, 38–43.
Newman, R.C., Wong, W.P. & Garner, A. (1986). *Corrosion*, **42**, 489–91.
Ogura, K. & Takesue, N. (1980). *Corrosion*, **36**, 487–92.
Pope, D.H. (1986). In *Proceedings of the Argentine–USA Workshop on Biodeterioration (CONICET-NSF)*, published by Aquatec Quimica S.A., Sao Paulo, Brazil, pp. 105–18.
Rickard, D.T. (1969). *Contribution to Geology*, **20**, 67–72.
Salvarezza, R.C. & Videla. H.A. (1980). *Corrosion*, **36**, 550–4.
Salvarezza, R.C. & Videla, H.A. (1984). *Proceedings of the 6th International Congress on Marine Corrosion and Fouling*, Athens, pp. 429–41.
Salvarezza, R.C., Videla, H.A. & Arvia, A.J. (1982). *Corrosion Science*, **22**, 815–29.
Salvarezza, R.C. & Videla, H.A. & Arvia, A.J. (1983). *Corrosion Science*, **23**, 717–32.
Scotto, V., DiCintio, R. & Marcenaro, G. (1985). *Corrosion Science*, **25**, 185–94.
Stoecker, J. (1986). In *Biologically Induced Corrosion*, NACE-8 *International Corrosion Conference Series*, NACE, Houston, Texas, pp. 324–9.
Tatnall, R.E. (1981). *Materials Performance*, **20**, 41–8.
Tiller, A.K. (1986). In *Biologically Induced Corrosion*, NACE-8, *International Corrosion Conference Series*, NACE, Houston, Texas, pp. 8–29.
Vasquez Moll, D.V., Salvarezza, R.C., Videla, H.A. & Arvia, A.J. (1984). *Corrosion Science*, **24**, 751–67.
Videla, H.A. (1986*a*). In *Biologically Induced Corrosion*, NACE-8, *International Conference Series*, NACE, Houston, Texas, pp. 162–70.
Videla, H.A. (1986*b*). *Proceedings of the Argentine–USA Workshop on Biodeterioration* (CONICET-NSF), published by Aquatic Quimica S.A., Sao Paulo, Brazil, pp. 43–63.
Videla, H.A., de Mele, M.F.L. & Brankevich, G. (1987). *Corrosion 87*, paper no. 365, NACE, Houston.
von Wolzogen Kuhr, G.A.H. & van der Vlugt, L.R. (1934). In *Corrosion*, **17**, 293–9 (1961).
Westlake, D.W.S., Semple, K.M. & Obuekwe, C.O. (1986). In *Biologically Induced Corrosion*, NACE-8, ed. S.C. Dexter, NACE, Houston, Texas.

The Antimicrobial Effect of Pigments in Corrosion Protective Paints

Maria Stranger-Johannessen

Center for Industrial Research, N-0314 Oslo 3, Norway

ABSTRACT

Microorganisms have been shown to deteriorate anti-corrosive paints, causing loss of adhesion and blistering of the film. The antimicrobial effect of zinc chromate, zinc borate and zinc phosphate was investigated in order to demonstrate that good corrosion inhibiting properties may be due to the prevention of microbiological attack on the coatings. An agar diffusion test of the pure pigments and incubation of painted steel plates in humidity cabinets together with aggressive fungi showed that zinc chromate and zinc borate inhibited the growth of fungi and prevented the deterioration of the coatings. Zinc phosphate was less effective against microorganisms and paints containing this pigment were heavily deteriorated in the test. The antimicrobial properties of the pigments correspond with the corrosion protective properties experienced in practical service and testing.

INTRODUCTION

Microorganisms have been found to be the cause for disbonding and blistering of corrosion protective coatings in various service situations (Stranger-Johannessen, 1980, 1986*b*). The contribution of microorganisms to these failures is still not generally recognized.

Some publications report the presence of large numbers of microorganisms in areas of coating deterioration. The authors presuppose that in order to cause loss of adhesion, they must act between the metal surface and the coating. Efforts are described to study how microorganisms may have penetrated the coating layer, without giving conclusive results (e.g. Boggs, 1969). Kemp *et al.* (1966), report that inner coatings of fuel storage containers performed well when they contained organic biocides.

This supports the postulate that the coating material itself is attacked by microorganisms from outside and deteriorated in such a way that the physical reactions leading to disbonding and blistering can take place (Stranger-Johannessen, 1980, 1986*b*). An antimicrobial effect of the material should prohibit this type of deterioration.

Certain corrosion inhibiting pigments which are used in anti-corrosive paints are known to effect the performance of the coatings in corrosion tests and in practical use. It is noteworthy that the effective corrosion inhibitors are often presumptive antimicrobials.

The aim of this work was to see if the corrosion inhibiting properties could be ascribed to the antimicrobial effect of the pigments rather than to physical and chemical processes.

METHODS

Incubation

Painted steel plates, about 200 × 70 mm, were inoculated with test organisms and incubated in humidity cabinets as described by Bravery *et al.* (1983). The cabinets were at room temperature and the water on the bottom was heated to 2–5°C above room temperature. The water was maintained in a sterile condition by the addition of quaternary ammonium compounds.

The test plates were sterilised by dipping in 0·1% hydrogen peroxide prior to inoculation.

Test organisms

No information was available on microbial strains suitable for testing the loss of adhesion and blistering of protective coatings.

The deteriogenic action of bacteria and fungi on coatings in service could be reproduced by laboratory testing with the isolated strains (Stranger-Johannessen, 1980, 1986*b*). Those used were two *Bacillus* spp., *Scopulariopsis brevicaulis* and a *Botrytis* sp. Several other fungal strains have been isolated from deteriorated coatings. Amongst them, *Cladosporium herbarum* was particularly aggressive.

The selection of proper test organisms is important. Fungi capable of good surface growth but without deteriogenic ability may overgrow and suppress that of the deteriogenic fungi. Such growth may even protect the coatings from microbial attack.

To determine the deteriogenic activity, 25 fungi and 15 bacteria were screened on 13 representative paint systems. Most of the selected cultures had previously been isolated from deteriorated paints or similar materials and from paint testing chambers. The remainder were selected on grounds of other promising characteristics.

Small painted steel strips (5 × 50 mm) were inoculated with suspensions of the single cultures and incubated on mineral salts agar in petri dishes. Based on the appearance of growth and detectable deterioration, nine were selected for further tests. These were:

Cladosporium herbarum
Alternaria alternata
Scopulariopsis brevicaulis
Cladosporium resinae
Botrytis sp.
and four unidentified fungal strains, isolated from deteriorated coatings.

Suitability of test organisms for particular paints

The deteriogenic effect of the test organisms on particular paint formulations was investigated with a spot test.

Suspensions of the test organism were mixed with molten agar media (malt extract or nutrient agar) and drops of the mixture placed on the painted test plates. Sterile media were used as controls. The plates were incubated in a humidity cabinet. Growth of the fungi and deterioration of the paint film were determined after 4 and 8 weeks.

Test organisms with pronounced deteriogenic effects were used for further testing.

Spore suspensions for inoculation

A mixed spore suspension of the selected test fungi was sprayed onto the painted plates before incubation in the humidity chamber. A corresponding sterile solution was sprayed onto a duplicate set and incubated in another humidity cabinet under sterile conditions.

Experiments have shown that the amount of nutrients contained in the spore suspension has an influence on the speed of attack. When 10% nutrient broth and 10% malt extract broth were added to physiological salt solution, growth and deterioration were observed after 8 weeks. A tenth of this amount produced no visible growth or deterioration but after 16 weeks very small, tightly bound blisters and fungal hyphae could be observed with a stereo-microscope. In the absence of nutrients no effects were observed.

The deteriogenic fungi obviously need external nutrients to support their growth. In the practical situation attack mainly occurs in the presence of nutrients, either provided by the medium in contact with the painted surface or by soiling.

RESULTS AND DISCUSSION

The effect of zinc chromate, zinc borate and zinc phosphate on the stability of paints was tested, both under sterile and microbiologically aggressive conditions.

Initially the antimicrobial effect of the pure pigments was investigated in an agar diffusion test, against four strains of representative bacteria (*Escherichia coli, Staphylococcus aureus, Pseudomonas* sp. and *Bacillus subtilis*), and the nine test fungi. The pigments were moistened and placed in holes cut in the centre of inoculated agar medium in petri dishes.

Zinc chromate gave the widest zones of inhibition and zinc borate was effective. However, zinc phosphate produced a very weak effect and most of the test organisms were uninhibited.

The test plates of steel were received ready coated with both solvent and water based alkyd paints containing the three respective pigments.

In the spot test all the fungi grew well on the paints with zinc phosphate pigments and produced severe blistering, in most cases, after 3–4 weeks. Much less growth was observed on the zinc borate paints and only two of the test fungi grew slightly on the zinc chromate paints, even after 18 weeks incubation.

On the basis of the results with zinc phosphate containing paints, all nine fungi were used in the mixed spore suspension. The suspension was prepared with 10% nutrient broth and 10% malt extract broth in physiological salt solution.

The plates were sprayed with the spore suspension and incubated in the humidity cabinet for 8 weeks. Sterile controls were tested simultaneously. The results are shown in Table 1.

Fungal growth on the surface and severe blistering of the paint films containing zinc phosphate was observed after 8 weeks. The paints containing zinc borate were much less, or not at all, deteriorated. No growth or deterioration could be seen on the zinc chromate paints.

The solvent based alkyd paints appear more resistant than the water based paints. The latter contain organic emulsifiers and stabilizers as additives, and may account for the better growth of the test fungi.

No change in the coatings was observed on the steel plates which had been incubated under sterile conditions.

It has been shown that loss of adhesion and blistering of the anti-corrosive paints was caused by fungal growth. Paints with effective antimicrobial pigments showed no such failures, even in the presence of the deteriogenic fungi.

The results correspond with practical experience of corrosion inhibiting pigments in paints and with test results reported by Kossmann (1985) for example. In efforts to replace zinc chromate, which is an excellent corrosion inhibitor, he found zinc borate gave better results than zinc phosphate in salt-spray testing and natural weathering. Zinc chromate and zinc borate were outstanding amongst the corrosion inhibitors tested.

The work reported here confirms that the antimicrobial effect of the corrosion inhibiting pigments plays an important role in determining the durability of the coatings by prohibiting microbiological attack. Paint experts are often unaware of this phenomenon, and are puzzled by failures which cannot be explained by chemical and physical phenomena. A broader recognition of microbial deterioration would be beneficial in improving the durability of anti-corrosive protective coatings.

TABLE 1
Fungal Growth (Scale – to +++) and Blistering (Scale 0 to 3) on Painted Steel Plates After Incubation for 8 Weeks in Humidity Cabinets

Pigment	*Alkyd paint in*			
	White spirit		*Water*	
	Growth	*Blisters*	*Growth*	*Blisters*
Zinc chromate	–	0	–	0
Zinc borate	–	0	+	2
Zinc phosphate	++	3	+++	3

REFERENCES

Boggs, W.A. (1969). Changes in aircraft design, fabrication and finish techniques and maintenance requirement dictated by microbe infested jet fuel sources. NACE, 25th Annual conf., Paper 49, 15 pp.

Bravery, A.F., Barry, S. & Worley, W. (1983). *Journal of the Oil & Colour Chemists' Association,* **66**, 39–43.

Kemp, H.T., Cooper, C.W. & Kell, R.M. (1966). *Journal of Paint Technology,* **38**, 363–7.

Kossmann, H. (1985). *Farbe und Lack,* **91**, 588–94.

Stranger-Johannessen, M. (1980). In *Biodeterioration,* ed. T.A. Oxley, G. Becker & D. Allsopp. Pitman, London, pp. 143–7.

Stranger-Johannessen, M. (1986*a*). In *Biodeterioration 6,* ed. S. Barry & D.R. Houghton. C.A.B. International Mycological Institute, The Biodeterioration Society, pp. 218–23.

Stranger-Johannessen, M. (1986*b*). In *XVIII. FATIPEC-Congress (Proceedings) Vol. 3,* Venezia, pp. 1–13.

Involvement of Bacterial Exopolymers in Biodeterioration of Metals

T.E. Ford, J.S. Maki & R. Mitchell

Harvard University, Cambridge, Massachusetts, USA

ABSTRACT

Microbial degradation of metals is associated with a wide range of biochemical processes, including acid metabolite production, galvanic coupling and differential aeration cell formation, in all of which the role of bacterial exopolymers is crucial. We obtained exopolymer from a number of pure cultures associated with metal surfaces and measured their binding characteristics with a number of metal ions (Fe, Cu, Mn, Ni). All of the exopolymers strongly bound the metals, but there was variation in maximum binding abilities and stability constants. Differential binding within the biofilm would result in preferential oxidation of particular metal species, e.g. concentration of specific ions by the exopolymer of the biofilm may result in oxidation of less noble ions in the metal lattice. Using electrical and polarization resistance techniques the effect of exopolymer-producing bacteria on metal biodeterioration rates was studied. This report discusses the relationship between exopolymer binding and corrosion kinetics.

INTRODUCTION

Production of bacterial extracellular materials is central to all processes of biodeterioration, both aerobic and anaerobic. They provide sites for aeration cells (Little *et al.*, 1986), ion concentration cells (Geesey *et al.*, 1986), and are important in providing a protective matrix against biocides and cleaning techniques (Pope *et al.*, 1983). Geesey *et al.* (1986)

and Ford *et al.* (1987*b*) have examined the role of bacterial exopolymers in metal binding. The ability of an exopolymer to bind specific metal ions will strongly influence its adhesion to metal surfaces and its ability to concentrate metal ions both from those surfaces and from the bulk media. Mittelman & Geesey (1985) have shown that exopolymer produced by a river sediment bacterium binds copper ions in solution. They suggest that when this bacterium is attached to copper alloy surfaces its exopolymer has the capacity to form copper concentration cells. Copper binding by exopolymers produced by a thermophilic bacterium isolated from corroded nickel tubing has been measured (Ford *et al.*, 1987*a*). It was shown that this thermophile accelerates corrosion of various metals and have obtained evidence that copper concentration cells are responsible for the corrosion (Little *et al.*, 1986).

It is important for the understanding of corrosion processes to determine the ability of exopolymers to bind different metal ions in order to establish potential galvanic cells, both in terms of ions within the metal lattice and free ions in solution.

This paper describes results of a number of binding studies using different bacterial exopolymer preparations. We also describe some initial experiments to quantify the effect of these organisms and their exopolymers on biocorrosion rates.

MATERIALS AND METHODS

Bacteria

Deleya marina (ATCC 25374) was obtained from the American Type Culture Collection (Rockville, MD). Cultures were grown in a defined minimal carbon medium with glucose as the carbon source.

Thermophilic microorganisms were isolated from corroding hot water systems. Species isolated belonged to the genera *Thermus* and *Bacillus* and were grown and maintained on dilute Castenholz medium containing 0·025% yeast extract and 0·025% tryptone. Isolates were maintained at 65°C.

Polymer isolation and metal binding

Polymer was isolated from batch cultures that had reached stationary phase.

Binding experiments were conducted using the Mittelman & Geesey

(1985) protocol. Solutions of exopolymer (0·1 mg/ml) were prepared at pH 6 and dialysed against iron, copper, nickel and manganese solutions also adjusted to pH 6 ranging from 50 to 800 μg/litre. Exopolymer solutions were dialysed for at least 36 h against metal solutions. Binding constants were calculated from the Langmuir linear expression $c/x = 1/k_1k_c + c/k_1$ (c is the free metal concentration, x is moles of metal bound per unit weight of metal-binding component, k_1 is the maximum binding ability (MBA), and k_c is the conditional stability constant).

Corrosion experiments

To quantify the effect of exopolymer on biodeterioration rates we compared rate of corrosion of mild steel (1010) in the presence of bacterial cultures and exopolymer preparations. Corrosion rates were determined using a simple polarization cell described in Ford *et al.* (1987*a*). Polarization resistance measurements were made using an AIS potentiostat (Model V-2LR-D), scanning between 20 mV negative to the measured corrosion potential (E_{corr}) and 20 mV positive to E_{corr} at a rate of 0·72 V/h until steady-state potential-current behaviour was observed. Electrical resistance measurements were made to estimate corrosion rates in stirred or flowing systems. Mild steel (1010) wire (30 gauge) was placed in stirred flasks or in flow-through columns. Wire resistance was continuously monitored, and a corrosion rate calculated from change in resistance through time.

RESULTS AND DISCUSSION

In initial experiments the exopolymer was extracted from surface bacteria, with potential involvement in biocorrosion reactions, and the metal binding capacities determined. Maximum binding abilities and conditional stability constants for a number of different microorganisms and metals are shown in Table 1. We observed considerable variation between different exopolymers and different metals; specific results are as follows.

(1) Overall, manganese displayed the highest maximum binding abilities.
(2) The strongest stability constant was between *Pedomicrobium manganicum* exopolymer and Fe^{3+}, despite a very low binding ability. At the pH value at which these experiments were conducted (pH 6), the solubility of Fe^{3+} is very low.

(3) Exopolymer obtained from suspended *Thermus* bound considerably less copper than exopolymer from the same bacterium growing on surfaces.

Interaction between metal ion and exopolymer is clearly complex and from these results it is apparent that environmental parameters (e.g. pH) are extremely important. *D. marina* is the only microorganism reported on here that has been previously studied for metal binding. Comparison of exopolymer–metal binding with trace metal assimilation order (Jones *et al.*, 1975) has been discussed by Ford *et al.* (1987*b*). Trace metal assimilation for whole cells is reported as Fe > Cu > Ni = Mn; in contrast, order of maximum binding ability of exopolymer is Mn > Cu > Ni > Fe. Concentration of metals intracellularly is unlikely to be influenced by extracellular binding by polymers. Conversely, it is likely that binding performs the important function of immobilizing toxic metals, preventing their entry into the cell.

All binding experiments were conducted using a primary exopolymer preparation. A secondary exopolymer obtained from *D. marina* after several deproteination steps displayed minimal binding capability (Ford *et al.*, 1987*b*), suggesting that protein groups may be important as binding moieties.

The effect of biofilms of *Deleya* on the corrosion rate of mild steel 1010 using measurement of resistance as an indicator of corrosion was analysed. The data (Table 2) indicate that in both stirred and flow-

TABLE 1
Maximum Binding Abilities and Conditional Stability Constants of Exopolymer/Metal Complexes

Exopolymer sample	*Metal*	*MBA (nmol/mg)*	K_c ($\times 10^8$)
[a]*Thermus* sp. (suspended)	Cu	9	0·7
[b]*Thermus* sp. (attached)	Cu	85	0·09
[b]*Deleya marina*	Cu	263	24
[b]*Deleya marina*	Mn	556	9
[b]*Deleya marina*	Fe	39	1
[b]*Deleya marina*	Ni	435	14
[b]*Pedomicrobium manganicum*	Fe	13	750
[b]*Pedomicrobium manganicum*	Mn	184	19
[b]*P. ferrugineum*	Mn	409	2·9

[a]Ford *et al.* (1986).
[b]Ford *et al.* (1987*b*).
[c]Black *et al.* (1987).
MBA: maximum binding ability.
K_c: conditional stability constant.

through systems biofilms of this bacterium strongly inhibited mild steel corrosion. Even at the very high corrosion rate observed in the stirred culture, *D. marina* biofilms acted as a strong corrosion inhibitor. This surprising result was supported by polarization resistance data (Table 3A). In this series of experiments, measurements were made in a static environment. Passivation of the metal by the bacterium was almost complete, reducing the corrosion rate by 94%.

The effect of biofilms of three thermophilic bacteria on corrosion rates was also tested, using polarization resistance as a measurement of corrosion. The data (Table 3B) show clearly that one of the bacteria tested (T3), tentatively identified as a *Bacillus* sp., behaved similarly to *D. marina*. It almost completely inhibited corrosion. In contrast, the two other thermophilic isolates tested (tentatively *Thermus* sp.) stimulated corrosion. Isolate T1 was particularly effective in the stimulation of corrosion, increasing the rate by 135% (Table 3B).

There are considerable difficulties in equating polymer production with biodeterioration rates. Many different processes occur within a biofilm. Working with pure cultures strong evidence has been provided that a microorganism may form a protective film on the metal surface. Alternatively, other acid producing bacteria, such as the *Thermus* species we have reported on here and elsewhere (Little *et al.*, 1986; Ford *et al.*, 1987*a*), accelerate corrosion in pure culture. In a mixed biofilm, microorganisms will interact, usually enhancing rates of metabolic processes. Within a consortium, species that apparently passivate metals in pure culture may enhance corrosion by providing a protective film for, for example, acid producers.

Bacterial exopolymer is central to all aspects of biofilm formation and function. Binding of metals may be important in both passivation and

TABLE 2
Corrosion Rates Measured as Change in Electrical Resistance through Time (ohms/day) of Mild Steel (1010) in the Presence of *Deleya marina*

	Resistance change (ohms/day)	*Corrosion rate (mils per year)*	*Reduction in corrosion rates compared to uninoculated control*
Experiment 1 (stirred)			
Uninoculated Control	0·044	46·7	
Inoculated	0·007	7·0	85%
Experiment 2 (flow through)			
Uninoculated Control	0·011	20·7	
Inoculated	0·002	3·1	85%

TABLE 3

Corrosion Rates Measured as Polarization Resistance in Volts/Milliamp (V/mA) of Mild Steel (1010) in the Presence of Different Bacterial Cultures

	Polarization (V/mA)	*Corrosion rate (mils per year)*	*Reduction or increase in corrosion rate compared to uninoculated controls*
(A) Uninoculated control (20°C)	4·7	3·1	
D. marina	64·3	0·2	−94%
(B) Uninoculated control (65°C)	1·9	7·7	
T1	0·8	18·1	+135%
T2	1·6	8·9	+16%
T3	160·0	0·09	−99%

T1–3 are three different thermophilic isolates.

activation reactions, particularly in establishment of galvanic cells. The biological importance of the exopolymer has been discussed extensively in the literature (e.g. Costerton *et al.*, 1978) and includes attachment, nutrient adsorption, protection, immobilization of toxic compounds, and a matrix for exoenzyme function. The role of exopolymers in biodeterioration of metals is extremely complex. Our preliminary data suggest that they may mediate competing electrochemical processes. The interaction between exopolymer chemistry and specific corrosion reactions is being studied.

ACKNOWLEDGEMENTS

This work was supported by Office of Naval Research Contract No. N00014-81-K-0624 to Harvard University.

REFERENCES

Black, J.P., Ford, T.E., Maki, J.S. & Mitchell, R. (1987). In American Society of Microbiology Annual Meeting, 1987, Atlanta, USA, Abstracts, p. 190.

Costerton, J.W., Geesey, G.G. & Cheng, K.J. (1978). *Scientific American,* **238**, 86–95.

Ford, T.E., Maki, J.S., Black, J.P. & Mitchell, R. (1986). In American Society of Microbiology Annual Meeting, Washington, USA, Abstracts, p. 170.

Ford, T.E., Walch, M. & Mitchell, R. (1987*a*). *Materials Performance*, **26**, 35–9.

Ford, T.E., Maki, J.S. & Mitchell, R. (1987*b*). *Corrosion/87*, National Association of Corrosion Engineers, Paper No. 380.

Geesey, G.G., Mittelman, M.W., Iwaoka, T. & Griffiths, P.R. (1986). *Materials Performance*, **25**, 37–40.
Jones, G.E., Royle, L.G. & Murray, L. (1975). *Proceedings of the Third International Biodegradation Symposium*, ed. J.M. Sharpley & A.M. Kaplan. Elsevier Applied Science, London, p. 889.
Little, B., Wagner, P., Gerchakov, S.M., Walch, M. & Mitchell, R. (1986). *Corrosion*, **42**, 533–6.
Mittelman, M.W. & Geesey, G.G. (1985). *Applied and Environmental Microbiology*, **49**, 846–51.
Pope, D.H., Duquette, D.J., Johannes, A.H. & Wayner, P.C. (1983). *Corrosion/83*, National Association of Corrosion Engineers, Paper No. 247.

An extended bibliography for this paper may be obtained from the senior author.

The Effect of Marine Fouling on Fatigue and Corrosion-Fatigue of Offshore Structures

R.G.J. Edyvean, C.J. Thomas & R. Brook

School of Materials, The University of Sheffield, Mappin St, Sheffield S1 3JD, UK

ABSTRACT

Corrosion-fatigue, the combination of a corrosive environment and cyclic stresses, can produce failures in a structure, such as an offshore platform, by the development and growth of cracks. Seawater provides the corrosive environment and, through wave action, the source of cyclic, fatigue loading. Marine fouling can enhance both factors, first by increasing the diameter and surface roughness of platform legs and bracing members and secondly by promoting corrosion. In relation to corrosion-fatigue this enhancement is mainly due to the production of H_2S *by sulphate-reducing bacteria. The effects of both enhanced loading and of hydrogen sulphide may not be alleviated by anti-corrosion measures and thus need to be quantified and incorporated into the design life of the structure.*

Data are presented on corrosion-fatigue crack growth in seawater containing various levels of both biologically produced and abiotic H_2S*. Crack growth rates are found to be enhanced, even at low levels, by biologically generated* H_2S*.*

INTRODUCTION

Corrosion-fatigue is the combination of a corrosive environment and applied cyclic stress to produce failure of a metal by the development and growth of cracks. Offshore structures, especially in deep water areas such as the northern North Sea, are in a hostile environment experiencing the effects of wind and storm action, constant wave loading of varying aggressiveness and continual immersion in a corrosive environment. Marine fouling on an offshore structure can have serious consequences, enhancing both the corrosion and stress components of corrosion-fatigue. First, the diameter of the tubular members of the structure is effectively increased by trapping a layer of water (Kingsbury, 1981). Secondly, the surface roughness, and hence the drag coefficient of the members, is increased (by a factor of three over a smooth cylinder for kelps (Wolfram & Theophanatos, 1985)). It has been calculated that a layer of fouling 150 mm thick increases the loading by 42·5%, the fatigue damage by 62% and decreases the predicted life by 54% (Heaf, 1979). Minor structural failures have been attributed to the reduction of fatigue life by marine fouling (Anon., 1978). Thirdly, biological activity can enhance corrosion by providing an environment and source of nutrients for sulphate-reducing bacteria.

The three stages of fatigue are: crack initiation, crack growth and final fracture. Processes such as differential aeration and the production of aggressive metabolites lead to crevice and pitting corrosion which may initiate fatigue cracks (Cragnolino & Tuovinen, 1984). It is assumed that, in structures as large and complex as offshore platforms, there will be flaws (e.g. in welds) which will considerably reduce the crack initiation time. A crack may be growing from early in the life of the structure. The rate of growth under corrosion-fatigue is controlled by the rate of dissolution of metal at the crack tip and by hydrogen embrittlement, which lowers the loads necessary to cause fracture (Bristoll & Roeleveld, 1978). Any external supply is likely to increase hydrogen embrittlement and thus crack growth rate. A particularly aggressive metabolic product is H_2S (as H_2S, HS^- or S^-, depending on conditions). Sulphide ions both stimulate anodic dissolution and poison the combination of hydrogen atoms to the gaseous molecule and make it more readily available for embrittlement. Cathodic protection may not counter the effects of H_2S as it has been shown to be both detrimental and protective in corrosion-fatigue, depending on conditions (Austen & Walker, 1984).

The main control of hydrodynamic loading and corrosion-fatigue is good design. It can reduce the risk of high stress areas and take into account the added forces due to fouling. However, it is only recently that reasonable predictions of the amount and effects of fouling have become available and data, especially on the effects of hydrogen sulphide, are still being gathered. Recent work has shown significant enhancement of crack growth at levels of H_2S produced by sulphate-reducing bacteria (Cowling & Appleton, 1986; Thomas *et al.*, in press).

To provide data from which design criteria can be developed, specimens of steel are subjected to known levels of cyclic stress in a corrosive environment. To reflect 'worst case' conditions a 'precracked' specimen is used for testing and crack growth studies. Providing specimens are designed to strict geometrical proportions, fracture mechanics enables laboratory experiments to be related to a structure by a stress intensity factor K (Ashby & Jones, 1980). Under cyclic loading, the characteristic change in K, ΔK, is a function of load range, the geometry of the specimen and the crack length, hence, as the load range and specimen geometry are fixed, ΔK will change as the crack grows. This allows crack growth rate data for the complete range of ΔK up to catastrophic failure to be gathered from one specimen (Thomas *et al.*, in press).

Dramatic increases in crack growth rate have been found in H_2S saturated environments when compared to the results found in air or seawater (Austen & Walker, 1984) but, for more realistic predictions, testing at lower levels of H_2S is required. However, there are few data for the levels of H_2S produced biologically. Reported values range from more than 2000 ppm in pure cultures to between 15 and 100 ppm in the interior of platform legs and levels of up to 600 ppm can be produced from decomposing algae in seawater without nutrient additions (Edyvean *et al.*, 1986).

MATERIALS AND METHODS

Biologically-active H_2S environments were created by the decomposition of marine algae (*Enteromorpha* spp., *Porphyra* sp. and *Pelvetia* sp.) in seawater (Thomas *et al.*, in press). Samples of the cultures (50 ml) were removed at intervals for pH and H_2S analysis and when a desired level of H_2S was obtained, that culture was used as a corrosion-fatigue test environment. Abiological H_2S environments were created by bubbling H_2S gas through natural seawater until the desired level was reached. Corrosion-fatigue crack growth measurements were made using 13 mm thick compact tension specimens (BS 5447, 1979) machined from steel grades RQT 501 and RQT 701 manufactured by British Steel Corporation. (Nominal composition RQT 501 wt% C = 0·18, Si = 0·5, Mn = 1·5, V = 0·08, P = 0·035, S = 0·015, Mo = 0·2, B = 0·003, N = 0·06; RQT 701 has the same composition except for carbon at 0·2 wt%) (Thomas *et al.*, in press). A crack was induced in each specimen prior to testing and tests were carried out in perspex chambers under conditions of free corrosion at the simulated wave frequency of 0·167 Hz and a stress ratio R (P_{min}/P_{max}) of 0·7. Analysis of the results enables values of the crack growth rate, da/dN, and of the range in stress intensity factor, ΔK, to be calculated.

RESULTS AND DISCUSSION

Data on crack growth rates at three levels of ΔK (12, 20 and 30 MPa/m) in various environments are given in Table 1. The data are presented as enhancement factors on the rates found in air and seawater. While the presence of seawater can increase crack growth rates by a factor of up to 7 over those in air (Austen & Walker, 1984), the additional presence of H_2S in the solution, even at low levels, considerably enhances crack growth (for example the rate of crack growth in 501 steel in 73 ppm H_2S at $\Delta K = 20$ is 11 times faster in seawater). This enhancement of crack growth is also shown in abiological H_2S environments. There is some evidence of a greater susceptibility to the effects of H_2S in abiological environments than similar levels produced biologically. For example at $\Delta K = 12$, 501 steel in 138 ppm abiotic H_2S shows crack growth enhancement of 25 times that of seawater but in 290 ppm H_2S produced biologically the enhancement is only 5 times.

Crack growth rates in 73–450 ppm H_2S levels generated biologically fall in a very narrow range for these steels (Thomas *et al.*, in press). This is not indicative of an upper limit for the effect of H_2S. The dissociation, adsorption and absorption of hydrogen into a metal to produce embrittlement is catalysed by an atomically 'clean' crack tip. Fouling of the region, and indeed the metal surface generally, by the adsorption of organic molecules, bacterial mucilage and the growth of bacteria, may result in a retardation of crack growth due to a slowing down of the embrittlement process. Thus, as an environment becomes more aggressive, due to greater biologically produced H_2S, there is more organic material available to limit its effects.

The results show that corrosion-fatigue can be considerably enhanced by biologically produced hydrogen sulphide even at relatively low levels. The consequences are potentially serious. While conditions generating high H_2S levels are likely to be transitory due to eventual nutrient limitation (for example, the decomposition of trapped algal or invertebrate material or the exhaustion of nutrient supply in the interior of platform legs) and diffusion, areas producing low levels of H_2S will be common. Laboratory results suggest that rapid failures could be predicted for offshore structures. However, this has not been the case, probably because the test assumes that a fatigue crack is already present in the structure and that the greatest crack-growth enhancement is at quite high ΔK values. For conditions offshore to be conducive to crack growth, H_2S production must coincide with a crack experiencing a critical range of stress for a sufficient length of time for the effects to become apparent. From the lack of reported failures it appears that such a combination of circumstances does not often occur. However, as the structures grow

TABLE 1

Crack Growth Enhancement over Air and Seawater for RQT 501 and 701 Steels in Seawater Containing Various Levels of Hydrogen Sulphide Generated Biologically (Biotic) and Abiologically (Abiotic)

H_2S (ppm)	*Environment*	*Crack growth enhancement K 12*		*Crack growth enhancement K 20*		*Crack growth enhancement K 30*	
		Air	*Seawater*	*Air*	*Seawater*	*Air*	*Seawater*
			STEEL 501				
73	Biotic	29	10	40	11	10	1
290	Biotic	13	5	30	9	20	3
450	Biotic	—	—	40	11	10	F
86	Abiotic	9	3	25	7	33	5
138	Abiotic	71	25	40	11	F	F
390	Abiotic	129	45	200	57	F	F
520	Abiotic	114	40	100	11	F	F
			STEEL 701				
130	Biotic	—	—	5	2	3	3
287	Biotic	25	10	40	14	5	4
422	Biotic	38	15	50	20	F	F
184	Abiotic	50	20	100	40	F	F
598	Abiotic	125	50	500	200	F	F
976	Abiotic	125	50	F	F	F	F

F = specimen failed (fast fracture).

older, corrosion promoting forces, both biological and abiological, will produce pits and crevices in susceptible areas, creating more cracks and thus increasing the risks of corrosion-fatigue failure.

ACKNOWLEDGEMENTS

The authors would like to thank I.M. Austen of the British Steel Corporation Research Organisation, Rotherham, UK, for his help and support of this work. CJT is funded by SERC. RGJE is the holder of the Sorby Research Fellowship of the Royal Society.

REFERENCES

Anon. (1978). *Ocean Industry,* August, 67–8.

Ashby, M.F. & Jones, D.R.H. (1980). *Engineering Materials.* Pergamon Press, Oxford.

Austen, I.M. & Walker, E.F. (1984). In *'Fatigue-84' Proc. 2nd Int. Conf. on Fatigue Thresholds,* Vol. 3, ed. C.J. Beevers. EMAS pub., pp. 1457–71.

Bristoll, P. & Roeleveld, J.A. (1978). *European offshore steels research project seminar.* Welding Institute, Cambridge. Preprints, Vol. 2, paper 18, pp. 439–58.
Cowling, M.J. & Appleton, R.J. (1986). Corrosion-fatigue of a C-Mn steel in seawater solutions. Paper C145/86, *Institute of Mechanical Engineering,* London, pp. 77–92.
Cragnolino, G. & Tuovinen, O.H. (1984), *International Biodeterioration,* **20**, 9–26.
Edyvean, R.G.J., Thomas, C.J., Brook, R. & Austen, I.M. (1986). In *Biologically Induced Corrosion,* ed. S.C. Dexter. NACE, Houston, pp. 254–67.
Heaf, N.J. (1979). In *11th Annual Offshore Technology Conference,* Vol. 1, Houston, Texas. pp. 255–6.
Kingsbury, R.W.S.M. (1981). In *Marine Fouling of Offshore Structures,* Vol. I. Society for Underwater Technology, London.
Thomas, C.J., Edyvean, R.G.J. & Brook, R. (In press). *Biofouling.*
Wolfram, J. & Theophanatos, A. (1985). OTC 4954. *17th Offshore Technology Conference,* Houston, May 1985, pp. 517-26.

Hydrogen Absorption During the Microbial Corrosion of Steel

C.H.J. Parker, K.J. Seal & M.J. Robinson

Biotechnology Centre & School of Industrial Science, Cranfield Institute of Technology, Cranfield, Beds, UK

ABSTRACT

Biogenic sulphide production in marine environments is known to encourage the embrittlement of steels due to an increased susceptibility to hydrogen absorption. However, it remains unclear to what extent a sustained SRB presence is necessary for enhanced hydrogen absorption to occur. By electrochemically measuring the permeation of hydrogen through steel membranes, changes in rates of hydrogen absorption have been monitored for periods of up to 10 days, using steel foils in cultures of Desulfovibrio desulfuricans *(Norway-4). Simultaneous tests have been performed using pairs of steel foils, one exposed to SRB, and one from which SRB are excluded, using a dialysis membrane. Shortly after inoculation with SRB, both sterile and colonised steel foils gave rise to hydrogen permeation peaks. Also, there was little difference between the peak hydrogen permeation rates attained by sterile foils, and foils exposed directly to the SRB. Such peaks were found to coincide with a rapid increase in hydrogen sulphide concentration, and a positive shift in the free corrosion potential, suggesting the establishment of a protective film at each of the steel surfaces. The nature and properties of these films was found to be independent of the presence of an SRB film.*

INTRODUCTION

In anaerobic, neutral pH environments, the activities of the sulphate-reducing bacteria (SRB) are widely held responsible for the removal of polarising cathodic hydrogen from steel surfaces, causing increased corrosion rates (Tiller, 1983). Such cathodic depolarisation has been attributed, with varying degrees of importance, to both the direct uptake of cathodic hydrogen by the SRB, as electron donor in dissimilatory sulphate reduction (Booth & Tiller, 1960), and also to the removal and absorption of hydrogen by the various forms of sulphide films which may develop, and which may also act as efficient cathode surfaces for the hydrogen evolution reaction (Smith & Miller, 1975). A further effect of cathodic depolarisation in soured environments, is the poisoning by H_2S on the interconversion of atomic and molecular forms of hydrogen (McCright, 1973). Consequently, in addition to enhancing the gross production of hydrogen, sulphide also encourages a greater proportion of hydrogen to exist in atomic form, which may subsequently enter the metal, and thus promote embrittlement and cracking. Using a development of the hydrogen permeation technique established by Devanathan & Stachurski (1962), it is possible to assess the significance of the hydrogen concentrations established within anaerobically corroding steel when exposed to an active SRB population. It may also help to indicate the importance of metabolically active SRB films present at the steel surface, compared to equally sour, but sterile conditions, for increased hydrogen absorption.

MATERIALS AND METHODS

Hydrogen permeation experiments were performed using a 'Double Chamber Bioreactor' apparatus (Fig. 1), essentially comprising two hydrogen detectors (Devanathan & Stachurski, 1962) mounted at each end of a central chamber, with a dialysis membrane partition (Medicell International Ltd).

Prior to insertion into the sterilised (121°C/20 min) Bioreactor assembly, identical shimm steel discs (0·17% carbon, 57 μm thick) were polished (600 grade silicon carbide paper), then ultrasonically cleaned and sterilised in isopropanol (30 min). The outer faces of the steel were held anodically (+150 mV) with respect to an anodised nickel cathode, such that diffused cathodic hydrogen (from microbial corrosion occurring at the inner face of the steel) oxidised as it emerged, and became reduced at the cathode. The small hydrogen permeation currents thus generated were measured with a zero-resistance ammeter. The use of an alkaline electrolyte rendered the steel passive to corrosion, thereby ensuring the measurement of exclusively diffused hydrogen.

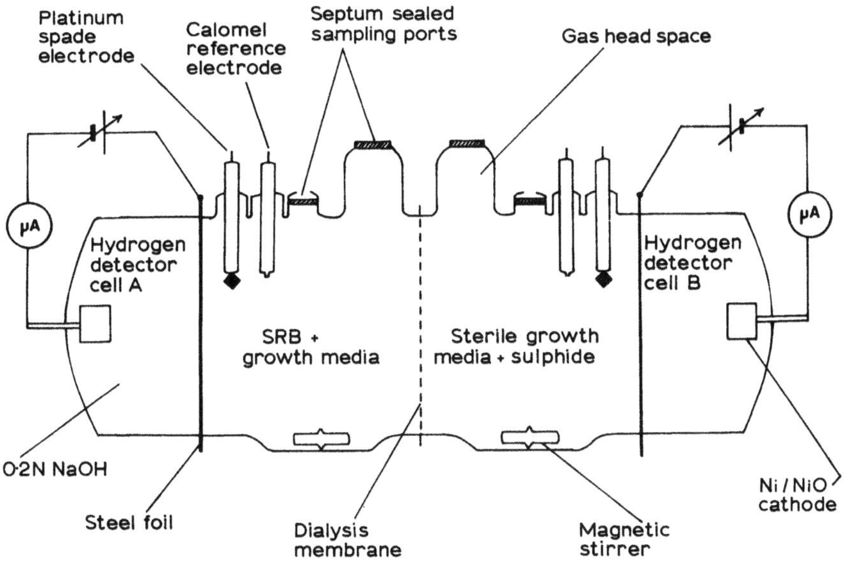

Fig. 1. Double chamber bioreactor used in hydrogen permeation tests (schematic).

Following the electrolytic depletion of hydrogen from the steel foils (2–3 days at 30°C), sterile Postgate's Medium 'C' (Postgate, 1984) was introduced simultaneously into each of the two bioreactor chambers, and after a further period of equilibrium (2–3 days), one chamber was inoculated by syringe with a triple-washed suspension of *Desulfovibrio desulfuricans* (Norway-4), giving a starting cell density of approximately 10^6/ml. Hydrogen permeation current was then monitored continuously for up to 10 days. Duplicate electrode ports facilitated the measurement of media redox potential and pH, and also the measurement of the corrosion potential for each steel specimen. Determinations of viable SRB numbers (Postgate, 1984) and sulphide concentration (gas chromatography) were possible using septum-sealed gas and media sampling ports.

RESULTS AND DISCUSSION

Earlier work involving comparisons of sour corrosion in the presence and absence of the SRB has attached various levels of importance to the role of bacterial film formation on steel, compared to the influence of biogenic H_2S alone (Booth *et al.*, 1968; Gaylarde & Johnston, 1980).

Hydrogen permeation

In this work, hydrogen permeation transients obtained for steel exposed to active SRB cultures demonstrate that, compared to steel exposed to a

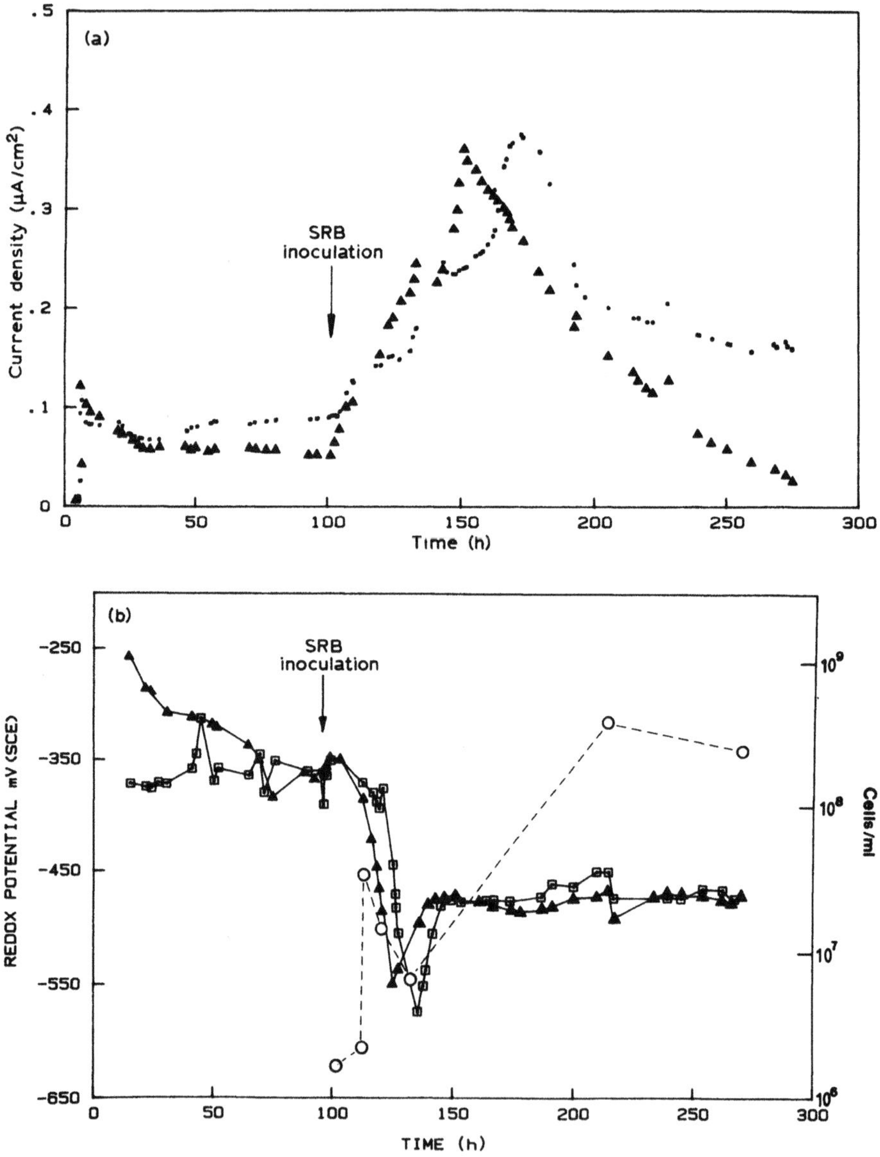

Fig. 2. Transients of (a) hydrogen permeation current; (b) redox potential and cell numbers; (c) pH; and (d) corrosion potential for duplicate shimm steel foils and culture chambers using a batch culture of *Desulfovibrio desulfuricans* (Norway-4) in Postgate's Medium 'C' at 30°C, with SRB present (▲) and absent (■) from the steel surface.

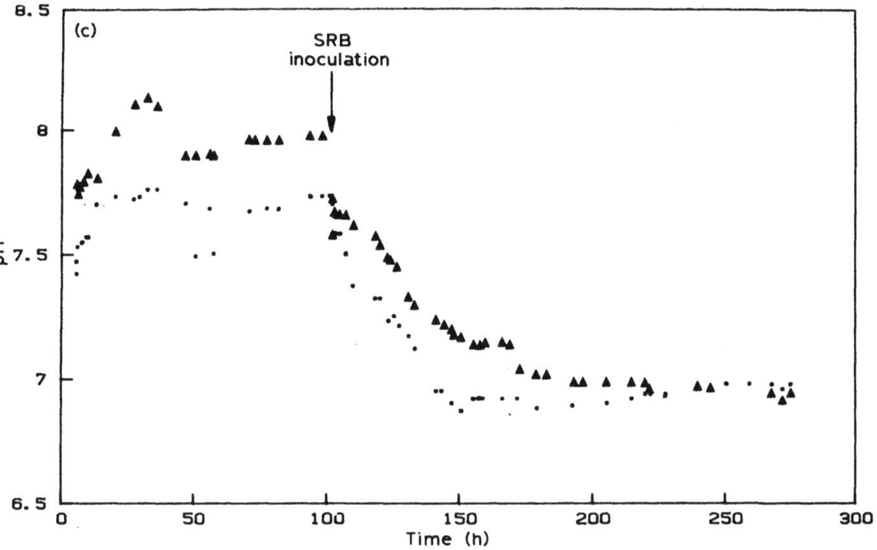

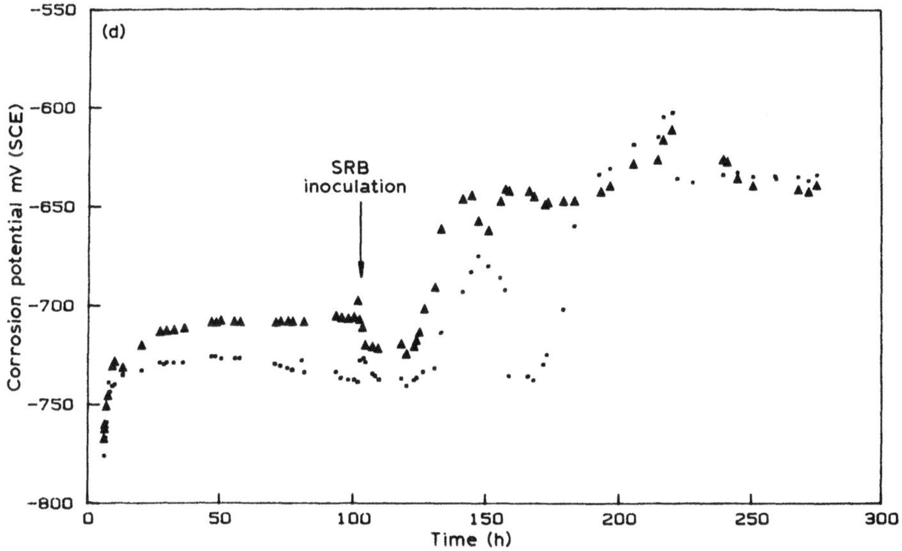

Fig. 2—*contd.*

bacteria-free but soured environment (containing SRB metabolites of up to 10 000 molecular weight), the development of a bacterial film has no additional effect on the generation and absorption of cathodic hydrogen at the steel surface (Fig. 2(a)). The similar and consistent peak hydrogen permeation currents obtained here, and in similar work described elsewhere (Robinson *et al.*, 1987), indicate the establishment of equally aggressive conditions on either side of the dialysis membrane with respect to pH and the concentration of corrosive metabolites (especially H_2S) generated by the SRB. Assuming that sulphide is the chief corrosive agent, then the delay shown between hydrogen permeation peaks is probably a function of the diffusion rate of H_2S across the dialysis membrane between the SRB and sterile bioreactor chambers.

Redox potential and pH

Rapid increases in hydrogen permeation current were coincident with a blackening of the culture medium and a sustained decrease in the redox potential by 130 mV (Fig. 2(b)). This is consistent with the rapid production of H_2S by the SRB, which itself partially dissociates to HS^- and H^+ ions, thereby lowering the pH (Fig. 2(c)).

SRB numbers

Viable counts (Fig. 2(b)) and a macroscopic examination of the SRB culture chamber, revealed the maintenance of a viable and highly motile population of SRB throughout the duration of the test, both in the bulk medium, and within the voluminous black film present at the vessel and steel surfaces.

Corrosion potential

The initial decline in free corrosion potential following SRB inoculation (Fig. 2(d)) suggests an anodic stimulation, coincident with H_2S evolution by the SRB at the newly colonised steel surface. The subsequent shift to a more noble corrosion potential (after about 24 h), is consistent with the development of a protective film on the steel surface. This anodic effect manifests itself similarly whether in the presence or absence of the SRB, and an examination of rinsed corroded steel foils revealed the presence of similar tenacious, uniform black sulphide films, with no evidence of underlying pitting. However, it remains to be seen whether such films remain protective over longer time periods, and to what extent film development depends upon other factors reported elsewhere, such as initial pH and H_2S concentration (Smith & Miller, 1975), and the concentration of ferrous ions (King *et al.*, 1976), where subsequent fluctuations in these factors may also influence the degree of protection or corrosiveness afforded by sulphide films.

CONCLUSIONS

As a measure of corrosion rate and the aggressiveness of corrosive media, the monitoring of cathodic hydrogen permeation using a double-chamber bioreactor which incorporates a bacterial filter has demonstrated that the presence of a growing SRB film has no additional effect on steel corrosion above that induced simultaneously by a cell-free cocktail of SRB metabolites. The system described has also shown the value of monitoring various corrosion and SRB growth parameters simultaneously, whilst also enabling direct comparisons between the corrosion of colonised and sterile steel surfaces.

REFERENCES

Booth, G.H. & Tiller, A.K. (1960). *Transactions of the Faraday Society*, **56**, 1689-96.

Booth, G.H., Elford, L. & Wakerley, D.S. (1968). *British Corrosion Journal*, **3**, 242-5.

Devanathan, M.A.V. & Stachurski, Z. (1962). *Proc. Roy. Soc.*, **A270**, 90-102.

Gaylarde, C.C. & Johnston, J.M. (1980). In *Microbial Adhesion to Surfaces*, ed. R.C.W. Berkeley & J.M. Lynch *et al.*, Ellis Horwood, pp. 511-13.

King, R.A., Dittmer, C.K. & Miller, J.D.A. (1976). *British Corrosion Journal*, **11**, 105-7.

McCright, R.D. (1973). In *Stress Corrosion Cracking and Hydrogen Embrittlement of Iron Alloys*, NACE, pp. 306-25.

Postgate, J.R. (1984). *The Sulphate-Reducing Bacteria*. Cambridge University Press, Cambridge.

Robinson, M.J., Parker, C.H.J. & Seal, K.J. (1987). *Proceedings of Corrosion 87* (in press).

Smith, J.S. & Miller, J.D.A. (1975). *British Corrosion Journal*, **10**, 136-43.

Tiller, A.K. (1983). In *Microbial Corrosion*. National Physical Laboratory, March 1983, Metals Society, pp. 54-64.

Thermophilic Sulphide Generating Bacteria Causing Corrosion in High Temperature Oilfield Systems

P.F. Sanders, M.J. Mosley & D.M. Holt

Micran Ltd, Unit A1, Wellheads Crescent, Wellheads Industrial Estate, Dyce, Aberdeen, AB2 0GA, UK

ABSTRACT

The increasing concentrations of H_2S in fluids of many North Sea hydrocarbon reservoirs is a major concern amongst oilfield operators because of sulphide induced corrosion. The combination of increasingly sour reservoirs and preliminary indications of thermophilic sulphate-reducing bacteria (t-SRB) growth in high temperature production facilities provided the stimulus to study microbial growth in these systems and to determine their role in reservoir souring and production plant corrosion. Initial isolation of sulphate-reducing bacteria capable of growth at 70° C from oil production systems indicated the possibility of t-SRB involvement in the souring and corrosion processes. A description is given here of studies designed to determine the extent of contamination of oil production plant and hydrocarbon reservoirs.

A model is presented which describes the involvement of t-SRB in reservoir souring. Specific isolation and enumeration methodology have been developed which permit the consistent recovery of t-SRB from such high temperature/high pressure systems. Results indicate a potential involvement of t-SRB in reservoir souring processes and subsequent contamination and corrosion of production plant.

The current state of knowledge of these bacteria is limited, but does permit a postulation for their interaction with environments associated with North Sea hydrocarbon production. The core of this model is that seawater flooding modifies the balance of sulphate, temperature and organic nutrient status in the reservoir.

INTRODUCTION

Hydrogen sulphide production from North Sea oil reservoirs is an increasing concern amongst operators, principally because of corrosion to production plant and processing installations. The souring of oilfields has been attributed to a variety of mechanisms, including sulphide generation by bacteria growing under reservoir temperatures and pressures. In a recent survey of possible souring mechanisms by Stott (1987) many of the proposed thermochemical processes were discounted on the grounds of the environmental conditions occurring in North Sea reservoirs.

The work of Sanders & Hamilton (1983) indicated that mesophilic sulphate-reducing bacteria (m-SRB) have contributed to hydrogen sulphide generation and corrosion problems in oilfield production facilities, especially in water injection systems. It is possible that souring derives from m-SRB activity close to the well bore, although nutrients would be a limiting factor.

Herbert & Stott (1983) were unable to isolate thermophilic SRB from North Sea water and could not acclimatise m-SRB to higher temperatures. They suggested that m-SRB could not survive the rapid changes in temperature and pressure that occur when seawater is injected into reservoirs during secondary oil recovery. Reports indicating the presence and growth of SRB at pressures of 500 bar and temperatures up to 104°C in oil bearing reservoirs by Zobell (1957) have not been substantiated.

In addition to sulphate reduction, sulphide generation through the dissimilatory reduction of sulphur by methanogenic bacteria has been suggested to be responsible for the formation of H_2S in habitats where sulphur is present (Stetter & Gaag, 1983). Thermophilic isolates of methanogenic bacteria have been obtained from volcanic habitats and it is possible that this type of organism may be responsible for reservoir souring by reduction of sulphur compounds in the formation.

Increasing levels of hydrogen sulphide combined with indications of thermophilic SRB growth in high temperature production facilities, were sufficient stimulus to initiate these studies into confirming the existence of t-SRB in these systems and defining their role in reservoir souring processes and production plant corrosion.

METHODS

Although the initial methodology was somewhat crude, being based upon well established m-SRB methods, the results of early surveys revealed thermophilic bacteria growing on internal surfaces of production vessels and associated pipework, even at temperatures above 80°C. During the past 2 years, more sophisticated sampling techniques, including high pressure filtration and pressure bomb sampling, have been developed to minimise explosive decompression of the samples taken from systems operating at pressures of up to 5000 psi. In addition, specific isolation media have been developed in which the composition is tailored to the conditions of osmolarity, nutrient status, etc., from which the organisms are isolated. Consistent recovery of viable thermophilic bacteria from oilfield systems is now routine.

RESULTS AND DISCUSSION

Table 1 shows that t-SRB have been detected from a variety of locations within oil production platforms. In addition, other thermophilic, non-SRB were commonly isolated. Surveys of a number of platforms serving souring reservoirs present a consistent pattern.

Characterisation of single isolates has shown that these are novel organisms. The rod shaped cells are variable in size and, depending on the age of culture, form chains, contain inclusions and may produce spores. They grow only under very reduced conditions utilising acetate, propionate, butyrate, benzoate, and reducing sulphate to sulphide.

TABLE 1

Occurrence of Thermophilic Sulphate-reducing Bacteria in Samples Taken from Locations Throughout Oil Production Platform Systems. The Exact Distribution of t-SRB is Site-specific, Depending upon Degree of Seawater Breakthrough, Operational Conditions, etc.

Site	*t-SRB*
Seawater in vicinity of production platform	—
Injection water	—
Bottom hole sample from water injection well	+
Connate water prior to waterflood	—
Seawater injection well backflow	+/—
Individual production wellhead water	+/—
Coupons from oil/water separation systems	+
Water from oil/water separation systems	+
Production water discharge	+

— t-SRB not detected; +/— variable detection; + consistent detection.

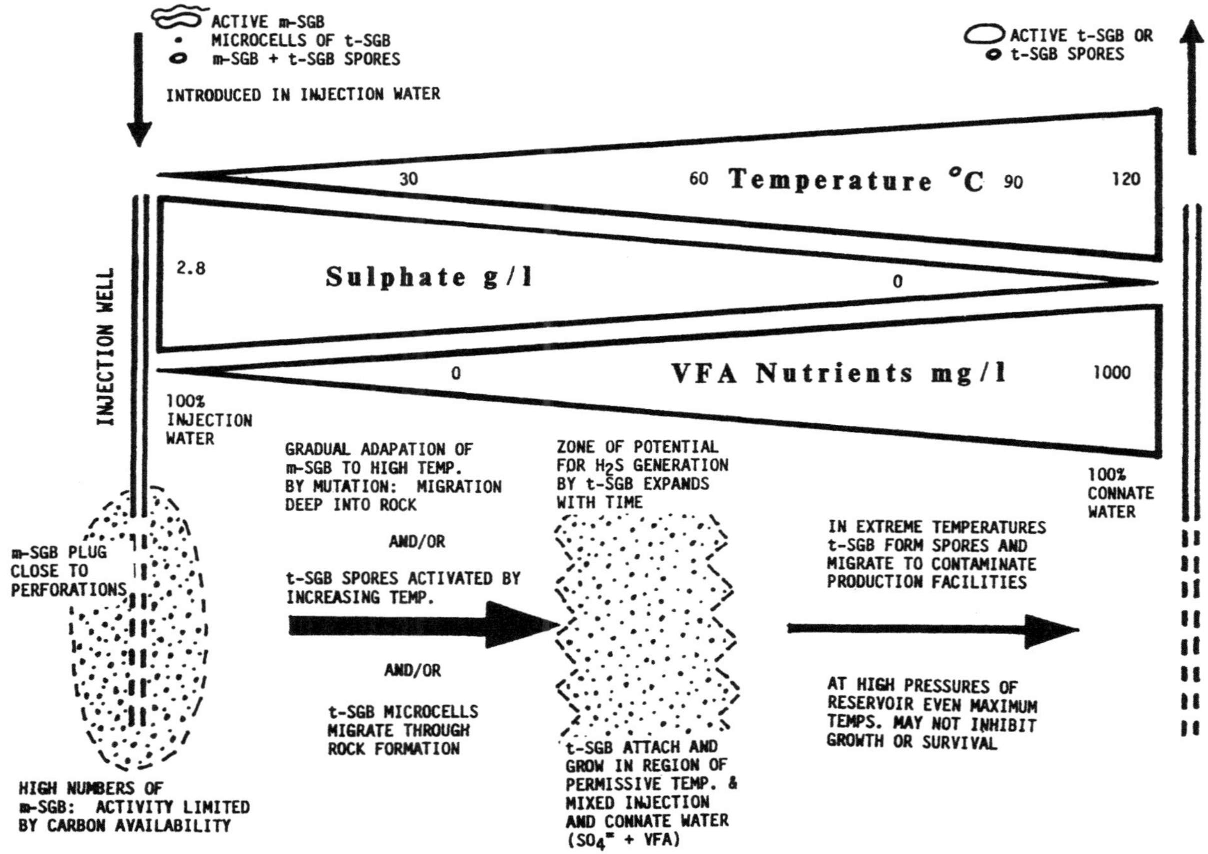

Fig. 1. Proposed theoretical model to explain microbiological souring of hydrocarbon reservoirs during extensive seawater flooding operations. Note that SGB (sulphate generating bacteria) is identical to SRB (sulphate reducing bacteria).

Maximum growth occurs at 60–70 °C at atmospheric pressure but limited growth occurs down to 45°C or up to 85°C.

Laboratory studies of the rates of sulphate reduction by thermophilic isolates derived from North Sea production facilities indicates that the amount of sulphide generated in reservoirs can be explained by t-SRB mediated sulphate reduction in a relatively small volume of rock. However, it is at present not possible to discount dissimilatory sulphur reduction by methanogenic bacteria using the appreciable quantities of sulphur present in North Sea hydrocarbon formations. We have isolated several types of thermophilic bacteria other than sulphate reducers from high temperature production systems and these have not yet been characterised.

The presence of t-SRB in water from production wells and from water injection well backflow samples indicates the extensive occurrence of these organisms within hydrocarbon reservoirs. Figure 1 shows a proposed scenario for the growth of both m-SRB and t-SRB within the gradations of physical and chemical conditions caused by water flooding.

No thermophilic SRB have as yet been detected in North Sea water samples, even using large filtered water samples. They may, however, be present in very low numbers, perhaps as ultra microbacteria (Kjelleberg *et al.*, 1982). Despite efficient control of water injection treatment systems, m-SRB are often present in significant numbers in treated water and are thus introduced into the formation. Injected seawater also introduces sulphate and produces a cooling effect in the vicinity of the well perforations. Gradations of sulphate concentration, temperature and volatile fatty acid (VFA) nutrient sources contained in connate water (Herbert *et al.*, 1985) are established due to mixing of connate and injection water. After some time of operation, seawater breaks through into production wells. The environment permissive to SRB growth will expand with time as cooling and dilution effects develop. Gradual adaptation by the organisms to extremes of temperature could occur, or dormant t-SRB in injection water could become active, allowing sulphide generation and contamination of the reservoir to increase. Eventually, t-SRB pass topside with produced fluids to contaminate production systems. Colonisation of vessel and pipework surfaces will add to the corrosive effect of sour produced fluids.

This model therefore provides a basis for developing an understanding of the souring phenomenon. It also has significant implications for the likely effectiveness of biocide strategies designed to combat downhole sulphide production.

REFERENCES

Herbert, B.N. & Stott, J.F.D. (1983). In *Microbial Corrosion*, ed. A.K. Tiller. Metals Society, London.

Herbert, B.N., Gilbert, P.D., Stockdale, H. & Watkinson, R.J. (1985). Factors controlling the activity of sulphate-reducing bacteria in reservoirs during water injection. SPE paper 13978.

Kjelleberg, S., Humphrey, B.A. & Marshall, K.C. (1982). *Applied and Environmental Microbiology*, **43**, 1166–72.

Sanders, P.F. & Hamilton, W.A. (1983). In *Conference Proceedings, Biologically Induced Corrosion*, ed. S. Dexter. Metals Society, London.

Stetter, K.O. & Gaag, G. (1983). *Nature (London)*, **305**, 309–11.

Stott, J.F.D. (1987). In *Sour Service in the Oil, Gas and Chemical Industry*. Symposium Publication, CAPCIS & IBC Technology, 17 pp.

Zobell, C.E.C. (1957). In *Sulphate-reducing Bacteria — their Relationship to the Secondary Recovery of Oil*. Symposium publication, St. Bonaventure University.

Methods for Insight into Mechanisms of Microbially Influenced Metal Corrosion

Nicholas J.E. Dowling,[a] Jean Guezennec[b] & David C. White[a]

[a]Institute for Applied Microbiology, University of Tennessee, Knoxville, TN 37932-2567, USA
[b]IFREMER Centre de Brest, BP337, 29273, Brest Cedex, France

ABSTRACT

Microbially influenced corrosion (MIC) produces areas of preferential metal loss and microbes exacerbate this process by generating acids, chelating metals, and inducing areas of differential cathodic activity. Ultrasensitive methods for determining the microbial biomass, community structure and nutritional status based on signature biomarker patterns of polar lipid fatty acids can be used to identify specific groups of microbes important in MIC. Monocultures compared for corrosion activity in a compartmented test apparatus permit experiments where the microbial metabolic byproducts are automatically taken into account.

INTRODUCTION

Microorganisms influence corrosion in the environment through complex consortia that form biofilms. The combination of physiologically different microbial types in close association greatly increases the metabolic versatility of the complex. This versatility probably determines the principal type of corrosion observed since it appears that most known 'corroders' grow best on the relatively simple substrates which are provided by other bacteria. The problem in understanding the activities of these consortia comes from the inability of classical microbiological methods to define the biofilm community structure and metabolic activities. Viable plate count methods require quantitative recovery from the substratum and media that universally allow microbial growth. The first is difficult and the second nonexistent. Direct microscopic examinations using immunological probes allow identification of specific microbes and have helped significantly in understanding biofilm dynamics leading to MIC.

METHODS

In 'signature' biomarker analysis of biofilms cell wall or membrane components restricted to specific subsets of the microbial community can be utilized to define the microbial community structure. Phospholipids are found in membranes of all cells and are quantitatively extracted without the necessity of recovery of the microbes from the substratum. Phospholipids are not utilized as storage molecules in eubacteria and break down rapidly with the loss of cell viability (Nichols *et al.,* 1985). Analysis of phospholipid ester-linked fatty acids (PLFA) by gas chromatography and mass spectrometry (GC/MS), particularly noting the position and configuration of double bonds, branching and cyclopropane ring formation, allows identification of specific groups of microbes at levels of a few hundred cells (Odham *et al.,* 1985). A sampling area of at least 1 cm^2 is required for extraction when the film biomass is low. Endogenous storage polymers such as poly beta-hydroxy alkanoate (PHA) or extracellular polysaccharide glycocalyx, allow an estimation of starvation stress (White, 1986). The redox level for facultative anaerobic bacteria can be correlated with the ratio of the respiratory benzo- to naphtho-quinones (Hedrick & White, 1986).

Unfortunately these methods are destructive to the biofilm and provide information of the average composition of the biofilm at the specific time of extraction. Nondestructive biofilm analysis by Fourier transform infrared spectrometry (FT/IR) using attenuated total reflectance (ATR) attachments allows nondestructive examination of the living biofilms. With the microscopic probe, areas of biofilm 10–20 μm in diameter can be analysed thereby providing spatial resolution.

Corrosion is an electrochemical process where the dissolution of metal is controlled not only by charge transfer but diffusive processes. The analyses of such processes have concentrated mainly upon weight loss and polarization methods. Weight loss experiments are problematic in that removal of corrosion products from carbon steel surfaces is difficult to reproduce. No mechanistic interpretation is possible. Polarization analyses, on the other hand, allow a limited description of the corrosion processes and a simple calculation to obtain the corrosion rate in the form of the current density I_{corr}. Usually this follows the Stearn-Geary equation:

$$I_{corr} = \frac{b_a b_c}{2{\cdot}303(b_a + b_c)R_p}$$

where b_a and b_c are the anodic and cathodic Tafel slopes respectively, and R_p is the polarization resistance. Since R_p is defined as the slope of a current/voltage plot when $I = 0$, a scan through the corrosion potential gives a value for R_p. However, direct current (DC) plots may be expected to modify both biological and chemical surface films due to the drastic change in potential which is usually in the order of several hundred millivolts. To counter this, relatively nondestructive techniques such as AC impedance analysis, galvanostatic coupling and 'noise' monitoring were developed.

The AC impedance technique applies low amplitude sinusoidal waveforms (potentiostatic or galvanostatic) over a range of frequencies to a sample under test. The phase shifts of current to potential may be calculated in terms of the impedance. A value for R_p may be obtained from low frequency impedance data (Gabrielli, 1984) and in conjunction with other frequency impedances may be interpreted to give an idea of the predominating type(s) of overall corrosion mechanism.

In galvanostatic coupling two identical electrodes are separated by a 0·2 μM membrane which permits the passage of metabolites but not bacteria. The electrodes are connected via a 'zero' resistance ammeter (Little *et al.*, 1986) and provide a value for the net current generated by the difference in rate of dissolution of metal.

Electrochemical noise is a technique where the oscillations in potential of a corroding system are monitored as a passivation layer is sequentially broken and repassivated. The larger amplitude oscillations imply that pitting is a significant process (Hepburn *et al.*, 1986) and is a common phenomenon during MIC.

RESULTS AND DISCUSSION

An obvious mechanism of MIC is the generation of strong mineral acids by microbial metabolism. Concrete sewers in which volatile reduced sulphur compounds diffuse into aerated areas encourage the growth of the sulphuric acid producing *Thiobacilli.* These organisms generate sulphuric acid and the concrete degenerates. The bacteria are difficult to culture, grow slowly, and may be present in small numbers. Fortunately they contain a wealth of unique PLFA which can be readily detected and rapidly identified. The use of PLFA biomarkers analysed by GC/MS has shown excellent correlations with the degree of biodeterioration of concrete both in test devices and in the field (Kerger *et al.*, 1987).

Studies with organic acid producing facultative heterotrophic microbes which produce acids like acetic acid when grown under conditions of limiting oxygen, have shown MIC estimated by both DC potentiodynamic and AC impedance methods. Traces of these organisms on the surface of metal coupons were detected by analysis of PLFA. Other studies carried out at IFREMER have shown that cathodic protection even at high impressed currents cannot prevent corrosion from occurring in environments with high numbers of SRB. Figure 1 shows a complex plane impedance plot after being cathodically protected for 2 months. The frequency decades are also listed. Figure 2 shows evolution of a similar plot with over 1 month. Thereafter the cathodic protection was removed and the corrosion rate accelerated until a passivation film was formed.

Microbes in natural biofilms occur in consortia. Gaylarde & Johnson

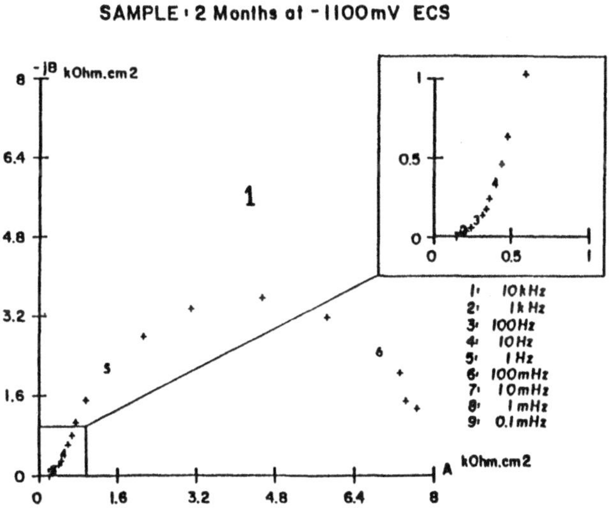

Fig. 1. This is an example of a complex plane impedance diagram. Here a 6 × 6 cm E24 carbon steel coupon was immersed in mud with high numbers of sulphate-reducing bacteria and cathodically protected at −1100 mV/Ag.AgCl. The data were obtained after an immersion time of 2 months. The semicircle describes a single capacitive loop to approximately 10 mHz. The extrapolated low frequency intercept on the descending (right arm) of the semicircle to the *X*-axis gives a value for R_p, while the high frequency intercept (left arm) provides the uncompensated resistance. This shows that there is still a significant rate of corrosion which is dominated by a single process, probably the transfer charge resistance. No diffusion phenomena are evinced which would imply that the cathodic protection has successfully prevented the formation of a passive film.

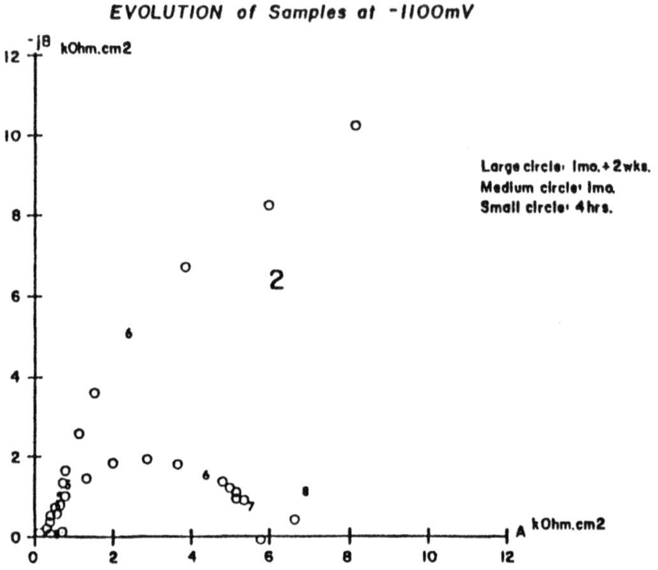

Fig. 2. Three plots are shown of coupons similar to those of Fig. 1 where cathodic protection is at −1100 mV/ECS. After 4 h immersion a very small semicircle (fast corrosion) was demonstrated. After 1 month the corrosion is slower (R_p 6 kOhms.cm^2) however when the cathodic protection is cut, corrosion occurs quickly and forms a passivation film after a further 2 weeks and R_p increases massively (large semicircle).

(1982) showed a combination of *Vibrio anguillarum* and *Desulfovibrio vulgaris,* under anaerobic conditions, induced greater weight loss than the sum of losses from either organism alone. Similar experiments in which aerated seawater was used in a continuous flow system confirmed the increased corrosion of the mixed culture and showed that the combination of heterotrophic, facultative acid-producing *Vibrio* could generate an anaerobic niche that SRB require for growth. The addition of bacteria with different physiological activities to artificial systems, with the coupons exposed to media that is continuously flowing, offer an excellent experimental tool to help define the mechanisms of MIC. Corrosion in these systems could be monitored by non-destructive techniques like AC impedance or electrochemical noise.

The ability to define accurately the microbes constituting a biofilm involved in MIC can be important in defining the mechanism. Careful analysis of SRB has shown that the lactate-utilizing *Desulfovibrio* contain significant quantities of unusual PLFA (Edland *et al.*, 1985) that can be readily differentiated from the acetate utilizing *Desulfobacter* type of sulfate-reducing bacteria (Taylor & Parkes, 1983). These biomarkers have been found in the biofilms formed on titanium and stainless steel pipes which have been exposed to pumped Atlantic seawater.

Desulfovibrio contains hydrogenases and is believed to facilitate corrosion by the classical mechanism of cathodic depolarization. *Desulfobacter* does not contain hydrogenase yet in co-culture with facultative heterotrophs shows facilitation of corrosion. Clearly there are other mechanisms besides cathodic depolarization involved in MIC with the SRBs. There is evidence however, that monocultures of *Desulfovibrio* probably facilitate corrosion to a greater degree than monocultures of *Desulfobacter*.

Many bacteria produce extracellular polymers that condition the microenvironment at surfaces and induce irreversible adhesion. Microscopic examination of biofilms of the heterotrophic *Vibrio natriegens* showed that increased corrosion rates of stainless steel corresponded to the elaboration of extracellular polysaccharide polymer by the biofilm. The examination of the coupons with FT/IR correlated bacterial growth with the appearance of adsorption at 1660 cm^{-1} (Amide I of the bacterial proteins) whereas the exopolymer maximal adsorption at 1440 cm^{-1} was found to be a calcium-polysaccharide. Rapid increases in the corrosion rate corresponded with increases in the 1440 cm^{-1} absorption not the bacteria themselves (1660 cm^{-1}). Removing the bacteria by sonication had little effect on the corrosion rate whereas removing the exopolymer decreased the corrosion rate to control levels. In these experiments it proved possible to show reversible MIC could result from nonuniform deposition of extracellular polymer.

The direct demonstration of facilitation of corrosion by bacterial polymers has been demonstrated elegantly (Geesey *et al.*, 1986). It is possible to follow the chemical changes in living biofilms by using the ATR attachment. In experiments using metal sputter-coated crystals (ATR) the application of acidic extracellular polymers purified from the bacteria were shown to increase significantly the rate of metal loss. Further studies by Geesey and colleagues with surface chemical techniques like Auger and PMS spectroscopy have shown that copper becomes oxidized as it is bound by bacterial extracellular polymers.

The study of MIC is obviously still in its infancy. As new nondisturbing techniques such as FT/IR, biomarkers and transitory-type electrochemical analyses become more widely used the mechanisms involved in MIC will be better explained.

REFERENCES

Edland, A., Nichols, P.D., Roffey, R. & White, D.C. (1985). *J. Lipid Res.*, **26**, 982–8.

Gabrielli, C. (1984). Identification of Electrochemical Processes by Frequency Responce Analysis. Technical report number 004/83. Solartron Instruments, Farnborough, UK.

Gaylarde, C.C. & Johnson, J.M. (1982). *Int. Biodeterior. Bull.,* **18**(4), 111–16.
Geesey, G.G., Mittleman, M.W., Iwaoka, T. & Griffiths, P.R. (1986). *Materials Performance,* **25**, 37.
Hedrick, D.B. & White, D.C. (1986). *J. Microbiol. Meth.,* **5**, 243–54.
Hepburn, B.J., Gowers, K.R. & Scantlebury, J.D. (1986). *Br. Corr. J.,* **21**(2), 105–8.
Kerger, B.D., Nichols, P.D., Sand, W., Bock, E. & White, D.C. (1987). *J. Industrial Micro.,* **2**, 63–9.
Little, B., Wagner, P., Gerchakov, S.M., Walch, M. & Mitchell, R. (1986). *Corrosion,* **42**(9), 533.
Nichols, P.D. & White, D.C. (1985). *J. Environ. Geolog. Water Sci.,* in press.
Odham, G., Tunlid, A., Westerdahl, G., Larsson, L., Guckert, J.B. & White, D.C. (1985). *J. Microbiol. Meth.,* **3**, 331–44.
Taylor, J. & Parkes, R.J. (1983). *J. Gen. Microbiol.,* **129**, 3303–9.
White, D.C. (1986). *Proceedings of the International Conference on Marine Biodeterioration,* Goa, India. January, 1986. Oxford and IBH Publishing Co., New Delhi, India.

An extended bibliography for this paper may be obtained from the senior author.

Monitoring Sulphate Reduction Activity in the Field Using Radiorespirometry

Stephen Maxwell

Media Supplies, Blair Cottage, Fintray, Aberdeen AB2 0JL, UK

INTRODUCTION

The methods employed for bacterial analysis in the oilfields are generally of a standard far below that expected in a high technology industry. Recently, more qualified microbiologists have been working in the field and with this a greater understanding of the nature of bacterial growth in offshore systems has developed. This has led to a demand for more accurate and meaningful techniques for assessing bacterial growth and the nature of the problem they present.

The predominant organisms associated with bacterial corrosion are the sulphate-reducing bacteria (SRB). Many different areas may be prone to attack, including water injection and produced water systems; oil storage cells; sea water ballasted legs; and the external structures of the platform itself.

In general, SRB are monitored in the field using cultural methods. They are being improved by the introduction of Most Probable Number (MPN) techniques; better sampling and detailed monitoring of sessile bacteria. Other techniques are being tested including epifluorescence microscopy, immunology and radiorespirometry. At present, however, judgements on the need to treat a system are still often based on the numbers of SRB counted using a single serial dilution count of bacteria in a bulk phase sample.

For several years sophisticated techniques employing radioactive [^{35}S]sulphate have been used to determine the activity of SRB in natural sediments (Jorgensen, 1977, 1978, 1982). These techniques were modified so they could be used for monitoring the activity of SRB in oilfield situations (Rosser & Hamilton, 1983). As bacterial corrosion is normally associated with surface phenomena, the technique was further modified to allow the determination of SRB activity in biofilms on metal surfaces (Maxwell & Hamilton, 1986). Two field trials were performed to show the applicability of these techniques in the field.

MATERIALS AND METHODS

Sampling sites

Two areas where SRB control measures (i.e. biocide addition) are not normally implemented are in ballast legs and on the external structures of the platform. SRB activity in these areas would, therefore, be expected to be high. For this reason these areas were chosen to illustrate how radiorespirometry can be used to improve our understanding of bacterially related corrosion in the field.

Thistle platform

Many drill muds used in drilling operations are organically rich. The cuttings produced will, therefore, be contaminated with organic material. On the Thistle jacket, cuttings have built up on some of the horizontal members. Active microbial populations, including SRB, can develop within them and initiate sulphide corrosion processes.

Dulplicate core samples of the cuttings piles on members 42 494 (−424 ft (−129 m)) and 52 494 (−524 ft (−160 m)) were taken by divers. Subsamples from one core were assayed immediately for sulphate reduction activity using the method of Rosser & Hamilton (1983). Numbers of SRB were determined using the triplicate Most Probable Number (MPN) technique. The duplicate core was transported back to the laboratory for chemical analysis of the mud and pore water. A profile was made of the two cuttings piles, in terms of the rate of sulphate reduction, numbers of SRB, sulphide and sulphate concentrations, water and organic content.

Brent Bravo platform

Entrained sea water within the drill leg of the Brent Bravo platform can become contaminated with organic material, e.g. waste oil, drilling mud, etc. The water is stagnant and the activity of aerobic bacteria can lead to a reduction in the dissolved oxygen content of the water. Where organic contamination is sufficiently large, the aerobic bacteria can reduce the dissolved oxygen concentration to the point where anaerobiosis occurs.

SRB may then become active with the concomitant production of H_2S (Wilkinson, 1982). The production of H_2S in these systems, in addition to sulphide film formation at metal surfaces and possible corrosion problems, can also lead to release into the atmosphere, causing personnel safety problems.

Corrosion coupons (3·0 × 1·5 × 0·5 cm) of 50D mild steel attached to polypropylene plates were immersed in the entrained sea water within the west drill leg for 73 days. Sulphate reduction rates were determined on the corrosion coupons and in 10 ml samples of the bulk phase using the method of Maxwell & Hamilton (1986). Sessile and planktonic SRB numbers were determined using the MPN technique. Sulphate concentration in the entrained water was measured, together with soluble sulphide in the bulk phase and acid volatile sulphide on the metal coupons.

RESULTS

The result of the field surveys on the Thistle and Brent Bravo platforms are presented in Tables 1 and 2 respectively.

DISCUSSION

Thistle survey

Sulphate reduction rates

In both cores the rate of sulphate reduction decreased through the core until, at the metal surface, sulphate reduction rates were minimal. In the absence of further sulphide production at the metal surface it is possible that any sulphide film formed will be protective to the underlying steel. When cathodic protection systems are in use, however, sulphide films may force the cathodic hydrogen back into the steel. In this case, the potential corrosion mechanism is not of sulphide pitting but of hydrogen embrittlement.

Sulphate and sulphide concentration

The sulphate concentration in the pore water of both cores decreased significantly with depth. This suggests that conditions in the deeper part of the pile have become limiting with regard to sulphate. The sulphide concentration might be expected to be inversely proportional to sulphate.

TABLE 1
Results from the Thistle Survey

	Cuttings					Pore water		
Member	*Depth of cuttings (cm)*	*Sulphate reduction rate*[a] *(nmol/g/d)*	*Number of sulphate reducers (cells/g)*	*Water content (%)*	*Organic content (%)*	*pH*	*$[S^{2-}]$ (mg/litre)*	*$[SO_4^{2-}]$ (g/litre)*
42494	+17 to +11	49·8	0	32·08	5·15	8·7	69	1·25
	+10 to +6	12·0	10^2	35·36	4·85	8·3	34	0·17
	+5 to 0	0·6	0	37·67	3·91	8·4	0	0·20
52494	+17 to +12	134·2	ND[b]	ND	ND	ND	ND	ND
	+11 to +6	55·1	10^2	32·31	4·44	8·9	10	0·27
	+5 to 0	0·7	0	32·94	6·01	7·3	83	0·36

[a] Sulphate reduction rates and numbers of sulphate-reducing bacteria are expressed per gram wet weight of cuttings.
[b] ND = not determined.

Significant levels of sulphide were found but no clear pattern could be seen when the two cores were compared. In the core taken at −424 ft (−129 m), no sulphide was found directly above the metal surface. In the core from −524 ft (−160 m), however, a sulphide concentration of 83 mg/litre was measured directly above the metal surface. This inconsistency may be due to the loss of soluble sulphides from the pore water via volatilisation, or reaction with the metal ions present to form insoluble metal sulphides.

Organic content of cuttings
In both cores significant levels of organic matter could be determined throughout the cuttings pile. Although it is not known whether this organic material is directly available as nutrients for SRB, it is suggested that carbon limitation in this situation is unlikely.

TABLE 2
Results from the Brent Bravo Survey

	Sulphate reduction rate ($mg/m^2/d$)	*SRB numbers*	*pH*	*$[S^{2-}]$ (mg/m^2)*	*$[SO_4^{2-}]$ (g/litre)*
Metal surface	47·98	10^8 cells/m^2	ND	2223	ND
Bulk phase	Below detection	10^3 cells/litre	7·2	0	2·10

Numbers of SRB
SRB were present in low numbers: 10^2/g wet weight of cuttings, and only in the mid-section of the cores. Counts of SRB showed no clear pattern through the cuttings pile and numbers did not correlate with the activity measurements in either core.

In summary, the data on sulphate reduction rates, when coupled with measurements of sulphate concentration in the pore water, indicate that SRB are sulphate-limited in the depths of the cuttings piles and that no significant sulphide production is occurring in the area near the metal surface. Bacterial counts alone reveal nothing of the biological processes taking place.

Brent Bravo survey

Sulphate reduction rates
Sulphate reduction rates measured in the biofilms were the highest recorded by the author in any system using this technique. Sulphate reduction was not detected in 10 ml samples of the bulk phase. It is clear, therefore, that in this situation the potential for sulphate reduction on surfaces is far greater than in the bulk phase.

Sulphate and sulphide concentration
Sulphate depletion was observed in the bulk phase; the sulphate concentration being 2·10 g/litre, whereas in open sea water the concentration is normally around 2·50–2·80 g/litre. Although the level of acid-volatile sulphide on the metal coupons was high, the level of soluble sulphide in the water column was below detection. Only an approximate value for acid-volatile sulphide can be given as this amount of sulphide will affect the efficiency of the sulphide assay. The same is true for the trapping of [^{35}S] sulphide in the radiorespirometric assay. Sulphide concentration and sulphate reduction rates reported for the metal coupons are, therefore, minimum values for both concentration and rate.

Although no soluble sulphide was observed in the bulk phase, the depleted sulphate concentration measured is indicative of sulphate reduction activity. Because H_2S is sparingly soluble in water it will partition into the atmosphere in a ratio in the order of 200 to 1 (Wilkinson, 1982). This partitioning effect may explain why sulphide cannot be detected in the bulk phase. The results suggest that the majority of H_2S produced on the surface of the coupon is precipitated as metal sulphides.

In the Brent Bravo drill leg oxygen sparging has been used in the past to control the levels of soluble sulphide in the entrained water. Oxidation of sulphide films may produce conditions which are extremely aggressive towards the steel, leading to accelerated corrosion rates (Hardy & Bown, 1984).

Organic content of system

In the drill leg the introduction of organic material into the water is intermittent and, therefore, conditions may become carbon-limited.

SRB

Numbers of SRB were 10^8 cells/m^2 on the metal surface and 10^3 cells/litre in the bulk phase.

The efficiency of enumeration techniques was estimated using the data obtained from pure culture studies, as described by Jorgensen (1978). It was found that the number of SRB determined on a lactate based medium could underestimate the actual numbers by 2–3 orders of magnitude.

CONCLUSIONS

In the Thistle survey activity measurements give a much more detailed picture of the conditions within the cuttings pile than do the bacterial counts. The poor counts cannot be completely explained but this type of erratic result is by no means uncommon.

In the Brent Bravo survey measurement of sulphate reduction activity allows a clearer understanding of the biology of the system. Radiorespirometric analysis of the corrosion coupons illustrates that sulphide is being produced at metal surfaces and precautions against corrosion may have to be taken.

The results emphasise that monitoring of the bulk phase does not allow an accurate assessment of the conditions at metal surfaces. The increased awareness that surface measurements are essential has already led to the development of special tools to allow monitoring of bacterially fouled surfaces, i.e. bioprobes for water injection systems.

ACKNOWLEDGEMENTS

I should like to acknowledge the help of Britoil plc and Shell Exploration and Production (UK) Ltd in making their facilities available for this work to be carried out. The research was supported by a SERC CASE award with Shell Exploration and Production (UK) Ltd, and was carried out in the Department of Microbiology at Aberdeen University, Scotland.

REFERENCES

Hardy, J.A. & Bown, J.L. (1984). In *Proceedings of UK Corrosion 1984 Conference*, London, November, 1984, pp. 135–8.

Jorgensen, B.B. (1977). *Marine Biology*, **41**, 7–17.

Jorgensen, B.B. (1978). *Geomicrobiology Journal*, **1**, 11–27.

Jorgensen, B.B. (1982). *Nature*, **296**, 643–5.

Maxwell, S. & Hamilton, W.A. (1986). *Journal of Microbiological Methods*, **5**, 83–91.

Rosser, H.R. & Hamilton, W.A. (1983). *Applied and Environmental Microbiology*, **45**, 1956–9.

Wilkinson, T. (1982). *Chemistry and Industry*, **4**, 115–23.

MUSEUMS

Actinomycetes and Biodeterioration in the Field of Fine Art

C. Giacobini, M.A. De Cicco, I. Tiglie & G. Accardo

Istituto Centrale per il Restauro, Ministero per i Beni Culturali ed Ambientali, Pizza S. Francesco di Paola 9, 00184 Roma RM, Italy

ABSTRACT

The paper examines the problems concerning the Actinomycetes *biodeteriogens on frescoes of underground environments. It illustrates: (1) the serious damage suffered by a complete series of famous frescoes; (2) the techniques of non-destructive sampling and (3) the media used for isolating both on site and in the laboratory. Two new aspects are presented: the study of the relation between the microclimate of the environment where the frescoes are situated and the humidity required by* Actinomycetes. *This has highlighted the close correlation between thermohygrometric values necessary for the development of* Actinomycetes *and the actual values. The isolated strains are classified for the first time, as those which attack mural paintings. They are:* Streptomyces rectus flexibilis, S. griseoulus, S. cinereoruber, S. vinaceus, S. albus *and* Nocardia *sp.*

INTRODUCTION

Among the biodeteriogen groups which can destroy a work of art, *Actinomycetes* are, without doubt, the most peculiar. In works of art these have never been isolated except from mural paintings (frescoes in particular) and only when they are located in crypts, grottos and tombs, etc. *Actinomycetes* produce chemical changes, and remarkable amounts of mineral salts are often mixed with their mycelium.

The first evidence that *Actinomycetes* were biodeteriogen is attributed to the ICR (Central Institute for Restoration) 25 years ago. Since then research has continued and the basic results have been recorded (C. Giacobini, 1963, 1985). A publication issued by the CNR (1986) concerns some studies on the biodeterioration, in St Clement's in Rome, and contains the classification of some *Actinomycetes* from the group of frescoes of that basilica (G. Agarossi *et al.*, 1986). The present paper provides an overview of the achievements attained through the study of various problems concerning the growth of *Actinomycetes*. Besides making proposals concerning standard methods of research this paper gives the classification of the actinomycetes that were responsible for attacking several groups of frescoes situated in different places and belonging to different centuries.

MATERIALS AND METHODS

Excavations, crypts, tombs and rupestrian grottos

A large number of sites were studied which contained paintings of great historical importance and artistic quality. An answer to the problem of their protection from the attacks of *Actinomycetes* is urgently required.

Measurement of the environmental parameters

Biological research cannot be separated from the microclimate of the sites where the paintings are located. Previously the microclimatic parameters have been established wherever possible, from data from former scientific studies, or from one-off measurements of the temperature and relative humidity. In the present study, they have been obtained from the regular measurements made by the Department of Physics ICR during microbiological research.

Microbiological analysis

'In situ'

The microbiological observations which must be made '*in situ*' are a visual examination of the species of *Actinomycetes* present by skilled experts and samples of the mycelium taken by scalpel or a swab. In some cases the presence of *Actinomycetes* can be assessed through the simple technique of the exposed plate, as in the present work, or through instruments such as the Andersen sampler and the surface air system (PBI).

Laboratory analysis

In the laboratory the samples are examined under the microscope to determine the gravity of the attack. From the collected material, which usually consists of about 20–50 mg, dilutions from 10^{-3} to 10^{-5} are prepared, and 1 ml aliquots inoculated into three different kinds of media (Mycological Agar, Starch Casein medium and Glycerol Arginine medium). After the plates have been incubated for 21 days at 28°C the colonies are counted and the severity of the infection assessed. The colonies are transferred onto Bacto I.S.P. Medium 5 (Bacto-Glycerol Asparagine Agar Base) to obtain pure cultures for identification, particularly those colonies which appear most frequently. Finally the species are identified. Professor R. Locci of the Institute of Patologia Vegetale of the University of Milan, Dr F. Parente and Dr G. Berretta of the 'Lepetit' Institute in Milan have undertaken the identification for the present work.

RESULTS AND DISCUSSION

The growth of *Actinomycetes* on the frescoes produces a whitish powdery mycelium, or colourless patinas that are very light, uniform and sometimes scattered with small light grey grains. The colours of the frescoes are faded and dusty in appearance. In particular, the state of conservation of the tomb frescoes usually shows slight but significant differences with respect to crypt and excavation frescoes. In the tombs, in fact, the frescoes appear to be either covered with a whitish patina or hidden by various types of efflorescence. On the other hand frescoes that are excavated are covered with a very slight, impalpable and shiny powder under which the colour has a dusty appearance. The powder changes into a thicker and more homogeneous layer, which is soft at first, but becomes harder and harder. Coverings half-way between the two have often been observed on crypt frescoes. As to the propagation time of the mycelium on tomb and crypt frescoes, it is interesting to note that when frescoes are restored without modifying the environment they appear to have recovered the original chromatic equilibrium, but a few years later they look 'dusted' again or are covered with a thick white layer that makes it impossible to read the particulars. The attack from *Actinomycetes* on excavated frescoes appear, instead, within about 2 months from the exposure of the painting to the air. After restoration and control of the ambient conditions the growth is usually impeded. If the excavation work is stopped because of the usual summer or winter pause,

TABLE 1
Distribution of the *Actinomycetes* Identified in the Different Sites

Genera and species	*Sites and localities* Capitolium BRESCIA	Domus Aurea ROMA	St Clement's ROMA	St Margareth MELFI	SS John and Paul ROMA	St Margareth MOTTOLA	Of the Ducks VEIO	Bartoccini TARQUINIA	Of the Bell VEIO	Of the Hannibal's warrior TARQUINIA	Of Hescanas ORVIETO	Of the Ogre TARQUINIA	Of the Relief TARQUINIA	*Total numbers for species* Surface	*Air
	Excavs		*Crypts*			*Grotto*	*Tombs*								
Streptomyces:															
S. albidoflavus					1									1	
S. albidus							*							—	1
S. albus		1	2					1	1					5	
S. antibioticus							1							1	
S. aurantiacus									1					1	
S. aureus			1											1	
S. catenulae							**							—	2
S. cinereoruber		2	1			1							1	5	
S. galbus									1					1	
S. griseoaurantiacus									*					—	1
S. griseolus							3**		2*					5	3
S. intermedius			1											1	
S. limosus							1		*					1	1
S. odorifer				1										1	
S. rectus flexibilis							2		2		6			10	
S. retinaculum											2			2	
S. roseogriseus							1		1*	1				3	1
S. spira							1		1		1			3	
S. vinaceus				2		1			1	1				5	
Nocardia:															
Nocardia sp.	1	1	2									1		5	
Total numbers for sites — Surface	1	4	7	3	1	2	9	1	10	2	9	1	1	51	
Total numbers for sites — *Air							5		4						9

the attack is usually ruinous by the next spring or autumn. In frescoes from excavations the actinomycetes often occur with bacteria, fungi and algae.

Studies on the microclimate of the sites have proved that it is usually characterised by a high degree of constant relative humidity, between 90% and 100%. The temperature varies between 6 and 7°C in winter to 16 and 18°C in summer. The thermohygrometric dynamics, which are affected by the sun, favour water exchange between the material and the environment through a process of absorption, adsorption, evaporation, condensation and formation of a liquid/water film on the surface of the painting.

It is possible to obviate such phenomena by modifying the microclimate to obtain values which do not favour the growth of the *Actinomycetes.*

Direct examination of the samples under an optical microscope only provides an indication of the gravity of the attack. More useful is examination under a scanning microscope, where the characteristics of the mycelium and spores can be observed.

Culturing gives positive results if several culture media are used simultaneously.

The number of *Actinomycetes* cells per gram of sample varies between 5000 and 1 000 000 from one fresco to another and from one spot to another.

About 200 strains of *Actinomycetes* have been isolated, sixty of which have been identified as far as the species. Their distribution in the different sites is shown in Table 1.

The strains which have been most frequently identified and therefore appear to be the most widespread are *Streptomyces rectus flexibilis, S. albus, S. cinereoruber, S. griseolus, S. vinaceus* and the *Nocardia* genus. The *S. rectus flexibilis* has been found ten times, *S. albus* fives times whilst *S. cinereoruber* has also been found five times.

S. griseolus has been found only twice, as has *S. vinaceus.*

S. roseogriseus has been found four times and *Nocardia* five.

CONCLUSIONS

The limitations of this work are numerous. Only sixty of the 200 strains that have been isolated have been classified due the difficulties in obtaining an accurate classification.

Another obvious limitation is the incompleteness of the research on *Actinomycetes*, the damage produced and the relation between the metabolic processes and the simultaneous presence of nitrates, chlorides and, chiefly, sulphates, on the frescoes.

Another difficulty is the need to collaborate with research chemists, which has not been possible.

Art historians and archaeologists, although in agreement on the need, are not always able to consent to the modification in environmental factors and the biologist cannot always advise on biocide products, for two reasons. First, experimental evidence is lacking and second, a global view on the dynamics and the developments of the phenomenon is still not available. However, general information is available suitable to control the development of these organisms in the stages which are most dangerous for the works of art.

ACKNOWLEDGEMENTS

We are greatly indebted to Professor R. Locci, Drs F. Parente and G. Berretta for their identification of the *Actinomycetes* species. We are also very grateful to Mr C. Cacace and Dr R. Rinaldi for their invaluable collaboration in the research and to Dr M.R. Giuliani, Mr B. Mazzone and Mr M. Spinucci for their assistance.

REFERENCES

Agarossi, G., Ferrari, R. & Monte, M. (1986). In *Scientific Methodologies Applied to Works of Art*, ed. P.L. Parrini, pp. 52–6.

Giacobini, C. (1963). In *Recent Advances in Conservation*, Butterworths, London, pp. 62–4.

Giacobini, C. (1985). In *Science Biology, Enc. I.* Geographical Institute De Agostini, Novara, pp. 323–34.

An extended bibliography for this paper may be obtained from the senior author.

The Effect of Microbial Growth on Synthetic Polymers Used on Works of Art

O. Salvadori

Laboratorio Scientifico della Soprintendenza ai Beni Artistici e Storici di Venezia, Cannaregio 3553, 30131 Venezia, Italy

&

M.P. Nugari

Istituto Centrale per il Restauro, Via di San Michele 22, 00153 Roma, Italy

ABSTRACT

Biological growth has been found on some synthetic polymers used in the treatment of monuments. The present work investigated suitable methods for evaluating the ability of these materials to sustain microbiological growth, and to determine if consequently the efficacy of the polymers was altered. Preliminary tests were carried out on some preservatives and consolidants. More extensive tests were made on samples treated with these products then incubated in a climatic cabinet, or submitted to soil burial testing. The possible change in the efficacy of the polymers after biological attack was evaluated by measuring contact angle and water absorption properties.

INTRODUCTION

The use of synthetic polymers in the field of the conservation of works of art has grown in the last few years and the materials suggested for use are increasingly numerous. The problem arises of their suitability for the consolidation and protection of outdoor monuments. In the context of suitability would come their long-term biological stability.

It was also necessary to study the possible interaction between biological growth and the treated stone surfaces.

On the basis of preliminary tests carried out on acrylic resin films, our research was extended to cover a larger group of products, including non-film-forming resins and products applied directly onto the stone.

Although atmospheric pollution problems provide information on the biodeterioration of synthetic polymers, it is not always easy to interpret these findings. The present work included a study of the validity of the internationally-accepted test methods when applied to materials used in restoration. A preliminary test was carried out using five resins in contact with active mud (industrial waste from factories producing synthetic resins) to see whether there was any interaction with the microflora. Subsequently testing was carried out using the soil burial method and climatic chamber exposure on stone samples treated with two different resins.

EXPERIMENTAL

A. Microbiological tests with polymeric materials

Five different resins were used: Silirain 50 (a methyl methoxy silane) and Paraloid B72 (an acrylate-methacrylate copolymer), dissolved in trichloroethane, EP 2101 (an epoxy resin) dissolved in isopropanol and toluene, Fomblin Y Met (a perfluoropolyether) dissolved in trichloro-trifluoroethane and Tegosivin HL100 (a siloxane derivative) dissolved in hydrocarbons.

Biologically active mud (50 ml) in sterile containers was treated with 5 ml of the particular resin in solution. The microbial content at zero time, 1 and 4 months was determined by the method of step dilution, inoculating Plate Count Agar with 1 ml of suspension at every dilution.

B. Microbiological test with treated stone samples

Stone samples, of three different types: Carrara marble, a sandstone and Lecce limestone, 5 × 5 × 2 cm, were treated by capillary rise with Tegosivin HL100 and another siloxane, Wacker 290L.

Half of the treated samples were subjected to the soil burial test, according to the British Standard 6058 (1981), section 2, appropriately modified.

The other half was exposed for one month to atmospheric pollutants then incubated in a climatic chamber at 28°C and 95% relative humidity.

After 30 days of incubation, growth on the samples was evaluated by a 0–5 scale which includes both the extent of cover and density of growth

(0 represents no growth, 1 slight growth and 5 total cover of dense growth) using visual and microscopical examination.

The determination of water repellency was assessed by contact angle measurement, using the Swedish Lorentz and Wettre instrument which permits a direct reading of both the height and diameter of the drops. Readings were taken 15 seconds after the drop was applied to the surface.

Capillarity absorption was measured using the method described in Normal 11/82 (1983), with the results expressed in % efficacy.

RESULTS AND DISCUSSION

The most striking result was that provided by tests on the active mud. All resin-inoculated samples showed an increase in the microbial flora in comparison with the control, except for the epoxy and perfluoropolyether resins. For these exceptions it was necessary to see whether the reduction was due to the solvents or whether these polymers were resistant to biological attack. Tests using only the solvents are now in progress.

The stone samples gave different results depending on which test method was employed. In those subjected to the burial test, growth was not easy to estimate, either visually or under a stereo-microscope, because of the soil particles clinging to the surface. On those in the climatic chamber the microbiological growth was clearly evident. No differences were observed between the two resins, nor between the stone samples.

The most frequently occurring species are shown in Table 1.

Data obtained by measurement of contact angle and capillarity absorption of the resin-treated sample show no distinction between the control and the biologically-treated samples. Only with Wacker 290L and soil burial testing was there a slight reduction in protective ability and water repellency with respect to the controls.

TABLE 1
Fungi Isolated from Specimens Treated with Tegosivin HL100 and Wacker 290L

1. *Alternaria alternata*
2. *Alternaria* sp.
3. *Cladosporium cladosporoides*
4. *Myrothecium verrucaria*
5. *Trichothecium roseum*
6. *Ulocladium atrum*

Comparing the results obtained with the two methods, soil burial is more severe.

CONCLUSIONS

Biological growth on the samples exposed to climatic chamber testing and the increase in the concentration on the active mud suggests adaptation of some microorganisms to these synthetic substrates, especially when an additional source of nutrient is present. This occurs in practice, when outdoor monuments accumulate atmospheric and earth particles.

The two test methods appear equally valid for an evaluation of possible interference with microbiological growth by the synthetic resins.

Both incubation methods were very severe and even the controls were affected by the temperature, relative humidity and soil burial.

Extended periods of incubation should now be employed, in order to see whether the metabolic products of the fungi can bring about an alteration in the polymers, over and above the simple deterioration caused by the physico-chemical test conditions. It is considered that even under the most favourable growth conditions, the microorganisms on the resin substrate would require an adaptation period before producing, when possible, adaptive enzymes capable of hydrolysing the polymer molecules.

REFERENCES

BSI.6085 (1981). Method of test for determination of the resistance of textiles to microbiological deterioration. Section two: soil burial test. BSI, London.

Normal 11/82 (1983), Assorbimento d'acqua per capillaritá-coefficiente di assorbimento capillare. CNA and ICR, Roma.

An extended bibliography for this paper may be obtained from the senior author.

Microbial Deterioration of Historic Glass Plate Negatives

B.J. Zyska, Z.T. Cieplik, A.R. Wójcik & Renata Kozłowska

Central Mining Institute, Plac Gwarków 1, 40-951 Katowice, Poland

ABSTRACT

The Krieger collection, comprising 12 482 glass plate negatives made in the years 1880–1926, has been stored from 1968 in the Historical Museum of the City of Cracow. Among 5414 glass negatives examined, 20 were found to be contaminated by fungal growth. Among 13 species of fungi Aspergillus nidulans, A. versicolor, Cladosporium cladosporioides, Penicillium cyclopium *and* P. janthinellum *were the most frequent. At 63% RH we confirmed a scarcely visible growth, on shellac and collodion, but not on dammar.*

INTRODUCTION

Since the earliest days of glass plate negatives thousands and thousands have accumulated in museums, archives, libraries and private collections all over the world (Iliński, 1970; Zyska, 1987). Photographic records, including glass plate negatives, represent an item of increasing importance in library and museum acquisitions, especially in view of the unusual reflection they give of our cultural and historical heritage (Williams, 1970).

This paper presents the results of investigations on the microbial deterioration of glass plate negatives from a study of certain factors governing the correct storage of such collections.

MATERIALS AND METHODS

The collection of glass plate negatives known as the 'Krieger' collection, has since 1968 been stored in the Historical Museum of the City of

Cracow. The collection comprises 12 482 glass negatives of which 5414 have been catalogued and examined. Of these, *c*.1400 were produced by the wet collodion process and 4000 by the bromised gelatine process. They were made in the years between 1880 and 1926. Some of the negatives are coated with varnishes. Chromatographic studies reveal that shellac, dammar and collodion have been used. The collection has miraculously survived two world wars.

The damage suffered is both chemical and mechanical. On some partial destruction and peeling of emulsion and wrinkling of the image surface were observed. Some were stuck together and others displayed numerous holes in the surface. On 50 glass negatives, especially on those coated with varnishes, colonies of different species of fungi were observed as:

—concentric circles on varnishes;
—growth on the collodion layer;
—light coloured, branched colonies strongly adhering to the glass.

Of the 50 negatives, 20 showed distinct fungal growth and were subjected to microbiological examination. The fungi were isolated by three methods, i.e.

—direct peeling of the mycelium from the glass by a sterile lancet into sterile, distilled water;
—swabbing with cotton wool from glass, collodion, gelatine and varnish;
—impression by pieces of sterile transparent, self-adhesive sealing tape taken from collodion, gelatine and varnish.

The following media were used: malt extract agar, Sabouraud agar, for fungi, carboxy-methyl cellulose agar and potato-carrot medium. All the plates were incubated at $25 \pm 1°C$ for 3 weeks and pure cultures identified.

In the humidity chambers a series of negatives was prepared to model the formulation of the museum specimens. The plates were coated by bromised gelatine and one of three varnishes, i.e. shellac, dammar or collodion. They were sprayed with a suspension in sterile water, of fungi isolated from the museum negatives. The mixture included *Penicillium canescens*, *P. cyaneum*, *P. cyclopium*, *P. janthinellum*, *Aspergillus repens*, *A. nidulans*, *A. versicolor*, *Paecilomyces lilacinus*. The samples were placed in their respective humidity chambers at 35, 63, 65, 76, 86, 92% RH. Non-infected control samples were kept at 35 and 92% RH. The tests lasted for 6 weeks. After which the fungi were cultured on malt extract agar and identified.

TABLE 1
Fungi Isolated from 20 Glass Plate Negatives in the Historical Museum of the City of Cracow

	Fungus species	*Number of times isolated*	*Percentage occurrence*					
			Glass plate	*Carrier*		*Varnish*		
				Bromised gelatine	*Collodion*	*Collodion*	*Dammar*	*Shellac*
1	*2*	*3*	*4*	*5*	*6*	*7*	*8*	*9*
1.	*Aspergillus nidulans* (Eidam) Winter	8	87·5	0	0	12·5	0	0
2.	*Aspergillus repens* (Corda) Saccardo	6	16·6	0	0	50·2	0	33·2
3.	*Aspergillus versicolor* (Vuillemin) Tiraboschi	8	62·5	12·5	0	12·5	12·5	0
4.	*Cladosporium chladosporioides* (Fresenius) de Vries	4	100·0	0	0	0	0	0

5. *Mycelia sterilia*	2	50·0	0	0	0	50·0	0
6. *Paecilomyces lilacinus* (Thom) Samson	1	100·0	0	0	0	0	0
7. *Paecilomyces variotti* Bainer	1	100·0	0	0	0	0	0
8 *Penicillium canescens* Sopp	1	100·0	0	0	0	0	0
9. *Penicillium cyaneum* (Bainier & Sartory) Biourge	1	100·0	0	0	0	0	0
10. *Penicillium cyclopium* Westling	5	80·0	0	0	20·0	0	0
11. *Penicillium expansum* Link ex S.F. Gray	1	100·0	0	0	0	0	0
12. *Penicillium janthinellum* Biourge	3	100·0	0	0	0	0	0
13. *Rhodotorula* sp	1	100·0	0	0	0	0	0

TABLE 2
Colonisation of Varnish Surfaces by Fungi After 6 weeks in the Humidity Chamber

RH (%)	*Dammar*		*Shellac*		*Collodion*	
	Growth estimation	*Species*	*Growth estimation*	*Species*	*Growth estimation*	*Species*
1	*2*	*3*	*4*	*5*	*6*	*7*
92%	4 / +++	*A. nidulans* *A. versicolor* *P. janthinellum*	4 / +++	*A. nidulans* *A. versicolor* *P. janthinellum*	4 / +++	*A. nidulans* *A. versicolor* *P. cyclopium* *P. janthinellum*
86%	4 / +++	*A. nidulans* *P. janthinellum*	4 / +++	*A. nidulans* *A. versicolor* *P. janthinellum*	4 / +++	*A. nidulans* *A. versicolor* *P. cyclopium*
76%	1 / ++	*A. repens* *A. versicolor*	3 / ++	*A. versicolor*	3 / ++	*A. versicolor* *P. cyclopium*

65%	1 / ++	*A. nidulans* *A. repens*	2 / ++	*A. versicolor*	2 / ++	*A. nidulans* *A. versicolor* *P. janthinellum*
63%	01 / –	–	1 / +	Single hypha	1 / +	Single hypha
35%	0 / –	–	0 / –	–	0 / –	–
Control samples of varnishes—not artificially infected						
92%	0 / –	–	4 / ++	*A. nidulans* *P. frequentans*	1 / ++	*A. nidulans* *A. versicolor* *P. cyclopium*
35%	0 / –	–	0 / –	–	0 / –	–

0, growth absent, not visible even under a microscope; 1, weak growth, under 25% of surface; 2, growth over 25–50% of surface; 3, growth over 50–75% of surface; 4, extensive growth over 75–100% of surface.
–, growth absent; +, somatic hypha; ++, partial sporogenesis; +++, sporogenesis in total.

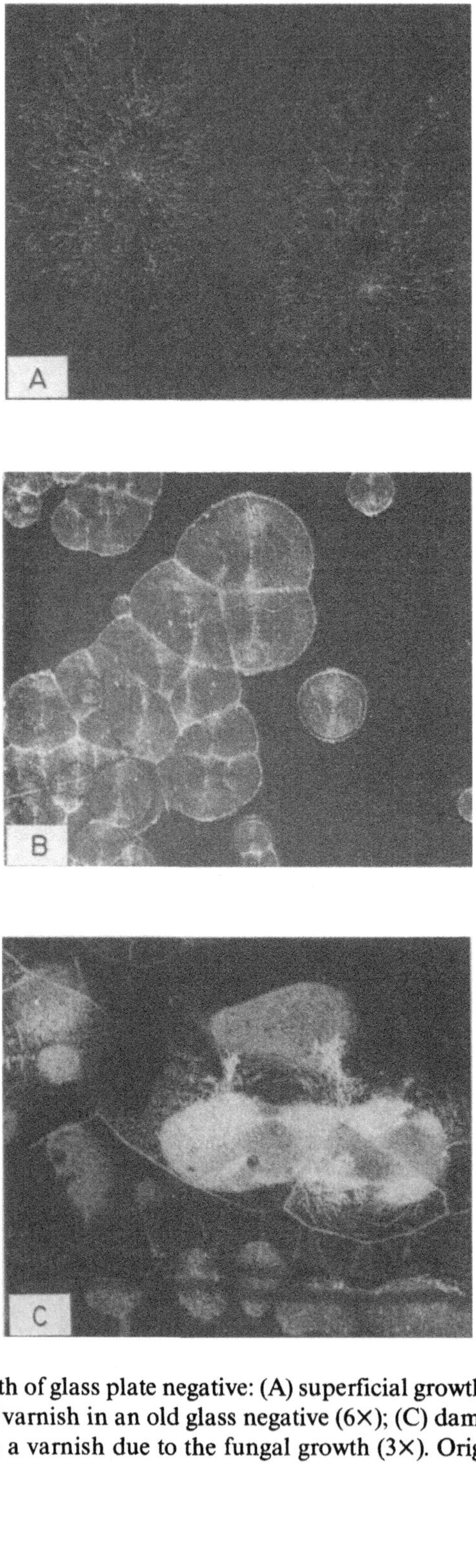

Fig. 1. Fungal growth of glass plate negative: (A) superficial growth on the glass surface (6×); (B) damage of varnish in an old glass negative (6×); (C) damage on the bromised gelatine coated with a varnish due to the fungal growth (3×). Original magnifications.

RESULTS AND DISCUSSION

Of the 20 glass plates the most advanced stage of deterioration was shown by negative no. 263/K, made in 1902. The emulsion of this negative is bromised gelatine dammar. The growth of mycelium is shown in Fig. 1(A). Figure 1(B) shows the deterioration of the varnish in the form of concentric circles. Nine species of fungi were isolated, i.e. *Cladosporium cladosporioides, Aspergillus versicolor, A. nidulans, Penicillium expansum, P. canescens, P. cyclopium, Paecilomyces lilacinus* and *Mycelia sterilia*. Only *Aspergillus versicolor* was isolated from the dammar. Table 1 lists the frequency of the 13 species of fungi, isolated from the 20 glass negatives. Table 2 shows the colonisation of varnish surfaces by fungi after six weeks in chambers of different RH. Figure 1(C) shows the growth of *Aspergillus versicolor* and *A. repens*, on the dammar varnish after 6 weeks at 76% RH. It can be seen that the fungi caused cavities similar to those on the glass plate negative presented in Fig. 1(B). The present work has shown, for the first time a broad range of fungi growing on glass plate negatives. The colonisation of the negatives by fungi at 63% RH was scarcely visible. The recommended RH is 20–30% for collodion plates and under 50% for gelatine plates (Gillet *et al.*, 1986). It is noteworthy that the strains isolated from glass plate negatives also include xerophilic species. Their growth will be studied further in relation to the combined effects of pH, solutes and water activity, following the procedure of Avari & Allsopp (1983).

REFERENCES

Avari, G.P. & Allsopp, D. (1983). In *Biodeterioration-5*, ed. T.A. Oxley & S. Barry. John Wiley, Chichester, pp. 548–56.

Cameron, D. (1968). *Museum News*, **46**, 17–21.

Gillet, M., Garnier, Ch. & Flieder, F. (1986). *Restaurator*, **7**, 49–80.

Iliński, M. (1970). *Black and White Photographic Materials*, 2nd edn. Wydawnictwo Artystyczne i Filmowe, Warsaw (in Polish).

Williams, E.E. (1970). In *Deterioration and Preservation of Library Materials*, ed. H.W. Winger & R.H. Smith. The University of Chicago Press, Chicago, pp. 3–17.

Zyska, B.J. (1987). *Conservation of Library Materials*, Vol. 1. Silesian University, Katowice, in press (in Polish).

Biologically Induced Corrosion of Natural Stones—Strong Contamination of Monuments with Nitrifying Organisms

Eberhard Bock, Wolfgang Sand, Michael Meincke, Brunhilde Wolters, Beate Ahlers, Carsten Meyer & Frank Sameluck

[a]Institut für Allgemeine Botanik, Abt. Mikrobiologie, Ohnhorststr.18, D-2000 Hamburg 52, FRG

ABSTRACT

Famous buildings all over the world are endangered by atmospheric pollution. The aggressive components are mainly of an acidic nature. The weathering is attributed to chemical and physical processes. However, reports about the involvement of microorganisms are either not available or their involvement is denied. Nevertheless in recent years it has become evident that lithoautotrophic nitrifying bacteria were involved. Due to the excretion of their metabolic end-product, nitric acid, concrete is degraded. The natural process could be modelled by means of simulation experiments.

INTRODUCTION

It is quite obvious even to the most casual observer that historic buildings and statues exhibit more and more corrosion. In the last 40 years the corrosion has increased significantly. During this time there has been a large increase in coal and oil usage by industry and a more intensive use of fertilizers for agriculture. These activities have changed the biosphere. Millions of tons of polluting gases are released into the atmosphere each year. In combination with industrial particulate material including adsorbed salts they react in the presence of oxygen, light and humidity. The best known industrial pollutant is SO_2. It has been known for a long time that this compound, which is soluble in water to form H_2SO_3, dissolves limestone (Braun & Wilson, 1970). The second most common pollutant is NO_x, i.e. $NO + NO_2$ (Galloway & Likens, 1981). These gases mainly originate from the emission of automobiles and from the soil. When NO_2 reacts with water, nitrous and nitric acids are produced according to the equation

$$2NO_2 + H_2O \rightarrow 2H^+ + NO_2^- + NO_3^-$$

These gases, SO_2 and NO_x, cause chemical corrosion. Little is known about the role of microorganisms, especially nitrifiers, in the deterioration of stone (Kaufmann, 1952). Here are presented the first results of a scientific research program funded by the Bundesminister für Forschung und Technologie (BMFT, FRG), which have been part published elsewhere (Bock, 1987).

METHODS

Samples were taken from the Kölner Dom (FRG), the Regensburger Dom (FRG) and the Alte Pinakothek in München (FRG). Scrape samples were taken from the surface to a depth of 5 mm by means of a hammer and a sterile chisel or by drilling cores of 90 mm diameter. The cores were divided by saw. All samples were cooled to 4°C and transferred within 6 h to the laboratory for further processing. The samples were homogenized by mortar and pestle and subsequently by means of a Virtis 45 homogenizer (USA) for 30 s at 18·000 rpm. For each operation 1 g of sample was processed. The suspension was diluted in 10-fold steps. For the determination of cell counts a three-tube most-probable-number technique using selective nutrients for ammonia oxidizers, nitrite oxidizers, heterotrophs, and fungi (Milde *et al.,* 1983) was used. Heterotrophs and fungi were evaluated after 2 weeks and ammonia and nitrite oxidizers after 6 weeks. Protein, as a measure of the biomass, was determined according to the method of Bradford (1976).

RESULTS AND DISCUSSION

How do nitrifying organisms infect buildings? It seems most likely that these bacteria are deposited on the surface in dust particles immediately prior to rain or a thunderstorm. Since they are very sensitive to light they must be moved rapidly from the surface to the inner part of the stone by moisture. Capillary action is responsible in porous stone.

Nitrifiers consist of two groups, the ammonia oxidizers and the nitrite oxidizers. The first group converts ammonia to nitrous acid, the second nitrous acid to nitric acid. The acid attacks and dissolves the alkaline binding material of the stone. Ammonia, the only substrate of ammonia oxidizers, is carried by dust as e.g. ammonium sulphate or ammonium nitrite salt. Nitrile as a substrate for nitrite oxidizers originates in part from the emission of automobiles, soil, industry, and from the metabolic activity of the ammonia oxidizers. Dissolved in water these compounds also penetrate the stone. Inside the stone the two groups of nitrifiers sometimes exist separately, sometimes together, but in some locations they are totally absent. This distribution is poorly understood. The ammonia oxidizers excrete nitrous acid and in the absence of nitrite oxidizers would explain the presence of high amounts of nitrite in some areas.

In the outermost layers to a depth of 5 mm the sandstone of the Kölner and Regensburger Dom are heavily contaminated with nitrifying bacteria. In Köln the numbers of nitrifiers are as high as those of the fungi. However it is well known that the MPN technique gives cell numbers which are greatly depressed. Therefore the direct method of TEM was used to enumerate the nitrite oxidizers of the genus *Nitrobacter*. Their ultrastructure makes them easy to identify. The total cells were separated from the stone by homogenization followed by differential centrifugation. Ultra thin sections of 1000 cells were examined and the percentage of nitrite oxidizers determined. By this method the nitrifiers are shown to be the most abundant microorganisms in the Schlaitdorfer Sandstone of the Kölner Dom. Although the total number of microorganisms decreases with increasing depth of the stone, the percentage of nitrifyers increases to about 80% of the total population at 5 cm depth. The pH-value at the surface is 5 and increases to 8·5 at 5 cm depth. The alkaline binding material—the dolomite—is used up by acid attack at the surface but not in the interior of the stone.

The ammonia oxidizing organisms isolated were *Nitrosovibrio, Nitrosospira* and *Nitrosomonas. Nitrosovibrio* was present at all locations. At the Kölner Dom it was the only ammonia oxidizer in 160 samples. Therefore *Nitrosovibrio* is the most abundant organism in natural stones. It should be mentioned that until now this nitrifier has only been isolated in Wales, Hawaii and in the acid tea soils of Asia. The strains we have in pure culture are characterized by carboxysomes and intracytoplasmic membranes. Furthermore, one of these strains is capable of resisting extremely high concentrations of its substrate ammonium sulphate (5 g/litre). Up to now only one strain has been characterized. The nitrite oxidizers all belong to the genus *Nitrobacter*. Most of the isolated strains are obligate chemolithoautotrophs. Their involvement in the corrosion of concrete was examined in a simulation chamber at the optimal growth conditions of 28°C and a RH of 95%. Ammonium sulphate was supplied continuously in the form of an aerosol. Different types of concrete were provided in the form of test blocks, 60 cm in height, 11 cm in depth, and 7 cm in width. They were sawn to provide 1·8 cm cubes for microbiological and physical tests. Pure cultures of *Nitrosomonas* and *Nitrobacter*, isolated from the cooling tower of the electric power station of Niederaußem near Köln (Kirstein *et al.*, 1986), were usd to inoculate the concrete blocks. Under these conditions the nitrifiers produced about 14 ml of 65% 'concentrated' nitric acid per concrete block per year. This acid converted the binding material of the concrete to calcium nitrate. On a labile test block the original surface has been totally corroded and sand particles could easily be removed from the concrete. Control experiments without nitrifying organisms did not cause corrosion. It is assumed that the same takes place in natural sandstones, when calcium or magnesium carbonates are the binding material. However under the naturally occurring conditions the corrosion process will be much slower.

Nitrifying organisms exist on the surface and within the pores of the sandstone in the form of a biofilm. They can survive freezing, thawing and drought. The biofilm is composed primarily of polysaccharides secreted by the organisms. These heteropolysaccharides form a gel and are hygroscopic. They accumulate water from the air and release the water slowly under conditions of low relative humidity. The overall process can be summarized as follows: The glycocalyx producing bacteria maintain a high moisture content in the stone. They also reduce the porosity. The nitrous acid produced by the ammonia oxidizers directly dissolves the calcareous binding material. It also chemically catalyses the conversion of sulphite to sulphuric acid. The sulphite is produced as a result of the solution of SO_2 in water. Usually on an acid surface without nitrous acid SO_2 will only be slowly dissolved and sulphite and sulphuric acid would not be produced. The nitrous acid is further oxidized by *Nitrobacter* to nitric acid.

From experiments it is concluded that nitrous and nitric acids produced from ammonia and not NO_x are the corrosive agents. Furthermore the conversion of SO_2 to sulphite and subsequently to sulphate is catalysed by nitrite. It is suggested that the overall process be known as 'nitrification-induced corrosion'.

ACKNOWLEDGEMENTS

This work was funded by the BMFT. The authors especially wish to thank H. Schulze (BMFT) for his generous support.

REFERENCES

Bock, E. (1987). *Befall mit Nitrifikanten. Bautenschutz, Bausanierung,* **10**, 42–5.
Bradford, M.M. (1976). *Analytical Biochemistry,* **72**, 248–54.
Braun, R.C. & Wilson, M.J.G. (1970). *Atmospheric Environment,* **4**, 371–8.
Galloway, J.N. & Likens, G.E. (1981). *Atmospheric Environment,* **15**, 1081–5.
Kauffmann, M.J. (1952). *Comptes Rendus Academie Sciences, Paris,* **234**, 2395–7.
Kirstein, K.O., Stiller, W. & Bock, E. (1986). *Beton- und Stahlbetonbau,* **81**, 202–5.
Milde, K., Sand, W., Wolff, W. & Bock, E. (1983). *Journal of General Microbiology,* **129**, 1327–33.

Status Report: Ongoing Studies of the Susceptibility of Stone Consolidants to Microbiologically Induced Deterioration

R.J. Koestler,[a] E.D. Santoro,[b] J. Druzik,[c] F. Preusser,[c] L. Koepp,[b] & M. Derrick[c]

[a] Dept. of Objects Conservation, The Metropolitan Museum of Art, 82nd St and 5th Ave., New York, NY 10028, USA
[b] Dept. Biology, Bloomfield College, Bloomfield, NJ 07003, USA
[c] The Getty Conservation Institute, 4503B Glencoe Ave., Marina del Ray, CA 90292–6537, USA

ABSTRACT

Over a 35-day period a testing methodology was employed to assess the biodeterioration susceptibility of thin films of 15 stone consolidants that had been or are being used in the stone conservation field. The program utilized inoculation of mixed fungal cultures onto thin films of the consolidants alone, without added substrates. A visual scoring regimen was employed to assess the biodeterioration of each consolidant. The initial results of our ongoing investigations are discussed in the context of determining the most microbiologically resistant of the consolidants studied. The utility of the testing methodology employed as an initial screening tool prior to a full scale in-situ test is discussed.

INTRODUCTION

The study of the susceptibility of stone monuments to biodeterioration has received some attention in recent years (Koestler *et al.*, 1985) while that of consolidants has received very little. Most of the attention focused on consolidants has been to study aspects of chemical or mechanical deterioration, as indeed our previous studies have done, e.g. see Tucci *et al.* (1985).

It is also necessary to evaluate and, accordingly, rank several of the more widely used consolidants as to their susceptibility to microbiological attack. In order to reduce visual bias it is necessary to evaluate the materials in a manner which is consistent with the deterioration assessment. Assessing the resistance to microbiological attack can be a lengthy process. The experimental design must be reproducible and effective. It is contended that studies of this type should be done in two phases, the first phase is to determine whether or not a consolidant can support microbial growth. A preliminary study was undertaken prior to the study reported herein (Koestler *et al.*, in press).

When designing such a screening study it needs to be of short duration, maximum utility and statistically reproducible. In addition, it should be conservative, in that it over-estimates the biodeterioration effect.

The present paper delineates the method used in the research and presents some preliminary data.

MATERIALS AND METHODS

Overview of methodology

Multiple samples of each consolidant, prepared as a thin film on glass slides, were inoculated with two different concentrations of fungal spore mixtures. Cultures were grown in a temperature and humidity controlled chamber for several months, assayed using grid matrix and random field procedures for signs of visual growths of fungi, for weight loss after fungal action, and by Fourier Transform Spectroscopy (FTIR) for alterations in the consolidant's chemical bonds.

Particle count methods

Prior to inoculating, particle counting of the spore suspensions was performed with an in-situ liquid particle counter, Model ILI 1000 (Spectrex Corp.). A focused laser beam is directed through a sample solution and particles are detected when they pass through the laser in the focus area. Light is scattered from the particle and a photo-electric cell detects that portion of the light reflected. The light pulse is analysed by a detection unit. Particle densities (1000/ml) are kept low so there is little chance of coincidence (shielding of a particle) as particles are crossing the laser path. Variation between replicates is reported to be less than 5% (Chung, 1982; Clayton; Blanchard *et al.*, 1976). Using this technique, spore concentrations of 10^4 and 10^8 were obtained and inoculated onto the consolidants.

Consolidant preparation

Consolidants were applied on glass substrates (5 × 20 cm) by airbrushing (Paasche) through an aluminium template; three or four rectangle samples per plates, ranging in size from approximately 1430 to 1660 mm^2. Concentrations and spraying techniques were balanced to achieve as consistent a film thickness between different materials as possible. Solution concentrations ranged from 2% for the PVAs, 5% for the acrylics, 40% solids for acrylic emulsions, on up to 75–96% for silanes.

Film-forming consolidants produced even layers immediately, however for nonfilm-formers as even a layer as possible was applied and allowed to polymerize. Under magnification this appeared as small hard droplets on the glass. Subsequent coats served to 'fill in' between previously applied coats and an approximation to a thin film was built up.

After the final application and air-drying, a 2–3 mm rag board mat was cut and glued to the edges of the substrate and a second glass plate taped over the mat. Groups of ten plates were wrapped in aluminium foil to protect them from light until inoculated.

The materials tested are presented in Table 1, many are consolidants currently used in conservation along with others that could be in the future. For example, the inclusion of polyvinyl butyral was included to represent Robert Feller's stability class B polymers. A few other natural resins were included for comparison, such as Dammar and Shellac.

Isolation of fungi

Fungi were isolated from cultures obtained from a Carrara marble statue that had been exposed to the semi-tropical conditions of Florida for some 80 years.

From earlier work (Koestler *et al.*, in press*a*; Koestler *et al.*, in press*b) it* became apparent that for a more accurate and complete isolation of fungi, an antibiotic must be incorporated into the primary isolation medium to reduce or eliminate bacterial growth which previously interfered with fungal growth. Cultures were examined routinely as young cultures of no more than 7–10 days old, since colonial morphology changes greatly as cultures get older.

The primary isolation medium consisted of Sabouraud's agar (SAB) to which Achromycin (Tetracycline - HCl, Lederle Lot No. 117–488 Exp. 5/90) was added in 35 μg/ml concentration. The tetracycline was prepared separately and added just before the agar was poured into Petri dishes. One gram of tetracycline was added to 150 ml of sterile distilled water. Before plates were poured, 0·05 ml of this solution were added to 10 ml of agar medium.

Cultures from all original samples were inoculated into the SAB tetracycline medium. Plates were examined daily for fungal growth. Plates had mould as well as yeast growth, while bacterial growth was inhibited by the antibiotic. The different mould colonies were isolated and transferred to the centre of a SAB plate by touching the surface of the colony with a flame-sterilized needle cooled in sterile medium. Yeast colonies were isolated on SAB using the simple streak plate technique used for bacterial isolation.

These techniques resulted in a higher variation of mould isolates as well as in the isolation of yeast which was absent in preliminary findings. The various mould and yeast isolates were kept refrigerated on SAB slants until used. Many of these isolates have been identified to the genus

TABLE 1
List of Consolidants Tested

Product name	*Area (mm²)*	*Manufacturer*	*Additives/contaminant*
Acryloid B-72	1657	Rohm and Haas	Information not available
Acryloid F-10	1550	Rohm and Haas	Information not available
AYAA	540	Union Carbide	No additives
AYAC	1575	Union Carbide	Acetic Acid < 0·05%
AYAF	1539	Union Carbide	No additives
AYAT	1596	Union Carbide	No additives
Mowital B-20-H	1564	American Hoechst	No additives; may be some by-products from the polymerization process such as PVAC, PVOH, butyraldehydes
Conservare H	1431	ProSoCo	Dibutyltin dilaurate catalyst
Conservare H40	1950	ProSoCo	Dibutyltin dilaurate catalyst
Conservare OH	1653	ProSoCo	Dibutyltin dilaurate catalyst
GE Silicon	1490	General Electric	100% silicone resin
Tegovakon V	1672	Goldschmidt	Solvent + alcohol; tin catalyst; silicic acid ester as a binder replacement
Imron 192S	1615	Dupont	HCl or triethylamine as an initiator; activator 192S has no additives
Dammar	1596	AF Suter	Natural resin
Shellac	1575	AF Suter	Natural resin

level using colony characteristics as well as microscopic examination.

From these cultures fungal spore suspensions were obtained by the method of Klausmeier (1972). Immediately prior to inoculation spore suspensions were counted via a Spectrex Corp. Particle Counter to assure 10^4 and 10^8 spores/ml were present in the inoculation.

Inoculation of fungi

Once in spore suspensions, 0·6 ml of the mixed spore suspension was inoculated onto each of the test samples. On each slide, one sample served as a control. The inoculated drop was spread across the sample surface with a sterile slide. Each inoculated test strip was incubated in a Powers incubator at 25°C±3°C, 90%±10% RH, under a 14/10 h light/dark cycle.

To ensure that an effect can be measured it is necessary to examine the consolidants statistically. Fifteen samples of each consolidant were prepared. These samples represent triplicate runs over five sample intervals of 7, 14, 21, 28 and 35 days exposure. Triplicate samples provide the minimum number for statistical assessment of the observable effect. On removal from the incubation chamber a series of observations were made by two persons, as described below:

1. Percentage of organism coverage was ascertained visually as a percentage of organism cover as a function of test strip discoloration.
2. Microscopic random field viewing. The sample was separated into 100 evenly spaced and numbered boxes approximately 2·1 mm^2 in size. Five grids were chosen for viewing by a random number table. Using a light microscope at low power (32×), the presence or absence of organisms on the grids were recorded.
3. Higher magnification (400×) viewing to describe hyphal length, presence of fruiting bodies, colony formation, etc.
4. As a further comparison, the sporulation scale presented in Smith and Nadim (1983) was also assessed in this study.

Upon the completion of these procedures the samples were cleaned using a 15% H_2O_2 solution.

RESULTS

Over the initial period, growth was not observed in consolidants exposed to 10^4 spores/ml. Conversely, growth was observed in most of the consolidants exposed to 10^8 spores/ml.

Initial comparisons of the random field procedure (RFP) to both visual and optical viewing techniques revealed that the RFP was more precise than the visual scoring procedure and provided a statistically valid assessment, especially when compared to higher magnification viewing. In many cases visual scoring proved inaccurate due to discoloration of the sample from the sterile water droplet rather than from actual fungal growth.

Organism growth

For this study the following classifications were used for assessment of growth:

Visual assessment —0, 25, 50, 75, 100% coverage.
Random fields —0–100% of the fields showing growth.
Magnification viewing —No growth, budding, moderate extension, sporulation, colony formation.

Assessment of the sporulation scale, after Smith and Nadim (1983) correlated closely with our RFP and high magnification viewing (see Table 2).

Using RFP and high magnification viewing the consolidants were ranked as to their ability to support fungal growth, as is presented in

TABLE 2
Sporulation Assessment

Consolidant	*Description*	*Sporulation scale*[a]
GE	No budding	0
Tegovakon V	Lt growth/med fruiting bodies (FB)	2
Dammar	Short hyphae/med FB	2
Mowital	Short-med hyphae/med FB	2
Conservare H	Extensive FB	3
B-72	Moderate-extensive FB	3
AYAF	Numerous FB	3
Conservare OH	No FB	0
F-10	Many FB	3
Conservare H-40	Many FB	3
Shellac	Numerous spores, no FB	0
AYAT	Many FB	3
AYAC	No FB	0
AYAA	No FB	0
Imron	Many FB	3

[a] Scale as defined by Smith & Nadim (1983). Scale compared to 0, no fruiting bodies; 1, poor fruiting bodies; 2, average fruiting bodies; 3, strong fruiting bodies.

Table 3. General Electric Silicone resin and Conservare OH showed virtually no growth after 1 week of incubation. Of the remaining resins, Tegovakon V and AYAC showed approximately 25% growth; Mowital showed approximately 50% growth; Dammar, B-72, F-10, and Shellac had approximately 75%; Conservare H, AYAF, Conservare H-40, AYAT, AYAA, and Imron all had greater than 75% fungal growth after the first week of incubation.

DISCUSSION

A reproducible assessment technique for evaluation of an effect is difficult and fraught with several types of bias. In the present project an attempt has been made to 'screen' consolidants for their ability to support fungal growth. The screening test outlined here is easy to set up, run, and rapid.

In the broader context of durability of a consolidant, factors other than chemical and mechanical stability have to be addressed. The potential for a consolidant to act as a food source for microbes must be understood before an intelligent choice can be made for a given application. This is especially true when an object resides in an environment that is high in

TABLE 3
Preliminary Ranking of Consolidants[a]

Consolidant	*% Growth*				
	0	*25*	*50*	*75*	*> 75*
GE	*				
Tegovakon V		*			
Dammar				*	
Mowital			*		
Conservare H					*
B-72				*	
AYAF					*
Conservare OH	*				
F-10				*	
Conservare H-40					*
Shellac				*	
AYAT					*
AYAC		*			
AYAA					*
Imron					*

[a] Based on a combination of the random field procedure and optical viewing technique.

relative humidity, such as the topics or sub-tropics. A half understood selection of a consolidant may succeed in minimizing physical and chemical deterioration while enhancing biological attack. This paper presents details of the experimental design and the results of the first week of data.

REFERENCES

Blanchard, J., Thomson, C. & Schwartz, J. (1976). *Comparison of Methods for Detection of Particulate Matter in Large-volume Parenterals*. American Society of Hospital Pharmacists.

Chung, Y. (1982). *Proceedings of Geological Society of China.*

Clayton, Z.A. (no date). *An Optical Method of Size Analysis for Silt Size Particles.* Publication by Spectrex Corp.

Klausmeier, R.E. (1972). *Biodeterioration of Materials, Vol. 2*, ed. A.H. Walters & E.H. Hueck-Van Plas. Halsted Press, Division of John Wiley & Sons, New York.

Koestler, R.J., Charola, A.E., Wypyski, M. & Lee, J.J. (1985). In *Vth International Congress on Deterioration and Conservation of Stone*. Lausanne, 25–27 Sept. 1985, ed. V. Furlan. Presses Polytechnique Romandes, Lausanne, pp. 617–26.

Koestler, R.J., Santoro, E.D., Preusser, F. & Rodarte, A. (in press *a*). In *Proceedings of the 1st Pan-American Biodeterioration Society Meeting*, Washington, D.C. 17–19 July 1986, ed. C. O'Rear & G.C. Llewellyn. Plenum Press, New York.

Koestler, R.J., Santoro, E.D., Ransick, L., Brill, R.H. & Lynn, M. (in press *b*). In *Proceedings of the 1st Pan-American Biodeterioration Society Meeting*, Washington, D.C., 17–19 July 1986, ed. C. O'Rear & G.C. Llewellyn. Plenum Press, New York.

Smith, R.N. & Nadim, L.M. (1983). In *Biodeterioration 5*, ed. T.A. Oxley & S. Barry. John Wiley, New York, pp. 538–47.

Tucci, A., Koestler, R.J., Charola, A.E. & Rossi-Manaresi, R. (1985). In *Proceedings of the Vth International Congress on Deterioration and Conservation of Stone*, Lausanne, 25–27 Sept. 1985, ed. V. Furlan. Presses Polytechnique Romandes, Lausanne, pp. 891–8.

Comparison of Possible Chemical and Microbial Factors Influencing Paper Decay by Iron-Gall Inks

K. Messner, L. Alberighi, G. Banik, E. Srebotnik, W. Sobotka & A. Mairinger

Institut für Biochemische Technologie und Mikrobiologie Abt. Mykologie TU-Wien, A-1060 Vienna, Getreidemarkt 9, Austria

ABSTRACT

Both chemical and microbial factors may be the causes of ink eating. Deteriogens produced by microbial action are organic acids, cellulase and H_2O_2. The fungi growing most frequently on ink were isolated and investigated for the production of deteriogens. The deteriogenic effect of the possible biological factors on paper was determined as a decrease in the folding endurance and compared with the effect of chemical factors (acidity, Fe^{2+}). Fungal growth in ink was rather poor. Cellulase and organic acids have to be taken into account as biological deteriogens but to a lesser extent than the radical mechanisms catalysed by Fe^{2+}.

INTRODUCTION

Iron-gallo tannate was the standard writing ink of the Middle Ages. A recipe by P. Canneparius (1660) contains 2 parts vitriol ($FeSO_4$), 3 parts galls, 1 part gum arabic, 30 parts water. An irreversible degradation of paper and parchment caused by iron-gall inks or by other mixed inks containing other transition metal ions is called ink eating. The degradation process of the substrate comprises several stages, the first being brownish discolorations in the immediate vicinity of the characters. This spreads diffusely over the substrate and also comes through on adjacent sheets. Finally, the destruction of the substrate (paper and parchment) manifests itself in the partial fracturing of individual characters, but can also cause losses of entire lines. The consequence is partial disastrous damage to manuscripts, documents and autographs.

In the present study an attempt has been made to clarify the role of microorganisms in ink eating, and their effect compared with that of chemical factors.

Concerning the growth of microorganisms do iron-gall inks provide sufficient nutrient in themselves? A quantity of nutrient salts is available in water. The chief carbon source contained in iron-gall inks is tannin (approx. 10%). Tannins are polyphenols consisting substantially of pentadigalloglucose. Moulds, such as *Aspergillus niger* and *Penicillium glaucum,* produce tannases which split tannins into glucose and gallic acid. Furthermore, gum arabic and sometimes vinegar was added to some inks. The cellulose fibres transferred from the paper to the ink might also be considered as an additional carbon source. Nitrogen is present in iron-gall inks originally, mostly in the form of impurities in the natural products. But considering that the paper was surface-sized with protein glues, a considerable amount could have been transferred into the ink, which could enhance the growth of microorganisms.

As most recipes produced inks which were acidic, bacterial growth was largely prevented, but the growth of fungi is stimulated. Since fungal growth cannot be excluded, it is necessary to clarify the depolymerizing effects that fungi may have on paper.

1. Organic acids: some fungi (e.g. *Aspergillus niger*) are capable of acidifying the medium by excreting organic acids. This would have an effect only with buffered inks with a neutral pH.
2. Cellulase: some fungi are equipped with a cellulase system which might be stimulated during growth in the ink.
3. H_2O_2: oxidases enable some microorganisms to produce and excrete hydrogen peroxide. This has a depolymerizing effect on cellulose and may start a Fenton-type reaction with Fe^{2+} ions, producing OH radicals largely depolymerizing cellulose.

Concerning the chemical factors: unbuffered iron-gall inks have a low pH value which depolymerizes cellulose. An additional, possibly even stronger depolymerizing effect may be produced by Fe^{2+} ions.

During the oxidation of Fe^{2+} to Fe^{3+} unstable complexes between metal ions and molecular oxygen are formed, which may result in the formation of different radicals which are strong oxidants and may cause numerous oxidations leading again to the depolymerization of cellulose.

Similar processes take place when paper ages, even if far more slowly because of the low concentration of the Fe^{2+} catalyst.

METHODS

Old recipes were used to make the following inks as substrates for studying the growth of microorganisms:

A: iron sulphate, 6 g; tannin, 9 g; gum arabic, 3 g; tap water 90 ml; pH = 2.

B: iron sulphate, 10 g; tannin, 10 g; gum arabic, 10 g; Na_2CO_3, 50 g; HNO_3, 10 ml; acetic acid (5%), 100 ml; tap water, 250 ml; pH = 8.

To isolate microorganisms, the inks were exposed for 6 weeks in a concentrated form, diluted and solidified with agar.

The 4 species occurring most frequently and 5 additional strains in ink have been reported previously (Kowalik, 1984); *Penicillium roquefortii, Aspergillus niger, A. fumigatus, A. flavus, Epicoccum nigrum,* as well as 2 cellulolytic fungi: *Trichoderma* sp., *Trichoderma pseudokoningii*, were cultivated at 25°C on different dilutions of the two inks which had been solidified with agar, and growth was quantified at pH = 2, 5 and 8.

To determine if fungi can reduce the pH value, 3 flasks with ink (pH = 8) and 3 flasks with ink mixed with Czapek medium were inoculated with a spore suspension of one of the 11 fungi and the pH measured after 0, 1, 2 and 3 weeks.

For the strains isolated from the ink, the cellulase activities formed under optimal growth conditions were determined as filter paper activity units (1 FPU = 1 μmol glucose/min ml) (Mandels *et al.,* 1976) and compared with the activities of *Trichoderma* strains.

H_2O_2 produced by microorganisms can be detected according to the method of Müller (1984) down to a detection limit of 0·0002% in the plate test.

In order to simulate the mechanisms of the effect of ink eating, the following procedure was chosen: 3 types of paper (A: archive copying paper; B: rag paper, surface-sized; C: rag paper, handmade) were cut into strips, immersed in the test solutions, placed for 15 min in the desiccator under vacuum, and subsequently dried at room temperature. This procedure corresponds to writing on the paper. After approximately 2 weeks the test strips were subjected to accelerated ageing in a climatic cabinet at 70°C and 90% RH for a week. The effect of the deteriogens was determined as a reduction of the folding endurance by determining the number of double folds according to Schopper. (Details of the methods are given in Alberighi *et al.*, in preparation.)

RESULTS AND DISCUSSION

Growth of fungi and production of possible deteriogens

Isolation of fungi: A larger number of fungal colonies developed on the exposed diluted inks than on the concentrated inks. Four species of fungi occurred most frequently on the exposed inks: *Penicillium spinulosum Thom, P. verrucosum Dierckx, P. raistrickii G. Smith, Paecilomyces variotii Bainier.*

Growth on inks: Some of the fungi investigated were capable of forming large colonies on ink agar at the 3 pH levels. The biomass formed was small, presumably because of the low nitrogen supply. Bacterial growth was observed only in inks with a pH of 8.

Acidification of inks by organic acids: Almost all fungi reduced the pH value from pH 8 by one to three units in ink-Czapek medium. Only *Aspergillus niger* was capable of reducing the medium to pH 1·7. None of the fungi reduced the pH of ink which did not contain Czapek medium. It can be concluded that the nutrients available in the inks did not suffice to form organic acids and that additional nutrients are necessary.

Cellulase activity: Only *Paecilomyces variotii* and *Penicillium verrucosum* showed average activities of 0·15 and 0·11 FPU, compared with 0·5 FPU for *Trichoderma viride.* It has to be borne in mind that these activities were achieved under optimal conditions for cellulase production which are never present in ink.

Formation of H_2O_2: All strains isolated from the ink gave negative results in the H_2O_2 plate test.

Effect of the deteriogens on paper

The control values (paper without additional treatment) showed an increased number of double folds (+30%) after 1 week of ageing (Table 1). This effect was confirmed for different papers in the first stage of ageing (Sobotka, 1983). Iron-gall inks with an excess of iron sulphate always result in a low pH and certain damage to the paper unless they are buffered. If under natural conditions the nutrient supply should suffice, the paper might be damaged by organic acids, as is shown by the results.

When assessing the reduction of the number of double folds caused by

TABLE 1
Reduction of Double Folds after Ageing (%)

Types of paper	*A*	*B*	*C*
Untreated paper	+32	+29	+38
Citric acid pH = 3	−55	−55	−78
Sulphuric acid pH = 3	−17	nt	−31
Acetate buffer pH = 4·5	nt	nt	−48
Cellulase pH = 3	−100	−92	nt
Cellulase pH = 4·5	−90	−25	nt
Iron sulphate pH = 7	−100[a]	−100[a]	nt
Iron sulphate pH = 3	−100[a]	−100[a]	nt
Ink (half concentrated) pH = 2	−100[a]	−100[a]	−100[a]

nt: no test.
[a]Numbers of double folds greatly reduced after treatment, before ageing.

the cellulase, one has to take into account that a cumulative effect occurs between acidity and cellulase activity. At pH 3 the number of double folds was reduced by a culture filtrate of *Trichoderma viride* by 100% and 92%, respectively. Both values are above those caused by the acids so that one may conclude that the cellulase solution has a depolymerizing effect under test conditions.

A 0·001% H_2O_2 solution, corresponding approximately to a biological concentration, showed no effect. The increase of double folds by 25% is equal to the value for untreated paper.

The number of double folds is reduced most rapidly by ferrous sulphate. All strips which had been immersed in samples containing ferrous sulphate showed greatly reduced numbers of double folds after brief storage in ambient conditions, prior to accelerated ageing. After ageing none of the samples could be double folded. This was confirmed also with ferrous sulphate solutions which were buffered to pH 7 with an acetate buffer to exclude the cumulative effect of acidity.

Summarizing, of the possible causes of ink eating, iron sulphate has the strongest depolymerizing effect on paper, the radical mechanisms being of the greatest importance. Moreover, at low pH the acidity of the ink to a lesser extent may have a destructive effect. Of the biological factors, cellulase and organic acids have to be considered. However, due to the rather poor growth of fungi in ink, high cellulase activities and acid concentrations cannot be expected. Ink eating therefore has to be regarded as a usual ageing process which takes place more rapidly because of the much higher concentration of Fe^{2+} ions and the acidic environment.

ACKNOWLEDGEMENTS

The research project Untersuchungen der desruktiven Wirkungen von Tinten auf Schriftträgermaterialien, sowie Entwicklungen geeigneter konservatorischen Maßnahmen, is being supported by the Stiftung Volkswaagenwerk. Collaborating institutes are the österreichische Nationalbibliothek, Vienna, and the Staatsbibliothek Preussischer Kulturbesitz, Berlin (West). The authors thank the Stiftung Volkswagenwerk for the support which made the present work possible.

REFERENCES

Kowalik, R. (1984). *Restaurator*, **6**, 75–114.
Mandels, M., Andreotti, R. & Roche, C. (1976). *Biotechnology and Bioengineering Symposium*, No. 6, pp. 35–53.
Müller, H.E. (1984). *Journal of Microbial Methods*, **2**, 101–2.
Sobotka, W. (1983). *Das österreichische Papier*, **20**(6), 27–34.

An extended bibliography for this paper may be obtained from the senior author.

Effect of Cleaning on the Distribution of Microorganisms on Rock Surfaces

Thomas Warscheid, Karin Petersen & Wolfgang E. Krumbein

Geomicrobiology Division, University of Oldenburg, Postbox 2503, 2900 Oldenburg, FRG

ABSTRACT

Bacteria, fungi, algae and actinomycetes were quantitatively determined in the uppermost layers of sandstones of several monuments within Germany before and after cleaning the stone surface by using high pressure water blasting. Dilution and inoculation for counts was performed using a spiral plater technique. The results of the study indicate that, depending on the original distribution of microorganisms, the microflora of sandstones immediately after treatment is not drastically altered, especially with regard to specific groups of microorganisms, which were isolated and counted under oligotrophic conditions. After a short period of time microorganisms again reach their previous number or even exceed it. Cleaning by the conventional methods cannot be regarded as a procedure which reduces microbial colonisation indefinitely.

INTRODUCTION

In recent years it has been widely demonstrated how microorganisms in addition to physical and chemical influences take part in the weathering of rocks (Webley *et al.*, 1963), as well as in the destruction of stone in buildings and historical monuments.

The pioneering inhabitants of freshly exposed rocks devoid of organic matter are usually photoautotrophic microorganisms accumulating photosynthetically synthesised organic carbon for copiotrophic (in nature zymogenous) and oligotrophic heterotrophs, which in turn supply sulphur bacteria and nitrifying bacteria with energy sources, such as hydrogen sulphide and ammonia (Krumbein, 1973).

However, only in distinct microhabitats are microorganisms able to attack the mineral-surface more vigorously than abiotic aggressive compounds. Beneath the excretion of enzymes, protons, inorganic and organic acids and complex forming agents microorganisms accumulate dissolved cations in their cells leading to a continuously steady state in the weathering process which ends in the destruction of the rock-forming mineral (Beveridge, 1984).

Nevertheless the mere occurrence of microorganisms on stone surfaces does not automatically imply destructive action. But these rock-populations may be involved in enhancing abiotic processes or lead to the formation of algae and microbial mats, which may in turn build surface crusts, to the aesthetic detriment of historic monuments.

In conservation it is common practice to clean stone surfaces by high pressure water blasting in order to remove biogenic layers and crusts and to prepare the stone for further treatment, such as rendering it hydrophobic.

This report represents results concerning the effectiveness of this procedure to reduce the numbers of microorganisms attached to the stone surface and subsurface and its influence on the distribution and re-settlement of these microorganisms.

MATERIALS AND METHODS

Samples

Specimens with typical deterioration characteristics of the following monuments were analysed:

Marien-Church, Gelnhausen—Quartzitic Lower Triassic-Sandstone. S 1—with algae layer and some crustose lichens on the solid stone surface; S 3—exfoliating stone surface covered with crustose lichens; S 2 and S 4—same as S 1, S 3 reinspected immediately after washing.

St. Lorenz-Church, Nuernberg—Quartzitic 'Burg'-Sandstone. S 5—Black crust on solid stone surface; S 7—thick algae carpet; some crustose lichens on solid stone surface; S 6 and S 8—same as S 5, S 7 reinspected between 1 and 2 weeks after washing.

St. Matthias-Church, Trier—Quartzitic Lower Triassic-Sandstone. S 9 with black-shaped algae layer on solid stone surface; S 10—sample site comparable to S 9 about 5 weeks after washing.

'Alte Pinakothek', Munich—carbonaceous Cretaceous-Sandstone ('Gruensandstein'). S 11—black crust with underlying crumbly stone surface; S 12—sample site comparable to S 11 about 8 weeks after washing.

Sampling

Incisions were cut about 2–3 mm deep in an area of 3 × 3 cm of the stone surface using a drill with a diamond wheel. Samples were removed with a stone chisel. Approximately 10 g of material was collected in sterile petri-dishes and analysed one day after collection.

Culture media

Copiotrophic, chemoorganotrophic bacteria: DEV Gelatine-Agar (Merck, Darmstadt W.G., Nr. 10685).
Oligotrophic, chemoorganotrophic bacteria: DEV Gelatine-Agar (same as above, but diluted 1/50) and Bunt & Rovira's medium.
Nitrifying bacteria: media submitted by Bock (Personal communication).
Copiotrophic fungi: Sabauraud 4% Maltose-Agar (Merck, Darmstadt W.G., Nr. 5439).
Oligotrophic fungi: Sabauraud 4% Maltose-Agar (as above, but diluted 1/50) and Czapek-Dox-Agar (Merck, Darmstadt W.G., Nr 5460 but diluted 1/50).
Algae and cyanobacteria: BG 11-Medium.
Actinomycetes: Glycerol-nitrate-casein-agar (but diluted 1/50).

Inoculation, incubation and enumeration

Sample material was carefully powdered in a hand mortar, suspended in sterile 0·001% Tween 80-solution (10%; w/v) and shaken continuously for 1 h. Chemoorganotrophic bacteria and fungi as well as actinomycetes were inoculated using the modified spiral plater technique of Lewis *et al.* (1985). Nitrifying bacteria, algae and cyanobacteria were inoculated as prescribed for the MPN method. The inoculated samples were incubated at 28°C in the case of chemoorganotrophic bacteria, fungi, actinomycetes and nitrifying bacteria and at 16°C in light (about 1000 lux) in the case of algae and cyanobacteria. Counting, of the chemoorganotrophic bacteria, fungi and actinomycetes according to the Spiral Plater procedure was carried out 2 weeks after inoculation. Nitrifying bacteria were enumerated 6 weeks, algae and cyanobacteria 8 weeks after inoculation. Counts were computed as CFU/g stone.

RESULTS AND DISCUSSION

Counts of isolated microorganisms from the different samples are summarised in Table 1. It can be shown that immediately after the high pressure water blasting treatment on the stone surface (S 1–S 4) a short-term impairment of the colonisation of microorganisms is obvious.

The species-composition however, remained extensively stable. Oligotrophic bacteria predominate to a greater extent compared with the copiotrophic ones. Fungi, whose differentiation with respect to copiotrophic and oligotrophic characteristics, in this case are less distinct, remain dominant in sample S 1/S 2 and inferior in sample S 3/S 4.

Nitrifying bacteria were found preferentially on the sample-site which was highly infected by a chemoorganotrophic microflora. This site is characterised by microsoil-conditions, arising behind the exfoliating stone surface, including the presence of actinomycetes (Krumbein, 1973; Strzelczyk, 1981).

Shortly after the washing procedure (S 5–S 8) the numbers of microorganisms develop differently, depending on their previous colonisation before treatment.

The species-composition of organisms remained unchanged to this point, but the diminution of their number on the slightly algae colonised black crust (S 5/S 6) was more lastingly preserved, especially in the case of nitrifying bacteria. Nevertheless the growth of algae is stimulated by the introduced water of the cleaning procedure.

In comparison the originally dense algae overgrown sample site S 7 shows a revival of the chemoorganotrophic bacterial flora after washing, which may be due to the fresh and unweathered stone surface, promoting new attachment of microorganisms or the increased algae growth stimulated by humidity.

Additionally, the partial conservation of the accumulated organic compounds on the stone, supporting microbial growth, by washing them into the interior of the stone has also to be considered regarding the coarse-grained character of the sandstone of this site, compared with the fine-porous character of the sandstone in the samples S 1–S 4. Coarse-grained sandstones possess larger pores allowing a high influx and efflux of nutrient enriched solutions thereby increasing the growth rate of colonising microorganisms.

Some weeks after the washing procedure (S 9/S 10) microorganisms were reactivated even on the fine-grained lower Triassic-Sandstone whereby oligotrophic bacteria again dominate.

More drastic was the recurrence of stone surface-infecting microorganisms on highly calcareous, water absorbing and porous Cretaceous-Sandstone about 8 weeks after the cleaning (S 11/S 12).

TABLE 1

Numbers of Microorganisms on Rock Surfaces (CFU/g stone) on Various Specimens Taken from Several Monuments within Germany before and after Cleaning Procedure with High Pressure Water Blasting

Sample	*Bacteria*			*Fungi*		*Algae*	*Actinomycetes*
	Copiotrophic	*Oligotrophic*	*Nitrifying*	*Copiotrophic*	*Oligotrophic*		
S 1	900	1 200	0	75 000	65 500	24 000	0
S 2	180	785	40	34 000	23 000	430	0
S 3	310 000	1 800 000	2 400	4 400	3 450	110 000	4 300
S 4	110 000	195 000	430	0	0	0	0
S 5	11 000	10 600	150 000	5 100	27 000	0	0
S 6	450	3 000	0	3 400	5 900	7 500	0
S 7	320 000	990 000	390	3 200	39 500	5 500	30 000
S 8	670 000	1 300 000	430	0	6 250	550 000	0
S 9	79 700	103 000	90	3 500	5 700	120 000	0
S 10	12 000	225 000	0	6 100	2 790	0	0
S 11	1 200	1 300	0	1 700	3 435	200	0
S 12	820 000	285 000	4 400	5 800	6 600	7 500	5 400

All groups of microorganisms analysed reach a large increase in numbers, which may also be due to the freshly exposed stone surface after cleaning as well as to the high porosity of the material offering optimal colonisation conditions for the organisms. Therefore it can be stated that high pressure water blasting only temporarily affects the number of microorganisms on stone surfaces.

After a short period, microorganisms recover and either reach their previous number or exceed it drastically, depending on the kind of sandstone in question.

After the initial cleaning process, additional conserving treatments, such as rendering the stone hydrophobic, are necessary to enable successful and enduring restoration.

Nevertheless it can be seen in recent analyses that even the protective skin of the hydrophobic medium offers new possibilities for the colonisation by different microorganisms.

In order to reach a better understanding of the attachment and distribution of microorganisms on stone surfaces further studies will be undertaken.

ACKNOWLEDGEMENTS

The Federal Minister of Science and Technology has largely contributed to our work through generous grants. D. Janssen has been of great help in our work with technical assistance.

REFERENCES

Beveridge. T.J. (1984). *Proceedings of the 3rd International Symposium on Microbial Ecology*, Michigan State University. American Society for Microbia, Washington D.C., pp. 601-7.

Krumbein, W.E. (1973). *Deutsche Kunst- und Denkmalpflege*, **31**, 54–71.

Krumbein, W.E. & Schönborn-Krumbein, C.E. (1987). *Bautenschutz und Bautensanierung*, **10**, 14–23; 110–17.

Lewis, F., May, E. & Bravery, A.F. (1985). *V'e Congrés international sur l'Alteration et la conservation de la Pierre*, Lausanne. Presses Polytechniques Roman des, Lausanne, pp. 633–42.

Strzelczyk, A.B. (1981). *Stone. Microbial Biodeterioration*, ed. A.M. Rose. Academic Press, New York, pp. 61–80.

Webley, D.M., Henderson, M.E.K. & Taylor, I.F. (1963). *Journal of Soil Science*, **14** (1), 102–12.

An extended bibliography for this paper may be obtained from the senior author.

Deterioration of Hair by Airborne Microorganisms: Implications for Museum Biological Collections

Catharine A. Hawks

Section of Mammals, The Carnegie Museum of Natural History, 4400 Forbes Avenue, Pittsburgh, Pennsylvania 15213, USA

&

Walter F. Rowe

Department of Forensic Sciences, The George Washington University, Washington, DC 20052, USA

ABSTRACT

When exposed to relative humidities of 80–90%, human and animal hair was rapidly attacked by airborne microorganisms. Damage included exfoliation of cuticular scales, distortion of the cortex and medulla, dramatic colour changes in some samples, and eventually, complete disintegration of portions of the hair shaft. Degreasing with diethyl ether briefly delayed the onset of attack, but had no other discernible effect on the process of biodeterioration. The hair was severely damaged long before the presence of microorganisms could be detected with the unaided eye. These findings suggest that museum collections of mammals, especially those from tropical environments, should be stored at relative humidities below 50% to prevent regeneration of mould spores.

INTRODUCTION

Museum collections of mammalian skins are an important scientific resource. Information about hair growth, colour and morphology are used for species identification. Steps are taken during field collection and preparation to minimize the deleterious effects of biodeteriogens on mammalian skins, especially in the tropics. To prevent colonization by fungi, they are removed from the animals, dried as quickly as possible and frequently degreased by immersion in organic solvents.

The research reported here replicated the conditions conducive to fungal infestation that occur when mammalian skins are collected or stored in humid environments. The rate of development of fungal infestation and the effect of degreasing on the rate of infestation were examined. The effects of fungal infestation on the features of mammalian hair were studied, particularly hair colour, cuticular scale pattern and medullary morphology. Finally, because most study skins are examined only with the naked eye or at low-power magnification, the degree of infestation at which the presence of fungi becomes overt was assessed.

METHODS AND MATERIALS

The substrates used in these experiments were human and domestic animal hair taken from living hosts. The samples of human head hair were cut from a blond Caucasian female (age 9 years) and a blond Caucasian male (age 18 years). The animal hairs were dark guard hairs plucked from the dorsal ridge of a mature domestic cat and white guard hairs and underfur plucked from the neck of a mature domestic dog (Collie). Domestic animal hairs were chosen because Connole (1965) found an absence of any quantity of kerotinolytic fungi on cat and dog hair and suggested that most of them were contaminants from other sources. Samples were placed in polyethylene envelopes until used in the experiments. To avoid denaturization of the keratin, hairs were neither autoclaved nor chemically sterilized; however, one sample of the blond hair and one of dog hair were cleaned of surface grease by two 10 minute washes in diethyl ether. Experimental samples were exposed briefly to the air in the Division of Mammals, United States National Museum of Natural History, and to the air in the laboratories of the Department of Forensic Sciences, The George Washington University. They were placed in closed humidity chambers and incubated at high relative humidity. Individual hairs were removed at about weekly intervals and examined microscopically at magnifications up to 500×, first as dry mounts, then stained with lactophenol trypan blue. Humidity chambers were constructed from new, unused aluminium food containers with plastic lids and from new polyethylene food storage containers. White nylon grids were suspended in the chambers to support the specimens. Both the containers and nylon grids were washed, then rinsed with distilled water and acetone. The bottom of each container was lined with a layer of silica gel saturated with distilled water. A small thermohygrometer was inserted in each chamber from time to time to verify that the humidity levels remained in the 80–90% range. The temperatures inside the chambers were the same as the laboratory temperatures, which varied from 20 to 31°C during the course of the experiments. The chambers were covered with cloth to reduce, but not eliminate, light.

RESULTS

All samples eventually developed fungal growth (Table 1). The shortest time observed was 9–14 days for uncleaned dog hair. Female human head hair was infected within 33–45 days in a polyethylene chamber, but required 49–58 days in an aluminium chamber. The results suggest that the relative humidity may have been held more consistently in the polyethylene chambers, possibly because there were more frequent air exchanges in the aluminium containers. Female human head hair placed in the chamber in which fungal growth had already developed on cat hair was more rapidly infected than that exposed in an empty chamber.

The microorganisms identified on the hair samples are given in Table 2. The type of damage and the sequences in which damage occurred varied greatly among individual hairs. However, alterations common to all samples included roughening of the hair surface as cuticular scales lifted and began to exfoliate, erosion of the cuticle and cortex, a tendency for the hairs to stick together as hyphae formed networks between hair

TABLE 1
Time Required for Fungal Activity to be Observed

Chamber[a]	*Sample*	*Detection of fungal activity*[b]
C-I	Male human head hair	51–56 days
	Female human head hair	49–58 days
C-II	Cat hair	21–28 days
	Female human head hair	21–26 days
C-III	Cleaned female human head hair	51–63 days
C-IV	Female human head hair	33–35 days
C-V	Cleaned dog hair	23–36 days
	Dog hair	9–14 days

[a]Chambers C-I and C-II were aluminium; the remainder were polyethylene.
[b]Based on microscopic detection of fungal hyphae.

shafts, penetration of the shaft by perforating structures, and changes in the birefringence of the anisotropic portions of the hair.

The lifting of cuticular scales was often the first microscopic indication of the activity of microorganisms and this was apparent only when the samples were stained. After 6 months exposure to high relative humidity, cleaned and uncleaned dog hair suffered a complete loss of cuticular scale pattern. Cat hair seemed to be particularly prone to erosion of the cuticle and cortex by microorganisms and developed areas where erosion progressed nearly through the shaft or, where the shaft was reduced, to a few thin fibres. In dry mounts, the damage often resembled the results of insect infestation and it was only after staining that fungal activity was apparent. Minor erosion of the exterior of the hair shaft occurred in all the samples at various times.

The hair shafts were penetrated by two types of perforating structures, both attacking at right angles to the longitudinal axis. These were thin, filamentous hyphae and broad cone-shaped hyphae, both of which seem to have the outer mycelial sheath described by English (1969). Deacon (1980) suggests that the more massive organs are characteristic of keratinolytic fungi, while the non-keratinolytic fungi produced the narrow, boring hyphae. The hyphae produced longitudinal erosion in the cortex and when viewed with crossed polars, it could be seen that they had altered the birefringence of the cortex at some distance from the microorganism. This suggests that the cortex had been denatured to a less structured, more amorphous state, possibly via enzymatic digestion. Animal hair was also attacked by penetrating hyphae extending from the medulla into the cortex. The medulla appeared to be filled with masses of filamentous hyphae before wide, cone-shaped penetrating organs attacked the cortex. In both human and animal hair, the medulla sometimes became extremely swollen, eventually splitting the hair shaft.

A dramatic colour change occurred in both samples of white dog hair after 96 days exposure to high relative humidity (83 days after fungal

TABLE 2
Microorganisms Identified

Microorganism	*Sample*
Chaetomium sp.	Female human head hair Cat Dog
Microsporum sp.	Dog
Penicillium sp.	Male human head hair Female human head hair

attack was confirmed). The white hair became distinctly yellow-orange and the intensity of the colour increased over time. It was noted that the hair degreased with diethyl ether lagged slightly behind the uncleaned hair both in rapidity of the colour change and fungal infestation. The colour appeared to be associated with a proliferation of brown hyphae and spores on the hair surface. Attempts to remove the microorganisms by washing the hair with ethanol or acetone were unsuccessful. Washing with water and a phosphate detergent removed some of the surface hyphae and spores, but the coloured hyphae persisted within the hair structure. It was not possible to determine whether the colour change is a function of the microorganism and its waste products as suggested by Allsopp (1985), or a function of accelerated oxidation of fatty acids due to exposure to high relative humidity. The colour change also may be related to physical changes in the medulla that alter its optical properties.

DISCUSSION

The results suggest that lack of climatic control for haired specimens in natural history collections may lead to deterioration over fairly brief periods of time. Storage of uninfected material with infected material can hasten fungal attack on uninfected specimens. If the environment favours the growth of microorganisms, even uninfected material stored in separate cabinets eventually would be attacked via fungal spores from the ambient environment. Degreasing of the hair surface slowed the onset of fungal attack, probably because it removed fats and oils which attract and hold hygroscopic airborne particulates. It did not prevent fungal attack or mitigate the damage from the fungi.

Much of the damage which occurred was invisible to the unaided eye and could be observed microscopically only after staining. Hair often appeared, to the naked eye, to be intact long after the structure had been weakened and infiltration by the microorganisms was well advanced. Visual inspection is therefore unlikely to provide sufficient warning before biodeterioration damages collections.

REFERENCES

Allsopp, D. (1985). *Journal of the Society of Archivists,* 7, 530–3.

Connole, M.D. (1965). *Sabouraudia,* **4**, 45–8.

Deacon, J.W. (1980). *Introduction to Modern Mycology.* Halsted Press, New York, pp. 170–2.

English, M.P. (1969). *Transactions of the British Mycological Society,* **55**, 247–55.

PAINT

Biodeterioration of Paint—a State-of-the-Art Comment

A.F. Bravery

Department of the Environment, Building Research Establishment, Princes Risborough Laboratory, Princes Risborough, Bucks, UK

INTRODUCTION

There is now a high level of acceptance amongst biodeterioration biologists and technologists in the paints and biocides industries, that deterioration of paints through biological attack is a fact of life. This view is not widely understood or accepted by users of paints nor by some producers of paint products. Many specifiers of paints also remain unconvinced that the service life might be extended by improving resistance to microbial attack.

Paints and painted surfaces provide in themselves a particularly wide range of different ecological niches which micro-organisms have succeeded in exploiting in a variety of intriguing ways. The individual components of paints often provide biodegradable substrates, the diluents and fluid additives are often susceptible, and in water-borne paints, the formulated product can be attacked during production, in pre-sale storage and as the re-usable residues. In service all paint firms can be vulnerable, especially when applied to substrates which are themselves susceptible to biodeterioration. Painted substrates indoors are exposed to a wide range of conditions from kitchens to cold stores, food production plants, brewing environments, etc., and outdoors from arctic to tropical, sheltered and shaded to full weather exposure. Small wonder that so many varied ways exist in which micro-organisms bring about biodeterioration.

This paper attempts to review the range of organisms implicated in paint biodeterioration, their interaction with paint and substrates to which it is applied, and to comment on some of the needs for future attention. As a preliminary the increasingly complex and varied range of components and formulations encompassed by the term 'paint' will be considered first.

THE COMPOSITION OF PAINT

'Paint' is essentially a suspension of finely divided solid material in a liquid matrix which dries or hardens after application to yield a continuous film which adheres to the underlying substrate. However the range of compositions varies considerably in order to confer specific desirable properties (Gray & Miller, 1986) and in recent years adaptations specifically for wood coatings have yielded a particularly diverse range of 'stain' type products (Bufkin & Wildman, 1980; BSI 1986*a*).

All paint compositions have in common 3 primary components:

—the binder medium or matrix which confers most of the properties;
—the pigments and extenders which confer colour and build;
—the solvents which are either organic (solvent-borne paints) or water (water-borne or emulsion paints).

There is also a huge range of additives comprising perhaps 5% or less of the total system. These include thickeners (often cellulose based), emulsifiers, plasticisers, colloids, all of which can be highly susceptible to microbial attack, and driers (usually organic metal soaps of manganese, cobalt, iron and lead). Variation in the choice and relative proportions of the primary components and additives confers the differing physical properties to the paint, whilst their relative susceptibilities individually and in combination determine the biodegradability of the formulated product or the dry film.

The medium, together with any organic additives, is generally the most vulnerable component. Formerly highly vulnerable materials such as linseed oil and other vegetable oils were popular as a medium. These now have been largely superseded by solutions of synthetic resins (alkyd, acrylic, polyurethane and epoxy) and often combinations of them are used which vary considerably in their biodegradability. Formulation of a semi-colloidal suspension of a resin in a water medium produces a latex (emulsion) which forms the basis of most water-borne paints. Another medium for water-thinned paints is water-soluble proteinaceous materials such as casein and glue, as well as water soluble synthetic resins. Many of the media used for water-thinned paints are highly susceptible to microbial attack.

Extenders, which might be regarded as additives, generally are inorganic particulates which do not affect the opacity of the paint. They are mainly resistant to micro-organisms with the exception of some slurry forms (zinc sulphide, lithopone).

Most solvents and pigments are not biodegradable though there are exceptions in that some organic pigments and dyes are vulnerable especially if formulated as pastes rather than dry powders.

The variety of components and of formulations has increased greatly recently due to advancement in the technology of paint production, in the types of substrates coated and the variety aesthetically of decorative and protective effects being sought. In particular the development and use of water-thinned products has increased markedly as a result of environmental and health and safety initiatives to find alternatives to oil-based coatings. Clearly the range of constituents and variety of formulations make them potentially susceptible to biodeterioration both as liquids and dry films.

ENVIRONMENTAL FACTORS

The conditions for growth of microbes clearly must be satisfied before attack can occur under any particular circumstances. Consideration of these factors reveals just how frequently minimal requirements can be met in practice.

Source of infection

Air-borne spores and propagules are always present except in clinically clean atmospheres. Sources of infection are more abundant wherever there are extraneous sources of nutrient such as decaying vegetation, kitchen, brewing and dairy products, fertilisers, foodstuffs, farm produce and rural airborne soiling.

Oxygen

Under most service conditions oxygen is never limiting, minimum requirements being met by normal levels of aeration. Where oxygen is limited, for example in liquid products, anaerobic species exploit the niche.

Food

Many constituents and formulated products contain a source of carbon which often is easily available to the enzymatic processes of microbes. Even if availability is limited, traces of organic matter soiling paint surfaces can permit growth through co-metabolism.

Water

The growth requirement most likely to be limiting, especially in oil-based liquid paints and in interior service situations, is water. As liquids, solvent-based paints generally contain insufficient water to support growth and even if contaminated by leakage or condensate, oil solvents tend to be sufficiently inhibitory to prevent significant microbial deterioration of the product. Dry films of all paints may become infected if wetted outdoors by rain or indoors by condensation, rising damp, splash, leakage, wash-down water, etc.

Temperature

The temperature at which paint constituents and formulations are prepared, and those at which paints are stored, are rarely limiting for microbial growth. As many of the constituents may be damaged by heat, possibilities for preventing or sterilising growths by heating are limited. In fact temperatures involved in many production facilities are almost ideal for the growth of microbes.

In service, paint films are exposed to a wide range of temperatures. Whilst repeated cycles of freezing outdoors or in cold stores may have some sterilising effect, in general, prevailing environmental conditions will tend to select out tolerant species. Heat can be inhibitory more through its drying effects than through elevated temperature alone, at least below 55–60°C. However, the severest conditions for microbial growth on painted surfaces outdoors are encountered in tropical regions where it is both warm and wet.

pH

Paint degrading organisms are tolerant of a very wide pH range from 3·5 to 10·0. Bacterial species implicated in liquid paint spoilage degrade a variety of carbohydrates over this range (Goll, 1958). Whilst most bacteria, yeasts and fungi have optima between pH 4·0 and 6·0, algae and lichens prefer more alkaline conditions (pH 7·0–9·0).

CAUSAL ORGANISMS

Four main groups of micro-organisms cause deterioration of paints:

1. bacteria;
2. yeasts;
3. fungi;
4. algae (and lichens)

Two main categories of microbiological deterioration of paints have been suggested by Goll (1958), namely attack during production and storage, and in service as dried films. A schematic summary of the way these organisms interact with liquid paint, the substrate to which it is to be applied and the dried film is given in Fig. 1.

Bacteria and yeasts

Liquid paints

The rôle of bacteria and yeasts in the fermentative putrefaction ('gassing') of liquid paints is now well established (Hess, 1965). Symptoms of biodeterioration include gas formation (often causing bulging or exploding cans), foul odours, loss of viscosity, coagulation and settling out of solids, reduced pH causing can corrosion and discolorations. Infections arise from a variety of sources, most particularly

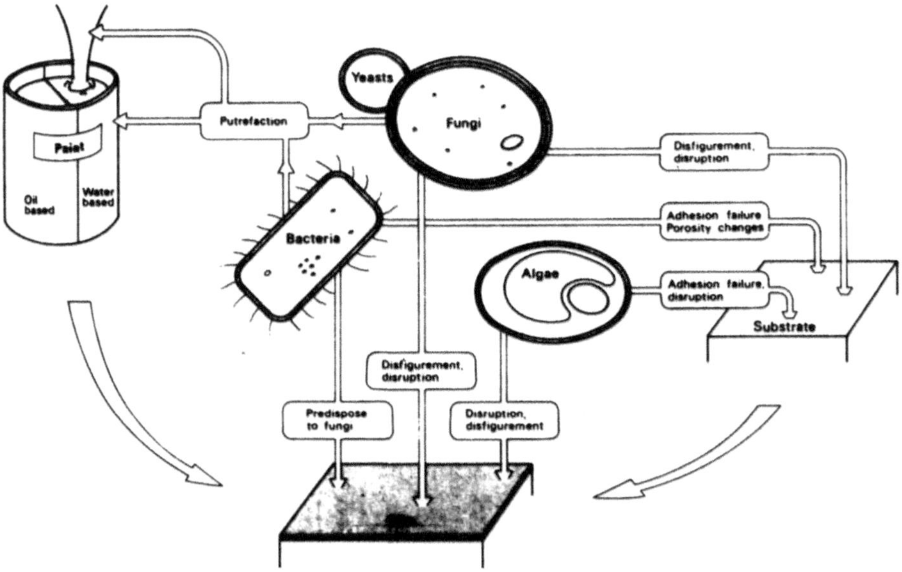

Fig. 1. Microbial interactions with paint.

from the atmosphere of the production plant though raw materials may be contaminated as may mixing waters.

Both anaerobic as well as aerobic species have been implicated (Opperman & Goll, 1984; Opperman, 1986) and all can be active over a wide pH from 3·5 to 10·0 (Goll, 1958). They appear to act on a range of components including sugars, carbohydrates, colloidal thickeners, carriers for water-thinned paints, emulsion stabilisers, viscosity control agents, dispersants, levelling agents, etc. Particular problems of loss of viscosity centre on proteinaceous and cellulosic thickeners.

Substrates

Bacteria can also interact with substrates to produce conditions which adversely affect paint performance. Infection of timber can arise in the growing tree through wounding, but infection introduced during water storage, flotation or sprinkling of logs or freshly sawn timber is more likely (Ellwood & Eklund, 1959). Bacteria can invade certain wood products in service; for example in exterior joinery as a consequence of water penetration of joints (Carey, 1980, 1982). The damaging effects come mainly from attack of ray parenchyma and pit membranes. As a result the porosity of the timber is greatly increased which increases its ability not only to absorb and retain water but also preservative solutions during treatment of components. Subsequent bleeding of solvents from this 'over-absorbent' timber causes retarded drying of paints, discoloration and film softening. Bacteria may develop too at the wood/paint interface (Duncan, 1963) or on exposed surfaces of badly maintained timber and affect adhesion of newly applied paints.

Interactions between bacteria and masonry materials can occur also. Heterotrophic and autotrophic bacteria grow in the surface layers of natural building stone and concrete (Paine *et al.*, 1933; Krumbein, 1983; Lewis *et al.*, 1985). Acidification especially by sulphur oxidising bacteria, produces an unstable, friable layer and adversely affects adhesion of masonry coating.

Paint films

Bacteria and yeasts colonise paint films especially where very high conditions of moisture prevail (Olson *et al.*, 1986). Pink, cream or yellow discoloration often correlate with yeast growth on walls of breweries and bakeries and on interior painted domestic wooden window frames subject to high condensation (Kapyla, 1985). Work in America has indicated that bacteria may be important, if not essential precursors for subsequent invasion of external paint films by microfungi by aiding adhesion and germination of fungal spores (Schmitt, 1974; Winters *et al.*,

1975). There is evidence too that bacteria and yeasts may be more tolerant of some paint biocides and may appear as primary colonisers of certain fungicidal coatings (Grant *et al.*, 1986)

Fungi

Liquid paints

Microfungi and actinomycetes are implicated with bacteria and yeasts in the deterioration of liquid water-borne paints. Ecologically they may occupy different niches, for oxygen tensions are often too low in the paints themselves. Fungi more commonly grow on the surfaces of liquid paint and on residues adhering to the sides of mixing vessels, stirring equipment, holding and distribution plant. Though the consequences for the paint products are essentially the same, the mechanism of the deterioration may be different in that it is the enzymes the fungi release into the paint rather than the growth itself which cause the deterioration (Jakubowski, 1969).

Substrates

Fungi interact with various substrates to interfere with the performance of applied paints. Invasion may occur either prior to application of paint, by penetration of applied films or by other routes which circumvent the coating itself. Fungal growth on paint films is highly dependent on the nature of the substrate and the ability of the substrate to retain and supply moisture and nutrients is critical (Brand & Kemp, 1973). Plastics materials, organic fibre composites, plaster and masonry can become colonised, though metals, glass and glazed ceramics are generally resistant unless heavily soiled. Wood and wood products, however, present the most significantly interactive substrate for fungal invasion.

The range of different types of fungi implicated in the degradation of painted wood is particularly wide.

Sap-stain and mould fungi. Early colonisers of green fresh-felled logs and sawn timber (sap-stain fungi) can offer a source of prior infection of timber destined to be painted. Sap-stain fungi of the genera *Ceratocystis* and *Leptographium* do not survive drying of the timber and application of oil-based paints and are not associated with paint failures. Other sap-stain species such as *Trichoderma*, *Cladosporium* and *Aureobasidium* have been isolated from disrupted coatings. However, some of these species are more often regarded as part of a second group of fungi disfiguring wood, those causing the so-called 'blue-stain in service'.

Blue-stain in service. Aureobasidium pullulans is the principal species involved, though many others including species of *Cladosporium*, *Diplodia, Alternaria, Schlerophoma* and *Stemphyllium* have been identified These fungi will invade wood prior to painting if it becomes re-wetted in storage or in service. Once the substrate is wet enough for hyphae to grow they penetrate rapidly along the wood cells, proliferating in the rays and larger-pored cells (Dickinson, 1971; Bravery & Miller, 1980). They are very damaging because the hyphae are darkly pigmented and because growth under subsequently applied paint films causes disruption and disfigurement.

Wood decaying fungi. Soft rot and basidiomycete types of fungi can affect painted wood by causing disruption and disintegration of the film as the wood is digested beneath. There is no evidence that they can invade an intact paint film by germinating spores. They arise as (a) infection in the wood from over mature or wounded standing trees, (b) infection of logs or sawn timber or (c) in service, by germination from spores washed into unpainted cracks and crevices in woody components. Their activity is dependent on the wood moisture content exceeding 20% for prolonged periods which in service can arise from rain or condensation penetrating at joints or behind putty. Once painted wood becomes wet, the rate of drying even with so-called 'breathing' paints, is normally insufficient to reduce the wood moisture below 20% often or long enough to eliminate the risk of decay. The influence of wood destroying fungi on the substrate is generally so extreme that deterioration of any paint is more useful in detecting the presence of rot than a problem in its own right.

Paint films

The main fungi capable of invading intact paint films are a mixed and varied group of phycomycetes, actinomycetes and fungi imperfecti. Interior and exterior paint films of both oil and water-borne types can be colonised provided sufficient moisture is available; co-metabolism of soiling nutrients will often permit colonisation and deterioration of paints otherwise resistant to fungal attack. Prior colonisation by bacteria may be important with exterior paints. The most serious problem with fungi on interior paints is the development of mould (predominantly species of *Penicillium, Cladosporium, Aspergillus, Stemphyllium, Alternaria* and *Stachybotrys*) on damp walls and ceilings.

Emulsion paints are more seriously affected than gloss paints and condensation is the most significant and widespread source of the dampness (Bravery, 1984). The most serious problem with fungi on exterior paints is that of blue-black disfigurement of light coloured and natural or transparent coatings. The worst problem, and one that is impossible to rectify, arises when the wood substrate beneath the finishes becomes so blue-black that their use must be abandoned in favour of opaque dark coloured coatings.

The mechanisms of colonisation and degradation of paint films have been the subject of numerous studies (Winters *et al.*, 1975; Craig & Pechak, 1978, 1979; Schmitt, 1978; Zabel & Terracina, 1978). There are two main possibilities for initiation of invasion, (1) attachment and germination of spores on the film surface to produce mycelial colonies and (2) germination of spores in or on fibres protruding through the film or in cracks, crevices or other imperfections in the film allowing mycelial development at the interface between paint layers or between paint film and substrate.

It has been suggested that certain fungi can germinate and grow on paint films without using any component of the film as nutrients. The rate of development is believed to be controlled by properties of the films (composition, hardness, porosity) and by environmental factors (substrate compounds, geographical location, climate, aspect). Low gloss, softer and more permeable films are more prone to attachment of spores and to physical penetration by hyphae. Enzymatic, hydrolytic and hydrostatic mechanisms are all implicated in the process of film penetration and degradation (Winters, 1978).

Aureobasidium is the most significant fungus, having been identified in over 95% of cases of fungus defacement of wood coatings. It is very variable in form and physiological responses (Craig & Pechak, 1978, 1979; Zabel & Terracina, 1978) which gives a clue to its success in exploiting the often harsh environment of a paint film.

Algae and lichens

Lichens are symbiotic associations of algae and fungi and can develop not only from propagules containing elements of both partners but also by 'lichenisation' of an algal matt if it is invaded by lichenising fungi. On paint films algal matts become lichenised under appropriate conditions. Algae are sometimes encountered on damp interior walls if sufficient sunlight is available.

Liquid paints

There is no published evidence that algae have been implicated in biodeterioration of liquid paints and it is hard to envisage circumstances under which they could be.

Substrates

Colonisation of the surfaces of wood and masonry materials by algae and lichens is common when moist conditions prevail (Wee & Lee, 1980; Grant, 1982; Olson *et al.*, 1986). Untreated wooden fences, cladding and garden furniture often reveal bright to dark green or black growths. Paving slabs, concrete or rendered facades and some natural building stones can be similarly colonised. Only water and a minimal supply of mineral salts is required. Inoculation is the more rapid where there is adjacent or overhanging vegetation and development is further accelerated if bird droppings or agricultural fertilisers introduce additional nitrogen. Algae and lichens tend to predominate over fungi on surfaces where the pH is 8 or above. Rough or porous surfaces are more vulnerable in facilitating attachment both of air-borne propagules and of accumulations of nutrient enriching soiling material. Surfaces worst affected are those apt to remain wetter longer being orientated away from direct sunshine and from the warmer drier prevailing winds. Peculiarly variable local patterns of growth are often observed (Wee & Lee, 1980). Surfaces which are smooth and highly exposed to driving rain and down wash may be self-cleaning and remain clear of colonisation except under protruding features such as cills and string courses which give shelter and break the flow. Conversely these features can sometimes give enough protection to reduce wetting to that growths do not occur under them but do develop elsewhere on the facades.

The algae implicated in fouling of terrestrial substrates have been reviewed by Grant (1982). The commonest species encountered are *Stichococcus*, *Chlorococcus*, *Trebouxia* and *Pleurococcus* and in tropical regions species of the blue-green algae *Scytonema* and *Oscillatoria* are also common. *Trentepohlia* is often encountered giving a characteristically rusty brown or reddish colouration. A very large number of lichen species have been recorded from masonry surfaces including *Lecanora dispersa*, *L. muralis*, *L. atra*, *Candelariella citrina*, *Xanthoria*, *Lecidia*, *Rhinodina*, *Protoblastinia* and various species of *Physcia*. In rural areas yellow and green coloured species can be prolific whilst in urban areas the grey and black species of *Lecanora* and *Rhinodina* predominate, sometimes as almost pure cultures.

Colonisation of substrates by algae and lichens can give serious problems for subsequent painting operations. Although it is customary to specify cleaning down prior to application of paints (BSI, 1982) lichen growth in particular can be very tenacious. Rhizoids penetrate into even dense, hard types of rock and stone (Bech-Andersen, 1981) which makes complete removal of the thallus very difficult. Even stiff wire brushing following an application of toxic washes may not prevent subsequent re-growth and disruption of coatings applied to affected concrete (Whiteley & Bravery, 1982). Algal growths are more amenable to physical methods of cleaning and more susceptible to toxic washes but detachment of masonry coatings in service often correlates with algae embedded in the first applied coat. Problems are more common with porous, inflexible coatings.

Paint films

The susceptibility of paint films to biodeterioration by algae and lichens is governed primarily by the same environmental factors as have been discussed already. The properties of the paint film itself can however enhance or reduce the risk (Whiteley, 1973; Rusch *et al.*, 1977).

Textured masonry paints present serious risks and low gloss characteristics also encourage attachment and entrapment of propagules. Paints which exhibit 'chalking' appear to vary in susceptibility to growths (Wright, 1986). The rate of chalking is important for whilst it produces an unstable surface less suitable for initiation of colonisation it also provides a matt surface more prone to attachment of propagules. Hard paint films and strongly water repellent ones, such as are produced by oil based alkyd paints, are more resistant to growth. However these surface characteristics may be transient, so durability is critical too. In general algae and lichens are more prolific on basic alkaline surfaces than acidic ones and therefore alkaline paint films are likely to be more susceptible. More permeable films, such as latex types, can be more vulnerable because they permit water transfer of alkalinity from the substrate (Drisko & Crilly, 1974). Algal fouling can be encouraged by the presence of nutrient enriching compounds such as increased levels of phosphate and nitrogen in some latex paints (Rusch *et al.*, 1977).

Little has been published on the mechanisms of colonisation and destruction of paint films by algae and lichens. Evidence of physical or biochemical erosion by algae is tenuous so biodeterioration is probably more aesthetic than structural. However it seems likely that penetration of masonry coatings by lichen rhizoids can occur with consequent disruption and detachment of the paint film. Simple fluctuations in the moisture content of algal films and lichen thalli will induce swelling and shrinkage which in turn can disrupt paint films. Anything which increases water penetration into and through paint films is potentially very damaging to their performance and to the integrity and stability of the substrate.

CONTROL OF PAINT BIODETERIORATION

The basic principle of control of biodeterioration which holds for any material is to limit one or other of the essential growth promoting factors. The options for environmental control are set out in Table 1. Clearly where it is not possible, practicable or economically viable to control the environment the material needs to be rendered inhospitable by the inclusion of control agents.

Liquid paints

With liquid paints there is a possibility of controlling biodeterioration by bacteria, yeasts and fungi by limiting the risk of infection. Air-borne inocula will always be present but the local concentrations can be kept low by good hygiene and plant practice. Mixing and storage vessels, distribution pipework and metering/filling equipment should be cleaned regularly to remove residues. Prolonged holding times bring an increased risk of infection if there are abnormal production or filling delays. A new production run can become inoculated with infected paint, even in small quantity, with thinning and mixing water or washing residues, as well as with infected raw materials. Components supplied as pastes or slurries are particularly at risk.

In so far as it is within the control of the manufacturer and compatible with the properties and economics of the particular product, there is some scope for formulating paints with resistant components rather than those which are inherently susceptible.

Paint films

Only the environmental factors of food and water offer realistic possibilities for control (Table 1). However, in exterior situations it is doubtful whether there is a possibility of controlling wetting, for most paints are part of the total system for protecting buildings, structures and materials from the weather. There are possibilities in the design and construction stages for providing features which shed water well clear of facades, which protect facades from wetting and which do not focus water onto particular zones. Painted timber joinery and other structures are more prone to fungal and algal colonisation if water is not shed quickly from otherwise horizontal surfaces.

Indoors, wetting may be inevitable as for example in swimming pools, dairies, kitchens and breweries where there is a high risk of condensation and much water is used in local operations. In cellars there may often be rising or penetrating damp as well as condensation. Whilst there is scope for reducing moisture levels, sometimes the costs and practicability prevent it as a total solution. In some dwellings where there are special difficulties of sufficient heat input, insulation and ventilation, condensation wetting cannot be adequately controlled and mould growth occurs.

There is some scope for controlling bioderioration by controlling the food source. Good standards of maintenance and hygiene reduce the levels of surface contamination by nutrient soiling. Both externally and internally, occasional cleansing of painted surfaces can reduce the risk of colonisation and the rate of development of organic growths but practical and economic factors often militate against this option.

The second possibility for controlling food source depends upon the extent to which product formulation can be based on components which in the dried films are resistant to biodeterioration. There have been many

TABLE 1

Options for the Environmental Control of Biodeterioration of Paints

Causal organisms	*Liquid paints*	*Paint films*		*Environmental factors*
		Interior	*Exterior*	
Bacteria and yeasts	√?	x	x	Infection
	x	x	x	Oxygen
	√?	√	√	Food
	x	√	x?	Water
	x	x	x	Temp.
	x	x	x	pH
Fungi	√?	x	x	Infection
	x	x	x	Oxygen
	√?	√	√	Food
	x	√	x?	Water
	x	x	x	Temp.
	x	x	x	pH
Algae and lichens	Not relevant	x	x	Infection
		x	x	Oxygen
		x	√?	Food
		√	x?	Temp.
		x	x	pH
		x	x	

x, not possible/practicable.
√, possible.
√?, limited possibilities.
x?, doubtful.

studies to identify the biodegradable components of paint formulations (Brand & Kemp, 1973). However the factors which determine the biodegradability of paint films have been less well studied and knowledge gained has yet to be used in a rational way to produce biologically resistant films. As long ago as 1951, Leonard & Pitman indicated that resistance to microbiological attack could be enhanced by selecting vehicles or binders that are non-nutrients, form the hardest possible films consistent with the flexibility needed in service, have as low moisture absorption as possible and are resistant to pore formation in the film.

More recently Becker *et al.* (1986) have demonstrated a direct correlation between porosity and mould susceptibility in latex paints. However, pigments and additives can vary markedly and be very specific in their effects on the biodegradability of a dried film (Brand & Kemp, 1973). The complexity of modern paint formulations and the variability in proclivity of components to biodegradation presents great difficulties in predicting susceptibility of films to biodeterioration. Therefore it must be determined for specific products at the development stage if the need for resistance is to be taken seriously and presented as a marketing attribute.

Biocidal additives

Unless and until inherently biologically resistant films can be developed, biocidal additives are the only option in situations where there is a demonstrable risk of biodeterioration yet no scope for environmental control.

Whether for preserving liquid paints or the dried film the biocide must meet the criteria.

(1) effective against the appropriate range of organisms;
(2) compatible with the paint;
(3) without effect on paint colour or other properties;
(4) without corrosive effects;
(5) odourless;
(6) non-volatile;
(7) stable;
(8) safe in handling and in use;
(9) cheap.

Despite such demanding requirements a recent review reveals no less than 147 candidate products (Hale & Springle, 1986).

Although a particular biocide may have a broad spectrum of activity against all combinations of target organisms it is highly improbable that any one could be formulated successfully for all types of paint and all purposes. Biocides suitable for and effective in liquid paints do not normally confer protection to a dried film at the same concentration (Hoffman & Saracz, 1967). Selection of biocides to preserve paint films requires careful development and proving. The wide range of paint formulations presents technical difficulties and there are also commercial, marketing disincentives. However there is an increasingly comprehensive range of fungicidal and algicidal products. A very wide range of exterior wood stains are available from low viscosity/low build preservative types to the high build paint-like products. Most of the specialist producers have accepted the need in these products for high resistance to biodeterioration and biocides such as dichlofluanid, tetrachloroisophthalonitrile, iodopropynylbutylcarbamate (IPBC), tetramethylthiuramdisulphide (TMTD) and thiocyanomethylthiobenzothiazole (TCMTD) are among those most commonly used (Pendleton *et al.*, 1987). These and other biocides are also used to confer fungal resistance in the much smaller range of wood paints available for exterior or interior use. Recent problems of mould in dwellings have stimulated the development of a range of interior fungicidal paints based on an increasing number of biocides. Earlier formulations relied heavily on Captan and TMTD which provided a performance standard for these products. The quantities of active ingredient incorporated varies markedly according to the formulation, the price and the performance standard adopted by the formulator. At 1987 prices, good quality fungicidal interior paints in the UK tend to cost on average 1·25–1·5 times that of an equivalent quality non-fungicidal grade.

Among more innovative approaches, pioneered in the USA, are those in which fungicides are chemically reacted with resins (O'Neill & Drisko, 1980) or bound into polymer films (Pittman *et al.*, 1978; Stahl & Pittman, 1978; Pittman & Lawyer, 1982). These biologically active polymers show considerable promise but have yet to be exploited commercially.

ASSESSMENT OF RESISTANCE TO BIODETERIORATION

Liquid paints

Despite studies on causal organisms and mechanisms of biodeterioration of liquid paints and the adoption of a standard in the USA (ASTM, 1986*a*) there is still no general agreement on standardised methods of test for preservative additives. Approaches can be based on isolation techniques to quantify viable propagules, though this fails to detect dead cells or the effects of enzymes. Other approaches involve detecting the consequences of microbial growth in the form of viscosity loss, enzyme activity, gas evolution, etc. Methods have to be adapted depending upon whether anaerobic or aerobic organisms are likely to be involved (Opperman & Goll, 1984; Opperman, 1986) and allowance made for the possible chemical mechanisms of thinning. With such a variety of approaches and techniques results of the tests are greatly influenced by variations in the precise testing protocols.

Within the Paints Group of the International Biodeterioration Research Group (IBRG) efforts have been directed at defining specific strains of test organisms and substrates in an attempt to develop an objective rationale and a consensus view.

Paint films

Laboratory techniques are only capable of a screening test function and field performance can only be judged by exposure of the paint to the full range of factors in a particular environment. However, laboratory methods have distinct advantages and can give useful results especially if reference products whose field performance is known are included for comparison. A wide variety of panel, fence and siding field-tests have been used together with patch trials on test walls or buildings. Laboratory tests also have been the subject of extensive studies and there is now a good deal of common ground internationally.

Early workers employed filter paper as the substrate, painted and placed onto agar in petri-dishes and inoculated with a single or mixed species spore suspension (Vicklund & Manowitz, 1947; Jakubowski, 1971). More recently humidity cabinets have been adopted in order to test under more realistic conditions and realistic substrates (O'Neil & Skinner, 1965; Skinner, 1970, 1971; ASTM, 1982, 1986*b*; Smith, 1978). Following co-operative studies within the IBRG, initially the glass tube test of Skinner and later a humidity chamber/panel test were validated for testing the mould resistance of paint films and the efficacy of paint film fungicides (Barry *et al.*, 1977; Bravery *et al.*, 1983). The panel test has now been adopted in the UK by the British Standards Institution and is shortly to be published as Part G6 of BS 3900 (BSI, 1986*b*). Also favoured are environmental chamber tests (Greenberger *et al.*, 1985; Grant *et al.*, 1986) which provide realistic conditions without limitation on the size and form of test samples.

Standardised laboratory test methodology is now available within Europe to permit assessment of the efficacy of fungicides against blue-stain fungi in service (BSI, 1987).

Laboratory methods of test for algicidal efficacy have been reviewed recently (Grant & Bravery, 1981*a*) and a more realistic alternative proposed in the form of a vermiculite bed technique suitable for a wide range of painted substrates (Grant & Bravery, 1981*b*, 1985).

STATUS OF PAINT BIODETERIORATION AND FUTURE DEVELOPMENTS

The foregoing has attempted to review the very wide range of interactions which can occur between paints, the environments during production and use of the multiplicity of organisms which can exploit the special ecological niches which arise. The requirements in service and developments in their formulation have increased rather than decreased the potential for biodeterioration in production, storage and as dry films.

Environmental pressures have caused the range of water-borne paints to increase considerably, so that wet-state biodeterioration has assumed greater significance. The range of organisms involved remains the subject of active study and though the rôle and significance of anaerobic species is better understood, that of yeasts and microfungi still remain uncertain. The relative significance of microbes, the enzymes and by-products of their metabolism and non-biological , chemical interactions which cause similar symptoms has yet to be satisfactorily resolved.

Large numbers of biocides have been developed and are available to control biodeterioration of liquid paints. Many are used successfully in commercial practice but they are tested and adopted in an *ad hoc* fashion. There still is no satisfactory way of dealing with thinning of latex paints where biocidal control of the growth of organisms is not successful. The complexity of the microbial ecology and biochemistry seems largely responsible for the fact that adoption of common methods of test has still to be achieved.

Much is now known of the types of fungi and algae which colonise paint films and *Aureobasidium pullulans* in particular has received considerable attention as the single most important fungal species. The rôle of bacteria as specific precursors for subsequent fungal invasion has been highlighted yet pure culture studies suggest that such relationships are not necessarily essential prerequisites.

The rôle of the substrate and its influence on micro-organisms colonising paint has been observed extensively in field trials and a significant relationship deduced. However there has been remarkably little rational and objective laboratory study of these interactions with the exception perhaps for fungi on woody substrates. Furthermore, though American work on the properties necessary for paint films to be resistant to micro-organisms has been pre-eminent, there is little evidence of specifically bio-resistant coatings being developed. Choice of paint constituents affects susceptibility to biodeterioration but the cataloguing of compounds is incomplete and the practical and economic possibilities have yet to be exploited. Despite considerable effort in the USA, the technology of biologically active polymers or of polymer-bound fungicides has still to be developed commercially.

The development of accepted standardised methods of test for mould resistance of paint films provides an opportunity for co-ordinated advance on a number of fronts using realistic substrates and relevant fungal species. Agreed methods are still needed however which relate specifically to the roles of yeasts, bacteria and indeed algae.

It seems ironic that in an era of heightened environmental awareness that there are nearly 150 paint biocides available and many good quality fungicidal and algicidal paints on the market, yet none, to the author's knowledge, which claim to be formulated as bioresistant and non-toxic. This is not to imply that paints containing fungicides or algicides are more hazardous than any other domestic pest control agent— only that there are perhaps possibilities for synthesising resistant paints, as with plastics polymers, which do not exist for many other biodegradable materials.

If the research efforts of biologists and chemists has still not delivered the goods we must ask 'why'? Maybe the need for paints which are adequately resistant to biodeterioration has not been established, perhaps because the rôle of biological agencies in the sub-standard or limited performance of paints has not been demonstrated convincingly enough to those who control both the demand and the resources needed to carry out the research.

REFERENCES

ASTM (1982). Standard D3274–82. American Society for Testing and Materials, Philadelphia, USA.

ASTM (1986*a*). Standard D2574–86. American Society for Testing and Materials, Philadelphia, USA.

ASTM (1986*b*). Standard D3273–86. American Society for Testing and Materials, Philadelphia, USA.

Barry, S., Bravery, A.F. & Coleman, L.J. (1977). *Int. Biodeter. Bull.*, **13** (2) 51–7.

Bech-Andersen, J. (1981). In *Biodeterioration 5*, ed. T.A. Oxley & S. Barry. John Wiley, New York.

Becker, R., Puterman, M. & Laks, J. (1986). *Dur. Building Mats*, **3**, 369–80.

Brand, B.G. & Kemp, H.T. (1973). *Mildew Defacement of Organic Coatings—a Review of Literature*. Paint Research Institute, Philadelphia, August 1973.
Bravery, A.F. (1984). In Holzschutz Forschung und Praxis, *Proc. Symposium publ. Holz Zentralblatt.* DRW-Verlag, Germany, pp. 78–80.
Bravery, A.F. & Miller, E.R. (1980). *Rec. Ann. Conv. Brit. Wood Preserv. Assoc.* (1980) 14–23.
Bravery, A.F., Barry, S. & Worley, W. (1983). *J. Oil. Col. Chem. Assoc.*, **66** (2), 39–43.
British Standards Institution (1982). *Code of Practice for Cleaning and Surface Repair of Buildings* Pt 1. *Natural Stone*. Pt 2. *Concrete rendered finishes*. BS 6270: Part 1:1982.
British Standards Institution (1986*a*). Draft British Standard. *Guide to Classification and Selection of Exterior Wood Coatings Systems* Draft for public comment, 86/53537 November 1986.
British Standards Institution (1986*b*). *Methods of test for paints*: Pt G6 *Assessment of resistance to fungal growth*. British Standard 3900: Pt G6. Doc No 86/52086. Draft for public comment.
British Standards Institution (1987). *Test methods for wood preservatives. Laboratory method for determining the preventive effectiveness of a preservative treatment against blue stain in service.* pr EN 152 Pts 1 and 2. Draft for public comment.
Bufkin, B.G. & Wildman, G.C. (1980). *Forest Prod. J.*, **30** (10), 37–44.
Carey, J.K. (1980). The mechanism of infection and decay of window joinery. Ph.D. Thesis, University of London, pp. 426.
Carey, J.K. (1982). *Holz als Roh- u Werk*, **40**, 269–74.
Craig, R.E. & Pechak, D.G. (1978). *J. Coatings Technol.*, **50** (639), 36–42.
Craig, R.E. & Pechak, D.G. (1979). *J. Coatings Technol.*, **51** (656), 33–6.
Dickinson, D.J. (1971). *Rec. Ann. Conv. Brit. Wood Preserv. Assoc.* (1971), 151–69.
Duncan, C.G. (1963). FPL Madison. USDA, *Official Digest*, **35** (465), 1003–12.
Ellwood, E.L. & Ecklund, B.A. (1959). *Forest Prod. J.*, **9**, 283–92.
Goll, M. (1958). *Official Digest*, **30**, 399, April.
Grant, C. (1982). *Int. Biodeter. Bull.*, **18** (3), 57–65.
Grant, C. & Bravery, A.F. (1981*a*). *Int. Biodeter. Bull.*, **17**, 113–23.
Grant, C. & Bravery, A.F. (1981*b*). *Int. Biodeter. Bull.*, **17**, 125–37.
Grant, C. & Bravery, A.F. (1985). *Int. Biodeter.*, **21** (4), 285–93.
Grant, C., Bravery, A.F., Springle, W.R. & Worley, W. (1986). *Int. Biodeter.*, **22** (3), 179–94.
Gray, R. & Miller, E.R. (1986). *Architects J.* Suppl. No. 16, *Third Series*, 3–38.
Greenberger, M., Kaplan, A.M. & Wiley, B.J. (1985). *J. Coatings Technol.*, **59** (731), 59–65.
Hale, S.C. & Springle, W.R. (1986). Evaluation of biocidal masonry coatings and guide to paint film fungicides. Paint RA Tech. Report TR/1/86, March.
Hess, M. (1965). *Paint Film Defects*. Reinhold, New York, 1965, 604 pp.
Hoffman, E. & Saracz, A. (1967). *J. Oil Col. Chem. Assoc.*, **50**, 425–40.
Jakubowski, J. (1969). *Am. Paint J.*, **53** (55), 70–4.
Jakubowski, J. (1971). *Am. Paint J.*, **56** (9), 56–76.
Kapyla, M. (1985). *Allergy*, **40**, 558–64.
Krumbein, W.E. (1983). *Microbial Geochemistry*. Blackwell, Oxford.
Leonard, J.M. & Pitman, A.L. (1951). *Ind. Engng. Chem.*, **43** (10), 2338–411.

Lewis, F.J., May, E. & Bravery, A.F. (1985). *Proc. Vth Int. Cong. Deter. and Conserv. Stone, Vol. 2,* pp. 633–42.
Olson, G.J., Iverson, W.P. & Brinckman,. F.E. (1986). In *Biodeterioration 6, Proc. 6th Int. Biodet. Symp.*, CAB International Mycology Institute, pp. 622–30.
O'Neil, L.A. & Skinner, C.E. (1965). *Soc. Chem. Ind. Monograph*, No. 23, pp. 170–8, April.
O'Neill, T.B. & Drisko, R.W. (1980). *Dev. Ind. Microbiol.*, **21**, 199–210.
Opperman, R.A. (1986). In *Biodeterioration 6, Proc. 6th Int. Biodet. Symp.*, CAB International Mycology Institute, pp. 610–16.
Opperman, R.A. & Goll, M. (1984). *J. Coatings Technol.*, **56** (712), 51–6.
Paine, S.G., Linggood, F.V., Schimmer, F. & Thrupp, T.C. (1933). *Phil. Trans. Roy. Soc.*, **222**, 97–127
Pendleton, D.E., O'Neill. T.B. & Drisko, R.W. (1987). *Proc. Steelstruct. Paint. Council.*, 5th Technical Symposium, pp. 136–64.
Pittman, C.U. & Lawyer, K.R. (1982). *J. Coatings Technol,* **54** (690), 41–6.
Pittman, C.U., Stahl, G.A. & Winters, H. (1978). *J. Coatings Technol.*, **50** (636), 49–56.
Rusch, T.E., Pritchett, E.A., Smith, R.A. & Abir, F.W. (1977). *Canad. J. Paint Finishing*, **51**, 84–8.
Schmitt, J.A. (1974). *J. Paint. Technol.*, **46** (599), 59–64.
Schmitt, J.A. (1978). *J. Coatings Technol.*, **50** (639), 51–3.
Skinner, C.E. (1970). *Paint Oil Colour J.*, **157**, 177–80.
Skinner, C.E. (1971). In *Biodeterioration of Materials*, ed. A.H. Walters & Hueck-van de Plas. Elsevier Applied Science, London, pp. 346–54.
Smith, R.A. (1978). *J. Coatings Technol,* **50** (639), 58–61.
Stahl, G.A. & Pittman, C.U. (1978). *J. Coatings Technol,* **50** (630), 62–5.
Vicklund, R.E. & Manowitz, M. (1947). *Am. Paint J.*, **32** (10), 67–89.
Wee, Y.C. & Lee, K.B. (1980). *Int. Biodeter. Bull.*, **16** (4), 113–17.
Whiteley, P. (1973). *J. Oil Col. Chem. Assoc.*, **56**, 382–7.
Whiteley, P. & Bravery, A.F. (1982). *J. Oil Col. Chem. Assoc.*, **65** (1), 25–7.
Winters, H. (1978). *J. Coatings Technol.*, **50** (638), 48–50.
Winters, H., Isquith, I.R. & Goll, M. (1975). *Dev. Ind. Microbiol.*, **17**, 167–71.
Wright, I.C. (1986). In *Biodeterioration 6, Proc. 6th Int. Biodet. Symp.*, CAB International Mycology Institute, pp. 637–43.
Zabel, R.A. & Terracina, F. (1978). *J. Coatings Technol,* **50** (639), 43–7.

Fungal Colonization of Organic Coatings in Cotton Mills

B.J. Zyska, Z.T. Cieplik, Danuta Kwiatkowska, Helena M. Wichary & Renata Kozłowska

Central Mining Institute, Plac Gwarków 1, 40-951 Katowice, Poland

ABSTRACT

Microbial deterioration of epoxy resin coatings in cotton mills is a constant problem. Studies were undertaken in two Polish mills to determine the factors governing the severe damage to chemicoresistant coatings such as epoxy resins. A total of 28 species of fungi was isolated from the air and coatings in the bleacher and dresser rooms. The differences between the mycoflora isolated from coatings in the mills are discussed. The microbial resistance of epoxy resin coatings in the environment is regarded as unsatisfactory.

INTRODUCTION

Mould growth on painted surfaces is a very serious problem world-wide and occurs in both exterior and indoor environments. Since Massee's (1911) description of *Phoma pigmentivora* as a contaminant of a paint film more than 500 papers have been published on the microbiology of paints, organic coatings and painted surfaces. Painted surfaces in cotton mills have on at least two occasions, in 1937, and 1948, been the subject of microbiological studies, as described by Turner (1967) and Zyska (1977). Workers in these mills frequently suffer from occupational asthma. The problem among weavers of cotton cloth, caused by the inhalation of fungal materials, was described by Collins (1913). The airborne micro-organisms of cotton mills are still being investigated (Van der Werff, 1958; Cinkotai *et al.*, 1977).

In studies of the microorganisms on paint films and organic coatings the ecological aspects cannot be neglected (Winters *et al.*, 1976). New materials are constantly employed in organic coatings technology invoking a whole new series of microbial problems (Winters, 1981). Reported here are the isolation and identification of fungi colonizing epoxy resin chemicoresistant coatings on surfaces in certain workshops of two Polish cotton mills. Investigations were supplemented by the estimation of viable numbers of airborne fungi in the same workshops.

MATERIALS AND METHODS

Investigations were conducted in cotton mills in Bialystok and Ozorków. The Bialystok mill was established in 1965. The yearly throughput of raw cotton averages 25 000 tonnes. The Ozorków cotton mill was founded in 1807, but the bleacher and dresser rooms were built in 1974. The yearly throughput of raw cotton averages 18 000 tonnes. In both mills the microbiological studies of air and organic coatings were undertaken in the bleacher and dresser rooms. In both places severely damaged epoxy resin coatings were analysed. The presence of a chlorine compound used in the production technology caused chemical corrosion of concrete elements of the building, resulting in lowered pH at the surface. Inadequate sealing during wet production processes and insufficient ventilation meant that the ceiling and walls were continuously wet. An additional factor promoting the surface growth of fungi was the formation of cotton and starch dust on the coatings.

The isolation of fungi from the organic coatings was carried out by three methods:

(1) Direct streaking of micro-organisms from discoloured coating surfaces into sterile physiological salt solution in test tubes.
(2) Swabbing with cotton wool from a 25 cm^2 surface (Upsher, 1977); (samples were used for quantitative estimation of fungi per cm^2).
(3) Direct sampling of coatings, including the plaster layer; the samples were shaken in sterile water, diluted and plated on appropriate agar. All the plates were incubated at $25 \pm 1\,°C$ for 3 weeks. The following media were used; Sabouraud's agar for fungi, malt extract agar and carboxy-methyl cellulose agar (Booth, 1971).

Airborne microorganisms were collected using the Polish 'Barbara 3A' air sampler on cellulose membrane filters of 1 μm pore size and the filters were then placed on Sabouraud's agar medium. The developing colonies of fungi were counted after 5 days of incubation under the conditions described above. In both mills 40 samples of air and 84 samples of coatings were taken.

RESULTS AND DISCUSSION

Table 1 gives data for the maximum number of colony forming units of fungi in both cotton mills. The frequency of airborne fungi and fungi colonizing organic coatings in the two mills is presented in Table 2. In the bleacher room at Bialystok seven species were identified on coating samples and 20 at Ozorków. In the dresser room, however, only four species were identified at Bialystok and nine at Ozorków. The difference in number of species of fungi isolated may be explained by the age of the coatings varying from 7–8 months in Bialystok to 2–3 months in Ozorków. On older coatings three typical fungi, i.e. *Cladosporium, Aureobasidium* and *Phoma* represented 58% of the total in the bleacher room and 64% of the total in the dresser room. On the 2–3 month old coatings the mycoflora was more differentiated but of the three particular species only *Cladosporium* was present. The composition of the mycoflora of the coatings in the Bialystok and Ozorków cotton mills agrees with the results of previous laboratory studies by Winters *et al.* (1976) who showed that *Aureobasidium pullulans* is unable to grow on vinyl acrylic or any other paint film not previously colonized by other fungi. This suggests that further studies on the ecological succession on epoxy resin coatings by strains of fungi isolated in the two mills would be of interest. It is noteworthy that the 2–3 months old coatings at Ozorków cotton mill are colonized by many fungi, almost the same genera as those isolated by Winters *et al.* (1976), but *Aureobasidium pullulans* is absent from this early stage of colonization.

Attention is drawn to the agreement between the fungi isolated from the air and from organic coatings determined in eleven cases for *Aspergillus stellatus, Aureobasidium pullulans, Cladosporium herbarum,*

TABLE 1
Colony Forming Units (cfu) of Fungi in Samples of Air and Epoxy Resin Coatings in Cotton Mills

Site	*cfu in air per m^3*		*cfu on coatings per cm^2*	
	Bleacher	*Dresser*	*Bleacher*	*Dresser*
Bialystok	1.8×10^3	1.7×10^4	4.2×10^5	3.7×10^5
Ozorków	2.6×10^3	1.3×10^3	1.0×10^7	3.2×10^5

TABLE 2
Frequency of Airborne Fungi and Fungi Colonizing Organic Coatings in Cotton Mills (%)

	Species of fungi	*Bialystok cotton mill*				*Ozorków cotton mill*			
		Bleacher room		*Dresser room*		*Bleacher room*		*Dresser room*	
		12 Air samples	*18 Coating samples*	*8 Air samples*	*10 Coating samples*	*16 Air samples*	*42 Coating samples*	*4 Air samples*	*14 Coating samples*
1	2	3	4	5	6	7	8	9	10
1.	*Acremonium strictum* W. Gams						1·5		
2.	*Alternaria alternata* (Fries) Keissler						1·5		
3.	*Alternaria mouchaccae* Simmons						6·3		
4.	*Aspergillus flavus* Link ex Fries						1·5		
5.	*Aspergillus glanulosus* Raper and Thom								5·3
6.	*Aspergillus niger* Van Tieghem						1·5		
7.	*Aspergillus sclerotiorum* Huber						1·5		
8.	*Aspergillus stellatus* Curzi	18·2	11·5			7·7	7·7		
9.	*Aureobasidium pullulans* (de Bary) Arnaud	27·2	23·1		27·3				
10.	*Chromelosporium* state of *Peziza ostracoderma* Korf					15·4			
11.	*Cladosporium herbarum* (Persoon) Link ex S.F. Gray	9·1	34·6		36·3	46·2	17·2	100·0	26·2
12.	*Cunninghamella* sp. Matruchot						1·5		
13.	*Fusarium solani* (Martius) Saccardo		11·5				4·1		10·5
14.	*Gliocladium roseum* Bainier						1·5		

(continued)

TABLE 2—*contd.*

	Species of fungi	*Bialystok cotton mill*				*Ozorków cotton mill*			
		Bleacher room		*Dresser room*		*Bleacher room*		*Dresser room*	
		12 Air samples	*18 Coating samples*	*8 Air samples*	*10 Coating samples*	*16 Air samples*	*42 Coating samples*	*4 Air samples*	*14 Coating samples*
1	2	3	4	5	6	7	8	9	10
15.	*Mortierella verticillata* Linnemann					7·7			
16.	*Mucor racemosus* Fresenius	9·1		20·0		7·7	7·7		
17.	Mycelia sterilia			20·0			1·5		5·3
18.	*Mycotypha microspora* Fenner								5·3
19.	*Nodulisporium* sp. Preuss		3·9						
20.	*Paecilomyces variotii* Bainier						4·1		5·3
21.	*Paecilomyces fumosoroseus* (Wize) Brown and Smith						1·5		
22.	*Penicillium* sp. Links ex Fries	36·4	7·7	20·0	18·2	15·3	10·1		15·8
23.	*Phoma violacea* (Bertel) Eveleigh			40·0	18·2				
24.	*Rhizopus stolonifer* (Ehrenberg ex Fries) Lind						5·3		
25.	*Rhodotorula* sp. Harrison		7·7						
26.	*Syncephalastrum racemosum* Cohn ex Schröter						1·5		
27.	*Trichoderma viride* Persoon ex S.F. Gray						5·3		10·5
28.	Yeast-like fungi						17·2		15·8

Penicillium sp., *Mucor racemosus* and *Phoma violacea* (Table 2). The coatings studied in these cotton mills were all epoxy resin types. The microbial resistance of epoxy resins was tested by Schwartz (1963) who reported that *Aspergillus* sp. penetrated the structure of this resin. This was confirmed to some extent in our studies.

The Catalogue of the Culture Collection of the Commonwealth Mycological Institute (1982) proved to be a very valuable reference work for information on the range of substrates colonized by different fungi. Of seven species of fungi present in the air of the mills four were confirmed as airborne fungi, i.e. *Aspergillus stellatus*, *Aureobasidium pullulans*, *Mucor racemosus* and species of *Penicillium*. Of 25 species of fungi isolated from the epoxy resin coatings, 12 were listed as isolated from the soil, i.e. *Alternaria alternata*, *A. mouchaccae*, *Aspergillus granulosus*, *A. stellatus*, *A. niger*, *A. sclerotium*, *Cunninghamella* sp., *Gliocladium roseum*, *Nodulisporium* sp., *Paecilomyces fumoseroseus*, *Syncephalastrum racemosum*, *Trichoderma viride*, and 9 were listed as isolated from cotton or other cellulosic materials. Based on our investigation of the fungal colonization of epoxy resin coatings in two cotton mills, their microbial resistance must be regarded as unsatisfactory.

Organic coatings in cotton mills provide a very interesting subject for further studies to investigate the ecological succession of microorganisms on an industrial scale. More tests are needed to develop effective fungicides for epoxy resin coatings.

REFERENCES

Booth, C. (1971). In *Methods in Microbiology, Vol. 4*, ed. C. Booth. Academic Press, London, pp. 51–94.

Catalogue of the Culture Collection of the Commonwealth Mycological Institute (1982). Commonwealth Mycological Institute.

Cinkotai, F.F., Lockwood, M.G. & Rylander, R. (1977). *American Industrial Hygiene Association Journal*, **38**, 554–9.

Collins, E.L. (1913). Unusual illness among weavers of cotton cloth. *Factories and Workshops. Annual report for the year 1912.* London.

Massee, G. (1911). In Catherine E. Skinner, *Laboratory Test Methods for Biocidal Paints.* In *Biodeterioration of Materials, Vol. 2*, ed. A.H. Walters & E.H. Heuck-Van der Plas. Elsevier Applied Science, London, (1972), pp. 346–54.

Schwartz, A. (1963). *Klasse für Chemie, Geologie und Biologie, No. 5.* Abhandlungen Deutscher Akademie der Wissenschaften Berlin.

Turner, J.N. (1967). *The Microbiology of Fabricated Materials.* J. and A. Churchill Ltd, London.

Upsher, F.J. (1977). In *Biodeterioration Investigation Techniques*, ed. A.H. Walters. Elsevier Applied Science, London, pp. 1–21.

Van der Werff, P.J. (1958). *Mould Fungi and Bronchial Asthma.* H.E. Stenfert Kroese N.V., Leiden.
Winters, H. (1981). In *Microbial Biodeterioration,* Ed. A.H. Rose. Academic Press, London, pp. 307–21.
Winters, H., Isquith, I.R. & Goll, M. (1976). In *Developments in Industrial Microbiology,* Vol. 17. American Institute of Biological Sciences, Washington, pp. 167–71.
Zyska, B. (1977). *Wydawnictwa Naukowo-Techniczne,* Warsaw (in Polish).

PAPER AND PACKAGING

Microbiology of Paper and Board Manufacture

D.M. Holt

Micran Ltd, Unit A1, Wellheads Crescent, Wellheads Industrial Estate, Dyce, Aberdeen, AB2 0GA, UK

ABSTRACT

There have been a number of significant changes in the UK paper and board industry over the last decade, in particular the increase in waste paper usage and the reduction in water consumption of mills per tonne of product. This review paper presents details of how these changes have affected the microbiological aspects of paper and board manufacture. Emphasis is directed to the reduction in water consumption 'closing up' in mill systems and methods of controlling microbial slime deposits. Details of microbial content of pulps, fibres and final products are also discussed.

INTRODUCTION

The growth of microorganisms in paper mill systems is well documented and considerable research has been carried out to identify the organisms responsible (Prendergast, 1948; Appling *et al.,* 1955; Wang, 1965) and to control deposit formation (Hughes, 1968; Conkey & Carlson, 1969; Kirchen & Louche, 1975). However in recent years, there have been a number of significant changes in the paper and board industry, in particular, the increase in waste paper usage and the reduction in water consumption of mills per tonne of product. The paper and board division at Pira have undertaken a number of major research projects on microbiology for the UK as these changes have occurred. This review presents a small proportion of the data obtained and outlines the problems caused by microorganisms and methods for control.

The microbiology of the pulp and paper industries can be conveniently divided into three broad sections, pulps, paper and board manufacture and final products. In the UK, the manufacturing aspects of microbiological degradation are by far the most important.

PULP MICROBIOLOGY

The degradation of wood fibres during the production of pulp is of little concern to the UK paper and board manufacturer. The vast majority of pulp used in the UK is either imported virgin fibre or derived from recycled waste papers.

When pulp is manufactured for export in Scandinavia, Canada, etc., it is partially dried and shipped in laps to the paper mill for rehydration and conversion into paper. The effect of microorganisms on the pulp is mainly dependent on moisture content and to a lesser extent temperature. Little or no growth is expected at less than 12% moisture. Fungi start to grow at around 14–20% moisture with an optimum of about 25–30%. Two principal types of fungal attack on pulps are recognised, cellulolytic, a process that weakens the fibres mechanically, and disfigurative, mainly caused by staining fungi. Common fungi responsible for fungal attack are given in Table 1 (from Weir, 1982). The majority of these fungi are also common contaminants of waste papers (Table 1). Many of the organisms found in both pulps and waste paper are soil or airborne microorganisms which do not normally affect the strength of the fibres during storage. However, severe contamination of pulp materials can lead to problems during paper and board manufacture. Higher moisture levels will allow the growth and proliferation of bacteria.

For reasons of economy in transport and handling costs, lap pulp is normally shipped as dry as possible. Thus the development of flash drying in the mid-1960s has been extremely effective in reducing biodeterioration. After mechanical removal of water, the pulp is dried in a fluffed state by a stream of hot air and combustion gases; thus the water in the fibre evaporates instantaneously. Flash-dried pulp can be easily compressed into bales or pelleted and its low moisture content (<10%) ensures resistance against microbiological degradation. As virgin fibres command a high price, care is normally taken to store the pulp bales undercover prior to use.

The increase in waste paper recycling in the UK appears to have caused few problems with regard to microbiological degradation.

TABLE 1
Cellulolytic and Staining Fungi Isolated from Mechanical and Chemical Pulps and Waste Papers

Pulps		*Waste paper*
Cellulolytic	*Disfigurative*	
Alternaria spp.	*Aleurisma* spp.	*Alternaria* spp.
Aspergillus fumigatus	*Aspergillus niger*	*Aleurisma* spp.
Basidiomycetes spp.	*Botrytis* spp.	*Aspergillus* spp.
Ceratostomella spp.	*Mucor* spp.	*Cladosporium* spp.
Cladosporium spp.	*Paecilomyces* spp.	*Fusarium* spp.
Epicoccum spp.	*Penicillium cyclopium*	*Geotrichum* spp.
Fusarium spp.	*Penicillium roquefortii*	*Monilia* spp.
Geotrichum spp.	*Pullularia* spp.	*Mucor* spp.
Gliocladium spp.		*Paecilomyces* spp.
Monilia spp.		*Penicillium* spp.
Phoma spp.		*Phoma* spp.
Stemphylium spp.		*Rhizopus* spp.
Trichoderma spp.		*Syncephalastrum* spp.
		Trichoderma spp.

Therefore the main concern with contaminated waste papers is the introduction of high levels of microorganisms into the paper or board making process. Waste-paper merchants have developed efficient systems of collecting, storing and delivering supplies to manufacturers. At the mill, storage is mainly outdoors but the turnover is usually rapid so that there is little time for significant degradation by microorganisms. Studies at Pira (Weir, 1977) have shown that storage of waste can affect the numbers of bacteria and the types of fungi but they are not considered important in the short term. Of more importance is the type of waste and the way it is handled and stored whilst loose (Weir *et al.*, 1981). Thus clean shavings are expected to be less contaminated than mixed merchant waste (Table 2). Weir *et al.* (1981) examined the possibility of using refuse derived fibre for paper and board manufacture. Microbial contamination was found to be comparable with that of mixed waste papers of domestic and industrial origin (Table 2). However the use of dustbin waste in the 1980s creates a hazard of microbial contamination from blood and faeces as well as the more traditional rubbish. Weir *et al.* (1981) detected several pathogenic bacteria and fungi, *Clostridium perfringens, Escherichia coli, Salmonella emek, Absidia* spp. and *Aspergillus fumigatus,* from both the loose material and aerosols generated by the separation process of the reclamation plant and suggested that sanitation of waste should be investigated. Due to the economies of

TABLE 2
Aerobic Bacterial Counts on Fibres from Different Sources (mainly from Weir (1977) and Weir *et al.* (1981))

Fibres	*Bacteria/g*
Pulps	$<10^3$
Clean waste	10^2–10^3
Newsprint	10^3–10^4
Container waste	10^3–10^4
Mixed merchant waste	10^3–10^5
Mixed council waste	10^4–10^6
Mixed waste (other sources)	10^6–10^{10}
Refuse derived fibres	$<10^5$–10^9

producing refuse derived fibre it has not found any significant market in the UK paper and board industries.

MICROBIOLOGY OF PAPER AND BOARD MANUFACTURE

Problems caused by microorganisms

Paper and board mill water systems provide an excellent environment for the growth of microorganisms. Process water temperatures of around 25–45°C, pHs between 5 and 9, and an abundant supply of dissolved nutrients encourages the proliferation of a wide range of bacteria and fungi. Where these organisms accumulate on papermaking machinery, they form the characteristic grey/brown/black gelatinous masses typical of mill slimes. Microorganisms can also cause degradation of additives such as starch and rosin size. Thus a number of serious problems can result from microbial activity:

(i) Machine runnability problems caused by breaks at the wet end.
(ii) Machine runnability problems caused by pipe blockages, plugging of wires and felts or holes in suction boxes and couch rolls. This affects the efficiency of the machine operation.
(iii) Sheet defects, e.g. spots and windows, blotches or specks. These survive passage through the machine but are detected by mill staff or the customer.
(iv) Corrosion: this can be a serious problem which is often not recognised until too late.
(v) Odours in the mill and in the final product. This can effect the health and safety of workers and if carried over into the product can result in customer complaints.

(vi) Toxic gas: this particularly relates to hydrogen sulphide which is dangerous to the workers and can also cause corrosion.

(vii) Possible contamination of final product: in heavily infected mill water systems, a large number of bacterial spores may survive the drying process.

(viii) Degradation of additives: microbial activity often leads to changes in viscosity of starch solutions and deterioration of sizing chemicals.

In practice slimes are found to develop as relatively small pockets of growth in a variety of places dependent upon each individual mill system. Slimes growing in the vicinity of stock approaching the wire, for example, in headboxes or vats, are the most crucial for machine runnability. Regions particularly susceptible to slime formation are where the water flow is reduced, stagnant tanks, underneath suction boxes, above the process water line, specific parts of vats, pipes of intermittent usage and blanked off 'Tee' and 'Y' joints. It should also be stressed that slimes act as a reservoir for re-infecting the mill water system.

Types of microorganisms found in the papermill waters

Given the right conditions, microorganisms can grow almost anywhere in the water circuits of a paper machine. The system is continually infected by fungi and bacteria from several sources. Microorganisms enter into the papermaking system through contaminated raw materials, fibres, chemicals and water. Once introduced microbial growth is sustained and unless they are eliminated by biocides will remain in the system until removed in the final product, as effluent or as solid waste.

There are many organisms and materials which may be found in slime deposits making classification extremely difficult. However, by choosing a number of key components it is possible to identify different types of slimes. The first distinction should be made between chemical and biological slimes. This can normally be determined using a microscope. Simple distinctions between bacteria, fungal or mixed microbiological populations can be made at this level and certain types of bacteria, e.g. *Beggiatoa* and *Gallionella* can be identified. As bacterial slimes develop there appears to be a succession from cocci and short rods to longer rods, spirochaetes and a range of motile species. An indication of whether the slime is newly formed can be ascertained. Chemical slimes can also be identified if crystals or resins are involved.

Details of microorganisms most commonly isolated from process waters and slime samples are presented in Table 3 (based on Pira studies). Bacteria such as *Aeromonas, Enterobacter* and *Klebsiella* are

TABLE 3
Microorganisms Associated with Slime Deposits

Bacteria		*Fungi*	
Acinetobacter	*Micrococcus*	*Aspergillus*	*Phoma*
Aeromonas	*Pseudomonas*	*Cladosporium*	*Pullularia*
Alcaligenes	*Staphylococcus*	*Fusarium*	*Trichoderma*
Bacillus	*Beggiatoa*	*Geotrichum*	*Mucor*
Coryneforms	*Gallionella*	*Penicillium*	*Rhizopus*
Enterobacter	*Sphaerotilus*	*Phialophora*	*Basidiomycetes*
Flavobacteria	*Desulphovibrio*		
Klebsiella			
Others: Yeasts, protozoa and nematodes			

frequently associated with slime masses but they are also inhabitants of free flowing waters. The various genera of fungi are ubiquitous in air, soil and water samples and are often found in microbiological deposits. Some species, e.g. *Phialophora, Trichoderma* and *Humicola* also have the ability to degrade cellulose.

The types of microorganisms which have the potential for causing slime related problems in the papermill system can be divided into four groups, aerobic bacteria, anaerobic bacteria, sulphate-reducing bacteria (SRB) and fungi. Organisms from each of these groups can be detected using specific growth media. The mechanism of slime formation varies depending on the type of microorganism. Some aerobic and anaerobic bacteria have the ability to produce large quantities of extracellular polymers which adhere the bacteria to one another and to organic and inorganic particles. Great variation in the structure and chemical composition of the polymers has been reported. However, the majority are carbohydrate based, either homopolysaccharides (only one type of sugar, e.g. starch or cellulose) or heteropolysaccharides (two or more types of sugar, e.g. xanthan). Fungi cause problems due to their hyphal growth which results in entanglement with papermaking materials. Finally, SRB do not cause slimes but they are able to grow in anaerobic deposits and corrode metals (Vestola & Korhonen, 1976; Soimajarui *et al.,* 1978; Miller, 1981).

Microorganisms in paper systems with low water usage

Historically, the paper and board industry is one of the largest consumers of fresh water per tonne of product. When paper was originally made by hand between 1000 and 1400 m^3 were required for each tonne of rag pulp produced. At the turn of the century, the amount needed had been halved and by 1930, it was reduced further to around 86–273 m^3/t product. Present day values average approximately 40 m^3/t in the UK with some mills operating below 10 m^3. These changes in water consumption have progressed gradually over the years due to economic factors closely associated with environmental regulations. The classical argument for reducing water consumption was to prevent loss of stock fibres, however in today's paper industry benefits also include reduced costs associated with fresh water supply, reduced energy requirements and reduced effluent loads and treatment costs.

When mills started to close their water circuits the build-up of electrolytes and organic materials was extensively monitored (Brecht *et al.*, 1973, 1974; Streebin *et al.*, 1976; Hurst-Gaul, 1981). The data show that there is a rapid initial increase of dissolved substances in the process waters until an almost steady state level is reached. Normal concentrations have been quoted as ranging from 10 to 20 times for mills changing from an open to a closed system (Hurst-Gaul, 1981; Geller, 1982). However, the quantities of chemicals dissolved in mill process waters can vary considerably. Alexander & Dobbins (1977) have calculated that some electrolytes may be concentrated by a factor of up to 160 times in totally closed circuits.

Practical experience has revealed several problems associated with closed water systems arising from the build-up of dissolved organic and inorganic materials and an increase in temperature (e.g. Dalpke, 1975; Geller & Luttgen, 1981; Mobius & Welcher, 1981). Higher concentrations can cause operational difficulties due to the formation of precipitates which block showers and screens. Alternatively they may promote excessive microbiological activity which adversely affects paper production. Several authors have reported a change in the numbers and range of microorganisms in closed water systems (Jung, 1978; Jung & Kutzner, 1978; Geller, 1982). Anaerobic and SRB can cause serious corrosion problems reputedly due to hydrogen sulphide formation and production of organic acids (Mueller & Muhonen, 1972; Bowers, 1978; Thorpe, 1979). In addition the quality of paper can be affected. The final product may also retain the metabolic by-products of microbial activity such as organic acids and hydrogen sulphide giving rise to odours and taints.

Although the problems attributed to microbial activity in closed systems were recognised during the onset of system closure, very few studies have been undertaken to quantify the changes which occur. The most comprehensive investigation was carried out by Jung & Kutzner (1978), who monitored the development of the microflora in closed waste-paper reprocessing mills.

The most noticeable change in the microflora is the increase in number of anaerobic bacteria relative to aerobes. Observations by Jung & Kutzner (1978), Gissler-Weber *et al.* (1981) and Holt *et al.* (1983) are shown in Table 4. The anaerobic bacteria rapidly multiply in closed water systems until a steady state level is reached. This is illustrated quite clearly by Holt *et al.* (1983). After the annual summer shutdown and start up with fresh water, at a waste based mill, the anaerobic bacterial population developed to its normal level within 2 days of mill operations. In a study by Jung & Kutzner (1978), in which the water was totally replaced, in a closed system, the anaerobic bacteria reached steady state numbers in 15 days. Initially, aerobic bacteria were the dominant organisms (maximum 10^8 colonies/ml after 8 days) but anaerobes gradually increased in numbers to 10^8 colonies/ml after 15 days. Geller (1982) suggested the change is due to the removal of oxygen by aerobic bacteria. This allows the proliferation of specific types of anaerobic bacteria able to convert soluble organics into volatile fatty acids. These simple weak fatty acids give rise to odour and possibly corrosion problems. They also provide the conditions and nutrients for SRB. Once established SRB readily convert sulphate to sulphide and cause pitting (Thorpe, 1979; Miller, 1981). These conversions occur within slime deposits so the detection of even low levels of sulphide in waters indicates a potentially serious problem. Few workers have measured the levels of sulphide in process waters and they are normally below 1–2 ppm even in systems contaminated by SRB (Holt *et al.*, 1983). However values as high as 72 mg/litre (Brecht *et al.*, 1974) and 225 mg/litre (Dalpke, 1975) have

TABLE 4
Changes in Numbers of Aerobes, Anaerobes and Sulphate-reducing Bacteria in Relation to Water Consumption

Specific water consumption (m^3/t)	*Number of bacteria*		
	Aerobes	*Anaerobes*	*SRB*
0[a]	1×10^8	9×10^8	10^4
0[b]	1×10^8	1×10^8	8×10^3
0–1[c]	10^8–10^9	10^7–10^8	10^3–10^4
2–6[c]	10^9	10^8–10^9	10^4
5–10[b]	1×10^8	7×10^7	10^3–10^4
10[a]	1×10^9	4×10^7	10^3
15–25[c]	10^6–10^9	10^5–10^7	10^0–10^3
20[a]	3×10^8	1×10^7	10^3
25–30[c]	10^5–10^8	10^4–10^6	0–10^3

[a] Jung & Kutzner (1978).
[b] Gissler-Weber *et al.* (1981).
[c] Holt *et al.* (1983).

been reported. The measurement of volatile fatty acids and sulphide can be used to indicate if anaerobic conditions have developed (Holt *et al.*, 1983).

Slime control

There are two generally accepted methods of slime control, good housekeeping and the use of biocides (added to fresh water, machine water and bulk stored materials, e.g. fillers and starches). A good housekeeping policy should include hosing down machines, particularly in areas where slimes develop, flushing out pipes, cleaning the walls of tanks and frequent chemical washouts. Storage of broke and waste paper under 'clean' conditions also minimises the level of microorganisms entering the water system.

The types of biocide available to the papermaker can be broadly categorised into three groups: Oxidising—chlorine, chlorine dioxide, hydrogen peroxide, ozone; non-oxidising—a range of chemical compounds, e.g. methylene bis-thiocyanate; enzymes—biological catalysts which attack the matrix of the slimes. Both oxidising and non-oxidising biocides are widely used and at the correct concentration kill any bacteria and fungi present. The use of enzymes is new to slime control in the paper industry and prevents their formation.

The properties that an ideal biocide should possess are listed in Table 5. Many of the current biocides in use fulfil most of the criteria and perhaps the only hazards they cause are due to improper protection

TABLE 5
Properties of a Biocide

Effectiveness
- (i) Toxic to microorganisms in low concentration
- (ii) Broad spectrum of activity
- (iii) Ability to penetrate and break up microbial slimes

Compatibility
- (i) Non-reactive with organic substances
- (ii) Non-reactive with other additives
- (iii) Non-reactive with construction materials

Safety
- (i) Non-toxic to higher forms of life at concentrations used
- (ii) Non-persistent, i.e. ultimately biodegradable when discharged into sewage systems and rivers
- (iii) Safe to handle, not spontaneously flammable or explosive
- (iv) Low volatility and odour.

during handling of the concentrated chemicals. Although enzymes are considered non-toxic, care should still be taken when handling concentrated solutions.

Oxidising biocides are mainly used for the treatment of fresh water entering the mill system. Chlorine is widely employed and is popular because of its efficacy and low cost. Although the water is rendered sterile, if the concentrations used are high enough, application on the machine is limited due to the high chlorine demand of the fibre. Lacrymatory problems with workers can also occur. Chlorine dioxide is more expensive but it has more oxidising power, yet is less affected by the chemical composition of the water than is chlorine. It has been used successfully in treating clarified water returning to the papermill system and has also been employed on machines. Chlorine dioxide is used more commonly in France but there is one mill in the UK carrying out the practice. The use of ozone in the UK is limited by its expense. It is possible to use it where pulping and papermaking units are combined, and ozone generators or bleaching operations are on-site (such as in the United States).

Numerous compounds have been employed as non-oxidising biocides. Often they are blended together to provide a synergistic and more effective control agent. At present upwards of 60 compounds or blends are available from various suppliers in the UK; a number are in the biocide survey compiled by Allsopp & Allsopp (1983). Suppliers of non-oxidising biocides normally have a range of compounds, some being specific to particular organisms, others exhibiting a broad spectrum of activity. Some active ingredients such as methylene bisthiocyanate are widely used whereas others are unique to certain companies. In the UK most mills using biocides on paper machines choose the non-oxidising type. The mode of action of the various biocides differs in that the micro-organism is attacked in a number of regions, cell wall, cytoplasmic membrane and cellular constituents. Because of this, organisms can have varying resistance to the active ingredients.

Enzyme deposit control agents are unique in that they do not kill the bacteria and fungi. They are new to the Paper Industry and have created a great deal of interest. Enzymes are biological catalysts and help reactions to take place without themselves being affected. The mechanism by which they control deposits is through breaking down the polysaccharide matrix which helps microorganisms to adhere together and to surfaces.

The choice of biocides is very much dependent upon cost and compatibility with the system. A number of suppliers also provide a back up service. Selection of a biocide is dependent upon two factors: (1) compatibility of the biocide with the system, and (2) that the organisms to be controlled are affected by the biocide.

In order to ensure system compatibility the following requirements must be met:

(i) The pH must be within the active range of the biocide.
(ii) Physical and chemical parameters must not affect the active ingredients. For example, high temperatures may cause volatilisation of biocide components or reaction with papermaking additives may give rise to noxious substances.
(iii) The biocide must not affect the quality of the water if this is discharged to rivers and sewers.
(iv) Biocides must not colour or taint final products and have to be approved for food grade end use.

Several methods have been developed for testing the efficacy of biocides. The majority use agar plating techniques which can only serve as a rough guide to their comparative activity. More meaningful techniques must include the following criteria.

(i) Tests must be performed in the presence of paper fibre. Many very adequate biocides are absorbed into the surface of fibres and effectively removed from the system.
(ii) As far as practical, the additives normally used in the mill must be present in the test system.
(iii) The time of contact in the mill is an important parameter.
(iv) Paper mills employ highly agitated systems, hence test systems that are not agitated are not truly representative of mill conditions.
(v) Unquestionably, the best medium for a specific mill is the white water from that mill.

Although programmes for controlling microbial deposits in open water systems are well established, the application of slimicides to closed water systems is a relatively new proposition. Because of the changes in bacterial populations from aerobic to anaerobic it is essential to examine the efficacy of biocide formulations against anaerobic bacteria. Geller (1982) tested a number of biocides against both aerobic and anaerobic bacteria. Using both mixed cultures and pure cultures of *Bifidobacterium bifidum* and *Pseudomonas aeruginosa,* he found that the biocides had different effects dependent on the type of organism. In a recent study by Holt & Pear (1985) sixty biocides were tested against mixed populations of aerobic, anaerobic and SRB derived from paper mill slimes. The data clearly showed that SRB were the most susceptible to a range of biocides whilst anaerobic bacteria were the most resistant.

MICROBIOLOGY OF FINAL PRODUCTS

Approximately 50% of the fibrous raw material currently used in the UK Paper and Board Industry is derived from waste papers. Together with the increase of water recycling, this has led to significant levels of solubles being incorporated into the final product. Modern papers and boards are often surface sized with starches providing an additional nutrient source for microorganisms. The only essential nutrient unavailable to bacteria and fungi in finished paper is water. Thus the easiest and most common method of preservation is simply to keep the paper or board reasonably dry, i.e. less than 12% moisture.

Weir (1977) found that the drying sections of paper and board machines achieved greater than 90% reduction in total bacterial count and the elimination of fungi from headbox to dry reel samples. The

TABLE 6

Counts of Aerobic Bacteria from Reel Samples of Final Products Made from Pulp or Waste Based Fibres (mainly from Weir (1977))

Product type	*Pulp type*	*Bacterial count (colonies/g)*
Paper		
Food wraps	Pulp	$1{\cdot}5 \times 10^3$
	Pulp	5×10^2
Sweet wraps	Pulp	$2{\cdot}7 \times 10^3$
Facial tissues	Pulp	$1{\cdot}0 \times 10^2$
Toilet tissue	Pulp	$2{\cdot}0 \times 10^2$
Bags (unspecified use)	65% pulp 35% clean waste	$1{\cdot}0 \times 10^2$
Hand towels (white)	Waste	$9{\cdot}2 \times 10^1$
(coloured)	Waste	$1{\cdot}5 \times 10^3$
(coloured)	Waste	$8{\cdot}9 \times 10^3$
Industrial tissues	Waste	$5{\cdot}0 \times 10^3$
Boards		
Light-weight board	Pulp	$1{\cdot}7 \times 10^3$
	Broke and clean waste	$2{\cdot}3 \times 10^3$
Light-weight board	Mixed waste	$1{\cdot}5 \times 10^5$
	Mixed waste	$2{\cdot}9 \times 10^5$
Heavy-weight board	Mixed waste	$3{\cdot}1 \times 10^5$
	Mixed waste	$1{\cdot}1 \times 10^5$

reduction in microbial numbers is dependent on the temperature of the driers, the drying time, the thickness of the paper or board and the level of contamination. Most vegetative bacterial cells do not survive the drying process leaving more resistant organisms, e.g. *Corynebacteria* and spore formers as contaminants. The majority of isolates from reel products are *Bacillus* spp. Fungi are not heat resistant and are rarely isolated from reel samples unless airborne contamination has occurred.

After the drying, calendering and press operations in a paper mill, the product is essentially free from vegetative bacteria and fungi. This does not imply that the paper is sterile (Table 6) as heat resistant bacteria (normally bacilli in the form of spores) may survive the drying process. These bacteria may be important, particularly in paper used in contact with food, and the determination of the total bacterial count is a common laboratory operation. However, provided the paper is maintained in a dry condition or suitably coated to prevent moisture uptake, the spoilage from bacterial growth is negligible.

Fungal attack of paper and board products is a much more common problem, the degree of attack being governed by temperature and moisture under the conditions of usage. The majority of infecting organisms will thrive at temperatures of 10–30°C and a relative humidity above 75%. As the final product is free from fungi, any subsequent growth must be the result of contamination by airborne spores, handling, etc. Thus most of the moulds growing on paper products are common airborne spoilage organisms such as *Penicillium, Aspergillus* and *Cladosporium.* Unfortunately, control of temperature and moisture conditions may not be possible on economic grounds and fungicides have to be used.

Preservation by fungicides

Fungicides must be compatible with the final product, particularly its strength, colour and printability. Similarly, they must not impart any odour to the paper that would alter customer reactions. The quantities and types of preservatives varies according to the nature of the product and coating colours, sizes and adhesives must also be taken into consideration. Commonly used preservatives include dichlorophen, benzisothiazolone, benzothiazole, sodium metaborate, dithiocarbamates, Copper-8-quinolinolate and 2,4-dichlorobenzyl alcohol. Techniques for applying preservatives range from incorporating into the pulp at the beater stage to treatments of formed sheets by roll sizes or spraying. The addition of agents to coatings is also practised.

REFERENCES

Alexander, S.D. & Dobbins, R.J. (1977). *Tappi,* **60**, 117-20.
Allsopp, C. & Allsopp, D. (1983). *International Biodeterioration Bulletin,* **19**, 99-146.
Appling, J.W. *et al.* (1955). *Tappi,* monograph No. 15.
Bowers, D.F. (1978). *Tappi,* **61**, 57-61.
Brecht, W., Dalpke, H.L. & Borner, F. (1973). *Wochenblatt Fur Papierfabrikation,* **101**, 113-19.
Brecht, W., Dalpke, H.L. & Borner, F. (1974). *Wochenblatt Fur Papierfabrikation,* **102**, 223-34.
Conkey, J.H. & Carlson, J.A. (1969). *Tappi,* **52**, 2311-18.
Dalpke, H.L. (1975). *Papier,* **29**, 236-40.
Geller, A. (1982). Complete water system closure in paper mills—the experience of the Federal Republic of Germany. Paper presented at Pira Seminar *Economic Pollution Control,* Leatherhead, April 1982.
Geller, A. & Luttgen, W. (1981). *Papier,* **35**, 151-9.
Gissler-Weber, R., Breuer, H., Boeltcher, H. & Minne, S. (1981). *Tappi Environmental Conference,* 1981.
Holt, D.M. & Pear, J.D.M. (1985). Applications of microbiological research to papermaking systems—an awareness programme on corrosion. Pira report No. PB 7 (R).
Holt, D.M., Webb, L.J., Pear, J.D.M. & Tompsett, S. (1983). Microbiological control in closed water systems. Pira report no. PB 1 (R).
Hughes, R.L. (1968). *Proceedings of the 1st International Biodeterioration Symposium.* Elsevier, London, pp. 281-90.
Hurst-Gaul, E. (1981). *Korrespondenz Abwasser,* **26**, 642-5.
Jung, W.K. (1978). *Das Papier,* **32**, 550-4.
Jung, W.K. & Kutzner, H.J. (1978). *European Journal of Applied Microbiology and Biotechnology,* **5**, 215-24.
Kirchen, C.P. & Louche, Y. (1975). *Pulp and Paper International,* **7**(8), 45-50.
Miller, J.D.A. (1981). In *Microbial Biodeterioration,* ed. A.H. Rose *Economic Microbiology,* Vol. 6. Academic Press, New York, pp. 149-202.
Mobius, C.H. & Welcher, U. (1981). *Wochenblatt Fur Papierfabrikation,* **109**, 411-15.
Mueller, W.A. & Muhonen, J.M. (1972). *Tappi,* **55**, 589-98.
Prendergast, A.G. (1948). *Paper Maker,* **116**, No. 3 TS21, 6 September.
Soimajarui, J., Pursiainen, M. & Korhonen, J. (1978). *European Journal of Applied Microbiology and Biotechnology,* **51**, 87-93.
Streebin, L., Reid, G., Law, P., Hogan, C. & Ruppersberger, J. (1976). *Tappi,* **59**, 105-8.
Thorpe, P.H. (1979). Bacterial corrosion of stainless steel caused by water recycling. Australian Paper Manufacturers Ltd, *Corr. Australasia,* **4**, 12-14.
Vestola, J. & Korhonen, J. (1976). *Tappi,* **59**, 130-3.
Wang, C.J.K. (1965). Fungi of pulp and paper. State University College of Forestry at Syracuse University, Technical publication No. 87.
Weir, B. (1977). Sources of microbial content of paper and board with significant waste paper content. Pira report No. PB 8 (R)/1977.
Weir, B. (1982). In *Principles and Practice of Disinfection, Preservation and Sterilisation.* ed. A.D. Russel, W.B. Hugo & G.A.J. Ayliffe. Blackwell, Oxford.
Weir, B., Pear, J.D.M. & Beattie, H.W.P. (1981). An evaluation of pathogenic microorganisms in recycled fibre from garbage, with a view to assessing hazards to health. EEC document No. 017-79-4 RCP (UK).

Bioluminescent Assessment of Fungal Growth on Plastic Packaging Materials

P.S. Senior & B.J. McCarthy

Science Group, Wira Technology Group, Wira House, West Park Ring Road, Leeds, LS16 6QL, UK

ABSTRACT

The biodegradability of plastic materials containing different levels of a starch filler was assessed by a modified challenge test. The levels of fungal growth were assessed by ATP luminescent assay at intervals over 96 h incubation instead of the usual visual assessment after 28 days or more. The levels of ATP were found to increase with time and this increase was greater when higher levels of starch filler were included in the material. Other potential uses of ATP luminescence in monitoring degradation of plastics are discussed.

INTRODUCTION

The growth of fungi on plastic materials can be considered as both problematic and advantageous. The adverse effects of undesired growth can result in losses occurring in many different areas. Direct damage can occur when packaging films are stored under adverse conditions either prior to or actually in use. Fungal growth on the in-use films results not only in loss of the film itself but can also lead to the transfer of contamination to the stored goods. Such problems lead to financial losses estimated to be around £17·5M per year.

However, the rapid degradation of numerous packaging films is highly desirable particularly from an environmental aspect. Many base polymers such as polyethylene or polypropylene are resistant to microbial attack (Klausmeier & Andrews, 1981) while other polymer formulations such as plasticized polyvinylchloride or polyurethanes will often support microbial growth. The extent to which microbial growth will develop is related to the addition of agents intended to modify the appearance and physical properties of the end-product. These include plasticizers, fillers, pigments and stabilizers (Wales *et al.*, 1987). The inclusion of starch as a filler is often used to increase the rate of degradation and has also been shown to act as an attractant for woodlice to aid the process of macrobiodeterioration (Griffin & Tarverdi, 1983).

This paper presents the results of a study into the use of bioluminescence to examine fungal growth on plastic packaging materials. The modification of growth rates by the inclusion of various degrees of starch filler was also examined.

METHODS

Challenge test procedures

The following five test organisms were used for both studies and are recommended for use with BS.6085:1981 and BS.2011:Part 2.1J:1985:

Aspergillus niger
Aspergillus terreus
Paecilomyces variotii
Penicillium ochro-chloron
Trichoderma viride

Spore suspensions were prepared as outlined in BS.6085 and used to inoculate the surface of mineral salt agar plates. Samples were immersed in the appropriate spore suspension and transferred to the agar plates which were incubated at 28°C. The amount of growth was then assessed at intervals by ATP assay.

ATP luminescent assay

Samples were transferred to 5 ml of a 0·02% Tween 80 solution and subjected to ultrasound (25 kHz) to remove the growth into suspension. The ATP content of this suspension was assessed by standard procedures using LKB reagents.

Starch filler trials

Samples containing starch levels varying from a 0% control to 5·6% were challenged with a mixed spore suspension containing all five test organisms. Samples were obtained from St. Lawrence Group Ltd, Bath, UK and were composed of Low Density Polyethylene (LDPE) and Linear Low Density Polyethylene (LLDPE).

RESULTS AND DISCUSSION

Starch filler trials

It can be seen that the ATP assay can be successfully used in monitoring fungal growth on a variety of plastic materials. The results obtained from the starch filler trials show that growth can be detected within 48 h of initiating the challenge and this period of time can often be less than 24 h. Table 1 shows rapid increases in ATP levels ranging from 4 to 7 times the original levels inside 24 h. Distinct differences in the growth rates are also noticeable with a change in starch content. The increase in starch levels resulted in overall higher degrees of growth and greater peak ATP levels indicating that initial growth of the fungal inoculum is encouraged by the addition of starch to the plastic.

The samples used in this particular trial all had the same basic composition of LDPE and LLDPE and as such any modification of the growth rate can probably be attributed to the inclusion of the starch fillers. These differences were noticeable at a very early stage of the trial and as such the ATP assay has great potential for use as a screening procedure when formulating new materials. Any changes in base polymer composition and addition of pigments, fillers, etc., may change

TABLE 1
Growth of a Mixed Fungal Inoculum on Starch-filled Polyethylene as Measured by ATP Luminescence

Test tim (h)	*ATP concentration (pmol/0·1 g)*		
	0% starch	*4·2% starch*	*5·6% starch*
0	97·3	69·1	103·7
6	174·9	140·2	150·9
24	309·8	297·4	751·4
30	348·1	604·4	736·4
48	314·3	358·1	873·9
54	357·7	466·1	665·7
72	287·9	543·9	510·9
78	318·1	563·5	655·1
96	229·1	398·0	582·7

the materials' susceptibility to attack. Such changes would be detected rapidly by the use of ATP assays. Screening of the biodegradability of individual components of plastic formulations is another area where luminescence could prove useful. The growth of challenge organisms on the base polymers, plasticizers and stabilizers can be determined rapidly.

As well as using luminescence to screen new plastic formulations the ATP assay has potential use in many other areas of challenge test assessments of packaging materials. It can be used, as with this trial, to monitor the overall growth rates of the challenge inoculum on the test material and reduce the duration of the trial. It can also be used at the completion of the trial to quantify growth levels more accurately than the standard visual grading system. The efficacy of any boicides added to preserve materials can be also assessed by this procedure.

REFERENCES

BS 2011: Part 2.1J (1985). *Basic environmental testing procedures.* British Standards Institution, London.

BS 6085 (1981). *Methods for test for determination of the resistance of textiles to microbiological deterioration*. British Standards Institution, London.

Griffin, G.J.L. & Tarverdi, K. (1983). In *Biodeterioration 5*, ed. T.A. Oxley & S. Barry. Wiley, New York, pp. 557–67.

Klausmeier, R.E. & Andrews, C.C. (1981). In *Economic Microbiology 6*, ed. A.H. Rose. Academic Press, London, pp. 431–74.

Wales, D.S., Hamlyn, P.F., Todd, J.E. & Sagar, B.F. (1987). In *Preservatives in the Food, Pharmaceutical and Environmental Industries*, ed. R.G. Board, M.C. Allwood & J.G. Banks. Blackwell, Oxford, pp. 99–113.

PHYSICAL METHODS OF CONTROL

Physical Conditions and Microbial Growth: Some Implications for Biodeterioration

J.W. Hopton

9 Birmingham Road, Alvechurch, Birmingham B48 7TB, UK

ABSTRACT

Microbial growth occurs when the genetic constitution of an organism and the physico-chemical condition of its environment are compatible. The effects of physical conditions on the growth of populations have been explored for many years and latterly such work has been augmented by studies of the locus of action of physical agencies at the cell level. The properties of organisms isolated from extreme environments have received attention. The operation of many natural environments and contrived industrial systems is dependent on the action of consortia of micro-organisms. Knowledge of the mechanisms of interaction of cells with the physical status of the environment is valuable for the evaluation of strategies for the control of biodeterioration and biodegradation.

INTRODUCTION

The effects of physical agencies on the growth of populations of micro-organisms has a long history, as witness the work of Pasteur on heat treatment. Techniques developed latterly have permitted study of the behaviour of individual cells and subcellular structures, including their susceptibility or resistance to physical factors.

Manipulation of a physical characteristic of the environment offers a means of exerting destructive or stimulatory control over the progress of a microbial population. Hence understanding of the mode of action of specific physical factors on growth assists evaluation of alternative strategies for the curtailment or prevention of biodeterioration.

This paper aims to present a broad view of the effect of selected physical factors of environments on the growth and metabolism of microbial inhabitants and to catalogue evidence for their locus of action.

THE GROWING MICRO-ORGANISM AND ITS ENVIRONMENT

For a micro-organism to exploit an environment for growth, the environment must be physically and chemically conducive and the cells must possess a genetic constitution which can produce the enzymes necessary to transform chemical constituents in its surroundings to new cell substance (Figs 1 and 2).

The cell can be visualised as consisting of outer envelopes enclosing a fluid interior containing mainly macromolecular components. It is bathed by an external fluid containing chemical compounds. These compounds, some or all of which will function as nutrients, can be complex in structure or rather simple organic substances and salts. An

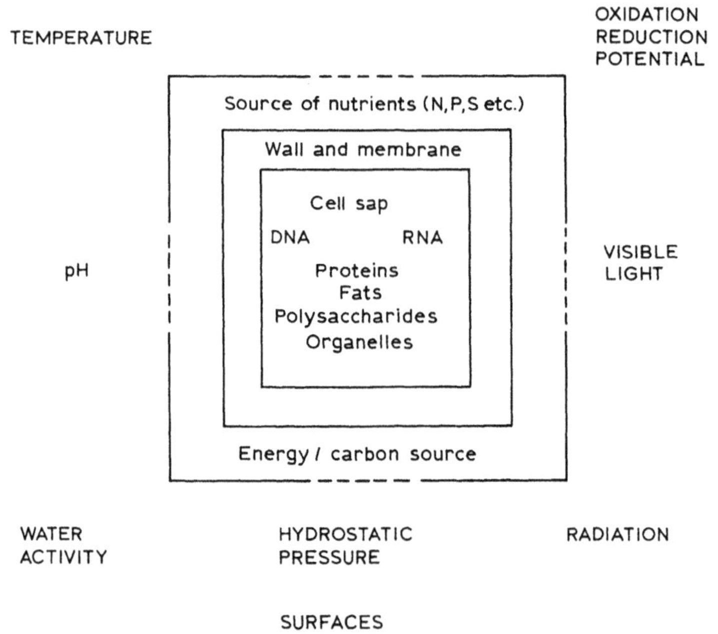

Fig. 1. The physical and chemical characteristics of a microbial cell and its surrounding environment. Capital letters signify physical characteristics. Dotted lines signify 'closed' systems.

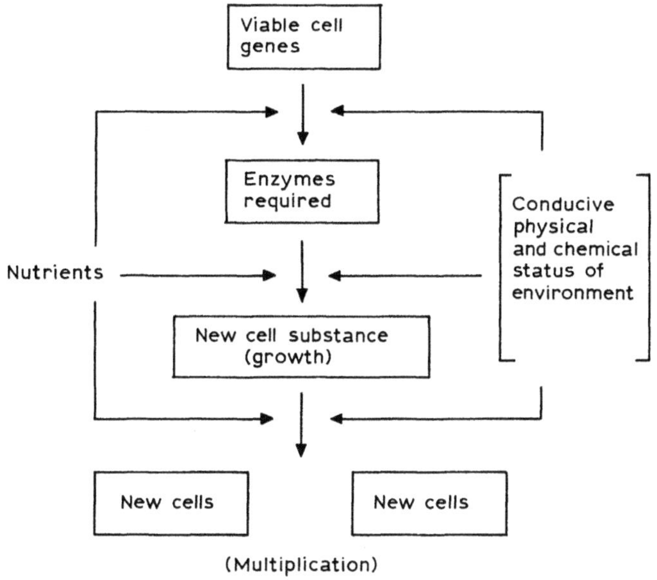

Fig. 2. Simplified representation of cell and population development.

absolute requirement is water, which participates in the establishment of the H^+ status of the environment.

Microbial systems can be 'closed' or 'open' (batch or continuous cultures) and harbour pure or mixed cultures. A system will at any one time have a battery of physical properties which will determine the rate and extent of growth of the population. Moreover these properties will change as a consequence of the growth itself but may also be altered by imposed regulatory control (Mandelstam *et al.*, 1982). Indeed such control provides a means of modifying microbial activity to encourage or discourage biodeterioration.

PHYSICAL PARAMETERS OF THE ENVIRONMENT AND MICROBIAL GROWTH

Hydrostatic pressure, visible light, radiation and oxidation-reduction potential

In some circumstances these environmental factors can be of overriding importance in determining the propensity of an item to microbiological deterioration. Conversely, imposition of a particular level and status of a factor can be the most satisfactory and economical means of alleviating deterioration.

Temperature, pH and substrate and solute concentration

The temperature/growth rate relationship is the same irrespective of the temperature range (psychrophilic, mesophilic, thermophilic) over which the organism grows but the effect of heat is abrupt at the upper end of the range (Fig. 3).

The growth response patterns for pH and substrate (solute) concentration are different to that for temperature. The symmetrical relationship for pH, with a broad or narrow optimum, is idealised in Fig. 3 but ability to grow at pH's outside the optimum, albeit at slower rates, has significance for slow biodeterioration.

The relationship between substrate concentration and growth rate encompasses the effect of high concentrations of solutes (water activity, a_W) at one extreme and the affinity characteristics of an organism for a substrate (as depicted by the Michaelis–Menten relationship and the magnitude of the saturation constant, K_s) at the other. Most organisms can grow at maximum rate at modest substrate concentrations but cannot tolerate a_W's lower than about 0·96. However, deterioration of sugary preserves and damage to materials in dry conditions (a_W's about 0·70) through microbial action is well known.

Growth of bacteria in ozonated drinking water is unwelcome evidence of utilisation of minute traces of organic compounds produced by ozone treatment (Van der Kooij & Hijnen, 1984).

Surfaces

Surfaces, exert an influence through physical interactions between the

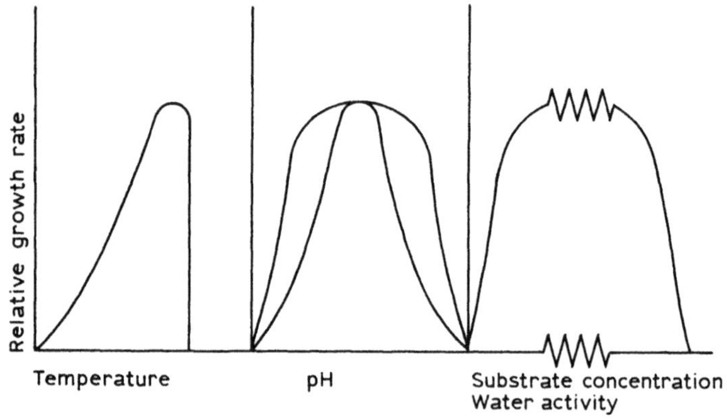

Fig. 3. Relationship between physical characteristics of the environment and growth rate of micro-organisms.

organism and the surface. Some surfaces can be rapidly colonised by micro-organisms and an organic matrix is frequently involved in the adhesion process (Costerton *et al.*, 1985). Curtailment of the effectiveness of some biocides against attached micro-organisms, has repercussions for biodeterioration control (Green *et al.*, 1987).

PHYSICAL EFFECTS IN MIXED CULTURE SYSTEMS

The functioning of many natural and contrived (industrially exploited) biological systems is attributable to the action of consortia of micro-organisms. The application of new discriminatory techniques to the study of such microbial interactions has shown how the microbial inhabitants progressively create new physical and nutritional states through their invididual activities. These in turn exert inhibitory or stimulatory pressures which differentially affect other microbial inhabitants. Eventually quasi-stable but nevertheless dynamic states become established (Bull & Slater, 1982).

THE SITE OF ACTION OF PHYSICAL EFFECTS

The information in Table 1 is from the study of micro-organisms which grow within normal physiological limits and of micro-organisms capable of colonising natural environments extreme in some physical properties (Kushner, 1980).

The response of cells to physical conditions can be related to the composition and structure of cellular constituents. Minor changes in the molecular composition of constituents with similar functions can bring about distinctive differences in their physiological properties. Regulatory devices such as the production of compatible solutes to counteract osmotic stress, involve metabolic switches.

PREVENTION OF BIODETERIORATION BY PHYSICAL MEANS

The information presented above indicates that there is a physiological range for any particular physical state over which an organism can develop. Conditions outside that range can damage the integrity of the organism and can constitute a potential means of control. Physical methods of control are mandatory under circumstances where biocide treatment or mechanical separation are ruled out.

TABLE 1
Locus of Action of Physical Agents on Microbial Cells

Physical agent	*Component of cell affected*	*Characteristics of resistant structures*
High temperature	Enzymes. Enzyme forming systems. General destruction/ denaturation	Heat stable enzymes and ribosomes. Modified structure of outer envelopes
Low temperature	Enzyme regulation. Membrane	Increase in unsaturated fatty acids. Stable regulatory genes
pH	Enzymes	Exoenzymes stable. Modified outer envelopes regulate H^+ entry—cell sap near neutral.
Water activity	Enzymes	Enzymes tolerant to high solute concentrations. Production of neutral compatible solutes which reduce water activity of cell sap
Radiation	DNA	Improved DNA repair systems

The choice of method will need to take account of the nature of the micro-organisms, the need or otherwise for absolute sterility, the sensitivity and fragility of the material to be treated, the bulk and weight of the material and the cost.

REFERENCES

Bull, A.T. & Slater. J.H. (1982). In *Microbial Interactions and Communities, Vol. 1*, ed. A.T. Bull & J.H. Slater. Academic Press, London, pp. 13–44.

Costerton, J.W., Marrie, T.J. & Cheng, K.J. (1985). In *Bacterial Adhesion. Mechanisms and Physiological Significance*, ed. D.C. Savage & M. Fletcher. Plenum Press, New York, pp. 3–43

Green, P.N., Bousfield, I.S. & Stones, A. (1987). In *Industrial Microbiological Testing*, ed. J.W. Hopton & E.C. Hill. Blackwell, Oxford, pp. 178–85.

Kushner, D.J. (1980). In *Contemporary Microbial Ecology*, ed. D.C. Ellwood, J.N. Hedger, M.J. Latham, J.M. Lynch & J.H. Slater. Academic Press, London, pp. 29–54.

Mandelstam, J., McQuillen, K. & Dawes, I. (1982). *Biochemistry of Bacterial Growth.* Blackwell, Oxford.

Van der Kooij, D. & Hijnen, W.A.M. (1984). *Applied and Environmental Microbiology*, **47**, 551–9.

Evaluation of Thermal Pasteurization for Control of Metalworking Fluid Biodeterioration

H.W. Rossmoore, L.A. Rossmoore & A.L. Kaiser

Department of Biological Sciences, Wayne State University, Detroit, MI 48202, USA and Biosan Laboratories, Ferndale, MI 48220, USA

ABSTRACT

A study was undertaken on metalworking fluids containing high levels of bacteria which were thermally pasteurized every other week for 10 weeks. Samples were prepared in duplicate so that following heat treatment, one was returned to its original biofouled container. Those fluids in the other set were returned to a thoroughly cleaned vessel. Heat treatment eventually proved less effective on the fluid from the 'dirty' container.

A shorter experiment (72 h) looked at the effect of biocide treatment of heated fluids. In both 'clean' and 'dirty' systems in the first experiment, bacterial counts were generally reduced > 99·9% but survivors returned to the original levels in 24 h. In the series using biocide, however population grow-back was prevented.

INTRODUCTION

Physical methods of controlling microbial populations have one thing in common; subsequent to treatment, there is no residual control of either the surviving population or post-treatment contamination. These methods have been limited to applications preventing contamination or survivor growth. Their use for highly contaminated industrial systems subject to recontamination is not commonplace because of lack of residual activity. Pulse treatment with LD_{50}s at intervals equal to generation times of the dominant population would be satisfactory but may not be technically or economically feasible (Rossmoore & Brazin, 1968).

Metalworking fluids (MWF) are of particular interest because of their propensity for contamination and their ability to support large populations of diverse microorganisms, which can potentially add to the burden of occupational hazards. Thus, any method for controlling microbial deterioration which reduces the burden in the workplace is to be welcomed.

A subsequent report (Heinrichs & Rossmoore, 1971) demonstrated that a regimen of 30 min at 50°C produced a significant decrease in the MWF population. More recently (Hill & Genner 1980, 1981; Hill & Elsmore, 1983), slipstream pasteurization in marine lubricating oil as well as in a MWF oil/water emulsion has been evaluated. Shipboard purifier/heat exchangers remove entrained water from the lubricating oil while increasing temperatures from 70 to 80°C for 10–15 s. Microbial levels in MWF appear to be similarly controlled in the laboratory.

The factors mitigating against success in the field are potentially larger systems, higher contamination levels, shorter generation times, and the need to control the temperature of the MWF below 30°C to maintain its cooling property.

Mechanical devices are employed in MWF systems to separate particulate and insoluble contaminants from the bulk fluid. Filtration of various kinds are found throughout the metalworking industry for removal of non-settlable metal fines or insoluble reaction products of the process.

Filters have been developed that remove bacteria (Symes & Cowap, 1976) but were limited to 10-gallon sumps, and attempts to improve flow rate and increase size have not been successful. Particulate and contaminant oil removed by centrifugation is common practice for small sumps and recycling systems (Rossmoore, 1986). A variation involves increasing temperature (to 71°C) with centrifugation (William & Geraghty, 1986) to increase separation efficiency and reduce microbial levels.

What effect the gross contamination has on the efficiency of pasteurization has never really been shown. It would be helpful if pasteurization could be used in combination with biocides at reduced doses or less frequent dosing to extend the utility of the latter while at the same time diminishing their toxicological impact.

The work presented in this paper evaluates the impact that cleanliness and biocide treatment have on the effectiveness of thermal pasteurization.

METHODS

Phase I

The first series of experiments evaluated the effect of sequential thermal pasteurization and centrifugation and the impact of the hygiene of the return vessel on microbial regrowth.

Four 4 litre systems contained the following:

(a) 35 g mild steel machining chips;
(b) 50 ml of tramp oil (25 ml hydraulic oil + 25 ml way oil);
(c) 350 ml of contaminated adapted soluble oil (inoculum); and
(d) 3150 ml of freshly made soluble oil (2·5% v/v soluble oil emulsion).

The systems were aerated continually for 5 days and left quiescent for 2 days. At this time, additions of water and fluid were made to compensate for 'evaporation'. A sample was removed for bacterial count. The aeration was continued for another 5 days. During the 10 working days, 50 ml of tramp oil was added daily. After the second weekend shutdown, each system was treated.

There were a total of four systems. System 1 was heated at 72°C for 15 s, centrifuged, and returned to its original container. System 2 was heated and centrifuged, but its container was thoroughly cleaned and new steel chips were added. System 3 was centrifuged only and then returned to its original 'dirty' container. System 4 was centrifuged, and returned to a 'clean' container with new steel chips.

Immediately after treatment a sample was removed for bacterial plate count. The study was carried out for 10 weeks. The separations were done in a bench model Alpha Laval high-speed disc bowl centrifuge with a throughput flow rate of 20 litres/h (Model No. 10CC Gyrotester). Fluids were heated and pulsed (Manostat Ministaltic Pump 72–100–000, American Scientific Products) through 5 m of aluminum tubing for 15 s in a water bath (Model BK-33, American Scientific Products) maintained at 80°C. An exit temperature of 72°C was reached from ambient (25°C) in 1–2 s and flowed directly from the water bath into the centrifuge without prior cooling.

Phase II

The same 5% soluble oil emulsion as described above but without tramp oil or metal chips was used for the study. Fluid was pulsed through the heated water bath for 15 s at (a) 72°C and (b) 60°C. Samples were collected in 100 ml aliquots in 180 ml bottles contained in an ice bath. After collection, biocide was added to them and to a control series at 25°C. Four biocides were tested. All samples were shaken at room temperature in a reciprocating shaker. At 18, 48 and 72 h aliquots were removed for bacterial counts.

RESULTS AND DISCUSSION

The cleanliness of the receiving container did not affect bacterial levels in the non-heated systems. However, there was a trend towards decreasing effectiveness of pasteurization, beginning at week 7, of the fluid from the so-called 'dirty' system (Table 1). In the 'dirty' system the same metal chips were retained and there was a build-up of biofilm. The accumulation of non-centrifugable tramp oil and dirt, together with biofilm and retained metal, may have contributed to the differential reduction noted.

The effect of biocide treatment of the heated fluid was examined (Table 2). Although the fluid was cooled in an ice bath immediately following heating, the population reduction was similar to that in previous experiments. In every case, the biocides prevented grow-back.

The demonstration of potentiation of thermal injury and/or chemical injury by combining the two was to be expected. Stressed organisms need recovery media to maintain viability. An excellent summary especially related to food (Jay, 1986) lists a very extensive number of injuries, including sublethal heat and the stress induced by several selective media. Additionally (Jay, 1983), a number of antibiotics used as food

TABLE 1
Effect of 72 C/15 s Pasteurization on Aerobic Bacteria in Metalworking Fluid

Pasteurized and centrifuged fluid returned to	*Week after inoculation*[a]								
	1	*2*	*3*	*4*	*5*	*6*	*7*	*8*	*9*
'Clean' system	−3·7	+3·0	−3·3	+4·5	−4·6	+6·0	−3·2	+2·3	−2·7
'Dirty' system	−2·8	+2·1	−2·9	+4·6	−1·3	+2·7	−0·5	+0·8	−1·0

[a]Inoculum: $3{\cdot}7 \times 10^8$ bacteria cfu/ml.
− = $\log_{10}$ reduction. + = $\log_{10}$ grow-back.
Pasteurization and centrifugation were on odd weeks; bacterial counts were done on odd and even weeks.

TABLE 2
Effect of Biocides on Pasteurized Metalworking Fluid Bacterial Survivors

Biocide/ppm[a]	*Log₁₀ cfu/ml aerobic bacteria*											
Temp (°C):	*Time zero*			*24 h after pasteurization*			*48 h after pasteurization*			*72 h after pasteurization*		
	25	*60*	*70*	*25*	*60*	*70*	*25*	*60*	*70*	*25*	*60*	*70*
Control	7	<3	<3	6·7	6·1	6·8	6	6	6·1	7·7	6·1	7·1
Bioban P-1487[b]/250				7·1	<3	<3	7·1	<0	<0	6·3	<0	<0
/500				4·8	<3	<3	4·8	<0	<0	6·8	<0	<0
/750				4·3	<3	<3	4·3	<0	<0	7·1	<0	<0
Kathon 886MW[c]/10				4·7	<3	<3	5	<0	<0	6·4	<0	<0
/25				<3	<3	<3	<0	<0	<0	1·3	<0	<0
/50				<3	<3	<3	<0	<0	<0	<0	<0	<0
Uconex 345[d]/250				7·4	<3	<3	6·7	4·1	>3	6	5·6	5
/500				7·2	<3	<3	5·4	3·0	1·4	6·1	5·4	6
/750				7·0	<3	<3	7·9	<0	<0	7·4	1·4	<0
Triadine 10[e]/250				7·2	<3	<3	5·8	2·4	1·6	4·7	5·5	1·6
/500				6·8	<3	<3	5·8	2·7	<0	4·0	<0	<0
/750				6·1	<3	<3	5·8	3·0	<0	5·9	5·6	<0

[a] Ppm of commercial product.
[b] 4,4′(2-ethyl-2-nitrotrimethylene) dimorpholine (20%) + 4-(2-nitrobutyl) morpholine (70%).
[c] 5-chloro-2-methyl-3(2H)-isothiazalone (11%) + 2-methyl-3(2H)-isothiazalone (3·5%).
[d] Glutaraldehyde (45%).
[e] 1,3,5-tris(2-hydroxyethyl) hexahydro-s-triazine (63·6%) + sodium 2-pyridinethiol-1-oxide (6·4%).

preservatives all reduce the level of heat required for food protection and pathogen destruction.

A prime target of thermal effects is probably the cell membrane. Although the specific site for action responsible for lethality is not known for any of the four biocides used here, a damaged cell membrane would probably facilitate the activity of all of them.

It can be concluded from these experiments that thermal pasteurization of MWF has a very limited short-term value since even with maximum population reduction, there is a grow-back to original population within 24 h. Since the practice in industry is to contrifuge/pasteurize biweekly, it would have little positive effect on the microorganisms. However, a combination of routine heat treatment and low level biocide dosing appears to have promise for field use.

REFERENCES

Heinrichs, T.F. & Rossmoore, H.W. (1971). *Developments in Industrial Microbiology*, **12**, 341–5.

Hill, E.C. & Elsmore, R. (1983). In *Proceedings of the 5th International Biodeterioration Symposium*. John Wiley, London, pp. 462–71.

Hill, E.C. & Genner, G. (1980). In *Proceedings of the 4th International Biodeterioration Symposium*. Pitman, London, pp. 37–43.

Hill, E.C. & Genner, G. (1981). *Tribology International*, April 1981, 62–74.

Jay, J.M. (1983). In *Food Microbiology*, ed. A.H. Rose. Academic Press, New York, pp. 117–43.

Jay, J.M. (1986). *Modern Food Microbiology*, 3rd edn, Chapter 5. Van Nostrand, New York, pp. 110–15.

Rossmoore, H.W. (1986). In *Comprehensive Biotechnology*, ed. Murray Moo-Young, Charles L. Cooney & Arthur E. Humphrey. Pergamon Press, New York, pp. 249–69.

Rossmoore, H.W. & Brazin, J.G. (1968). In *Biodeterioration of Materials: Microbiological and Allied Aspects.* Proceedings of the 1st International Biodeterioration Symposium, ed. A. Harry Walter & John J. Elphick. Elsevier, UK, pp. 386–402.

Symes, W.R. & Cowap, D. (1976). In *Proceedings of the Third International Biodegradation Symposium*, ed. J.M. Sharpley & A.M. Kaplan. Elsevier Applied Science, London, pp. 223–42.

William, G.F. & Geraghty, J.A. (1986). *Filtration News*, July–August, 10–13.

The Control of Microorganisms Using Ionising Radiation

F.J. Ley

Isotron plc, Elgin Estate, Swindon, Wilts SN2 6DU, UK

ABSTRACT

The lethal effect of ionising radiation on living cells is used for medical product sterilisation and for the control of microorganisms in raw materials for the pharmaceutical and cosmetic industries. Food irradiation is applied, in a limited way, in a number of countries, with applications involving the inactivation of bacteria, insect control and interference to biochemical mechanisms as in sprout inhibition of certain root crops. Gamma radiation is also used in other areas, for example, for the preservation of works of art and books and papers of historic interest.

The choice of dose will depend on the initial numbers of contaminating organisms and a knowledge of the resistance of the species involved. Furthermore, account must be taken of the radiation stability of materials. Gamma radiation facilities based on the use of cobalt-60 or electrons generated by electrical machines are widely available and control techniques including dose measurement well established.

INTRODUCTION

The development of nuclear power has provided the facility to produce radionuclides in sufficient quantity to allow industrial scale radiation processing. Cobalt-60 is a particularly attractive source with its 5·3 year half-life, gamma ray emission and ready production from cobalt-59 by

TABLE 1
Summary of Old and New (SI) Irradiation Units

	Absorbed dose units		*Radioactivity units*	
	Old	*New*	*Old*	*New*
Name	rad	gray	curie	becquerel
Symbol	rad	Gy	Ci	Bq
Units	100 erg/g	1 J/kg	$3{\cdot}7 \times 10^{10}$ dis/s	1 dis/s

neutron bombardment. A total of about 100 million curies (3·7 EBq) are now installed in some 150 plants throughout the world.

Gamma rays are a form of electromagnetic radiation and identical with X-rays. They do not induce radioactivity in the material being treated and there are no restrictions on the subsequent handling of material no matter how high the dose applied. However, they bring about chemical change by causing ionisation of molecules, directly or through the ionisation of water. In the case of living cells, the primary target is thought to be deoxyribonucleic acid (DNA) leading to interference with cell division. Undesirable damage to materials might be expected when using high doses and this is often associated with the crosslinking or degradation of polymers. The choice of dose for the control of micro-organisms, insects, etc., must be consistent with minimum chemical damage to the irradiated product.

The units of dose have now been brought in line with the International System of Units (SI); old and new units and their definitions are summarised in Table 1. The gray and the becquerel are now in common usage. However, the becquerel being such a small unit, radioactivity is also quoted in curies in this review.

APPLICATIONS

General

Many of the applications are based on the lethal effect of the radiation on living cells, cleverly exploited in medical therapy in relation to the inactivation of tumour cells. The combination of lethal effect with high penetrating power and small rise in temperature, makes the process attractive in other areas such as sterilisation and the control and inhibition of the biodeterioration of food and other products.

Table 2 summarises the areas of commercial application with respect to the control of microorganisms, parasites and insects. The levels of dose reflect the significant differences in radiation resistance. With microorganisms this difference is apparent between species and furthermore the choice of a treatment dose depends on the numbers of organisms involved, the influence of environmental conditions during irradiation and the level of inactivation required. In general, it is accepted that viruses are more resistant than bacterial spores, resistance

TABLE 2
Present Status of Various Radiation Processes

Application	*Dose range (kGy)*	*Commercial status*
Sterilisation of medical products, e.g. disposable devices, pharmaceuticals	15–35	Established—general
Sterilisation of packaging materials for food and medical aseptic filling	10–25	Established—general
Inactivation of *Anthrax* spores, e.g. in bales of hair and fur	20–25	Established—mainly Australia, UK
Inactivation of bacteria in raw materials, e.g. talc and kaolin powders	2·5–25	Established—general
Inactivation of pathogens in agricultural area, e.g. control of bee diseases	20–25	Established—mainly Australia, UK
Conservation of works of art and historical objects: disinfestation and disinfection	2–50	Small scale—France, UK
Radiation preservation of food		
Sterilisation of spices and herbs	10–30	Established—about 10 countries
Control of *Salmonella,* etc., in frozen prawns, chicken meat	2–7	Established—mainly Netherlands, Belgium
Control of mould, e.g. soft fruit, tropical fruits	2–3	Established—mainly South Africa, China
Disinfestation of fruits, e.g. control of fruit fly—mango, papaya	0·15–1	Recently introduced—US, South Africa
Disinfestation of grain	1–2	Large scale facility—USSR[a]
Inhibition of sprouting in root groups, e.g. potatoes, onions	0·1–0·2	Established—Japan, Italy, Hungary
Inactivation of parasites, e.g. *Trichinella spiralis*	0·3–1	Recently approved—US, China
To prolong 'chill' life (0–4°) of meat or fish	2–4	Experimental only

[a] Electron machines in use, otherwise gamma radiation.

increasing with decreasing particle size, and in turn spores are more resistant than yeasts and moulds. However, there are some exceptions, e.g. *Micrococcus radiodurans* which fortunately is non-pathogenic and uncommon. Figure 1 illustrates the dose/survival curves for different species irradiated in the same environment, plotted in a semi-logarithmic manner to convenience mathematical manipulation. The difference in resistance can be compared conveniently by observing that a 10^5 reduction in initial population covers a dose range of 1–16 kGy.

The influence of different factors on the resistance of a given species may be measured using the D_{10} value or decimal reduction dose which can be obtained from the linear part of the dose/survival curve. The role of oxygen in enhancing radiation damage is well established. Values of D_{10} for bacteria may be increased by between 2 and 4 times under anaerobic conditions. Temperature during irradiation has a striking effect which is noteworthy in relation to the possibility of the treatment of

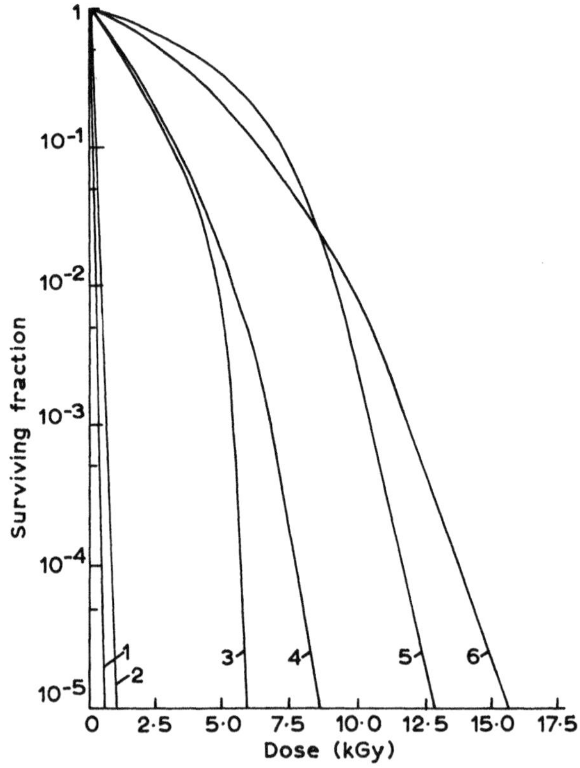

Fig. 1. Typical dose/survival curves for a number of bacterial species irradiated in buffer suspension in the presence of air. 1. *Pseudomonas* sp. 2. *Salmonella typhimurium*. 3. *Streptococcus faecium* A.1. 4. *Bacillus pumilis* spores. 5. *Clostridium botulinum* type A spores. 6. *Micrococcus radiodurans*.

frozen foods for control of pathogens. Microbes are more resistant to irradiation in the frozen state. Additionally, the very nature of the product itself is important in modifying radiation resistance.

The large differences are accounted for not only in terms of gaseous environment and temperature, but also by the presence of natural sensitising and protective agents and even by the water content of the organisms. A review has been made (Ley, 1973) of the influence of such factors, as well as those involving the radiation itself, such as dose rate and dose fractionation.

Food

Food irradiation is the most controversial area having aroused considerable consumer reaction mainly through misunderstanding of the nature of the process (Bruhn *et al.,* 1986). The question of the safety of irradiated food for consumption was reviewed by the UN (WHO, 1981) and the UK (UK Report, 1986). Both concluded that there are no hazards associated with the practical application of the process (up to 10 kGy) and no special nutritional or microbiological problems. However, no action to modify existing prohibitive legislation has been taken in the UK but a number of countries do allow specific processes. In Japan potatoes are irradiated for sprout inhibition and in the Netherlands frozen prawns are treated for the control of *Salmonella.* The low dose required for killing fruitfly does not effect the quality of mangoes and papaya and radiation will replace ethylene dibromide treatment for the fruit exported into the US from Puerto Rico and Hawaii. Irradiation of spices replaces the use of ethylene oxide gas.

In many instances the efficacy of the process is yet to be determined. Radiation plants involve high capital investment and are economic only with large throughput on a continuous basis. About 5% of worldwide radiation capacity is currently used for food but there is no doubt that there is sufficient potential for health authorities to investigate safety aspects and give guidance through legislation as recommended by WHO. The Codex Commission of the UN has already published a General Standard for Irradiated Food and a Code of Practice for operation of radiation facilities (Codex, 1983).

Conservation

Gamma radiation has already been used on a small but significant scale in a number of countries for the purpose of conserving objects of historical interest, particularly in France (Ramiere, 1981). A dose of

18 kGy was given to the mummy of Rameses II to control fungus and destroy insect larvae, the mummy remaining sealed in a sterile environment. Furniture, sculptures and maps, books and documents have also been treated. Another approach has been to impregnate wooden or stone objects with a monomer such as styrene then gamma irradiate to cause polymerisation thus consolidating the object and making it more resistant to decay. An 18th-century inlaid wood floor was successfully dismantled and treated at the Grenoble Nuclear Research Centre.

CONCLUSION

It is concluded that there are a number of areas where irradiation has a practical and advantageous application.

REFERENCES

Bruhn, Christine M., Sommer, R. & Schultz, H.G. (1986). *Journal of Industrial Irradiation Technology,* **4**(1), 1–20.

Codex (1983). *Codex Alimentarious Vol. XV.* Codex General Standard for Irradiated Foods (Codex Stan 106–1983) and Recommended International Code of Practice for the Operation of Radiation Facilities used for the Treatment of Foods.

Ley, F.J. (1973). The effect of ionising radiation on bacteria. Manual on Radiation Sterilisation of Medical and Biological Materials. IAEA Technical Report Serial No. 149. HMSO, London, pp. 37–63.

Ramiere, R. (1981). *International Conference on Industrial Application of Radioisotope and Radiation Technology,* IAEA, Grenoble, France.

UK Report (1986). *Report on The Safety and Wholesomeness of Irradiated Foods by the Advisory Committee on Irradiated and Novel Foods.* HMSO, London.

WHO (1981). Wholesomeness of Irradiated Food. Report of a Joint FAO/IAEA/WHO Expert Committee. WHO Technical Report, Serial No. 659. HMSO, London.

Control of Microbial Growth through the Exclusion of Air

G.W. Gould

Unilever Research Laboratory, Colworth House, Sharnbrook, Bedford MK44 1LQ, UK

ABSTRACT

Reduction in O_2 levels prevents the growth of strict aerobes, unless an alternative electron acceptor is available. It reduces the growth rate and yield of many facultative anaerobes; and encourages the growth of most strict anaerobes. Exclusion of air is synergistic with antimicrobial procedures, including reductions in pH value and water activity, and with the presence of some preservatives. Reduction may be achieved by vacuum packing and/or by inert gas flushing, but very low O_2 residuals are difficult to achieve and maintain without the use of expensive barrier films. Active O_2 absorbing films or sachets may help to overcome these difficulties. The specific antimicrobial effects of CO_2 have led to its increasing use, sometimes alone, but more often with nitrogen, with O_2, and with other gas mixtures in 'controlled atmosphere' storage and in 'modified atmosphere' packaging.

INTRODUCTION

Removal of oxygen (O_2) prevents the growth of all strictly aerobic microorganisms unless some other usable electron acceptor is present, but will not prevent the growth of facultative or strict anaerobes. Exclusion of oxygen by vacuum packaging or, more recently by nitrogen flushing, has been used to delay microbiologically catalysed deterioration, particularly of foodstuffs and also to prevent non-microbiological oxidative deterioration reactions. More recently, in addition to reducing the levels of O_2, preservation has been enhanced by raising the level of carbon dioxide (CO_2) because it has a specific inhibiting effect on some microorganisms. These procedures are known as 'controlled atmosphere storage' or 'modified atmosphere packaging'. In this paper the major aspects of the control of microbial growth through the maintenance of low O_2 levels and by raised levels of CO_2 has been summarised.

OXYGEN DEPENDENCE OF MICROBIAL GROWTH

In most practical deterioration situations, microorganisms are exposed to O_2 in solution and, whilst air contains about 21% O_2, its solubility in water is such that a typical medium, or moist food, etc., may contain only about 5 μg of O_2 per ml. Wimpenny (1969) listed apparent K_m values for O_2 uptake for a number of different microorganisms, from about 500-fold to about 25-fold less than the levels in air-saturated water. Consequently, substantial inhibition of O_2 dependent growth will occur only when levels are reduced to a few per cent or less of saturation levels at normal pressures and temperatures.

A second aspect of practical relevance is 'yield'. For facultative anaerobes that use O_2 when growing aerobically, the yield of cells obtained is lower in the absence of O_2 than in its presence, often by factors of about 2–6. However, electron acceptors other than O_2 may be utilised anaerobically and allow otherwise strict aerobes to grow. This may be important practically in allowing the growth of aerobes in vacuum packed foods such as cured meats that contain nitrate. Availability of alternative electron acceptors may also increase the yield, and therefore the deterioration potential, of facultative anaerobes. Hadjipetrou & Stouthamer (1965) found yields of *Aerobacter aerogenes*, grown on glucose, to rise from 26 g/mole anaerobically to 46 g/mole anaerobically plus nitrate, and to 73 g/mole aerobically.

OXYGEN RESIDUALS

Many moulds and many *Pseudomonas* spp. will still grow in levels of O_2 as low as 1% (Shaw & Nicol, 1969). Achievement of levels much below this is difficult to obtain without the use of techniques such as evacuation-inert gas flushing-evacuation-gas flushing cycles. It is here that additional levels of CO_2 can be beneficial by inhibiting oxidative organisms even though low O_2 residuals remain. In a sealed O_2 impermeable pack, however, the total amount of O_2 dependent growth possible will be limited by the total amount of O_2 present, rather than by its initial concentration.

An alternative to the removal of O_2 from packed products is to incorporate oxygen-absorbing materials within the closed packs. Oxidisable materials such as sulphite or ascorbic acid can be incorporated into sachets or, in some instances, into the product itself, O_2 is removed as the materials oxidise. Alternatively, O_2 absorbing packaging films containing palladium to catalyse the removal of O_2 and prevent excessive ingress from the air have been developed. In Japan, companies have been active, generating many patents on sachet systems which are especially effective in protecting from oxidative rancidity those foods that contain high levels of polyunsaturated fatty acids. The mechanisms of O_2 absorption vary but many of the most successful are based on the oxidation of iron prepared with particular additives, to have different total absorption capacities, rates of absorption, temperature characteristics, reaction to water vapour, etc.

EFFECTS ON SPOILAGE FLORA

Exclusion of O_2 greatly alters the spoilage flora of most materials but the detailed alterations will depend on factors such as the temperature, pH value, water activity, etc. Fresh meat, as normally prepared, contains a wide range of bacteria, yeasts and moulds. On chill storage in air almost all of them, are rapidly outgrown by the typical meat spoilage flora, consisting predominantly of strictly oxidative Gram-negative rods typified by *Pseudomonas* spp. and, to a lesser extent, *Acinetobacter*. On packaging in O_2-impermeable films, they still grow but, as the residual O_2 levels falls and CO_2 rises, a Gram-positive facultatively anaerobic flora is selected. Organisms such as *Brocothrix thermosphacta* form a substantial fraction of the population. If the O_2 level is deliberately reduced at the start of storage, and the CO_2 level raised, the Gram-negative flora does not develop, and *Brocothrix* predominates. If, in addition, the temperature is reduced to about $-1°C$, then *Brocothrix* is replaced by other slow growing Gram-positive facultative anaerobes such as the lactic acid bacterium, *Leuconostoc* and *Lactobacillus* species. A major shift in flora is brought about and changes from the putrid-slimy deterioration typical of the Gram-negative aerobes to a slow souring, due to the formation of lactic acid. Analogous flora changes occur on poultry (Mead, 1983). If the temperature is raised to 13°C or above, other lactic acid bacteria such as *Lactobacillus* and *Streptococcus* sp. grow most rapidly and there is the possibility of growth of the fermentative Gram-negative Enterobacteriaceae and of facultatively anaerobic species of the Gram-positive spore-forming genus, *Bacillus*, and of the Gram-positive spore-forming *Clostridium* species, wihch are strict anaerobes.

The vacuum packing of raw meats, such as bacon, that have been cured first protects the red pigment nitrosomyoglobin from oxidative fading. Secondly it reduces the rancidity resulting from oxidation of the unsaturated fatty acids present in pork fat. Thirdly it extends shelf life, by delaying microbiological spoilage caused by oxidative organisms such as *Micrococcus* species, from about one week to 3 or 4 weeks or more depending on the effectiveness of temperature control in the chill distribution chain (Gardner, 1983). The shelf life is greater than that of fresh meat due to the raised salt concentration, and consequently lower water activity, brought about during curing. Under conditions of chill storage, slow souring due to growth of lactobacilli normally takes place. However, if the salt level is particularly high and especially if the temperature is raised above 20°C, *Micrococcus* and *Staphylococcus* become dominant. As with fresh meat, a shift in flora and type of spoilage occurs, i.e. to slow lactic souring under well-controlled vacuum-packed chill storage, but to cheesy off-notes under high salt and warm conditions.

Many moulds readily grow on the surface of dried and smoked bacon and other cured meats, 'dry cured' hams, fermented-dried salamis, etc. *Aspergillus, Penicillium* and *Cladosporium* are amongst the most common. The strict O_2 dependence of moulds makes vacuum- or nitrogen-packaging very effective means for their inhibition.

CARBON DIOXIDE-ENRICHMENT

The benefits of CO_2-enrichment are such that the procedure is now widely used commercially, e.g. for 'controlled atmosphere' storage of bulk products or for 'modified atmosphere' storage and distribution of packed products. Some microorganisms (e.g. some *Pseudomonas* spp.; Gardner & Carson, 1967) are sensitive to levels as low as 3% CO_2 (at *c.* 4°C) whilst many are delayed or inhibited at 10–20%. Many others (lactic acid bacteria; many yeasts) are very CO_2 tolerant.

In some instances, CO_2 enrichment is applied without exclusion of O_2, or even with the level of O_2 increased above 21%. The optimal gas mixture for preservation of chill stored fresh meat is about 20% CO_2:80% O_2. The 20% CO_2 delays growth of the typical meat spoilage oxidative Gram-negative bacteria, whilst the high level of O_2 maintains the meat haem pigments in the bright red oxymyoglobin form that consumers prefer. Optimal mixtures of O_2, CO_2 and N_2 differ for different products, such as meat, cured meats, fish, vegetables, etc.

The antimicrobial activity of CO_2 is greatly enhanced with reduction in temperature. The reasons for the strong synergism with chill are not fully understood, but may include the increase in solubility of CO_2 with fall in temperature (Enfors & Molin, 1981).

SAFETY CONSIDERATIONS

Since O_2 free packaging may favour the growth of facultative (e.g. *Staphylococcus aureus*) and obligate (e.g. *Clostridium perfringens*; *C. botulinum*) food poisoning and toxinogenic anaerobes, care must be taken in the selection of materials to which low O_2 vacuum and modified atmosphere packaging techniques are applied. Some of the earliest uses of vacuum-packaging led to a number of incidents of food-poisoning resulting from growth of toxin formation by *C. botulinium* type E, which is capable of growth down to about 3·3°C. Nowadays the risk is minimised by various combinations of antimicrobial treatments, including heating, cooking and chill control, as well as reduction in pH and water activity, use of preservatives, hygienic manufacturing techniques, etc., so that vacuum or modified atmosphere packaging is often only one component in a multicomponent preservation system.

Indeed, with respect to the prevention of growth of many facultative pathogenic or spoilage microorganisms, an important general principle is that the exclusion of air reduces the energy generated from metabolism of the most commonly available substrates. This, in turn, reduces the ability of the cells to overcome other stresses that make additional energy demands. Such stresses result from a number of widely used antimicrobial procedures, including reduction in pH value, lowering of water activity, addition of antimicrobial agents, etc. (Gould *et al.*, 1983), so that, for optimal control of microbial growth by the removal of air, consideration should always be given to the practically-realistic additional stresses that can be imposed.

REFERENCES

Enfors, S.O. & Molin, G. (1981). *Can. J. Microbiol.*, **27**, 15–19.

Gardner, G.A. (1983). In *Food Microbiology Advances and Prospects*, Society for Applied Bacteriology Symposium Series No. 11, ed. T.A. Roberts & F.A. Skinner. Academic Press, London, pp. 179–202.

Gardner, G.A. & Carson, A.W. (1967). Relationship between carbon dioxide production and growth of pure strains of bacteria on porcine muscle. *J. Appl. Bacteriol.*, **30**, 500–10.
Gould, G.W., Brown, M.H. & Fletcher, B.C. (1983). In *Food Microbiology: Advances and Prospects*, Society for Applied Bacteriology Symposium Series No. 11, ed. T.A. Roberts & F.A. Skinner. Academic Press, London, pp. 67–84.
Hadjipetrou, L.P. & Stouthamer, A.H. (1965). *Journal of General Microbiology*, **38**, 29–38.
Mead, G.C. (1983). In *Food Microbiology: Advances and Prospects*, Society for Applied Bacteriology Symposium Series No. 11, ed. T.A. Roberts & F.A. Skinner. Academic Press, London, pp. 203–16.
Shaw, M.K. & Nicol, D.J. (1969). *Proceedings of the 15th European Meeting of Meat Research Workers,* Helsinki, pp. 226–32.
Suzuki, H., Wada, S., Hayakawa, S. & Tamura, S. (1985). *Journal of Food Science*, **50**, 358–60.
Wimpenny, J.W.T. (1969). In *Microbial Growth*, Nineteenth Symposium of the Society for General Microbiology, ed. P.M. Meadow & S.J. Pirt. The University Press, Cambridge, pp. 161–97.

An extended bibliography for this paper may be obtained from the author.

PLASTICS AND RUBBERS

Microbial Deterioration of Rubber

B.J. Zyska

Central Mining Institute, Plac Gwarków 1, 40–951 Katowice, Poland

ABSTRACT

During the past 40 years the production of natural rubber has increased three times and of synthetic rubbers 15 times, the current global production being 13 million tonnes per annum. The activity of microorganisms in the Hevea latex and its prevention is discussed. Studies on the mechanisms of microbial degradation in natural rubber are reviewed and the microbial resistance of synthetic rubbers, including their compounding ingredients, are discussed. The protection of rubber goods against microorganisms has advanced continuously during the last decade and numerous new microbiocides have been patented.

INTRODUCTION

Rubber consumption appears to be one of the characteristic indices indicating the development of industrialised countries. Figure 1 shows the increase in world production of natural rubber and synthetic rubbers in the period 1950–84. In 1984, world production of rubbers was $12{\cdot}31 \times 10^6$ t, that is $4{\cdot}29 \times 10^6$ t of natural rubber and $7{\cdot}16 \times 10^6$ t of synthetic rubbers. These figures do not include the production of synthetic rubbers in the USSR and China, which are $1{\cdot}95 \times 10^6$ and $0{\cdot}17 \times 10^6$ t, respectively. Hence world production may be estimated at a level of $13{\cdot}43 \times 10^6$ t. The main producers of natural rubber are Malaysia, Indonesia and Thailand, accounting for 77% of total production. On the other hand, the ten main producers of synthetic rubbers are the USA, USSR, Japan, FRG, France, United Kingdom, Italy, Holland, Brazil and Canada, and account for 81% of world production. In 1985 the world total of Gross National Products was 55% higher than in 1970. In the corresponding period the increase in natural rubber production was 47% and in synthetic rubbers 39% showing clearly the relation between overall production and rubber demand.

It is noteworthy that 70% of total rubber consumption is needed for tyre production, and each tyre needs up to 70% of natural rubber.

Still further increases in the production of natural rubber are anticipated, according to estimates made in the early 1980s the production of natural and synthetic rubbers may reach a level of 19 million by 1990. Of this, natural rubber production is estimated to reach a level of 5·5 million t, i.e. 1·5 million t more than in 1980 (Winkler, 1983). For rubber goods of every type there is a scheduled list of properties, which are achieved by compounding. For example, hardness and modulus are increased by the use of particulate fillers. Compounding contributing to a more tightly knit crosslinking system will enhance the elastic properties. Most of the properties required in rubber goods were determined a long

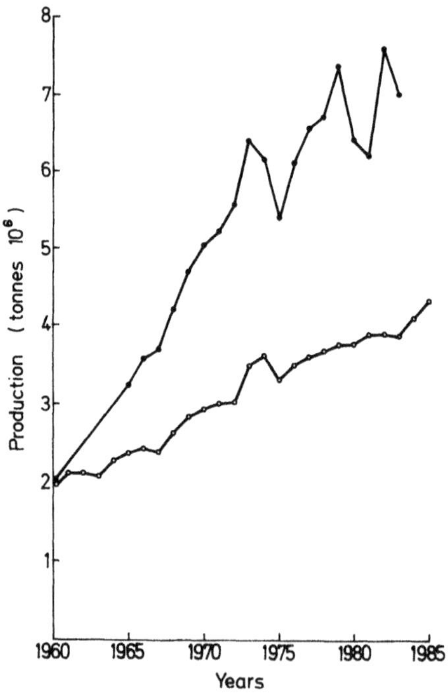

Fig. 1. World production of natural and synthetic rubbers for the period 1960–85. •—•, synthetic rubber; o—o, natural rubber.

time ago and broadly discussed in the literature, however, production of rubber goods resistant to microorganisms has seldom been dealt with in monographs on rubber technology and manufacture. Nevertheless for some rubber manufacturers and microbiologists this question has been of serious concern for many years. High resistance to microorganisms is one of the required properties of major importance in certain types of rubber goods, for example, when rubber is used in moist air, domestic water supply, sea water, sewage, soil, deep mines and in tropical climates. Extensive references on microbial deterioration of elastomers and rubbers have been given previously (Zyska, 1977, 1981). The present review mainly covers the last decade.

MICROBIAL DETERIORATION OF HEVEA LATEX AND ITS PRESERVATION

In the discussion on microbial degradation of natural rubber it is impossible to omit the microbiological problems of the latex collected from big estates or by smallholders. Latex is collected from the tree *Hevea brasiliensis* by daily or periodical tapping of the cambium to a depth of 1·0–1·5 mm. The annual yield of latex from newer plantations exceeds 2200 kg/ha/year (2000 lb/acre/year). In older plantations the average yield was only one fifth of this. Hevea latex is a dispersion of rubber particles in the water phase and contains 34–37% of rubber, 52–60% water. The total quantity of non-rubbers is approximately 4%, comprising mainly proteins, amino acids, lipids, quebrachitol, carbohydrates and other inorganic and organic constituents. The untapped rubber latex in the tree is sterile. After tapping it becomes heavily contaminated by microorganisms. The microbiology of Hevea latex has been investigated since 1915 (Eaton & Grantham, 1915) and references on this subject exceed 200 papers (John, 1978).

The collected latex is either treated to extract the rubber, or it is despatched to the consumer in liquid form. According to Le Bras (1965) only about 10% of the production is handled in liquid form. The direct use of Hevea latex was estimated in 1965 to be 200 000 t (expressed as dry rubber content). Concentration of rubber in latex to be transported is 60–62%. In the years 1966–78 extensive investigations on the microbiology of latex were undertaken at the Rubber Research Institute of Malaysia (John, 1978). Certain of these investigations have been reported (John & Verstraete, 1980). The main areas investigated were the microbial destabilisation of latex, its preservation and methods of coagulation. Detailed research included analysis of the activity of bacteria in latex, the yeast population, breakdown of amino acids, monosaccharides and quebrachitol by latex bacteria and the microbial destabilisation of fresh and ammoniated latex. Investigations on the preservation of latex included short-term stabilisation, continuous preservation, control of volatile fatty acid build-up and factors governing hygiene in factories preparing latex concentrate. Research included tests on different systems for the biological, chemical and physical coagulation of latex. The broad range of investigations gives evidence of the important role of micro-organisms in the process of rubber extraction from latex and the preparation of concentrated latex for use by the consumer all over the world. To keep Hevea latex fluid for a day or two, it is essential to protect the protein-lipid layer surrounding the rubber particle. This is easily achieved by the addition of small quantities of preservatives (anti-coagulants). If the latex is to be kept for longer periods, the normal practice is to treat the field latex with a strong preservative, which will either kill most of the acid-producing bacteria or prevent them from producing acids. Higher concentrations of preservatives, commonly called long-term preservatives, are required in the manufacture of latex concentrate to be transported to Europe or North America. Progress up to 1980 in preservation of Hevea latex was presented by Zyska (1981). It must be stressed that the combination of tetramethylthiuram disulphide and zinc oxide was found in 1978 to be more effective than all other long-term preservatives known and used earlier (John, 1978; John & Verstraete, 1980). A 50/50 combination of tetramethylthiuram disulphide and zinc oxide at a concentration of 0·025%, satisfactorily keeps the latex fluid for up to 20 days. The level of ammonia needs to be sustained at 0·2%. This invention is patented by the Rubber Research Institute of Malaysia (1977). In two Indian patents (Krishnaswamy *et al.*, 1981*a, b*) the formulation has been slightly modified but the need for retesting is suggested. The tetramethylthiuram disulphide–zinc oxide preservation system was analysed by Calvert *et al.* (1983). Two variants of ultra-low ammonia lattices were prepared. ULAZN latex, containing 0·1% NH_3 and KLAZN latex, containing 0·05% NH_3 and a small amount of KOH, on a 4000 litre scale and subjected to circulatory treatment to simulate bulk transfer. The properties and behaviour of both lattices were similar to those of well-preserved commercial low-ammonia latex. Because of its lower NH_3 content and small reserve of fixed alkali, KLAZN was the preferred ultra-low ammonia natural rubber latex. As Malaysia is the main exporter of natural rubber, including latex, the paper on standard and modified latex concentrates presented by Sum Ng Chiew (1978) is of particular interest. Six types of modified latex concentrates were considered and treated using different procedures and various preservatives to protect against microorganisms, to suit the requirements of further manufacture. Natural rubber latex is used for manufacturing medical and food-handling utensils. A series of four patents for bacteria-resistant latex compositions were taken out by the Japanese company Unitika Ltd. The protected bactericides for latex compositions are:

6,9- diamino-2-ethoxyacridine (Unitika Ltd, 1985*a*), benzyldimethylhexadecylammonium iodide (Unitika Ltd, 1985*b*);

1,6- bis 4-chlorophenylbiguanido hexane gluconate (Unitika Ltd, 1985*c*) and an amino acid hydrochloride (Unitika Ltd, 1986).

MICROBIAL RESISTANCE OF RAW NATURAL RUBBER AND SYNTHETIC RUBBERS

The microbiology of raw natural rubber has been investigated for more than 70 years, since Söhngen & Fol (1914) gave the first evidence of its deterioration by *Actinomyces elastica* and *A. fluscus.* Weight losses in natural rubber have been reinvestigated by Spence and van Niel (1936), Shaposhnikov *et al.* (1952*a, b*), Nette *et al.* (1959), Borel *et al.* (1982), Williams (1982) and Tsuchii *et al.* (1985). Of all these researchers only Tsuchii *et al.* (1985) found a microorganism which was able to degrade unvulcanised natural rubber up to 100% during 8 weeks of incubation. This was an actinomycete *Nocardia* sp. strain 835 A. Tsuchii *et al.* (1985) stressed that the taxonomy is very similar to that of *Proactinomyces ruber,* reported by Nette *et al.* (1959), which degraded natural rubber by 25·8% in 45 days. Apart from the laboratory tests on microbial deterioration of natural rubber, the most outstanding observations and investigations were made by Taysum (1966) at plantations of Hevea trees in Malaysia (Zyska, 1981). Taysum showed that the leaves of these trees contain 0·45% of natural rubber, calculated per dry mass of leaves. The annual leaf fall causes a total deposition of approximately 1·78 g of cis-1,4-polyisoprene/m^2. If this compound did not deteriorate, then considerable quantities of rubber should be found in the soil. The content of natural rubber in the soil of a 35-year-old plantation was found, however, to be only 0·12%, giving evidence of continuous degradation of natural rubber by soil microorganisms. The microbial resistance of raw unvulcanised synthetic rubbers has been evaluated over many years by estimating the intensity of fungal growth on the samples. Investigations by Schwartz (1963), Dubok *et al.* (1971) and Lazar (1973) have been compared by Zyska (1981). Recently the same method of evaluating the microbial resistance of unvulcanised synthetic rubbers was used by Williams (1982).

Synthetic rubbers contain certain ingredients such as emulsifiers, stabilisers, coagulants, oil and/or carbon black. Therefore, an exact description of the type and grade of the rubber polymer must be given to make research results valid for comparison. The numerical scale of evaluation of the intensity of fungal growth on rubber samples is as follows:

0 inert, no growth in medium or on material;
1 semi-resistant, very weak growth in medium, no growth on material;
2 sensitive, moderate growth in medium and on material;
3 very sensitive, heavy growth in medium and on material;
4 very sensitive, very heavy growth in medium and on material.

Table 1 lists the unvulcanised synthetic rubbers presented by Zyska

TABLE 1
Microbial Resistance of Unvulcanised Synthetic Rubbers

Type of rubber	*Number of evaluations and intensity of growth*	*Index of microbial resistance*
Styrene-butadiene	1 × 0	2·4
	1 × 2	Sensitive
	5 × 3	
Polyisoprene	1 × 2	3·0
	1 × 3	Very sensitive
	1 × 4	
Polybutadiene	1 × 1	2·0
	1 × 3	Sensitive
Isobutene-isoprene (butyl)	1 × 1	2·5
	1 × 2	Sensitive
	1 × 3	
	1 × 4	
Ethylen-propylene	1 × 0	1·7
	1 × 2	Sensitive
	1 × 3	
Chloroprene	4 × 0	0·9
	1 × 1	Semi-resistant
	1 × 2	
	1 × 3	
Acrylonitrile-butadiene	2 × 0	1·0
(nitrile)	2 × 2	Semi-resistant
	1 × 3	
Fluorocarbon	1 × 1	1·5
	1 × 2	Semi-resistant

(1981), and the synthetic rubbers tested by Williams (1982). Williams' results were evaluated by a numerical scale based on sample descriptions, after 9 months soil burial. Table 1 incorporates the results of five investigators published between 1963 and 1982 but evaluations of the microbial resistance of the rubbers are not uniform. Differences arise due to testing different grades and trade names and also to differences in test methods. Evaluation of the intensity of fungal growth usually gives only preliminary results, which may prove to be anomalous. A new way of evaluating microbial resistance of raw unvulcanised synthetic rubbers could be that of the Fermentation Research Institute in Japan (Tsuchii *et al.*, 1978, 1979, 1984, 1985). This team is conducting intensive research to find the microorganisms responsible for the degradation of elastomers. Studies on the microbial degradation of liquid polybutadiene showed that *Acinetobacter* sp. 351, isolated from soil, degraded a commercial sample of type cis-1,4 and M_n = 650 up to 30% in three days (Tsuchii *et al.*, 1978). From gel permeation chromatography of the substrate after degradation, the strain was shown to degrade butadiene oligomers with a degree of polymerisation up to about 10, but not higher oligomers. The strain also grows on *n*-alkanes with carbon numbers from 9 to 40, and on some alkenes and fatty acid methyl esters. In further studies Tsuchii *et al.* (1984) isolated a *Moraxella* strain from the soil utilising 1,4-type polybutadiene as sole C source. About 40% of the polybutadiene, having an average MW of 2350, was degraded and assimilated by the strain in 5 days but polybutadiene, with an average MW of 16 100, was hardly degraded at all. It is possible that the size of the polymer emulsion may influence growth rate and vinyl branches in the polybutadiene prevented its microbial degradation. The same authors Tsuchii *et al.* (1979) isolated an organism from the soil which grew well on isoprene oligomers and degraded 60% of a sample of synthetic cis-1,4-polyisoprene, with an average MW M_n = 940, in 4 days. This same organism hardly degraded a synthetic polyisoprene with a higher average MW of 2500. From gel permeation chromatography, oligomers with MW as high as 1000 were shown to be attacked. The residual substrate after degradation was shown by IR spectroscopy to have a higher quantity of 3,4-structure than the original. Apparently, 3,4-linkages in synthetic polymers prevent biodegradation of higher oligomers. The latest published results on the degradation of unvulcanised synthetic rubber presented by Tsuchii *et al.* (1985) again provide evidence of degradation of polyisoprenes, in this case by the actinomycete *Nocardia* sp. strain 835A. The isoprene rubbers investigated were:

—Isoprene rubber A, produced by Aldrich Chemical Co. Inc. with a high cis-1,4-content (97–98%).

—Isoprene rubber B, produced by Seitetsu Chemical Industries Co. Ltd, trade name Maxprene IR-900, with a cis-1,4-content (92–93%).

Nocardia sp. strain 835A degraded isoprene rubber A up to 100%, and the isoprene rubber B up to 50% in 8 weeks in a mineral salt liquid medium. Butadiene, chloroprene and styrene-butadiene rubbers were degraded by this organism on a very small scale, resulting in a weight loss of 1–4%. Reviewing investigations on degradation of raw unvulcanised synthetic rubbers during recent years it is clear that estimation of the superficial fungal growth on samples is of limited value. On the other hand investigation of the degradation processes of synthetic rubbers by selected microorganisms in liquid media needs to be studied in more detail to determine the mechanism of this phenomenon.

MICROBIAL RESISTANCE OF COMPOUNDING INGREDIENTS

The use of lattices, natural and synthetic rubbers in the raw state or as unvulcanised mixtures, accounts for only a very small proportion of the total consumption of these products. In the case of natural rubber it is less than 3%. The required properties of rubber vulcanisates can be achieved by physical mixing and cross-linking in the vulcanisation process of many individual constituent materials. Today, a technical vulcanisate is made up of a base-polymer or blend of polymers, cross-linking agents, accelerators of the crosslinking reaction and their modifiers, antioxidants, antiozonants, inhibitors of metal catalysed oxidation, protective waxes, reinforcing fillers, processing aids, softeners, plasticisers, dispersing aids, tackifiers, factice, lubricants, diluents (inert mineral fillers, organic materials, extending oils), organic and inorganic colouring materials, blowing agents, fungicides and fibrous materials.

It has been demonstrated that the microbial resistance of plastics or vulcanisates is related to the processing aids. For this reason the microbial resistance of compounding ingredients is of overall importance. Moreover, in compounding microbial resistant rubber goods some knowledge of the microbial resistance of ingredients and their compatibility with fungicides cannot be neglected. Hundreds of compounding ingredients are known in rubber technology and manufacture. Their microbial activity or resistance has been investigated for over 40 years. The fungicidal and bactericidal properties of accelerators have been known since the trials of Dimond & Horsfall (1943). Silicon rubbers and natural rubbers are used in a variety of applications in medicine and pharmacy. In such applications the sterility factor and their resistance to pathogenic bacteria are important. Winner (1957) tested certain accelerators and antidegradants as bacteriostats to *Staphylococcus aureus, Streptococcus pyogenes, Escherichia coli, Proteus vulgaris* and *Pseudomonas aeruginosa*. Diphenyl guanidine and β-naphthol were found to be effective against all five species. Further work on microbial resistance of compounding ingredients has been carried out by Ritzinger (1959), Petrujova & Zanowa (1960), Dubok *et al.* (1971), Lazar (1973) and Williams (1984). Carbon blacks, non-black fillers and colouring materials are resistant, but when in contact with edible substances, they are susceptible to fungal growth. Sulphur, as the vulcanising agent, appears to be slightly fungitoxic. Only a few of the plasticisers, softeners and extenders are fungicidal or semiresistant, most being very sensitive to fungal growth (Zyska, 1981). More work is required to evaluate the interactions between the microorganisms and the hundreds of compounding ingredients in use.

MICROBIAL RESISTANCE OF VULCANISATES OF NATURAL RUBBER

The role of microorganisms in causing the degradation of rubber goods manufactured from natural rubber was investigated in the years 1942–49, although the first report of mould damage to an electrical insulation was made in 1911, by an anonymous technician in USA (Anon., 1911). Scott (1920) reported that in India natural rubber vulcanisate is attacked by the fungus *Stemphylium macrosporoideum*. The efforts made to elucidate the processes involved in the microbial deterioration of natural rubber vulcanisates in the years since Rook (1955) published his paper on the microbial deterioration of rubber gaskets, to when Cundell & Mulcock (1975) and Kwiatkowska *et al.* (1980) presented their results based on a long series of tests have been reviewed several times (Cundell & Mulcock, 1972; Zyska, 1976, 1977, 1981). Changes in natural rubber vulcanisates due to microbial degradation can be summarised as follows:

—increase in the equilibrium degree of swelling in toluene;
—decrease in network chain density;
—in the infrared spectrum decline in prominence of $\genfrac{}{}{0pt}{}{R}{R}\!\!>\!C{=}CH_2$- peak at 890 cm^{-1} and appearance of a broad peak in 1000 cm^{-1};
—decrease of calorific value;
—weight loss due to soil microorganisms (up to 40% during 91 days in natural rubber vulcanisates without carbon black and only 15% in vulcanisates with a carbon black content of 45 pphr);
—decrease in tensile strength;
—decrease in elasticity.

Since the last review (Zyska, 1981) the contributions of Tsuchii *et al.* (1985), Williams (1986) and Kwiatkowska & Zyska (1987) represent further contributions to the knowledge of the relationship between natural rubber vulcanisates and microorganisms. From studies by Tsuchii *et al.* (1985) on the degradation of natural rubber vulcanisates by *Nocardia* sp. strain 835A, the following effects may be stated:

—weight loss up to 95% after an 8-week cultivation period;
—degradation products of natural rubber vulcanisates are found as oligomers with molecular weights from 10^4 to 10^3 reaching a maximum during the 9th and 12th day of incubation and then gradually decreasing;
—from the ^{1}H-NMR spectra the purified oligomers are expected to be cis-1,4-polyisoprene mixtures with structure

$$OHC{-}CH_2{-}[{-}CH_2{-}C({-}CH_3){=}CH{-}CH_2{-}]_n$$
$$-CH_2{-}C({=}O){-}CH_3;$$

—*Nocardia* sp. strain 835A cleaves natural rubber at the double bond;
—natural rubber may be degraded by an oxygenase reaction of *Nocardia* sp., though the mechanism of chain scissors requires further studies.

At the Central Mining Institute, Katowice, the relationship between natural rubber vulcanisates and microorganisms has been investigated for many years with reference to changes in the network chain density (Zyska *et al.*, 1971, 1973; Reszka *et al.*, 1975; Fudalej *et al.*, 1976; Kwiatkowska *et al.* 1980).

The results have been reviewed by Zyska (1981) and the results of further studies have been presented (Kwiatkowska & Zyska, 1987). The microbial degradation of vulcanisates of natural rubber was assessed relative to the carbon black loading rate. In the early stage of degradation structural changes are shown as a decrease in network chain density. This degradation may occur in a number of situations, e.g. due to microorganisms in soil burial, airborne microorganisms in the return air of a coal mine and in pure culture laboratory tests with *Aspergillus* sp. strains, isolated from rubber, exposed in the mine air. All data are statistically significant at the confidence level 0·95.

The microbial deterioration of natural rubber vulcanisates was studied by Williams (1986) with reference to the effect of stearic acid, zinc oxide, sulphur, accelerators, antioxidants, fillers and processing aids in the vulcanised rubber compositions. It was found that soil microorganisms attack preferentially the stearic acid component of vulcanised natural rubber. Zinc oxide has very little effect on soil microorganisms and high sulphur concentrations could protect the rubber. The presence of antioxidants in the rubber did not prevent attack by soil microorganisms. High levels of carbon black in the vulcanisate of natural rubber could protect the rubber from microbial attack. Paraffin wax increases susceptibility to microorganisms, but increasing concentrations of accelerator may inhibit this process. With reference to these studies by Williams (1986) some remarks are necessary. In the 6 months soil burial the weight losses varied from 0·5 to 4·5%. In similar soil burial tests for natural rubber vulcanisate sheet of thickness 0·07 mm Kwiatkowska *et al.* (1980) reported a decrease of weight up to 90% in a three month test. This gives evidence of considerable activity by the soil microorganisms. In these studies the soil activity was analysed before the investigations started and thickness of rubber samples was chosen carefully.

PROTECTION OF RUBBER GOODS AGAINST MICROORGANISMS

The early reports of rubber goods being colonised by microorganisms are listed in Table 2. In 1911 when it was indicated that fungi are involved in the deterioration of rubber insulation the annual world consumption of natural rubber was 94 000 t and did not reach 2 million t until 1950 (Le Bras, 1965). The microbial deterioration of certain rubber goods was intensively studied in the period 1945–80 with special reference to ring joints in gas, sewage and water pipelines, electric cables, telephone and electronic equipment, fuel storage-tank linings and aircraft fuel-tank coatings (for details see Zyska (1981)). In the last decade attention has been focused mainly on the protection of rubber goods. A growing number of patents have been taken out for agents offering microbial resistance for rubber. In the period 1977–87 in *Chemical Abstracts* 51 patents were published. It is significant that 45% of these were Japanese, 16% were from USA and the rest from 10 other countries. Publications in the last decade can be classified as follows:

—protection of various rubber compositions irrespective of intended application;
—paint and coating materials;
—sealing compositions;
—medical and articles of hygiene;
—antimycotic footwear.

In Soviet publications on the protection of rubber against micro-

TABLE 2
Early Literature Concerning Rubber Goods Colonised by Microorganisms

Rubber goods	*References*
Rubber insulation	Anon. (1911) Blake & Kitchin (1949)
Fire hoses	Thaysen *et al.* (1945)
Gaskets	Englerth (1945)
Ring joints in water pipelines	Rook (1955)
Polysulphide rubber lining	Allen & Fore (1953)
Natural rubber vulcanisates	Scott (1920)
Natural rubber stoppers used for determining the BOD of water samples	ZoBell & Grant (1942)

organisms sodium pentachlorophenate (Shcherba *et al.*, 1976) and thiuram with Captax (Mazur, 1976) are proposed as fungicides. For nitrile rubber, ethylene-propene rubber, silicon rubber and fluorocarbon rubber Gerashchenko & Prakazova (1986) recommend diffusion saturation with 4–8% thiuram or trilan. Resistance to biodegradability was found to be increased for more than 6 months.

Research work on rubber deterioration has been undertaken for many years at the Research Institute of the Electro-technical Industry in Bucharest. Results of the fungitoxic action of some chemicals used for protection of certain industrial products have been reported (Ionita *et al.*, 1976). Antimykotikum, *ortho-* and *para*-benzyl phenol in mixtures with salicylanilide and tetramethylthiuram disulphide, were retested and found to give good protection for rubber. Earlier results on this product were reviewed by Zyska (1981). A protection system for natural rubber vulcanisates against thermal deterioration and microbial attack has been published by Stanescu & Vladoiu (1976). The use of urethane vulcanising agents in combination with sulphur vulcanisation additives gave products resistant to degradation by heat, oxidation and micro-organisms. Macesanu *et al.* (1982) in studies on the behaviour of certain rubber mixtures in laboratory accelerated climatic tests stressed that pentachlorophenol gave the best fungicidal protection of the three compounds investigated. Lacatusu *et al.* (1982) analysed the antifungal protection of certain rubber mixtures for application in tropical climates. The experiments were carried out at 28°C and 95–100% RH, using *Aspergillus niger, Chaetomium globosum, Trichoderma viride* and other fungi. The addition of 3% pentachlorophenol, sodium pentachlorophenolate or β-naphthol protected rubber against fungi in laboratory tests for more than 28 days. Antimycoticum proved to be less effective.

In patents dealing with the protection of rubber against micro-organisms claims were made for the following compounds:

—cyanuric chloride (Kyowa Gas Chemical Industry Co., 1977);
—2-trichloromethyl-4-nitrobenzenesulphenic acid derivatives (BASF A.G., 1978);
—powdered sulphur (Miki Fimisuke, 1981);
—mixtures of zinc salts and additive or reactive organic halides (VEB Synthesewerk, 1984);
—β-naphthol (Interprinderea, 1986).

Good resistance to microbial degradation of rubber packings of 10 μm was achieved by chlorination over 90% of the surface (Sekisui Chemical Co., 1985). A rubber coating composition for protecting rollers in the textile industry from mildew in tropical and continental temperate climates contains diphenylguanidine and di-*o*-tolylguanidine (Interprinderea, 1978). For polysilicone rubbers the following compounds are claimed as microbiocides:

—2-(4-thiazolyl) benzimidazole (Torey Silicon Co., 1979);
—2-[3-(trimethoxysilyl) propyl]-1,1,3,3-tetramethylguanidine (Shin-Etsu Chemical Industries Co., 1981)
—diiodomethyl *p*-tolyl sulphone (General Electric Co., 1981).

Chlorinated elastomers are often used as coating materials. Schoenen *et al.* (1978) investigated the resistance to slime of chlorinated elastomers in drinking water basins and tanks. They found that after 10 months slime growth is stopped. Tomescu *et al.* (1982) used an organotin compound and Preventol A4, a Bayer product, to prevent degradation of chlorinated rubber paint film due to attack by microorganisms, without changing the physicomechanical properties of the film. An organotin compound provided microbial resistance to a chlorinated rubber formulation for use on masonry and concrete surfaces (Drisko *et al.*, 1977). It has been known for some years that sealing rubber compositions are sensitive to microbial growth. It was found that neoprene, nitrile and EPDM rubbers covulcanised with tributyltin acrylate in the presence of peroxides give vulcanisates having a powerful and long-lasting fungicidal action. Microbial degradation of pipe-connecting materials and natural rubber was discussed in a paper by Tsuru Shin'ya (1978). Fungicides for sealing compositions were developed for use with butyl-rubber-based compounds and silicone rubber sealants. In the butyl-rubber mix for underground pipe joints tannin and an unnamed bactericide are claimed as microbiocides by the Hayakawa Rubber Company (1981). Microbiocides for silicone rubber sealants were Me N-[1-(methylcarbamoyl)benzimidazol-2-yl] carbamate (Toshiba Silicone Co., 1981), methoxytributyltin (Toshiba Silicone Co., 1982), oxybis (tributyltin) (Shin-Etsu Chemical Industry Co., 1982) and thiuram disulphides (Rhone-Poulenc Specialités Chemique, 1985). For medical and hygiene articles the following compounds are claimed to be microbiocidal:

—2-(4-thiazolyl) benzimidazole (Kanebo Ltd, 1984);
—4,5-dichloro-N-cyclo-hexylisothiazalin-3-one (Rohm & Haas Co., 1985);
—1,6-bis (4-chlorophenylbiguanido) hexane-HCl (Unitika Ltd, 1984);
—iodine (Le Veen, 1983).

The Food and Drug Administration in USA permit the use of 4-(diiodomethylsulphonyl) toluene as a fungicide in rubber sealants and caulking agents, intended for repeated use in contact with food. The maximum level allowed is 0·3% (Food and Drug Administration, 1979).

Silicone rubber, used in hydrocephalus shunts, was tested for antimicrobial activity when impregnated with clindamycin-HCl, Na fusidate, and diethanolamine fusidate. Silicon rubber prepared in this way prevents colonisation of the cerebrospinal fluid shunt by coagulate-negative *Staphylococcus albus* (Bayston & Milner, 1981). Rubber articles for medical use must comply with several stringent requirements. It was found, for instance, that common eye drop preservatives, such as benzalkonium chloride, phenylmercuric borate, chlorhexidine digluconate and phenylethanol are sorbed by most rubbers and plastics used for eye drop bottles. These include silicone and butyl rubber. Only chlorobutyl rubber did not sorb the preservatives in detectable quantities (Boer & Nijland, 1981).

Colonisation of rubber products by pathogenic microorganisms has been researched in detail (Zyska, 1981). To prevent dermatophytosis of the foot, shoe insoles containing flexible polyurethane foams are prepared with microbiocides, such as 1-hydroxy-2(1H) pyridinethiones or their salts and 2,2′-dithiobispyridine-1,1′-dioxide and its adducts alkaline earth metal salts (Olin Corp., 1983). Balabanov *et al.* (1980) have reported eight new, but not identified, antimycotic dithiocarbamides, sulphonamides, benzothiazoles and thiuram disulphides, used for rubber impregnation before vulcanisation. The microbiocide laboratory code 3A41, gave antimycotic properties to rubber shoes for a year. It was found that the rubber handrails of subway escalators may propagate pathogenic microorganisms. Treatment of rubber handrails with 0·1% peracetic or 1% of either dezam, chlorcin or amphosept inhibited 80–100% of the microflora. Similar results were obtained by exposure of the rubber handrails to UV light for 1–10 min (Pisarenko *et al.,* 1981). Ouchi Koji *et al.* (Hitachi Cable Ltd, 1977) claimed that copper oxide dispersed on the surface of rubber inhibits the generation of microorganisms enabling the production of hygienic rubber articles.

During the last few years the chief incentive for the investigation of elastomers was the need to prevent or retard attack by microorganisms. Many studies dealt with the choice of bioresistant additives. Research has also been directed towards providing protection for the susceptible ingredients or the elastomer itself by the use of fungicides or bactericides.

In synthetic polymer chemistry the development of biodegradable polymers is one of the main aims (Kumar *et al.*, 1983). Elastomer chemists are also attempting to develop biodegradable synthetic rubbers. Voegeli & Cousminer (1978) showed that with the addition of certain nitrogen compounds butadiene-acrylonitrile rubber could be a source for protein production, but the industrial application is a long way off. Research must be directed towards creating a clean environment for the future. The chemist and the biologist can contribute by developing new biodegradable elastomers.

REFERENCES

Allen, F.H. & Fore, D. (1953). *Industrial and Engineering Chemistry*, **45**, 374–7.

Anon. (1911). *Electrical Review*, **58**, 646.

Balabanov, V., Mircheva, V. & Uzunov, P. (1980). *Dermatol. Venerol.* (Sofia), **19**, 230–4.

BASF A.G. (1978). 2-trichloromethyl-4-nitrobenzenesulfenic acid derivatives. *Federal Republic of Germany Patent Ger. Offen.*, 2 721 917.

Bayston, R. & Milner, R.D.G. (1981). *Journal of Clinical Pathology*, **34**, 1057–62.

Blake, J.T. & Kitchin, D.W. (1949). *Industrial and Engineering Chemistry*, **41**, 1630–41.

Boer, Y. & Nijland, C.J. (1981). *Pharm. Weekbl.*, **116**, 371–4.

Borel, M., Kergomard, A. & Renard, M.F. (1982). *Agricultural and Biological Chemistry*, **46**, 877–81.

Calvert, K.O., Tan, T.Y., Prichards, R.J. & Sundaram, P. (1983). *Plast. Rubber Process. Appl.*, **3**, 137–43.

Cundell, A.M. & Mulcock, A.P. (1972). *International Biodeterioration Bulletin*, **8**, 119–25.

Cundell, A.M. & Mulcock, A.P. (1975). *Developments in Industrial Microbiology*, **16**, 88–96.

Dimond, A.E. & Horsfall, J.G. (1943). *Science*, **97**, 144–5.

Drisko, R.W., O'Neill, T.B. & Schwab, L.K. (1977). *Government Report Announcement Index* (U.S.), **77**, 157.

Dubok, N.N., Angert, L.G. & Ruban, G.I. (1971). *Soviet Rubber Technology*, **30**, 17–20.

Dunn, P. & Oldfield, D. (1978). *Government Report Announcement Index* (U.S.), **78**, 176.

Eaton, B.J. & Grantham, J. (1915). *Agricultural Bulletin F.S.M.*, **4**, 26.
Englerth, G.H. (1945). *Mould Deterioration of Rubber Gaskets.* US Department of Agriculture, Washington.
Food and Drug Administration (1979). *Federal Register,* 7 September 1979, **44**, 52–189.
Fudalej, P.S., Zyska, B.J., Fudalej, D.S. & Kuczera, M.H. (1976). In *Proceedings of the Third International Biodegradation Symposium,* ed. J.M. Sharpley & A.M. Kaplan, Elsevier Applied Science, London, pp. 347–55.
General Electric Co. (1981). Mildew-resistant RTV silicone compositions. *European Patent Application,* EP 34 877.
Gerashchenko, E.I. & Prokazova, E.V. (1986). *Khim. Tekhnol. (Kiev),* **2**, 21–5.
Hayakawa Rubber Co. (1981). Anticorrosive sealing compositions for underground pipe joints. *Japanese Patent,* JP 81 111 659.
Hitachi Cable Ltd (1977). Hygienic rubber and plastic articles. *Japanese Patent,* JP 7736 142.
Interprinderea de Utilaj Chimic (1978). Rubber composition for climatic protection. *Romanian Patent,* RO 63 507.
Interprinderea de Articole Tehnice din Cauciuc (1986). Rubber compositions resistant to tropical climates. *Romanian Patent,* RO 88 587 B1.
Ionita, I., Lacatusu, V. & Budulan, F. (1976). In *Comunicarile Stiintifice ale celui de al VI-lea simpozion de biodeteriorare, climatizare, navalizare, Vol. 1,* pp. 151–8.
John, C.K. (1978). Microbiological Studies of Hevea Latex with Particular Reference to its Keeping Quality and Coagulation. D. Sc. thesis, University of Ghent, Belgium.
John, C.K. & Verstraete, W. (1980). In *Biodeterioration,* ed. T.A. Oxley, G. Becker & D. Allsopp. Pitman, London, pp. 127–34.
Kanebo Ltd. (1984). Antimicrobial sponge rubbers. *Japanese Patent,* JP 84 102 933.
Krishnaswamy, Chittur Subramania Iyer & Subramaniam, Chittur Krishnaswamy (1981*a*). Preserving natural rubber latex from deterioration. *Indian Patent,* IN 148 997.
Krishnaswamy, Chittur Subramania Iyer & Subramaniam, Chittur Krishnaswamy (1981*b*). Preserving rubber latex. *Indian Patent,* IN 148 998.
Kumar, G.S., Kalpagam, V. & Nandi, U.S. (1982–1983). *Journal of Macromolecular Science-Reviews* in *Macromolecular Chemistry and Physics,* **C22**, 225–60.
Kwiatkowska, D., Zyska, B.J. & Zankowicz, L.P. (1980). In *Biodeterioration,* ed. T.A. Oxley, G. Becker & D. Allsopp. Pitman, London, pp. 135–41.
Kyowa Gas Chemical Industry Co. (1977). Bactericidal elastomers. *Japanese Patent,* JP 7 738 600.
Lacatusu, V., Tutos, G., Canepa, M. & Marinescu, M. (1982). In *Lucr. de al IX-lea simpozion de biodeteriorare climatizare,* **2**, 556–61.
Lazar, V. (1973). In *Modelling of Environmental Effects on Electrical and Mechanical Equipment, 4th International Symposium,* Czechoslovak Academy of Sciences, Praha, pp. 261–5.
Le Bras, J. (1965). *Introduction to Rubber.* Maclaren, London.
Le Veen, H.H. (1983). Thermoplastic polyurethane article treated with iodine for antibacterial use. *United States Patent,* US 4 381 380.
Macesanu, L., Popa, I. & Canepa, M. (1982). In *Lucr. de al IX-lea simpozion de biodeteriorare climatizare,* **2**, 347–61.
Mazur, F.F. (1976). In *Biokorroz., Biopovrezhd-eniya, Obrastanija, Mater. Vses. Shk.,* Starostin I.V., pp. 27–33.

Miki Fumisuke (1981). Fungicide for natural or synthetic rubbers. *Japanese Patent,* JP 8 122 351.

Nette, I.T., Pomortseva, N.V. & Kozlova, E.I. (1959). *Mikrobiologiya,* **28**, 881-6.

Olin Corp. (1983). Shoe insole having antibacterial and antifungal properties. *United States Patent,* US 4 401 770.

Petrujova, A. & Zanowa, V. (1960). *Soviet Rubber Technology,* **19**, 16-17.

Pisarenko, G. Ya., Lebedeva, N.S. & Satdykov, Sh. I. (1981). *Gigiena i Sanitarija,* **4**, 70-1.

Reszka, J., Zyska, B.J., Fudalej, P.S. & Reszka, K.R. (1975). *International Biodeterioration Bulletin,* **11**, 71-7.

Rhone-Poulenc Specialités Chemique (1985). Monocomponent organopolysiloxane compositions resistant to microorganisms. *French Patent,* FR 2 550 541 A1.

Ritzinger, G.B. (1959). *Rubber and Plastic Age,* **40**, 1067, 1069.

Rohm and Haas Co. (1985). Microbicidal or antimicrobial biomedical devices. *European Patent Application,* EP 147 970 A1.

Rook, J.J. (1955). *Applied Microbiology,* **3**, 302-8.

Rubber Research Institute of Malaysia (1977). Preserved natural rubber latex concentrate. *Brazilian Patent,* PI 7 605 454.

Schoenen, D., Thofern, E. & Dott, W. (1978). *Gesundheits-Ingenieur,* **99**, 122-8.

Schwartz, A. (1963). *Klasse für Chemie, Geologie und Biologie,* Nr 5, Akademie-Verlag, Berlin, DDR, 138 pp.

Scott, J. (1920). *India Rubber Journal,* **60**, 410-12.

Sekisui Chemical Co. (1985). Rubber packings having good resistance to microbial degradation. *Japanese Patent,* JP 8 550 212 B4.

Shaposhnikov, V.N., Robotnova, I.L., Yarmola, G.A. & Kuznetsova, V.M. (1952*a*). *Mikrobiologiya,* **21**, 280-2.

Shaposhnikov, V.N., Rabotnova, I.L., Yarmola, G.A., Kuznetsova, V.M. & Mozokhina-Porshnyakova, N.N. (1952*b*). *Mikrobiologiya,* **21**, 146-54.

Shcherba, V.P., Ilin, I.A., Ruban, G.I., Dovner, N.A. & Zakharov, N.D. (1976). In *Biokorroz., Biopovrezhdeniya, Obrastaniya, Mater. Vses. Shk.,* ed. I.V. Starostin, pp. 23-26.

Shin-Etsu Chemical Industry Co. (1981). Guanidyl-containing organosilicon compounds. *United States Patent,* US 4 248 992.

Shin-Etsu Chemical Industry Co. (1982). Silicone rubber sealants. *Japanese Patent,* JP 82 133 150.

Söhngen, N.L. & Fol, J.G. (1914). *Centralblatt für Bakteriologie.,* II, *Abteilung,* **40**, 87-95.

Spence, D. & van Niel, C.B. (1936). *Industrial and Engineering Chemistry,* **28**, 847-50.

Stanescu, C. & Vladoiu, N. (1976). In *Comunicarile Stiintifice ale celui de al VI-lea simpozion de biodeteriorare, climatizare, navalizare.* Vol. 1, pp. 166-72.

Sum, Ng Chiew (1978). Standard and modified latex concentrate. *Lecture Notes.* Rubber Research Institute of Malaysia, Kuala Lumpur, pp. 114-28.

Taysum, D.H. (1966). In *Microbiological Deterioration in the Tropics,* ed. N.J. Butler & H.O.W. Eggins, Monograph No. 23, Society of Chemical Industry, London, pp. 105-20.

Thaysen, A.C., Bunker, H.J. & Adams, M.E. (1945). *Nature,* **155**, 322-5.

Tomescu, I., Ievreinov, M., Ionescu-Homoriceanu, S., Beschea, T. & Popescu, A. (1982). *Constructii (Bucharest),* **6**, 25-7.

Torey Silicone Co. (1979). Mold- and mildew-resistant organopoly-siloxane compositions. *Federal Republic of Germany Patent,* Ger. Offen. 2 910 739.
Toshiba Silicone Co. (1981). Fungicides for moisture-curable silicone rubber sealants. *Japanese Patent,* JP 8 138 348.
Toshiba Silicone Co. (1982). Fungicides for room temperature-curable silicone rubber sealing compositions. *Japanese Patent* JP 8 296 044.
Tsuchii, A., Suzuki, T. & Takahara, Y. (1978). *Agricultural and Biological Chemistry,* **42**, 1217–22.
Tsuchii, A., Suzuki, T. & Takahara, Y. (1979). *Agricultural and Biological Chemistry,* **43,** 2441–6.
Tsuchii, A., Suzuki, T. & Fukuoka, S. (1984). *Agricultural and Biological Chemistry,* **48**, 621–5.
Tsuchii, A., Suzuki, T. & Takeda, K. (1985). *Applied and Environmental Microbiology,* **50**, 965–70.
Tsuru, Shin'ya (1978). Microbial degradation of pipe-connecting materials and natural rubber. *Bokin Bobai* **6**, 274–8. In *Chemical Abstracts* (1979), **91**, 108831.
Unitika Ltd (1984). Antibacterial urinary catheters. *Japanese Patent,* JP 84 228 856 A2.
Unitika Ltd (1985*a*). Bacteria-resistant latex compositions. *Japanese Patent,* JP 8 540 139 A2.
Unitika Ltd (1985*b*). Antimicrobial latex compositions. *Japanese Patent,* JP 8 581 232 A2.
Unitika Ltd (1985*c*). Bactericide-containing rubber latex. *Japanese Patent,* JP 8 594 460 A2.
Unitika Ltd (1986). Antibacterial latex compositions. *Japanese Patent,* JP 86 145 235 A2.
VEB Synthesewerk (1984). Microbe-resistant finished polyurethane elastomers. *German Democratic Republik Patent,* DD 208 160 A1.
Voegeli, H.E. & Cousminer, J.J. (1978). *International Biodeterioration Bulletin,* **14**, 119–22.
Williams, G.R. (1982). *International Biodeterioration Bulletin,* **18**, 31–6.
Williams, G.R. (1984). *International Biodeterioration,* **20**, 173–5.
Williams, G.R. (1986). *International Biodeterioration,* **22**, 307–11.
Winkler, H. (1983). *Weltressourcen.* Urania Verlag, Leipzig.
Winner, H.I. (1957). *Journal of Applied Bacteriology,* **20**, 88–9.
ZoBell, C.E. & Grant, C.W. (1942). *Science,* **96**, 379–80.
Zyska, B.J. (1976). In *Proceedings of the Third International Biodegradation Symposium,* ed. J.M. Sharpley & A.M. Kaplan. Elsevier Applied Science, London, pp. 331–41.
Zyska, B.J. (1977). Microbial corrosion. *Wydawnictwa Naukowo-Techniczne,* Warsaw (in Polish).
Zyska, B.J. (1981). In *Economic Microbiology,* Vol. 6, *Microbial Biodeterioration,* ed. A.H. Rose. Academic Press, London, pp. 323–85.
Zyska, B.J., Fudalej, P.S. & Rytych, B.J. (1971). *International Biodeterioration Bulletin,* **7**, 155–9.
Zyska, B.J., Fudalej, P.S., Rytych, B.J. & Fudalej, D.S. (1973). *International Biodeterioration Bulletin,* **9**, 23–7.

Microbial Degradation of Synthetic Polymers

J.A. Cameron,[a] C.L. Bunch,[a] & S.J. Huang[b]

[a]Department of Molecular and Cell Biology and [b]Department of Chemistry, The University of Connecticut, Storrs, Connecticut 06268, USA

ABSTRACT

Degradable synthetic polymers have generally been those which have linkages found in natural polymers. These include amide, urea, urethane and ester bonds. The polyester, polycaprolactone (PCL) has been used as a model compound to examine some of the biochemical events involved in the process. A variety of microorganisms have been isolated capable of degrading PCL and several selected for further study. The extracellular enzyme system of the yeast Cryptococcus laurentii *consists of a constitutive enzyme with a molecular weight (MW) of 23 000 and a low MW cofactor bound to it by non-covalent bonds. It is inactivated at 42°C, has a broad pH optimum between pH 7 and 11, and an isolectric point of pH 7·3. The extracellular enzyme system of* Pseudomonas aeruginosa *consists of a constitutive enzyme with a MW of 20 000 with an aggregate or larger form possible. The pH optimum is sharp at 7·3 and the isoelectric point is pH 5·5.*

INTRODUCTION

The use of synthetic polymers has shown a continual increase since they were first developed. World production of plastics and resins reached 115 billion pounds in 1985. With increased use has come a variety of problems associated with the long life of these polymers. They constitute a significant (8–10%) portion of solid waste in some areas and pose a significant pollution problem in marine as well as terrestrial environments. The problems associated with the disposal of polymers have led to a number of proposals to legislate the use of certain polymer products by banning or taxing them.

The development of polymers which are biodegradable or the development of systems to enhance biodegradation are designed to resolve the disposal problem. Griffin (1985) has reported on the use of starch as a biodegradable filler in polyethylene and polystyrene. Most efforts, have been directed toward polymers with hydrolysable linkages, primarily polyesters and polyurethanes.

Potts *et al.* (1973) indicated that only polymers with molecular weights (MW) less than 450 supported the growth of fungi. Later reports, however, have shown that higher MW polymers are broken down (e.g. Tokiwa *et al.*, 1974; Haines & Alexander, 1976; Cook *et al.*, 1981; Benedict *et al.*, 1983*a*).

The biodegradability of polyurethane was established by the work of Osserfort & Testroet (1966). Kaplan *et al.* (1968) and Darby & Kaplan (1968) reported that polyester polyurethanes are more susceptible to fungal degradation than polyether polyurethanes. The ability of fungi to degrade polyurethane in the marine environment was demonstrated by Jones & Le Campion-Alsumard (1970). Filip (1978), based on infra-red spectroscopy studies, has proposed the following sequence of events in polyurethane degradation by complex mixtures of organisms: (1) degradation of free isocyanate groups in the preparation; (2) splitting of urea and amide groups; (3) cleavage of the urethane groups; and (4) cleavage of the isocyanuric acid rings. The role of extracellular enzymes in the degradative process has been established by the work of Pathirana & Seal (1983) and Wales & Sagar (1985). Huang *et al.* (1981) reported that the degradation of polyester polyurethanes by filamentous fungi and the yeast *Cryptococcus laurentii* was enhanced by the addition of nutritional nitrogen in the environment. It was also found that biodegradation increased with increasing ester chain length.

The most widely studied polyester is polycaprolactone. Fukamura (1966) reported two bacterial enzymes that were capable of degrading a polycaprolactone. Potts *et al.* (1973) reported the breakdown of a polycaprolactone with a MW of 30 000 during soil burial tests. Fields *et al.* (1974) showed that a low MW polycaprolactone was degraded by *Pullularia pullulans*, a *Penicillium* sp. and a mixture of soil bacteria in the presence of other nutrient sources. Tokiwa *et al.* (1976) demonstrated the degradation of a polycaprolactone with a MW of 25 000 by a *Penicillium* species. Tokiwa & Suzuki (1977*a*) demonstrated that isolated lipases from hog pancreas and from filamentous fungi are capable of degrading polycaprolactone. An aromatic polyester was not hydrolysed by these enzymes. They also isolated and characterized a lipase enzyme from a *Penicillium* isolate that was capable of degrading a variety of polyesters. The enzyme was not affected by the presence of end groups, such as phthalate, but was unable to degrade polymers with a 'few hybrid bonds' in the chain Tokiwa & Suzuki (1977*b*). Kavelman & Kendrick (1978) demonstrated the fungal breakdown of polycaprolactones by changes in the tensile strength of the polymer. Kinoshita *et al.* (1977, 1981) demonstrated two bacterial hydrolases from a *Flavobacterium* which were capable of cleaving 6-amino hexanoic acid cyclic dimers and linear oligomers. One enzyme was capable of splitting the cyclic dimer to a linear dimer, the other split the linear dimer and was able to cleave linear oligomers up to icosamers, but not hectamers. Negoro *et al.* (1983) was able to transfer the plasmid for these enzymes into *Escherichia coli* and study the genetic apparatus involved.

The biodegradative process utilizing polycaprolactone as a model degradable polymer has been examined. Using enrichment techniques a variety of fungi, bacteria and yeasts capable of degrading PCL of varying MW have been isolated (Cook *et al.*, 1981; Benedict *et al.*, 1983*a*, *b*). Cell associated enzymes have been demonstrated for these organisms as well as extracellular enzymes. A number of bacterial isolates and the yeast *Cryptococcus laurentii* have been shown to require an additional cofactor (Jarrett *et al.*, 1984; Cameron *et al.*, 1986). This cofactor is common to all the isolates that require it. Yeast cofactor will function with the enzymes from the bacteria and bacterial cofactor will function with yeast enzyme. The cofactor has not been found in strains which lack the enzyme and it does not enhance the activity of enzymes that do not require it (unpublished data). No filamentous fungi have been found to require this cofactor.

The enzyme attacks the amorphous regions of the polymer first, then the more crystalline regions (Cook *et al.*, 1981). The breakdown products of the extracellular enzymes examined so far have been primarily trimers (Jarrett *et al.*, 1984). This is in contrast to the report of Tokiwa *et al.* (1976) who reported that the degradation product was epsilon hydroxy caproic acid.

A more detailed examination of the enzyme systems of several of the degrading organisms has been started. This paper describes the preliminary characterization of extracellular enzymes from two organisms, *C. laurentii* and *Pseudomonas aeruginosa*.

METHODS

C. laurentii working stock cultures were carried on a solid medium consisting of basal mineral salts broth (BMS, Benedict *et al.*, 1983*a*), 1% noble agar (Difco), and a suspension of particles of PCL as the carbon source. A number of strains of *P. aeruginosa* were screened for polycaprolactone depolymerase activity and a strain from the culture collection of A.H. Romano, of this department, was selected for further study. It was carried on the same medium as *C. laurentii*.

Four PCL preparations were used in this work; PCL 700, with a MW of 35 000, PCL 300, with a MW of 18 000, PCL 1250, a polycaprolactone diol with a MW of 1250, and PCL 530, another polycaprolactone, with a MW of 530. PCL 700 and 300 were used primarily with *C. laurentii* while PCL 1250 and 530 were used with *P. aeruginosa*. Starter cultures were 25 or 50 ml volumes of trypticase soy broth (Baltimore Biological Laboratories) in 250 or 500 ml flasks. The flasks were inoculated from plates and incubated on a shaker overnight at 20°C. Enzyme was routinely produced in BMS broth at pH 7·0 in 1 litre flasks containing 90 ml. *P. aeruginosa* cultures were found to grow more readily and produce more enzyme with the addition of 0·5% trypticase soy broth. Flasks were inoculated with 1–3 ml of starter culture and shaken at 20°C. Flasks were monitored at intervals and harvested when the optical density exceeded 1·0 and the rate of depolymerization of substrate by the supernatant was over 1 unit. A unit is defined as a change of 1 O.D. unit/ml/min $\times 10^{-3}$. Cultures were pooled and the cells were removed by centrifugation. Sodium azide (0·02%) was added to the supernatant to suppress microbial contamination. Some supernatants were further concentrated by ultrafiltration using an ultrafilter with a MW cut off of 10 000.

A densitometric assay was used to determine depolymerase activity. The substrate consisted of fine particles of PCL prepared by precipitating the polymer from an acetone solution with water and filtering the resulting suspension to remove large aggregates (Jarrett *et al.*, 1984). The reaction mixture consisted of 1·3 ml of chocolate buffer (pH 7·0), 0·2 ml of PCL substrate and 1 ml of enzyme preparation. The tube was mixed and the optical density read at 580 nm at 0 time and at 5 min intervals for a minimum of 30 min. Reaction mixtures were routinely held at room temperature except for the determination of temperature optimum. Appropriate controls were followed simultaneously.

The production of enzyme during growth was followed by sampling the flasks periodically and determining the optical density of the culture and the enzyme activity of the supernatant.

Esterase activity was measured as an increase in absorbance at 420 nm caused by the cleavage of ester bonds in the substrate, *p*-nitrophenyl caproate (PNP), releasing *p*-nitrophenol. The reaction mixture contained: 0·5 ml of enzyme, 1·75 ml cholate buffer (as above) and 0·25 ml PNP (in cholate buffer). Absorbance was recorded every 30 s for 5 min and activity expressed as change in $O.D._{420}$/min/ml × 10^{-3}.

The optimum temperature for enzyme activity was determined by prewarming all of the components of the reaction mixture for 10 min at the appropriate temperature, mixing and holding them at that temperature, except when making the densitometric reading. Temperatures used were 25, 31, 36, 41 and 47°C. Temperature stability was determined by holding 2 ml aliquots of the enzyme preparation at the desired temperature for 30 min, followed by chilling in an ice bath. The time of heating was monitored with a control tube containing a thermometer in 2 ml of water. After chilling, the enzyme was allowed to warm to room temperature before assaying for residual activity. Temperatures used were 24, 30, 35, 41 and 47°C.

The optimum pH of the depolymerase reaction was determined by using buffers at increments of 0·5 pH units between pH 5·5 and 12·5. Sorenson's phosphate buffer was used between pH 5·5 and 7·0, tris(hydroxymethyl)aminomethane-HCl buffer was used between pH 7·5 and 8·5, and Sorenson's glycine buffer was used above pH 8·5. The pH of the reaction mixtures was verified after mixing.

The MW of the depolymerase was determined by gel permeation chromatography on Sephadex G-100 or G-75 (Pharmacia). The gel was equilibrated and gravity packed in a 2·5 × 36 cm column. Following washing with 0·05 M phosphate buffer, pH 7·0, 5 ml samples were chromatographed using the same buffer. Fractions of 70 drops were collected and assayed for depolymerase and esterase activity. Standards used were: blue dextran 2000 (Pharmacia), arginase, bovine serum albumin, trypsin and cytochrome C (Sigma Chemical Co.).

Chromatofocusing was employed to determine the isoelectric point of the enzymes. Polybuffer exchanger 94 resin and Polybuffers 74 and 96 (Pharmacia) were used, following the directions of the manufacturer.

RESULTS AND DISCUSSION

Polyester depolymerase was released into the fluid portion of the culture generally parallel to the growth of the organism and reached its maximum shortly after the maximum stationary phase of growth had been reached. Although the turbidity of the culture remained high after reaching maximum stationary phase, the enzyme level soon dropped, presumably due to the utilization of extracellular proteins for maintenance.

In the case of the *Pseudomonas* culture the esterase enzyme, as determined with PNP-caproate, continued to increase after the

TABLE 1
Characteristics of some Polymer Degrading Enzymes

Organism	*Substrate*	*MW × 10³*	*pH opt.*	*Heat[a] stability*	*Temp.opt.*	*Reference*
C. laurentii	Polyester	23	7–11	47	42	This paper
P. aeruginosa	Polyester	20	7·3	58	50	This paper
Flavobacterium	6-amino hexanoate cyclic	110	7·3	50	33	Kinoshita *et al.* (1977)
Flavobacterium	6-amino hexanoate linear	87	8–9	ND	40	Kinoshita *et al* (1981)
Penicillium	Polyester	25	4·5	50	45	Tokiwa *et al.* (1976)
Aspergillus terreus	Polyurethane	160, 140	ND	ND	ND	Wales & Sagar (1985)
Chaetomium globosum	Polyurethane	135	ND	ND	ND	Wales & Sagar (1985)
Rhizopus stolonifer	Polyurethane	150	ND	ND	ND	Wales & Sagar (1985)
Trichoderma viride	Polyurethane	160	ND	ND	ND	Wales & Sagar (1985)

[a]Temperature at which significant loss of activity was reported.

depolymerase activity reached its maximum, indicating the possibility of two esterolytic enzymes in the culture.

The enzyme reaction rate of both enzymes increased with temperature until denaturation of the enzyme began. In the case of the *Cryptococcal* enzyme this occurred at approximately 42°C, after which it dropped rapidly. At 47°C 90% of the activity was lost. In the case of *Pseudomonas* denaturation began at about 50°C. In the case of the *Cryptococcus* enzyme the pH optimum showed a broad plateau between pH 7 and 11, with a slight peak at pH 8·5. In the case of the polyester depolymerase of *P. aeruginosa* a distinct optimum was seen at pH 7·3.

The MW of the *Cryptococcus* enzyme, as determined on G-100, is about 23 000. It eluted immediately after trypsin (MW 23 281). The MW of the *Pseudomonas* enzyme, determined using G-75, showed two peaks of activity, one at the void volume of the column, indicating a MW of over 45 000, and one at 20 000. The high MW peak is probably an aggregate of the 20 000 MW enzyme. Since only one sharp pH optimum and only one peak of activity were seen on chromatofocusing it seems unlikely that it was a higher MW species. The preparation used had been concentrated by ultrafiltration. Preliminary studies with gel electrophoresis support the presumption. The low MW reported here are in sharp contrast to the MW reported by Wales & Sagar (1985) who reported MW between 104 000 and 160 000 for esterases isolated from filamentous fungi.

The pI of the *Cryptococcal* enzyme was 7·5, whilst that of the *Pseudomonas* enzyme was 5·5. The *Pseudomonas* enzyme did not require a cofactor to function and its activity was not enhanced by the *Cryptococcus* cofactor.

It is apparent that while the enzymes described here have a similar activity and size, they differ in many important respects. The difference in cofactor requirement, pI, heat stability and pH optimum indicate significant structural differences. These enzymes also differ from other esterolytic enzymes reported by other laboratories. Table 1 compares some of the properties of the enzymes reported here with those reported in the literature. The fact that they differ is, of course, not surprising since they are derived from different organisms which may have differing microenvironments and probably evolved for the utilization of different natural substrates. They may, in the long run, provide a useful means of dealing with new and useful polymers in the environment.

ACKNOWLEDGEMENT

This paper was supported in part by the US Army Research Office.

REFERENCES

Benedict, C.V., Cook, W.J., Jarrett, P., Cameron, J.A., Huang, S.J. & Bell, J.P. (1983*a*). *Journal of Applied Polymer Science*, **28**, 327–34.

Benedict, C.V., Cameron, J.A. & Huang, S.J. (1983*b*). *Journal of Applied Polymer Science*, **28**, 335–42.

Cameron, J.A., Benedict, C.V., Jarrett, P., Bell, J.P. & Huang, S.J. (1986). In *Biodeterioration 6*, ed. C.C. Llewellyn & C.E. O'Rear. CAB International Mycological Institute, Slough, pp. 601–5.

Cook, W.J., Cameron, J.A., Bell, J.P. & Huang, S.J. (1981). *Journal of Polymer Science, Polymer Letters Edition*, **19**, 159–65.

Darby, R.T. & Kaplan, A.M. (1968). *Applied Microbiology*, **16**, 900–5.

Embry, W.L., Gaylinn, S. Buchanan, S.G., Allen, J. & Baratz, M.S. (1986). *1986 CBR Commodity Year Book*. Commodity Research Bureau, Publishers, Jersey City, p. 191.

Fields, R.D., Rodriguez, F. & Finn, R.K. (1974). *Journal of Applied Polymer Science*, **18**, 3571–9.

Filip, Z. (1978). *European Journal of Applied Microbiology and Biotechnology*, **5**, 225–31.

Fukumara, T. (1966). *Journal of Biochemistry*, **59**, 537–44.

Griffin, G.J.L. (1985). In *Biodeterioration and Biodegradation of Plastics and Polymers*, ed. K.J. Seal. Biodeterioration Society, Kew, pp. 11–26.

Haines, J.R. & Alexander, M. (1974). *Applied Microbiology*, **28**, 1084–5.

Huang, S.J., Macri, C., Roby, M., Benedict, C.V. & Cameron, J.A. (1981). In *Polyurethane Chemistry and Applications*, ed. K.N. Edwards. American Chemical Society, Washington, pp. 471–87.

Jarrett, P., Benedict, C.V., Bell, J.P., Cameron, J.A. & Huang, S.J. (1984). In *Biomedical Applications of Polymers*, ed. S.W. Hoffmann *et al.* Plenum Publishing, New York, pp. 181–192.

Jones, E.B.G. & Le Campion-Alsumard, T. (1970). *International Biodeterioration Bulletin*, **6**, 119–24.

Kaplan, A.M., Darby, R.T., Greenberger, M. & Rogers, M.R. (1968). *Developments in Industrial Microbiology*, **9**, 201–17.

Kavelman, R. & Kendrick, B. (1978). *Mycologia*, **70**, 87–103.

Kinoshita, S., Negoro, S., Muramatsu, M., Bisaria, V.S., Sawada, S. & Okada, H. (1977). *European Journal of Biochemistry*, **80**, 489–95.

Kinoshita, S., Terada, T., Taniguchi, T., Takene, Y., Masuda, S., Matsunaga, M. & Okada, H. (1981). *European Journal of Biochemistry*, **116**, 547–51.

Negoro, S., Taniguchi, T., Kanoaka, M., Kimura, H. & Okada, H. (1983). *Journal of Bacteriology*, **155**, 22–31.

Osserfort, Z.T. & Testroet, F.B. (1966). *Rubber Chemistry and Technology*, **39**, 1308–27.

Pathirana, R.A. & Seal, K.J. (1983). In *Biodeterioration 5*, ed. T.A. Oxley & S. Barry. John Wiley, Chichester, pp. 679–89.

Potts, J.E., Clendening, R.A., Ackart, W.B. & Niegisch, W.D. (1973). In *Polymers and Ecological Problems*, ed. S. Guillet. Plenum Press, New York, pp. 51–80.

Tokiwa, Y. & Suzuki, T. (1977*a*). *Nature*, **270**, 76–8.

Tokiwa, Y. & Suzuki, T. (1977*b*). *Agricultural and Biological Chemistry*, **41**, 265–74.

Tokiwa, Y., Ando, T. & Suzuki, T. (1976). *Journal of Fermentation Technology*, **54**, 603–8.

Wales, D.S. & Sagar, B.F. (1985). In *Biodeterioration and Biodegradation of Plastics and Polymers*, ed. K.J. Seal. Biodegradation Society, Kew, pp. 56–69.

Novel Test Methods for the Microbial Deterioration of Polyester Polyurethanes

R.H. Bentham,[a] L.H.G. Morton,[a] & N.G. Allen[b]

[a]School of Applied Biology, [b]School of Mechanical and Production Engineering, Lancashire Polytechnic, Preston, Lancashire PR1 2TQ, UK

ABSTRACT

An account is presented of experimental procedures designed to evaluate rapidly the susceptibility of polyurethane formulations to microbial attack.

The effect of stress upon the susceptibility of polyester polyurethane degradation is examined and evidence is presented which suggests that stressing increases the rate of degradation significantly.

In vitro tests are outlined which give rapid qualitative assays of the susceptibility of polyurethane formulations to microbial attack.

INTRODUCTION

The early development of standard test methods for plastics was based primarily on results from plasticised PVC (Hazeu, 1965; Hitz *et al.*, 1967; Hitz & Zinkernagel, 1967; Klausmeier, 1972). There are an increasing number of formulations for plastics, many of which are used in environments where more traditional materials, e.g. rubber, were previously employed. This is because they have suitable physical and mechanical properties (Seal & Morton, 1986). Plastics may be exposed to soil or to humid warm environments often rich in biologically active wastes. Many failures have occurred because the material employed had not been suitably tested in biologically hostile environments. Polyester type polyurethanes have been shown to be susceptible to microbial attack, whereas polyether type polyurethanes are relatively resistant (Darby & Kaplan, 1968).

Seal & Pantke (1986) concluded from the International Biodeterioration Research Group (IBRG) co-operative tests on plastics that visual assessment should only be used as a support to other assessment procedures of deterioration, and that 28 days was not sufficient to demonstrate the susceptibility of a known polyurethane using the test protocol. Therefore rapid assessments of deterioration using novel test protocols are desirable.

This paper describes soil burial experiments and in vitro nutrient supplemented laboratory test methods. Deterioration was assessed by tensile stress and percentage elongation estimations. Visual assessment was modified to include a procedure using a fluorescein diacetate to detect the penetration of esterase enzymes through polyester polyurethane thin films.

It is in response to work of the type described by Shuttleworth & Seal (1986) that the following investigations were motivated.

MATERIALS AND METHODS

Test material

Two similar formulations of polyurethane foam were investigated, supplied as dumbbell-shaped test pieces. They were both polyester polyurethanes, one formulation, however, contained VINYZENE B.P.5–2 PG (0·11 parts per thousand of 10,10′-oxybis-phenoarsine) as the biocide.

Test soil

Bainbridge Brothers 'Cal-Val' John Innes No. 2 compost (a loam, sphagnum moss, lime and lime-free grit mixture) was used as the test soil at 60% water holding capacity.

Rapid soil burial method

Sets of 12 replicate test pieces of the materials were buried for 28 days in a stressed and unstressed attitude, in clean plastic boxes containing the compost. The unstressed dumbbells were buried horizontally at a depth of 2·5 cm below the surface. Stressing the dumbbells was achieved by placing them in a PVC clip which held their opposite ends together.

Evaluation of physical properties

Tensile stress at break (N/mm^2) and percentage elongation at break of the test pieces were measured on a vertical Tensometer with a 2000 N load cell and a test speed of 100 mm/min.

Isolation of biodeteriogenic fungi

Where test pieces showed obvious signs of decay and discoloration they were gently washed free of adhering soil particles and portions of the test material were plated out onto 3% malt agar and incubated at 25°C. The fungi isolated were brought into pure culture and identified where possible. Cultures of all isolates which were able to clear polycaprolactone (polyester) agar plates (Pathirana & Seal, 1984) have been sent to the Commonwealth Mycologyical Institute for identification.

Isolates were maintained on a medium of the following composition g/litre: $NaNO_3$ 2·0; KH_2PO_4 0·7; K_2HPO_4 0·3; KCl 0·5; $MgSO_4.7H_2O$ 0·5; $FeSO_4.7H_2O$ 0·01; Glucose 10·0. The pH was adjusted to between 6·0 and 6·5 with 0·01 M sodium hydroxide.

Portions of a low density open foam measuring 40 × 12 × 12 mm were placed into tubes containing 15 ml of the liquid culture medium.

The tubes were autoclaved and when cool inoculated with spore suspensions of the species isolated from the previous soil burial experiments and left to incubate at 25°C.

Rapid test methods using an in vitro technique with polyester polyurethane thin films

Thin polyurethane film samples (0·07 mm) containing no biocide were cleaned in acetone and surface sterilised in 70% ethanol solution for 4 h (Shuttleworth & Seal, 1986). The thin films were then placed on agar plates of the liquid medium with the inclusion of 10 g/litre agar and 0·02 g/litre fluorescein diacetate.

Agar discs (1 cm dia.) were cut from 7 day malt agar cultures of the fungi isolated above, placed with the inoculum face downwards on the centre of the thin film and incubated at 25°C. Controls omitted the polyurethane thin films.

Assessment of film degradation

The petri dishes were viewed in ultra-violet light (254 nm) after 7 days incubation. Fluorescence was indicative of esterase activity.

RESULTS

Physical properties of polyurethane foam after soil burial

Table 1 presents the results for tensile stress evaluation of the various test pieces. They may be summarised as follows:

1. Untreated and unexposed control material showed no significant difference in tensile stress or percentage elongation at break whether they contained biocide or not (A, A1).
2. (i) Exposed specimens showed that: the presence of biocide had a marked effect upon the tensile stress and percentage elongation at break of test material. The test pieces containing biocide showed no significant signs of decay whether they were stressed or not (B, C);
   (ii) the test pieces which did not contain biocide showed a marked loss of strength in both the stressed and unstressed condition (B1, C1).
3. The most marked decrease in tensile stress and percentage elongation at break was recorded in the stressed samples containing no biocide (B1). The stressed samples showed a significant decrease in both compared to the unstressed samples.

TABLE 1
The Mean Tensile Stress at Break and Percentage Elongation of the Foam Samples After 28 Days Soil Burial at 20°C

Sample designation	*Biocide in formulation*	*Tensile stress (N/mm²)*	*%Elongation*
A	+	$9{\cdot}6 \pm 0{\cdot}5^a$	875 ± 50
A1	−	$9{\cdot}5 \pm 0{\cdot}5$	886 ± 27
B	+	$8{\cdot}7 \pm 0{\cdot}9$	819 ± 54
B1	−	$4{\cdot}1 \pm 1{\cdot}5$	480 ± 160
C	+	$8{\cdot}7 \pm 0{\cdot}8$	826 ± 44
C1	−	$5{\cdot}9 \pm 2{\cdot}1$	624 ± 143

A = Unexposed controls with boicide.
A1 = Unexposed controls without biocide.
B = Test pieces with biocide subject to stress.
B1 = Test pieces without biocide subject to stress.
C = Test pieces with biocide not subject to stress.
C1 = Test pieces without biocide not subject to stress.
+ = present.
− = absent.
^{a} = Standard deviation.

Biodeteriogenic fungi in buried test pieces

Twenty isolates from the test pieces were shown to bring about clearing of the polycaprolactone agar. The fungi isolated were *Aspergillus fischeri* (I.M.I. 301256), *Aspergillus niger* (I.M.I. 314389), *Aspergillus ustus* (I.M.I. 31482), *Aspergillus versicolor* (I.M.I. 314386). *Emericella nidulans* (I.M.I. 314388), *Fusarium culmorum* (I.M.I. 314383), *Fusarium solani* (I.M.I. 314385), *Penicillium chrysogenum* (I.M.I. 314384, 314387), *Trichoderma harzianum* (I.M.I. 314381), *Aspergillus* strains 1–6, *Chaetomium* strain 1, *Penicillium* strains 1–3.

Fungal culture on unpreserved polyurethane open foam

The growth of test fungi in tubes containing unpreserved polyurethane foam was apparent after 2 days. Preliminary work has shown that cultures remain viable and morphologically unchanged for up to 6 months.

Rapid test using thin films

The growth of all the test fungi placed on the thin films caused penetration and the release of fluorescein within 4–7 days of incubation. Controls without polyurethane showed no release of fluorescein.

DISCUSSION

The results show that, over the test period, stress alone had no effect upon the physical strength of the polyurethane samples.

Stressed material in the presence of a microbial flora showed a greater degree of degradation than unstressed material, both visually and by physical measurements.

Samples containing biocide were considered to be representative of an abiotic control, as the biocide would prevent microbial attack but not chemical or physical deterioration.

Deterioration of stressed material was reduced significantly when a biocide was included.

The production of fungal esterases causes the penetration of polyurethane thin films (Shuttleworth & Seal, 1986). In the method described the penetration of the thin film will release the esterase enzyme into the agar which will cause liberation of fluorescein from fluorescein diacetate. Glucose in the agar medium will ensure that if penetration of the fungi through existing holes occurs liberation of fluorescein will be prevented as the glucose will be preferentially metabolised. The controls showed that the esterase enzyme was not constitutive to the fungi tested but was induced in the presence of the polyurethane.

CONCLUSION

The rapid test methods outlined above may be of value when assessing the efficacy of a biocide which has been included in a polyurethane formulation.

As a result of this investigation cultures of fungi known to colonise polyurethane are available which may be used to screen prospective biocide formulations.

REFERENCES

Darby, R.T. & Kaplan, A.M. (1968). *Applied Microbiology*, **16**, 900–5.

Hazeu, I.W. (1965). *T.N.O. Nieuws*, **21**, 323, 15–18.

Hitz, H.R. & Zinkernagel, R. (1967). *International Biodeterioration Bulletin*, **3**, 21-3.

Hitz, H.R., Merz, A. & Zinkernagel, R. (1967). *Material und Organismen*, **2**, 271–96.

Klausmeier, R.E. (1972). *International Biodeterioration Bulletin*, **8**, 3–7.

Pathirana, R.A. & Seal, K.J. (1984). *International Biodeterioration*, **20**, 229–35.

Seal, K.J. & Morton, L.H.G. (1986). In: *Biotechnology 8*, Chap. 18, VCH, Frankfurt, pp. 584–606.

Seal, K.J. & Morton, L.H.G. (1986). In *Biotechnology 8*, Chap. 18, VCH,

Shuttleworth, W.A. & Seal, K.J. (1986). *Applied Microbiology and Biotechnology*, **23**, 407–9.

An extended bibliography for this paper may be obtained from the senior author.

Airborne Fungi in Coal Mines and Their Relation to the Colonization of Certain Materials

Z.T. Cieplik & B.J. Zyska

Central Mining Institute, Plac Gwarków 1, 40-951 Katowice, Poland

ABSTRACT

In 1983 a research programme on the mycological aspects of mine air microbiology was initiated in two Polish and a Chinese coal mine. Results obtained are for one level 300 m below the surface of the Katowice coal mine. The quantity of airborne fungi estimated just behind the longwall was 24 000–32 000 colony forming units per m^3. Of 55 species of fungi, 28 were airborne and in the materials, 19 were only airborne and 8 were only in the materials. Ten different materials used in the mine were analysed for fungal colonization.

INTRODUCTION

For many years the microbiology of deep coal mines has been studied at the Central Mining Institute. The microbial deterioration of materials in deep mines has been discussed previously (Zyska, 1968; Zyska *et al.*, 1972). The role of various airborne fungi in the biodeterioration of materials in the tropics has for many years been investigated by Upsher (1972, 1984). As stressed earlier by Zyska (1968) mine workings have a unique climate, exhibiting small temperature amplitude, often high humidity and controlled air velocity. Conditions in most European deep mines approximate to a humid tropical climate with a mean monthly temperature above 18°C throughout the year and monthly absolute humidity averages of 20 g/m^3 for at least 6 months of the year.

In such an environment a significant relationship would be expected between the airborne microorganisms and microbial deterioration of materials used in the mining of coal. In this area, Osborne & Throver (1966), investigating the deterioration of timber in zinc ore mines, identified 22 species of fungi in the atmosphere. However, quantitative estimation of airborne fungi was not presented. Gregory (1973) reported the occurrence of microorganisms in the air of caves and subways and Baross & Morita (1978), studying microbial life at low temperatures, reviewed the presence of bacteria in subterranean and glaciated caves. Presented here are the results of studies on airborne fungi and fungi colonizing ten materials in a Polish coal mine.

MATERIALS AND METHODS

Investigations were conducted in the Katowice coal mine at the 300 m level. The mine was established in 1822 and has a daily production of 8000 t. Airborne fungi were sampled at four sites, i.e. the surface, a stone drift 50 m from the pit bottom in the fresh air, 30 m behind the longwall and in the return air 30 m from the outtake shaft. Each result quoted is the average of five measurements. The samples were taken in January, April and June. The Polish Barbara 3A air sampler on cellulose membrane filters of pore size 1 μm were used for sampling. After sampling the filters were placed on Sabouraud's agar medium for fungi. At the same sites the open plate method was used for collecting a broad spectrum of airborne fungi with four media, i.e. malt extract agar, Martin's medium agar, carboxy methyl cellulose agar, Sabouraud's agar for fungi. Fungi were isolated from an electric rubber cable, polypropylene and jute—polypropylene fabric used in hydraulic stowing, PVC coated polyamide brattice cloth, rubber and asbestos-resin packing, paper and jute sacks, rubber and PVC conveyor belting. The samples were taken in the return air, behind the longwall. Fungi were isolated by a procedure presented earlier (Zyska *et al.*, 1972). The following media were used: malt extract agar and Czapek-Dox agar.

RESULTS AND DISCUSSION

The following estimations of airborne fungi in the mine were found at level 300 m:

—surface, near the shaft, 200–2000 cfu/m^3
—stone drift, 1600–2800 cfu/m^3
—behind the longwall, 24 000–32 000 cfu/m^3
—return air, 16 600–18 400 cfu/m^3

TABLE 1
Occurrence of Airborne Fungi and Fungi Colonizing Materials in the Katowice Coal Mine, Poland

Species of fungi	*Air sampling*		*Materials*									
	Membrane filters	*Dishes*	*Rubber electric cable*	*Polypropylene fabric*	*Jute polypropylene fabric*	*PVC coated polyamide brattice cloth*	*Rubber packing*	*Asbestos resin packing*	*Paper sacks*	*Jute sacks*	*Rubber conveyor belting*	*PVC conveyor belting*
Alternaria alternata (Fries) Keissler	+	+	+									
Arthrobotrys oligospora Fresenius		+									+	
Aspergillus flavus Link ex Fries	+						+			+		
Aspergillus fumigatus Fresenius	+		+						+	+		
Aspergillus giganteus Wehmer	+											
Aspergillus nidulans (Eidam) Winter		+										
Aspergillus niger Van Tieghem				+					+			
Aspergillus ochraceus Wilhelm	+	+	+				+	+				+
Aspergillus restrictus. G. Smith	+	+	+	+								

Aspergillus sclerotiorum Huber								+	+			
Aspergillus sulphureus (Fresenius) Thom and Church	+	+										
Aspergillus tamarii Kita				+								
Aspergillus ustus (Bainier) Thom and Church		+										
Aspergillus versicolor (Vuillemin) Tiraboschi	+	+	+	+			+		+	+	+	
Aspergillus wentii Wehmer		+										
Basidiomycotina sp.	+	+										
Botrytis cinerea Persoon ex Persoon	+	+										
Cephalosporium acremonium Corda		+	+									
Chaetomium globosum Kunze ex Steudel											+	+
Circinella muscae (Sorokin) Berlese and de Toni		+										
Cladosporium herbarum (Persoon) Link ex S.F. Gray	+	+		+						+		
Cladosporium cladosporioides (Fresenius) de Vries	+											
Doratomyces stemonitis (Persoon ex Fries) Morton and G. Smith	+	+	+				+					
Fusarium solani (Martius) Saccardo	+	+	+	+	+	+	+	+	+	+		+
Fusarium sporotrichioides Sherbakoff		+										

contd.

TABLE 1—*contd.*

Species of fungi	*Air sampling*		*Materials*									
	Membrane filters	*Dishes*	*Rubber electric cable*	*Polypropylene fabric*	*Jute polypropylene fabric*	*PVC coated polyamide brattice cloth*	*Rubber packing*	*Asbestos resin packing*	*Paper sacks*	*Jute sacks*	*Rubber conveyor belting*	*PVC conveyor belting*
Fusarium sp Link					+	+		+				
Gliocladium catenulatum Gilman and Abbott		+				+						
Gliocladium roseum Bainier	+	+		+	+				+			
Gliomastix murorum var felina (Marchal) Hughes		+						+				
Mortierella sp. Coemans				+								
Mucor fragilis Bainier		+										
Mucor globosus Fischer	+	+							+			
Mucor racemosus Fresenius s.l.		+			+		+					
Oidiodendron maius Barron	+	+										
Paecilomyces fumosoreus (Wize) Brown and Smith		+										
Paecilomyces variotii Bainier	+	+							+		+	
Penicillium chermesinum Biourge	+				+		+					
Penicillium chrysogenum Thom	+					+						

Penicillium citrinum Thom			+									
Penicillium cyclopium Westling	+	+				+			+			+
Penicillium decumbens Thom	+			+								
Penicillium fellutanum Biourge	+	+		+		+						
Penicillium frequentans Westling		+		+					+			
Penicillium funiculosum Thom	+				+					+		+
Penicillium janthinellum Biourge	+		+								+	+
Penicillium luteum Zukal												+
Penicillium notatum Westling		+										
Penicillium rugulosum Thom	+	+										
Penicillium variabile Sopp	+											
Penicillium viridicatum Westling		+										
Rhizopus nigricans Ehrenberg		+										
Rhizopus stolonifer (Ehrenberg ex Fries) Lind	+											
Sporothrix schenckii Hektoen and Perkins		+										
Stachybotrys atra Corda		+		+	+		+	+	+			
Trichoderma viride Persoon ex S.F. Gray	+	+	+	+	+	+	+	+	+	+	+	
Tritirachium roseum Van Beyma	+	+										+

The quantity of airborne fungi shows a large increase towards the longwall. The differences in quantity of airborne fungi between workings of various types were confirmed in another Polish coal mine and in a Chinese coal mine. The results will be published elsewhere.

Table 1 shows the occurrence of fungal species in the mine air and in the materials investigated. Of 55 species of fungi, 28 were found in the air and the materials, 19 were only airborne and 8 only in the materials. A strong correlation between the airborne fungi and the fungi colonizing the materials was established. Further studies on the airborne microorganisms in deep mines currently in progress should provide results of significant scientific, health and economic value.

REFERENCES

Baross, J.A. & Morita, R.Y. (1978). In *Microbial Life in Extreme Environments*, ed. D.J. Kushner. Academic Press, London, pp. 9–104.

Gregory, P.H. (1973). *Microbiology of the Atmosphere*, 2nd edn. Leonard Hill, Aylesbury.

Osborne, L.D. & Throver, L.B. (1966). *Holzforschung*, **20**, 160–4.

Upsher, F.J. (1972). In *Biodeterioration of Materials*, Vol. 2, ed. A.H. Walters & E.H. Hueck-van der Plas. Elsevier Applied Science, London, pp. 27–34.

Upsher, F.J. (1984). *International Biodeterioration*, **20**, 73–8.

Zyska, B.J. (1968). In *Biodeterioration of Materials*, ed. A.H. Walters & J.J. Elphick. Elsevier, Amsterdam, pp. 111–19.

Zyska, B.J., Rytych, B.J., Zankowicz, L.P. & Fudalej, D.S. (1972). In *Biodeterioration of Materials*, Vol. 2, ed. A.H. Walters & E.H. Hueck-van der Plas. Elsevier Applied Science, London, pp. 256–67.

Changes in Natural Rubber Vulcanizates Due to Microbial Degradation

Danuta Kwiatkowska & B.J. Zyska

Central Mining Institute, Plac Gwarków 1, 40-951 Katowice, Poland

ABSTRACT

The microbial degradation of vulcanizates of natural rubber (NR) was assessed relative to carbon black loading. The structural changes, in the early stage of degradation, are a decrease in network chain density, due to the activity of microorganisms in soil burial, airborne microorganisms in the return air of the Katowice coal mine and Aspergillus *sp. strains isolated from rubber exposed in the mine air. All data are statistically significant at the confidence level 0·95.*

INTRODUCTION

The processes of microbial deterioration of natural and synthetic rubbers have, for many years, been the subject of investigations. Current knowledge of these processes was reviewed in 1981 (Zyska, 1981). Since then Tsuchii *et al.* (1985) have studied the microbial degradation of various kinds of unvulcanized rubber and vulcanized natural rubber products by *Nocardia* sp. strain 835 A. Based on ^{1}H-NMR spectra of the purified oligomers they produced evidence that this strain cleaved natural rubber at the double bond. Williams (1982, 1986) has studied the microbial breakdown of rubber polymers and vulcanized rubbers.

The present study examined the microbial deterioration of vulcanizates of NR with reference to changes in structure, relative to the carbon black loading. Samples were exposed to soil microorganisms and to airborne microorganisms in the return air of the coal mine. The first part of this series of tests has been presented previously (Kwiatkowska *et al.*, 1980). Examination of network chain density, in the early stages of microbial deterioration, appears to give the most convincing proof of the destructive action of microorganisms in NR vulcanizate sheet.

MATERIALS AND METHODS

NR vulcanizate sheet of 0·07 mm thickness was prepared from NR latex. The carbon black loading was 0, 5, 15 and 45 pphr. Strips of vulcanized NR 45 × 45 × 0·07 mm were fastened in a polyethylene frame and exposed to soil microorganisms, to airborne microorganisms in a coal mine and to three different strains of *Aspergillus* sp., isolated from NR samples after exposure in the coal mine. The preparation of NR vulcanizate sheet, estimation of NR sheet weight loss, evaluation of the equilibrium, degree of swelling and calculation of the NR network chain density has previously been reported (Kwiatkowska *et al.*, 1980; Zyska *et al.*, 1972).

RESULTS AND DISCUSSION

The changes in network chain density with carbon black loading of 0, 5, 15 and 45 pphr as a function of duration of soil burial, during spring, are given in Fig. 1. Soil burial in autumn always gave slightly different results. The network chain density in NR vulcanizate sheet as a function of duration of soil burial was calculated based on a selected least squares curve of the form:

TABLE 1

Value of the Coefficients a_0, a_1, a_2 and of the Index — in eqn. 1

Carbon black loading (pphr)	a_0	a_1	a_2	m
0	1·54	0·067	−0·15	0·65
5	1·89	0·15	−0·44	0·67
15	1·85	0·091	−0·29	0·65
45	2·09	0·11	−0·26	0·65

$$y_1 = a_0 + a_1 t + a_2 t^m \quad (1)$$

Values of a_0, a_1, a_2 and m for curves in Fig. 1 are given in Table 1.

The maximum relative error for average values of network density presented in Fig. 1 is 15·7%. The correlation coefficient is 0·98 for unfilled and 0·91–0·99 for filled NR vulcanizate sheet.

Weight losses for the exposure period in the soil burial test are presented in Table 2.

As can be seen in Fig. 1 the decrease in the network chain density appears before the 7th day, when the first weight loss was measured.

The observed increase in network chain density which follows between the 10th and 21st day of soil burial, during which there is further evident weight loss, needs to be investigated in further tests. Tsuchii *et al.* (1985) extracted the NR vulcanizate samples by acetone and $CHCl_3$ prior to soil burial. Our procedure, i.e. not extracting samples with a solvent, was based on the principle of not treating NR vulcanizate samples with any chemical compound before exposure to microorganisms. Table 3 shows the weight losses and changes in network chain density of NR vulcanizate sheet after 28 days of exposure to the airborne microorganisms in the Katowice coal mine and to three *Aspergillus* sp. strains. All the data are statistically significant at the confidence level of 0·95. These tests confirmed our earlier findings on the colonization of rubber cables by microorganisms, performed in four deep coal mines. *Aspergillus* spp. were shown to be among the predominating fungi in the rubber cable insulation layers and caused remarkable changes in the capacitance, leakage, electrical strength and fatigue strength of rubber cables (Zyska *et al.*, 1972). This confirmation of our previous results enables development of substantially new formulations for cable insulation rubber to minimize microbial degradation of the insulation layers.

TABLE 2
Natural Rubber Vulcanizate Sheet Weight Losses in the Soil Burial Test (%)

Duration of the test (days)	*Carbon black loading (pphr)*		
	0	*5*	*45*
7	14·7	12·8	2·8
14	17·8	18·6	9·0
28	20·6	21·9	9·3

TABLE 3

Percentage Weight Loss and Changes in NR Vulcanizate Sheet Structure Due to the Activity of Microorganisms after 28 Days

Carbon black loading (pphr)	*Airborne microorganisms in the coal mine*		Aspergillus *sp. strains isolated from NR exposed in the coal mine air*					
	Weight loss	*Network chain density decrease*	*I*		*II*		*III*	
			Weight	*Network chain density decrease*	*Weight*	*Network chain density decrease*	*Weight*	*Network chain density decrease*
0	nd	5·3	2·8	16·6	3·2	15·9	3·3	15·3
5	nd	12·0	2·3	11·7	2·1	19·5	8·5	15·6
45	nd	1·0	2·4	0	1·8	3·4	2·2	6·8

nd, not determined.

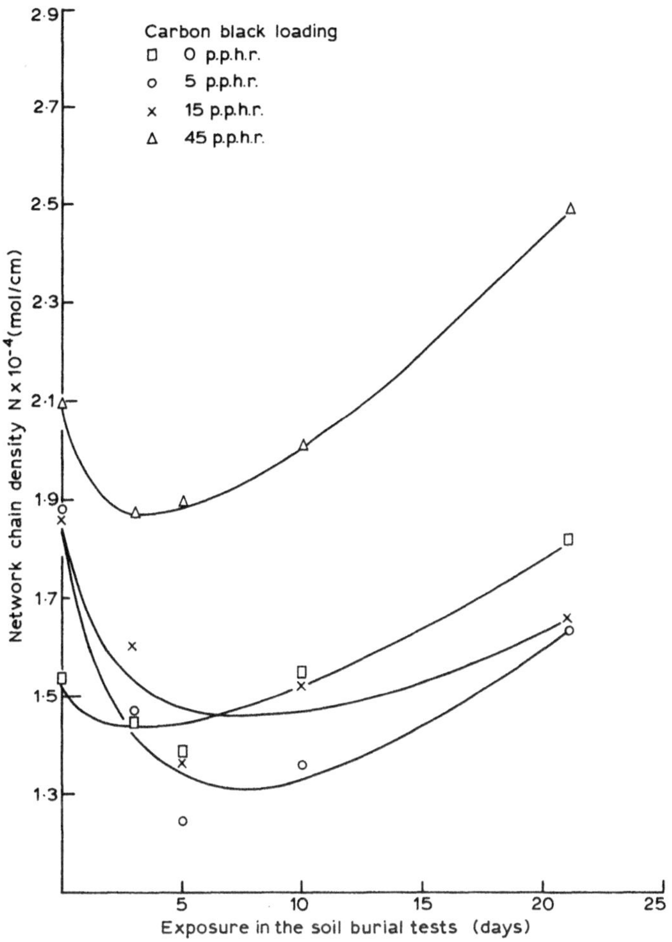

Fig. 1. Changes in network chain density as a function of duration of soil burial.

REFERENCES

Kwiatkowska, D., Zyska, B.J. & Zankowicz, L.P. (1980). In *Proceedings of the 4th International Biodeterioration Symposium*, Berlin. Pitman, London, pp. 135–41.

Tsuchii, A., Suzuki, T. & Takeda, K. (1985). *Applied and Environmental Microbiology*, **50**, 965–70.

Williams, G.R. (1982). *International Biodeterioration Bulletin*, **18**, 31–6.

Williams, G.R. (1986). *International Biodeterioration*, **22**, 307–11.

Zyska, B.J., Rytych, B.J., Zankowicz, L.P. & Fudalej, D.S. (1972). In *Biodeterioration of Materials*, Vol. 2, ed. A.H. Walters & E.H. Hueck-van der Plas. Elsevier Applied Science, London, pp. 256–67.

Zyska, B.J. (1981). In *Economic Microbiology*, Vol. 6, *Microbial Biodeterioration*, ed. A.H. Rose. Academic Press, London, pp. 323–85.

The Study of Fungal Contamination in the Field of Electronics

Mayumi Inoue

Inoue Biodeterioration Laboratory, 26–10, Sekimachiminami 2-Chome, Nerima-ku(177), Tokyo, Japan

ABSTRACT

Fungal contamination of electronic equipment has been recognized as a serious problem in the past few years. Many plastic materials such as epoxy, polyethylene and polypropylene are components of electronic equipment. The deterioration of these materials resulting from fungal growth is a very important problem. The predominant fungi isolated from deteriorated printed circuit, IC and silicon wafer plate were species of Trichoderma viride, Alternaria alternata, Cladosporium sphaeropermum, Fusarim semitectum *and* Pestalotia adusta. *The results of tests carried out are described and discussed.*

INTRODUCTION

This study is concerned with fungal contamination in the field of electronics. Because the climate in Japan has a high humidity, it is suitable for the growth of various kinds of fungi. They can grow not only on foods but on aluminium (Inoue, 1966), plastic materials (Inoue, 1983) and even electronic components.

EXPERIMENTAL METHOD

Samples were collected from the surface of printed circuits, electric cables, telephone systems, automatic control equipment, IC and silicon wafers.

Potato dextrose agar plate medium was used for the isolation of the fungi. Small pieces of the samples were spread on the surface of the agar plate and kept at 24–27°C, 90–95 per cent RH for 5–7 days. When growth occurred, single colonies of fungi were transferred to a new agar plate and incubated further. The species of fungi were identified according to the method of Gilman (1957), Ainsworth (1965) and Barnett (1972).

RESULTS AND DISCUSSION

The fungal contamination of the various samples are set out below:

Printed circuits

The following species of fungi were isolated from the surface of printed circuits:

Cladosporium cladosporioides
Fusarium solani
Chaetomium globosum
Penicillium frequentans
Penicillium stekkii

One of the contaminated printed circuits is shown in Fig. 1.

Fig. 1. Fungus growth on printed circuit.

Electric cables

Deterioration of electric cables from the underground railway, underground road, underground shopping centre and tunnel were principally caused by fungal growth. Extensive damage was found in the tunnel beneath the strait of Seikan (between Aomori and Hakodate).

Telegraph cables were also deteriorated by a range of fungi (*Fusarium semitectum, Rhizopus nigricans* and *Penicillium stekkii*).

About 10 years ago, the deterioration of electric cables covered with polyvinyl chloride (PVC) was observed. The fungi isolated from the PVC were

Fusarium semitectum
Cladosporium spherospermum
Alternaria alternata
Pestalotia adusta

Recently, cable coated with polyethylene was badly attacked by fungal growth.

Silicon wafer

An important observation is that fungi can grow on the surface of a silicon wafer despite the fact that it has a mirror-like polish (Fig. 2). The following fungi were isolated from the surface:

Rhizopus nigricans
Fusarium semitectum
Alternaria alternata
Nigrospora oryzae
Paecilomyces varioti

The reason for the high contamination of silicon wafers is that the same species of fungi isolated from the surface were found on the ceiling, walls and floor of the clean room used for silicon wafer manufacture. The HEPA filter was heavily contaminated by black sporing fungi and the following species were isolated from the surface of the glass fibre of the filter:

Fusarium semitectum
Paecilomyces varioti
Aspergillus flavus
Cladosporium sphaerospermum

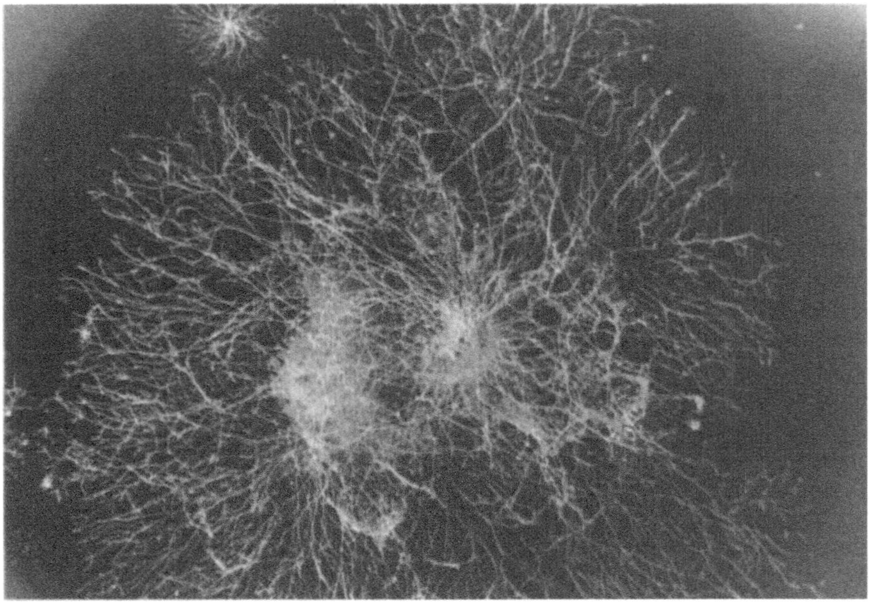

Fig. 2. Fungal growth on a silicon wafer.

Alternaria alternata
Rhizopus nigricans
Aureobasidium pullulans

It was confirmed that spores of these fungi passed through the HEPA filter into the clean room.

The floor of a biological clean room of the IC factory was buckled and peeling. It is made from aluminium panels about 80 mm thick to which the PVC flooring is attached by an adhesive.

Detailed investigation revealed that the various kinds of fungi grew between the aluminium panels and PVC flooring causing corrosion of the aluminium and deterioration of the PVC. This is also an important source of contamination of the biologically clean room. However, the main source of fungal infection was a contaminated environment caused by fungi attaching to and growing on the walls, ceiling and floor.

It is recognized that manufacture of articles in a microbiologically clean environment improves the performance, accuracy, reliability and yield of the article.

Instead of using a HEPA filter, a study is being made of an electrostatic method, the Bio Clean Air Unit. The results of which will be reported later.

REFERENCES

Ainsworth, G.C. (1965). *The Fungi, an Advanced Treatise.* Academic Press, New York.

Barnett, H.L. (1972). *Illustrated Genera of Imperfect Fungi.* Burgess Publishing Company.

Gilman, J.C. (1957). *A Manual of Soil Fungi.* The Iowa State University Press, Iowa.

Inoue, Mayumi (1966). *The ninth Japan Congress on Testing Materials,* 88.

Inoue, Mayumi (1983). *Biodeterioration 5.* John Wiley, New York, p. 535.

POST-HARVEST DECAY

A Review of the Nature and Causes of Post-harvest Deterioration in Fruits and Vegetables, with Especial Reference to those in International Trade

Anna L. Snowdon

Department of Applied Biology, University of Cambridge, Pembroke Street, Cambridge CB2 3DX, UK

ABSTRACT

Harvested fruits and vegetables are subject to a wide range of diseases (incited by infectious micro-organisms) and disorders (induced by disruption of normal physiological processes). Consideration is given to the biology and control of the major post-harvest diseases and disorders affecting pome, stone, berry and soft fruits; tropical and sub-tropical fruits; cucurbits and solanaceous fruits; legumes; flower, stem and leaf vegetables; bulbs; temperate roots and tubers; tropical roots and tubers. Contributory factors, which together influence the nature and extent of deterioration, include choice of cultivar; weather during the growing season; crop husbandry; harvesting and handling techniques; post-harvest treatments; pre-cooling and packaging. The storage environment is crucial and, if produce is exported, important factors include design and function of ship, container or truck; pattern of stowage; carriage instructions and policy; and transit time. In a 15-year study it was found that in 39% of cargoes deterioration resulted from adverse pre-shipment factor(s) only, in 42% from adverse shipboard factor(s) only, and in 19% from both pre-shipment and shipboard factors. Analysis of causes can help to reduce future losses.

INTRODUCTION

The most important characteristic of freshly harvested fruits and vegetables is that they are still alive and respiring. On the basis of respiration pattern, horticultural products can be divided into two main groups. Vegetables and some fruits (e.g. citrus, cherries, grapes, pineapples) have a respiration rate which, at a given temperature, remains fairly constant (or even falls slightly) during the post-harvest period (Hardenburg *et al.*, 1986). The second group includes fruits such as apples, pears, plums, tomatoes, avocados, bananas, mangoes and many others. They are characterised by a sudden surge in respiration rate (usually after harvest) known as the climacteric rise. This is 'triggered' by ethylene gas, produced in minute quantities by the fruits themselves (Blanpied, 1985). Over a period of several days (at moderate temperature) a characteristic skin colour becomes apparent, the flesh softens, starch is converted into sugar, and flavour and aroma develop. This is the stage of prime eating quality or 'ripeness'. Subsequently, as the rate of respiration declines, a process of senescence sets in, off-flavours develop, and the dying tissues become increasingly susceptible to attack by decay organisms (Burton, 1982).

Vegetables and non-climacteric fruits begin to decline as soon as they are harvested, the rate depending partly on their intrinsic storage potential and partly on the environmental conditions. Deteriorative changes are hastened by the presence of ethylene, whether from the produce itself or from an external source (Knee *et al.*, 1985). The most characteristic aspect of senescence in green vegetables is the loss of chlorophyll and resultant yellowing of tissue. Internal features of senescence in vegetables may include toughening of the tissues and deterioration in flavour and aroma (Lieberman, 1983; Weichmann, 1987).

Another relevant physiological process characteristic of plants is transpiration, which is the loss of water by evaporation from the plant surface. When the plant is in the ground, moisture lost by evaporation from the leaves can be replaced by water drawn up through the roots from the soil. Once harvested, the plant organ is deprived of the means of replenishment, and is prone to weight loss, wilting and shrivelling (Wills *et al.*, 1981).

Besides the 'normal' deteriorative changes described above, there are other changes which occur only under certain circumstances. For example, low temperatures (whether before or after harvest) may disrupt the complex sequence of biochemical reactions taking place in plant tissue and cause a disorder known as 'chilling injury' (Watada, 1982). Disease organisms (chiefly fungi and bacteria) may cause infection before, during or after harvest (Dennis, 1983; Moline, 1984; Burchill & Maude, 1986). Each horticultural commodity is subject to a wide range of diseases and disorders (Snowdon, 1988), of which only the most important or topical are considered here.

Pome fruits

The storage of apples and pears is now very sophisticated, with accurate control of oxygen and carbon dioxide concentrations, and removal of ethylene in order to maximise storage life (Knee *et al.*, 1985). Control of post-harvest diseases and disorders begins in the orchard. It may be necessary to use a fungicide spray at blossom time to prevent infection by *Botrytis cinerea*, which would otherwise be manifest after harvest as a calyx-end rot of the fruit. Spraying is also necessary against *Gloeosporium* spp., which invade developing fruits via the lenticels. These fungi typically remain quiescent until after harvest; as the fruits ripen, their resistance declines, allowing the fungi to initiate decay. Nutritional sprays (e.g. calcium chloride) and the application of a growth regulator (e.g. paclobutrazol) can prevent or delay the onset of physiological disorders such as bitter pit and senescent breakdown (Porritt *et al.*, 1982; Sharples, 1984). Such treatments can also confer resistance to post-harvest fungal invasion. *Penicillium expansum* (blue mould) (Cappellini *et al.*, 1987) and *Monilinia* spp. (the brown rot fungi) tend to invade via wounds such as insect punctures or, more especially, injuries sustained during harvesting and handling. Control involves strict hygiene in orchard and packing house in order to minimise the build-up of spores. A post-harvest fungicide dip or drench can be very effective, especially when applied in conjunction with calcium treatment, but the time of application is critical and delays of a few hours can result in large increases in subsequent decay.

Two diseases, mucor rot and phytophthora rot, have become important in recent years. Investigations in North America and the UK have established that infection by *Mucor piriformis* occurs from sporangiospores in soil entering the packing house on the boxes used for harvesting (Michaelides & Spotts, 1986). Control involves improvements in hygiene, and refinements in the application of post-harvest treatments. Also soil borne are *Phytophthora cactorum* and *P. syringae* (Harris, 1985). During wet conditions in the orchard, zoospores may be splashed on to the fruit and can cause direct infection. Lesions may not be obvious at harvest time, but development continues in store, mould spreading into adjacent sound fruits and forming 'nests' of decay. The increased incidence of phytophthora rot is probably related to the use of dwarfing rootstocks and pruning systems which encourage production of fruit near the ground. If the soil under the trees is kept bare it should be drenched with fungicide, and the fruit should be protected by orchard sprays and a post-harvest dip or drench.During wet weather particular care should be taken to avoid leaving harvested fruit in the orchard.

Other diseases which may cause post-harvest losses in apples and pears include alternaria rot, bitter rot (*Glomerella cingulata*), botryosphaeria rot, core rot and mouldy core (numerous species), cylindrocarpon rot (*Nectria galligena*), pink mould rot (*Trichothecium roseum*), phacidiopycnis rot, rhizopus rot, scab (*Venturia* spp.) and side rot (*Phialophora malorum*).

Stone fruits

Orchard spraying is necessary to control *Monilinia* spp. and *Botrytis cinerea* (grey mould), which are important pathogens of peaches, nectarines, apricots, plums and cherries. As well as causing blossom blight and latent infection of developing fruits, these fungi can also invade harvested fruits via injuries. Wounds are a prerequisite for infection by *Rhizopus stolonifer*. Strategies for preventing post-harvest decay depend on the fruit. For example, peaches and nectarines can be given a fungicide treatment (Fourie & Holz, 1987) and heat treatment to inactivate incipient infections, followed by hydro-cooling. It is essential to chlorinate the water in the hydro-cooler and to renew it frequently in order to prevent a build-up of spores (Kader *et al*., 1985).

In view of problems associated with the use of fungicides (fungal insensitivity, undesirable residues, etc.), recent research efforts have been directed towards biological control. Promising results have been obtained by dipping the fruit in a suspension of bacteria antagonistic to the fungi but otherwise harmless (Wilson & Pusey, 1985). Other diseases of stone fruits include alternaria rot, bacterial spot (*Xanthomonas campestris* pv. *pruni*), blue mould rot, cladosporium rot, pustular spot (*Stigmina carpophila*) and rust (*Tranzschelia discolor*).

Soft fruits and berry fruits

Grey mould rot caused by *Botrytis cinerea* is the most important disease of strawberries (Maas, 1984), raspberries, grapes and kiwifruits. It is especially prevalent during wet seasons. Since the fungus typically infects flowers, resulting in latent infection of the fruits, timely field spraying is an essential part of control. Harvested fruits should be cooled to 1°C as rapidly as possible. Grapes can be protected from post-harvest infection by means of sulphur dioxide, which acts as a fungistat (Kokkalos, 1986). The gas is released into the store or, alternatively, it is generated by sodium metabisulphite contained in sachets included in each pack of grapes. With regard to soft fruits, storage in a modified atmosphere (15% carbon dioxide) inhibits the growth of grey mould, and this technique has proved especially beneficial in consignments which have received inadequate refrigeration.

Another important disease of soft fruits is 'leak' caused by species of *Mucor* and *Rhizopus*. The fungi are common soil inhabitants, and one control strategy involves the use of a mulch to prevent infested soil particles from being splashed on to fruits. Few fungicides are effective against these fungi. Most *Rhizopus* spp. are completely inhibited by temperatures below 5°C, but *M. piriformis* can make progress even at 0°C (Dennis, 1983).

Anthracnose (black spot) of strawberries is caused by species of *Colletotrichum* and *Gloeosporium*. The disease has been recorded in the USA, Australia, South Africa and Italy. In an attempt to prevent *C. acutatum* from becoming established in the UK, there are quarantine regulations which require the destruction of any infected imports of strawberries.

The main post-harvest problem of kiwifruit is premature ripening, when the fruits may lose firmness and deteriorate in taste and texture before arrival in the market. Softening is precipitated by ethylene gas, produced by the fruits themselves or from an external source (Ben-Arie & Sonego, 1985). Current research is on pre-harvest as well as post-harvest physiology, in order to elucidate the factors which may predispose the crop to premature ripening.

Tropical and sub-tropical fruits

Avocados, bananas, mangoes and papayas are climacteric fruits which are harvested mature but unripe. They are subject to latent infection by the anthracnose fungus *Glomerella cingulata* (conidial state: *Colletotrichum gloeosporioides*). Infection occurs in developing fruits, but the fungus remains quiescent until the onset of post-harvest ripening, when the host tissue loses its resistance. Symptoms consist of numerous dark circular spots, which mar the appearance of the ripe fruit. The same fungus is responsible for lens-shaped lesions on bananas resulting from wound-infection; these can develop even on green fruit. Control of anthracnose must start in the plantation with fungicide sprays (Zentmyer, 1984). To reduce the likelihood of the development of tolerant fungal strains, an unrelated fungicide should be used after harvest (Eckert & Ogawa, 1985). Post-harvest treatments for mangoes and papayas may include immersion for several minutes in a hot water bath containing a fungicide; time and temperature are critical if the disease is to be controlled without heat injury to the fruit (Alvarez & Nishijima, 1987). Irradiation is sometimes used as a supplementary treatment (Spalding & Reeder, 1986). Post-harvest treatments for avocados may include a fungicide dip and the use of controlled atmosphere storage. Post-harvest measures against banana anthracnose consist of a fungicide dip, prompt cooling and rapid artificial ripening before distribution.

Currently, the most serious type of decay in bananas is 'crown rot' arising from infection of the tissue exposed when the 'hand' is severed from the bunch (Slabaugh & Grove, 1982). It is a disease complex, caused by several fungi and bacteria acting together (often including species of *Colletotrichum, Fusarium* and *Verticillium*). The traditional means of control is to subject the hands to a fungicide dip (or spray) in the packing station. An alternative method is to sever the hands in the plantation as soon as the bunch has been cut, and to apply a fungicide-impregnated pad to the newly exposed crown tissue. The pad also serves to absorb latex which exudes from the cut surface. Regardless of the method used to protect the crown, it is of great importance to minimise the time between cutting and cooling the fruit, especially if a long voyage is scheduled. Rapid reduction of temperature to about 13°C serves to slow down development of decay.

Pineapples are non-climacteric fruits and must therefore be allowed to develop most of their eating qualities whilst still attached to the plant. An important post-harvest disease is black rot caused by *Ceratocystis paradoxa* (conidial state: *Thielaviopsis paradoxa*). Infection may occur in the field, but the fungus more usually invades via the cut stem shortly after harvest. Control involves hygiene in field and packing station, and prompt fungicide treatment of harvested fruits (Rohrbach & Apt, 1986). Fruitlet core rot may be caused by species of *Fusarium*, *Penicillium*, *Erwinia* and *Pseudomonas*, alone or in combination. The organisms survive on debris in the soil, and are carried by mites and mealybugs into the floral cavities of the plant. Thus the chief means of reducing losses is the use of an acaricide in the field.

The most common causes of deterioration in citrus fruits are green mould (*Penicillium digitatum*) and blue mould (*P. italicum*). Initial infection is via wounds. Once established, blue mould is capable of spreading directly into uninjured healthy fruits in contact, causing 'nesting', and this can have serious consequences in long-term storage. In contrast, green mould is generally unable to infect adjacent fruits unless they are injured. Nevertheless, even straightforward 'soiling' by clouds of spores can render sound fruit unmarketable. Furthermore, ethylene is evolved both by the fungus and by infected fruit, and this has the effect of hastening senescence in the rest of the fruit.

Citrus grown in areas having a substantial summer rainfall is subject to infection by the stem-end rot fungi, *Botryodiplodia theobromae* and *Diaporthe citri* (*Phomopsis citri*). These organisms lodge under the calyx and cause infection which remains quiescent until after harvest. Fungal advance is possible only after senescence of the 'button' tissue. This may occur after extended storage or, prematurely, if the fruit is subjected to ethylene degreening.

Alternaria citri also causes a latent infection of the stem-end, but tends to progress into the centre of the fruit (causing a black or brown discoloration) without being apparent externally. Because of the difficulty of detection, alternaria rot causes problems in the processing industry; even low levels of infection can result in off-flavoured juice, with the further hazard of mycotoxin formation (Stinson *et al*., 1981).

Geotrichum candidum, a wound pathogen, is the cause of sour rot, in which cell wall-degrading enzymes cause complete disintegration of tissue. Spore-laden fluid leaks out, resulting in infection of nearby fruits.

Control measures against citrus diseases include orchard sprays (against fruit-piercing insects as well as fungi), care in harvesting and handling, attention to the degreening regime if used, application of a growth-regulator (to maintain 'button' vitality), and prompt treatment with post-harvest fungicides. It is only quite recently that effective chemicals have been developed for the control of alternaria rot and sour rot (Eckert & Ogawa, 1985). With regard to control of penicillium rots, recent studies (Gardner *et al*., 1986) have shown that simple changes in packhouse layout (in particular, enclosure of the 'dump' area) can be very effective in reducing the aerial spore population and lowering the incidence of decay. Hygienic practices and the rational use of different fungicides are also essential in delaying the emergence of insensitive fungal strains. 'Seal-packaging' of individual fruits has the advantage of isolating diseased fruits from their neighbours (Shewfelt, 1986).

The use of low temperature to retard deterioration of tropical and subtropical fruits is severely limited by their susceptibility to chilling injury, characterised by external and/or internal discoloration and (in the climacteric fruits) failure to ripen normally. The critical temperature, below which injury may occur, varies with commodity, cultivar, season and maturity (Brown, 1986). Various strategies (depending on commodity) have been devised to reduce susceptibility to chilling injury: pre-treatment with ethylene; use of certain fungicide treatments, minimisation of moisture loss by means of wax coating or semi-permeable packaging; 'cold-conditioning' by stepwise cooling; and controlled atmosphere storage (Watada, 1982).

Cucurbits

One of the most important diseases of cucurbits is black rot caused by *Didymella bryoniae*. Also common is grey mould rot caused by *Botrytis cinerea*. In both diseases fruits are invaded via the dying flowers, through the cut stem at harvest time, and through wounds. Control must begin with the use of clean or treated seed (van Steekelenburg, 1986). Field hygiene is essential, since the fungi persist in crop debris; in high-value protected crops it may be feasible to carry out soil sterilisation before planting. Irrigation and ventilation must be well-controlled to prevent the build-up of excessive moisture, and fungicide sprays should be carefully timed. Some fungicides used against *B. cinerea* remain effective for only one or two seasons, owing to the emergence of new fungal strains tolerant of the chemical. An alternative strategy is reported from Japan; spore production can be inhibited in greenhouses made from special plastic film which cuts out certain UV wavelengths (Honda & Yunoki, 1980).

Melons also suffer from these two diseases but, in hot countries, more important causes of loss are alternaria rot, botryodiplodia rot, charcoal rot (*Macrophomina phaseolina*), cottony leak (*Pythium* spp.), fusarium rot and phytophthora rot. General methods of control include seed treatment, fungicide spraying, attention to soil moisture, harvesting at optimal maturity and gentle handling. Stem-end rots may be controlled by the use of a fungicidal paste applied to the cut stem. Control by the use of refrigeration is limited by the susceptibility of cucurbits to chilling injury, the critical temperature for cucumbers being about 9°C and for melons about 5°C.

Other diseases leading to post-harvest losses of cucurbits are anthracnose, blue mould rot, choanephora rot, cladosporium rot, myrothecium rot, pink mould rot (*Trichothecium roseum*), rhizopus rot, sclerotium rot (*Corticium rolfsii*), soil rot (*Rhizoctonia solani*), sour rot (*Geotrichum candidum*) and watery soft rot (*Sclerotinia* spp.). Bacterial soft rot, caused by *Pseudomonas* sp., is particularly difficult to control (Risse *et al.*, 1985).

Solanaceous fruits

Tomatoes, peppers and eggplants are frequently attacked by *Botrytis cinerea*, and control methods are similar to those described for cucurbits. *Alternaria alternata* is the cause of black spot. This fungus is generally regarded as only mildly pathogenic, but it can rapidly colonise tissue which has been weakened, for example by unfavourable growing conditions or by post-harvest chilling injury. A toxin is produced, necessitating scrupulous quality control of fruits destined for processing (Stinson *et al.*, 1981).

Bacterial soft rot may be caused by species of *Bacillus, Erwinia* and *Pseudomonas*. Infection occurs via wounds or through the cut stem at harvest time. Post-harvest washing can cause infection in de-stemmed tomatoes by forcing bacteria into the stem-scar through hydrostatic pressure (Bartz, 1982). Thus control depends on gentle handling, regular renewal and chlorination of wash-water, brief immersion times, and drying of the fruit surface before packing.

Soil rot is caused by *Rhizoctonia solani*. It is particularly prevalent on tomatoes intended for processing, since these crops are often grown without the aid of stakes; fruits may therefore be in contact with the ground and, furthermore, they are not harvested until fully ripe. Efforts are being made to breed resistant cultivars but, meanwhile, control measures include seed treatment, mulching and staking. Ethephon treatment is beneficial, not only because it promotes uniform ripening (permitting prompt harvesting) but also because it causes leaf drop, resulting in a reduction of humidity and consequently less likelihood of infection.

Other diseases of solanaceous fruits include anthracnose, cladosporium rot, cottony leak (*Pythium* spp.), fulvia rot, fusarium rot, phoma rot, phomopsis rot, phytophthora rot, pleospora rot, rhizopus rot, sclerotium rot (*Corticium rolfsii*), sour rot, and watery soft rot.

Legumes

Peas and beans are subject to pod-spotting caused by various species of *Ascochyta*. To avoid these diseases it is desirable to obtain clean seed, or to treat the seed with a fungicide before sowing (Maude *et al.*, 1986). Crop debris should be destroyed, and it is also important to control weeds, several of which can act as alternative hosts for these fungi. For control of chocolate spot of broad beans caused by *Botrytis fabae*, and grey mould rot of peas and beans caused by *B. cinerea*, there are certain cultural practices which help to reduce disease incidence. Such measures include correct irrigation and fertilisation, wide spacing and suitable orientation of rows to permit effective natural aeration and to prevent accumulation of moisture within the canopy.

Similar practices are used against watery soft rot caused by *Sclerotinia* spp. The less bushy cultivars of peas and beans are more likely to escape these diseases. Otherwise, reliance must be placed on a fungicide spraying programme to protect the crop at all stages from bloom to harvest (Sherf & MacNab, 1986).

Other diseases of peas and beans include alternaria blight, angular leaf spot (also affecting pods), anthracnose, various bacterial blights, cottony leak (*Pythium* spp.), downy mildew (*Peronospora* spp.), phomopsis blight, powdery mildew (*Erysiphe* spp.), phytophthora rot, rust (*Uromyces* spp.), sclerotium rot (*Corticium rolfsii*) and soil rot (*Rhizoctonia solani*). Frequently, however, the primary cause of post-harvest loss is wilting and mechanical damage (Ceponis & Butterfield, 1984).

Flower, stem and leaf vegetables

Cabbage can be stored for several months at 1°C, and its storage life can be extended by the use of controlled atmosphere (Geeson & Browne, 1980). Losses through decay are mainly caused by *B. cinerea, Alternaria* spp. and *Mycosphaerella brassicicola*. Good control may be achieved by removing the outer leaves (often infected) at harvest time, and applying a fungicide treatment (drench, dip or fog) to the trimmed heads (Geeson *et al.*, 1982). Cooling should be as rapid as possible. Bacterial rotting can be a problem even under refrigeration; *Pseudomonas marginalis* can cause decay at temperatures as low as 0°C (Liao & Wells, 1987).

Cauliflowers, Brussels sprouts and broccoli are affected by similar organisms. It has been suggested that fungal disease problems in the vegetable brassicas have increased in the UK as a result of the widespread planting of oil-seed rape, a related host. The evidence is not clear-cut, however, and investigations continue (Wafford *et al.*, 1986).

Celery, fennel, lettuce, endive and chicory are all subject to watery soft rot caused by *Sclerotinia minor* or *S. sclerotiorum*. These fungi persist in the soil as sclerotia which are difficult to eradicate. Partial control may be achieved by deep ploughing or by flooding the land for several weeks before planting, in order to kill the sclerotia. In countries with a hot dry season 'solarisation' may be feasible. A clear plastic tarpaulin is laid over the ground for several weeks before the next crop is planted; the soil is thereby heated to a temperature which is lethal to sclerotia (Stapleton & DeVay, 1986). Current research is directed towards improving the prospects for biological control, in which conditions are established that encourage the breakdown of sclerotia by hyperparasitic (but otherwise harmless) soil organisms (Patterson *et al.*, 1986).

Other diseases affecting these crops are bacterial soft rot, cercospora spot, grey mould rot, licorice rot (*Mycocentrospora acerina*), rhizoctonia rot and septoria spot. Besides measures taken in the field, prompt cooling and storage near 0°C help to delay the onset of rotting.

Salad onions and leeks are prone to white rot caused by *Sclerotium cepivorum*. This disease is becoming increasingly prevalent and can cause heavy losses, both pre-harvest and post-harvest, chiefly in temperate growing areas but also in warmer countries such as Egypt. Chemical and biological methods have been investigated. A novel approach is to stimulate the sclerotia into premature germination (in the absence of the host crop) by the addition of onion or garlic extract to the soil; by the time the crop is planted, most of the sclerotia have decayed. Results are not always satisfactory but the technique shows promise (Coley-Smith & Parfitt, 1986).

Bulbs

Although *S. cepivorum* was initially a problem only in salad onions, it is causing increasing losses in bulb onions.

Black mould, *Aspergillus niger*, thrives at temperatures of 30°C and above, and is common on onions grown in the tropics (Abdel-Rahim & Arbab, 1985). In recent years it has appeared in UK onions, as a result of a change in post-harvest handling techniques. Whereas previously onions were dried and cured in the field, they are now harvested directly and subjected to high temperature drying in store. Such conditions may be conducive to proliferation of black mould, and care must be taken to ensure that 'curing' temperatures are not excessive.

In contrast, neck rot (caused by *Botrytis allii* and other species) is less prevalent than in the past, following the discovery that the disease is often seedborne. Current practice involves seed treatment, and careful drying of harvested bulbs (Burchill & Maude, 1986).

Other diseases of onions and garlic are bacterial soft rot (species of *Erwinia, Pseudomonas* and *Lactobacillus*), blue mould rot, fusarium basal rot, purple blotch (*Alternaria porri*), sclerotium rot (*Corticium rolfsii*) and smudge (*Colletotrichum circinans*).

Temperate roots and tubers

With regard to long-term storage of carrots, *B. cinerea* and *Mycocentrospora acerina* are the major pathogens in temperate countries. The latter (which causes licorice rot) increased substantially with the advent of mechanical harvesting and consequent injury to the roots. A post-harvest fungicide dip or drench can be an effective control measure. However, it may not be active against crater rot caused by *Rhizoctonia carotae*, now recognised as an important cause of post-harvest loss (Punja, 1987). This fungus may be carried in soil particles adhering to wooden storage boxes. Hence one method of control is to steam-clean the boxes or line them with plastic film, which also serves to keep the roots turgid.

Sclerotinia sclerotiorum occasionally causes serious losses, and there is no 'lag phase' before rotting begins in store (Snowdon, 1983). This fungus has also been responsible for rapid deterioration in cargoes of carrots shipped under inadequate refrigeration.

Chalaropsis thielavioides and *Thielaviopsis basicola* thrive in a warm moist environment, and have been found on carrots packed in plastic film at ambient temperature during distribution.

Cavity spot has received much attention in recent years, since it is a disfiguring blemish which reduces market value. Formerly believed to be a physiological disorder, it has now been linked with infection by *Pythium* spp. and possibly other organisms, and can be controlled by treating seed and soil with an appropriate fungicide (White, 1986). Other diseases of carrots are bacterial soft rot, black rot (*Alternaria* spp.), rhizopus rot, sclerotium rot (*Corticium rolfsii*), sour rot (*Geotrichum candidum*) and violet root rot (*Helicobasidium purpureum*).

Important pathogens of potatoes include *Phytophthora infestans* (the cause of late blight), *Phoma exigua* (gangrene), *Fusarium* spp. (dry rot) and *Erwinia* spp. (bacterial soft rot) (Nolte *et al.*, 1987). With increased sale of pre-packed washed tubers, superficial skin blemishes have assumed greater importance, for example black dot (*Colletotrichum coccodes*), black scurf (*Rhizoctonia solani*), common scab (*Streptomyces scabies*), silver scurf (*Helminthosporium solani*), and skin spot (*Polyscytalum pustulans*) (Cayley *et al.*, 1983).

Essential measures for control of potato diseases include the use of clean or treated seed tubers (Jeffs, 1986); field hygiene and timely spraying with fungicide; destruction of foliage at least 2 weeks before harvest; treatment of tubers with a fungicide dust, dip or spray immediately after harvest; provision for efficient 'curing', followed by cool storage.

Bacterial soft rot can be a serious problem, both in bulk-stored potatoes and in potato cargoes shipped in ventilated space at ambient temperature. Applying bactericidal chemicals has generally proved ineffective (Wyatt & Lund, 1981), and control relies on efficient ventilation and the prevention of condensation on the tuber surface.

Tropical roots and tubers

Tropical food crops include cassava, yams, sweet potatoes, tannias and taros. Much work has been done on the phenomenon of vascular streaking in cassava, a physiological disorder which is thought to be a result of stress following moisture loss. The blue-black discoloration can occur within a day or two of harvest (Rickard, 1981). Consequently cassava for local consumption is left in the ground until required. For exported roots, control measures include prompt application of wax, or alternatively plastic film packaging and cool storage. The other commodities must be kept at moderate temperatures because of their susceptibility to chilling injury, and may suffer heavy losses from *Botryodiplodia theobromae*, *Corticium rolfsii*, and species of *Aspergillus*, *Penicillium*, *Fusarium* and *Rhizopus* (Ikotun, 1983). The chief means of control are cultural methods, efficient 'curing' and the use of a post-harvest fungicide (Linnemann, 1981).

FACTORS INFLUENCING DETERIORATION

It may be seen from the foregoing that deterioration can take place during handling, storage, transport or marketing, and that numerous factors, both pre-harvest and post-harvest, influence the nature and extent of deterioration (Snowdon, 1983). Causes of deterioration in fruit and vegetable cargoes (shipped by sea or overland) were investigated, over a period of 15 years, in the UK, the Netherlands, Belgium and France. Methods have been described in an earlier paper (Snowdon, 1979). Information was obtained either during out-turn surveys (on behalf of cargo receivers, insurers, ship-owners or charterers) or during the study of claims documents (on behalf of lawyers acting for one of these parties). For 462 of the cargoes (of which 327 were surveyed) the available information permitted a diagnosis of the underlying causes of deterioration, and for each of these cargoes the causes were assigned to one or more of ten categories (Fig. 1). In over half of the cargoes deterioration resulted from more than one cause (Fig. 2). In 39% of the cargoes deterioration resulted from adverse pre-shipment factor(s) only, in 42% from adverse shipboard factor(s) only, and in 19% from both pre-shipment and shipboard factors.

It was found that the selected means of transport was sometimes intrinsically unsuited to the task, for example general cargo vessels used for onions and ill-designed refrigerated ships used for unpre-cooled carrots or bananas. The commonest fault in ship stowage (Snowdon, 1981*b*) was failure to appreciate the risk in continuing the stow up into the hatchcoaming. With regard to container stowage (Snowdon, 1981*a*) there was a lack of awareness of the special needs of cargo presented at ambient temperature. Shippers chartering vessels did not always

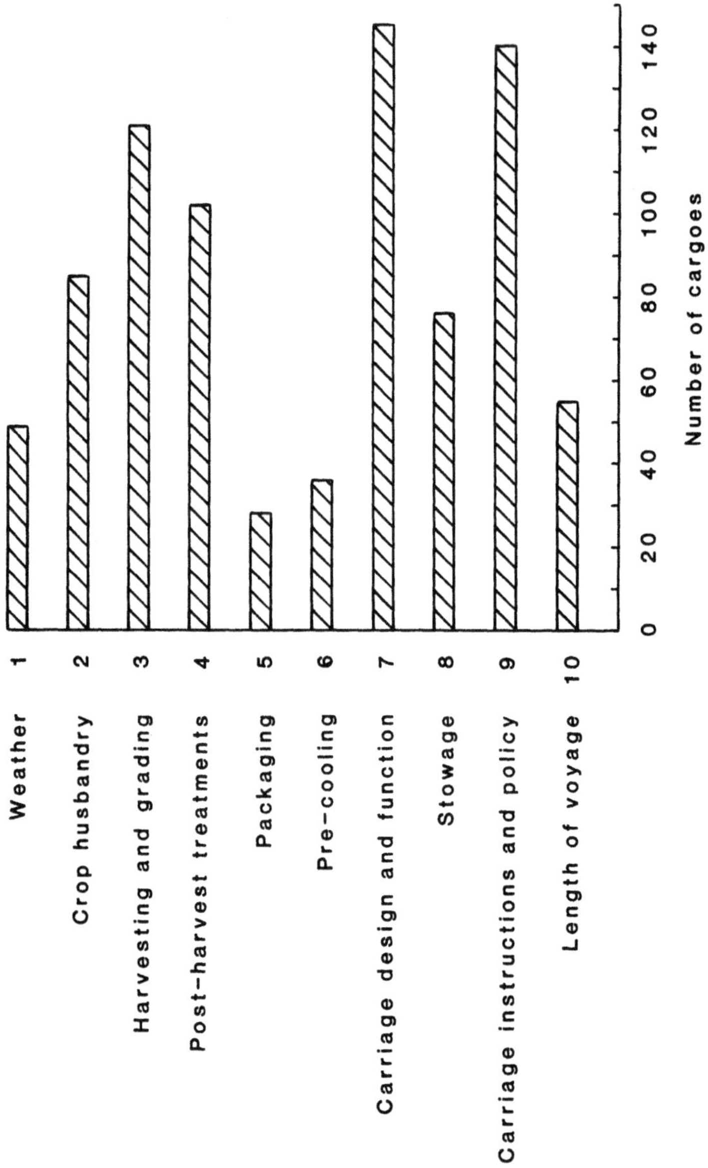

Fig. 1. Factors influencing deterioration in cargoes.

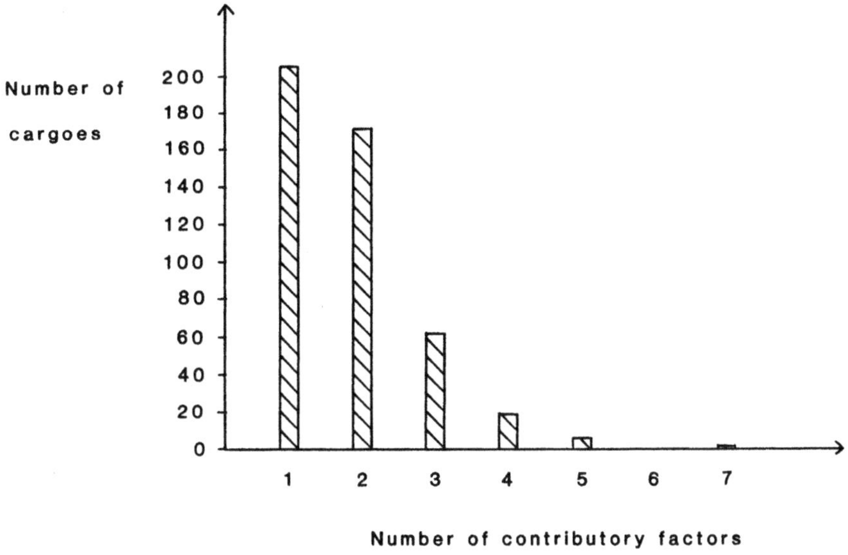

Fig. 2. Number of contributory causes.

understand how the holds functioned, and gave instructions which were difficult to carry out or which had a deleterious effect on their cargoes. Shipboard policies were sometimes misguided, for example deliberate slow cooling of vegetable cargoes. The effect of a delayed voyage depended on the commodity, but tended to be especially serious in ventilated cargoes. The results represent only a very small proportion of cargoes in international trade, but nevertheless give some idea of the reasons for losses. A general rule is that deterioration is frequently the result of a combination of adverse factors.

DISCUSSION

Each type of fruit and vegetable is subject to a range of diseases and disorders (Cappellini *et al.*, 1987) and research is being carried out on the influence of both pre- and post-harvest factors, in attempts to control post-harvest deterioration and so prolong storage- and shelf-life (Sharples, 1984; Kader *et al.*, 1985; Shewfelt, 1986). For most commodities optimal storage conditions have been defined (Snowdon & Ahmed, 1981; Ryall & Pentzer, 1982; Hardenburg *et al.*, 1986) and store operators are generally familiar with the requirements of local produce. Even so, for a variety of reasons there may be serious losses in store (Burchill & Maude, 1986). For goods which are transported from one area to another (and sometimes half way round the globe) there are additional problems which are not always appreciated or understood (Isenberg, 1984). Although the vast majority of consignments are delivered in good condition, there are many which undergo partial or complete deterioration owing to a lack of awareness (on the part of the importer, exporter or carrier) of fundamental principles concerning the nature and causes of post-harvest change. There is a need for wider dissemination of practical information (Harvey, 1981) on the requirements of the various commodities (including the less familiar 'exotic' fruits and vegetables which are being traded in increasing quantities), together with improved collaboration between scientific, technical and commercial personnel engaged in international trade in fresh fruits and vegetables.

REFERENCES

Abdel-Rahim, A.M. & Arbab, H.A. (1985) *Mycopathologia*, **89**, 75–9.

Alvarez, A.M. & Nishijima, W.T. (1987). *Plant Disease*, **71**, 681–6.

Bartz, J.A. (1982). *Plant Disease*, **66**, 302–6.

Ben-Arie, R. & Sonego, L. (1985). *Scientia Horticulturae*, **27**, 263–73.

Blanpied, G.D. (Ed.) (1985). *HortScience*, **20**, 39–60.

Brown, B.I. (1986). *Acta Horticulturae*, **175**, 339–42.

Burchill, R.T. & Maude, R.B. (1986). *Outlook on Agriculture*, **15**, 160–6.

Burton, W.G. (1982). *Post-harvest Physiology of Food Crops*. Longman, London, 339 pp.

Cappellini, R.A., Ceponis, M.J. & Lightner, G.W. (1987). *Plant Disease*, **71**, 852–6.

Cayley, G.R., Hide, G.A., Read, P.J. & Dunne, Y. (1983). *Potato Research*, **26**, 163–73.

Ceponis, M.J. & Butterfield, J.E. (1984). *HortScience*, **19**, 571–2.

Coley-Smith, J.R. & Parfitt, D. (1986). *Pesticide Science*, **17**, 587–94.

Dennis, C. (Ed.) (1983). *Post-harvest Pathology of Fruits and Vegetables*. Academic Press, London, 264 pp.

Eckert, J.W. & Ogawa, J.M. (1985). *Annual Review of Phytopathology*, **23**, 421–54.

Fourie, J.F. & Holz, G. (1987). *Phytophylactica*, **19**, 73–8.

Gardner, P.D., Eckert, J.W., Baritelle, J.L. & Bancroft, M.N. (1986). *Crop Protection*, **5**, 26–32.

Geeson, J.D. & Browne, K.M. (1980). *Annals of Applied Biology*, **95**, 267–72.

Geeson, J.D., Robinson, R.C. & Williams, D.J. (1982). *Tests of Agrochemicals and Cultivars (Annals of Applied Biology*, **100**, Supplement) No. 3, 36–7.

Hardenburg, R.E., Watada, A.E. & Wang, C.Y. (1986). The commercial storage of fruits, vegetables, and florist and nursery stocks. *Agriculture Handbook of the United States Department of Agriculture*, No. 66, 130 pp.

Harris, D.C. (1985). *Transactions of the British Mycological Society*, **85**, 153–5.

Harvey, J.M. (1981). *International Journal of Refrigeration*, **4**, 293–8.

Honda, Y. & Yunoki, T. (1980). *Japan Agricultural Research Quarterly*, **14**, 78–83.

Ikotun, T. (1983). *Fitopatologia Brasileira*, **8**, 1–7.

Isenberg, F.M.R. (1984). *Acta Horticulturae*, **157**, 169–76.

Jeffs, K.A. (Ed.) (1986). *Seed treatment*, 2nd edn. British Crop Protection Council, Thornton Heath, 332 pp.

Kader, A.A., Kasmire, R.F., Mitchell, F.G., Reid, M.S., Sommer, N.F. & Thompson, J.F. (1985). *Postharvest Technology of Horticultural Crops*. Berkeley, University of California (Division of Agriculture and Natural Resources) Special Publication 3311, 192 pp.

Knee, M., Proctor, F.J. & Dover, C.J. (1985). *Annals of Applied Biology*, **107**, 581–95.

Kokkalos, T.I. (1986). *American Journal of Enology and Viticulture*, **37**, 149–51.

Liao, C.H. & Wells, J.M. (1987). *Phytopathology*, **77**, 673–7.

Lieberman, M. (Ed.) (1983). *Post-harvest Physiology and Crop Preservation*. Plenum Press, New York, 572 pp.

Linnemann, A.R. (1981). *Abstracts on Tropical Agriculture*, **7**, 9–20.

Maas, J.L. (Ed.) (1984). *Compendium of Strawberry Diseases*. American Phytopathological Society, St Paul, Minnesota, 138 pp.

Maude, R.B., Bambridge, J.M. & Spencer, A. (1986). *Tests of Agrochemicals and Cultivars (Annals of Applied Biology*, **108**, Supplement), No. 7, 70–1

Michaelides, T.J. & Spotts, R.A. (1986). *Plant Disease*, **70**, 1060–3.

Moline, H.E. (ed.) (1984). *Postharvest Pathology of Fruits and Vegetables: Postharvest Losses in Perishable Crops*. Berkeley, University of California Publication NE-87 (UC Bulletin 1914), 80 pp.

Nolte, P., Secor, G.A. & Gudmestad, N.C. (1987). *American Potato Journal*, **64**, 1–9.

Patterson, C.L., Grogan, R.G. & Campbell, R.N. (1986). *Plant Disease*, **70**, 982–7.

Porritt, S.W., Meheriuk, M. & Lidster, P.D. (1982). Postharvest disorders of apples and pears. Agriculture Canada Publication 1737, 66 pp.

Punja, Z.K. (1987). *Canadian Journal of Plant Pathology*, **9**, 24–31.

Rickard, J. (1981). *Tropical Science*, **23**, 235–7.

Risse, L.A., Miller, W.R. & Chun, D. (1985). *Proceedings of Florida State Horticultural Society*, **98**, 189–91.

Rohrbach, K.G. & Apt, W.J. (1986). *Plant Disease*, **70**, 81–7.

Ryall, A.L. & Pentzer, W.T. (1982). *Handling, Transportation, and Storage of Fruits and Vegetables, Vol. 2*, Fruits and Tree Nuts, 2nd edn. AVI Publishing Co., Westport, Connecticut, 610 pp.

Sharples, R.O. (1984). *Acta Horticulturae*, **157**, 93–104.

Sherf, A.F. & MacNab, A.A. (1986). *Vegetable Diseases and their Control*, 2nd edn. John Wiley, New York, 728 pp.

Shewfelt, R.L. (1986). *Food Technology*, **40**, 70–80, 89.

Slabaugh, W.R. & Grove, M.D. (1982). *Plant Disease*, **66**, 746–50.

Snowdon, A.L. (1979). *Annals of Applied Biology*, **91**, 404–9.

Snowdon, A.L. (1981*a*). *Container Technology*, **3**, 139–48.

Snowdon, A.L. (1981*b*). *Transactions of the Institute of Marine Engineers, Series C*, **94**, 33–7.

Snowdon, A.L. (1983). *Proceedings of the Institute of Refrigeration*, 1982–83, 7.1–7.7

Snowdon, A.L. (1988). *A Colour Atlas of Post-harvest Diseases and Disorders of Fruits and Vegetables*. Wolfe Publishing, London (in press).
Snowdon, A.L. & Ahmed, A.H.M. (1981). *The Storage and Transport of Fresh Fruit and Vegetables*. National Institute of Fresh Produce, London, 32 pp.
Spalding, D.H. & Reeder, W.F. (1986). *Plant Disease*, **70**, 1149–51.
Stapleton, J.J. & DeVay, J.E. (1986). *Crop Protection*, **5**, 190–8.
Steekelenburg, N.A.M. van (1986). *Netherlands Journal of Plant Pathology*, **92**, 81–91.
Stinson, E.E., Osman, S.F., Heisler, E.G., Siciliano, J. & Bills, D.D. (1981). *Journal of Agricultural and Food Chemistry*, **29**, 790–2.
Wafford, J.D., Gladders, P. & McPherson, G.M. (1986). *Aspects of Applied Biology*, **13**, 1–12.
Watada, A.E. (1982). *HortScience*, **17**, 159–86.
Weichmann, J. (Ed.) (1987). *Postharvest Physiology of Vegetables.* Marcel Dekker, New York, 616 pp.
White, J.G. (1986). *Annals of Applied Biology*, **108**, 265–73.
Wills, R.H.H., Lee, T.H., Graham, D., McGlasson, W.B. & Hall, E.G. (1981). *Postharvest: an Introduction to the Physiology and Handling of Fruit and Vegetables.* Granada, London, 150 pp.
Wilson, C.L. & Pusey, P.L. (1985). *Plant Disease*, **69**, 375–8.
Wyatt, G.M. & Lund, B.M. (1981). *Potato Research*, **24**, 315–29.
Zentmyer, G.A. (1984). *Tropical Pest Management*, **30**, 388–400.

Stimulation of Aflatoxin Formation in Maize Seeds

J.E. Smith

Department of Bioscience and Biotechnology, University of Strathclyde, Glasgow, UK

R.G. Cuero

USDA-ARS Southern Regional Research Center, New Orleans, USA

&

J. Lacey

Institute of Arable Crops Research, Rothamsted Experimental Station, Harpenden, Herts, UK

ABSTRACT

Aflatoxin production by Aspergillus flavus *was measured when cultivated alone and when paired with either* A. niger, A. oryzae, Penicillium viridicatum, Fusarium graminearum, Hyphopichia burtonii *or* Bacillus amyloliquefaciens *in irradiation-sterilised maize seeds at three water activities (0·98, 0·95 and 0·90* a_w*) and 25°C. Compared to pure culture, aflatoxin production in mixed culture with filamentous fungi was decreased at 0·98* a_w *but was enhanced at 0·90* a_w*. Both* H. burtonii *and* B. amyloliquefaciens *markedly stimulated aflatoxin production, especially at 0·95 and 0·98* a_w*. The importance of microbial interaction in toxin production in living seeds is discussed.*

INTRODUCTION

Under natural conditions, mycotoxins are usually formed in living seeds of cereals in the presence of other microbial contaminants. How these microorganisms affect the formation in cereal grains is not well understood. In the past mycotoxin production has usually been studied in monocultures in rich organic media under controlled conditions. There have been few experiments that examined mycotoxin formation in dual or mixed microbial culture and these have usually used liquid media or autoclaved (non-viable) grain (Lillehoj *et al.,* 1967; Boller & Schroeder, 1974; Wicklow & Hesseltine, 1980). Neither substrate is comparable with viable cereal grains. This paper describes the production of aflatoxin by *A. flavus* when grown in combination with other toxigenic or non-toxigenic filamentous fungi, a yeast or a bacterium on irradiation-sterilised viable maize seeds at different controlled temperatures and water activities.

METHODS

Organisms and inoculum preparation

Aspergillus flavus CMI 102566 was obtained from the Commonwealth Mycological Institute, Kew, *Penicillium viridicatum* PIL 5574 from the Ministry of Agriculture Fisheries and Food Slough Laboratory, Slough, Berks, UK; *A. oryzae* and *A. niger* from the University of Strathclyde Culture Collection; and *Fusarium graminearum, Hyphopichia burtonii* and *Bacillus amyloliquefaciens* were isolated as natural contaminants of the maize used in these investigations. Spore suspensions used to inoculate grain were freshly prepared or stored at 4°C for no more than 2 days.

Source and treatment of maize

North American hybrid maize seeds (Pioneer 3369a) were obtained from the Southern Regional Research Center, New Orleans, USA. They were separately cracked in an electronic pilot plant machine (Universal Dryer, Richard Simon & Sons, Nottingham, UK) to simulate mechanical damage or insect attack and to aid invasion by the microorganisms. The initial a_w of maize seeds was determined at 25°C using an electronic dew point meter (Protimeter DP 680) and adjusted with sterile water to 0·98, 0·95 or 0·90 a_w. The grain was then sterilised by gamma irradiation at 25°C in a ^{60}Co source at the Scottish Reactor Centre, National Engineering Laboratory, East Kilbride, Glasgow (Cuero *et al.,* 1986). A dose of 1200 Krad was measured with a Perspex dosemeter (IAEA, 1977).

Microbial culture on cereals

After irradiation, 50 g of sterilised cracked maize seeds, with a_w adjusted, were placed in a sterile beaker and inoculated with suspensions of the paired microorganisms. The inoculated seeds were transferred to microporous film bags (Cuero *et al.*, 1985), sealed and incubated in Fison Environmental Cabinets at 25°C, and 90, 95 or 98% relative humidity as appropriate. The bags were removed daily and shaken to avoid mycelial compaction.

Mycotoxin extraction and quantification

After 12 days the contents of the bags were removed and aflatoxin extracted using standard methods (Gorst-Allerman & Steyn, 1979; AOAC, 1980). Aflatoxin was quantified by high performance liquid chromatography using a Pye Unicam Model LC-XPD instrument.

RESULTS AND DISCUSSION

In single culture at 25°C most aflatoxin was produced at 0·98 a_w (1020 ng/g), slightly less at 0.95 a_w and very little at 0.90 a_w (100 ng/g) (Table 1). In paired filamentous fungal cultures, aflatoxin formation at 25°C showed three distinct patterns related to water activity (Table 1). At 0·98 a_w, where *A. flavus* grew profusely, aflatoxin production was considerably decreased in all pairings. At 0·95 a_w, the ability of *A. flavus* to produce the aflatoxins was little affected by any pairing, while at 0·90 a_w all pairings gave considerably enhanced production of the aflatoxins compared with pure cultures of *A. flavus.*

Decreased aflatoxin production has often been found in paired cultures of *A. flavus* and *A. niger* utilising autoclaved substrates at high water activity (Wicklow *et al.*, 1980; Horn & Wicklow, 1983). It appears that at high water activities in both autoclaved and irradiation-sterilised cereals *A. flavus* is less able to produce aflatoxins in competition with other fungi than when growing alone. At 0·95 aflatoxin production was little affected by any fungal pairing. However, the most interesting observation at 25°C was the stimulation of aflatoxin production at 0·90 a_w when *A. flavus* was co-cultivated with any of the test fungi. No biochemical or physiological explanations for these changes in aflatoxin production can be given but they undoubtedly reflect the important role of water availability on fungal metabolism.

When *A. flavus* and *H. burtonii* were grown in dual culture at 25°C aflatoxin production increased under all conditions of water activity. The highest total production was at 0·95 a_w (2009 ng/g), a two-fold increase over single cultures, at 0·90 a_w the increase in aflatoxin production in dual culture was at least ten-fold (Table 1) (Cuero *et al.*, 1987).

TABLE 1
Aflatoxin Production in Single and Dual Culture on Maize at 25°C

Water activity	*0·98*	*0·95*	*0·90*
Single cultures			
A. flavus (total aflatoxins)	1 020·0[a] (28·4[b])	989·0 (63·9)	100·2 (12·6)
Dual cultures			
A. flavus + *A. niger*	552·9 (66·0)	936·0 (110)	534·0 (41·7)
A. flavus + *A. oryzae*	291·0 (62·0)	896·0 (100)	219·0 (60·0)
A. flavus + *P. viridicatum*	231·0 (64·0)	738·0 (93·0)	1 443·0 (167·0)
A. flavus + *F. graminearum*	578·0 (80·1)	879·0 (100·1)	1 095·0 (120·2)
A. flavus + *H. burtonii*	1 321·6 (55·40)	2 009·0 (87·52)	1 185·3 (64·90)
A. flavus + *B. amyloliquefaciens*	2 131·6 (207·03)	931·0 (91·2)	1 302·6 (157·60)

[a] ng toxin/gm maize.
[b] Standard error of mean.

Dual cultures of *A. Flavus* and *B. amyloliquefaciens* produced most aflatoxin (2132 ng/g) at 0·98 a_w. At 0·95 a_w, production differed little from single *A. flavus* cultures, but at 0·90 it was at least ten times greater (Table 1).

CONCLUSION

Since both *H. burtonii and B. amyloliquefaciens* were found to be common contaminants of maize seeds they may play an important role in determining mycotoxin production, particularly at temperatures and water activities that are suboptimal for aflatoxin production by pure cultures of *Aspergillus flavus.*

REFERENCES

AOAC (1980). *Official Method of Analysis, 13th edn.* Arlington, Va., AOAC.

Boller, R.A. & Schroeder, H.W. (1974). *Phytopathology*, **64**, 121–3.
Cuero, R.G., Smith, J.E. & Lacey, J. (1985). *Biotechnology Letters*, **7**, 463–6.
Cuero, R.G., Smith, J.E. & Lacey, J. (1986). *Food Microbiology*, **3**, 107–13.
Cuero, R.G., Smith, J.E. & Lacey, J.E. (1987). *Applied and Environmental Microbiology*, **53**, 1142–6.
Gorst-Allerman, C.P. & Steyn, P.A. (1979). *Journal of Chromatography*, **175**, 325–31.
Horn, B.W. & Wicklow, D.T. (1983). *Canadian Journal of Microbiology*, **29**, 1087–91.
International Atomic Energy Agency (IAEA) (1977). *Manual of food irradiation dosimetry*. IAEA Technical Reports Series No. 178. International Atomic Energy Agency, Vienna.
Lillehoj, E.B., Ciegler, A. & Hall, H.H. (1967). *Journal of Bacteriology*, **83**, 464–8.
Wicklow, D.T. & Hesseltine, C.W. (1980). *Phytopathology*, **69**, 589–91.
Wicklow, D.T., Hesseltine, C.W., Shotwell, O.L. & Adams, G.L. (1980). *Phytopathology*, **70**, 761–4.

Involvement of Trichothecenes in the Pathogenicity of *Myrothecium roridum*

G.A. Bean, J. Kuti & T. Ng

Departments of Botany and Horticulture, University of Maryland, USA

ABSTRACT

Trichothecenes are secondary metabolites produced in vitro *by plant pathogenic fungi including* Myrothecium *spp. Their role in the pathogenicity of fungi has been investigated in five cultivars of muskmelon (*Cucumis melo *L.) inoculated with* Myrothecium roridum *(ATCC 52485) which produces roridin E. Addition of roridin E to a spore suspension of* M. roridum *caused a significant increase in lesion size and the amount of sporulation in muskmelon leaf tissue. The degree of wilting and the severity of phytotoxicity in 6-week-old seedlings were also correlated with the response of cultivars inoculated with* M. roridum. *Increasing concentrations of roridin E caused increasing electrolytic leakage from muskmelon leaf cells; the level of leakage was cultivar dependent. Roridin E produced by* M. roridum *appears to be involved in pathogenicity of this fungus.*

INTRODUCTION

Myrothecium roridum Tode ex Fries (ATCC No. 52485) causes severe leaf spot and fruit rot of muskmelon. It is also known to produce the macrocyclic trichothecene roridin E (Bean *et al.*, 1984).

Roridin E is a secondary metabolite produced by several genera of fungi (Ueno, 1977). These compounds are also found naturally in plant systems as 'baccharanoids' (Kupchan *et al.*, 1977). It has been suggested that their origin is from fungi in the soil (Jarvis *et al.*, 1981).

The trichothecenes are divided into non-macrocyclic trichothecenes, e.g. T-2 toxin and diacetoxyscirpenol and macrocyclic trichothecenes, e.g. verrucarins and roridins (roridin A, D, E and H). Myrotoxins (A and B) are also macrocyclic trichothecenes.

The biological activities of non-macrocyclic trichothecenes have been investigated extensively. Only a few studies have been made of macrocyclic trichothecenes (Cutler & Jarvis, 1985). The cytotoxic properties of some macrocyclic trichothecenes have led to studies on their potential as anti-leukaemic agents.

The objective of this study was to examine the effects of roridin E on the pathogenicity of *M. roridum* to muskmelon.

MATERIALS AND METHODS

Source of trichothecenes

Isolation and characterization of the macrocyclic trichothecenes have been described elsewhere (Bean *et al.*, 1984). Partially purified crystalline forms of roridin E isolated from muskmelon (ATCC No. 52485), myrotoxin isolated from tomato (ATCC No. 60379), and roridin A isolated from soil (ATCC No. 20685) were used. Aqueous 1% methanol (MeOH) was used as a solvent to prepare the toxin solutions because of its low toxicity to plant tissue and fungal cells.

Plant material

Plant material included seedlings and fully expanded leaves of five muskmelon genotypes (*Cucumis melo* L.): 'Amarelo', 'Hales Best', 'Iroquois', 'Perlita' and 'PRM45'.

Role of trichothecenes in pathogenicity

Fully expanded leaves were inoculated with 25 μl of a solution of 5 μg/ml roridin A, roridin E and myrotoxin. Leaves were kept in Petri dishes lined with moistened filter paper for 7 days after which lesion diameter (both necrotic and chlorotic areas) was measured. To assess the role of trichothecene in the pathogenicity of *M. roridum* 25 μl droplets of a spore suspension, with or without mycotoxins, were placed on detached leaves of 'Hales Best' (resistant) and 'Iroquois' (susceptible). After 7 days, lesion diameter of necrotic plus chlorotic tissue was measured and the level of sporulation estimated.

Muskmelon seedling assay

Six-week-old seedlings were washed in tap water, and the roots immersed in either 1% MeOH or 5 μg/ml of roridin E for 2 h. The seedlings were washed and immersed in tap water and examined for phytotoxic symptoms, which were ranked on a scale of 0–5: 0 = no visible symptom and 5 = severe discoloration.

Roots of 'Amarelo' seedlings were also immersed in solutions containing 0, 5, 10, 25, 50 μg/ml of roridin E for 2, 4 and 8 h, and observed for phytotoxicity. In a third experiment, they were treated with 0, 5, 10, 25, or 50 μg/ml of myrotoxin A+B, roridin A and roridin E for 2 h, then transferred to tap water and ranked for wilting on a 0–5 scale, 0 = no symptoms, 1 = mild leaf roll and 5 = severe wilting and tissue desiccation.

Electrolyte leakage studies

To assess the effects of the roridin E on muskmelon leaf membrane stability, leaf discs (18 mm diam.) were excised from healthy leaves, agitated in 25 ml of deionized water for 30 min and floated on 15 ml of a solution containing 0, 5, 10, 25, 50 and 100 μg/ml of roridin E. The leaves were incubated on an orbital shaker for 2 h, after which 5 ml aliquots were collected and tested for electrolytic conductivity (μmhos) with a conductivity bridge (YSI model 31). Electrolytic leakage (T1) was determined by comparing conductivity results with stock roridin E solution. Leaf discs were then boiled in deionized water for 15 min to obtain the remaining electrolyte content (T2). Leaf discs, of the five genotypes, were also incubated with 5 μg/ml of roridin E for 2, 4, and 8 h.

For each experiment, leaf discs were obtained from 10 healthy leaves of each genotype-treatment combination and percentage injury was calculated by dividing the electrolyte content of the treatment solution (T1) by total electrolyte content (T1+T2) multiplied by 100.

RESULTS AND DISCUSSION

Trichothecenes produced necrotic lesions on all muskmelon genotypes. Necrosis first appeared 24 h after inoculation. The rate of chlorotic and necrotic spot development differed with the trichothecene; roridin A caused larger lesions than roridin E while myrotoxin induced the smallest lesion size. All muskmelon genotypes responded in the same manner to roridin A and myrotoxin but had differential responses to roridin E. The responses to roridin E were highly correlated ($r = 0{\cdot}99$,

$p = 0{\cdot}01$) with genotype responses following inoculation with *M. roridum*, whereas responses to roridin A and myrotoxin were not ($r = 0{\cdot}68$, ns and $r = 0{\cdot}21$, ns respectively). These results are important in that roridin E, the only trichothecene in this study produced by the muskmelon isolate of *M. roridum*, appears to possess selectivity for identifying genotype resistance in the host.

Addition of roridin E to the three *M. roridum* isolates at inoculation caused a significant increase in lesion size only when added to the muskmelon isolate on both resistant and susceptible cultivars. Sporulation of the muskmelon isolate was increased when roridin E was added at inoculation; the amount of sporulation increase was greatest on the susceptible cultivar. The addition of roridin E to the other two *M. roridum* isolates did not affect either lesion size or the amount of sporulation.

In the seedling assay, muskmelon genotypes responded differently when immersed in 5 μg/ml roridin E. Cotyledon tissue was the most sensitive whereas only 'Amarelo' and 'Iroquois', the most susceptible genotypes, showed seedling injury in both roots and hypocotyls.

'Amarelo' was the only genotype to show damage in the leaf tissue, indicating translocation of roridin E into leaves. When 'Amarelo' was exposed to increasing concentrations and exposure time to roridin E, injury in all tissues increased. The rates at which trichothecenes are taken up in the root systems by different plant species have been studied by several researchers (Bean *et al.*, 1984). Jarvis *et al.* (1981) reported that *Baccharis megapotomica* was able to absorb roridins and verrucarrins via the root system and translocate the toxins to the upper part of the plant within a few hours. Seedlings of tomatoes, peppers and artichokes also showed leaf tissue damage after a 72 h exposure. Bean *et al.* (1984) reported that muskmelon seedlings absorbed, translocated and metabolized roridin A to 8 β-hydroxyroridin A and showed severe wilting response 24 h after exposure.

Roridins A, E and myrotoxin also induced wilting in 'Amarelo' seedlings. Roridins A and E caused wilting at a concentration of 5 μg/ml while seedlings were able to tolerate myrotoxin up to 10 μg/ml with no evidence of wilting. Severity of wilting in seedlings increased as the concentration increased. This may have resulted from disruption of cell membranes (Novacky, 1980) or a reduction of flow of water either by mechanical occlusion or an increase in viscosity of the xylem sap (Sutton & Williams, 1970).

Electrolyte leakage studies

Roridin E induced cellular leakage of muskmelon leaf tissue as evidenced by changes in electrical conductance. This leakage increased linearly with increasing toxin concentration (Fig. 1).

Muskmelon genotype responses were consistent with those following inoculation with *M. roridum*. 'Hales Best', the most resistant genotype, showed little injury while the susceptible 'Iroquois' showed an increase in membrane injury with increasing concentration. The injury observed may be due to a number of factors such as non-selective increases in permeability or in ion transport across the membrane (Novacky, 1980). Specific ion analysis of the leachates following exposure to roridin E would be necessary to clarify the mode of toxic action on the membrane.

The differences in the effect of roridin E on the genotype membranes may be due to the presence of toxin receptors or binding sites in susceptible genotypes and their absence from resistant genotypes. Evidence for these receptors on plant cell membranes has been reported by researchers working with host selective toxins (Kenfield & Strobel, 1981). Although this study has shown that cell membranes were affected by treatment with roridin E, the main site of action may lie elsewhere in the plant cell. Other studies have shown that they act as protein synthesis inhibitors by selectively binding to the 60S ribosomal subunit of the cells (Cundliffe *et al.*, 1974). Nivalenol, T-2 and verrucarins are inhibitors of polypeptide chain initiation whereas trichodermin inhibits the elongation and/or termination of protein synthesis (Cundliffe *et al.*, 1974).

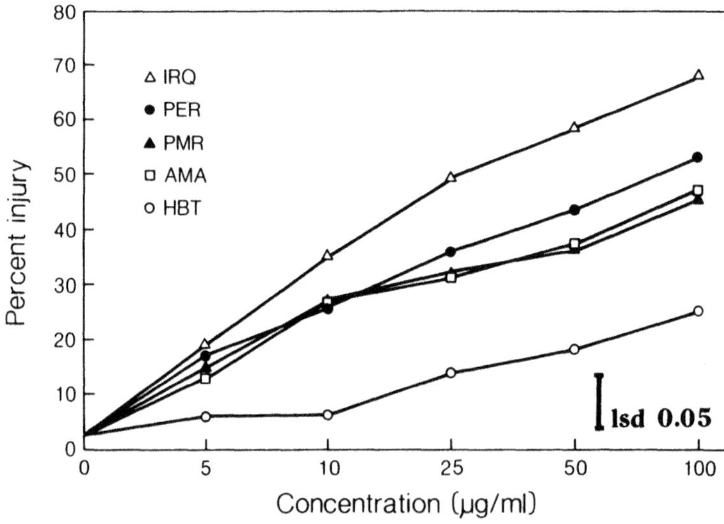

Fig. 1. Effect of increasing roridin E concentration on leaf membrane damage in 5 muskmelon genotypes.

CONCLUSION

Trichothecenes produced by the plant pathogen *M. roridum* induced various phytotoxic effects in muskmelon genotypes that were correlated with the response of the genotypes after inoculation with the fungus. Although there is no evidence of a direct involvement of roridin E in pathogenicity of *M. roridum,* this study indicates that roridin E may be playing a secondary role in pathogenicity by acting as a virulence factor in the invasion of muskmelon tissue.

REFERENCES

Bean, G.A., Fernando, T.A., Jarvis, B.B. & Bruton, B.D. (1984). *Journal of Natural Products,* **47**, 727-9.

Cundliffe, E., Cannon, M. & Davies, J. (1974). *Proceedings of the National Academy of Science, USA,* **71**, 30-4.

Cutler, H.G. & Jarvis, B.B. (1985). *Environmental and Experimental Botany,* **25**, 115-28.

Jarvis, B.B., Midiwo, J.O. Tuthill, D. & Bean, G.A. (1981). *Science,* **214**, 460-2.

Kenfield, D.S. & Strobel, G.A. (1981). *Plant Physiology,* **67**, 1174-80.

Kupchan, S.M., Streelman, D.R., Jarvis, B.B., Dailey, R.G., Jr & Sneden, A.T. (1977). *Journal of Organic Chemistry,* **23**, 4221-5.

Novacky, A. (1980). In *Plant Membrane Transport; Current Conceptual Issues,* ed. R.M. Spanswick, W.J. Ivias & J. Dainly. Elsevier North-Holland, Amsterdam, pp. 369-78.

Sutton, J.C. & Williams, P.H. (1970). *Canadian Journal of Botany,* **48**, 391-401.

Ueno, Y. (1977). In *Advances in Nutritional Research,* ed. H.H. Draper. Plenum, New York, pp. 301-53.

An extended bibliography for this paper may be obtained from the senior author.

Grain Storage: the Management of Ecological Change

J. Lacey

Institute of Arable Crops Research, Rothamsted Experimental Station, Harpenden, Herts AL5 2JQ, UK

ABSTRACT

Fungi colonizing grain before harvest have to contend with extremes of temperature, water availability and radiation. Field fungi, e.g. Cladosporium *and* Alternaria, *are tolerant of these conditions but can seldom develop in stored grain where conditions are usually more stable and there is competition from better adapted storage fungi, e.g.* Aspergillus *and* Penicillium. *Such storage fungi are found in only small numbers before harvest but develop abundantly afterwards, especially if there is spontaneous heating. The conditions in which grain is stored should control the ecological succession of microorganisms and arthropods, and maintain the grain in good condition with minimum loss of quality. This can be done by drying, chilling or modification of the gaseous composition of the intergranular atmosphere, either naturally or with the artificial addition of gases, or by chemical treatment. Grain stored in these ways is often in an unstable condition and equipment failure, insect or mite infestation, or poor management can lead to rapid colonization by fungi and consequent deterioration. This paper discusses how different storage environments affect the colonization of grain by fungi and the consequences of failure to control them adequately. The future development of storage methods integrating two or more ways of preventing deterioration is considered.*

INTRODUCTION

Ecology is the study of the relationships of organisms with one another and with their environment. The complex of substrate, environment and organisms forms an ecosystem which is characteristically subject to change. Any new substrate will be colonized by a succession of microorganisms as nutrient availability changes, but this pattern may be altered by changes in the physical environment. Grain stored in bulk forms a man-made ecosystem with a limited, non-regenerating energy supply. The grain is a rich source of nutrients, not only for the growth of fungi and other microorganisms but also for insects and mites. Compared with many natural ecosystems, it provides a relatively large concentration of energy per unit of biomass. However, because of frequent interference from the physical environment and from man it must be considered an immature unstable ecosystem, with a relatively simple structure and composed of species with high growth and reproductive rates and limited specialization. Colonization by microorganisms and arthropods decreases the energy content of the ecosystem and its value as human and animal food (Sinha, 1973, 1979). The purpose of all grain storage systems is to maintain the grain in an unstable equilibrium so that deterioration is minimised. This may be achieved by modifying environmental conditions in the store so that they become unfavourable for the growth of fungi, insects and mites. This is usually done by drying or chilling the grain, by modifying the intergranular gaseous environment or by chemical treatment.

Microbial colonization of grain commences soon after ear emergence and may continue until the grain is utilized for feed, milling, malting or seed. A major disturbance of the ecosystem occurs as the grain is harvested and placed in store, often after drying, with a change from the environmental conditions occurring in the field to those in store and a dramatic effect on the microflora. During storage, the extent of ecological change is controlled by the nature of the storage system and its management. Change may be gradual or sudden depending on the instability of the storage system and whether the changes are due to equipment failure, insect or mite infestation, degradation of preservatives or poor management.

THE FIELD ENVIRONMENT

Fungi colonizing the aerial surfaces of plants are subject to extremes of environmental conditions. Although leaf temperatures, at the level of maximum leaf area index, are usually greater by day and lower by night than in the ear, the ear is exposed to lower vapour pressures and greater wind speeds and radiation than other parts of the plant, especially by day. Water availability to colonizing microorganisms is determined, at least in part, by the water potential of the underlying tissue while it is still green but, as the grain ripens and dries, it is increasingly controlled by rainfall and atmospheric humidity. The availability of nutrients also changes. Before ripening, surface moisture may contain solutes from plant cells, microorganisms and pollutants. These can either provide nutrients or be inhibitory, and may change the microflora. With ripening and senescence, chlorophyll and proteins are lost, degradative enzyme activity increases and, as membranes lose their integrity, nutrient losses from the cells increase and become more readily available to surface microorganisms.

HARVEST

Harvest marks the transition from the field to the storage environment and has an influence on subsequent microbial development that is seldom taken into account. Apart from perhaps ensuring that the grain is as dry as possible, little thought is given to the effect of harvesting conditions on storage. However, Flannigan (1978) has shown that considerable contamination of the grain with storage fungi, especially *Penicillium* spp. can occur at this stage. After combine harvesting, grain contained 250 times more propagules than grain from the same source threshed by hand, although numbers of *Eurotium* spp. differed little between the two samples. Such contamination may come from fungi growing on residues retained within the harvester and from soil, and more attention paid to the cleanliness of the harvester might decrease this inoculum.

Grain is often dried in hot air before it is placed in store. In maize, and perhaps also in other grains, this can alter the relationship between water availability and water content and lead to greater susceptibility to moulding at a given water content (Tuite & Foster, 1963).

THE STORAGE ENVIRONMENT

Conditions in stored grain are more stable than in the field. Because of the insulating effect of the grain, diurnal temperature changes within the bulk are less than those of the ambient air while humidity of the intergranular air is largely controlled by the availability of water in the grain. Air flows only by convection or by artificial ventilation.

The availability of water in the grain is important in determining not only the vigour of microbial growth but also the species that can grow. It is determined by the water content of the grain and is measured in terms of equilibrium relative humidity (ERH, %), water activity (a_w) or water potential (ψ, MPa). a_w is widely used in storage and food microbiology and ψ in soil microbiology and microbial and plant physiology. There is pressure to use ψ universally as a measure of water availability and in this paper both figures will be given.

Grain temperature changes in response to changes both in the ambient air temperature and in microbial activity. Energy is released by microbial respiration and when it is produced faster than it can escape, spontaneous heating occurs. Heating usually occurs rapidly when grain is first placed in store, reaching a maximum within a few days after which gradual cooling occurs over a few weeks, as water is driven off and nutrient reserves are depleted. Water availability determines the rate of microbial growth and respiration and hence determines the extent of spontaneous heating. Temperatures up to 35°C occur at about 0·7 a_w (−49·1 MPa) to about 65–70°C at 1·0 a_w (0 MPa) (Lacey, 1980). Heating also leads to convection currents in the grain bulk and, if these strike cold air above the grain, to condensation at the surface which allows further moulding.

FIELD AND STORAGE FUNGI

Commonly, fungi have been classified into *field* and *storage* fungi (Christensen, 1974). Field fungi are those that grow on the grain before harvest but which rarely develop further in store. They include species of *Cladosporium, Alternaria, Epicoccum* and the yeast-like *Aureobasidium pullulans*, often with dark hyphae and spores to protect them from harmful ultra-violet radiation. If the grain was not harvested and placed in store, field fungi would continue to develop with the appearance of new species in a typical succession (Hudson, 1965). Because of the different environmental conditions in field and store, the microflora of grain before harvest is quite different from that afterwards. Storage fungi are few on grain before harvest but typically become much more abundant during storage; typical examples of storage fungi are *Aspergillus* and *Penicillium* spp. However, the concept of field and storage fungi was derived in temperate regions and is sometimes difficult to apply in warmer humid climates where, for instance, *Aspergillus flavus* and *Penicillium* spp. may be common on maize and other seeds before harvest (Hill *et al.*, 1985). Also typical field fungi, such as *Fusarium* spp., can sometimes grow in store. Christensen (1974) proposed an intermediate category for the field fungi able to grow in store but storage fungi sometimes found in the field were not included. Indeed, it may be questioned whether it would not be more informative to classify the organisms on their ecological requirements.

The temperature and water requirements of some field and storage fungi are given in Table 1. Typically, field fungi grow within temperature ranges of about 0 to 30–35°C and require at least 0·85 a_w (−22·4 MPa) for germination, 0·88 a_w (−17·6 MPa) for growth and 0·90 a_w (−14·5 MPa) for sporulation. By contrast, storage fungi differ greatly in their requirements. Some *Penicillium* spp. grow between −4 and 35°C and *Aspergillus* spp. between 10 and 55°C while thermophilic fungi, e.g. *Humicola lanuginosa*, grow only between 30 and 60°C. Storage fungi can also tolerate lower a_w than field fungi, with minima for germination ranging between 0·70 and 0·85 a_w (−49·1 to −22·4 MPa) and for sporulation between 0·75 and 0·95 (−39·6 to −7·1 MPa). Usually, the more thermotolerant or thermophilic the species, the greater the water requirement. The intermediate fungi may have requirements close to the borderline between the two groups. For example, *A. flavus* has a minimum a_w for germination of 0·80 (−30·7 MPa) and for sporulation of 0·85 a_w (−22·4 MPa), and a temperature range for growth of 6–8 to

TABLE 1

Minimum Water Activities Allowing Germination, Growth and Sporulation of Field and Storage Fungi and Minimum and Maximum Temperatures for Growth (Magan & Lacey, 1984*a*)

Species	*Minimum a_w allowing:*			*Temperature (°C)*	
	Germination	*Growth*	*Sporulation*	*Min.*	*Max.*
Alternaria alternata	0·85	0·88	0·90	0	34
Cladosporium herbarum	0·85	0·90	0·90	−5	30–32
C. cladosporioides	0·86	0·88	0·90	2	30
Epicoccum nigrum	0·86	0·88	0·90	3	32
Fusarium culmorum	0·87	0·90	0·92	3	36
Aspergillus candidus	0·75	0·80	0·83	10–15	50–55
A. flavus	0·80	0·80	0·85	6–8	42–45
A. fumigatus	0·86	0·94	0·95	10–12	52–55
A. nidulans	0·80	0·80	0·80	6–10	46–51
A. versicolor	0·75	0·78	0·80	5	40
Eurotium repens	0·72	0·75	0·78	5	42
Penicillium aurantiogriseum	0·80	0·83	0·85	−4	35
P. brevicompactum	0·80	0·82	0·85	−2	30
P. piceum	0·79	0·85	0·89	12	48
P. roquefortii	0·83	0·83	0·83	4	35

42–45°C. *Fusarium culmorum* has a minimum for growth of 0·90 a_w (−14·5 MPa) and a temperature range of 3–36°C and occurs chiefly in store in wet grain maintained at a low temperature by ventilation or chilling (Magan & Lacey, 1984*a*).

CONTROL OF THE FIELD ENVIRONMENT

The only practical method of modifying the field environment in order to modify subsequent colonization of the grain by storage fungi is through the use of fungicides. However, they have only transient effects on field fungi, so that populations in fungicide-treated plots have often returned to the levels in untreated plots within 4–6 weeks, while storage fungi are little affected. Thus in one experiment, although *Cladosporium* spp. populations were significantly smaller at harvest in barley grain treated previously with fungicide than in untreated, differences in the microflora after storage at 0·95 a_w (−7·1 MPa) were few. At 5–10°C, populations of *Penicillium aurantiogriseum* were slightly smaller in the fungicide-treated grain, but not significantly so, while at 15°C *Trichothecium roseum* was increased by some fungicides. At 20°C, all treatments were heavily moulded (Hill & Lacey, 1983*a*, *b*; Magan & Lacey, 1986).

CONTROL OF THE STORAGE ENVIRONMENT

The stored grain ecosystem can be modified by changing several abiotic and biotic factors to decrease microbial growth. Drying is the method most frequently used to control the storage environment but temperature can be lowered by chilling or by ventilation with ambient air and the intergranular atmosphere may be modified by restricting aeration or by adding gases.

Water availability

Decreasing the availability of water for microbial growth decreases the number of species that are able to grow while their ability to germinate, grow and sporulate is retarded. Consequently, the degree of drying determines how long grain can be stored without significant colonization and deterioration. As can be seen from Table 1, the most xerophilic storage fungi, chiefly *Eurotium* spp., can grow down to about 0·70 a_w (−49·1 MPa). As a_w is increased, so more species are able to grow and the extent of spontaneous heating increases. These species include a range of

Aspergillus spp. and, at higher a_w, *Penicillium* spp., miscellaneous other fungi and actinomycetes, often including thermophiles. However, fungi are not necessarily most numerous at the temperatures and a_w at which they grow best. Often, the conditions are those at which they survive or compete best with other species. Thus, *Penicillium aurantiogriseum* (optima 25°C and 1·0 a_w, 0 MPa) is most numerous in grain at about 0°C and 1·0 a_w and *Eurotium* spp. (optima 30°C and 0·9 a_w; −14·5 MPa) at 30°C and 0·7 a_w (Hill & Lacey, 1983*b*). Despite this, the predominant species can be good indicators of the conditions in which the grain has been stored (Table 2; see also Lacey, 1986). At low a_w, a small change can have a large effect on the activity of storage fungi. Thus an increase from 14·4 to 15·3% water content (approximately 0·65 to 0·70 a_w, −59·3 to −49·1 MPa) increased populations of *Aspergillus restrictus*, in hard red winter wheat after 36 days' storage at 20–25°C, by 100 times. Over 100 days, water released by metabolism increased the water content of all treatments by 0·2–0·3% and allowed growth of *Eurotium repens* in samples stored with an initial water content of 14·9 and 15·3%. At the same time, the number of surface-sterilized kernels yielding *A. restrictus* declined from 96%, in grain initially at 14·4% which developed no *E. repens*, to 45% in grain at 15·5%, of which 55% yielded *E. repens* (Christensen, 1963). Up to 15% contamination with *E. repens* occurred before there was a significant decline in *A. restrictus* and then only at 14·9% water content or above.

Snow *et al.* (1944) considered that drying to 0·72 a_w (−45·2 MPa) was adequate for storage up to 3 months but that 0·65 a_w (−59·3 MPa) was

TABLE 2
Predominant Microorganisms in Grain Stored at Different Water Activities and Water Potentials with Different Degrees of Spontaneous Heating

Predominant species	a_w	ψ (−MPa)	*Max. temp. (°C)*
Field fungi	< 0·65	< 59·3	ambient
Aspergillus restrictus	0·7–0·8	49·1–30·7	ambient
Eurotium spp.	0·8–0·9	30·7–14·5	35
A. versicolor, Scopulariopsis spp. *Streptomyces griseus*	0·9–0·95	14·5–7·1	40
A. candidus, Penicillium spp.	0·95	7·1	45
A, nidulans, Absidia spp. *Streptomyces albus*	0·98	2·8	50–55
A. fumigatus, Rhizomucor pusillus Humicola lanuginosa, Talaromyces thermophilus, Faenia rectivirgula Thermoactinomyces spp., *Hordeomyces gregorii*	1·00	0	60–65

necessary for longer periods, up to 2–3 years. These a_w are often below those normal in commercial trading and in grading standards (Table 3). For instance, the specified water contents for grades 2 and 3 of soya beans are, respectively, 14·0 and 16·0% and for the same grades of maize grain, 15·5 and 17·5% (Christensen, 1974). These are not safe for storage and can lead to abuses through the mixing of wet grain, that might have deteriorated, with drier grain to achieve the desired water content. Prior invasion of part of the grain with storage fungi may render the whole more liable to deterioration as the fungi can continue to grow at a slower rate after mixing. Moulding may also alter the water content/a_w relationship, making the grain more susceptible to further moulding than indicated by the water content, especially if the grain is not well mixed (Christensen, 1974). The importance of uniform drying at harvest is emphasized by studies of grain samples containing many unripe grains (with 60% water content). Samples containing green grains were more susceptible to heating and moulding than uniformly mature samples with the same mean water content (Hill & Lacey, 1983*b*).

Another phenomenon that can lead to moulding, if not counteracted, is moisture migration, i.e. movement of water vapour from a warm area to a cooler area where it condenses, increasing a_w and the risk of moulding and heating. Temperature differences between the grain and the ambient air may lead to convection currents. If the grain is warmer than the ambient air, these may carry warm moist air to the surface where condensation may occur. Conversely, if ambient temperatures are greater than those in the grain bulk, the circulation is reversed, with warm air rising near the walls and descending at the centre to give condensation within the grain bulk. Warming of one side of a silo by the sun may also cause migration of moisture to the cooler north side. Where

TABLE 3

Water Contents for Safe Storage of Different Grains for Short (3 month) and Long (2–3 year) Periods at 15–20°C (after Snow *et al.*, 1944)

Grain type	*Water content (%) for storage for:*	
	Short periods *(0·72a_w) (−45·2 MPa)*	*Long periods* *(0·65a_w) (−59·3 MPa)*
Wheat	15·7	14·6
Maize	14·8	13·7
Barley	14·8	13·6
Oats	14·5	13·4
Soybeans	13·3	11·4

condensation occurs, spontaneous heating may lead to the formation of a hot spot.

The ecology of both fungal- and insect-induced hot spots has been studied in detail by Wallace and Sinha in the Canadian Prairies (Wallace & Sinha, 1962; Sinha & Wallace, 1965, 1966). Fungal-induced hot spots can develop rapidly in stored grain, even when the temperatures are below freezing, and may be marked by sprouting of the grain at the surface above the hot spot where water has condensed. In naturally occurring hot spots, temperatures ranged from 20 to 45°C when temperatures in unheated grain were as low as −5–0°C. A typical succession of fungi was defined in an artificial hot spot created by inserting 27 kg of grain at 23% water content (about 0·9 a_w, −14·5 MPa) in a 45 cm diameter pocket, with its centre 60 cm below the surface of a bulk of grain with a water content of 14%. Heating was only initiated about 4 months after inserting the wet grain but then within 4 weeks the temperature increased from −6–+8°C to 64°C. It then cooled again to ambient over the following 3 weeks. Immediately prior to the start of heating, numbers of *Penicillium* spp., especially *P. aurantiogriseum*, increased greatly. This was followed successively by the development of *P. funiculosum, Aspergillus flavus, A. versicolor, Absidia* spp. and *Streptomyces* spp. Colonization by different species overlapped but sometimes the disappearance of a species occurred suddenly, as with *A. flavus* when the temperature approached 64°C. *Alternaria* spp., numerous in unheated grain, disappeared rapidly when heating commenced. Heating can occur much more rapidly than is evident from studies of Canadian elevators. With the higher ambient temperatures after harvest in England (15–20°C), damp grain can heat to 60°C within 7–10 days and may remain above ambient for 40–100 days. At high temperatures, large populations of thermophilic fungi and actinomycetes can develop (Hill & Lacey, 1983*b*).

The effects of temperature and a_w on microbial colonization of grain may be achieved not only by the regulation of germination and growth rate but also by modifying competitive ability and by regulating the production of metabolites. The competitive ability of both field and storage fungi changes with temperature, a_w and substrate but results from agar culture appear to be poor indicators of behaviour in grain (Magan & Lacey, 1984*b*, 1985). For example, although *A. versicolor* competed poorly in culture, it was predominant in wheat grain inoculated with a mixture of *Aspergillus* spp. at 0·90 and 0·95 a_w (−14·5 and −7·1 MPa) at 15, 25 and 30°C. *A. candidus* and *A. repens* were also numerous at 15°C although only *A. candidus* competed well in culture. At 25°C, *A. nidulans* and *A. fumigatus* increased in abundance and *A. repens* declined. Neither *A.*

repens nor *A. candidus* could be detected at 30°C but *A. nidulans* was as numerous as *A. versicolor*. In mixtures of *Penicillium* spp., *P. hordei* predominated, except at 30°C where it was replaced by *P. piceum*. *P. aurantiogriseum, P. brevicompactum* and *P. roquefortii* all grew better at 15°C than at higher temperatures.

Multivariate analyses of data on populations of grain storage fungi in Canada suggested that grain storage fungi had a marked detrimental effect on *Alternaria* but little effect on one another (Wallace *et al.*, 1983). It was suggested that this was possibly a consequence of their requirement for different environmental conditions. In other Canadian studies (Wallace & Sinha, 1981), principal component analysis was used to determine the main constituents, any underlying structure and the order in the system, in large bodies of data from random samples. The principal component analysis utilized correlation matrices to identify the variables or principal components that accounted for portions of the total variance and to measure the contribution of each variate to them. As much of the total variance as possible is accounted for by the first principal component, as much as possible of the remainder by the second component and so on. None of the principal components was correlated with another. Water availability was an important variate in each of the first four principal components, that together accounted for 39·5% of the total variability. Temperature also contributed to three of these and storage fungi were major contributors to the first, that accounted for 19·7% variability, and third, that accounted for 6·8% variability. In the first, storage fungi were inversely related to field fungi but in the third they were, surprisingly, inversely related to moisture and temperature, perhaps resulting from interactions with insect and mite infestations (see below).

Production by *E. rubrum* of the enzymes, malate dehydrogenase and isocitrate dehydrogenase, and protein production declined as a_w was increased from 0·93 to 0·99. However, it is not clear whether the decreased protein production was the cause or result of slower growth or whether it was caused by a more rapid turnover of protein (Stevens *et al.*, 1983). Mycotoxin production also changes with a_w and temperature. Greatest production of aflatoxin on irradiated maize occurred at 0·95–0·98 a_w (−7·1 to −2·8 MPa) and 25°C, the highest temperature tested, of zearalenone at 0·98 a_w (−2·8 MPa) and 25°C or at 0·95 a_w (−7·1 MPa) and 16°C while ochratoxin production at 25°C appeared insensitive to a_w between 0·90 and 0·98 a_w (−14·5 and −2·8 MPa). However, the presence of other microorganisms can also affect mycotoxin production. For instance, *Hyphopichia burtonii* caused aflatoxin production by *A. flavus* to almost double at 0·95 a_w (−7·1 MPa) and 25°C while *Bacillus amyloliquefaciens* had a similar effect at 0·98 a_w (−2·8 MPa) and 25°C. At 0·98 a_w and 16°C, although aflatoxin production by *A. flavus* in both single and paired cultures was much smaller, the stimulation by *H. burtonii* and *B. amyloliquefaciens* was proportionately much greater. On irradiated rice, aflatoxin was only produced in paired cultures of *A. flavus* with these two organisms growing alone (see also Smith *et al.*, this volume). Filamentous fungi also stimulated aflatoxin production by *A. flavus* under some conditions of temperature and a_w but suppressed it in others (Cuero *et al.*, 1987*a*, *b*, *c*). Changes in mycotoxin production with temperature, a_w and substrate may partly be a consequence of changes in the competitive ability of storage and field fungi but this does not seem to be the complete explanation.

Temperature

Although decreasing water availability contributes to control of temperature, it can also be controlled actively to decrease the deterioration of damp grain. Air that has been passed over refrigeration coils and then dried by raising the temperature 1–2°C can be blown through to chill the grain bulk. This is particularly useful in countries where ambient temperatures are insufficiently low to give natural chilling. The temperature required to limit fungal deterioration changes markedly with water availability but insect infestations are only likely to occur at temperatures above about 17°C and mite infestations above 3°C with a_w above about 0·4 (−126 MPa) (Christensen, 1974).

Field fungi can survive at low temperatures, and may even grow actively if at high a_w, together with some storage fungi (Christensen, 1974). In particular, *Fusarium*, *Cladosporium* and *Penicillium* can grow well at −6 to −8°C and can sometimes sporulate and produce mycotoxins at these temperatures. As demonstrated in Canada, heating can still occur in such cold grain, although its commencement may be delayed until *Penicillium* populations have reached a critical level, depending on both a_w and temperature. Thus at 16% water content (about 0·82 a_w, −27·3 MPa), temperatures below 15°C may allow mould-free storage for more than 100 weeks while at 18% (about 0·87 a_w, −19·2 MPa), temperatures below 0°C are required which at 20% water content (about 0·90 a_w, −14·5 MPa) allow only 32 weeks' storage. Grain stored at 3·8°C all showed visible moulding at 26–29% water content (about 0·96–0·98 a_w, −5·6–2.8 MPa) after 180–225 days and at 24% (about 0·94 a_w, −8·5 MPa) aftr 330 days while only 25% were affected at 22% (about 0·93 a_w, −10·0 MPa) and 5% at 20% (about 0·90 a_w, −14·5 MPa) after 420 days storage. Once mould growth had started, the rate of increase could be comparatively rapid, with increases from 20 to 100% of grains moulded within 50 days, after taking 130–180 days for 20% of the grain to mould (Christensen, 1974).

Ventilation

Low rate ventilation with ambient air may be used in cool climates to create uniform conditions throughout a grain bulk by removing respiratory heat, CO_2 and water vapour, so preventing moisture migration. The object is to cool the grain below 10°C so that mould development is slowed. Thus, grain containing 18% water (about 0·87 a_w, −19·2 MPa), cooled to 5–10°C, will keep for up to 8 months in Britain. In warmer climates, drying to 12–13% (about 0·52–0·60 a_w, −90·0 to −70·3 MPa) before ventilation may be necessary to prevent fungal growth and care may be necessary to prevent the addition of water to the grain during ventilation. This may be achieved by manual or automatic control of the fan's operation so that grain is not ventilated in humid weather or by ensuring that the ventilation system warms the air by 1–2° to decrease its humidity (Christensen, 1974). However, even in cool climates musty odours may still occur during ventilation, suggesting that some moulding has occurred. Studies of moulding during ventilation of grain have invariably assessed visible moulding rather than total fungal development, so that it is possible that growth could still occur within damp grain but sporulation be inhibited by air movement or unfavourable humidities.

Intergranular gas composition

The fungi that cause moulding of grain are usually considered obligate aerobes but the partial pressure of O_2 necessary for growth is often overestimated. With many fungi, it is necessary to decrease the concentration of O_2 to $< 0{\cdot}14\%$ to decrease linear growth by half. However, species differ much in their tolerance of both low O_2 and high CO_2 concentrations. Both interact with one another so that, for instance, a fungus may be more sensitive to high CO_2 when O_2 concentrations are low, than when they are close to ambient levels. Fungal responses are also affected by temperature and a_w. Growth is sometimes stimulated by concentrations of 5–10% CO_2, especially at high a_w and often >15% CO_2 is required to halve growth rates. Indeed, *P. roquefortii* is reported to grow with up to 75% CO_2 although it appears more sensitive to high CO_2 at 0·90 a_w (−14·5 MPa) than at 0·95–0·98 a_w (−7·1 to −2·8 MPa) (Goulding, 1940; Magan & Lacey, 1984*c*).

The sensitivity of fungi to high CO_2 and low O_2 concentrations is utilized in the storage of damp grain, especially barley and maize for animal feed, in sealed and unsealed silos (Burmeister *et al.*, 1965; Messer *et al.*, 1965; Lacey, 1971; Clarke & Hill, 1981) and also in controlled-atmosphere storage of grain (Shejbal, 1980). Sealed storage of barley grain with 19–22% water content (about 0·89–0·93 a_w, −16·0 to −10·0 MPa) and unsealed storage, usually of grain with 20–40% water (about 0·9–1·0 a_w, −14·5 to 0 MPa), both rely on the replacement of O_2 by CO_2 through the respiration of the grain and its associated microflora. This can give peak CO_2 concentrations up to 60–85% and minimum O_2 concentrations of < 1% within the first 7 days of storage. Artificially controlled atmospheres are achieved by adding CO_2 or N_2 gases to grain in sealed bins giving more rapid anaerobic conditions than would otherwise be possible with dry grain but the regimes proposed have usually been designed for insect control and are often insufficient to prevent fungal colonization.

Control of moulding depends on the maintenance of suitable modified atmospheres throughout the storage period. However, N_2 treatment of grain above 14·5% water content (about 0·75 a_w, −39·6 MPa) in Italy was limited by mould growth. The presence of only 0·2% O_2 in the atmosphere allowed slow moulding which was only prevented if O_2 was totally excluded from the atmosphere (Shejbal, 1980). In Australia, the recommendations for wheat grain with 12% water content (about 0·5 a_w, −90·0 MPa) are >70% CO_2 initially, with >35% for 10 days, or <1%O_2 in N_2 for at least 6 weeks. In practice, welded steel bins mostly lost 3–6% CO_2 per day, from an initial concentration of 60–75%, while concrete bins lost 10–20% per day. The storage periods in both types of bin extended up to 12 months although, at the end of this period, some insect infestations were found (Shejbal, 1980). Fungal contamination was not assessed.

Without added gases, CO_2 concentrations in hermatically sealed welded steel silos could reach 95% with wheat of 23% water content (about 0·93 a_w, −10·0 MPa) or 30% with 17% water content (about 0·85 a_w, −22·4 MPa), then remaining constant for up to 100 days. However in commercial practice, with bolted metal silos, the maximum concentration attained was usually less than 50% after 30 days with grain at 20–22% water content (about 0·90–0·92a_w, −14·5 to 11·5 MPa), declining to about 15% CO_2 after 200 days. The losses are due to air entering the silos in response to diurnal temperature fluctuations and to removal of grain from the silos. When the silos are nearly empty, respiratory activity is insufficient to maintain the high CO_2/low O_2 atmosphere, and microbial development with spontaneous heating may result (Christensen, 1974). In unsealed silos, the initially high CO_2 concentration declines to about

15% after 100 days storage and O_2 concentrations increased to 4·5% in the same period (Messer *et al.*, 1965). Although this was sufficient to limit microbial development to bacteria and to prevent moulding in the bulk of the grain, air exchange occurred at the exposed surface with moulding and heating (Lacey, 1971).

The nature and extent of moulding in the unsealed silos depended on grain water content, the type and thickness of the top seal during the initial storage period and, subsequently, on the rate at which grain was removed from the exposed surface. Moulding could occur to 30 cm depth during storage but was less under wilted grass that ensiled than under straw or polyethylene sheet either alone or together. Moulding could be heavy, with evidence of spontaneous heating and colonization by *A. flavus*, thermophilic fungi and actinomycetes (Table 2). Heating was controlled during the unloading period by the removal of at least 7·5 cm daily from the exposed surface of the grain. As rate of use declined below 7·5 cm/day, a characteristic succession of microorganisms developed with increasing heating. First colonizers, with temperatures remaining close to ambient, were yeasts, especially *Hyphopichia burtonii* and *Hansenula anomala*, followed by *Penicillium roquefortii*. With heating to 25–30°C, these were replaced by *P. rugulosum*, *Eurotium* spp. and *Aspergillus candidus*. At 30–35°C, *Absidia corymbifera* and *Rhizomucor pusillus* predominated. *Aspergillus fumigatus* appeared at 40°C and continued to increase with increasing temperature. The thermophilic fungus, *Humicola lanuginosa*, and actinomycetes, *Faenia rectivirgula, Hordeomyces gregorii* and *Thermoactinomyces* spp., developed to give maximum numbers at 50–60°C, sometimes in discrete zones in the grain (Lacey, 1971). A similar succession has been found in sealed silos as the O_2 concentration increases in spring (Clarke & Hill, 1981).

Chemicals

Volatile fatty acids have been used for the preservation of grains for animal feeds and a large number of other chemicals have been tested for this purpose. Mostly, the chemicals have been based on propionic acid, although formic and sorbic acids have also been used, and they have been fungistatic rather than fungicidal. Storage fungi differ in their susceptibility to inhibition by propionic acid. *Paecilomyces variotii* and *Eurotium* spp. can tolerate concentrations up to 1% in agar media, when other species are inhibited by $< 0·2\%$, and can also degrade the propionate. Spore germination appears more susceptible to inhibition than mycelial growth. Consequently, application of the chemical must be uniform and adequate for the water content if control is to be maintained. Untreated or undertreated pockets can lead to the development first of tolerant species that metabolize the acid allowing subsequent development of less tolerant species and a succession characteristic of untreated grain at the same water content (Lord *et al.*, 1981; Lacey *et al.*, 1983). Sometimes *A. flavus* has been a major colonizer of grain treated inadequately with propionic acid and has formed aflatoxins (Hacking & Biggs, 1979), perhaps stimulated by the propionic acid (Al Hilli & Smith, 1979) or by other microorganisms colonizing the grain (Cuero *et al.*, 1987*a*, *c*).

Insects and mites

Insects and mites are important in their own right as agents of deterioration of cereal grains but they are also important because of their interactions with fungi and their control contributes to the control of microorganisms. Insects require at least 17°C to infest grain but their occurrence is independent of a_w in the range usually found in stored grains. Mite infestations can occur between 3 and 30°C and above 12% water content (about 0·52 a_w, −90·0 MPa). The metabolic activity of insects and mites can lead to spontaneous heating to 40°C, even in grain too dry for fungal heating. Their respiration increases the a_w, so favouring fungal activity, and their faecal material may provide a substrate for growth. Fungi may provide food for insects and mites but can also be pathogenic to some of them.

Interactions between microbial colonization and insects infesting stored grain can take several forms, as illustrated by the multivariate study of Canadian wheat stores referred to above (Wallace & Sinha, 1981). The second principal component, accounting for 7·9% of the variation, referred only to arthropods, showing that they were more affected by temperature than by water content and that fumigation also has a marked effect. Component III (6·8% of variation) included both mite and fungal variables. Increasing temperature and, to a lesser extent, water content favoured mites but there was an unexpected inverse relationship of fungi with these factors. Two alternative explanations were given. Either the fungi in this component were superficial contaminants on the grain that grew and sporulated only on the isolation plates or *Acarus soir* fed and bred on *Penicillium*, especially *P. aurantiogriseum*, while *A. gracilis* fed on both *Penicillium* and *A. versicolor*. Other evidence of mites feeding on fungi in stored grain comes from English studies although the possibility of an inhibitory secretion was also suggested. *Eurotium* spp. were fewer in grain infested with *Acarus siro, Lepidoglyphus destructor* and *Tyrophagus longior* while *Penicillium* spp. were fewer only in the presence of *A. siro* (Armitage & George, 1986).

Component VI (4·3% of variation) includes two genera of fungivorous beetles, *Cryptophagus* and *Ahasverus*, and the mites, *Lepidoglyphus*, *Tydeus* and *Tarsonemus* that scavenge decomposed microorganisms and grain debris. Components IX, X and XI, each accounting for about 3% of variation, all included *Rhizopus* with the beetles *Cryptolestes*, *Cryptophagus* and *Ahasverus* either alone or together. These components indicated that *Cryptophagus* and *Ahasverus* deterred *Rhizopus*, perhaps through feeding on it, but that *Cryptolestes* favoured it by damaging the grain embryo, facilitating invasion. However, when all three insects were present, their roles were reversed although the reasons were unclear. It was also evident that the beetles favoured *Streptomyces*, probably because their excreta form a good substrate. Armitage & George (1986) also found increases in the populations of some fungi in the presence of mites on which they were pathogenic. Thus, *A. restrictus* was more numerous in the presence of *L. destructor* and *Wallemia sebi* with both *A. Siro* and *L. destructor*.

Less evident in these studies was the role of insects in initiating hot spots in grain drier than 14% (about 0·65 a_w, −59·3 MPa. For instance, *Oryzaephilus surinamensis* can initiate heating in grain of 12–14% water content (about 0·5–0·65 a_w, −90 to −59·3 MPa), giving heating to 25–35°C, before being replaced by *Cryptolestes ferrugineus*, which is itself killed when the temperature reaches near 40°C. By their respiration, the insects increase the water content of the grain by up to 3% giving conditions that favour the growth of *Penicillium* and *Streptomyces* (Sinha & Wallace, 1966).

There have been many studies on the growth and reproduction of insects and mites on diets consisting only of fungi, mostly by Sinha and his colleagues in Winnipeg. These demonstrate that many fungi provide adequate nutrition for a wide range of grain storage arthropods. Some examples are shown in Table 4. In this study (Sinha, 1968), field fungi such as *Alternaria alternata* and *Cladosporium cladosporioides* were most favourable for growth and reproduction while some storage fungi, e.g. *A. fumigatus, A. niger* as well as *Streptomyces griseus* supported very few of the species tested. However, the beetles, *C. ferrugineus* and *O. surinamensis* could grow on 10 of the 24 species of fungi tested and *Tribolium castaneum* and *T. confusum* on eight and seven, respectively. Of the mites, *A. siro* grew on 17 and *L. destructor* on 12. Other studies have shown that insects can be inhibited by fungi with, for instance, *A. candidus, A. flavus, A. niger* and *A. ochraceus* inhibiting egg deposition by *Ahasverus advena* while it shows a preference for *P. citrinum* and *Eurotium amstelodami* (Bulla *et al.*, 1978). *Penicillium* spp., and sometimes even isolates of the same species, differ in their attractiveness to *T. confusum* and in their ability to support growth and reproduction. *P. viridicatum, P. citrinum, P. chrysogenum* and *P. aurantiogriseum* (*P. cyclopium*) were attractive to *T. confusum*, while *P. crustosum*, *P. urticae* and *P. oxalicum* were neutral and *P. aurantiogriseum* (*P. martensii*), another isolate of *P. citrinum* and *P. purpurogenum* were repellent. Most neutral species and even some of those that were attractive to *T. confusum* inhibited larval development while *P. purpurogenum* caused high mortality of adult beetles (Wright *et al.*, 1980*a, b*). Some of these effects could be caused by mycotoxins such as patulin and citrinin although only small amounts were detected. In other experiments, T-2 toxin at 100 ppm increased the fecundity of *T. confusum* but also decreased the hatch of eggs by 10–12%, retarded larval development and caused some mortality of larvae. Zearalenone had little effect apart from prolonging the reproductive life of *T. confusum* (Wright *et al.*, 1976).

TABLE 4

Ability of Fungi to Support Growth and Reproduction of Grain Storage Insects and Mites

Species	*Fungi*																
	Field						*Storage*										
	1	*2*	*3*	*4*	*5*	*6*	*7*	*8*	*9*	*10*	*11*	*12*	*13*	*14*	*15*	*16*	*17*
Insects																	
Cryptolestes ferrugineus	+		+	+	+	+				+	+					+	
Oryzaephilus surinamensis		+	+		+	+				+	+			+	+	+	
Tribolium castaneum		+	+	+	+					+					+		
T. confusum	+	+	+		+	+				+							
Mites																	
Acarus siro	+	+	+		+	+	+	+	+	+	+		+	+	+	+	
A. gracilis	+	+	+		+	+	+		+	+	+		+		+	+	
Glycyphagus destructor	+	+	+	+	+	+				+	+	+			+		
Caloglyphus redikorzevi	+	+	+	+	+	+	+	+	+	+	+	+		+	+	+	+

Only *Tyrophagus casei* and *T. putrescentiae* grew on *Aspergillus fumigatus* and *A. niger*.
Key to column headings: 1, *Alternaria alternata*; 2, *Cladosporium cladosporioides*; 3, *Curvularia tetramera*; 4, *Helminthosporium sativum*; 5, *Nigrospora sphaerica*; 6, *Stemphylium botryosum*; 7, *Aspergillus flavus*; 8, *A. ochraceus*; 9, *A. versicolor*; 10, *Mucor sphaerosporus*; 11, *Penicillium aurantiogriseum*; 12, *P. funiculosum*; 13, *P. terrestre*; 14, *Rhizopus arrhizus*; 15, *Scopulariopsis brevicaulis*; 16, *Trichothecium roseum*; 17, *Streptomyces griseus*.

Other factors

The stored grain ecosystem can also be modified by controlling the content of broken grains and foreign material that enters the store or by irradiation of the grain. Broken grains and foreign material often separate into discrete zones when the store is filled, interfering with ventilation, perhaps differing in a_w from the grain and providing a focus for fungal and insect infestation. Damaged grain is more susceptible to fungal colonization and may also show more visible moulding. In barley, at least, *Penicillium* spp. are favoured by grain damage more than *Aspergillus* spp. With maize, germ damage increases susceptibility to moulding more than damage to other parts of the grain (Lacey, 1986).

Irradiation can be used to sterilize grain before storage but large doses are necessary. *Fusarium*, yeasts and *Bacillus* are particularly resistant to irradiation, requiring up to 1200 Krad to ensure sterility. However, *Aspergillus* and *Penicillium* are killed by 120 Krad and Mucoraceae by 600 Krad (Cuero *et al.*, 1986).

CONCLUSION: PROSPECTS FOR INTEGRATED CONTROL

Fungal colonization of grain can be modified and controlled by modifying the storage ecosystem in a variety of ways. However, each of these seeks to maintain an unstable state that could easily change as a result of changes in environmental conditions, deficiencies in management or of equipment failures. Some methods of controlling infestation and moulding are expensive, impractical, or difficult with a given storage system. Large quantities of expensive fungistatic chemicals are required to prevent moulding. For controlled-atmosphere storage, sealed stores are necessary to maintain large concentrations of CO_2 or N_2 and with the latter, at least, complete removal of O_2. While irradiation requires impractically large doses for complete sterility. Almost all investigations of the control of the grain ecosystem have concentrated on one aspect, neglecting the possibility of interactions between different factors and possible synergism. There is a need to investigate the possible use of integrated methods of control that rely on the modification of several parameters to obtain savings in costs without diminishing the efficiency of control. Possible combinations could include the use of low levels of irradiation, small CO_2 concentrations and small applications of fungistatic and/or insecticidal chemicals. Experimental work is necessary to define suitable combinations but it is possible that such an integrated approach could produce a more stable ecosystem at economic cost.

REFERENCES

Al-Hilli, A.L. & Smith, J.E. (1979). *FEMS Microbiology Letters*, **6**, 367–70.
Armitage, D.M. & George, C.L. (1986). *Experimental & Applied Acarology*, **2**, 111–24.
Bulla, L.A., Kramer, K.J. & Speirs, R.D. (1978). *Advances in Cereal Science and Technology*, **2**, 91–133.
Burmeister, H.R., Hartman, P.A. & Saul, R.A. (1965). *Applied Microbiology*, **14**, 31–4.
Christensen, C.M. (1963). *Cereal Chemistry*, **40**, 385–90.
Christensen, C.M. (Ed.) 1974). *Storage of Cereal Grain and its Products.* American Association of Cereal Chemists, St. Paul.
Clarke, J.H. & Hill, S.T. (1981). *Transactions of the British Mycological Society*, **77**, 557–65.
Cuero, R.G., Smith, J.E. & Lacey, J. (1986). *Food Microbiology*, **3**, 107–13.
Cuero, R.G., Smith, J.E. & Lacey, J. (1987*a*). *Applied and Environmental Microbiology*, **53**, 1142–6.
Cuero, R.G., Smith, J.E. & Lacey, J. (1987*b*). *Transactions of the British Mycological Society*, **89**, 221–6.
Cuero, R.G., Smith, J.E. & Lacey, J. (1987*c*). *Journal of Food Protection* (in press).
Flannigan, B. (1978). *Transactions of the British Mycological Society*, **71**, 37–42.
Golding, N.S. (1940). *Journal of Dairy Science*, **23**, 891–8.
Hacking, A. & Biggs, N.R. (1979). *Nature, London*, **282**, 128.
Hill, R.A. & Lacey, J. (1983*a*). *Annals of Applied Biology*, **102**, 455–65.
Hill, R.A. & Lacey, J. (1983*b*). *Annals of Applied Biology*, **102**, 467–83.
Hill, R.A., Wilson, D.M., McMillan, W.W., Cole, R.J., Sanders, T.H. & Blankenship, P.D. (1985). In *Trichothecenes and other Mycotoxins*, ed. J. Lacey. John Wiley, Chichester, pp. 79–95.
Hudson, H.J. (1965). *New Phytologist*, **67**, 837–74.
Lacey, J. (1971). *Annals of Applied Biology*, **69**, 187–212.
Lacey, J. (1980). In *Microbial Growth in Extreme Environments*, ed. J.A.S. Corry & R.G. Board, Society of Applied Bacteriology Technical Series No. 15. Academic Press, London, pp. 53–70.
Lacey, J. (1986). In *Spoilage and Mycotoxins of Cereals and other Stored Products*, ed. B. Flannigan. CAB International, Farnham Royal, pp. 29–34.
Lacey, J., Lord, K.A., Cayley, G.R., Holden, M.R. & Sneath, R.W. (1983). *Animal Feed Science and Technology*, **8**, 283–301.
Lord, K.A., Lacey, J. & Cayley, G.R. (1981). *Transactions of the British Mycological Society*, **77**, 41–5.
Magan, N. & Lacey, J. (1984*a*). *Transactions of the British Mycological Society*, **82**, 71–81.
Magan, N. & Lacey, J. (1984*b*). *Transactions of the British Mycological Society*, **82**, 83–93.
Magan, N. & Lacey, J. (1984*c*). *Transactions of the British Mycological Society*, **82**, 305–14.
Magan, N. & Lacey, J. (1985). *Transactions of the British Mycological Society*, **85**, 29–37.
Magan, N. & Lacey, J. (1986). *Annals of Applied Biology*, **109**, 117–28.

Messer, H.J.M., Hill, J.M., Whittenbury, R. & Lacey, J. (1965). The use of concrete staved silos for storing high-moisture barley. Agricultural Research Council Experimental Buildings Report No. 4.

Shejbal, J. (Ed.) (1980). *Controlled Atmosphere Storage of Grains*. Elsevier, Amsterdam.

Sinha, R.N. (1968). *Evolution*, **22**, 785–98.

Sinha, R.N. (1973). *Annales de Technologie Agricole*, **22**, 351–69.

Sinha, R.N. (1979). *Annales de Technologie Agricole*, **28**, 191–209.

Sinha, R.N. & Wallace, H.A.H. (1965). *Canadian Journal of Plant Science*, **45**, 48–59.

Sinha, R.N. & Wallace, H.A.H. (1966). *Researches in Population Ecology*, **8**, 107–32.

Smith, J.E., Cuero, R.G. & Lacey, J. This volume, pp. 603–7.

Snow, D., Crichton, M.H.G. & Wright, N.C. (1944). *Annals of Applied Biology*, **31**, 111–16.

Stevens, L., Dix, N.J. & Thompstone, A. (1983). *Transactions of the British Mycological Society*, **80**, 527–9.

Tuite, J. & Foster, G.H. (1963). *Cereal Chemistry*, **40**, 630–7.

Wallace, H.A.H. & Sinha, R.N. (1962). *Canadian Journal of Plant Science*, **42**, 130–41.

Wallace, H.A.H. & Sinha, R.N. (1981). In *The Fungal Community: its Organization and Role in the Ecosystem*, ed. D.T. Wicklow & G.C. Carroll. Marcel Dekker, New York, pp. 233–47.

Wallace, H.A.H., Sholberg, P.L., Sinha, R.N. & Muir, W.E. (1983). *Mycopathologia*, **82**, 65–76.

Wright, V.F., De las Casas, E. & Harein, P.K. (1976). *Environmental Entomology*, **5**, 371–4.

Wright, V.F., De las Casas, E. & Harein, P.K. (1980*a*). *Environmental Entomology*, **9**, 204–12.

Wright, V.F., Harein, P.K. & Collins, N.A. (1980*b*). *Environmental Entomology*, **9**, 213–16.

Spices and Mycotoxins—a Review of Biodeterioration and Health Implications

G.C. Llewellyn, V.L. O'Dell

Bureau of Toxic Substances, Virginia Department of Health, Richmond, Virginia 23219, USA

&

B. Flannigan

Department of Brewing and Biological Sciences, Heriot-Watt University, Edinburgh, EH1 1HX, UK

ABSTRACT

This paper updates an earlier review and is concerned primarily with mycotoxicological aspects relevant to spices. Fungal counts in excess of 10^6 propagules/g continue to be reported for various spices, but gamma radiation has proved to be effective in destroying both bacteria and fungi in black pepper, cardamom, cinnamon, cloves and nutmeg. In India, aflatoxin contamination is highest at the level of the grower and the trader with poor drying, storage and handling being the main contributory factors. Heat treatment has proved more effective than light in detoxifying aflatoxin-contaminated black cumin and coriander. Experimental studies suggest that black pepper, cardamom and turmeric and poor substrates for fungal growth and aflatoxin production, but ginger and red pepper appear to be high risk commodities. Other experiments have demonstrated that cinnamaldehyde in cinnamon and eugenol in cloves are strongly antifungal. Ginger and some herbs and other flavouring materials have been shown to inhibit growth and/or aflatoxin production also, and addition of cinnamon or pepper at relatively high rates has been found to control fungal invasion in crushed maize stored at high moisture contents.

INTRODUCTION

The longstanding use of spices for flavourings, carminatives, preservatives and antioxidants, and for improved digestibility of food, has resulted in international trade which is worth more than $365 000 000 annually (Pruthi, 1980; Madhyastha & Bhat, 1985). Spices may therefore play an important part in the national economy of the countries producing and exporting them, and to a lesser degree in those importing them (Pruthi, 1980). Large quantities of spices are consumed locally and are not processed for use outside the region or country of origin. Often, material that is sold for export is of better quality than that retained for local consumption. Uptake of moisture from the environment may encourage mould growth and subsequent toxin production in such spices. The probability of mycotoxin development appears also to be increased during transportation and long-term storage may also contribute to such development.

Bearing these points in mind, the recent literature is reviewed with particular emphasis on mycotoxicological aspects, including the anti-aflatoxigenic potential (Llewellyn *et al.*, 1981, 1982). This updates an earlier review of the literature up to 1982 (Flannigan & Llewellyn, 1986) which dealt with bacterial as well as the fungal flora of spices and in which the emphasis was primarily on biodeterioration, although mycotoxicological aspects were also considered. The previous review (Flannigan & Llewellyn, 1986) excluded herbs, generally defined as the fresh or dried leafy tops of soft-stemmed plants and not the other dried aromatic parts (rhizomes, bark, flower buds, fruits, seeds and arils) known as spices, and other vegetable products used for flavouring (garlic, horseradish, vanilla pods, etc.). However, the term 'spice' is often used to cover all three types of flavouring materials and it is in this broader sense that the term is used in the present review.

MICROBIAL CONTAMINATION AND ITS CONTROL

The level of microbial contamination of spices and measures to reduce numbers of contaminants continue to receive attention, particularly in India. The effects of environmental conditions during storage and retailing on the microbiological status of unpackaged cumin seeds and chilli powder has been examined by Bhat *et al.* (1987). They found high levels of bacterial and fungal contaminants in all samples of chilli powder taken from 18 shops in Bombay. Aflatoxigenic strains of *Aspergillus flavus* and *A. parasiticus* were detected in 88% of these samples.

In contrast, fungi could not be detected in 98% of cumin seed samples from the same premises, and where they were present the counts were much lower than in chilli. The findings of these workers indicate the antimicrobial effects of the spice as well as a definite correlation between environmental hygiene and the microbial burden. Another survey of the microbiological quality of whole black pepper, probably the most important spice in world trade (Rosegarten, 1981), and turmeric powder showed that there were high levels of contamination in samples of both spices for sale in Bombay (Geeta & Kulkarni, 1987). Mould counts ranged from $0{\cdot}6 \times 10^4$ to $1{\cdot}6 \times 10^6$/g of black pepper and from $0{\cdot}5 \times 10^3$ to $1{\cdot}1 \times 10^6$ for turmeric. The principal moulds were *Aspergillus* spp.: *A. niger*, *A. flavus* and *A. parasiticus* were identified in all samples of both spices. However, the counts did not show any definite correlation with the environment, possibly because most fungal contamination occurs before the spices reach the retail shops, with the fungal spores remaining dormant unless conditions are favourable for germination and growth.

Sharma *et al.* (1984) have examined gamma radiation as a means of decontaminating some common Indian spices, viz. black pepper, cardamom, cinnamon, cloves and nutmeg. Black pepper had the greatest microbial contamination and cloves the lowest counts of both bacteria and fungi. Radiation doses of 7·5 kGy were sufficient to kill all bacteria in cinnamon and cloves, but 10 kGy was required with black pepper, cardamom and nutmeg. All fungal spores in the spices were killed by a dose of 5 kGy.

TOXIGENIC FUNGI AND MYCOTOXINS

Since the World Health Organization (1976) published a report on 'Microbiological Aspects of Food Hygiene', which provided a series of microbiological safety specifications for foods and a system for laboratory quality assessment, Kurata & Hitokoto (1985) have proposed a modified system for the evaluation of quality. This is based on the presence of certain sets of mycological indicators. Kurata & Hitokoto (1985) support the microbiological evaluation of food as a means of predicting further deterioration by fungi and mycotoxin contamination of marketed foods. They examined a total of nearly 6500 samples of 26 categories of food from normal marketing channels. Spices were ranked either in the middle or towards the higher risk end of the list evaluated as potential hazards to health. Their assessment was based upon the fungal flora, including species of *Aspergillus, Penicillium* and *Fusarium*, and the water activity of the foods. Five groups of fungi were selected as an indicator set for biodeterioration and mycotoxin risk, and a three-indicator set of *A. flavus* (*sensu lata*), *A. ochraceus* and *A. versicolor* for risk from mycotoxigenesis. These organisms Kurata & Hitokoto (1985) consider to be extremely important toxigenic species, and they suggest that the test for these fungi is suitable for risk assessment in spices and other foods.

The most important mycotoxins likely to be encountered in spices are, of course, the aflatoxins (Flannigan & Hui, 1976) produced by the related species *A. flavus* and *A. parasiticus*, which regularly grow on a variety of agricultural commodities under favourable environmental conditions. Of the four parental aflatoxins in the known series of these secondary metabolites aflatoxin B_1 (AFB_1) is reported to be the most hepatocarcinogenic (Butler, 1965). Cell and tissue damage (Golumbic & Kulik, 1969) result primarily from its toxicity and carcinogenicity but it is also mutagenic (Maher & Summers, 1970) and teratogenic (Dipaolo *et al.*, 1967). Aflatoxins have been reported to be involved in the aetiology of human liver cancer in certain areas of the world (Denizel & Kosker, 1972; Karapinar, 1986), and aflatoxicosis has also occurred where man has been exposed to substantial quantities of the mycotoxins in his diet (Deger, 1976; Jewers, 1982). The widespread use of spices makes it important, therefore, to control contamination by aflatoxigenic and other mycotoxin-producing fungi.

The Indian Council of Medical Research (1985) has concluded that the risk of aflatoxin contamination of spices is highest at the level of the grower and trader. The observed methods of drying and storing spices by the small grower and the insanitary methods of handling spices by many small traders appear to be the cause of much aflatoxin contamination. In a survey to determine the dietary intake of aflatoxin via the major spices, it was found that the intake of chilli was higher than that of any other spice and that spice consumption in general was greater in rural India than in the urban areas. The actual ingestion of aflatoxin in spices was very low in relation to that taken in with staples such as cereals. Nevertheless, efforts are still being directed to reducing the amounts of aflatoxins in spices. Bilgrami *et al.* (1984) have investigated the effects of light and heat on levels in coriander and black cumin. These spices have previously been noted as good substrates for aflatoxin production and were subjected to sunlight, fluorescent tube light, tungsten bulb light, dry heat and autoclaving. In general, heat was more effective in detoxification than light. Autoclaving for 1 h at 121°C reduced aflatoxin levels by more than 50%.

ANTIFUNGAL ACTIVITY OF SPICES

In examining the growth of *A. parasiticus* on five types of spice, Madhyastha & Bhat (1985) observed that growth and aflatoxin production were well supported by ginger and red pepper (Table 1). However, black pepper (whole and ground), cardamom and turmeric were poor substrates for growth and toxigenesis. Ether and chloroform extracts of cardamom entirely inhibited growth and toxin production, whilst those of turmeric were less inhibitory to growth than to toxin production, which was affected only moderately. When the effects of the oils were examined, cardamom had a dose-dependent inhibitory effect on both functions, but although high concentrations of turmeric inhibited toxin production almost entirely some lower concentrations stimulated growth of the mould. Sharma *et al.* (1984) extended their previously mentioned study of the effects of gamma radiation on the microflora of spices to encompass the effects of radiation on the antimicrobial properties of spices themselves (Table 1). In cloves, the component which is inhibitory to growth and aflatoxin production in *A. parasiticus*, i.e. eugenol, was apparently unaffected by exposure to 10 kGy irradiation. Extracts of irradiated seeds still inhibited growth of *A. parasiticus*, and when flasks containing irradiated cloves were inoculated with the organism there was no visible growth even after 1 month. There was no growth either in corresponding flasks containing irradiated cinnamon, but the efficacy of the inhibitory compound in extracts of cinnamon, viz. cinnamaldehyde, was reduced by approximately 45–62% in various irradiated samples.

Tiwari *et al.* (1983) also studied the effects of essential oils on *A. parasiticus* (Table 1). In addition to anise, eucalyptus, orange bitter, rosemary, sandalwood and thymol (present in oil of thyme), the oils of cinnamon, ginger and nutmeg all decreased production of aflatoxin, which in controls was present in greatest amounts at 7 days. Anise had the greatest inhibitory effect and nutmeg, rosemary, eucalyptus and sandalwood oils the least effect on toxin production, but cinnamon, ginger, orange bitter and thymol reduced production to less than 25% of controls. When Karapinar (1986) examined growth and aflatoxin production of *A. parasiticus* in liquid medium they found that when the medium was amended by addition by both thyme or either orange or lemon oil, growth was inhibited. The organism grew well in the presence of aniseed, bay leaves, mint or red pepper. Since esters of sucrose, which are approved food additives for use as emulsifiers, are also known to inhibit the growth of some bacteria and moulds, Marshall & Bullerman (1986) tested combinations of ester with cinnamon and other mould inhibitors for additive or synergistic effects (Table 1). At concentrations

TABLE 1

A Summary Evaluating the Influence of Various Spices on Mould Growth and Aflatoxin Production

Species	*Inhibition of fungal growth and aflatoxin production*[a]	*Related observations*	*References*
Black pepper	+++		Madhyastha & Bhat (1985)
Cardamom	+++		
Ginger	0		
Red pepper	0		
Tumeric (curcumin)			
Aniseed	0		Karapinar (1986)
Bay leaves	0		
Mint	0		
Red pepper	0		
Thyme	+++		
Anise	++++		Tiwari *et al.* (1983)
Cinnamon	+++	Less than 25% of control value at peak	
Eucalyptus	+		
Ginger	+++	Less than 25% of control value at peak	
Nutmeg	+		
Orange bitter	+++	Less than 25% of control value at peak	
Rosemary	+		
Sandalwood	+		
Thymol	+++	Less than 25% of control value at peak	
Cinnamon (0·1% at pH 4)	−−−		Marshall & Bullerman (1986)
Cinnamon (0·1% at pH 6)	+++		
Cinnamon (1·0% at pH 4)	+++		
Cinnamon (1·0% at pH 6)	+++		
Cinnamon and sucrose ester	+++	At pH 4 and pH 6	
Black pepper	+++	At moisture levels of 16·61 and 20·04%, if added at sufficient rates: cinnamon at 1·5%, black pepper at 2·5%	Abdelhamid *et al.* (1985)
Cinnamon	+++		

[a] ++++, substantial inhibition; +++, moderate inhibition; ++, low levels of inhibition; +, minimal inhibition; 0, no inhibition; −−−, stimulated growth.

of 0·1%, sucrose ester significantly inhibited growth of *A. parasiticus* at pH 4 and 6, while cinnamon inhibited at pH 6 and stimulated at pH 4. When the ester and the spice were combined, there was no increase relative to the effect of the ester alone. However, when the concentration of cinnamon was increased to 1·0%, complete inhibition of growth was achieved at both pH levels by the spice, whether alone or in combination with 1·0% sucrose ester. The effect of cinnamon on growth and toxin production has also been studied in crushed maize stored at different moisture levels (Abdelhamid *et al.*, 1985). Even at high moisture levels (Table 1), both this spice and black pepper were found to inhibit fungal growth if added at sufficient rates, i.e. 1·5% for cinnamon and 2·5% for black pepper.

CONCLUSION

This short, updated review of the literature has revealed in the first instance that there is a continuing problem in relation to mycotoxins, which is posed by contamination of spices and spice-containing foods. Secondly, substantial strides have been made in evaluating the mycotoxigenic potential of certain spices via various fungal indicator sets. It is evident that if a quality assurance system employing such indicators were initiated there would be a gradual and eventually substantial reduction of fungal problems in spices and spiced foods. Thirdly, it is evident that scientists from around the world are continuing to recognize this problem of contamination as they evaluate microbial profiles of various spice and herbal products. Fourthly, in an effort to attack this problem scientists have, on an international basis, studied a multiplicity of approaches. These research evaluations range from ionizing irradiation, which appears to show marked potential as a control mechanism, to heat and light treatments and the use of the chemical components of spices themselves to control or inhibit fungal growth and toxin production. However, although substantial inhibition has been found with a variety of spices (Table 1), studies on ginger show that conflicting results may be obtained, presumably as a consequence of differences in methodology, fungal strains and the parameters studied (Table 1).

Despite the advances that have been made, it is evident that satisfactory solutions to the problem of moulds and their mycotoxins in spices have not yet been found. Additional research in this area is needed to gain even slight benefits in making more efficient use of our foodstuffs, as well as reducing the risk to the consumer.

REFERENCES

Abdelhamid, A.M., Sadik, E.A. & Fayzella, E.A. (1985). *Acta Phytopathologica Academiae Scientiarum Hungaricae*, **20**, 309–20.

Bhat, R., Geeta, H. & Kulkarni, P.R. (1987). *Journal of Food Protection*, **50**, 418–19.

Bilgrami, K.S., Singh, A. & Ranjan, K.S. (1984). *National Academy Science Letters, India*, **7**, 273–4.

Butler, W.J. (1965). In *Mycotoxins in Foodstuffs*, MIT Press, Cambridge, MA, p. 175.

Deger, G.E. (1976). *Annals of Internal Medicine*, **85**, 204.

Denizel, T. & Kosker, O. (1972). A mycological survey of various kinds of edible nuts commercially available in the U.K. with reference to mycotoxins. University of Ankara, *Yearbook of the Faculty of Agriculture*, pp. 168–99.

Diapaolo, J.A., Elis, A. & Erwin, H. (1967). *Nature, London*, **215**, 638.

Flannigan, B. & Hui, S.C. (1976). *Journal of Applied Bacteriology*, **41**, 411–18.

Flannigan, B. & Llewellyn, G.C. (1986). In *Biodeterioration 6*. Commonwealth Agricultural Bureau International Mycological Institute, United Kingdom, pp. 273–9.

Geeta, H. & Kulkarni, P.R. (1987). *Journal of Food Protection*, **50**, 401–3.

Golumbic, C. & Kulik, M. (1969). In *Aflatoxin—Scientific Background, Control and Implications*, ed. L.A. Goldblatt. Academic Press, New York, pp. 307–27.

Indian Council of Medical Research (1985). Effect of cardamon oil and curcumin on growth and aflatoxin production by *Aspergillus parasiticus*. National Institute of Nutrition, Hyderabad-500007, India, pp. 183–7.

Jewers, K. (1982). *Royal Society Health Journal*, **102**, 114–18.

Karapinar, M. (1986). *International Journal of Food Microbiology*, **2**, 239–45.

Kurata, H. & Hitokoto, H. (1985). In *Filamentous Microorganisms: Biomedical Aspects*, ed. T. Arai. Japan Scientific Societies Press, Tokyo, Japan, pp. 165–74.

Llewellyn, G.C., Burkett, M.L. & Eadie, T. (1981). *Journal of the Association of Official Analytical Chemists*, **64**, 955–60.

Llewellyn, G.C., Gentry, C.L., Mayo, E.S. & Dashek, W.V. (1982). *Zeitschrift für Lebensmitteluntersuchung und-Forschung*, **174**, 18–22.

Madhyastha, M.S. & Bhat, R.V. (1985). *Journal of Food Science*, **50**, 376–8.

Maher, V. & Summers, W. (1970). *Nature, London*, **225**, 68.

Marshall, D.L. & Bullerman, L.B. (1986). *Journal of Food Protection*, **49**, 378–82, 388.

Pruthi, J.S. (1980). *Advances in Food Research, Suppl. 4*, Academic Press, N.Y., p. 13.

Rosengarten, F. Jr (1981). *The Book of Spices*. Jove Publications, Inc. New York, New York.

Sharma, A., Ghanekar, A.S., Padwal-Desai, S.R. & Nadkarni, G.B. (1984). *Journal of Agriculture and Food Chemistry*, **32**, 1061–3.

Tiwari, R., Dikshit, R.P., Chandan, N.C., Saxena, A., Gupta, K.G. & Vadehra, D.E. (1983). *Journal of Food Science and Technology*, **20**, 131–2.

Wogan, G.N. (1966). *Bacteriology Review*, **30**, 460–70.

World Health Organization (1976). World Health Organization Technical Report Series, No. 598. WHO, New York.

Post-harvest Losses Caused by Anthracnose (*Colletotrichum gloeosporioides*) of Tropical Fruits and Vegetables

M.J. Jeger & R.A. Plumbley

Overseas Development Natural Resources Institute, 56–62 Gray's Inn Road, London WC1X 8LU, UK

ABSTRACT

Severe post-harvest losses in tropical fruits and vegetables are caused by anthracnose. The disease is a major constraint on the expansion of export trade in fruits such as mango. The situation can be improved by a better understanding of the quiescent nature of anthracnose and the epidemiology of the disease pre-harvest. Management of the disease can best be effected by an integrated approach incorporating pre- and post-harvest treatments and biological and environmental components of control.

INTRODUCTION

Anthracnose, caused by the fungus *Colletotrichum gloeosporioides* (Penz.) Sacc., is a major disease of tropical fruits and vegetables throughout the world, especially under warm moist conditions. *C. gloeosporioides*, the anamorph of *Glomerella cingulata* (Stonem.) Spaulding & Schrenk (Mordue, 1971), infects a large number of hosts and plant parts, including stems, leaves, flowers and fruits. Of the major tropical fruits. *C. gloeosporioides* attacks mango, papaya, avocado, passion fruit and guava. Many solanaceous vegetables are attacked, including aubergine and *Capsicum*. The taxonomic status of *Colletotrichum* spp. is generally problematical (Baxter *et al.*, 1985). Although the basis for distinguishing at the species level exists, the extent of intraspecific variation is large and 'stains' are poorly defined (Geiger *et al.*, 1980). In this paper most attention will be given to mango anthracnose, although much of the discussion applies to anthracnoses of other fruit and vegetables. In mango the anthracnose pathogen has been named as *C. gloeosporioides* var *minor*, but this convention has not been universally followed or accepted.

LOSSES CAUSED BY *C. gloeosporioides*

Losses in yield of mango are caused by pre-harvest infection and disease development on leaves and blossoms (Prior & Ryder, 1987). Fruit set can be reduced markedly and fruitlets often fall prematurely. Infection occurs at any time from fruit set but appressoria formed remain latent or quiescent until the fruits ripen post-harvest. The epidemiology of quiescent infection of fruits in the field has not been investigated for *C. gloeosporioides* on mango, although some data are available for avocado and for another pathogen causing a quiescent disease, *Alternaria alternata*, on mango and persimmon. A consequence of quiescence is that fruits may be totally unblemished by disease lesions at harvest and a considerable effort may be invested in marketing the crop. Depending upon the length of time involved, almost total crop loss can occur when the fruits ripen, although these losses are often anecdotal in terms of % fruits infected. Carefully-designed quantitative surveys on post-harvest losses due to anthracnose have rarely been undertaken. In one such survey in Mexico City, Noon (1984) estimated that up to 25% of total market losses of mangoes were caused by anthracnose, which was by far the most prevalent pathogen found.

The consequences of quiescence are most felt when mangoes are exported. There is a substantial world market for tropical horticultural fruits and sea frieght would be one means of reducing costs. Typical shipment times are 19–21 days from South Africa and 16 days from Israel to Europe. Even if temperatures are reduced during shipment, post-harvest disease development can cause 100% losses when fruits are ripened. None of the treatments discussed later in this paper gives reliable control of anthracnose over such an extended storage period. The situation can be further exacerbated by other pathogens that occur in conjunction with *C. gloeosporioides* on mango and other tropical fruits.

POST-HARVEST DISEASE DEVELOPMENT

The physiology of quiescence has been studied in great detail for tropical fruits (reviewed by Swinburne, 1983), although most work has been done with *C. musae* on banana rather than with *C. gloeosporioides*. There is a diverse range of mechanisms by which quiescence can be maintained in unripe, immature fruits; these include pre-formed fungitoxic compounds in unripe but not ripe fruits, an inappropriate substrate for the nutritional and energy requirements of the pathogen, an inadequate enzyme potential to colonise unripe fruit, and the production of phytoalexins in unripe fruit following infection. Mechanisms of quiescence have not adequately been examined for *C. gloeosporioides*, although the morphological nature of quiescence has been established for avocado, papaya and mango. Subcuticular but arrested hyphae of *C. gloeosporioides* have been observed in immature mango and papaya fruit, but not in avocado. Prusky *et al.* (1982) found a preformed antifungal diene in unripe avocado which inhibited growth of *C. gloeosporioides*, and subsequently proposed a mechanism by which growth of the fungus was subsequently allowed.

Surprisingly little work has been done on the gross post-harvest physiological changes that occur following quiescent infection. It has been shown that the production of cell-wall-degrading enzymes is linear with disease development of *C. gloeosporioides* in *Capsicum annuum*, and activity of a range of enzymes was greater in infected fruits. There were also marked changes in ascorbic acid, proteins, phenols, sugars, organic acids and amino acids in infected fruits.

The quantitative relationship between post-harvest disease development and pre-harvest infection for *C. gloeosporioides* has not been reported as for *Alternaria alternata* in mango, although its importance has been recognised. It would be of great value to determine these relationships to enable better decisions on the timing of the spray treatments and post-harvest dips described in the next section.

CONTROL OF POST-HARVEST ANTHRACNOSE

Two strategies can be distinguished in controlling post-harvest anthracnose development, whether by chemicals, host resistance or other management practices. If the numbers of appressoria developing on fruits can be restricted pre-harvest then clearly there is less opportunity for extensive lesion development when fruits ripen. One problem is that it is not known whether all quiescent appressoria of *C. gloeosporioides* (e.g. early, late or both) on fruits actually lead to post-harvest symptoms. Conversely, control can be achieved by preventing appressoria from developing into lesions post-harvest. This approach would certainly be facilitated by a knowledge of the relationship between pre- and post-harvest anthracnose alluded to in the previous section. In practice, effective control will depend upon both pre- and post-harvest treatments to prevent post-harvest lesion development.

Pre-harvest sprays of copper fungicides can give good control of the blossom blight stage of the disease in mangoes (Thompson, 1987). Improved fungicide schedules have been achieved by incorporating surfactants, polyisobutene stickers to increase retention (Prior & Ryder, 1987), and new systemic fungicides (Thompson, 1987).

Several problems remain, however, if pre-harvest foliar sprays are to prevent post-harvest anthracnose. There is the necessity to maintain effective control from flowering to harvest, which could mean up to 20–25 sprays, and the associated risk of insensitivity developing in the pathogen population. One means of reducing these problems is to develop epidemiologically-based forecasting models for scheduling fungicide sprays.

Post-harvest treatments of tropical fruits against anthracnose has had a long history, and usually involve some combination of hot water and a fungicide such as benomyl (Hatton & Reeder, 1964). Despite this, no universally-accepted optimal treatment has evolved for all varieties and scales of operation. Examples of the effectiveness of these combination treatments are to be found in many recent reports (e.g. Muirhead, 1976). Most recently the use of combination hot water, fungicide and irradiation treatment has been reported and recommended (Spalding & Reeder, 1986). Such treatments will probably be premature for the majority of developing countries, where simple technology for applying post-harvest treatments has not fully been worked out (Martin & Thompson, 1984).

CONCLUSIONS: AVOIDING POST-HARVEST LOSSES FROM ANTHRACNOSE

Improvements in control and a concomitant reduction in yield and quality loss are necessary prerequisites for expanding export trade in mangoes and other tropical fruits. Improvements may arise from the purely technical development of more effective fungicides and apparatus, but sustained long-term advances must come from a better understanding of the basic physiology and epidemiology of quiescence. Some areas have been almost totally neglected: host resistance, controlled atmosphere storage effects on anthracnose and biological control, whether by indirect (Lim & Khor, 1982) or direct (Khetmalas *et al.*, 1984) means. The development of integrated systems of control would be further facilitated by forecasting schemes and threshold criteria for determining the need for fungicide sprays and post-harvest treatments. It may then prove possible to alleviate the large post-harvest losses caused by *C. gloeosporioides*, to the benefit of producer, exporter and consumer.

REFERENCES

Baxter, A.P., Van der Westhuizen, G.C.A. & Eicker, A. (1985). *Phytophylactica,* **17**, 15–18.

Geiger, J.P., Lourd, M. & Huguenin, B. (1980). *Annales de Phytopathologie,* **12**, 177–91.

Hatton, T.T. & Reeder, W.F. (1964). *Proceedings, Caribbean Region, American Society of Horticultural Science,* **8**, 76–84.

Khetmalas, M.B., Hasabnis, S.N., Sardeshpande, J.S. & Diwakar, M.P. (1984). *Current Science,* **53**, 862–3.

Lim, T.K. & Khor, H.T. (1982). *Zeitschrift für Pflanzenkrankheiten und Pflanzenschutz,* **89**, 125–31.

Martin, D.A. & Thompson, A.K. (1984). Modification and testing of a mango dipping machine in Jamaica. Tropical Development and Research Institute Report R1200(L), 24 pp.

Mordue, J.E.M. (1971). *Glomerella cingulata.* CMI Descriptions of Pathogenic Fungi and Bacteria, No 315. Commonwealth Mycological Institute, Kew.

Muirhead, I.F. (1976). *Australian Journal of Experimental Agriculture and Animal Husbandry,* **16**, 601–3.

Noon, R.A. (1984). In *Plant Diseases: Infection Damage and Loss*, eds R.K.S. Wood & G.H. Jellis. Blackwell, Oxford, pp. 299–310.

Prior, C. & Ryder, K. (1987). *Tropical Pest Management* (in press).

Prusky, D., Keen, N.T., Sims, J.J. & Midland, S.L. (1982). *Phytopathology,* **72**, 1578–82.

Spalding, D.H. & Reeder, W.F. (1986). *Plant Disease,* **70**, 1149–51.

Swinburne, T.R. (1983). In *Post-harvest Pathology of Fruits and Vegetables,* ed. C. Dennis. Academic Press, London, pp. 1-21.

Thompson, A.K. (1987). *Proceedings, Commonwealth Secretariat Workshop, Mangoes in the Caribbean, Dominica, March 1987* (in press).

An extended bibliography for this paper may be obtained from the senior author.

RAPID METHODS

Rapid Methods and Automation in Microbiology for Biomass Estimation

Daniel Y.C. Fung

Department of Animal Sciences and Industry, Kansas State University, Manhattan, Kansas 66506, USA

ABSTRACT

*In the past 15 years a series of international symposia was held to discuss the growing field of Rapid Methods and Automation in Microbiology: Stockholm (1973), Cambridge, UK (1976), Washington, D.C. (1981) and Berlin (1984). The fifth symposium will be held in Florence, Italy, in November, 1987. Proceedings of these symposia were published by the following authors: Heden & Illeni (1975*a,b*), Johnson & Newsom (1976), Tilton (1982) and Habermehl (1985). An excellent review of rapid methods for food-borne microorganisms and their toxins was produced by Pierson & Stern (1986). An international workshop focusing on practical application of rapid methods was initiated by the author at Kansas State University in 1981. The seventh workshop was held in July of 1987.*

The purposes of these symposia and workshops were to study all aspects of detection, isolation, identification, and characterization of microorganisms in clinical, food, environmental and industrial specimens. In the Sixth International Biodeterioration Symposium the author presented a review of rapid methods in enumeration, estimation, and identification of microorganisms (Fung et al.*, 1986). The purpose of the present paper is to up-date developments in this area.*

MINIATURIZATION OF CONVENTIONAL PROCEDURES AND DEVELOPMENT OF DIAGNOSTIC KITS

The author has miniaturized a large number of microbiological testing procedures since 1968 (Fung & Hartman, 1975). The methods were used successfully in bacteriological testings of aerobic bacteria (Fung & Miller, 1970, 1971, 1973; Fung & Neimeic, 1977). The entire concept and procedure was used to study anaerobic bacteria by Wilkins & Walker (1975) and Wilkins *et al.* (1975). More recently these methods were used successfully in studying food yeasts in the author's laboratory (Lin & Fung, 1985, 1987*a*,*b*). Currently these methods are being tested for food mould cultivations.

On the commercial side, API, Enterotube, R/B system, Minitek system, MicroID, and IDS have been extensively tested and reviewed (Cox *et al*, 1977; Fung & Cox, 1981). The conclusion is that these diagnostic kits provide *c.* 90–95% accuracy compared with conventional methods. Their usefulness in clinical and food microbiological laboratories will continue to be important. A new system called Spectrum 10 has been recently introduced to the market. It gives 95–96% accuracy compared with MicroID and API systems for stock cultures and for fresh isolates, 93% accuracy to the genus level, and 82% accuracy to the species level (Cox *et al.*, 1985).

IMPROVEMENTS OF VIABLE CELL COUNT PROCEDURE

Standard plate count is still the 'standard' for counting live microorganisms in food and environmental samples. This method is time consuming both in terms of operation and collection of data. Many methods have been tried to improve the efficiency of the standard plate count method. The Spiral Plating Method was developed to spread a liquid sample on the surface of pre-poured agar. The principle is to deposit a thin film of liquid sample in a spiral fashion at a decreasing rate on a rotating plate. The system has been tested extensively in the US (Schakowsky, 1986). One of the problems of this system is the clogging of the dispensing tube by food particles or other solids. To overcome this problem Konuma & Kurata (1982) modified the stomacher bag by placing a filter in it to prevent solid particles coming into the spiral plating system.

Another improved method is the Isogrid Method. This involves filtering samples through a hydrophobic grid and placing the grid on cultivating agar and later to observe growth of colonies of bacteria, yeast or mould. The hydrophobic grids prevent colonies from growing further than the square grid. This facilitates counting of the colonies both manually and electronically. The system has been successfully used for viable cell counts of a variety of food products (Entis, 1983, 1986). Lin *et al.* (1984) developed a trypan blue agar for use in conjunction with the Isogrid system to enumerate yeast and mould from foods.

Petrifilm SM System and Redigel are two new modifications of the standard plate count method. The Petrifilm SM system uses a dry rehydratable film with nutrients imbedded in it. Liquid (1 ml) samples are added directly to the dry medium. The plates are then incubated at the desired temperature and time before counting the colonies. This method has been successfully used to make aerobic plate counts of milk and freshly ground beef (Ginn *et al.*, 1986; Smith *et al.*, 1986). Data from the author's laboratory (Table 1) indicated a 0·99 correlation coefficient

TABLE 1

Comparison of the Standard Plate Count Method and the Petrifilm Method for Viable Cell Counts of Shrimp, Perch, Cod and Whiting (data from Fung *et al.* (1987))

Seafood		*Colony forming units/g*[a]	
		Standard plate count	*Petrifilm SM*
Shrimp	sample 1	$2{\cdot}1 \times 10^4$	$0{\cdot}9 \times 10^4$
	sample 2	$9{\cdot}3 \times 10^3$	$6{\cdot}8 \times 10^3$
	sample 3	$1{\cdot}5 \times 10^4$	$1{\cdot}2 \times 10^4$
	sample 4	$4{\cdot}4 \times 10^4$	$2{\cdot}0 \times 10^4$
	sample 5	$2{\cdot}0 \times 10^4$	$0{\cdot}9 \times 10^4$
Perch	sample 1	$1{\cdot}4 \times 10^3$	$1{\cdot}2 \times 10^3$
	sample 2	$5{\cdot}4 \times 10^2$	$3{\cdot}7 \times 10^2$
	sample 3	$8{\cdot}0 \times 10^2$	$4{\cdot}8 \times 10^2$
	sample 4	$5{\cdot}0 \times 10^2$	$3{\cdot}2 \times 10^2$
	sample 5	$1{\cdot}0 \times 10^3$	$1{\cdot}2 \times 10^3$
Cod	sample 1	$1{\cdot}2 \times 10^2$	$1{\cdot}1 \times 10^2$
	sample 2	$2{\cdot}2 \times 10^2$	$2{\cdot}4 \times 10^2$
	sample 3	$1{\cdot}0 \times 10^4$	$1{\cdot}0 \times 10^4$
	sample 4	$2{\cdot}1 \times 10^4$	$2{\cdot}3 \times 10^4$
	sample 5	$1{\cdot}2 \times 10^4$	$1{\cdot}2 \times 10^4$
Whiting	sample 1	$3{\cdot}0 \times 10^2$	$3{\cdot}1 \times 10^2$
	sample 2	$5{\cdot}8 \times 10^2$	$5{\cdot}6 \times 10^2$
	sample 3	$1{\cdot}2 \times 10^3$	$1{\cdot}1 \times 10^3$
	sample 4	$5{\cdot}4 \times 10^2$	$5{\cdot}2 \times 10^2$
	sample 5	$8{\cdot}8 \times 10^2$	$8{\cdot}4 \times 10^2$

[a] Samples were massaged in a stomacher for 1 min in sterile diluent. Viable cell counts were made according to a standard method. Incubation time was 48 h at 32°C. All samples were done in duplicate. Correlation coefficient between the two methods is $r = 0{\cdot}99$.

comparing Petrifilm SM with the conventional viable cell count method. The Redigel System consists of a tube containing sterile nutrient in liquid form. A sample (e.g. 1 ml of slurry of food) is placed in this medium. After mixing, the sample is poured into a special petri dish previously coated with a gelating material. When the liquid comes in contact with the gelating, it forms a complex that swells to resemble the conventional agar. After incubation for an appropriate time and at an appropriate temperature, colonies can be counted. Both systems are prepackaged and no sterilization is needed. These systems are ideal for on site testing. Data from the author's laboratory indicate high correlation between the Redigel system and the conventional method. Currently it is under evaluation through the US AOAC procedure.

A fast method called the DEFT test was developed to evaluate rapidly viable cells under UV microscopy. They are filtered through a special filter and then stained with acridine orange before observation under the UV microscope. Viable cells fluoresce orange and dead cells fluoresce green. Pettipher and colleagues studied this technique extensively and concluded that the method is comparable to the viable cell method for milk samples (Pettipher *et al.*, 1980; Pettipher & Rodrigues, 1980).

DEVELOPMENT OF NEW METHODS FOR ESTIMATION OF MICROBIAL POPULATIONS

These approaches utilize new ideas and instruments. All living things contain ATP. In the presence of a firefly enzyme system and ATP, light will be produced. Several companies have marketed instruments and procedures for bioluminescence. The most recent data by Littel *et al.* (1986) indicated that the ATP procedure was able to predict bacterial levels with $0{\cdot}5\ \log_{10}$ of the actual count for beef and chicken samples. Minimum sensitivity is 5×10^4 colony forming units/g of meat sample. Ward *et al.* (1986) also found positive correlation between the ATP method and the conventional method in evaluating fish samples. Bioluminescence ATP assay was also successfully used in winery quality control (Pavelka, 1987).

Impedance measurements rely on the fact that as bacteria grow they change the electrical conductivity of the liquid around them. By measuring the changes in impedance and/or conductance microbial activities can be monitored. The Bactomatic system was designed to monitor electrical impedance and has been used for a variety of microbial samples (Eden & Eden, 1985). The Malthus Microbial Analyser is designed to measure conductance. Gibson & Ogden (1980) used this method to study the quality of fish. A great deal of development work in this system is being done by the Malthus Instruments Company, Stoke-on-Trent, UK.

A radiometric method is based on the fact that many bacteria generate carbon dioxide during metabolism of carbohydrates. A very low level of radioactivity is introduced to the carbon molecule of glucose or other sugars. As the organisms grow they metabolize and release radioactive CO_2 into the head space of the container. An instrument (Bactec) is used to measure the radioactivity of the gas. The amount of radioactivity detected is proportional to the number of bacteria in the sample. This method has been used to detect bacteria in foods (Limpi *et al.*, 1974; Rowley *et al.*, 1979). Radiometric methods work well in hospitals but have not been used extensively in other microbiology areas.

Microcalorimetry depends on heat generated by bacterial growth. Instruments have been designed to detect small temperature changes in samples. Studies on ground meat were recently made by Gram & Sogaard (1986) who reported that within the range of 10^5–10^8 colony forming units/g an instrument called BioActivity Monitor was a promising analytical tool for estimating bacteria number.

The Limulus Amoebocyte Lysate Method, which is the most sensitive method for measuring endotoxin or lipopolysaccharides of Gram-negative cell walls, has been used to estimate the quality of ground beef by Jay (1986). The principle involves a reaction of the endotoxin material with a lysate of the amoebocytes of *Limulus*. A gel is formed in the presence of endotoxins within 1 h.

A relatively new procedure utilizing the enzyme catalase has been tested for estimating bacteria in foods. The instrument designed for this purpose is called a Catalasemeter. The principle is based on the flotation time of a paper disc containing catalase (from a bacterial sample, for example) in a tube containing H_2O_2. A biochemical reaction between catalase and H_2O_2 generates molecular oxygen, which causes the paper disc to float from the bottom of the tube to the surface. In the presence of a high level of catalase (indicating high level catalase positive microorganisms), the flotation time(s) will be short. Conversely, the flotation time will be long (100–1000 s) in the presence of a low catalase concentration. No flotation will occur in the absence of catalase. Since most commercial foods are cold stored under aerobic conditions, the spoilage organisms are mainly psychrotrophs and since most are catalase positive, it is possible to use this method to estimate the population in certain foods. Wang & Fung (1986*a*) made a detailed review of the significance of catalase in food microbiology and reported the feasibility of using catalase activity as an index of microbial loads on chicken surfaces (Wang & Fung, 1986*b*). Figure 1 shows the relationship between number of bacteria (*Micrococcus luteus*) and flotation time in the catalasemeter. There is a inverse relationship (i.e. the higher the number of bacteria, the shorter the flotation time). The limit of sensitivity is about 10^5 for this organism. Through filtration this limit can be decreased. On the same figure a simple gas column method is also shown (Fung *et al.*, 1986). Although the gas column is less sensitive that the catalasemeter it is extremely simple to use and it does provide a trend showing that in the presence of high levels of catalase, a large column of gas is generated in the capillary tube whereas in the absence of catalase, no gas is produced.

Figure 2 shows the result of comparing flotation time against psychrotroph population on the surface of cold stored chicken. In this study a filtration step is added and the test medium is acidified, since it was found that blood can interfere with this test and a pH of 3·3 will minimise the activity of non-bacterial catalase. From the figure, it can be seen that a flotation time of less than 100 s indicates a high number of bacteria on the chicken surface. At flotation times around 1000 s (log 3 s),

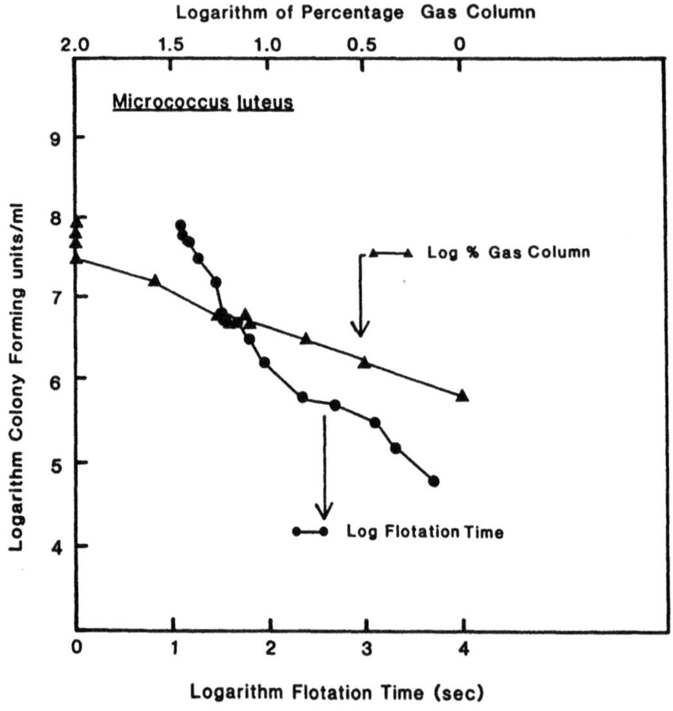

Fig. 1. Measurement of catalase activity of *Micrococcus luteus* by catalasemeter and gas column method. Reproduced with permission from *J. Food Sci.* (Wang & Fung, 1986*b*).

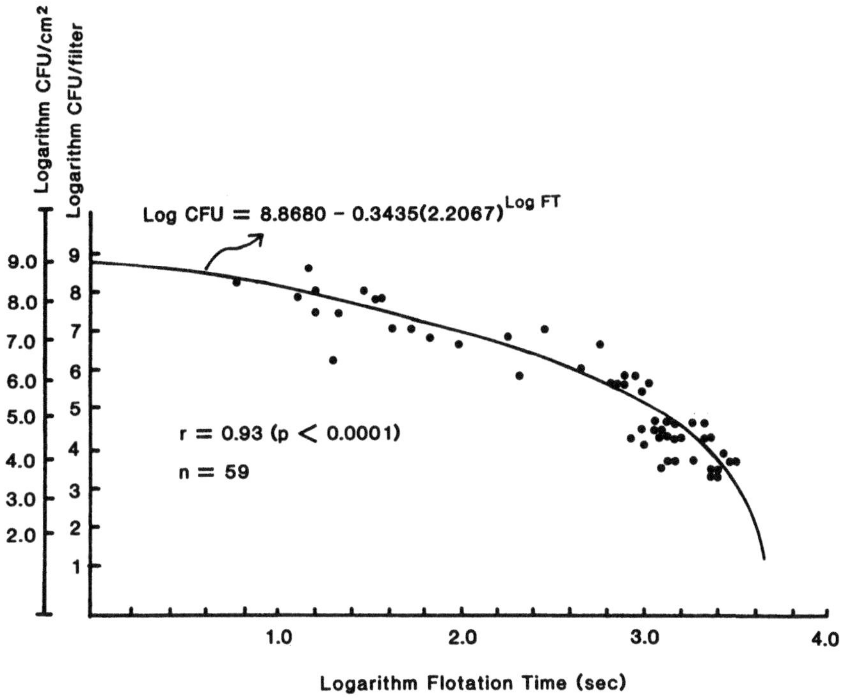

Fig. 2. Catalase activity of psychrotroph population from cold stored (7°C) chicken samples in pH 3·3 phosphate buffer using a catalasemeter with membrane filter method. Reproduced with permission from *J. Food Sci.* (Wang & Fung, 1986*b*).

the curve seems to show a break indicating that one can make some judgement of acceptability or rejection of a food by flotation time, which reflects the microbial count of the chicken. Although more research and development are needed for the catalasemeter to be successful, the speed of the test warrants continued investigation.

IDENTIFICATION OF MICROBES BY SOPHISTICATED INSTRUMENTS AND PROCEDURES

Although the primary purpose of this paper is to review biomass estimation it should be pointed out that there is a great need for better and faster methods of identification of microbes in foods, clinical specimens and environmental samples. Some of the more promising ones include the AMS System (Vitek), DNA Probe (Gene-Trak), EIA (Organon Teknika), *Salmonella* One-Two test (BioControl), Protein profiling (AMBIS System) and cell composition profiling (Hewitt-

Packard System). These systems and methods are designed to identify specific pathogens or organisms with the minimum of interpretation.

One of the questions frequently asked by potential users of the above-mentioned concepts and instruments is 'Which is the best?' Table 2 is a summary of the author's view on the ideal automated microbiology assay system.

In conclusion this paper gives a synopsis of some of the more generally recognized methods for estimating biomass as well as identifying microbes from food, clinical and environmental samples. The field of rapid methods and automation in microbiology is exciting and will continue to flourish since applied microbiologists will always need to find more sensitive, efficient and better methods to study microorganisms.

TABLE 2

Attributes for an Ideal Automated Microbiology Assay System

1. Accuracy for the intended purpose
   —sensitivity: minimal detectable limits
   —specificity of test system
   —versatility: potential applications
   —comparison to referenced methods
2. Speed-productivity
   —in obtaining results
   —number of samples processed per run; per day
3. Cost
   —initial, per test, reagents, others
4. Acceptability
   —by scientific community
   —by regulatory agencies
5. Simplicity of operation
   —sample preparation
   —operation of test equipment
   —computer versatility
6. Training
   —on site; how long
   —quality of training personnel
7. Reagents
   —reagent preparation—stability—availability—consistency
8. Company reputation
9. Technical service
   —speed and availability
   —scope of technical background
   —cost
10. Utility and space requirements

REFERENCES

Cox, N.A., McHan, F. & Fung, D.Y.C. (1977). *J. Food Protect.*, **40**, 866–72.

Cox, N.A., Van Wart, M., Bailey, J.S. & Thomson, J.E. (1985). *J. Food Protect.*, **48**, 76–9.

Eden, R. & Eden, G. (1985). *Impedance Microbiology*. John Wiley, New York.

Entis, P. (1983). *Assoc. Off. Anal. Chem.*, **66**, 897–904.

Entis, P. (1986). In *Foodborne Microorganisms and Their Toxins: Developing Methodology*, ed. M.D. Pierson & N.J. Stern. Marcel Dekker, New York, pp. 91–106.

Fung, D.Y.C. & Miller, R.D. (1970). *Appl. Microbiol.*, **20**, 527–8.

Fung, D.Y.C. & Miller, R.D. (1971). *J. Milk Food Technol.*, **35**, 328–9.

Fung, D.Y.C. & Miller, R.D. (1973). *Appl. Microbiol.*, **25**, 793–9.

Fung, D.Y.C. & Hartman, P.A. (1975). In *New Approaches to the Identification of Microorganisms*, ed. C.G. Heden & T. Illeni. John Wiley, New York, pp. 347–70.

Fung, D.Y.C. & Neimeic, M. (1977). *Lab Health Sci.*, **14**, 273–8.

Fung, D.Y.C. & Cox, N.A. (1981) *J. Food Protect.*, **44**, 877–80.

Fung, D.Y.C., Wang, G.I.J., Bailey, J.S. & Cox, N.A. (1986). In *Biodeterioration VI*, ed. S. Barry, D.R. Houghton, G.C. Llewellyn, & C.E. O'Rear. The Biodeterioration Society, UK, pp. 433–8.

Fung, D.Y.C., Hart, R. & Chain, V. (1987). *Proc. International Symp. Seafood Quality Determination*, Alaska, in press.

Gibson, D.M. & Ogden, I.D. (1980). *J. Appl. Bacteriol.*, **49**, 12–18.

Ginn, R.E., Packard, V.S. & Fox, T.L. (1986). *J. Assoc. Off. Anal. Chem.*, **69**, 527–71.

Gram, L. & Sogaard, H. (1986). *J. Food Protect.*, **48**, 341–5.

Habermehl, K.O. (Ed.) (1985). *Rapid Methods and Automation in Microbiology and Immunology*. Springer-Verlag, Berlin.

Heden. C.G. & Illeni, T. (Eds) (1975*a*). *Automation in Microbiology and Immunology*. John Wiley, New York.

Heden, C.G. & Illeni, T. (Eds) (1975*b*). *New Approaches to the Identification of Microorganisms*. John Wiley, New York.

Jay, J.M. (1986). *Foodborne Microorganisms and Their Toxins: Developing Methodology*, ed. M.D. Pierson & N.J. Stern. Marcel Dekker, New York, pp. 219–40.

Johnson, H.H. & Newsom, S.W.B. (Eds) (1976). *Rapid Methods and Automation in Microbiology*. Learned Information Ltd, Oxford.

Konuma, H.A. & Kurata, H. (1982). *Appl. Environ. Microbiol.*, **44**, 765–9.

Limpi. R.A., Mikelson, D.A., Rowley, D.B., Previte, J.J. & Wells, R.E. (1974). *Food Technol.*, **28**, 52–8.

Lin, C.C.S. & Fung, D.Y.C. (1985). *J. Food Sci.*, **50**, 241–4.

Lin, C.C.S. & Fung, D.Y.C. (1987*a*). *CRC Press*, **14** (4), 273–89.

Lin, C.C.S. & Fung, D.Y.C. (1987*b*). *CRC Critical Review*, in press.

Lin, C.C.S., Fung, D.Y.C. & Entis, P. (1984). *Can. J. Microbiol.*, **30**, 1405–7.

Littel, K.J., Pikelis, S. & Spurgash, A. (1986). *J. Food Protect.*, **49**, 18–22.

Pavelka, R. (1987). In *Rapid Microbiological Methods*, ed. D.L. Downing & Y.D. Hang. New York Agr. Exp. Station, Cornell University, No. 60 p. 7.

Pettipher, G.L. & Rodrigues, U.M. (1980). *J. Appl. Bacteriol.*, **50**, 157–66.

Pettipher, G.L., Mansell, R., McKinnon, C.H. & Cousins, C.M. (1980). *J. Appl. Environ. Microbiol.*, **39**, 423–9.

Pierson, M.D. & Stern, N.J. (eds) (1986). *Foodborne Microorganisms and Their Toxins: Developing Methodology*. Marcel Dekker, New York.

Rowley, D.R., Vandemark, P., Johnson, D. & Shatluck, E. (1979). *J. Food Protect.*, **42**, 335–41.

Schakowsky, S. (1986). In *Foodborne Microorganisms and their Toxins: Developing Methodology*, ed. M.D. Pierson & N.J. Stern. Marcel Dekker, New York.

Smith, L.B., Fox, T.L. & Busta, F.F. (1986). *J. Food Protect.*, **48**, 1044–5.

Tilton, R.C. (Ed.) (1982). *Rapid Methods and Automation in Microbiology*. American Society for Microbiology, Washington, D.C.

Wang. G.I.J. & Fung, D.Y.C. (1986*a*). *J. Food Safety*, **8**, 47–67.

Wang, G.I.J. & Fung, D.Y.C. (1986*b*). *J. Food Sci.*, **51**, 1442–4.

Ward, D.R., LaRocco, K.A. & Hopson, D.J. (1986). *J. Food Protect.*, **49**, 647–50.

Wilkins, T.D. & Walker, C.B. (1975). *Appl. Microbiol.*, **30**, 825–30.

Wilkins, T.D., Walker, C.B. & Moore, W.E.C. (1975). *Appl. Microbiol.*, **30**, 831–7.

Rapid Methods for Sulphate-reducing Bacteria

Christine Gaylarde & Paul Cook

Department of Biological Sciences, City of London Polytechnic, Old Castle St, London E1 7NT, UK

ABSTRACT

Methods routinely used for the detection and enumeration of SRB are lengthy and insensitive. A number of more rapid techniques are suggested. The agar tube method requires little operator time and is inexpensive. The microtitre plate growth assay allows easy replication and hence increases sensitivity whilst remaining inexpensive. Immunological techniques such as the ELISA are extremely rapid and sensitive but require skilled personnel, as does the radiorespirometric technique. Agglutination titrations in microtitre plates provide a rapid method of identification if the appropriate antisera are available.

INTRODUCTION

The sulphate-reducing bacteria (SRB) are a somewhat heterogeneous group of Gram-negative anaerobic bacteria which are linked together by one major characteristic — the use of inorganic sulphate as a final electron acceptor in respiration. The net result of their metabolism in moist environments is the production of hydrogen sulphide. Since these organisms are ubiquitous, there are diverse problems resulting from their activities, not the least of which are the corrosion of metal structures and the foul odours of stagnant fluid systems.

In spite of their common occurrence in nature, the laboratory culture and maintenance of this group of organisms is difficult. They grow slowly on the specialised media necessary for their peculiar requirements and often do not grow at all. Pure cultures give particular problems. Thus standard techniques for their detection and enumeration are unreliable and lengthy, involving prolonged incubation periods of 3 days to 4 weeks. Identification of the species within the group is unsophisticated, employing morphological, physiological, nutritional and biochemical studies (Postgate, 1984). Simpler and more rapid techniques are required for detection, enumeration and identification of this industrially important group of organisms and also for the testing of biocides aimed at combating the problems caused by them.

RAPID METHODS AVAILABLE

An improved technique for the determination of SRB activity has been reported by Hamilton and his co-workers (Rosser & Hamilton, 1983; Maxwell & Hamilton, 1986). This involves incubation with radiolabelled sulphate and measurement of the radioactive sulphide produced. Although not suitable for use in many situations, it is a very rapid method and is recommended for the measurement of SRB activity providing suitable equipment, containment and trained staff are available. We have used this technique with great success for evaluating biocide activities against SRB. Figure 1 shows the effects of different concentrations of a biocide on sulphate reduction by *Desulfovibrio desulfuricans*. Even at 20 ppm it is obvious that a high level of inhibition is produced and this result is obtained within 60 min.

If the sophisticated equipment needed for this 'radiorespirometry' technique is unavailable, there are alternatives.

An extremely simple test, which may be classed as 'rapid' for SRB, is the agar tube method. Variations of this test have been developed by Hill and by the authors. The latter use cylindrical plastic tubes made by removing the tapered ends from 10 ml pipettes. Alternatively, the pipettes may be used without modification. They are filled with an agar medium based on Postgate's Medium B (Postgate, 1984), a known volume of the sample is layered on to the agar and sterile liquid paraffin is placed above the sample to exclude air. Living SRB present in the sample produce hydrogen sulphide, which diffuses through the column reacting with the ferrous sulphate present in the medium to form black ferrous sulphide. The depth of blackening can be read off on the calibrated tube and used to determine the concentration of SRB cells. Including extra iron in the test, in the form of a steel wire, helps to reduce the redox potential and increase the sensitivity. Figure 2 shows the calibration curve after 5 days incubation with and without steel wire. Figure 3 shows that the optimum

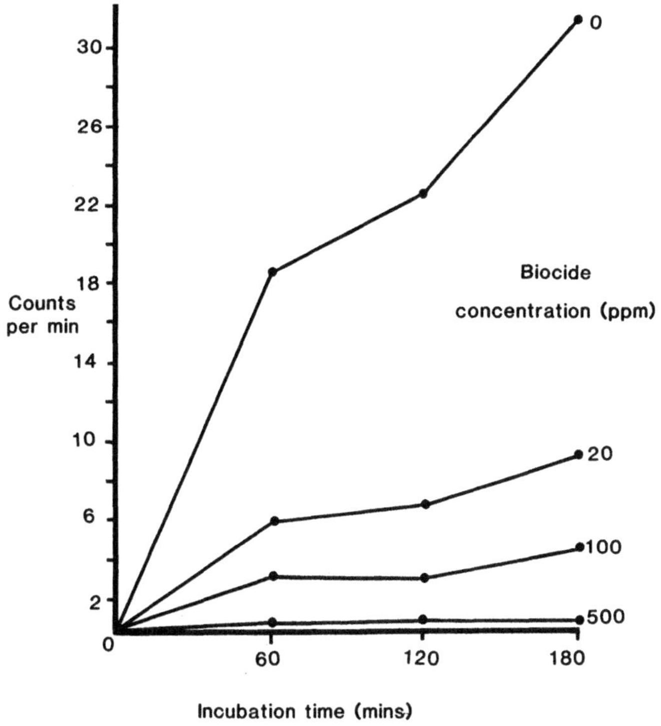

Fig. 1. ^{35}Sulphide produced by *D. desulphuricans* incubated with varying concentrations of biocide.

agar concentration is 0·25% and indicates that results can be obtained within 3 days using this technique.

Although easy to perform and quite inexpensive, the agar tube method is fairly wasteful of incubator space, especially if a number of replicates are to be performed. The microtitre plate assay developed at City of London Polytechnic overcomes this problem. Here, the sample under investigation is diluted in a suitable liquid medium (usually Postgate's medium C) in the wells of a microtitre plate. Small volumes (typically 200 μl per well) reduce costs and the multichannel pipettes available for use with microtitre plates allow a high level of replication. Miniaturised MPN determinations can be performed in these plates and have been found to give good and reproducible results. Since SRB are strictly anaerobic in their growth, it is necessary either to incubate the plates in an artificial anaerobic atmosphere or to seal the prepared plates with paraffin wax. This can be done by making a 'former' around the edges of the plate with autoclave tape and pouring the molten wax into this slowly and evenly. Results are normally obtained within 3–5 days, but this can be reduced by the addition of iron powder to the wells of the plate. This

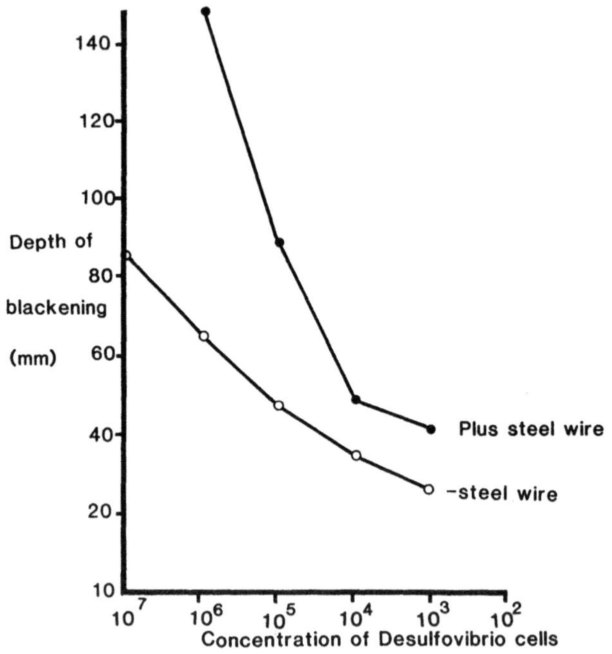

Fig. 2. Calibration curve for agar tube method.

modification allows results to be obtained more rapidly, and makes available a large surface area within each well which permits the formation of a large area of biofilm. Such miniaturised systems can be very useful for biocide testing. We have shown that biocide MICs determined by this method are increased at least twofold, and often much more, when powdered iron is included in the medium.

The last two methods are undemanding of apparatus and skill and are rapidly set up but they still require incubation. Since SRB are slow growing organisms, any really rapid test will not require cultivation. The obvious choice is an immunological method. In the USA, Pope (1986) has described the use of fluorescent antibodies in the detection of SRB, but this method is difficult to apply to environmental samples which are heavily contaminated with particulate matter. More readily applicable is the enzyme-linked immunosorbent assay (ELISA), versions of which are widely used in clinical diagnosis. Gaylarde & Cook have described three variations of the basic ELISA technique which can be used for SRB. The direct ELISA is a very specific assay, capable of detecting 10^1–10^2 cells/ml, but unsuitable for the determination of total SRB numbers. The indirect ELISA, equally sensitive, is less species specific and is therefore an acceptable method where it is desired to estimate total SRB (usually

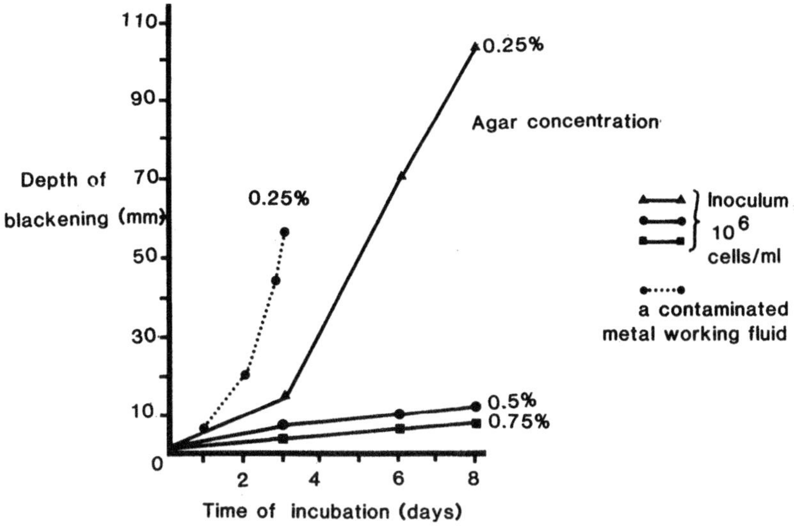

Fig. 3. Effect of agar concentration on speed of blackening in agar tube test.

the case in industrial situations). For maximum sensitivity, the indirect ELISA requires a total time of 48 h. This is reduced by the use of titanous hydroxide in the microtitre plate wells — the 'modified indirect ELISA'. Using this method, a result can be obtained within 6 h, but at the expense of some sensitivity. The modified indirect ELISA can only reliably detect 10^3–10^4 cells/ml.

All the methods discussed so far may be used to enumerate SRB. The immunological techniques are obviously also suitable for rapid identification. The direct ELISA, with its high specificity, may be used for this purpose, but a much simpler technique is cell agglutination. Using standard suspensions of bacterial cells, pipetted in 200 μl volumes into the wells of microtitre plates and tested with various antisera raised against species and strains of *Desulfovibrio*, it can be shown that high antibody titres are obtained only where the antiserum and cells are homologous (Table 1). This degree of specificity is quite unexpected when one considers the similarities found between protein patterns of the various strains (Bradley, 1985) and suggests that the antigenic determinants may be the lipopolysaccharides on the cell surfaces rather than the surface proteins. In spite of this unexpected result, the fact remains that agglutination with specific antisera is an effective and extremely rapid method of identifying SRB.

The lack of cross reaction between *Desulfovibrio* antisera and bacteria of genera other than the SRB has been demonstrated by agglutination (results not shown) and by the ELISA (Gaylarde & Cook, 1987).

TABLE 1
Titres of Species-specific Antisera against Various Strains of SRB (as reciprocals of serum dilutions)

	D. vulgaris (Woolwich) (1)	16 384				
Antiserum	*D. desulfuricans* (Essex) (2)	8	8192			
raised	*D. desulfuricans* (Sylt3) (3)	16	4	65 536		
against:	*D. vulgaris* (Hildenborough) (4)	1024	8	16	16 384	
	D. salexigens (California) (5)	16	16	16	32	4 096
		(1)	(2)	(3)	(4)	(5)
		Antigen (whole cells)				

CONCLUSIONS AND FUTURE DEVELOPMENTS

For biocide assays, the radiorespirometric technique and the microtitre plate growth assay both offer certain advantages over traditional methods. The choice between them will depend on the preferences of the laboratory and availability of equipment.

The ELISA is the most rapid technique presently available for the detection of total SRB numbers. This assay measures both live and dead cells and so is possibly not suitable alone for biocide testing, although dead cells do eventually lyse and become undetectable. For rapid identification of SRB strains, the direct ELISA or simple agglutination tests are of use. Other methods at present available include the detection of fatty acids, specific to some genera (Boon *et al.*, 1977; Taylor & Parkes, 1983; Dowling *et al.*, 1985). However, the use of relatively sophisticated and expensive apparatus limits this approach for the time being. It seems likely that a gene probe will be produced for the SRB in the near future. Such probes are already available for other genera, some of which are of industrial importance (e.g. the *Pseudomonas fluorescens* probe made by Schleifer's group in Munich; Festl *et al.*, 1986). A gene probe for the SRB, whilst utilising rather sophisticated methodology, would allow the detection of these organisms rapidly and with high specificity even when they were present as a minor component of a population. Such a goal is well within our sights.

REFERENCES

Boon, J.J., De Leeuw, J.W., Hoek, G.J. & Vosjan, J.H. (1977). *J. Bacteriol.*, **129**, 1183–91.
Bradley, G. (1985). Ph.D. thesis, City of London Polytechnic.
Dowling, N.J.E., Widdel, F. & White, D.C. (1985). *J. Gen. Microbiol.*, **132**, 1815–29.

Festl, H., Ludwig, W. & Schleifer, K.H. (1986). *Appl. Env. Microbiol.,* **52**, 1190–4.
Gaylarde, C.C. & Cook, P.E. (1987) ELISA techniques for the detection of sulphate-reducing bacteria. SAB Technical Series, in press.
Maxwell, S. & Hamilton, W.A. (1986). *J. Microbiol. Methods,* **4**, 83–91.
Pope, D. (1986). In *Biologically Induced Corrosion*, NACE-8, International Corrosion Conference Series, NACE, Houston, Texas.
Postgate, J.R. (1984). *The Sulphate-reducing Bacteria*. Cambridge University Press, Cambridge.
Rosser, H.R. & Hamilton, W.A. (1983). *Appl. Env. Microbiol.,* **45**, 1956–9.
Taylor, J. & Parkes, R.J. (1983). *J. Gen. Microbiol.,* **129**, 3303–9.

Rapid Microbiology: the Use of Luminescence and Adenosine Triphosphate (ATP) for Enumerating Microbes and Checking Effectiveness of Biocides: Present Status and Future Prospects

Philip E. Stanley

Consultant Scientist, 48 Glisson Rd, Cambridge CB1 2HF, UK

ABSTRACT

All living cells contain ATP and typically a bacterial cell contains about a femtogram (10^{-15}g) of ATP. The enzyme firefly luciferase reacts specifically with ATP to produce a photon of light for each ATP molecule hydrolysed. It is possible to assay as little as 10^{-13}g ATP by measuring the light produced using a sensitive luminometer. The technique is suited for rapid enumeration of microbes in a range of samples. Once the ATP has been extracted from the microorganism the light producing reaction takes less than a minute to perform. Efforts are being made to improve the means by which ATP is extracted from a range of microbial cells and spores. Extractants must be effective and compatible with the firefly luciferase and not cause inhibition or colour quenching. Most real samples also contain free ATP and some cellular somatic ATP (e.g. from animal or plant cells) and methods need improvement for removal of this ATP prior to extraction and measurement of microbial ATP. Better specifications are also expected in the sensitivity of luminometers and available luciferase and accompanying reagents. Future improvements in the technology will allow the ATP luminescence technique to be used to enumerate at least some individual species of microorganisms.

Many biocides cause microorganisms to leak ATP or have a dramatic effect on cellular ATP status. The ATP luminescence technology can be applied in checking the effectiveness of biocides and in fundamental studies of their actions.

INTRODUCTION

The unrestricted growth of microorganisms in certain industrial products and processes is a serious problem which is sometimes difficult to detect and/or control. Traditional assay procedures for microbiological testing are often time consuming and labour intensive; a typical assay requiring 1-3 days to complete. In the food industry, for example, this length of time is often unacceptable not only from the viewpoint of spoilage but also from storage as well as financial and marketing considerations. Most modern techniques for 'rapid microbiology' involve amplification of cell numbers by a period of microbial growth followed by enumeration using instrumentation. Such techniques reduce the amount of labour but many hours are still needed for the growth phase. The effectiveness of protocols involving a growth period is dependent upon the microbe multiplying to a level sufficient for enumeration. However, if conditions are unfavourable or the organism is stressed and fails to recover in time, their presence may be overlooked.

Because the ATP level in microbial cells stays more or less constant it is a useful indicator since it can be measured luminometrically using firefly luciferase down to subpicogram (10^{-12} g) levels. Most workers agree that dead microbial cells do not contain ATP.

The main advantages are:

1. there is generally no period of growth;
2. all cells contain ATP;
3. stressed cells and spores are detected;
4. there is a wide dynamic range, and rarely a need for dilution series;
5. at present it can measure down to one thousand cells per assay.

The main disadvantages are:

1. the presence of non-microbial ATP in most samples;
2. protocols are not well established for a wide range of samples;
3. the means and methods for extracting ATP from cells need improvement;
4. automation is required for sample preparation.

Commercial reagents and instrumentation are available which cover a range of requirements but there is a general need for improvement in sensitivity (below 1000 microbes) and a requirement for species specificity.

ATP is the key energy currency molecule of all cells and any disturbance to their metabolism results in a variation in ATP level. Thus when a biocide is applied the ATP status provides an indication of biocidal effectiveness. ATP may leak from the cell or the cell level may show a steady decline with time. The use of the firefly luciferase technique to monitor ATP is thus a practical means of assessing and/or screening biocide efficacy.

A COMMENTARY ON THE ATP AND LUCIFERASE TECHNIQUE

This methodology for assaying ATP has been available since the mid-1950s but its use for enumerating microbes in simple cultures was only reported in the mid-1960s whilst its application to more complex samples such as food, took until the early 1970s. The reasons are several fold. First, there was no commercial reagents and preparations of firefly luciferase often left a good deal to be desired in terms of reproducibility, background (blank), sensitivity and specificity. Second, the instrumentation for measuring the small amounts of light was of a research type used by the physical sciences. Third, the techniques for selective extraction of ATP and removal of unwanted ATP (somatic and free ATP) were slow to be developed. Fourth, much of the development in analytical luminescence was carried out in chemistry and biochemistry laboratories and not microbiology laboratories. As a result until a few years ago, in the area of microbiology, those new or potential users have often been less than encouraged with the technique when they have tried it for themselves.

The situation is changing and a number of initiatives are under way, the most interesting of which is that furthered by the Biotechnology Unit of the Department of Trade and Industry in UK. This is bringing together researchers, users and commercial companies into consortia which, with government funds, are developing and proving this technology. Improvements are expected in the sensitivity of the technique achieved by better reagents and more sensitive instruments. An important aspect is the support in developing and using protocols for particular products and samples.

The reaction of ATP with firefly luciferase

Light production from luciferase occurs only in the presence of ATP and as little as a femtogram (10^{-15} g) may be measured which happens to be the amount of ATP typically found in an average bacterial cell. However, under practical conditions the smallest amount that can be detected is around a picogram (10^{-12} g) which is equivalent to about 1000 bacterial cells or 10 yeast cells. This decrease in sensitivity is caused by a number of factors including luciferase enzyme inhibition by the sample as well as the inherent background (blank) of the luciferase reagent and the instrument noise.

Present reagents are formulated to give a more or less constant light output for a matter of minutes following mixing. The presence of certain extractants for cellular ATP such as detergents and surfactants, do however, cause the light level to decrease rather rapidly causing difficulties when using certain types of internal standardisation with added ATP, as well as for time dependent measurements. For the highest sensitivity it is generally necessary to use large quantities of luciferase and use 'peak height' or 'flash kinetics' where the light is generated as a peak over a few seconds. Special instrumentation is required to record and measure the signal. In this situation ATP and luciferase must be mixed in the dark within the sample chamber of the instrument and in front of its light detector (photomultiplier). With the constant light output signal light emission may be initiated inside or outside the instrument.

The optimum temperature is about 20°C at a pH around 7·75. Magnesium ions are required as well as the cofactor D-luciferin and oxygen. Halide anions and certain cations e.g. cadmium, are inhibitory.

A typical protocol for a sample (e.g. a sample of meat) which also contains non-microbial ATP is as follows:

1. homogenise 10 g meat in 100 ml Ringer solution (ATP free);
2. centrifuge at 100 G for 2 min or allow to settle for 10 min;
3. remove 100 μl of supernatant from below fat layer into cuvette;
4. add Triton-X100 to lyse somatic cells and release their ATP;
5. treat mixture with an ATPase (e.g. potato apyrase) for 15 min to remove all traces of somatic and free ATP;
6. extract microbial ATP (using proprietary extractant (10 s) or trichloracetic acid plus neutralization (5 min);
7. add 100 μl firefly luciferase reagent and immediately measure light output over 10 s interval;
8. calibrate the system against meat samples measured by conventional plating procedures.

Variations may be employed to deal with textiles, paper, hygienic work surfaces, microbial based corrosion, plastic surfaces (protheses), cutting oils, liquid fuels, soils, beverages, water samples, cooling water, dairy products, etc.

The use of luminescence and ATP to monitor biocide efficacy

Luminescence and ATP may be utilised in two ways for monitoring biocides. First, it may be used on a routine basis to check that the microbial population has decreased to an acceptable level as might be the case after cleaning or perhaps when monitoring cooling or process water. This simply involves monitoring biomass taking care to deal with any free ATP which might have leaked from the affected microbial cells. Second, it may be used to monitor the ATP level when screening for biocidal effect. As ATP is a key metabolite, a major change in cell metabolism will occur at death, and be reflected in (usually) a rapid drop in cellular ATP level. Thus monitoring for 30 min at 5 min intervals following application of biocide should give a swift indication of biocide effectiveness.

REFERENCES

A list of papers for further reading may be obtained from the author.

Industrial Application of the Malthus Analyser

S.D. Abbiss

Malthus Instruments Ltd, William Clowes Street, Burslem, Stoke-on-Trent ST6 3AT, UK

ABSTRACT

Rapid microbiological methods in general, and electrical methods in particular, are the methods chosen by an increasing number of industrial laboratories. The Malthus microbiological analyser measures microbial activity by monitoring changes in conductance.

INTRODUCTION

Changes in the conductance of a culture medium arise from the metabolic activity of a growing microbial population (Allison *et al.*, 1938). When change in bacterial numbers is plotted against change in conductance, a linear relationship is obtained (Richards *et al.*, 1978). This is the fundamental relationship upon which conductance monitoring is based.

METHODOLOGY AND APPLICATIONS

A sterile medium produces no significant change in the conductance response; if organisms are present, a sigmoidal conductance curve is produced and the point at which the curve deviates from the baseline is the detection time (Gibson & Clark, 1987). Typically, this occurs when the number of organisms reaches 10^5–10^6 cells/ml, but varies depending on the conditions. The higher the number of organisms present initially, the shorter the detection time. In an industrial context, it is often necessary to compare detection times to plate counts. This is done by preparing a calibration curve with log numbers plotted against detection times. For any specified standard expressed in the form of counts, the corresponding detection time can be read from the calibration curve. Sample preparation is minimal. Liquid samples are pipetted directly into the cells, solid samples are homogenised first in a diluent. The sample cells are connected to the incubator via a bridge board. The test commences as soon as the electrical connection has been made and 256 cells can be monitored at one time.

Samples can be added to the incubator at any time without disturbing cells that are already running. This is because of the high thermal capacity of the water-filled incubators. At any time during the test, sample information, such as product names and codes can be entered into the computer and the current status of the cells checked. The results are automatically categorised as PASS, FAIL or SUSPECT according to limits which the user has defined. Information can be transferred directly to remote sites by using the Malthus computer as part of an on-line management system.

Variations on the standard Malthus method allow the system to be used for a wide range of industrial applications. One of the simplest is sterility testing. With sterile products, a pre-incubation stage is usually necessary to increase the level of any contaminants to a detectable level, then a sub-sample is tested. If the product is sterile, there is no significant change in conductance.

The simple presence or absence test can also be used to test for specific organisms or groups of organisms using an appropriate selective medium. For example, Malthus Coliform Broth can be used to analyse untreated water samples. The medium can detect one coliform organism in 100 ml water within 12 h. Non-coliform organisms are retarded beyond 12 h, even if they are present in large numbers.

A conductance medium for the detection of Enterobacteriaceae is also available. For any product, it is necessary to determine the critical detection time. Samples are analysed by the Malthus and conventional methods so that log numbers can be plotted against detection times. The statistical analysis is done by the Malthus statistics program which plots the regression line and confidence limits. The point at which the upper confidence limit intercepts the x axis is the detection time limit or the time required to detect one Enterobacteriaceae organism in this medium. Samples which produce responses within the time limit are classified as Enterobacteriaceae positive.

More recently, a medium has been optimised for the detection of salmonella by conductance monitoring (Easter & Gibson, 1985). It is based on the reduction of trimethylamine oxide to trimethylamine — a reaction which is carried out by salmonellae and yields a large conductance signal of over 600 μs. The method involves a pre-enrichment stage followed by subculture into the selective TMAO medium in Malthus cells. If salmonellae are present, a characteristic conductance response is obtained, generally within 24 h. Consequently the test is reduced by 2–3 days compared to traditional methods.

In the field of environmental monitoring surfaces can be swabbed and the swab placed directly into the Malthus cell. If microorganisms are present, a detection time will be obtained which will relate to the degree of contamination of the surface. Membrane filters can also be placed directly into Malthus cells, allowing large volumes of liquids, such as rinse waters, to be tested. The electrodes themselves can be dipped into liquids or semisolids and then transferred into the Malthus cells to provide a quick and simple environmental check.

Malthus systems can be used to assess biocide efficacy by challenging a series of cells containing different concentrations of the biocide with a standard inoculum. The detection time lengthens as the biocide concentration increases until total inhibition is demonstrated.

Preservative efficacy can also be tested. The organisms are inoculated into the preserved product and their decline is monitored over a defined period. Automation of challenge testing results in considerable savings in labour, time and materials.

The Malthus analyser can be used for monitoring process hygiene. McMurdo & Whyard (1984) demonstrated clear economic advantages of the system for hygiene management of a spray drier plant.

CONCLUSION

The major advantages of Malthus conductance microbiology are faster results and automation, which have significant benefits. Laboratory testing is simplified and greater numbers of samples can be analysed with less expenditure on media and consumables. Conductance monitoring provides a good 'early warning system' which can be used effectively as part of an environmental monitoring programme. The results are available in hours rather than days. In an industrial context, this can mean that positive microbiological clearance becomes feasible, thereby reducing wastage, storage costs and minimising the possibility of product recall.

REFERENCES

Allison, J.B., Anderson, J.A. & Cole, W.H. (1938). *Journal of Bacteriology*, **36**, 571–86.

Easter, M.C. & Gibson, D.M. (1985). *Journal of Hygiene, Cambridge*, **94**, 245–62.

Gibson, D.M. & Clark, A.E. (1987). *Laboratory Practice*, **36**(3), 47–8, 49.

McMurdo, I.H. & Whyard, S. (1984). *Journal of the Society of Dairy Technologists*, **37**(1), 4–9.

Richards, J.C.S., Jason, A.C., Hobbs, G., Gibson, D.M. & Christie, R.H. (1978). *Journal of Physics E: Scientific Instruments*, **11**, 560–8.

Real-time Screening of Microorganisms in Waters and Effluents

J. Scholefield

Centre for Bio-Medical Instrumentation, University of Strathclyde, 131 Albion Street, Glasgow G1 1SD, UK

ABSTRACT

A range of waters and effluents from food processing plants was microbiologically screened using Autotrak, an automated Direct Fluorescence Microscope system. This 'real-time' screening approach provides results in less that 30 s and will examine 140 specimens/h. Compared with standard cultural procedures correlation values of 0·748–0·920 were obtained for the Autotrak evaluation of process waters and effluents in the range 10^4–10^8 CFU/ml.

INTRODUCTION

Microbiological screening is routinely performed using cultural procedures which provide results following incubation for 18–72 h. Together with dissolved organic material and particulate matter, process waters and effluents often carry high 'microbial loads' of food spoilage and public health significance. Unknown, stressed or fastidious microorganisms may not be recovered by standard cultural techniques and results following incubation may be of only retrospective significance so far as control action is concerned. Efforts in 'cost-containment' have resulted in the use of the 'dip-slide', a coarse cultural screen, and automated media and sample preparation devices, together with the 'Spiral Plater' (Gilchrist *et al.*, 1973) are now widely used. Incubation times may be reduced using the hydrophobic grid-membrane technique for the enumeration of micro colonies (Sharpe, 1975).

Accelerated techniques including oxidation–reduction methods have been applied for milk, effluent and shellfish screening. Results may be obtained in 30 min to 6 h but precision is poor for other than highly contaminated samples.

Automated systems based on the detection of early growth response or of microbial metabolites represent a major area of development in the last 20 years. Such instrumentation is based on the detection by optical (Autobac), carbon dioxide (Bactec) or impedance (Bactometer) changes. These systems are batch operated, provide results in 3–18 h depending on the initial 'microbial load' and require calibration with respect to relevant isolates and specimen composition. Wood & Gibbs (1982) have reviewed their application in the food industry. Such systems are primarily of use for identification rather than primary screening. While it may be convenient to assume a simple relationship between the initial 'microbial load' and the 'end point', it may be shown that variation in the microbial population, their state of metabolism and specimen composition may affect the duration of the 'lag-phase' and thereby the accuracy of an early growth response 'end point'.

'Real-time' screening must be based on the direct determination of the microbial cells or of some enzymic or compositional property. The photometric determination of microbial ATP based on bioluminescence has been applied to milk and other foods (Wood & Gibbs, 1982). Specimens require 30–45 min preparation and to date the technique has not been fully automated. Reagents are expensive and of variable quality and shelf-life. Gram-negative bacteria may be estimated using the Limulus Lysate Assay. Particle analysis (Coulter) has been used for microbial screening but the non-microbial particulate content of foods, waters and effluents causes flow-path blocking and malfunction.

Despite the widespread use of the microscope for the qualitative assessment of specimens and cultures, quantitative microscopy has not been developed for primary screening. The Breed Smear test was developed for milk screening and has been used for other foods although the technique is laborious and subjective. However, the semiquantitative microscopical examination of urine specimens is part of the routine procedure used in many clinical microbiology laboratories.

Early studies by Strugger (1948) demonstrated the potential for Direct Fluorescence Microscopy and the development of incident illumination by Ploem (1967) resolved the problem of progressive fluorescence fading of transmission microscope preparations. Studies of the reactions between fluorochromes and different microbial species in a range of menstrua (Scholefield *et al.*, 1976) led to the development of real-time screening of microorganisms in urines (Scholefield *et al.*, 1981), milk, waters, effluents and on carcases (Scholefield *et al.* 1985, 1986; Harran & Scholefield, 1987). Autotrak, an automated Direct Fluorescence Microscope (DFM) system was developed in the University of Strathclyde from 1983 to 1986 for the microbiological screening of urine, milk and other liquid menstrua.

METHODS

Specimens

A total of 87 process water, effluent and sewage specimens from food manufacturing sources were examined using DFM and cultural procedures.

Cultural procedure

Serial dilutions of specimens were prepared in quarter strength Ringers solution or Mueller-Hinton Broth (Oxoid Ltd, UK). Surface plates of Nutrient or Plate Count Agar (Oxoid Ltd, UK.) were inoculated in triplicate and colony counts made after 1, 3 and 7 days incubation at 30°C.

Autotrak procedure

The DFM system comprises a microprocessor controlled epifluorescence microscope with autosampler and operates using a consumable kit of tape, stain-reagent, etc., for 1000 analyses. Replicate filtered or unfiltered water and effluent samples (1–2 ml) were loaded into the autosampler. Up to 50 samples may be loaded at any time and up to 140 samples examined per hour. A sampling probe takes up aliquots of sample which are applied to a coated moving tape. The applied sample stripe is heat fixed and the contained microorganisms are stained with an acridine orange reagent during passage of the tape through a process block. Following washing and drying the tape passes under the epifluorescence microscope head. On excitation at *c.* 490 nm, transient fluorescent signals are emitted by the microbial cells which are detected by a photomultiplier tube. Analogue-to-digital signal processing results are printed out within 30 s as the number of Colony Forming Units per ml of sample. Results may be stored or transmitted through an RS232 interface.

TABLE 1
Comparison of Autotrak Counts (y) and Colony Counts (x); 14 Vegetable Process Effluent Samples

	Unfiltered	*Filtered*
Regression (y)	$0{\cdot}906 + 0{\cdot}747x$	$0{\cdot}781 + 0{\cdot}793x$
Correlation Coefficient (r-value)	0·787	0·903

RESULTS AND DISCUSSION

Vegetable process effluents

Table 1 shows the relationship between direct Autotrak counts and colony counts after 48 h incubation at 30°C for 14 unfiltered specimens. Following filtration of samples through a 100 μm metal filter to remove gross particulate matter the results indicate an improvement in the Correlation Coefficient from 0·787 to 0·903 without significant loss of recovery of microorganisms.

Meat plant effluents

Forty-one specimens were obtained from abattoir floor drains, carcase washings and food contact and non-contact surfaces on meat handling lines. Figure 1 shows the Autotrak versus Colony Count regression analysis. The precision of the DFM determinations is below 10^4 CFU per

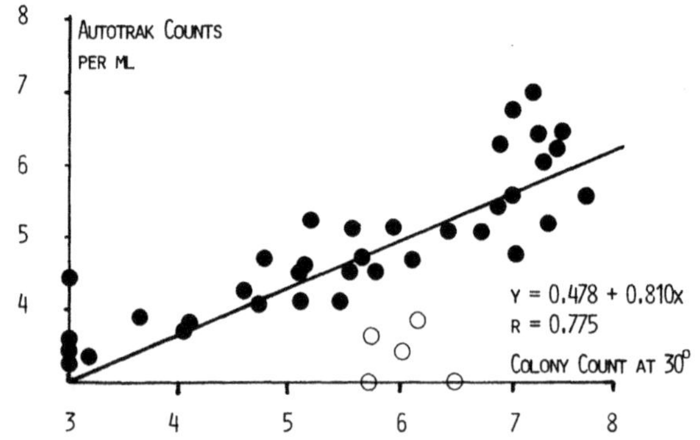

Fig. 1. Comparison of Autotrak (y) and colony counts (x): 14 meat process effluents.

TABLE 2
Comparison of Autotrak Counts (y) and Colony Counts (x) at 24 and 72 h and 7 days: 13 Meat Plant Effluents

	24 h	*72 h*	*7 days*
Regression (y)	$0{\cdot}350 + 0{\cdot}802x$	$0{\cdot}518 + 0{\cdot}804x$	$0{\cdot}866 + 0{\cdot}767x$
Correlation Coefficient (r-value)	0·748	0·770	0·756

ml, which is low. A sub-group of 5 samples from a carcase scalding tank gave low DFM counts. This may be due to the effect of residual chlorine in the water. The presence of high levels of fat in other samples dictated the addition of a counterstaining solution (Nigrosine) to samples for DFM examination.

Effect of incubation time

DFM counts were performed on 13 non-filtered samples of sewage from the drain outlet of a meat processing plant. Colony counts were determined at 30°C following 24 h, 72 h and 7 days incubation. The results in Table 2 demonstrate a progressive improvement in the regression characteristics with an increase in the correlation coefficient from 0·748 to 0·770 between 24 and 72 h incubation.

These preliminary studies demonstrate the potential value of the Autotrak-DFM system for rapid primary screening of process waters and effluents in the range 10^4–10^8 CFU/ml. With the development of specific and appropriate protocols, it is apparent that this instrument would provide ‘real-time’ information of value for microbiological quality and process control in the food industry.

REFERENCES

Gilchrist, J.E., Campbell, J., Donnelly, C.B., Peeler, J.T. & Delaney, J.M. (1973). *Applied Microbiol.*, **25**, 244.

Harran, D. & Scholefield, J. (1987). *Proc. 33rd Int. Congress of Meat Science and Technology*, Helsinki, Finland.

Ploem, J.S. (1967). *Z. Wiss. Mikroskopie*, **68** (3), 129–42.

Scholefield, J. (1986). *Proc. Conf. Emerging Medical Technologies,* November, Medica, Dusseldorf, FRG, p. 4.

Scholefield, J., Manson, R. & Johnston, R.J. (1976). *Proc. 2nd Int. Symp. 'Rapid Methods and Autom. in Microbiol.* Cambridge, p. 253.

Scholefield, J., Manson, R., Johnston, R.J., Scott, R. & Spinell, M. (1981). *Proc. 3rd*

Int. Symp 'Rapid Methods and Autom. in Microbiol. Washington D.C., USA, p. 1979.
Scholefield, J., Manson, R., Johnston, R.J. & Scott, R. (1985). *Urol Res.*, **13**, 141.
Sharpe, A.N. (1975). *Applied Microbiology*, **30** (4), 519–24.
Strugger, S. (1948). *Can. J. Res. Serc.*, **26**, 188.
Wood, J.M. & Gibbs, P.A. (1982). *Devel. Food Microbiol.*, ed. R. Davies. Elsevier Applied Science, London, p. 183.

Flow Micro-fluorimetry as an Aid to DNA Determinations in the Genus *Penicillium*

P.D. Bridge

CAB International Mycological Institute, Ferry Lane, Kew, Surrey TW9 3AF, UK

ABSTRACT

Flow micro-fluorimetric studies on a number of closely related fasciculate species of Penicillium *have been carried out using Feulgen-acriflavine stained conidial suspensions. Conidial DNA determinations using samples of 10^4 conidia have shown that the amount of DNA can vary both between strains and between different phenotypic lines of the same strain. The differences were reproducible and showed a good correlation with variations in conidium size. Some inter-strain comparisons showed differences corresponding to a doubling of the DNA content, indicating a possible variation in ploidy level. The results demonstrate that flow cytometry may provide a useful, rapid aid to DNA determinations in conidial fungi.*

INTRODUCTION

Flow cytometry is a technique which has found many applications in immunology, particularly for separating cells. Samples of particulate materials, such as cell suspensions, are injected into a continuous flow carrier liquid which flows through a small 'window', illuminated by a tuneable laser. Light that is reflected, refracted or emitted by the particles is detected by photomultipliers and this information is monitored by a computer. The varying degrees of reflection and refraction can be used to give an estimate of the size and topography of the particles, and one aspect of the equipment is computer controlled deflection plates that allow particles with different characteristics to be separated (Crissman *et al.*, 1978; Muirhead *et al.*, 1985). As the system is computer controlled and involves a continuous flow, large numbers of particles may be examined in a very short time span. Inclusion of filters and fluorescent stains allows quantitative comparisons to be made on the basis of staining intensity. A number of different fluorescent stains are available that will bind specifically with DNA or with A–T base pairs and this has been used to separate mixed populations of bacteria with different mol% G+C values (van Dilla *et al.*, 1983). This paper considers recent studies on variation in DNA content of fluorescent Feulgen stained conidia from isolates of *Penicillium*.

MATERIALS AND METHODS

Cultures of *Penicillium viridicatum* (IMI 39758ii), *P. martensii* (IMI 297891) and *P. crustosum* (IMI 291543) were maintained on slopes of Czapek agar (CZ; Smith & Onions, 1983). Up to 50 single conidium isolates were made by isolation from conidial suspensions plated onto Tap Water agar (TWA; Smith & Onions, 1983) after 36 h incubation. Strains for flow cytometry were prepared from single conidium lines grown on CZ slopes for 7 days. Suspensions were prepared from the slopes in 0.2% (v/v) aqueous Tween 80 and washed twice by centrifugation. Conidia were fixed in Carnoy's fluid and Feulgen stained with acriflavine using the method of Bridge *et al.* (1986).

Flow micro-fluorimetry was performed using an EPICS V instrument with an argon ion laser (300 mW at 488 nm) followed by analysis for forward angle light scatter (FALS) and emitted fluorescence (with 515 nm long-pass interference and absorbance filters (Bridge *et al.*, 1986; 1987)).

RESULTS AND DISCUSSION

Measurement of FALS for each sample showed a single, normally distributed peak that represented the conidia. Using the size range of this peak as limits, fluorescence was measured as the log integral green fluorescent light (LIGFL). In each sample this gave two normally distributed peaks, one large peak and one smaller peak which was equivalent to double that of the first. This doubling factor was seen in all samples although the relative numbers of conidia corresponding with each peak varied. In general, younger samples showed a large first peak

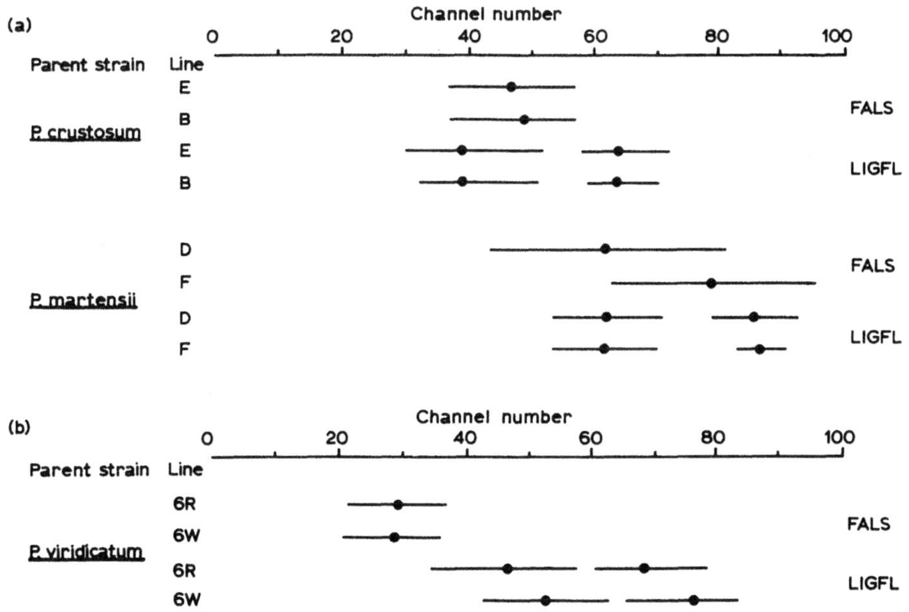

Fig. 1. Conidial size (FALS) and DNA content (LIGFL) in lines of (a) *Penicillium crustosum* and *Penicillium martensii* (derived from Bridge *et al.*, 1987) and (b) *Penicillium viridicatum* (derived from Bridge *et al.*, 1986).

and a very small second peak, and this was reversed in older samples. It seems likely that this second peak is due to the small number of binucleate conidia seen in each sample and as the ratio varies with time, nuclear replication prior to the onset of germination may also be involved.

The peak positions, and so the relative DNA contents, for the conidia from representative single conidial lines of *P. crustosum* and *P. martensii* are shown in Fig. 1(a). Lines of *P. crustosum* showed similar conidial sizes and DNA contents. Lines of *P. martensii* showed large variable conidia with almost exactly double the DNA contents of the *P. crustosum* lines. The isolate of *P. martensii* that these lines were derived from is known to produce both sectors and sterile mycelium in culture, both of which are features suggestive, but not indicative, of an aneuploid or diploid culture. This is apparently supported by the DNA contents and the larger conidium size.

Lines of *P. viridicatum* fell into two groups (Fig. 1(b)) which matched previously reported morphological variations. There was no significant difference in conidium size but there was a difference of approximately 17% in DNA content. The two morphological distinct lines were interconvertible, in that isolates of the w-line could occasionally give rise to sectors identical to the r-line. These also showed the reduced DNA content of the other r-lines. Similar variations have been reported in a number of higher plants and although the exact mechanism of these changes in DNA content are not clear, possible mechanisms include aneuploidy, part chromosome duplication and amplification. Flow cytometry provides a rapid and simple method for screening particulate samples and can have applications where changes in ploidy level are suspected, as in isolates showing highly variable characteristics (Tolmsoff, 1983) or high yielding industrial strains (Pathak & Elander, 1971).

ACKNOWLEDGEMENTS

This work was undertaken in conjunction with Science and Engineering Research Council contract SO/17/84. Flow cytometric analyses were undertaken by Professor L. Hudson, Department of Immunology, St George's Hospital Medical School.

REFERENCES

Bridge, P.D., Hudson, L., Hawksworth, D.L. & Bridge, D.A. (1986). *FEMS Microbiology Letters*, **37**, 241–4.

Bridge, P.D., Hudson, L., Kozakiewicz, Z., Onions, A.H.S. & Paterson, R.R.M. (1987). *Journal of General Microbiology*, **133**, 995–1004.

Crissman, H.A., Stevenson, A.P., Kissane, R.J. & Tobey, R.A. (1978). In *Flow Cytometry and Sorting*, eds R. Melamed., P.F. Mullaney & M.L. Mendelsohn. John Wiley, Chichester.

van Dilla, M.A., Langlois, R.G., Pinkel, D., Yajko, D. & Hadley, W.K. (1983). *Science*, **220**, 620–2.

Muirhead, K.A., Horan, P.K. & Poste, G. (1985). *Biotechnology*, **3**, 337–56.

Pathak, S.G. & Elander, R.P. (1971). *Applied Microbiology*, **22**, 366–71.

Smith, D. & Onions, A.H.S. (1983). *The Preservation and Maintenance of Living Fungi*. Commonwealth Mycological Institute, Kew, UK.

Tolmsoff, W.J. (1983). *Annual Review of Phytopathology*, **21**, 317–40.

TEXTILES

Biodeterioration of Textile Materials and Textile Preservation

B.F. Sagar

Shirley Institute, Didsbury, Manchester M20 8RX, UK

ABSTRACT

Biodeterioration of textile materials is not just confined to mildew and rotting of cellulosic products—micro-organisms interfere with processes such as water-jet weaving and they attack many textile auxiliaries including conventional sizing agents, fibre yarn lubricants, textile adhesives, back fillings, and latex emulsions, together with a variety of coated and laminated textile materials.

Synthetic textile materials (with the exception of certain polyurethanes) are inherently resistant—nevertheless they are frequently susceptible to microbial spoilage because of the presence of the biodegradable processing and finishing agents. Unsightly mildew growth on synthetic textiles such as those derived from polyamides, polyester, polypropylene or polyacrylonitrile can be just as serious as the degradation of cellulosics because of the loss of aesthetic appeal.

A review of textile biodeterioration is presented, along with methods of preventing microbial attack, detection of the presence of micro-organisms, and techniques for assessing propensity to mildew and biodeterioration and the efficacy of anti-fungal treatments.

PROFILE OF THE TEXTILE INDUSTRY

The textile industry contributed an estimated £2914 million to the UK national economy in 1986. Taking into account the closely related clothing industry, the combined contribution (£4830 million) was more than that of the whole motor vehicle industry (£4440 million), 1·6 times that of the areospace industry (£2990 million), almost three times as much as office machinery and data processing equipment (£1661 million) and over six times as much as shipbuilding (£757 million) (Anon, 1987).

Almost 10% of the total UK employment in manufacturing industry in 1986 was in the textile and clothing industries, far outweighing that of any of the above industries. Only the engineering and the food, drink and tobacco industries provided more manufacturing jobs than the 507 000 employed in textile and clothing manufacture. Output of textiles rose in 1986 for the fourth successive year. It is now 10% above the 1982 low point but still almost 20% lower than the pre-recession 1979 level. Output per worker in the textile industry has risen by 40% since 1980.

The long-term trend is for consumption of textiles and clothing in the UK to grow in volume terms by nearly 1% per annum. UK textile imports rose by 7% in 1986, and sadly we are now importing nearly twice the weight of textiles that we imported 10 years ago; imports now exceed exports by a factor of more than three.

DURABILITY OF TEXTILE MATERIALS

Textile materials fail in service when their physical properties deteriorate to some critically unacceptable level. Deterioration may be brought about by various mechanisms; purely physical effects associated with mechanical processes such as tensile stress, abrasion, flexing, etc., and four processes of ageing associated with either atmospheric oxygen (thermal- or photo-autoxidation) or hydrolysis (induced either chemically or enzymatically). The natural and synthetic polymers involved in the textile industry differ greatly in their susceptibility to these four modes of chemical/biochemical breakdown. For example, polyester textile fibres are much more resistant to photo-oxidation than polyamides, hence the predominance of polyester net curtains.

Quite severe conditions of pH and temperature are required to hydrolyse the polymers that constitute textile fibres; obviously the polymer must be able to withstand the conditions of wet processing (scouring, bleaching, dyeing, etc.) and subsequent laundering. But many micro-organisms elaborate enzymes that catalyse the hydrolysis of several important textile fibres at ambient temperature and at a pH close to neutrality.

BIODETERIORATION/BIODEGRADATION OF TEXTILES

Micro-organisms cause considerable losses both to the textile industry and to the consumer. This wastage, which must amount to many millions of pounds a year, is especially caused by the carbon-heterotrophous micro-organisms. Not all breakdown of materials by micro-organisms is undesirable. Depending on the circumstances, one and the same process may be beneficial or detrimental to man; the damage caused by saprophytic micro-organisms is far outweighed by the vital role they play as the regulators of the dynamic equilibrium of nature. About 3×10^{10} t of carbon, in the form of CO_2, are transformed yearly into plant material by the photosynthetic process and biodegradation of the ensuing dead vegetation ensures the ultimate return of all of this CO_2 to the atmosphere. Unfortunately, micro-organisms cannot distinguish between materials that are useful to man and plant debris and man's rubbish, and we have to spend large sums to prevent the biodeterioration of textile materials.

We must bear in mind the distinction between biodeterioration and biodegradation. The former has been defined as '*any* undesirable change in the properties of a material caused by the vital activities of organisms' (Hueck, 1968). Biodegradation on the other hand includes those activities of organisms which result in the actual breakdown of a material, either to man's detriment or good; in the latter sense by producing new substances of value.

MATERIALS AT RISK

Since bacteria and micro-fungal spores are omnipresent in any non-sterile environment, all textile materials become contaminated during, or soon after manufacture, unless special precautions are taken to maintain sterility, as with surgical dressings. Textile materials containing any kind of nutrient are at risk therefore whenever they are exposed to a humid atmosphere or stored under damp conditions conducive to the germination and growth of these ubiquitous spores. Besides requiring water and a metabolisable source of carbon all deteriogens need additional nutrients, particularly a source of nitrogen, phosphorus, and other essential elements. It is for this reason that micro-organisms do not grow so readily on clean, scoured and bleached cotton, which contains virtually no nitrogen. Micro-organisms differ in their ability to grow and cause damage under conditions of physiological drought. It must be remembered, however, that the total amount of water in a material is of little biological importance; it is how tightly the water is bound to the substrate that matters. A measure of the amount of water available to the organism is given by the 'water activity', which is related to the ratio of the partial vapour pressure of the water in the material to that of pure water at the same temperature and pressure and is numerically equivalent to the equilibrium relative humidity. Provided the water activity is kept sufficiently low in the micro-environment around the individual fibres, micro-organisms will not grow.

With the notable exception of polyurethanes, most wholly synthetic, man-made fibres are inherently resistant to degradation of their polymer structure by micro-organisms. Nevertheless, all fibres are susceptible to microbial spoilage if they contain biodegradable impurities, processing aids, or finishing agents. Unsightly mildew growth on synthetic textiles such as those derived from polyamides, polyester, polypropylene, or polyacrylonitrile fibres can be just as serious as the biodegradation of textile materials containing cellulosic fibres (cotton, viscose, flax, jute, hemp, ramie) because of the loss of aesthetic appeal.

Depending on the circumstances and the fibre type, the biodeterioration of textile materials can take several forms, including loss of strength, actual loss of material, staining, dyeing defects, and malodours. Particularly striking effects are the serious loss of strength of cellulosic materials (which can be complete within a few days if conditions are conducive to the rapid growth of cellulolytic microfungi) and the damage to woollen materials by insects; both of these deteriogens can cause complete destruction of the textile. Less dramatic but equally serious effects include permanent staining associated with the pigments produced by many micro-organisms and dyeing defects brought about by the presence of organisms which perhaps were not detected before dyeing.

Whether a textile material needs to be protected by treatment with a biocide is governed by its end-use. At serious risk are materials containing cellulosic fibres and likely to come into contact with moist soil, which inevitably contains a diverse range of cellulolytic micro-organisms. A significant proportion of all textile products containing cotton, viscose, flax, jute, ramie and hemp intended for use as canvas, tarpaulins, tentage, awnings, flags, shop blinds, boat covers, carpet backing, ropes, cordage, webbings, sacking, nets, filters, pressed felts, and coated textiles is therefore treated with biocides.

BIODETERIORATION DURING TEXTILE PROCESSING

Biodeterioration of fibre- and yarn-lubricants

Lubricants are employed for several purposes in textile processing and may lead to mildew problems if the lubricated yarn is stored under conditions conducive to the growth of micro-fungi. Applications of lubricants to fibres and yarns include spin finishes, knitting oils, texturing oils, raising and sueding oils, and spinning conditioning agents. Several components of these fibre- and yarn-lubricants are biodegradable and may therefore provide nutrient for microbial growth at any intermediate stage during processing. Since many of these lubricants are applied from aqueous emulsions the yarn may inadvertently be stored in a damp state, facilitating the growth of micro-organisms. Synthetic fibres are just as prone to mildew attack in this way as are cellulosic fibres.

Growth of mildew on packaged yarn before final processing leads to wastage of production. Moreover, if the lubricant is not removed during fabric finishing, the final product may also be at risk. Effective removal may require an additional process involving scouring, although some raising oils are designed to be removed simply by heating when the relatively low boiling lubricants are volatilised. If these biodegradable materials are not removed during processing, biocides may have to be incorporated in the original lubricant to guard against the possibility of the final textile product being susceptible to mildew.

Biodeterioration of warp sizes and loomstate fabric

Starch is still widely used for sizing, although less biodegradable synthetic size materials are finding increasing use in the weaving industry. Micro-organisms quickly grow in a starch size bath. It was common practice in earlier years to mix large batches of size preparation to last for several days and protection of this size material with biocide was essential. Modern techniques of size preparation allow small batches to be economically mixed and the need for biocides in this area has declined. Likewise there is less demand for protection of the sized warp sheet and loomstate fabric; export of loomstate fabrics to tropical countries has diminished greatly and better awareness of the importance of correct storage of yarns and fabrics containing size has mitigated this problem.

Biocidal agents used in size applications should be effective at low concentrations and be compatible with the size mixture. They should not cause staining of the yarn and they should be easily removed during desizing and scouring. If protection of the size mix itself is the main consideration then the most suitable biocide is formaldehyde. Since this compound is volatile it is quickly lost after weaving and other biocides must be used to protect the warp sheet and loomstate fabric. Suitable agents for this purpose include: salicylanilide ('Shirlan'), sodium *o*-phenyl phenate dichlorophen, and bis(tributyl tin) oxide.

Water-jet weaving

Weaving by water-jet loom has grown in importance during recent years. This method of weaving, where weft insertion is by a jet of water, is restricted to yarns composed entirely of man-made fibres which do not swell in water, e.g. polyamide, polyester, acrylic, and cellulose acetate fibres. Whilst these fibres are inherently resistant to degradation by micro-organisms they may contain low molecular weight monomers or oligomers (e.g. caprolactum in nylon 6), or lubricants and anti-static agents in the spin-finish applied by the fibre manufacturer immediately after extrusion to assist in processing the yarn, e.g. during winding; many of these materials are biodegradable and will support the growth of micro-organisms. Likewise the size material applied to the warp yarns to assist weaving, although chosen for its poor water-solubility at ambient temperature, may contain components which support microbial growth.

There are two reasons for the applications of biocides during water-jet weaving; firstly, to prevent growth of micro-organisms (principally bacteria) in the water, which may contain substantial amounts of nutrients due to extraction of some of the organic material described above from the yarn. Severe development of bacteria in the water can lead to blockages and consequent loom stoppages, to foul odours, and to entrapment of the microbial sludge in the fabric. Secondly, if adequate precautions are not taken to dry the fabric as it comes off the loom, mildew growth can occur on the damp fabric after weaving.

It is a common practice, therefore, to treat the water system with a suitable biocide. The problem is complicated by the fact that some strains of micro-organisms develop resistance to an initially successful biocide after a period of use. The biocide must have sufficient water-solubility to allow effective additions to be made to the water supply. It must not produce staining or odours on the fabric nor be incompatible with subsequent processes, including dyeing. Only compounds which are comparatively non-toxic to man can be used since there is inevitably some spray generated by the water-jet loom. Biocides which have been used with varying degrees of success include: *o*-benzyl *p*-chloro-phenol, 5,5′-dichloro-2,2′-dihydroxydiphenyl sulphide, 2,4,4′-trichloro-2′-hydroxydiphenylether, sodium hypochlorite, and sodium dichloro-isocyanurate.

Biodeterioration of textile adhesives, back-fillings, and latex and synthetic emulsions

Considerable quantities of biodegradable materials are used in the back-coating of carpets, for back-filling specialised materials such as traditional book-binding fabrics, bed-tickings, abrasive cloths, and window-blind materials, for adhesives in laminating, and in latex and synthetic emulsions. The most susceptible materials (starch and glue) require protection from the action of micro-organisms both during processing and in the final products. Temporary protection of starch and glue mixes during back-filling operations is adequately and cheaply achieved by addition of formaldehyde. When protection of the final product (e.g. book-binding fabrics, window-blind material) is the objective, more permanent biocides are required but these do not have to withstand leaching with water; sodium *o*-phenyl phenate or dichlorophen are commonly used.

Similarly, several types of latex emulsion may need protection with biocide to prevent biodeterioration during storage and also in the final product. SBR latex emulsions, for example, can deteriorate in storage due to bacterial action, whilst the final product may be susceptible to fungal attack. Other natural and synthetic rubber latex emulsions may require little protection during storage because of their high ammonia content but they too may require the addition of a permanent biocide to give protection to the finished carpet if this is likely to be exposed to arduous environmental conditions. Sodium mercaptobenzothiazole is frequently used to protect SBR latex emulsions. This biocide has both anti-bacterial properties (for protection of the emulsion during storage) and anti-fungal properties (for protection of, for example, latex-backed carpets).

Biodeterioration during wet processing

Most textile materials are prone to microbial spoilage during wet processing (fabric preparation, dyeing and finishing). The likelihood of problems of this kind depend on a number of factors, including the fibre type, presence of impurities, sizes, or textile auxiliaries, yarn or fabric pH, temperature, water activity, atmospheric conditions, stage in processing, time delay/storage time, and whether a biocide is present.

BIODEGRADATION OF CELLULOSIC FIBRES

Some micro-organisms are able to utilise cellulosic fibres directly as a source of carbon. Many other more readily available sources of carbon are also utilised by these cellulolytic organisms and they often start their growth on residual textile auxiliaries or starch size, but in the presence of cellulose they quickly elaborate enzymes which catalyse the hydrolysis of the cellulose polymer to produce low molecular weight sugars and ultimately glucose. The glucose is assimilated by the micro-organism to provide its principal source of energy.

Several of the cellulolytic microfungi can be cultivated under controlled fermentation conditions to produce cell-free isolates capable of extensive degradation (under aseptic conditions) of the highly-ordered, crystalline forms of cellulose present in the natural fibres. Production of the complex mixture of extracellular cellulase enzymes requires the presence of an inducer (cellulose, cellobiose or sophorose) and is subject to catabolite repression. Several individual enzymes are involved; (i) a C_1 component, so named (Reese *et al.*, 1950) because it was believed to act first on crystalline cellulose, de-aggregating the cellulose chains in preparation for subsequent attack by the hydrolytic enzymes; (ii) the exo- and endo-β-1,4-glucanases (generally referred to as C_x enzymes, the x reflecting the fact that normally there are several of these components in culture filtrates of cellulolytic micro-fungi, which readily hydrolyse soluble derivatives of cellulose such as carboxymethyl cellulose, or swollen and partially degraded celluloses, but cannot attack highly-ordered substrates in the absence of C_1, and finally, (iii) the β-glucosidases, which hydrolyse cellobiose and short-chain cello-oligosaccharides to glucose (Fig. 1). The role of C_1 remains controversial; it seems to be characterised by its ability to catalyse the hydrolysis of β-1,4-glucan bonds in cellulose molecules situated in one pair of surfaces of the crystalline structural elements in native cellulose, where the spatial disposition of the hydroxyl groups available for hydrogen-bonding with an enzyme molecule is quite unique (Sagar, 1985). Figure 2 illustrates the disintegration of the structure of cotton fibres by the total extracellular cellulase complex elaborated by *Trichoderma viride*. Other producers of cellulase enzyme systems active against highly-ordered cellulose substrates include *T. koningii, Penicillium funiculosum, Fusarium solani* and *Sporotrichum pulverulentum*.

BIODETERIORATION OF COATED OR LAMINATED TEXTILE MATERIALS

Woven and knitted fabrics coated or laminated with polyvinylchloride, polyurethane, or natural rubber are widely used as waterproof materials

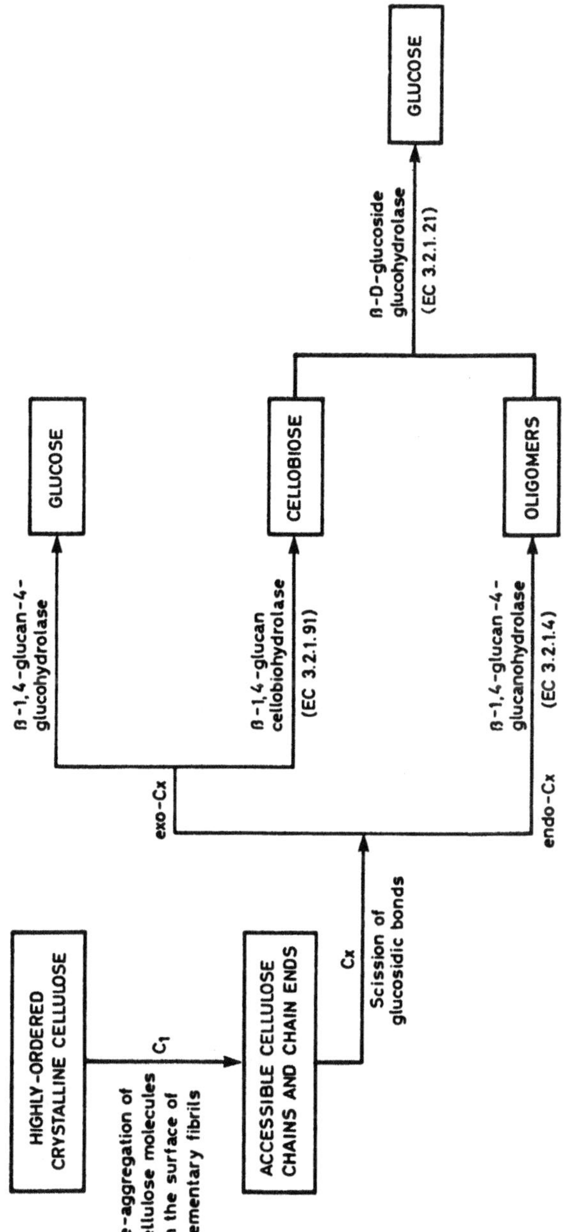

Fig. 1. Action of the individual components of the cellulase complex.

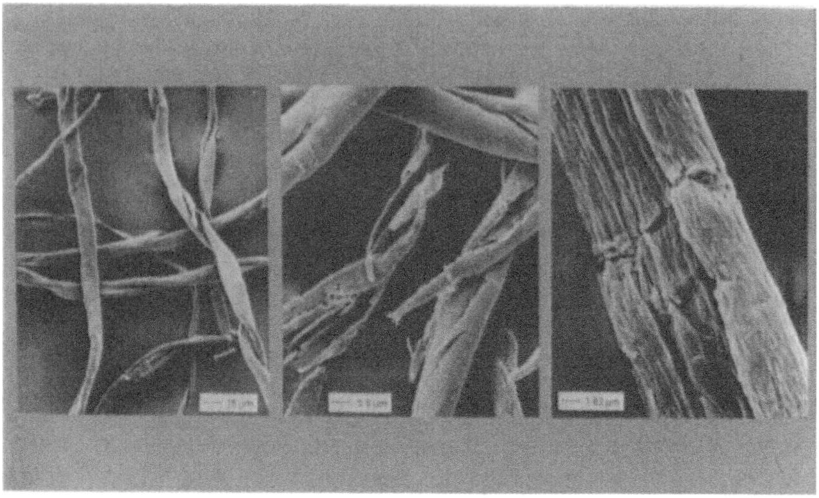

Fig. 2. SEM of cotton fibres treated with the cellulase complex from *T. viride*.

for fabricating a wide range of textile products such as tarpaulins, rain- and sports-wear, rubber dinghies, etc. The base fabric may be cellulosic but synthetic fibres are frequently used for increased strength and lower weight. Whilst the need for rot- and mildew-protection of the cellulosic-base fabrics is widely recognised, it is perhaps not so generally realised that the plastic or rubber coating itself is also susceptible to biodeterioration.

Several micro-organisms have been shown to attack natural rubber and degradation of rubber-coated fabrics can occur due to microbiological deterioration of the rubber; some accelerators and anti-oxidants, however, may serve the dual purpose of acting as biocides.

Although polymers such as polyvinylchloride do not act as a primary carbon source for bacteria and micro-fungi, these PVC plastic films and coatings may be susceptible to attack by micro-organisms depending on the amount and type of plasticisers, fillers, pigments, stabilisers, and processing aids incorporated in the plastic; most plasticisers, particularly those based on derivatives of azelaic acid, adipic acid, sebacic acid, succinic acid and epoxidised soya bean oil, are highly susceptible to microbiological attack and biocides are a necessary feature for such vinyl systems to protect the coating against organisms which affect the aesthetic qualities and to prevent consequent changes in mechanical properties of the plastic film due to destruction of the plasticisers.

Unlike most synthetic polymeric materials, polyurethanes are inherently susceptible to attack by micro-organisms. Polyurethanes are produced by addition reactions between polyisocyanates (difunctional

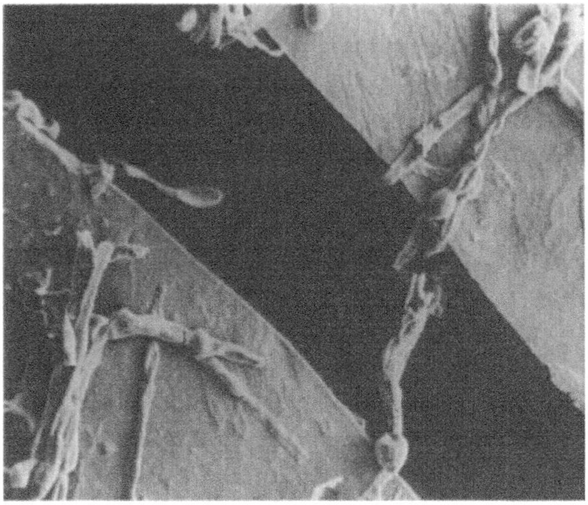

Fig. 3. SEM of biodegraded polyurethane film.

or higher) and polyhydroxy-compounds such as glycols, polyesters, and polyethers. Their susceptibilities to biodeterioration are dependent upon the polymer structure; polyether polyurethanes are moderately resistant to micro-fungal attack whereas most polyester polyurethanes are highly susceptible to this form of biodeterioration (Darby & Kaplan, 1968; Wales & Sagar, 1985). Compared with other synthetic polymer systems, such as PVC, where the polymer *per se* does not serve as a microbial nutrient and only additives (plasticisers, etc.) are metabolisable,

TABLE 1
Micro-organisms Shown to Attack Polyester-polyurethanes

Micro fungi	*Alternaria tenuis*
	Aureobasidium pullulans
	Aspergillus amstelodami
	Chaetomium globosum
	Penicillium chrysogenum
	Penicillium ochrochloron
	Trichoderma viride
	Stachybotrys atra
Bacteria	*Pseudomonas aeruginosa*
	Pseudomonas sp.
	Flavobacterium sp.
	Micrococcus sp.

polyurethanes are unique and pose special problems. Initial microbial attack of polyurethane-coated fabric is manifested by incipient cracking of the coating (Fig. 3). A wide range of micro-organisms (some examples are given in Table 1) elaborate the non-specific esterases implicated in the biodegradation of these polyester-based polyurethanes.

CONTROL OF BIODETERIORATION

Control of the environmental physical conditions is by far the most effective way of preventing the biodeterioration of textile materials. Textiles in the dry Egyptian desert protected from sunlight have withstood the ages, while wool and furs kept at temperatures below 10°C are practically invulnerable to insect damage. Recognising the impracticability of this approach a biocide is usually incorporated in materials susceptible to biodeterioration, especially if they are intended to be used under conditions conducive to microbial and insect proliferation. Control of biodeterioration by the application of a biocide is widely practised in the textile industry. The technical and patent literature refers to a profusion of very varied types of both organic and inorganic chemicals with biocidal activity for use during processing and as additives to protect the finished products (Allsopp & Allsopp, 1983). But many of these biocides are not in current use and reliable information as to the actual chemicals being used is often difficult to obtain. The current climate of opinion regarding the toxicological hazards of such chemicals makes acceptance of new compounds, even when supported by extensive (and expensive) toxicity testing data, a very slow process.

Microbial biocides are customarily considered to be fungistatic/ bacteriostatic (i.e. to inhibit further development of the micro-organism) or fungicidal/bacteriocidal (cause death of the micro-organism) but the dividing line is difficult to define and it is probably better simply to refer to biocides as anti-fungal or anti-bacterial agents.

Rot and mildew proofing

The attributes of an ideal biocide are listed in Table 2. Three principal types of biocide have found use for the rot- and mildew proofing of textile materials, viz. organic compounds, organo-metal compounds, and inorganic salts. Inorganic treatments are less widely used these days and many of the more toxic organo-metal biocides have naturally fallen out of favour in recent years and several have been justifiably banned in many countries, e.g. the organo-mercurials.

TABLE 2
Attributes of an Ideal Biocide

1.	Effective against wide range of micro-organisms
2.	Low mammalian toxicity and non-toxicity to humans
3.	Active during life of product (resistant to sunlight and leaching)
4.	Colourless and odourless
5.	Effective at low concentrations
6.	Inexpensive and easy to apply
7.	Without effect on fabric 'handle' or strength
8.	Must not sensitise fabric to damage by light or other influences
9.	Compatible with water-repellent and flame-retardant agents, dyes, and other textile auxiliaries
10.	No adverse effects on other materials in close proximity

Some commonly used mildew- and rot-proofing agents are listed in Table 3. Two of the most important chemical systems still in current use are pentachlorophenyl esters of mixed fatty acids of C_{12} average chain length (usually containing some free pentachlorophenol) and copper naphthenate. The latter is relatively cheap and is used whenever its green colour and characteristic, unpleasant smell are of no disadvantage. The effect of copper naphthenate on the susceptibility to degradation by sunlight is variable and is probably related to the purity of the product and the presence of inorganic copper salts in the system. Both agents are reasonably resistant to leaching by water in service—a major consideration for biocides in this field. The search for the 'ideal' biocide continues; none is completely effective against all organisms and without some

TABLE 3
Commonly Used Mildew- and Rot-proofing Agents

Phenolics	Sodium *o*-phenyl phenate (Mystox WFA)
Chlorinated phenols	Pentachlorophenyl laurate (Mystox LPL) 5,5′-Dichloro 2,2′-dihydroxy diphenyl methane (Panacide)
Organo-metallic compounds	Copper-8-hydroxy quinolinolate (Mystox 8) Bis(tri-*n*-butyltin)oxide (Irgarol B1)
Other organic compounds	N-(2-methyl-1-naphthyl) maleimide (Vinyzene T-129) N-trichloromethylthio-4-cyclohexene-1,2-dicarboximide (Vancide 89) 6-Acetoxy-2,4-dimethyl *m*-dioxane (Giv-Gard DXN)

disadvantage. The use of a single biocide has some fundamental limitations and it is often advantageous to apply two or more mutually compatible agents in a single treatment.

Biocides are often applied as aqueous emulsions along with other finishing agents. For example, an FWR finish on military tentage consists of a combination of fire-retardant, water repellent and rot-proofing agents. The amount of biocide must be carefully controlled and contracts for rot-proofed materials (especially for military end uses) generally follow the specifications in BS 2087, Part 1 (Anon., 1981*a*). Part 2 of this standard (Anon., 1981*b*) describes the methods of chemical analysis to be used.

Moth- and insect-proofing

All textile materials containing the natural fibre, wool, are susceptible to insect damage. In the United Kingdom the grub of the common clothes moth, *Tineola bisselliella*, is considered to be the most important insect that attacks materials made from wool or woolen blends, whilst in other parts of the world carpet beetles are important pests, in particular the black carpet beetle, *Attagenus piceus*, the furniture beetle, *Anthrenus flavipes*, and the woolly bear, *Anthrenus vorax*. It is common practice to blend wool with synthetic fibres such as polyamides, polyester, or acrylics but these wool/synthetic mixtures are equally susceptible to attack by insects and therefore also require to be protected with biocides. Although the clothes cloth grub consumes both nylon and wool indiscriminately in a blended material, the nylon is excreted unchanged and the presence of moth-proofing agent on the nylon is thought not to be toxic to the larvae; toxicity appears to be related to the concentrations of biocide on the wool and larger amounts may therefore have to be applied to wool/nylon blends using those biocides which have a greater affinity for nylon than for the wool component of the blend.

Biocides for moth- and insect-proofing are of two main types—contact insecticides are larval stomach poisons. The only contact insecticide to have found common use for this purpose was dieldrin. This material is substantive to wool, effective at extremely low concentrations, and resistant to washing. Because of its effectiveness at low concentrations it is the most economical method of moth-proofing but current concern about the toxicity of insecticides and their build-up in the environment has led to doubts as to the advisability of continuing to use dieldrin for moth-proofing. The principal larval stomach poisons used in Europe for moth-proofing are the proprietary produce Eulan WA/new (Bayer) and Mitin LP or FF (Ciba-Geigy).

Hygienic finishes

Application of anti-bacterial agents to textiles to prevent growth of bacteria in the interests of hygiene is proving to be of increasing interest. One of the principal objectives in applying hygiene or sanitised finishes is the elimination of the cause of unpleasant odours associated with bacterial decomposition of perspiration. Some protection against the micro-fungi (*Epidermophyton flocosum* and *Trichophyton mentagrophytes*) responsible for athlete's foot is also claimed for several of these agents and treatment of footwear and floor-coverings is carried out to prevent infection.

Hygienic finishes are applied to a wide range of materials including socks, stockings, linings for footwear, underwear, mattresses, pillows, suit linings, dress shields, shower curtains, towels, hospital blankets, floor-coverings and carpets. Biocides for most of these hygiene applications obviously must have low human toxicity. Special attention needs to be given to the possibilities of dermatitic effects and dermal absorption of the biocide into the bloodstream when treated articles are worn in contact with the skin. A particular problem with articles that require frequent laundering, e.g. socks, is the requirement for good washfastness of the hygienic finish; but some proprietary products claim resistance to repeated laundering and dry-cleaning.

DETECTION OF MICRO-ORGANISMS ON TEXTILE MATERIALS

Gross contamination is indicated by clearly visible pigmentation, fluorescence under UV illumination and the characteristic odour of mildew. Hyphae and spores of many strains of micro-fungi are easily recognised on textile materials under a good stereo-binocular microscope or when examined microscopically in a suitable mounting medium (e.g. lactophenol) at a magnification of about 150×. Bacteria and strains of micro-fungi with very fine hyphae are more difficult to detect unless the microbiological material is first stained. A Safranin/Picro Aniline Blue stain is very suitable to use with contaminated textile materials. The stain is made up in two parts: Stain A consists of 1% Safranin (water soluble) in distilled water; Stain B consists of Aniline Blue (water soluble) (0·25 g) dissolved in a saturated aqueous solution of picric acid (100 ml). The sample is immersed in Stain A for 1 min at room temperature, washed in water, blotted dry, heated gently in Stain B until the liquid boils, allowed to cool in the solution, rinsed thoroughly first in water and then in absolute alcohol, and finally dried thoroughly before mounting on a microscope slide in either cedar-wood oil or ALP2 immersion oil. Hyphae, spores and bacteria are stained deep blue. This staining technique is applicable to the detection of micro-organisms on all textile materials except silk which is itself stained blue. Alternatively, the micro-organisms can be readily visualised using a scanning electron microscope. A method for the detection of microbial activity at an early stage of textile spoilage, based on the firefly bioluminescence assay of adenosine triphosphate (ATP) has been described (McCarthy, 1983).

TESTING THE PROPENSITY TO MICROBIAL ATTACK AND EFFICACY OF ANTIMICROBIAL TREATMENTS

The general aims for any microbiological test are to induce and assess microbial growth under controlled conditions. Methods of microbiological testing of textiles have been recently reviewed (Wales *et al.*, 1987). Somewhat different microbiological test conditions may be necessary depending on whether the intention is to assess the inherent susceptibility of a textile material to micro-organisms or to evaluate the effectiveness of a biocide treatment in preventing attack. Most test procedures involve micro-fungi; whilst bacteria are essential for testing the efficacy of hygienic finishes it is regrettable that bacteria do not feature among the organisms specifically chosen for biodeterioration testing by any of the National Standards. Standard test methods are also not available for assessing the resistance of textile materials to actinomycetes or algae, although existing methods can be modified (Fermor & Eggins, 1981).

Three approaches to microbial testing have been adopted; screening tests, simulated end-use application tests, and field tests. Screening tests are used for the initial assessment of the susceptibility of a material to microbial attack, during the development of a new biocide, or for routine quality assurance testing. Agar plate tests are commonly used with pure cultures of either a single organism or more usually with a mixture of several different organisms. Table 4 lists the micro-fungi used in various National Standard test procedures. The original selection of the test organisms was largely governed by their perceived genetic stability, tolerance to antifungal agents, and type of material to be tested. Not all the micro-fungi listed are cellulolytic; some are known to attack plastics, lacquers, leather and rubber. Two types of agar plate tests are commonly used. If the test is simply directed to the assessment of the textile's propensity to mildew, the agar medium should contain only inorganic salts and the ability of the specimens of test fabric to serve as the sole source of carbon for growth of micro-fungi is assessed. However, if the

TABLE 4
Micro-fungi Used in National Standards

	BS 6085	*NFS 41–600*	*DIN 53931*	*SNV 195921*	*AATCC 30*	*ASTM G21*
Aspergillus amstelodami		√				
Aspergillus flavus		√	√			
Aspergillus niger	√	√			√	√
Aspergillus terreus	√					
Aureobasidium pullulans	√					√
Chaetomium globosum	√	√	√		√	√
Candida albicans				√		
Memnoniella echinata		√				
Myrothecium verrucaria		√				
Paecilomyces varioti	√	√				
Penicillium funiculosum	√	√	√			√
Penicillium ochro-chloron	√					
Scopulariopsis brevicaulis	√					
Stachybotrys atra		√				
Trichoderma viride	√	√	√			√
Trichophyton mentagrophytes				√		

purpose of the test is to evaluate the efficacy of a biocide treatment, a metabolisable carbon source is included in the mineral-salts agar and the resistance of the specimens to encroachment by a profuse growth of test organisms on the surrounding agar is assessed. It is customary to assess and express the extent of visual growth on the specimen according to the scale of 0–5 shown in Table 5.

End-use application tests are designed to simulate in the laboratory the conditions of practical use. Two such tests, widely used for the evaluation of textile materials, are exposure of inoculated (or sometimes

TABLE 5
Assessment of Extent of Micro-fungal Growth

Observed growth on specimen	*Rating*
None	0
Growth only apparent under the microscope	1
Traces of growth (less than 10%) visible to naked eye	2
Light growth (10–30%)	3
Medium growth (30–60%)	4
Heavy growth (> 60%)	5

uninoculated) test specimens to a controlled humid environment and soil burial. Exposure of samples to a laboratory environment simulating end-use conditions may involve no more than incubation in a chamber with the temperature kept constant (usually at 27°C) and with the relative humidity at 95%, but some assessments call for exposure of test specimens in a tropical chamber where the temperature and humidity are cycled to simulate the fluctuating diurnal conditions in the tropics. The test strips may be assessed visually on the same scale (Table 5) used in the agar plate tests, by ATP assay based on the firefly (*Photinus pyralis*) luciferin–luciferase reaction (McCarthy, 1983), or by measuring the change in some physical property, such as tensile strength.

Probably the most drastic laboratory test method is the soil burial test. For testing to BS 6085 (Anon., 1981*c*) specimens of fabric for test are buried to a total depth of 80–100 mm in soil (garden loam or naturally fertile top soil, free from pesticides with a moisture content between 20 and 30% and a pH of 5·5–7·0) in a suitable container. A control, cotton cloth is similarly prepared. The containers are then covered and incubated at 28 ± 1°C. After incubation for 7 days the control cloths are disinterred, cleaned and sterilised by soaking in 70% ethanol for 4 h at room temperature, dried and conditioned in the standard atmosphere for physical testing. To confirm the cellulolytic activity of the soil and therefore the validity of the test, the control specimens must have lost at least 80% of their original tensile strength (Anon., 1981*c*). Incubation of the test specimens is usually continued for up to 28 days. They are then removed from the soil and prepared for physical testing in the manner described above for the control specimens.

Finally it may be desirable to carry out a field test under actual conditions. Whilst a full-scale field-exposure evaluation is undoubtedly the crucial test of effectiveness of a biocide treatment, there are usually difficulties involved in carrying out such exposure studies and experimental conditions are rarely reproducible under the uncontrolled climatic conditions occurring in nature. An important aspect of field-exposure testing is the possible loss of antimicrobial compounds by leaching with rain or by actinic decomposition; it is usually necessary to simulate these effects in the laboratory when carrying out laboratory evaluations of biocide efficacy.

TESTING HYGIENIC FINISHES

The anti-bacterial properties of textile materials treated with a hygienic finish are generally assessed against the common Gram-positive bacterium *Staphylococcus aureus* and a representative Gram-negative bacterium *Escherichia coli*. Two methods are used; a qualitative agar plate inhibition test and a quantitative bacterial challenge test in which the numbers of viable bacteria recoverable from a treated textile are determined. In the former assessment samples of treated fabric are placed on tryptone soya agar plates which have been inoculated with cultures of the test organism. After incubation for 18–24 h at 37°C the plates are examined for zones of inhibition and for the presence or absence of growth under the specimen. In the bacterial challenge test circular swatches (*c*. 50 mm diameter) are cut from the test fabric. (The number of swatches used in each test container should be sufficient to absorb 1 ml of inoculum without leaving any free liquid in the container.) The swatches are sterilised either by autoclaving, or by exposure to ethylene oxide vapour. Specimens from a control fabric without any anti-bacterial properties should also be similarly prepared. Swatches of test and control fabrics are placed in sterile 8 oz wide-mouthed glass bottles (with screw tops). The swatches are inoculated with 1 ml of an appropriate dilution of a 24 h bacterial culture to give around $1{\cdot}5 \times 10^5$ colony forming units (cfu) per ml. Immediately after inoculation, 100 ml of a suitable neutraliser solution (e.g. Leathean broth) are added to half the control and test fabric swatches, and the jars shaken vigorously for 1 min. Serial dilutions of the neutraliser solution extracts are made and plate counts carried out on aliquots (0·1 ml) spread on tryptone soya agar to give the cfu recoverable after zero contact time. The remaining jars are incubated at 37°C for 18–24 h and the recoverable cfu determined as described above, to give the percentage reduction in the number of bacteria remaining viable in contact with the treated fabric.

REFERENCES

Allsopp, C. & Allsopp, D. (1983). *International Biodeterioration*, **19**, 99–146.

Anon. (1981*a*). Preservative treatments for textiles: Part 1. Specifications for treatments. *British Standard 2087, Part 1.* British Standards Institution, London.

Anon. (1981*b*). Preservative treatments for textiles: Part 2. Methods of test. *British Standard 2087, Part 2.* British Standards Institution, London.

Anon. (1981*c*). Determination of the resistance of textiles to microbiological deterioration. *British Standard 6085.* British Standards Institution, London.

Anon. (1987). *Review of 1986/87 and Annual Report for 1986.* British Textile Confederation, London.

Darby, R.T. & Kaplan, A.M. (1968). *Applied Microbiology*, **16** (6), 900–5.

Fermor, T.R. & Eggins, H.O.W. (1981). *International Biodeterioration Bulletin*, **17**, 15–17.

Hueck, H.J. (1968). In *Biodeterioration of Materials*, ed. A.H. Walters & J.J. Elphick. Elsevier, Amsterdam, pp. 6–12.

McCarthy, B.J. (1983). *International Biodeterioration Bulletin*, **19** (2), 53–7.

Reese, E.T., Siu, R.G.G. & Levinson, H.S. (1950). *Journal of Bacteriology*, **59**, 485.

Sagar, B.F. (1985). In *Cellulose and Its Derivatives*, ed. J.F. Kennedy, G.O. Phillips, D.J. Wedlock & P.A. Williams. Ellis Horwood, Chichester, pp. 199–207.

Wales, D.S. & Sagar, B.F. (1985). In *Biodeterioration and Biodegradation of Plastics and Polymers*, ed. K.G. Seal. Biodeterioration Society, pp. 56–69.

Wales, D.S., Hamlyn, P.F., Todd, J.E. & Sagar, B.F. (1987). In *S.A.B. Technical Series 22, Preservatives in the Food, Pharmaceutical, and Environmental Industries*, ed. R.G. Board, M.C. Allwood & J.G. Banks. Blackwell, Oxford, pp. 99–113.

TIMBER

The Importance of Microfungi and Bacteria in the Deterioration of Timber

B.O. Henningsson

Swedish University of Agricultural Sciences, Department of Forest Products, Box 7008, S-750 07 Uppsala, Sweden

ABSTRACT

The importance of microfungi and bacteria in man's utilization of timber is discussed based on our present knowledge. Both the early and less harmful colonizers of wood as well as the true wood destroying organisms are considered. The early colonizers include sapstain fungi, moulds, yeasts and certain bacteria which live mainly on easily available nutrients in the wood such as free sugars, low polymerized carbohydrates and proteins. The wood destroyers included are soft rot fungi, tunnelling, erosion and cavitation bacteria, which are able to attack cellulose, hemicelluloses and even lignin.

INTRODUCTION

Microfungi and bacteria play an enormous rôle in the biodeterioration of lignocellulosic material in nature and for keeping the carbon cycle running. Dead plants and trees, leaves, needles, dead branches and other cellulosic debris are partly taken care of by these organisms and converted to carbon dioxide. However, when timber constructions and other useful articles of wood or wood-based materials are attacked and degraded these organisms become noxious. They cause damage which is noticeable in economic terms. In this presentation mainly organisms and problems associated with man's utilization of the world's timber resources will be considered.

Some microfungi and bacteria are early colonizers of timber. They may invade the wood as soon as a tree has been felled or when a piece of wood is exposed to moist conditions, i.e. when they can come in close contact with a moist wood surface. This means that timber may be attacked by these early colonizers during the whole period of its processing and use. Bacteria, yeast fungi, sapstain (blue stain) fungi and mould fungi may therefore be referred to as the early colonizers. They are mainly utilizing easily available nutrients in the wood, such as sugars, low-polymerized carbohydrates and proteins mainly occurring in ray parenchyma and other cells that were living when the tree was cut (Butcher, 1968; Käärik, 1971).

However, microfungi and bacteria may also attack wood in later stages of the microbial succession, during the service life of the timber. These later colonizers are in many cases true wood destroyers. They are able, by the use of enzymes or other bioactive agents, to degrade the highly polymerized structural elements of the wood.

MICROFUNGI

Fungi are regarded as more important organisms than bacteria in the deterioration of wood. They are mostly well adapted to colonize the wood substrate by their hyphae and their spores can easily infect wood surfaces. When a tree has been felled it is immediately subject to infection by microfungi causing sapstain. The infection can be brought about either by air-borne spores landing on a moist wood surface or by insects (e.g. bark beetles) which make their galleries in the wood or in the layer between the inner bark and the wood. Frequently there is an association between the insect and the fungal species that it carries.

Insect-borne sapstain is a problem usually restricted to round timber before sawing, since they depend on fresh bark for their egg laying. Insect-borne sapstain may develop further after sawing as long as moisture and temperature conditions in the wood are favourable and it cannot be prevented or stopped by chemical treatment.

Some of the insect-borne sapstaining fungi have a very rapid growth and cause an intensive discoloration. When the discoloration is very dark it may completely obscure simultaneous attack by decay fungi. This can lead to quality problems since decay cannot be accepted in timber for constructional purposes.

Air-borne sapstain may occur on round timber as well as on sawn timber during drying, transport and storage. The genus *Ceratocystis* contains a number of sapstaining fungi of both the insect-borne and air-borne type. Air-borne blue stain may also attack timber in service, e.g.

window joinery, other exterior joinery, cladding, garden furniture etc.

Even if the staining fungi do not influence the strength of the wood, they may have other more or less undesirable effects. The discoloration in itself is not acceptable for certain uses, e.g. floors, wall panels and furniture.

When penetrating the wood, sapstain fungi break down certain structures: the tori of the bordered pits, various membranes in pores and other pits and certain cells of the wood rays. The permeability of the wood is increased leading to increased water uptake in, for example, external joinery and increased risk for attack by decay fungi. Timber with sapstain is accordingly not suitable for window joinery. The increased permeability may also lead to over-absorption of preservative solution when timber is treated. This is a problem especially in the double vacuum treatment of joinery timber.

Active blue stain fungi also have a capacity to retain moisture in the wood during drying. As a result stained parts of the wood may dry slower than surrounding sound wood. When timber is dried according to a standardised drying schedule sapstain may then lead to uneven moisture distribution and to patches with unacceptably high moisture contents in the dried wood.

The activity of the staining fungi causes chemical changes in the wood. There is a depletion of certain nutrients, e.g. free sugars, a degradation of extractives, a change in the wood's pH, etc. Knowledge on how such changes influence for instance the further microbial invasion and breakdown or later chemical processes such as the chemical fixation and persistence of wood preservatives is by no means complete.

Although mould fungi generally grow more superficially on the wood, they may bring about similar unfavourable changes as do the staining fungi. Heavily sporulating mould fungi may also cause health problems for people working with contaminated timber. Thermotolerant and thermophilic moulds, for instance, have occasionally caused great health problems in saw mills where kiln drying is practised (Henningsson, 1979). Also mould growth on wood or wood-based products in buildings has created health problems (allergies).

Microfungi include species with a real wood degrading ability. The most widespread type of attack on the wood fibres is erosion of the cell wall from the lumen surface. The hyphae then grow in the cell lumen of the wood cells. Usually this type of attack is fairly slow and less destructive than that caused by soft rot fungi which have the ability to grow within the wood cell walls. This attack leads to formation of the well known and typical cavities and causes a dangerous reduction in strength of the attacked timber (Hoffmeyer, 1976). Considerable research has been devoted to the soft rot fungi in the past 30 years since their importance was established (Savory, 1954). At the present time the biochemical and micromorphological mechanisms of soft rot decay are being studied in many laboratories around the world.

Soft rot frequently occurs in timber in ground contact such as poles, fence posts, pilings, railway sleepers, vineyard and orchard sticks, etc., and in timber in above ground situations (window joinery). Timber in freshwater as well as in sea-water may also be attacked by soft rot fungi. Soft rot in timber in water cooling towers is a classical example of severe damage to timber constructions in contact with water (Eaton, 1976).

Soft rot fungi have been shown to attack hardwood to a greater extent than softwood. This has been attributed to the lower lignin content, the chemically different types of lignin and the generally thicker fibre cell walls found in hardwood compared to softwood.

Many soft rot fungi show a higher resistance to certain conventional preservative treatments of wood than do brown and white rot fungi. This has allowed soft rot fungi to attack preservative treated timber. Serious problems have for instance been reported in CCA-treated (copper/chrome/arsenic) Eucalyptus poles in Australia, CCA-treated vineyard stakes in New Zealand and in BIS-treated (zinc/chrome/arsenic) utility poles in Sweden.

BACTERIA

Certain bacteria are early colonizers of timber in service. Studies on window joinery and on special test samples (L-joints) have shown that they invade the wood through end surfaces in the joints. Their activity increases the water permeability of the wood which makes it possible for microfungi and basidiomycete fungi to enter. Thus bacteria can play an important rôle in the establishment of decay in timber in service. It is known that certain bacteria are able to detoxify certain preservative chemicals. The significance of such bacterial activities in practice, however, is not sufficiently known.

Bacteria also play a part in the so called sick-house syndrome. They occur together with moulds and other microfungi on moist wood and wood-based materials in buildings. Actinomycetes are probably responsible for much of the bad odour in these instances.

It is common practice in many countries to store round timber before sawing for varying periods in water or under intermittent spray (wet storage). During such storage bacteria invade the wood. Ray cells, tori of the bordered pits and other pit membranes are degraded by the bacteria and the permeability of the invaded wood increases (Dunleavy & Fogarty, 1971). If the wet storage period is extended the bacterial attack and the increased permeability become a problem. Unacceptable water uptake in window joinery, uneven absorption of stains and paints in joinery and furniture and over absorption of preservative solution in treated timber may be the result of extended wet storage. Recommendations based on a series of studies have been issued in Sweden regarding the wet storage of round timber.

The activity of the bacteria, in addition to permeability changes, results in chemical changes in the wood, e.g. depletion of nutrients, formation of organic acids and occasionally unacceptable discoloration. The effects of such chemical changes for further microbial invasion and for the various uses of the timber are not known.

Certain bacteria are also able to degrade the structural elements of the wood-cellulose, hemicelluloses and lignin. Early indications of true bacterial breakdown of wood (Boutelje & Bravery, 1968) have recently been fully confirmed and clearly demonstrated in experiments (Daniel & Nilsson, 1985). By the use of modern electron microscopic and other techniques several different morphological types of bacterial attack on the wood structure have been observed. Three such types have recently been described: (i) erosion bacteria, (ii) cavitation bacteria and (iii) tunnelling bacteria. The last mentioned group is the most spectacular resulting in very intricate tunnels in the wood cell walls.

Wood degradation by bacteria has been found in terrestrial as well as in marine environments. Both untreated and preservative treated timber can be attacked and certain bacteria seem to be extremely resistant to conventional wood preservatives, although attack is mainly confined to the surface layers of the timber. The significance of bacteria in the deterioration of timber is not yet sufficiently understood.

REFERENCES

Butcher, J.A. (1968). In *Biodeterioration of Materials*, ed. A.H. Walters & J.J. Elphick, Applied Science Publishers, London, pp. 444–59.

Boutelje, J. & Bravery, A.F. (1968). *J. Inst. Wood Sci.*, **20**, 47–57.

Daniel, G. & Nilsson, T. (1985). Ultrastructural and T.E.M.-Edax studies on the degradation of CCA treated radiata pine by tunnelling bacteria. *Int. Res. Group Wood Preserv.*, Doc. No. IRG/WP/1260.

Dunleavy, J.A. & Fogarty, W.M. (1971). *Rec. Brit. Wood Preserv. Assoc. Ann. Conv.*, 1–28.

Eaton, R.A. (1976). In *Recent Advances in Aquatic Mycology*, ed. E.B.G. Jones. Elek Science, London, pp. 359–87.

Henningsson, B.O. (1979). *Int. J. Wood Preserv.*, **1**, 131-5.
Hoffmeyer, P. (1976). In *Soft Rot in Utility Poles Salt-treated in the Years 1940–1954.* Swedish Wood Preservation Institute Rep. No. 117E.
Käärik, A. (1971). *Mitt. Dt. Ges. Holzforsch.*, **57**, 23–34.
Savory, J.G. (1954). *Ann. Appl. Biol.*, **41**, 336–47.

An extended bibliography for this paper may be obtained from the author.

Immunological Methods for the Detection and Characterisation of Wood Decay Basidiomycetes

J.W. Palfreyman, A. Bruce, D. Button, H. Glancy, A. Vigrow & B. King

Wood Technology Group, Department of Molecular and Life Sciences,
Dundee College of Technology, Dundee, UK

ABSTRACT

The potential of immunological probes, viz. antisera, as specific detection systems for the basidiomycete organisms Lentinus lepideus *(a major cause of distribution pole failure) and* Coriolus versicolor *have been investigated. Using immunodot-blot techniques semi-quantitative detection systems have been developed for the former organism and immunocytochemical techniques for the latter, allowing its visualisation in wood. Specificity studies on the antisera produced indicate that cross reactivity is limited to basidiomycetes, and in particular specific brown and white rot organisms. Cross reactivity can be diminished by suitable absorption. Western blot studies on* L. lepideus *indicate that the organism contains a number of antigenic species. Immunological methods are appropriate to research in the biodeterioration of wood and will prove valuable in the detection and identification of decay microorganisms.*

INTRODUCTION

Studies on the organisms responsible for the biodeterioration of wood are hampered by the lack of reagents capable of identifying organisms colonising wood samples. Though it is possible by classical staining techniques to visualise fungal hyphae within wood tissue, such methods are incapable of identifying fungal isolates *in situ*. To obtain such information specific probes are required.

To be of use any probe, as well as being organism specific, would need to (i) be able to detect target organism in the presence of contaminating material, (ii) be applicable to a variety of techniques allowing information on diverse aspects of the organism to be obtained and (iii) be relatively easy to produce. Immunological probes are compatible with each of these criteria.

Immunological probes have recently been used as specific probes for fungi causing biodeterioration (Clarke *et al.*, 1986). However, they are ideal tools for the detection and identification of microorganisms since they can be designed to be highly specific and will detect their target (antigen) even when it is present in highly complex structures such as wood. The first application was developed by Benhamou *et al.* (1985) who produced monoclonal antibodies which would detect the Dutch Elm Disease pathogen, *Ophiostoma ulmi* within infected wood samples. Preliminary studies on the development of both antisera and monoclonal antibodies for the brown rot basidiomycete *Poria placenta* were reported in 1986 by Goodell & Jellison.

The present study has been directed towards the generation of immunological probes capable of detecting and identifying the wood rot basidiomycete organisms, *Lentinus lepideus* (a typical brown rot), and *Coriolus versicolor* (a typical white rot). Though initial studies involved the use of agar grown mycelia the objective was the production of systems for studying hyphae within their natural substrate, i.e. wood. In this paper some of the methods developed to approach this problem are discussed.

MATERIALS AND METHODS

Fungi were obtained from Dr A. Rayner of the School of Biological Sciences, University of Bath, UK or Dr A. Bravery of the Building Research Laboratory, Princes Risborough, UK. Antisera were prepared by Multiple immunisations of mycelial preparations suspended in Freunds complete or incomplete adjuvant. All antisera were raised in rabbits. The methodology used will be published elsewhere. Quantitative and qualitative techniques have been used. The major quantitative method was the immunodot blot (Towbin & Gordon, 1984), qualitative methods included immunocytochemistry using the PAP technique and Western blotting.

RESULTS AND DISCUSSION

Work has been concentrated on two organisms, *L. lepideus* and *C.*

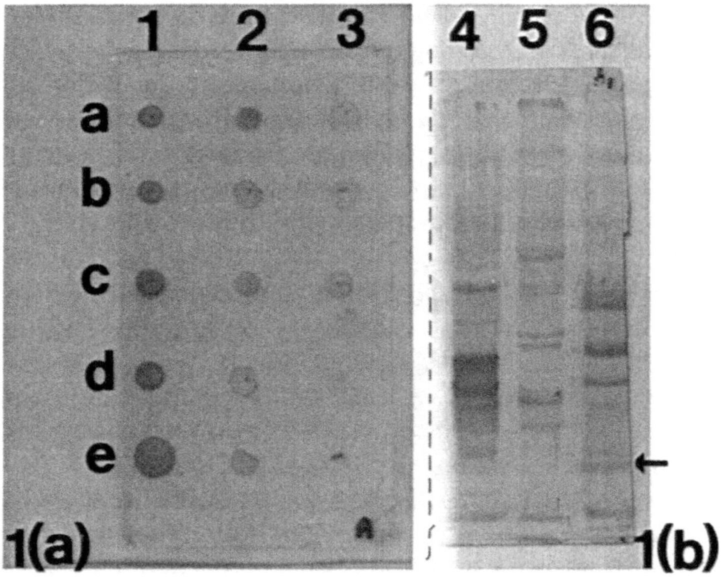

Fig. 1. Detection of *L. lipideus* antigens using dot blot immunoassay (a), and analysis of the antigens using the Western blotting technique (b). (a) Cross reaction of *L. lepideus* antiserum with different isolates of *L. lepideus* (1(a–e), 2(a)), other basidiomycetes (2(b–e), 3(a–c)) and 2 moulds (3(d,c)). (b) Reaction of antiserum with *L. lepideus* (track 6) and 2 *Poria* species (tracks 4 and 5). The specific *L. lepideus* band is arrowed.

versicolor. Studies with the dot blot assay indicated that although the initial antisera were not highly specific for *L. lepideus* (Fig. 1(a)), absorption of the antisera with the cross reacting fungi allowed the development of more specific reagents. Application of such antisera to wood blocks infected with *L. lepideus* indicated that the fungi could be detected using the dot blot assay, and that it could be detected before appreciable weight loss had occurred. The dots illustrated in Fig. 1 form the basis of a semi-quantitative method since their intensity can be estimated by scanning densitometry.

The nature of the antigens detected by the *L. lepideus* antisera was investigated by SDS polyacrylamide gel electrophoresis (SDS-PAGE) followed by Western blotting. A number of different molecular weight bands could be detected by this technique (Fig. 1(b)), one of which, having a molecular weight of around 17 000 daltons appears to be specific for *L. lepideus*. Efforts are currently directed towards purification of this antigen. Western blotting of wood extracted material indicated that it was also present when *L. lepideus* was grown in its natural substrate, i.e. wood.

During these studies it was noted that a number of different fungi showed some reaction with control antisera. The brown rot organism

Poria placenta was particularly reactive with such sera and the mycelia from this fungus also cross reacted strongly with both *L. lepideus* and *C. versicolor* antisera. It is possible that commercial preparations of control (non-immune) serum may be taken from animals which have inadvertently been exposed to cross-reacting fungal antigens. Alternatively hyphal preparations of *P. placenta* may contain elements, for example lectins, which bind antibodies non-specifically (Nordbring-Hertz & Chet, 1986).

Unlike the dot and Western blot methods, immunocytochemical techniques can be used to visualise intact structures by interacting labelled antibodies with tissue sections. Reaction of *C. versicolor* antisera with slide preparations of *C. versicolor* indicated that staining of mycelia could be attained by the PAP technique. Control experiments using non-immune rabbit serum gave negative results.

Application of the PAP method to wood sections indicated that non-specific reaction of antisera occurred with both infected and uninfected wood. All antisera used in the cytochemistry reactions were therefore preabsorbed with wood extracts prior to use. After appropriate preabsorption fungal hyphae growing in their natural substrate could be stained, confirming that mycelia grown in agar must share antigenic properties with mycelia grown in wood. Conventional staining by the safranin/picro-aniline blue method (Cartwright, 1929) gave similar results though this method is not organism specific. Considerably more material has been stained in the *C. versicolor* infected preparation which is consistent with the weight loss findings of the *C. versicolor* and *L. lepideus* infected blocks used in this experiment (20% and $< 2\%$ respectively).

Cross-reactivity studies indicated that the antisera could interact with a variety of fungi (in particular other basidiomycetes). By analysis of absorption experiments using the dot blot technique it was confirmed that the *C. versicolor* antiserum could be made organism specific by preincubation with two cross-reacting species, *S. sanguinolentum* and *L. lepideus*. Neither *C. versicolor* nor *L. lepideus* were stained when non-immune antiserum was applied to infected wood sections or if the antiserum was preabsorbed with the homologous antigen.

The staining patterns observed with both *L. lepideus* and *C. versicolor* indicate that the fungal hyphae are largely localised to the vessels within the wood structure. Furthermore the staining is localised to the hyphae itself. It might be considered that antigenic material, such as enzymes, would be released from actively growing hyphae and hence detected by immunocytochemistry. To test this hypothesis antisera are being produced to fungal hyphae grown in wood blocks, since the antigen used for initial immunisations might not be expected to contain extracellular components.

Work to date has been limited to two organisms, *L. lepideus* and *C. versicolor*, however application to other wood decay basidiomycetes is underway. Our current studies have utilised conventional antiserum; however, the potential specificity of immunological techniques could also be realised with monoclonal antibodies. Selected monoclonal antibodies would be specific for particular organisms and it might be possible to select those that did not interact with uninfected wood thus avoiding preabsorption stages. Alternatively individual monoclonal antibodies could be modified to prevent them from non-specifically interacting with wood.

CONCLUSION

It has been demonstrated that immunological probes can be used to detect and identify wood decay basidiomycetes growing in their natural substrate. Given the large contribution that such reagents have made to other fields of biological research it is suggested that the use of immunological probes and techniques should be more widely applied to the study of decay organisms and the decay process.

REFERENCES

Benhamou, N., Ouellette, G.B., Lafontaine, J.G. & Joly, J.R. (1985). *Can. J. Bot.*, **63**, 1177–84.

Cartwright, K.St.G. (1929). *Ann Bot.*, **43**, 412–3.

Clarke, J.H., MacNicoll, A.D. & Norman, J.A. (1986). *International Biodeterioration Supp.*, **22**, 123–30.

Goodell, B.S. & Jellison, J. (1986). Detection of a brown rot fungus using serological assays. International Research Group on Wood Preservation, Document No. IRG/WP/1305.

Nordbring-Hertz, B. & Chet, I. (1986). In *Microbial Lectins and Agglutinins*, ed. D. Mirelman. Wiley-Interscience, New York, pp. 393–408.

Towbin, H. & Gordon, J. (1984). *J. Immunol. Methods*, **72**, 313–40.

Recent Trends in Wood Preservation Research

John A. Butcher

Ministry of Forestry Forest Research Institute, Wood Technology Division, Rotorua, New Zealand

ABSTRACT

A generic review of literature was undertaken to establish the broad changes in research emphasis in wood preservation between 1980 and 1987. An analysis of world literature showed no major changes over this time. In contrast, the selective papers of the International Research Group on Wood Preservation displayed increasing emphasis on wood preservative systems. Possible future trends in processing and preservation systems are briefly discussed with reference to application systems for antisapstain chemicals, 'just-in-time' processing, boron-based preservatives, and chemical wood modification.

INTRODUCTION

In 1980, a symposium on 'Wood preservation from around the world' (American Wood Preservers' Association, 1980) addressed the current status, recent highlights and future possibilities of research by reference to ten countries.

At that time considerable research effort was focused on the problem of soft-rot control in CCA-treated hardwoods, both at the fundamental and applied level. Concern about environmental and health hazards of wood preservation had resulted in increased attention on the development of new (third-generation) wood preservatives, antisapstain chemicals, and insecticides. Test procedures to evaluate candidate fungicides and insecticides were being developed with the aim of allowing more precise predictions of field performance. Many other topics were covered, most of which displayed the regional problems and interests of the wood preservation industry.

Butcher (1983), in examining future international trends in wood preservation research, selected four topics for discussion:

—understanding the ecosystem of wood in soil contact;
—wood-preservative interactions;
—alternative chemicals for wood protection;
—prediction of preservative performance and residual service life in treated commodities.

These four topics were selected to represent areas of research expected to receive greater attention over the next decade. The validity of these predictions will be discussed later.

BROAD TRENDS 1980–87

The collected documents of the International Research Group on Wood Preservation (IRG) provide one data base for examining broad trends in research. The IRG is a very active organisation currently having a personal membership of 356 scientists from 58 countries. Documents tend to report research in progress, and can be interpreted as providing a progressive record of current research activity from around the world. Table 1 summarises research emphasis in some broad areas between the years 1980–87 inclusive.

An analysis of these data allows the following broad conclusions:

(1) Fundamental studies on organisms appear to be maintaining a constant level of activity, but no longer represent the major research activity of IRG. Minor trends were:
 —increasing interest in bacteria since 1983;
 —a decline in research on dry wood insects in favour of termites;
 —a continuing decline in marine organisms research;
 —a recent increase in studies of decay mechanisms.
(2) There has been a very large increase in research on preservatives, especially concerned with fundamental studies on CCA, organotin preservatives, and development of alternative preservative systems.
(3) Research on sapstain control has shown a steady increase that is being maintained.

TABLE 1
An Analysis of Research Activity in IRG (1980–87) (Number of Documents)

Subject area	*1980*	*1981*	*1982*	*1983*	*1984*	*1985*	*1986*	*1987*
Bacteria and fungi	11	4	5	6	12	7	12	2
Insects	6	6	14	19	6	15	12	6
Marine organisms	11	16	5	6	5	7	5	4
Decay mechanisms	2	2	1	5	3	2	9	11
Total	30	28	25	36	26	31	38	23
Preservatives	9	9	15	22	32	15	23	35
Insecticides	1	1	1	1	4	3	4	1
Treatment processes	2	1	2	4	3	0	2	4
Total	12	11	18	27	39	18	29	40
Sapstain control	2	3	4	4	8	3	7	7
Finishes/weathering	1	1	0	0	2	5	3	4

(4) Surface finishes and weathering of wood begin to attract more attention.

Another data base for examining broad trends in wood preservation research was Forest Products Abstracts (1983–86 inclusive). The numerous papers (*c.* 1000) in the survey cover work published between 1980 and 1987, but should not be regarded as all inclusive.

Table 2 summarises data in terms of allocating subject matter of papers to one or more specific areas. Broad conclusions are:

(1) Fundamental and applied research on fungi have been maintained and there is an indication of a slight decrease of research on insects.
(2) Preservatives represent the dominant research topic; especially CCA preservatives, pentachlorophenol, and creosote. Less than 10% of citations concerned new preservative compounds.
(3) Wood storage continues to be a major area of research interest; most attention is focused on storage of roundwood.
(4) Research on insecticides has been variable, and reflects existing compounds.
(5) Surface finishes and weathering of wood is an active area of research which shows a major downward trend in 1986.

As expected, the two data bases provide somewhat different perspectives regarding current research trends. The IRG (Table 1) is a dynamic organisation comprising the leading research workers in wood preservation. Research emphasis has changed significantly over the past 8 years especially in the balance between basic and applied research with a shift towards preservatives and processing. In contrast, Forest Products Abstracts (Table 2) records world literature and reflects the *status quo* rather than the leading edge of research.

Of particular interest was the very small amount of research into treatment processes recorded in either data base, and little of it displaying any novelty.

The broad trends outlined above can be compared with the predictions of Butcher (1983). The suggested increase in research to understand the ecosystem of wood in soil contact was not forthcoming, although such problems as the lack of performance of CCA-treated horticulture posts (Hedley & Drysdale, 1985) highlighted the need for an intensive research effort in this area.

Research on wood–preservative interactions, especially in relation to CCA-preservatives, remained very active with major contributions being made by Pizzi and co-workers at the Processing and Chemical Manufacturing Technology Division of the CSIR in South Africa. The problem of soft-rot control in CCA-treated hardwoods is now well understood and in consequence research effort is declining.

A safe prediction was that research on alternative wood preservation

TABLE 2
An Analysis of Research Topics Referenced in Forest Products Abstracts (1983–86) (Number of Papers Published Including Cross-referencing)

Subject area	*1983*	*1984*	*1985*	*1986*
Fungi	125	130	98	117
Drywood insects	67	53	93	46
Termites	143	110	77	67
Marine organisms	14	13	13	6
Decay mechanisms	39	57	60	43
Preservatives	165	210	130	150
Insecticides	50	72	75	28
Treatment processes	21	20	22	17
Wood storage	53	71	67	92
Surface finishes	23	39	61	25
Weathering	23	29	24	12

chemicals would continue over the next decade. However, one disturbing trend shown in the literature, by the recent IUFRO (1987) Symposium on storage of softwoods, and by numerous personal communications, is an increasing uniformity in research where most investigations are concerned with a similar, if not identical, range of candidate materials.

Finally, prediction of preservative performance has not as yet received the attention it deserves, although the climate for advances in this area is gradually changing. Problems of predicting residual service life of commodities has remained largely a regional (Scandinavian) problem.

POSSIBLE FUTURE TRENDS

Since the collected papers of this symposium consider mainly research on wood destroying organisms, attention will be focussed on the broad areas of treatment processes and wood protection systems.

(a) Application systems for antisapstain chemicals

Until recently, research to gain improvements in protection of stored timber and roundwood concentrated on the development of alternative chemical formulations. Such work still continues, but most candidate materials suffer from high cost and/or an incomplete spectrum of biological performance. Precise application of the biocide at minimum effective dosage on targeted wood surfaces is a mechanism for partially overcoming these deficiencies. At the same time, the minimal loss of chemical that occurs by run-off reduces environmental and health risk. There is now a growing awareness that ultra-low volume application systems represent the future technology in sapstain control rather than dependence on particular biocides. Key developments have been the invention of a foam-based application system in New Zealand, and the commitment to install a spray technology facility at the Western Forest Products Laboratory of Forintek, Canada to extend research on systems incorporating electrostatic charging of spray particles.

(b) 'Just-in-time' processing

Butcher (1987), presented an argument outlining the benefits of applying 'just-in-time' processing to the wood preservation industry. Essentially, this is processing undertaken in response to intermediate market demand in contrast to the established 'just-in-case' processing where wood is preservative-treated and placed in stock to await potential orders.

A necessary pre-requisite for 'just-in-time' processing is immediacy in treatment. For some years light organic solvent preservative (LOSP) treatment has been an isolated example of a treatment system applicable to this type of processing, but in many countries it has been applied conventionally as a distinct and separate part of processing, and with treated material being placed in stock.

In New Zealand, processes for pressure treatment of green wood with copper–chrome–arsenate preservatives, have been developed over many years. They have been widely adopted by the industry and have led to the elimination of drying before treatment, the ability to treat material rapidly in response to orders, and to a marked reduction in costly stock holding. Such processes have recently been introduced in Australia, and are now attracting attention in USA (Barnes, 1987) and Europe (Peek, 1987).

These green wood treatments are not fully compatible with the 'just-in-time' philosophy of processing because they do not have true immediacy, and are not fully integrated into the overall processing system. However, they do represent a major move in this direction and illustrate some of the benefits that can be gained.

Recent research now examines vapour-phase treatment of timber (Vinden *et al.*, 1985) and board products (Turner & Murphy, 1987) using trimethylborate. This represents a very exciting area of development because treatment can be undertaken in a matter of minutes and on demand. This may well signal a new direction in wood preservation processing technology and should provide the necessary stimulus for a much needed research effort.

(c) Boron-based wood preservative systems

Boron has been widely used for several decades as an insecticidal treatment for softwoods and hardwoods, as a base ingredient in antisapstain chemicals, in certain multisalt preservatives (e.g. copper–chrome–boron), and increasingly as a remedial treatment against decay. More widespread use as a fungicide has been limited by its leachability from treated wood.

Interest in boron as a fungicide/insecticide is now increasing worldwide. However, the true potential of this biocide will not be reached unless it can be 'fixed' in treated wood. It is believed that research is again active on this front and it is certainly worthy of concerted effort. Boron has well proven insecticidal and fungicidal activity and above all is deemed a biocide of low health and environmental risk.

(d) Chemical wood modification

Research on the chemical modification of wood has been active for many years with major emphasis on improving dimensional stability and water repellency in both solid wood and composite board materials. Although it is also well known that chemical modification can confer resistance to biological attack, its potential has never been fully examined. Butcher (1987) expressed the opinion that this technology may represent the major advance needed by the industry to find a long-term solution to the problems associated with use of toxicants. Rowell (USDA Forest Products Laboratory, Madison), considers that interest from the wood preservation industry is now increasing. In Japan, the Laboratory for Improved Wood of Kyoto University has an active programme investigating resistance of modified wood to fungal and insect attack, and similar programmes are being developed elsewhere.

REFERENCES

American Wood Preservers' Association (1980). Wood preservation research from around the world. Joint Symposium of the International Research Group on Wood Preservation and American Wood Preservers' Association. *Proc. A.W.P.A. Annual Meeting 1980*, Nashville, Tennessee.

Barnes, H.M. (1987). Treatment of partially seasoned southern pine with CCA. *Proc. A.W.P.A. Annual Meeting 1987*, Toronto, 20 pp. (in press).

Butcher, J.A. (1983). International trends in wood preservation research. Wood Preservation Symposium, National Timber Research Institute, CSIR, Pretoria, South Africa, March, 1983 (NZFS Reprint 1664).

Butcher, J.A. (1987). The perception and future perspectives of wood preservation. Reginald B. Colley Lecture. A.W.P.A. Annual Meeting 1987, Toronto (in press).

Hedley, M.E. & Drysdale, J. (1985). *Mat. u. Org.* **20** (1), 35–51.

IUFRO (1987). Storage of softwoods. IUFRO S.5.03.03 Symposium, Honey Harbour, Ontario, Canada, May 1987 (in press).

Peek, R.D. (1987). Application of oscillating pressure to improve treatability of refractory species. The International Research Group on Wood Preservation. Document No. IRG/WP/3449.

Turner, P. & Murphy, R.J. (1987). A novel process for the treatment of manufactured wood based board materials. IUFRO S.5.03.02 Symposium on the Protection of Wood-based Composites, Honey Harbour, Ontario, Canada, May 1987 (in press).

Vinden, P., Fenton, T. & Nasheri, K. (1985). Options for accelerated boron treatment. A practical review of alternatives. The International Research Group on Wood Preservation. Document No. IRG/WP/3329.

Termite Control: Changing Attitudes and Technologies

J.P. La Fage

Department of Entomology, Louisiana State University Agricultural Experiment Station, LSU Agricultural Center, Baton Rouge, Louisiana, 70803, USA

ABSTRACT

Termites account for nearly $2 billion per annum in losses to structures throughout the world. After nearly 40 years, chlorinated hydrocarbon termiticides are being replaced with less persistent and presumably safer soil pesticides. Changing attitudes are providing a foundation for experimentation and eventual implementation of alternative control strategies involving wood preservatives, baiting systems, biological control, and physical methods.

INTRODUCTION

Wood is one of our most valuable construction materials. While it possesses many favourable attributes including strength, widespread availability, natural beauty, and renewability, it is compromised by at least one serious deficiency. When not properly protected or when lacking natural durability, it is biodegradable. Despite a substantial body of knowledge of wood preservation, about one third of the world's timber production is lost annually to various agents of biodegradation (Edwards & Mill 1986). Although about 2200 species of termites (Isoptera) are known worldwide, with most restricted to the tropics, they figure prominently among the causal agents of wood destruction.

ECONOMIC IMPORTANCE

While termites are most generally recognized for their attacks upon structures, their proclivity for destroying cellulosic materials is not limited to timbers in buildings. Utility poles and railway sleepers are attacked throughout the tropics. In Louisiana (USA), even creosoted electric transmission poles are regularly attacked by the Formosan subterranean termite, *Coptotermes formosanus* Shiraki. Other items damaged include paper products, boats and ships, underground cables, agricultural crops, pasture grasses, and live trees in forests, orchards and urban areas. In New Orleans, Louisiana, and other coastal cities in the USA, *C. formosanus* is a serious pest in marine pilings treated with creosote.

Credible information on the economic losses by termites is difficult to obtain and virtually lacking for anything but damage to structures. Mauldin (1982) reported annual losses in the United States of $750 million resulting from prevention, remedial control and repair. About 95% of the loss is attributed to subterranean termites (*Reticulitermes, Heterotermes, Coptotermes* spp.). Based upon reports from the pest-control industry, set costs of termite treatments worldwide are $1·92 billion for 1985. However, not all termites are destructive. The vast majority of species actually serve a critical role in nutrient cycling in tropical and temperate ecosystems.

TERMITE CONTROL—PAST AND PRESENT

The remainder of this paper will focus on control procedures for subterranean termites. An excellent review of control methods for drywood termites is given by Edwards & Mill (1986).

Prior to World War II, subterranean termite control consisted largely of creosote or pentachlorophenol impregnations for wood and experimental soil treatments with pitch, tar, creosote, arsenic salts, orthodichlorobenzene, trichlorobenzene, and chlorinated naphthalenes. Many of these compounds were objectionable because of health hazards, odour, or phytotoxicity. In 1939, a laboratory was established by the USDA at Gulfport, Mississippi, dedicated to the study of termite problems. Long-term soil-treatment tests were installed there during 1943. The cyclodiene pesticides, discovered at about this time, were first incorporated into the testing programme with the inclusion of chlordane in 1948. In 1949, the related compounds, dieldrin and aldrin were included and heptachlor in 1952. Now known as the Forestry Sciences Laboratory, this group remains very active in the testing of candidate pesticides for soil treatments in the United States.

The four cyclodiene soil termiticides demonstrated exceptional efficacy providing 100% effectiveness in ground-board studies for at least 34 years.

With nearly 5 years of field efficacy, chlordane was marketed for use as a soil termiticide in 1952. Although several other compounds including lindane (gamma-HCH) and DDT, have been registered for soil treatment, none has achieved so widespread acceptance as the four cyclodienes. In addition to persistence in soil, they have been relatively inexpensive, widely available, and, until recently, considered safe.

In 1978, concern about possible health hazards led the US Environmental Protection Agency (EPA) to cancel all uses of chlordane except for subterranean termite control. A paper by Livingston & Jones (1981) which reported contamination of living spaces by chlordane was the first indication that the cyclodienes were in trouble. Subsequent reports on television, radio, and in the press eventually led to the banning of chlordane in New York and Massachusetts during 1984–85. Dieldrin and aldrin were removed from the market voluntarily by manufacturers. Even greater pressure to ban cyclodienes is apparent now as a subcommittee of the US House of Representatives considers an amendment to the federal Clean Air Act which would ban chlordane in the United States on the basis of potential health hazards. Saudi Arabia was the first country to ban cyclodienes for termite control (in 1984) and Japan has eliminated the use of these compounds.

As concerns over pesticide safety have increased, so has acceptance of more expensive and less persistent termiticides. In 1980, the organophosphate, chlorpyriphos, was registered as a soil termiticide and another Isofenphos, in 1982. More recently, a pyrethroid, permethrin, has become available for use in soil treatments. Several additional compounds are effective for termite control but are not yet registered or marketed. These include endosulfan, bromocyclen, fenvalerate, a particularly promising compound, cypermethrin and several others.

Nearly all termiticides today are applied as soil treatments. Nevertheless, concerns about human health and environmental contamination continue to mount. Nearly 500 million kg of pesticides are applied to *c.* 148 million hectares annually. While the total amount applied to agricultural lands far exceeds that in urban areas, the rate of application in or under structures is much higher than on farmland or forests. Chlordane is currently applied at a rate of 390 kg/h under and adjacent to buildings (La Fage, 1986). Data such as these have prompted discussion and initiation of research on alternative termite-control strategies.

FUTURE TRENDS IN TERMITE CONTROL

Future methods for subterranean termite control will likely fall into four categories: (1) pesticides in wood, (2) baiting systems, (3) biological control, and (4) physical methods (Edwards & Mill, 1986).

Pesticides in wood

Many wood preservatives are currently used to protect structural timbers against decay. Some of these, including CCA, ACAs, AACs, dimethylol compounds and borates, are also effective against termites (Williams & Amburgey, 1987). However, termites can crawl over treated products to reach unprotected wood. Among the organophosphorus insecticides, chlorpyriphos is effective as a wood preservative for protection against termites and is being marketed in a number of countries. Several pyrethroids also protect against termites when used as wood preservatives. Although the possibility of using natural wood extractives or their synthetic analogues has been widely researched (Carter & Beal, 1982), no commercial products have been developed to date. Research has also been conducted on the resistance to termite attack of wood impregnated with plastic resins and polyethylene glycol (De Groot & Esenther, 1982). Other studies have produced termite repellents and antifeedants (Scheffrahn & Rust, 1983).

Baiting systems

In the bait-block method of subterranean termite control, wooden baits impregnated with a slow-acting, non-repellent toxicant are placed in the foraging territory of termite colonies. Termites feed on them and return the toxicant to the central nest where it is shared with nestmates via trophallactic exchanges and grooming. The entire colony may be destroyed within a few weeks. Dechlorane (mirex), has been used effectively as a bait toxicant in the USA, Australia and China. Although the use of this insecticide has been banned by the EPA, no suitable alternative has yet been discovered. A number of insect growth regulators including methoprene (Jones, 1987) and fenoxycarb show promise in laboratory studies. The future of these compounds will not be known until field trials, now underway, are completed. Other toxicants that have been tested include the fluorolipids, chitin synthesis inhibitors and the protozoacide, chlortetracycline.

Biological control

In late 1983 the nematode, *Neoaplectana carpocapsae* F., was introduced in the USA as a biological control for subterranean termites. Subsequent research at Gulfport, failed to establish the efficacy of this organism and recent reports (Mix, 1986) have largely discounted its future use.

Among the fungi which attack termites, several species in the genera *Metarhizium* and *Beauveria* appear to have some potential as control agents for subterranean termites. Studies in Hawaii (Lai *et al.*, 1982) and Australia (Hänel & Watson, 1983) yielded inconclusive but promising results.

Physical control

Little research has been conducted on physical methods of controlling subterranean termites. Ebeling & Pence (1957) evaluated substrate particle size as a deterrent to substrate penetration by *Reticulitermes hesperus* (Banks). More recently, Tamashiro *et al.* (1987) demonstrated in laboratory trials that *C. formosanus* is incapable of tunnelling through a substrate containing basaltic particles 1·7–2·4 mm in diameter.

CONCLUSION

In many countries, the era of subterranean termite control with cyclodiene soil insecticides is nearing an end. Many new soil termiticides have been introduced and more will probably follow. Looking further into the future, it is likely that alternative control strategies, especially baiting systems, will be perfected and ultimately accepted by the pest-control industry and consumer.

REFERENCES

Carter, F.L. & Beal, R.H. (1982). *International Journal of Wood Preservation*, **2**, 185–91.

De Groot, R.C. & Esenther, G.R. (1982). In *Structural Use of Wood in Adverse Environments*, ed. R.W. Meyer & R.M. Kellogg. Van Nostrand Reinhold, New York, pp. 219–45.

Ebeling, W. & Pence, R.J. (1957). *Journal of Economic Entomology*, **50**, 690–2.

Edwards, R. & Mill, A.E. (1986). *Termites in Buildings*. Rentokil Ltd, Felcourt East Grinstead, W. Sussex, 261 pp.

Hänel, H. & Watson, J.A.L. (1983). *Bulletin of Entomological Research*, **73**, 305–13.

Jones, S.C. (1987). Effects of methoprene on *Coptotermes formosanus* (Isoptera: Rhinotermitidae). International Research Group on Wood Preservation, IRG/WP/1322.

La Fage, J.P. (1986). In *Proceedings of the National Conference on Urban Entomology*, ed. P.A. Zungoli. 24–27 February, 1986, College Park, MD, University of Maryland, pp. 45–57.
Lai, P-Y., Tamashiro, M. & Fujii, J. (1982). *Journal of Invertebrate Pathology*, **39**, 1-5.
Livingston, J.M. & Jones, C.R. (1981). *Bulletin of Environmental Contamination and Toxicology*, **27**, 406–11.
Mauldin, J.K. (1982). In *Biology of Social Insects*, ed. M.D. Breed, C.D. Michener & H.E. Evans. Proc. 9th Congress Int. Union Study Social Insects. Westview Press, Boulder, CO, pp. 138–41.
Mix, J. (1986). *Pest Control*, **54**, 48, 54.
Scheffrahn, R.H. & Rust, M.K. (1983). *Journal of Chemical Ecology*, **9**, 39–55.
Tamashiro, M., Yates, J.R. & Ebesu, R.H. (1987). The Formosan subterranean termite in Hawaii. Proceedings of a symposium on *The Biology, Ecology, and Control of the Formosan Subterranean Termite*, 25 June, 1985, Pacific Branch of the Entomological Society of America, Honolulu, HI (in press).
Williams, L.H. & Amburgey, T.L. (1987). *Forest Products Journal*, **37**, 10–17.

An extended bibliography for this paper may be obtained from the author.

The Wood-Boring Isopod, *Sphaeroma*: a Threat to Maritime Structures in Warm Waters

S.M. Cragg

School of Biological Sciences, King Henry Building, King Henry 1st Street, Portsmouth PO1 2DY, UK

ABSTRACT

Several species of Sphaeroma *burrow. Some can burrow into non-wood materials and into wood treated with copper–chrome–arsenic (CCA), creosote, combined CCA/creosote and other preservatives: hard, durable timbers are not immune to attack. These species are restricted to tropical and warm temperate waters, but tolerate a wide range of salinities. They are most active in the intertidal zone. There is no evidence that they ingest wood so conventional preservatives may not reach the digestive tract. Physical barriers or contact control agents may be more effective.*

INTRODUCTION

Wood below the high tide mark is subject to degradation by marine borers. In temperate waters, bivalve molluscs of the family Teredinidae and the isopod crustacean, *Limnoria* cause the most severe damage. In warmer waters, another isopod borer, *Sphaeroma* may also damage wood. This paper assesses the threat of *Sphaeroma* to maritime timbers and evaluates the control measures used to date.

THE PROBLEM OF *SPHAEROMA* CONTROL

Harrison & Holdich (1984) recognize 29 species of *Sphaeroma*, including the following wood borers: *S. annandalei annandalei, S. annandalei travancorensis*, *S. peruvianum*, *S. quoyanum*, *S. retrolaeve*, *S. sieboldii*, *S. terebans* and *S. triste*. Some also bore into non-wood substrates, for example: mudstone and sandstone (*S. quoyanum*: Mills, 1978); laterite and hard clay (*S. annandalei annandalei*: Dharmaraj & Nair, 1982); rope, carpet, styrofoam, limestone and sandstone (*S. terebrans*: Johnson *et al.*, 1987). Timbers with a reputation for resistance to other marine borers due to a high silica or extractive content, such as turpentine (*Syncarpia glomulifera*), greenheart (*Ocotea rodiaei*) and kasi kasi (*Xanthostemon* sp.), have been found to be vulnerable to *Sphaeroma* (Cragg & Icely, 1982; Barnacle *et al.*, 1986; Johnson *et al.*, 1987).

Table 1 summarizes evidence that preservatives commonly used to protect marine pilings are insufficient to prevent *Sphaeroma* attack. Figures 1 and 2 illustrate the damage that *Sphaeroma* can inflict on treated timber. A comparison of preservative loadings of timber attacked by *Sphaeroma* (Mills, 1978; Barnacle *et al.*, 1986; Johnson *et al.*, 1987) with the stringent requirements for marine pilings in the tropical waters of Queensland, Australia (Leightley & Kennedy, 1983) shows that treatment

TABLE 1
Records of Preservative Treated Wood Attacked by *Sphaeroma*

Preservative	*Timber*	*Borer*	*Site*	*Authors*[a]
Creosote	*Pinus longifolia*	*S. terebrans*	Cochin, India	a
	Eucalyptus maculata[b]	*S. terebrans*	Bowen, Australia	d
	Southern pine	*S. terebrans*	Florida, USA	e
Creosote/coal tar	*Pinus longifolia*	*S. terebrans*	Cochin, India	a
Creosote/CCA	*Pinus radiata*	*S. terebrans*	Bowen, Australia	d
	Eucalyptus maculata[b]	*S. terebrans*	Bowen, Australia	d
CCA	*Pinus radiata*	*S. quoyanum*	Tauranga, New Zealand	b
	Araucaria cunninghamii	*S. triste* and *S. terebrans*	Port Moresby, Papua New Guinea	c
	Tectona grandis[b]	*S. terebrans*	Port Moresby, Papua New Guinea	c
	Pinus elliottii	*S. terebrans*	Bowen, Australia	d
	Syncarpia glomulifera[b]	*S. terebrans*	Bowen, Australia	d
	Southern pine	*S. terebrans*	Florida, USA	e
Pentachlorophenol	*Larix decidua*	*S. quoyanum*	Tauranga, New Zealand	b

[a] Authors: a = John (1965); b = Mills (1978); c = Cragg & Levy (1979); d = Barnacle *et al.* (1986); e = Johnson *et al.*, (1987).
[b] Hardwood (broad-leaf) species.

Fig. 1. *Sphaeroma* burrows in a CCA-treated pile at Port Moresby, Papua New Guinea.

was not substandard. Nor is excessive leaching of preservative a likely explanation: Cragg & Icely (1982) found that samples taken from heavily burrowed wood at the site shown in Fig. 2 had retained the copper–chrome–arsenic preservative at levels in excess of the Queensland requirements. Increasing loadings is not a viable approach as attack has

Fig. 2. CCA-treated pilings at Port Moresby, Papua New Guinea, destroyed by *Sphaeroma*.

been detected in material with close to the practical maximum loading.

Treated wood particles may not be ingested—Krishnan (1973) found only algae in the gut of *S. terebrans*. Rotramel (1975) suggests that *S. quoyanum* derives its nutrition by filter-feeding rather than by ingesting wood. The setae on the legs and mouth-parts of other wood-boring *Sphaeroma* resemble those implicated in the filtering process described for *S. quoyanum* (Harrison & Holdich, 1984). Thus stomach-toxin type preservatives may not reach the target tissues.

Alternative preservatives have been tested by Johnson *et al.* (1987): chemical modification of wood by propylene oxide, butylene oxide or acetic anhydride did not prevent attack, but, for a 3 year period, chloropyrifos added to creosote was effective.

Annual applications of creosote by the floating-collar method has proved effective over a considerable period (Johnson, 1947). Barnacle *et al.* (1986) found that though creosote-treated wood is eventually attacked by *Sphaeroma* the onset of attack is delayed, and Johnson *et al.* (1987) report that higher loadings of creosote reduce attack. Creosote may act by being absorbed through the external cuticle rather than through the gut.

FACTORS AFFECTING THE DISTRIBUTION OF WOOD-BORING *SPHAEROMA*

S. annandalei ranges between India and Brazil, *S. terebrans* between S.W. USA and N.E. Australia, *S. triste* between India and N.E. Australia; *S. peruvianum* occurs in Peru, *S. retrolaeve* and *S. sieboldii* in Japan; *S. quoyanum* is found in Australia, New Zealand and on the Pacific coast of the USA and Mexico (Harrison & Holdich, 1984). Within the species range, distribution can be patchy and the occurrence of *Sphaeroma* as a pest may not correlate closely with its natural distribution. It is widespread and numerous in mangrove forests around the coast of Papua New Guinea (unpublished observation), but Shillinglaw & Moore (1947) found few timber wharves out of the many that they examined with *Sphaeroma* burrows.

Sphaeroma does not have a planktonic larval stage, but it can be transported over considerable distances in wood and so may appear as a pest in areas where it was previously unknown. Indeed, Rotramel (1972) suggests that the populations of *S. quoyanum* in northern California probably originated from those in Australia or New Zealand. Also, the population of *S. triste* causing the damage at the site in Fig. 2 has no large source population nearby. Borers may have been carried in wooden-hulled canoes infested from populations in mangroves adjacent to a village 150 km away (Cragg & Icely, 1982) with strong trading connections with the settlement at the site.

Temperature probably plays a role in determining the limits of distribution. *S. terebrans* is restricted to tropical and sub-tropical waters and *S. quoyanum* to sub-tropical and warm temperate waters.

Some species of *Sphaeroma* tolerate a wide range of salinity conditions. *S. terebrans* and *S. triste* occur at coastal sites in Papua New Guinea (Cragg & Levy, 1979) where the salinity rarely drops below 30 ppt, and also where salinity is always very low—as low as 0·2 ppt in the case of *S. triste*—in the Gulf of Papua (Cragg & Aruga, in review). *S. terebrans* is found in the backwaters of Cochin, India where the salinity fluctuates seasonally between less than 1 ppt and 30 ppt (Cheriyan, 1973). However, in Florida, it is common in estuarine sites, but not at coastal high-salinity sites (Estevez & Simon, 1975; Johnson *et al.*, 1987).

Burrowing activity, whether in mangroves (Estevez & Simon, 1975), wood samples (Dharmaraj & Nair, 1979), or pilings (Cragg & Levy, 1979; Barnacle *et al.*, 1986) is most intense in the intertidal zone, though Santhakumari & Nair (1984) report that during the lowered salinities of the monsoon, settlement of *Sphaeroma terebrans* may occur lower in the water column. To date the threat posed by *Sphaeroma* has been underestimated because marine tests of wood preservatives have generally been installed below the low water mark (Barnacle *et al.*, 1986).

Prior to the erection of wooden structures along the shoreline, the main habitat of *Sphaeroma* would have been mangrove forests. The prop roots of *Rhizophora* are a particularly favoured niche (Estevez & Simon, 1975).

CONCLUSIONS

Sphaeroma should be considered a potential threat to any wooden structure in the intertidal zone in warm temperate to tropical estuarine and marine waters. Conventional preservatives cannot prevent its attack. So in addition to protection against basidiomycetes and termites above the water-line and teredinids and *Limnoria* below, the intertidal zone at least may require protection against *Sphaeroma*. The floating collar method, though effective, may not prove widely acceptable in view of the increasing scrutiny of the use of chemicals in harbour areas. Plastic sheathing of polyurethane coatings may be appropriate where abrasion or impacts are unlikely to occur. Non-leachable preservatives which are contact toxins could perhaps provide damage-resistant protection.

REFERENCES

Barnacle, J.E., Cookson, L.J. & McEvoy, C.N. (1986). An appraisal of the vertical distribution of attack of untreated and treated wood by warm water sphaeromatids at some tropical sites—a discussion paper. Document IRG/WP/4124, 24 pp.

Cheriyan, P.V. (1973). *Forma et Functio,* **6**, 1-68.

Cragg, S.M. & Aruga, J. (in review). Intertidal and subtidal wood-boring faunas in the Gulf of Papua, Papua New Guinea.

Cragg, S.M. & Icely, J.D. (1982). An interim report on studies of the tolerance of Sphaeroma (Crustacea: Isopoda) of CCA-treated timber. Document IRG/WP/491, 26 pp.

Cragg, S.M. & Levy, C.R. (1979). *International Journal of Wood Preservation*, **1** (4), 161-8.

Dharmaraj, K. & Nair, N.B. (1979). *Aquatic Biology*, **4**, 49-71.

Dharmaraj, K. & Nair, N.B. (1982). *Indian Journal of Marine Science*, **11**, 93-5.

Estevez, E.D. & Simon, J.L. (1975). In *Proceedings of the International Symposium on Biology and Management of Mangroves* Vol. 1, ed. G.E. Walsh, S.C. Snedaker & H.J. Teas. University of Florida, Gainesville, pp. 286-304.

Harrison, K. & Holdich, D.M. (1984). *Zoological Journal of the Linnean Society*, **81**, 275-387

John, P.A. (1965). In *Symposium on Crustacea, Part IV, Ernakulam.* Marine Biological Association of India, pp. 1269-73.

Johnson, B.R., Estevez, E.D. & Rice, S.A. (1987). *Sphaeroma terebrans* Bate: a note on distribution and preservative tolerance in Florida coastal waters. Document IRG/WP/4135, 6 pp.

Johnson, R.A. (1947). *Port of Sydney Journal*, **1** (5), 148-50.

Krishnan, R.V. (1973). *Journal of the Timber Development Association of India*, **19**, 5-6.

Leightley, L.E. & Kennedy, M.J. (1983). Wood preservation requirement in Queensland, Australia. Document IRG/WP/3227, 16 pp.

Mills, P.E. (1978). *International Biodeterioration Bulletin*, **14**, 35-6.

Rotramel, G. (1972). *Crustaceana*, suppl. **III**, 193-7.

Rotramel, G. (1975). *Crustaceana*, **28** (1), 7-10.

Santhakumari, V. & Nair, N.B. (1984). *Fishery Technology*, **21**, 118-25.

Shillinglaw, A.W. & Moore, D.D. (1947). Report of the marine borer survey in New Guinean waters. Council for Scientific and Industrial Research, Melbourne, Bulletin No. 223, 40 pp.

In-service Decay of Timber Joinery and its Eradication

Janice K. Carey & A.F. Bravery

Princes Risborough Laboratory, Building Research Establishment, Princes Risborough, Bucks, UK

ABSTRACT

Premature decay of external softwood joinery in the UK reached serious proportions in about 1960; as a result requirements for preservative pre-treatment have been progressively introduced. To extend the useful service life of joinery installed either untreated or with minimal treatment two methods of applying preservative in situ *have been investigated. Organic solvent-based preservatives, applied under pressure gave complete protection to the more vulnerable sapwood close to the joint. Treatment of heartwood was less effective. Distribution of boron from solid preservative plugs was dependent on moisture content but largely independent of heartwood, sapwood or preservative plug type. At c. 50% moisture content decay was prevented in sapwood for approximately 90 mm along the grain but for only 40 mm at 30% moisture content under the test conditions. Greater protection was afforded to heartwood.*

INTRODUCTION

Problems of premature decay of exterior joinery in the UK were reviewed by Savory & Carey (1979). The introduction of preservative pretreatment, as a control measure, began as early as 1965 and specifications were prepared for suitable preservatives and treatments by the Forest Products Research Laboratory (FPRL) (1967), and introduced by the

National House-Building Council (1969, 1974, 1979), Property Services Agency (1968, 1971) and other specifiers.

A Code of Practice for Preservation of timber (BS 5589) was published in 1978 and is referred to in subsequent revisions of joinery standards (BS 1186: Part 1:1986). In recent years installation of untreated softwood exterior joinery has been rare but there is a legacy of joinery either untreated or which received the earliest, more minimal treatments. The cost of replacing decayed joinery is very high, interest therefore developed in extending its life by applying preservative treatment *in situ*. To date, the potential effectiveness of two types of systems have been assessed at PRL; pressure injection using organic solvent-based preservatives (Lea, 1980) and diffusion from solid preservative plugs containing boron (Carey & Bravery, 1987).

METHODS

Pressure injection

Details of the technique have been described fully by Lea (1980). Untreated 12 year old Scots pine opening lights were divided into eight groups equivalent in sapwood and decay content. Five specialist firms co-operated in the work by treating one group of frames each using their normal injection schedule and preservative through non-return valves located near joints; the three remaining groups were used as untreated controls.

After a minimum period of 1 month, the frame members were sawn up and assessed for discoloration and decay, the distribution of preservative, the incidence of decay fungi and the protection afforded against decay.

Boron diffusion

Details of the technique are described more fully in Carey & Bravery (1987). Essentially test blocks of Scots pine were pre-wetted to either 25% or 50% moisture content (sapwood) or 25% or 40% (heartwood) placed in resealable polythene bags and sterilised by gamma irradiation.

The blocks were infected with *Coniophora puteana* FPRL 11E from dowelling inserted in a hole down the centre of each block (Fig. 1) 1 week before a sample of solid wood preservative was inserted in a second hole at 90° to the dowel and 20 mm beyond its end. The hole was filled and the blocks incubated at 22°C for 4, 12, 18 or 24 weeks.

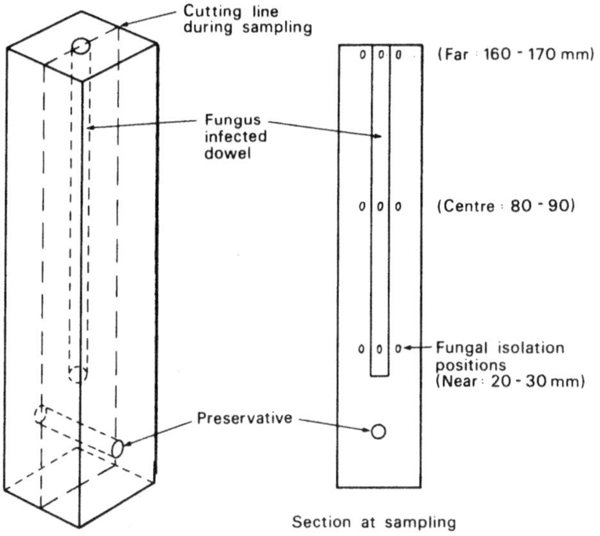

Fig. 1. Boron diffusion test method.

Each block was sawn lengthwise cutting across the preservative plug and the moisture content of one half established by oven drying. Nine sample chips were cut from the other half, planted on a selective medium and growth of the test fungus recorded. After air drying the newly cut face was sprayed with curcumin reagent to detect the presence of boron.

RESULTS AND DISCUSSION

Pressure injection

Old frames were used to assess eradicant efficacy of the treatments because they are naturally infected with decay fungi and pine sapwood wetted in service becomes more permeable so 'new' wood cannot provide an accurate model.

The results from all the tests were variable and there were no statistical differences between the treatments. Preservative penetration of sapwood was good, particularly between the injector and the joint; longitudinal penetration was better and less variable than cross-sectional penetration. Less penetration was obtained in heartwood.

The incidence of decay fungi isolated from the three untreated groups was variable. Therefore, although overall the treated members showed a lower incidence of live decay fungus than the untreated, indicating some

TABLE 1
Number of Isolation Positions Yielding Viable Test Fungus[a]

Type of timber (Target mc)	*Distance from preservative plug (mm)*	*Incubation period (weeks)*							
		Untreated				*Treated*[d]			
		4	*12*	*18*	*24*	*4*	*12*	*18*	*24*
Sapwood	20–30	3[b]	6[c]	9	9	4	6	1	0
(25%)	80–90	5[b]	6[c]	9	9	7	8	9	9
	120–130	7	6[c]	9	9	8	9	9	9
Sapwood	20–30	8	9	9	6[c]	0	0	0	0
(50%)	80–90	9	9	9	6[c]	7	2	1	1
	120–130	9	9	9	6[c]	9	6	9	9
Heartwood	20–30	2[b]	7	2	6	2	1	0	1
(25%)	80–90	2[b]	4	5	6	1	3	5	3
	120–130	6	5	7	7	4	7	6	7
Heartwood	20–30	0[b]	4	6	5	0	1	0	0[c]
(40%)	80–90	2[b]	2	5	3	3	3	1	2[c]
	120–130	6	3	8	8	7	6	7	5[c]

[a] Max 3 per replicate, i.e. 9 total.
[b] Artificially low due to only one infection dowel being inserted.
[c] Two replicates only (max 6).
[d] Three types of preservative plug were used. In general all gave similar results; data are presented for one type only, for simplicity.

eradicant action, more detailed interpretation was not possible.

Most sapwood was completely protected against further decay though there were exceptions particularly when the injection hole was completely within the heartwood. Heartwood protection was always variable.

Boron diffusion

After 4 weeks incubation, boron distribution varied with the moisture content in both heartwood and sapwood and irrespective of the type of preservative plug. After longer periods, presumably as a result of decay interfering with the curcumin dye reaction (Edlund, 1982), distribution could not be determined with certainty.

In sapwood, the test fungus established rapidly in untreated blocks (Table 1). At *c.* 25% all the treatments progressively reduced the incidence of viable fungus from the sample zone closest (20–30 mm) to the

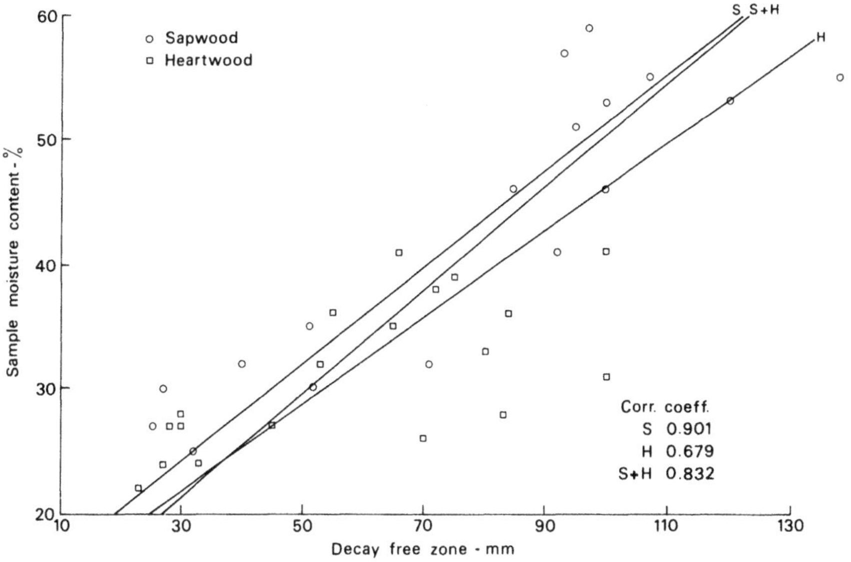

Fig. 2. Moisture content and decay free zone (12 weeks).

preservative plug. At *c.* 50%, all types of preservative plug had eradicated the fungus from the closest zone within 4 weeks and at the second zone (80–90 mm) the fungus was progressively eradicated; there was little evidence of any effect at 120–130 mm.

Colonisation of untreated heartwood was much slower than sapwood. Overall, the effectiveness of the preservative plugs in eradicating fungus was similar to sapwood at equivalent moisture contents.

After 12 weeks exposure decay was visible. The decay free zone, the minimum distance between the preservative plug and visibly decayed wood, varied with moisture content (Fig. 2) and in sapwood was almost identical to boron movement recorded after 4 weeks. The larger decay free zones in heartwood are due to the slower rate of colonisation and decay.

In a few instances live fungus was obtained from zones without visible decay, possibly because the fungus had been inactivated by sub-lethal doses of boron.

The method used provides a severe test for the preservative plugs in that the boron can move only by diffusion and not by mass flow of water. In practice, this condition would exist for only a short period after refurbishment of the finish (Miller & Boxall, 1984) until the paint seal across the joint is breached when water will enter and distribute the boron by mass flow. In addition, the test fungus used (*C. puteana* FPRL

11E) is particularly active and also tolerant of boron. In practice, fungi causing decay in joinery are likely to be rather less active than *C. puteana* and/or less tolerant of boron. It is concluded that implanting boron-containing preservative plugs in joinery *in situ* shows potential for both the eradication of existing infection and the provision of future protection against decay. Further diffusion rate studies are in progress and service trials are currently being monitored.

REFERENCES

British Standards Institution (1978). Code of practice for preservation of timber. BS 5589:1978.

British Standards Institution (1986). Timber for and workmanship in joinery Part I Specification for timber. BS 1186:Part 1:1986.

Carey, J.K. & Bravery, A.F. (1987). A technique for determining the efficacy of water diffusible preservative plugs for implanting in joinery in service. *International Research Group on Wood Preservation Document No.* IRG/WP/?, 12 pp.

Edlund, M-L. (1982). *International Journal of Wood Preservation*, **3** (1), 3–22.

Forest Products Research Laboratory (1967). Preservative treatments for external joinery. FPRL Technical Note 24. Princes Risborough, UK.

Lea, R.G. (1980). *British Wood Preserving Association Newsheet*, **162**, 4–7.

Miller, E.R. & Boxall, J. (1984). *Hols als Roh- und Werkstoff,* **42** (1), 27–34.

Savory, J.G. & Carey, J.K. (1979). *Journal Institute Wood Science*, **8** (4), 176–80.

Bacterial Attack of Wood Cell Walls

Thomas Nilsson & Geoffrey Daniel

Department of Forest Products, Wood Mycology Section, Swedish University of Agricultural Sciences, Box 7008, S-750 07 Uppsala, Sweden

ABSTRACT

This paper is a brief review of our present knowledge of bacterial attack of wood cell walls.

Pure cellulose is readily degraded by a large number of cellulolytic microorganisms, fungi as well as bacteria. Pure cellulose, is, however, rare in nature; most of it is associated with lignin and hemicelluloses. Lignified plant material is not degraded by ordinary cellulolytic microorganisms. The well known fungal decay categories: white rot, brown rot and soft rot represent groups of fungi which are specifically adapted for degrading lignified plant material. Wood-degrading bacteria represent the procaryotic counterpart of the decay fungi.

So far only one wood-degrading bacterium has been obtained in pure culture. It has tentatively been identified as an *Erwinia* sp. All attempts to isolate other wood-degrading bacteria have so far been unsuccessful. Thus the specific identity of the bacteria is unknown. All observations on bacterial degradation of wood cell walls have been carried out on wood samples exposed either in the field (Greaves, 1968; Holt *et al.*, 1979; Holt, 1981) or to the mixed microflora in soil brought to the laboratory. Bacterial consortia, free of other microorganisms, have also been extensively used in the laboratory for reproducing the different forms of bacterial attack.

Bacterial attack in wood is commonly observed in material where fungal activity is suppressed, either due to preservatives, toxic heartwood extractives or limited oxygen supply. Thus bacterial attack is often found in wood lying at the bottom of lakes or the sea. The most recent example is the extensive bacterial attack observed in wood samples from the Swedish warship 'Kronan' which during a battle in 1676 exploded and sank in the Baltic sea.

Using light microscopy, TEM and SEM, it has been possible to distinguish at least three different morphological forms of bacterial attack of wood cell walls.

The first form of attack is a type of erosion of the wood cell walls by Gram-negative rod-shaped bacteria. The bacteria grow in the wood cell lumina and attach themselves to the cell walls by means of glycocalyces (Daniel & Nilsson, 1986). Each individual bacterium erodes the wall in a groove-like manner (Figs 1 and 2). The combined activity of several bacteria results in a thinning of the cell wall which extends to the middle lamella which is apparently not degraded. Localised erosion by a group of bacteria often leads to large troughs extending deep into the cell wall (Singh & Butcher, 1986). Erosion bacteria have been observed to produce numerous membrane bound vesicles (Fig. 3) whose function is still unclear. They may contain packages of muralytic enzymes and may also aid in attachment of bacteria to the wood cell walls (Daniel & Nilsson, 1986). Erosion bacteria attack hardwoods as well as the more lignified softwoods.

The two other forms of attack are characterised by the fact that the bacteria grow and degrade the wood substance within the wood fibre wall itself. In one of the forms, cavitation, the attack results in the formation of discrete cavities in the fibre walls. The cavities sometimes have a form similar to cavities formed by soft rot fungi. Most of the bacteria observed within the cavities are long slender forms, but other types have also been noted. Cavitation has been found in preservative treated pine and eucalypt poles (Nilsson & Singh, 1984; Willoughby & Leightley, 1984).

The third form of attack is the most extensively studied. Here the bacteria actively penetrate into wood fibre walls and once inside commence tunnelling through the entire substrate. Branching of the tunnels occurs when recently divided individuals initiate additional tunnels in new directions. At later stages the network of tunnels coalesces to form a central very degraded area which is surrounded by a fringe of radiating tunnels.

A most typical feature of tunnelling bacteria is the concentric arrangement of wall-like structures left behind the bacteria in the tunnels (Fig. 4) (Daniel & Nilsson, 1985). These structures probably represent secreted muco-polysaccharides. The deposition of the structures suggests

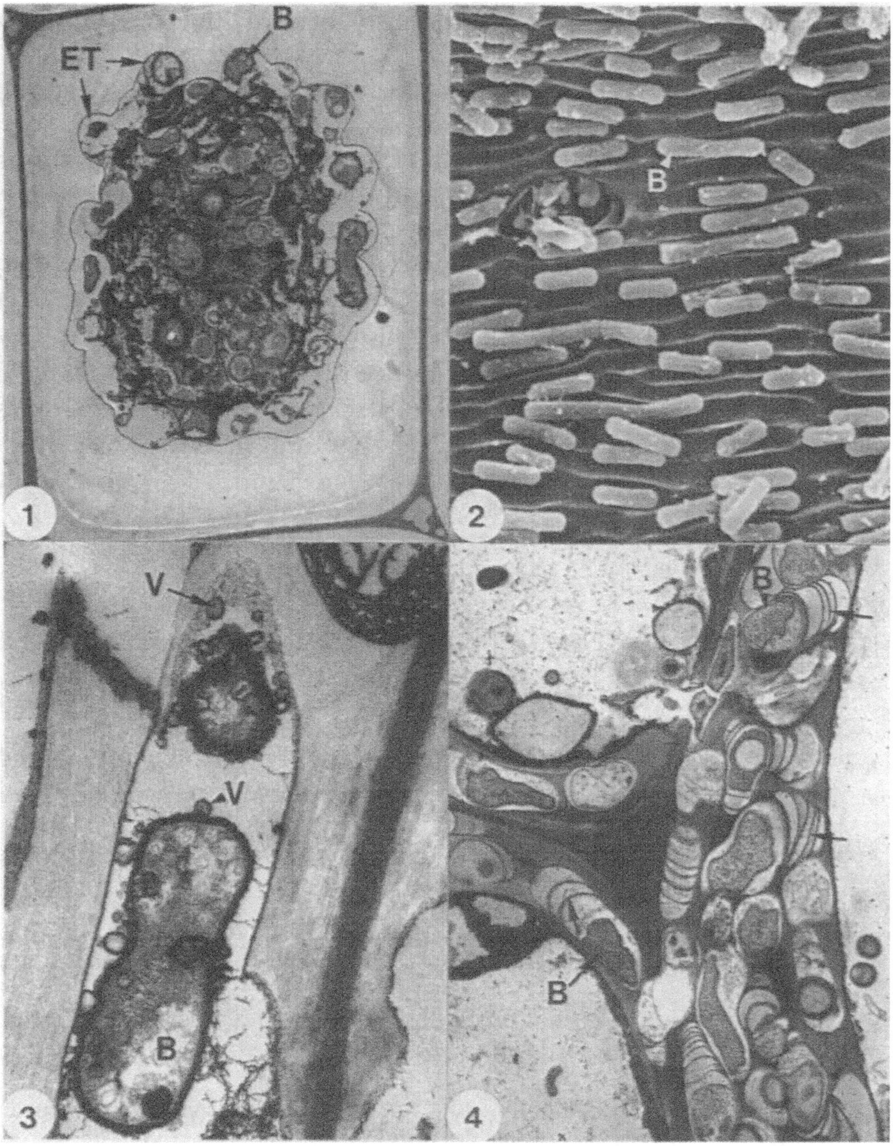

Figs 1–4. Wood degrading erosion and tunnelling bacteria.
1. TEM cross-section of a birch fibre showing well-defined erosion troughs (ET) produced by closely adhering bacteria (B). By progressive erosion, the S_2 wall layer is degraded. 2. Corresponding SEM to show erosion troughs and aligned rod-shaped bacteria. 3. TEM showing typical vesicular bodies (V) routinely observed associated with the bacteria and neighbouring degraded sites. 4. Active tunnelling bacterial attack of *Homalium foetidum* with bacteria enclosed within discrete tunnels. The tunnels are characterised by prominent semi-concentric layers (arrows) which are laid down behind the bacteria.

that the bacteria do not move continuously forward but rather by a type of stop, start action.

Tunnelling bacteria actively degrade a number of woody substrates which are very or almost completely resistant to fungal attack. Such substrates include timbers with high lignin content, timbers containing high levels of toxic extractives, chemically modified timbers and timbers which have been treated with copper–chrome–arsenic formulations.

Evolution of $^{14}CO_2$ from ^{14}C-labelled DHP's has demonstrated that erosion and tunnelling bacteria possess ligninolytic abilities (Nilsson & Daniel, 1986; Daniel *et al.*, 1987). Tunnelling bacteria also caused evolution of $^{14}CO_2$ from aspen and spruce wood containing ^{14}C-labelled lignin. Klason lignin analyses have also confirmed that tunnelling bacteria cause a substantial depletion of lignin from intact wood during the attack process.

REFERENCES

Daniel, G. & Nilsson, T. (1985). Ultrastructural and T.E.M.-E.D.A.X. studies on the degradation of CCA treated radiata pine by tunnelling bacteria. International Research Group on Wood Preservation, Document No. IRG/WP/1260, 33 pp.

Daniel, G.F. & Nilsson, T. (1986). Ultrastructural observations on wood degrading erosion bacteria. International Research Group on Wood Preservation, Document No. IRG/WP/1283, 28 pp.

Daniel, G.F., Nilsson, T. & Singh, A.P. (1987). *Canadian Journal of Microbiology* (in press).

Greaves, H. (1968). *Applied Microbiology*, **16**, 1599-601.

Holt, D.M. (1981). Bacterial breakdown of timber in aquatic habitats and their relationship with wood degrading fungi. Ph.D. thesis CNAA, Portsmouth Polytechnic.

Holt, D.M., Jones, E.B.G. & Furtado, S.E.I. (1979). Bacterial breakdown of wood in aquatic habitats. Record Annual Conference British Wood Preservation Association, pp. 13–24.

Nilsson, T. & Singh, A.P. (1984). Cavitation bacteria. International Research Group on Wood Preservation, Document No. IRG/WP/1235.

Nilsson, T. & Daniel, G.F. (1986). In *Proceedings Biotechnology in the Pulp and Paper Industry*, Stockholm. Published by Swedish Forest Products Research Laboratory and Swedish Association of Pulp and Paper Engineers, pp. 54–7.

Singh, A.P. & Butcher, J.A. (1986). *Journal Institute Wood Science*, **10**, 140–4.

Willoughby, G.A. & Leightley, L.E. (1984). Patterns of bacterial decay in preservative treated Eucalypt power transmission poles. International Research Group on Wood Preservation, Document No. IRG/WP/1223, 19 pp.

Cellulolytic Nitrogen-Fixing Bacteria in the Teredinidae (Mollusca: Bivalvia)

Ruth D. Turner

Museum of Comparative Zoology, Harvard University, Cambridge, MA 02138, USA

ABSTRACT

Isolation of a symbiotic bacterium from the gland of Deshayes in the gills of teredinids has led to an increased understanding of shipworm nutrition. It is the first bacterium known that is capable of digesting cellulose and fixing nitrogen. The symbiotic relationship allowing shipworms to use wood as their primary food source and to satisfy their nitrogen requirements has been a critical factor in the evolution of this unique and diverse family of bivalves.

INTRODUCTION

Teredinids are morphologically and physiologically among the most specialized of the Bivalvia. Research on their nutrition extends back to the early 1700s but intriguing problems remain. For abstracts of papers prior to 1954 see Clapp & Kenk (1963) and for reviews of later work the reports of Turner (1966), Nair and Saraswathy (1971), Morton (1978), Mann (1984, in press) and Turner (1984, in press). The mystery of teredinid nutrition is inexorably tied to the gland of Deshayes so a brief account of its history seems appropriate.

Deshayes (1848) noted a glandular tissue in the umbonal area of *Teredo navalis* (= *Nototeredo knoxi* (Bartsch)) and a similar tissue in the gills and the lining of the afferent branchial vein. He suggested that the umbonal portion might secrete a fluid to soften the wood in the formation of the burrow and the portion in the gills was a mucous gland

and served for the nutrition of viviparous embryos. Sigerfoos (1908) considered the three glands as all part of a single structure, the sizes of which varied with the species. He described the gland in detail and named it after Deshayes, stating that the umbonal portion was very large in *T. dilatata* Stimpson (= *Psiloteredo megatara* (Hanley)), small in *Xylotrya* (= *Bankia gouldi* (Bartsch)) and rudimentary in *Teredo navalis* Linnaeus. He noted that the anterior portion was present in newly settled larvae and was ducted anteriorly, that as the animal grew the two parts of the gland separated but were connected by a duct lying in the afferent branchial vein. He also noted spherical cells within the duct and the gland. Though Sigerfoos did not attribute any definite function to the gland he refuted the ideas of Deshayes stating that the anterior portion was too small in *T. navalis* to produce secretions to soften the wood and that the portion in the gills could not be for the nutrition of viviparous embryos because it was found in males and females and in species that did not retain larvae in the gills. Sigerfoos was not even able to guess 'the special function of this remarkable structure' but he suggested it might be 'for the elaboration of some internal secretion or other material for whose formation the presence of both blood and water is necessary'.

The gland of Deshayes was then essentially forgotten until Saraswathy & Nair (1971) described the anatomy of the gland from the gills of *Nausitora hedleyi* (Schepman) and *Teredo furcifera* von Martens. They noted that it was ducted into the oesophagus and that the gland and ducts contained spherical bodies for almost their entire length, observations that agreed with those of Sigerfoos. They also concluded that 'the function of these structures, unique to the Teredinidae, cannot even vaguely be guessed'. However, in their review of teredinid biology Nair & Saraswathy (1971) suggested that the gland of Deshayes might be involved in the digestion of cellulose, but on p. 445 reiterated Nair's earlier statement that 'cellulose splitting seemed to take place in two different sites, one extracellularly in the caecum and the other intracellularly in the vacuoles of the digestive diverticula which has a cellobiase powerful enough to complete digestion before absorption . . .,' and made no further reference to the gland of Deshayes.

A major break-through came when Popham & Dickson (1973) examined the gland of Deshayes in *Bankia australis* Calman using transmission electron microscopy (TEM). They described and figured the bacterium in the 'gland' and stated that it was not glandular tissue but an association of densely packed, rod-like, Gram-negative bacteria which were probably symbiotic. In a second paper Popham (1975) suggested that these symbiotic bacteria might provide the teredinids with vitamins and essential amino acids. Carpenter & Culliney (1975) used rates of acetylene reduction to assay for nitrogen fixation in four species of teredinids and concluded that nitrogen fixation was important in areas low in plankton and in young teredinids with gills too small to obtain adequate phytoplankton by filter feeding. They isolated from the caecum of *Lyrodus pedicellatus* (Quatrefages) a Gram-negative, rod-shaped, nitrogen-fixing bacterium and found that it grew under either aerobic or anaerobic conditions, reduced acetylene only under anaerobic conditions and liquified cellulose. While recognizing the work of Popham mentioned above, Carpenter and Culliney noted that 'no nitrogen fixing bacteria were obtained from the gill or the gland of Deshayes'. In the light of subsequent work by Waterbury *et al.* (1983) this is very surprising and almost suggests that, by some strange mishap, the cultures from the caecum and from the gland of Deshayes had been switched.

In his survey of feeding and digestion in teredinids Morton (1978) emphasized functional morphology and the digestion of cellulose. He made no mention of nitrogen nutrition and after discussing the work Popham & Dickson (1973), and others stated that 'There is little evidence to suggest that marine bacteria and fungi play any more than a minimal and fortuitous role in the digestion of wood in the shipworm gut'. Morton (1983) concluded his brief section on the Teredinidae (p. 119) with essentially the same statement. Trytek & Allen (1980) incubating isolated gills of *Bankia setacea* (Tryon) in ^{14}C glucose, found that most of the essential amino acids became labelled and inferred that the bacteria in the gland of Deshayes were capable of synthesizing them. They suggested that the bacterial symbionts might supply essential amino acids and vitamins when the food supply was inadequate but emphasized the need for experimental proof. The first step in testing this or any other hypothesis concerning the function of the bacterial symbionts was to isolate the bacterium in pure culture. This led to the collaborative research by Waterbury (microbiology), Calloway (morphology and histology) and Turner (anatomy and malacology).

METHODS

See Waterbury *et al.* (1983) for procedures in handling specimens, isolating the bacterium and maintaining stock cultures.

RESULTS AND DISCUSSION

The following description of the bacterial symbionts is largely condensed from Waterbury *et al.* noted above. The bacterial isolates from all species of teredinids studied are similar and probably represent a single species of bacterium. It is Gram-negative, rod-shaped, moves by means of a single polar flagellum, is an aerobic chemoheterotroph and obligately marine. It requires a source of combined nitrogen if vigorously aerated but fixes molecular nitrogen under microaerophilic conditions. Calloway suggested that the position of the 'gland' (i.e. lining the afferent branchial vein) places it in the least aerated portion of the circulatory system, and Waterbury noted that the close packing of the bacteria in clusters would insure that the rapid utilization of oxygen at the periphery of the clusters and promote microaerophilic conditions within (see Calloway (in) Turner (1984), p. 13, Figs. 3(a–d) for the morphology and histology of the area and the bacterium).

The bacterium is a new species, unique to the Teredinidae, and does not fit into any of the known genera. That the bacterium isolated is the same as the one in the gland of Deshayes is supported by the fact that: (1) Contamination was minimized during removal of the 'gland' and isolations were made only from serial dilutions of the homogenate at 10^7 or 10^8 dilutions. (2) It was possible to isolate repeatedly the same bacterium from different animals of the same species taken from the same or widely separated localities and from 23 different species representing 9 genera and world-wide collecting. (3) We have been unable to isolate the bacterium from the seawater tanks, from the surface of the wood in which the teredinids were boring or from any part of the animal except the gills. (4) The bacterial isolates are as similar morphologically to those in the gland of Deshayes as they are to each other. (5) This is the first bacterium known to both fix nitrogen and degrade cellulose, and the first example of a symbiotic cellulolytic bacterium occurring in an organ not a part of the digestive system.

Recent work by Kamicker *et al.* (1985, 1986) using rates of acetylene reduction to measure nitrogen fixation by the bacterial symbionts in the gills of *Bankia gouldi* showed that: (1) Shipworms grown in wood held in running seawater derived 5–15% of their nitrogen requirements from the bacterial symbionts. (2) Shipworms grown in wood held in filtered, autoclaved seawater depended on their symbionts for as much as 98% of their nitrogen. (3) Young shipworms receiving a phytoplankton supplement grew faster and fixed more nitrogen per gram dry weight than those that did not. It was suggested the phytoplankton supplement helped establish a vigorous effective symbiosis. They concluded that nitrogen fixed by the symbionts was excreted to the host based on the fact that there was no evidence of phagocytized or lysed bacteria in the gills and that the only site of nitrogenase activity in the shipworm, as determined by acetylene reduction, was in the gill. The rates of acetylene reduction for intact shipworms demonstrated that nitrogen is fixed according to need and this suggests that the bacterium functions in the shipworm as it does in culture. The important next step toward understanding nitrogen nutrition in shipworms is to determine the amount of fixed nitrogen incorporated by the shipworm. Experiments to examine this using standard ^{15}N procedures are in progress (Kamicker & Waterbury, personal communication). The cellulase activity of the bacterial symbionts in culture studied by Kamicker & Waterbury (1986) showed that: (1) They were similar to cellulolytic *Clostridium* and *Bacillus* species in degrading filter paper. (2) The cellulase was extracellular. (3) Reduction of cellobiose was intracellular. Whether the bacterium behaves in the shipworm as it does in culture is still unknown but the authors suggested that 'in shipworms, extracellular cellulases are probably released from the bacteria in the gills to the digestive system where they degrade the cellulose in the wood'. Sigerfoos (1908) and Saraswathy & Nair (1971) reported that the gland of Deshayes was ducted forward to the oesophagus. Calloway (personal communication), using light microscopy, has seen the ducts, but has been unable to observe the precise point of entrance into the digestive system. Further histological work is needed to document the path of the cellulase to the digestive system and to the site or sites where cellulose digestion actually occurs. The ability of teredinids to digest wood and to satisfy their nitrogen needs by the facultative use of phytoplankton, by nitrogen conservatism, by uptake of free amino acids and ammonia, and particularly by nitrogen fixation has allowed for the evolution of the great anatomical diversity exhibited by this unique family of bivalves. Though the general pattern of teredinid nutrition outlined here appears similar in all groups, the known variation in the anatomy of the gills, reproductive system and digestive tract of shipworms demands caution before making generalizations.

Another unsolved problem in the biology of the bacterial symbionts in shipworm gills is the means by which the bacterium is transferred from the parent to the young. For a discussion of this problem, the possibilities of transfer of the bacterium via the egg and for on-going research see Turner (in press). There is obviously a great deal of exciting research yet to be done in this area.

ACKNOWLEDGEMENTS

Thanks are extended to all those who have contributed to this collaborative research and particularly to the Office of Naval Research for their continued support.

REFERENCES

Carpenter, E.J. & Culliney, J.L. (1975). *Science*, **187**, 551-2.

Clapp, W.F. & Kenk, R. (1963). *Marine Borers, an Annotated Bibliography*. Office of Naval Research, Department of the Navy, Washington, D.C., pp. 10 + 1136

Deshayes, P. (1848). *Zoologie*, **1**, 35-76, 5-10.

Kamicker, B. & Waterbury, J.B. (1986). *Measurements of the Cellulase Activity of the Bacterial Symbiont from the Gills of Shipworms*. American Society of Microbiologists, Washington, D.C., March 1986 (abstract).

Kamicker, B., Gallager, S. & Waterbury, J.B. (1985). *The Contribution of Symbiotic Nitrogen Fixation to the Nitrogen Budget of Shipworms*. Sixth International Symposium on Nitrogen Fixation, Corvalis, Oregon, August 1985 (abstract).

Kamicker, B., Gallager, S. & Waterbury, J.B. (1986). *The Contribution of Symbiotic Nitrogen Fixation to the Nitrogen Budget of Shipworms*. American Society of Microbiologists, Washington, D.C., March 1986 (abstract).

Mann, R. (1984). *Marine Biodeterioration: an Interdisciplinary Study*, ed. J.B. Costlow & R.C. Tipper, Naval Institute Press, Annapolis, Maryland, pp. 24-9.

Mann, R. (in press). *Marine Biodeterioration as it pertains to the Indian Ocean*, ed. M.F. Thompson & R. Nagabushanam.

Morton, B. (1978) *Oceanography and Marine Biology, Annual Review*, **16**, 107-44.

Morton, B. (1983). *The Mollusca*, **5**, 117-19.

Nair, N.B. & Saraswathy, M. (1971). *Advances in Marine Biology*, **9**, 335-509 (section on Food and Digestion, pp. 441-6).

Popham, J.D. (1975). *Veliger*, **18** (1), 55-6, Figs 1-5.

Popham, J.D. & Dickson, M.R. (1973). *Marine Biology*, **19**, 330-40.

Saraswathy, M. & Nair, N.B. (1971). *Transactions Royal Society Edinburgh*, **68** (14), 507-69.

Sigerfoos, C.P. (1908). *Bulletin Bureau of Fisheries*, **27** (for 1907), 191-231.

Trytek, R.E. & Allen, W.V. (1980). *Comparative Biochemistry and Physiology*, **67A**, 419-27.

Turner, R.D. (1966). *A Survey and Illustrated Catalogue of the Teredinidae (Mollusca: Bivalvia)*. Special Publication, Museum of Comparative Zoology, Harvard University, Cambridge, MA, 265 pp.

Turner, R.D. (1984). In *Marine Biodeterioration: an Interdisciplinary Study*, ed. J.D. Costlow & R.C. Tipper. Naval Institute Press, Annapolis, Maryland, pp. 3-16.

Turner, R.D. (in press). In *Marine Science of the Arabian Sea*, ed. M.F. Thompson & N.M. Tirmizi. American Institute of Biological Sciences, Washington, D.C.

Waterbury, J., Calloway, C.C. & Turner, R.D. (1983). *Science*, **221**, 1401-3.

Soft-rot Decay of Treated Hardwoods

M.D.C. Hale

Department of Forestry and Wood Science, University College of North Wales, Bangor, Gwynedd LL57 2UW, UK

&

R.A. Eaton

School of Biological Sciences, Portsmouth Polytechnic, King Henry Building, King Henry 1 Street, Portsmouth PO1 2DY, UK

ABSTRACT

The lumen surface and S_2 wood cell wall areas of Copper–chrome–arsenic (CCA) preservative-treated wood decayed in a laboratory system by a preservative tolerant soft-rot fungus were investigated to examine relationships between preservative microdistribution and sites of decay. The results showed that decay initiation commonly occurred in the fibres adjacent to ray cells and that preservative elemental levels declined from the surface of the wood inwards. Despite reasonably high levels of chromium overall, low levels of copper were recorded even in unattacked fibres.

INTRODUCTION

Several interacting theories have been proposed to explain the ability of soft-rot fungi to decay some species of preservative-treated hardwoods, i.e. preservative tolerance and detoxification (Greaves, 1972), preservative microdistribution (tissue: Greaves, 1974; cell wall: Dickinson, 1974), cell wall lignin (Butcher & Nilsson, 1982; Nilsson, 1982) and the polyflavonoid tannin theory (Pizzi *et al.*, 1986).

Few attempts have been made to correlate the preservative microdistribution in the treated fibres with areas of decay and preferential colonisation. The work described here has attempted to achieve this with a laboratory decay system combined with analysis of preservative microdistribution (cell wall level) by energy-dispersive (EDS) X-ray microanalysis in TEM. Where sufficient data have been collected the profile of preservative levels in fibres along partial transects taken from the wood surface inwards have been constructed.

MATERIALS AND METHODS

Treatment and source details of the wood material and the fungal culture can be found in Hale & Eaton (1986).

RL sections (25 μm thick) of *Acacia mangium, Alstonia scholaris, Eucalyptus deglupta* and *Gmelina arborea* were infected over agar with *Lecythophora mutabilis* (Hale & Eaton, 1985). The sections were cut from untreated and CCA-treated sap- or outerwood of 20 × 20 mm thick stakes.

After incubation periods of 1–28 days sections were removed and examined microscopically. Transverse sections were subsequently prepared for TEM analysis. Sections were analysed in an EMMA 4 TEM fitted with a Kevex EDS detector. Rigorously standardised analysis conditions were used. P/B ratios of Cu (Kα) and Cr (Kα) were processed.

Elemental analyses of Cr and Cu were made of the cell wall layers across each section or transect. These were decay-rated by numbers of cavities per cell, presence or absence of hyphae in lumina and by excessive amorphous preservative lumen deposits. Central block portions only were analysed in treated *A. scholaris* (i.e. 10 mm in) as no decay was induced. In *G. arborea* regions 2 mm deep were analysed. Resin analyses were done in empty spaces within fibre lumina of all samples.

RESULTS AND DISCUSSION

Conventional microscopy

Light microscopy of all of the susceptible wood species, treated or control, showed that S_2 decay was initiated in fibres adjacent to the rays. TEM confirmed these observations and in the treated sections, intense electron dense deposits (EDD) were visible in the ray lumina (Fig. 1) and

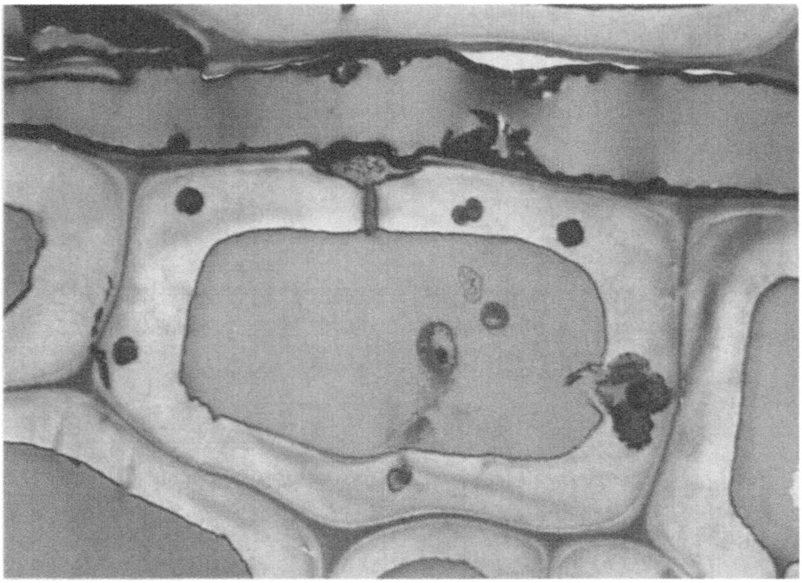

Fig. 1. Soft-rot adjacent to a well treated ray in a *E. deglupta* fibre with an electron dense lining.

in the outer 0–3 fibres. Further into the wood, narrower EDD and clumped amorphous material lined fibre lumina similar to those previously reported (Bravery, 1975). These luminal linings were continuous and visible in all cells examined. The presence of hyphae did not affect the appearance of EDD. Some of the cell lumina were completely filled with EDD.

Decay took longer to initiate in the treated wood sections and took longer in some wood species than others, i.e. treated: *A. mangium* (6 days), *G. arborea* (10 days), *E. deglupta* (14 days). After 28 days' incubation no decay occurred in *A. scholaris*. In the treated sections the extent of decay varied across the transect; in *G. arborea* sections, only areas corresponding to stake regions of 2 mm+ deep were decayed.

Microanalysis—general

The analytical results produced from the embedding resin alone were low; Cr: 0–0·11; Cu: 0–0·23. Values within these ranges were regarded as zero.

Microanalysis—transects

The microanalysis results (Fig. 2) of *A. mangium* and *E. deglupta* showed a decrease in Cr and Cu on lumen surfaces with increasing depth into the

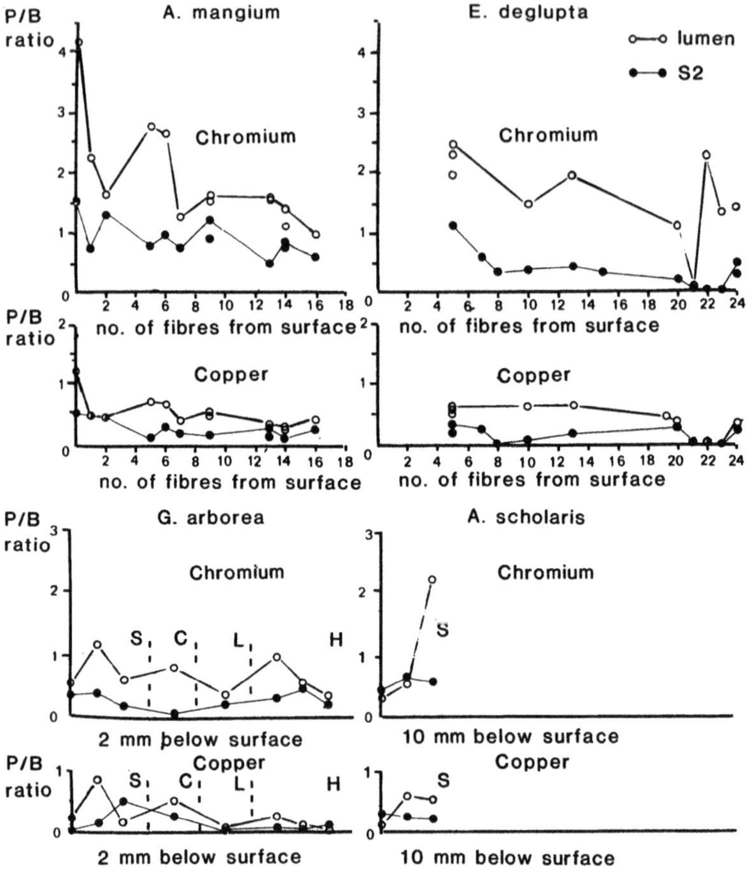

Fig. 2. P/B ratios of chromium and copper determined across sections of 4 hardwoods.

wood. The Cr levels on the lumen surfaces of inner cell walls of both wood species remained or stayed at values generally above 1 even at the maximum depths analysed, i.e. 16 and 24 cells respectively.

In *A. mangium*, Cu levels on lumen surfaces remained more or less the same across the section (0·26–0·5) apart form higher levels at the wood surface. Similar observations were made in *E. deglupta* but at a depth of 21 fibres. Cu levels on fibre surfaces adjacent to rays were close to zero in a number of the cells examined. This was despite plentiful ray deposits of CCA and Cr fibre levels >1. Even on those fibres with lumina filled with EDD the lumen values for Cu were low.

The Cr and Cu levels in S_2 cell wall layers declined with increased depth into the wood. Copper levels in the innermost fibres of *E. deglupta* were negligible despite cell lumina being filled with EDD.

Decay was noted at depths of 7 (*A. mangium*) and 10 (*E. deglupta*) cells deep. Beyond these depths a number of undecayed cells appeared to have Cr and Cu values similar to those of decaying cells despite all cells being subjected to similar biological exposure. This suggests that most fibres of these species would have failed given sufficient incubation. However, in *E. deglupta*, cell wall protection of uncolonised fibres and those fibres containing lumen hyphae was afforded by high Cr levels at the lumen surface.

Microanalysis—other observations

The values for Cr on the lumen surfaces of *A. scholaris* were highly variable while the Cu values showed less variation. The cell wall S_2 values of these inner fibres were low when compared with the surface values of the other two species (Cr 0·49–0·60, Cu 0·22–0·3) but considering the 10 mm depth origin of these specimens values are still appreciable. In *G. arborea* both Cr and Cu values were low on both the lumen surface and in the S_2.

Despite the quite high Cr levels encountered in *A. mangium* at fibre lumen surfaces these appear to be insufficient to prevent decay initiation. In *A. mangium* therefore low fibre Cu levels must partially contribute to poor performance although published values for Klason lignin contents of *Acacia* spp. (Pettersen, 1984) are low (19–20%). Butcher & Nilsson (1982) have suggested that low fibre Cu and lignin may lead to high susceptibility. Furthermore low copper levels were also recorded in *E. deglupta*, a species with a higher lignin content (26–32%; Pettersen, 1984), and performance was consequently better. Similarly *G. arborea* and *A. scholaris* both have moderately high (29%) and high (29–34%) lignin contents respectively (Levy, 1978; Butcher & Nilsson, 1982; Pettersen, 1984). Again better performance was observed in these species.

Reasons for preferential decay in regions surrounding rays may be connected to the microdistribution of nutrients in wood. Nitrogenous compounds were well supplied in the incubation medium but the presence of easily broken down polysaccharides may induce greater decay and induce greater preservative tolerance.

Many of the SEM microanalysis results on CCA treated hardwoods are in accordance with the above results. Low Cu values in fibres are typical while higher Cu levels are often found in vessels and rays (Greaves & Nilsson, 1982). Reasons for this disproportionation may include flow pathways within wood, low flow across hardwood ray/fibre pit membranes and rapid fixation onto wood tannin compounds.

The results presented above suggest that high preservative levels at the wood surface can only be regarded as giving limited protection against soft-rot decay. Decay may be initiated just a few cells subsurface. In addition results suggest that in this laboratory experiment even with poor microdistribution low copper may be tolerated if other soft-rot excluding factors like high lignin contents and adequate Cr levels occur. Unpublished results of a parallel field study with these wood species indicate that after 3 years soft rot failures are occurring with *A. mangium, E. deglupta* and *G. arborea*. Further fundamental work with more sensitive analytical systems needs to be undertaken to examine flow, distribution and fixation in hardwoods.

REFERENCES

Bravery, A.F. (1975). *Material und Organismen*, **10**, 31–41.

Butcher, J.A. & Nilsson, T. (1982). Influence of variable lignin content amongst hardwoods on soft-rot susceptibility and performance of CCA preservatives. The International Research Group on Wood Preservation, Document No. IRG/WP/1151, 16 pp.

Dickinson, D.J. (1974). *Material und Organismen*, **9**, 21–33.

Greaves, H. (1972). *Canadian Journal of Microbiology*, **18**, 1923–31.

Greaves, H. (1974). *Holzforschung*, **28**, 193–200.

Greaves, H. & Nilsson, T. (1982). *Holzforschung*, **36**, 207–13.

Hale, M. & Eaton, R.A. (1985). *Mycologia*, **77**, 447–63.

Hale, M. & Eaton, R.A. (1986). *Transactions of the British Mycological Society*, **86**, 585–90.

Levy, C.R. (1978). *Proceedings of the American Wood Preservers' Association*, **74**, 145–64.

Nilsson, T. (1982). Comments on soft-rot attack in timbers treated with CCA preservatives: a document for discussion. The International Research Group on Wood Preservation Document No. IRG/WP/1167, 10 pp.

Pettersen, R.C. (1984). *Advances in Chemistry Series*, 57–126.

Pizzi, A., Conradie, W.E. & Bariska, M. (1986). Polyflavonoid tannins—from a cause of CCA soft-rot failure to the "missing link" between lignin and microdistribution theories. The International Research Group on Wood Preservation. Document No. IRG/3359, 20 pp.

Studies on the Toxicity of Biocides towards Mould and Sapstain Fungi

G.R. Williams

Hickson Timber Products Ltd, Castleford, UK

&

R.A. Eaton

School of Biological Sciences, Portsmouth Polytechnic, King Henry Building, King Henry 1 Street, Portsmouth PO1 2DY, UK

ABSTRACT

Toxicity tests were carried out to determine the biocidal effectiveness of sodium pentachlorophenoxide (NaPCP) and methylene-bis-thiocyanate (MBT) on a range of mould and sapstain fungi. The results indicated that whilst the MIC values of the two biocides were similar in many cases, exposure to higher concentrations of the biocides in a simulated dipping test gave different results. It was found that exposure periods of between 24 and 48 h were required to prevent spore germination using a treatment solution of 1·0% NaPCP whilst the same effect could be obtained using a 0·24% solution of MBT in only 1 min, except with the tolerant mould Trichoderma viride.

INTRODUCTION

Tests to determine the relative performance of potential biocides for timber protection range from rapid screening methods using synthetic substrates to full scale field trials. For the evaluation of biocides to control post-harvest deterioration, many rapid screening tests have used a timber substrate which aims to simulate more closely the conditions present in the sawmill environment. In many of these tests, and even in some field test procedures, an artificial inoculum comprising various fungal genera are used in an attempt to standardise the test procedure. Such tests invariably produce a ranking of compounds in terms of activity to prevent the development of particular fungal groups, i.e. sapstain, mould and basidiomycetes. Whilst this type of data may ultimately be the most useful in determining the actual concentrations of active ingredient required to prevent defacement, it provides little information relating to the toxicity of the biocide to the individual target fungi. This paper describes the assessment of the toxicity of two biocides, methylene-bis-thiocyanate (MBT) and sodium pentachlorophenoxide (NaPCP) toward specific target fungi.

Two tests were carried out: the first to assess the toxicity of each biocide during the dipping phase of treatment when preservative concentration would be relatively high for a short period of time, the second to determine the equilibrium concentration of the biocide necessary to prevent longer term viability of fungal propagules at the timber surface (the minimum inhibition concentration—MIC value).

METHODS

The following fungi were used in all tests:

1. *Ceratocystis pilifera* (Fries) C. Moreau
2. *Ceratocystiopsis falcata* (Wright & Cain) Upadhyay Comb. Nov.
3. *Gliocladium roseum* (Link) Bainier
4. *Graphium aureum* Corda
5. *Leptographium lundbergii* (Lagerb. and Melin) Conant.
6. *Phialophora fastigiata* (Lagerb. and Melin)
7. *Trichoderma viride* Pers. ex Gray

The toxicity of MBT and NaPCP during the dipping phase of treatment

Fungal spore suspensions were prepared in sterile water and were used within 30 min of preparation. A standard Millipore filter apparatus was used to achieve brief exposure of the spores to the biocide solution. This comprised an upper reservoir, a filter screen onto which is placed a 47·0 mm diameter (0·45 μm pore size) cellulose nitrate filter to retain fungal spores, and a lower reservoir connected to vacuum. Spore suspensions of the test fungi were introduced into the upper reservoir and a vacuum drawn thereby settling the spores onto the filter surface.

The test solution was then introduced into the upper reservoir and left for different treatment periods. After exposure, the biocide solution was removed by re-applying the vacuum followed by washing with sterile distilled water to remove excess preservative from the spores and filter. The Millipore filter was removed and the spores transferred onto 2·5% malt agar by inverting the filter and 'printing' the spores onto the agar surface. Tests with *C. falcata* used 1·7% corn meal agar. Plates were incubated for 2 and 5 days and the percentage germination determined by microscopical examination. The biocide concentrations used were 0·09, 0·15 and 0·24% MBT and 1·0% NaPCP. Treatment times ranged from 10 s to 48 h.

Minimum inhibitory concentration values (MIC) for MBT and NaPCP

MIC values for MBT and NaPCP were determined in liquid culture using a modified Eggins and Pugh medium containing 1% glucose as the sole carbon source. Spore suspensions of the test fungi were prepared in this nutrient medium and 20 ml aliquots placed in universal bottles to which amounts of preservative solution were introduced in order to obtain pre-determined active ingredient concentrates ranging from 0·1 to 25 ppm of MBT and NaPCP. All bottles were incubated at 25°C for 10 days after which they were inspected and a visual assessment of growth made in comparison with untreated controls.

Williams (1987) describes a detailed methodology for both of these test procedures.

RESULTS AND DISCUSSION

A 10 s exposure to the 1·0% NaPCP solution had no effect on the spore viability of the fungi tested (Table 1). However, increasing the exposure period did affect spore viability of all the fungi, particularly after 48 h when no germination was recorded even after 5 days incubation. The two Ascomycetes, *C. falcata* and *C. pilifera* were relatively sensitive to NaPCP whilst the mould fungi were more resistant to the treatment. *Trichoderma viride* was the most tolerant of the test fungi throughout this test.

Exposure to test solutions of MBT had a greater inhibitory effect on spore viability. Even at the lowest test concentration of 0·09% and the shortest exposure period of 10 s, significant reduction in the viability of the fungal spores was recorded for all species except the mould fungi *G. roseum* and *T. viride* (Table 2). The sensitivity of many fungi to MBT was more pronounced at higher test concentrates and at 0·24% a.i., *T. viride*

TABLE 1
Percentage Germination of Seven Selected Fungi Incubated over Two Recovery Periods and Following Exposure to 1·0% NaPCP for 5 Different Time Periods

		Recovery periods									
		2 days					*5 days*				
		Exposure period									
Test fungi	*Control*	*10 s*	*60 min*	*18 h*	*24 h*	*48 h*	*10 s*	*60 min*	*18 h*	*24 h*	*48 h*
C. falcata	100	100	<1	ng[a]	ng	ng	+[b]	+	+	+	ng
C. pilifera	100	100	5	<1	<1	ng	+	+	+	+	ng
G. aureum	100	100	10	<1	<1	ng	+	+	+	+	ng
G. roseum	100	100	95	ng	ng	ng	+	+	+	+	ng
L. lundbergii	100	100	85	ng	ng	ng	+	+	+	+	ng
P. fastigiata	100	100	95	<1	<1	ng	+	+	+	+	ng
T. viride	100	100	100	40	<1	ng	+	+	+	+	ng

[a] ng = no germination recorded.
[b] + = 100% germination and significant hyphal growth.

was the only organism showing signs of growth after a 10 s treatment.

These results indicate that MBT is considerably more toxic than NaPCP towards fungal inocula during the dipping phase of treatment.

The results for the MIC determinations are presented in Table 3. The MIC values again show the variation in susceptibility of the various fungi to solutions of NaPCP and MBT. Although overall the results for the two compounds were essentially comparable the sapstaining fungi *L. lundbergii* and *C. falcata* were found to be the most sensitive to both compounds whilst the mould *T. viride* and the sapstain organism *P. fastigiata* were the most tolerant.

The fact that moulds such as *T. viride* and *G. roseum* had a much higher tolerance to the test biocides reinforces the observation that tolerant mould fungi such as *T. viride* are often the most problematical organisms on treated freshly-felled timber. More specifically, a number of authors including Butcher (1973), Butcher & Drysdale (1978) and Williams (1987) have recorded even more growth of *T. viride* on MBT-treated timber than on untreated controls. This apparent stimulation of mould growth may result from both an increased tolerance of the mould fungi to MBT and elimination of less tolerant fungal competitors.

The large difference in toxic effect between MBT and NaPCP was a surprising feature of the toxicity tests and may indicate a difference in the mode of action of the two biocides. Levi (1977) pointed out that in

TABLE 2

Percentage Germination of Seven Selected Fungi Incubated over Two Recovery Periods and Following Exposure to 0·09%, 0·15 over 0·24% a.i. solution of MBT for 2 Time Periods

		Treatment concentration (% a.i.)											
		0·09%				*0·15%*				*0·24%*			
		Recovery period				*Recovery period*				*Recovery period*			
		2 days		*5 days*		*2 days*		*5 days*		*2 days*		*5 days*	
		Exposure period											
Test fungi	*Control*	*10 s*	*1 min*	*10 s*	*1 min*	*10 s*	*1 min*	*10 s*	*1 min*	*10 s*	*1 min*	*10 s*	*1 min*
C. falcata	100	ng[a]	ng	ng	ng	ng	ng	ng	ng	ng	ng	ng	ng
C. pilifera	100	<1	ng	+[b]	+	ng	ng	+	ng	ng	ng	ng	ng
G. aureum	100	<1	ng	+	+	ng	ng	+	ng	ng	ng	ng	ng
G. roseum	100	85	ng	+	+	80	<1	+	+	ng	ng	ng	ng
L. lundbergii	100	25	ng	+	+	<1	ng	+	ng	ng	ng	ng	ng
P. fastigiata	100	ng	ng	ng	ng	ng	ng	ng	ng	ng	ng	ng	ng
T. viride	100	100	5	+	+	95	5	+	+	<1	ng	+	+

[a] ng = no germination recorded.
[b] + = 100% germination and significant hyphal growth.

TABLE 3
Minimum Inhibitory Concentrations (ppm) of Active Ingredient of NaPCP and MBT Tested Against 7 Sapstain and Mould Fungi

Fungi	*NaPCP*	*MBT*
C. falcata	0·5–1·0	<0·1
C. pilifera	5·0–10·0	0·5–1·0
G. aureum	5·0–10·0	1·0–5·0
G. roseum	1·0–5·0	1·0–5·0
L. lundbergii	0·1–0·5	<0·1
P. fastigiata	10·0–25·0	5·0–10·0
T. viride	10·0–25·0	10·0–25·0

growing cells, NaPCP primarily acts by uncoupling oxidative phosphorylation although studies by Crang & Robinson (1984) using *Aureobasidium pullulans* found no disruption of cell integrity at the ultrastructural level when spores were exposed to solutions of NaPCP. This might explain, in part, the high percentage of spore recovery recorded during the present study after washing spores to remove residual NaPCP.

The mode of action of MBT and other thiocyanates is not known although the rapid and often permanent effect on fungal spore viability might suggest either fungal spore membrane damage or intracellular disruption. Bearing in mind the often permanent inhibitory effect of MBT on the fungal spores, this might indicate a fungicidal effect by this biocide compared with a fungistatic effect by NaPCP.

Finally in defining the overall effectiveness of a candidate biocide as a prophylactic treatment, it is important to consider not only its short-term fungitoxic activity at the dipping phase of treatment but also its subsequent mobility in wood and the residual levels of biocide at the timber surface following storage (Lewis *et al.*, 1985).

REFERENCES

Butcher, J.A. (1973). *Material und Organismen*, **8**, 51–70.

Butcher, J.A. & Drysdale, J. (1978). *International Biodeterioration Bulletin*, **14**, 11–19.

Crang, R.E. & Robinson, S.G. (1984). *Plant Research Institute Proceedings*, No. 51.

Levi, M.P. (1977). In *Antifungal Compounds*, ed. M.R. Siegel & H.D. Sister. Marcel Dekker, New York, pp. 397–437.
Lewis, D.A., Williams, G.R. & Eaton, R.A. (1985). *Record of the 1985 Annual Convention of the British Wood Preserving Association*, pp. 14–28.
Williams, G.R. (1987). Sapstain and mould control of freshly felled timber. Ph.D. thesis, CNAA, Portsmouth Polytechnic.

Wood Preservative Efficacy Against Dry Wood Insects

M.M. Serment

Centre Technique du Bois et de l'Ameublement, 10 avenue de Saint-Mandé, 75012 Paris, France

ABSTRACT

The mode of action of different insecticides and the correlation between their thresholds and the commercial formulations in which they are used are considered in the light of laboratory test results. Though a parallel can be drawn between these toxic values and commercial formulations with preventive action, the same relation is not found with eradicant action. It seems that the toxicity of active ingredients is not the only factor to be taken into account. Beyond a certain point, the efficacy is not necessarily proportional to the amount of insecticide used.

INTRODUCTION

Wood destroying insects that damage wood used for building purposes belong to the order of Coleoptera, except for termites, and more specifically to the families of Cerambycidae, Anobiidae, Lyctidae and Bostrychidae. Depending on the country, different species abound depending on climatic conditions: in Germany and France, the number one enemy is the house longhorn beetle, *Hylotrupes bajulus*, while in the UK it is the furniture beetle, *Anobium punctatum*. The death watch beetle, *Xestobium rufovillosum* comes a close second but its presence is the result of its eating habits showing a distinct preference for rotten wood, rather than for climatic reasons. In temperate regions the Lyctidae, most commonly *Lyctus brunneus*, and the Bostrychidae, in tropical and subtropical latitudes are also important in the attack on dry wood.

PRESERVATION PRODUCTS

The currently available products on the market are of two types.

Waterborne

Salts with a fluorine or boron base have been used since the beginning of the century and combinations of certain salts can be fixed in wood. These are used in a waterborne solution at specific concentrations depending on their efficacy.

Organic

Since the end of World War II, mineral salts have met with competition from soluble synthetic substances in organic solvents. These formulations were and are still more appreciated for their penetrating properties. The addition of fixing resins provides a long-lasting effect.

Solvent costs have risen significantly in the last 8 years and this, coupled with ecological reasons, has resulted in these solvents being replaced by water. Concentrated organic pesticides are therefore found dispersed in emulsion form and are known as water-hydrodispersible.

Insecticides are divided into five main classes according to their chemical nature:

1. organometallic products;
2. organochlorinated products;
3. organophosphate products;
4. carbamates; and
5. a more recent addition to the market, pyrethroïds.

But new products are constantly being studied, e.g. isothiazolone, from the organosulphate product group.

Their efficacy varies according to the target insect and is species specific.

When examining a new product it is necessary to determine its threshold with regard to different species.

The most common species chosen as a reference, in Europe are *Hylotrupes bajulus, Anobium punctatum* and *Lyctus brunneus*. As it is not commercially feasible to produce a specific product for each insect, the toxic value is determined by the most resistant insect:

According to the class of insecticides and Table 1, it can generally be said that:

- organometallic products have a relatively low efficacy in comparison to the other classes;
- organochlorinated products are very effective against the house longhorn beetle but less so against the furniture beetle;
- organophosphate products are more effective in combating the furniture beetle;
- carbamates are less virulent than the preceding classes;
- pyrethroïds are as effective or even more so than organochlorinated products against *H. bajulus* and *A. punctatum*, but are less effective against *Lyctus*.

This explains why some of these classes of insecticide are used in sinergy.

For mineral salts, the toxic values determined are the exact reflection of their efficacy. The dose used is slightly higher than the toxic value obtained for the species to be combated. The quantities required are generally several hundred grams per cubic metre of wood.

Organic products, however, are effective at very low concentrations making it difficult to distribute the product evenly throughout the wood, especially because of the need to use them in a formulation.

The quantities of these products required will therefore vary according to:

- the nature of the solvent for substances in organic solvents;
- the emulsifying agents for emulsions;
- the type and quantity of fixing resins.

The active dosage required is several times the toxic values determined for an active ingredient. Because of the very different nature of formulations and even though the tests provide an approximate idea, this adjustment is not a simple correlation factor. It will vary according to the desired level of efficacy, preventive or curative, the latter requiring greater concentrations. It is therefore necessary to run additional tests on the insecticide so as to determine its efficacy on wood *in situ*. Tried and tested, standardised methods are available for this purpose.

EFFICACY

Up until the last few years, two types of effect were sought in practice for protecting wood.

TABLE 1
Effectiveness of some Used Insecticides Against Different Species of Insect

Type of insecticide	*Name of insecticide*	*Toxic values against* H. bajulus *nearly hatched larvae (EN 47) (g/m³)*	*Toxic values against* A. punctatum *by larvae transfer (EN 21) (g/m³)*
Organochlorine	Lindane	0·78< <2	20< <32[a]
	Dieldrine	0·1< <0·2[b]	52<83<140[b]
Organophosphate	Aethyl parathion	3<5<10[b]	8·4< <13[b]
	Phoxim	7< <12[b]	9<14<22[b]
Carbamates	Bassa	17<30<44[b]	400[b]
	Propoxur	18< <30[b]	1100< <2000[b]
Pyrethroids	Permethrin 25:75	5·5< 11·9[a]	42·4< <96·3[a]
	Permethrin 25:75	5< <8[b]	
	Permethrin cis	0·7< <2[b]	
	Permethrin trans	8[b]	
	Permethrin 75:25/50/50/40:60	2< <4[b]	
	Cypermethrin	1·1< <5·7	20·3< <39·3[a]
	Deltamethrin	0·0078< <0·078	4·6< <9·1[a]

[a] Data of BRE Princes Risborough Laboratory.
[b] Data of Desowag-Bayer Laboratory.

Preventive efficacy

Preventive efficacy consists in establishing a continuous barrier around the piece of wood in order to prevent egg laying. To be effective, it has to be intact.

Insecticides can act severally or jointly in three different ways:

— contact
— ingestion
— inhalation.

Laboratory tests have revealed that:

— mineral salts are only effective on ingestion. The newly hatched larvae will have to bore through the wood to be affected;
— products in organic solvents have a repellent effect on adults, especially during the first months because of the solvents' vapour phase;
— this repellent effect is, in most cases, coupled with a contact effect which has generally been the case up until now.

With the same active ingredient, it is stronger for organic formulations than for emulsions.

It decreases with time. This fact must be taken into account during laboratory testing. Before exposing adult *Lyctus* or *Anobium* as mentioned in standards EN 49 and 20, it is strongly recommended that test specimens be subjected to an evaporative ageing procedure.

However, pesticide losses are only superficial and only occur over a few tenths of a millimetre in depth. The degree of loss will depend on the species of wood, nature of the active ingredient and the initial loading.

Even so, the residue is sufficient to prevent newly-hatched larvae from growing by either contact or ingestion. Samples taken from in situ roof trusses treated 23 years earlier with organochlorinated products were shown to be free of infestation and there was still sufficient quantity of active ingredient in the wood.

It is difficult to evaluate precisely the preventative efficacy of inhalation of the active ingredient which seems to be overridden by the contact phenomenon.

Curative efficacy

The aim is to stop infestation as quickly as possible. All the larvae in the wood must be killed.

As mineral salts work by ingestion and are restricted to the surface layers they will only have a limited effect whatever the quantity applied.

The amounts of insecticide used must not only be toxic on ingestion but must also have an effect by inhalation and contact. This is particularly so for organochlorinated and synthetic pyrethroïd products and to a lesser extent for organophosphate products.

Eradicant action

In France, the main aim is immediate, or almost immediate, curative action. The main insect to be controlled is the house longhorn beetle and given the size of its larvae, damage can be very extensive over time notably in load-bearing structures. The insecticide must kill the larvae as fast as possible.

The general practice is to place the insecticide in a light hydrocarbon solvent. These solvents carry the insecticide deep into the wood and the toxicity of their vapour phase has a shock effect on the larvae.

The toxicity and inhalation effects of the active ingredient are increased by that of the resins which are more or less responsible for fixing the products in the wood. It is therefore difficult to know in advance the results that will be obtained. Hence, before any practical application of new formulae, it is necessary to carry out laboratory tests.

Remedial action

The oil crisis brought about the trend toward water-borne emulsions and was accompanied by a new approach to curative efficacy using the contact effect of insecticides to halt the adult emergence. The insects can no longer copulate nor, as in the case of *Anobium* or *Lyctus* lay their eggs in the existing larval galleries or pupal chambers. Infestation is self arresting. When such products are applied, a very low death rate has been observed of only 25–35%. The effect is slow and can only be used with difficulty for eradicating the large wood consumers, such as the house longhorn beetle.

The advantage of these curative products is that their effect is twofold as they also ensure preventive protection.

CONCLUSIONS—OUTLOOK

In addition to the foregoing, three very distinct, current trends should be pointed out:

(a) research into products which are toxic to insects whilst being environmentally acceptable, e.g. mimetic hormonal substances, and chitin inhibitors, such as diflubenzuron or triflumuron appear promising even if they are selective with regard to species;
(b) reduction in the use of solvents and the return to water-soluble or water-borne products;
(c) and finally, the quantity of active ingredients of the organic type used in commercial formulations is being reduced to its minimum level while still retaining a certain permanent effect. It should not be forgotten, however, that renewed treatment may be necessary in the event of reinfestation. This should be drawn to the attention of preservatives applicators with respect to guarantees.

REFERENCES

A bibliography for this paper may be obtained from the author.

Wood Degrading Activity of some Lignicolous Marine Fungi

L.L.P. Vrijmoed

Department of Applied Biology and Chemical Technology, Hong Kong Polytechnic, Hung Hom, Hong Kong

&

I.J. Hodgkiss

Department of Botany, University of Hong Kong, Pokfulam Road, Hong Kong

ABSTRACT

The wood degrading activity of six lignicolous marine fungi, namely, Antennospora quadricornuta, Ceriosporopsis halima, Corollospora maritima, Halosphaeriopsis mediosetigera, Lulworthia *sp. and* Periconia prolifica *was compared with that of a terrestrial species,* Chaetomium globosum. *The loss in weight of balsa wood (*Ochroma lagopus *Sw.) was estimated in a basal medium either incorporated with or without yeast extract in artificial seawater or in distilled water after 12 weeks of incubation at 25°C. With the exception of* A. quadricornuta, *all test fungi resulted in higher percentage weight loss in the presence of yeast extract. All species, except* A. quadricornuta *and* H. mediosetigera, *yielded higher percentage weight loss in artificial seawater medium.*

INTRODUCTION

Detection of wood degrading activities of marine fungi was first recorded by Barghoorn & Linder (1944) who discovered 'a conspicuous softening occurs in the outer parts of piling and submerged wood...' and histological study revealed that the action of fungi was the main cause of the decay. The pattern of ramification of hyphae within the wood cells, especially in the S2 layer was found to be identical to that observed by Bailey & Vestal (1937) within wood of the terrestrial environment. Savory (1954) named this type of surface deterioration 'soft rot' so as to distinguish this type of decay from the brown and white rots caused by wood-destroying Basidiomycetes. Jones (1971) first demonstrated weight loss of wood samples caused by marine fungi under laboratory conditions. The investigation described here aims at producing weight loss figures due to fungi which are common in the subtropical coastal waters of Hong Kong and which have been described previously (Vrijmoed *et al.*, 1982).

METHODS

The method used was adapted from Eaton and Jones (1971). Balsa wood blocks (*Ochroma lagopus* Sw.) of 1 × 1 × 1 cm were dried for 24 h at 105°C and cooled in a desiccator before weighing. Five blocks per 250 ml flask were then vacuum impregnated (172·24 kPa) for 3 days with 50 ml of the basal medium. Four different media were used for each test fungus, viz. basal medium with or without yeast extract in distilled or artificial seawater. After impregnation, 15 g of vermiculite was added to each flask to cover the wood blocks and the entire flask was subsequently autoclaved.

In order to estimate the amount of salt absorbed by a wood block after vacuum impregnation in nutrient solution, blocks from 3 replicate flasks of each of the 4 media were removed after the impregnation process, dried, cooled and reweighed. The difference in dry weights before and after impregnation provides an average value of the salt absorbed by each block.

Five lignicolous marine species common in Hong Kong coastal waters (Vrijmoed *et al.*, 1982), one temperate/ubiquitous species and one terrestrial species for comparison were tested (Table 1). Each flask was inoculated with five mycelial agar plugs from artificial seawater agar cultures. Controls without fungal inocula were included. They were incubated at 25°C in the dark. After 12 and 18 weeks, the wood blocks were cleansed of superficial mycelium or perithecia before drying, cooling and reweighing. The final weight losses were adjusted for nutrient salts absorbed by the block after vacuum impregnation as well as the weight changes in control blocks.

TABLE 1
List of Fungi Used in the Balsa Wood Degrading Ability Experiment

Fungus	*Source*	*Accession No.*	
		HKU[a]	*ATCC*[b]
Antennospora quadricornuta (Cribb et Cribb) Johnson	ATCC	—	18588
Ceriosporopsis halima Linder	Hong Kong	T18–78	48620
Chaetomium globosum Kunze ex Fries	ATCC	—	6205
Corollospora maritima Werdermann	Hong Kong	T19–79	48621
Halosphaeriopsis mediosetigera (Cribb et Cribb) Johnson	ATCC	—	16934
Lulworthia sp.	Hong Kong	T20–79	48622
Periconia prolifica Anastasiou	Hong Kong	S6–72	48623

[a] HKU = Botany Department, University of Hong Kong, Hong Kong.
[b] ATCC = American Type Culture Collection.

TABLE 2
Weight Losses[a] of Balsa Wood Blocks (%) Caused by Growth of Fungi in Pure Cultures after 12 Weeks of Incubation at 25°C

Fungus	*BM + DW*	*BM + Y + DW*	*BM + ASW*	*BM + Y + ASW*
Antennospora quadricornuta	0·23	+	+	2·93
Ceriosporopsis halima	3·67	8·33	9·33	13·57
Chaetomium globosum	1·97	9·09	10·05	14·25
Corollospora maritima	5·82	7·03	18·22	25·34
Halosphaeriopsis mediosetigera	5·47	11·00	2·73	7·19
Lulworthia sp.	1·87	6·22	20·33	23·94
Periconia prolifica	2·08	5·02	21·43	22·08
Control	−2·17	−1·22	+15·26	+26·49
Salt uptake after vacuum impregnation	0·3	0·48	10·75	12·68

[a] Average value from 4 balsa blocks, corrected for salt uptake after vacuum impregnation and natural changes during incubation.
BM, basal medium which consists of NH_4NO_3 3 g/litre, K_2HPO_4 2 g/litre, KH_2PO_4 2·5 g/litre, $MgSO_4.7H_2O$ 2 g/litre.
Y, yeast extract.
DW, distilled water.
ASW, artificial seawater, 30‰ (Lyman & Fleming, 1940).
+, increase in weight observed even after correction for salt uptake and natural gain in weight.

RESULTS AND DISCUSSION

The weight losses of the balsa blocks after 12 weeks of incubation at 25°C with the test fungi are presented in Table 2.

With the exception of *Antennospora quadricornuta*, all the test fungi resulted in a higher percentage weight loss in the presence of yeast extract in the medium. Artificial seawater yielded higher percentage weight loss when compared with distilled water, even for the terrestrial species *Chaetomium globosum*, but, with the exception of *A. quadricornuta* and *Halosphaeriopsis mediosetigera*. The higher weight loss figures in the seawater media are likely to be attributed to better mycelial growth in that medium as compared with distilled water.

Comparative weight loss figures are available for *Corollospora maritima* (Jones, 1971), *H. mediosetigera*, *Lulworthia* sp. (Jones, 1971; Henningsson, 1976) and *C. globosum* (Eaton & Jones, 1971), but not for the rest of the test fungi.

For *Corollospora maritima*, similar percentage weight loss of beech (Jones, 1971) and balsa were recorded though a higher temperature and longer incubation period was used for the beech block test. A recent study has shown that *C. maritima* consists of distinct physiological races which maintain characteristic temperature responses in the laboratory and different strains adapt to grow best near the temperatures at which they are found in nature (Bebout *et al.*, 1987). The British strain used on beech would not be at its optimum growth at 27°C whereas 25°C was found to be the optimum temperature for the Hong Kong strain (Vrijmoed, unpublished). This may explain why a high weight loss figure on balsa compared with that on beech was obtained. The general observation is that softwood is more resistant to attack by soft rot fungi than hardwood (Savory, 1954; Jones, 1971).

H. mediosetigera also produced comparable weight loss figures on birch and balsa blocks suggesting both the incubation temperature and the substrates favoured the growth of these isolates. The species used on balsa wood is not a local isolate but obtained from The American Type Culture Collection (Table 1) and originated from temperate waters. The very low values recorded for beech and Scots pine are, by contrast, likely to be due to the wood itself as fungi produced much less weight loss on both of these substrates as compared with balsa. Although *H. mediosetigera* is frequently collected in temperate waters and does occur in tropical regions (Kohlmeyer, 1984), if the same principle as described earlier for *C. maritima* applies, 28°C may be too high for the growth of the British isolate, and hence the low wood-degrading activity.

The local *Lulworthia* sp. has an ascospore length of between 250 and 400 μm which cannot be identified at present (Kohlmeyer, 1984). *L. medusa* and *L. purpurea* have been considered to belong to this group (Kohlmeyer & Kohlmeyer, 1979). Performance on balsa is again better than on birch and beech, especially in seawater media, in spite of the longer incubation period and slightly higher incubation temperature. Perhaps, strict comparison of these fungi may not be valid as they may be different species within the group (Kohlmeyer & Kohlmeyer, 1979). However, these results showed that various species of the *Lulworthia* group exhibit different wood degrading ability.

Chaetomium globosum has been shown to produce higher weight loss figures on beech compared with Scots pine and that similar results were obtained for media in distilled water and seawater. This is in contrast to results on balsa where seawater supports the highest weight loss. It is difficult to explain why the fungus can apparently produce better growth in a saline medium when it is essentially a non-marine species. The beech figures are two- to ten-fold higher than those of the balsa, suggesting that beech is a better substrate for *C. globosum*.

Antennospora quadricornuta produced little weight loss on balsa blocks (Table 2). Results from other treatments can be discarded because of abnormal increase in weight which cannot be explained. This fungus has however been observed to cause rapid decay of fishing craft in India (Becker & Kohlmeyer, 1958), thus establishing its wood degrading capability. Its low activity under laboratory conditions is likely to be due to the incubation temperature (25°C); its optimal temperature for mycelial growth is observed to be at 31°C (Vrijmoed, unpublished), being a tropical species (Kohlmeyer, 1984). Presumably, the temperature of the Indian waters (Kerala, India) where the rapid decay of fishing craft occurred was higher, thus favouring wood degrading activity.

Ceriosporopsis halima and *Periconia prolifica* showed similar results to *A. quadricornuta*, with saline conditions favouring higher percentage weight loss of wood (Table 2). The high weight loss figures of *P. prolifica* indicate that it is a soft rot fungus with high activity.

CONCLUSION

The present investigation has shown that wood degrading ability of marine fungi and possibly some terrestrial species is higher in saline conditions, although the type of wood and temperature of incubation are important. Weight loss figures on balsa have been provided for *C. maritima*, *H. mediosetigera*, *Lulworthia* sp. and *C. globosum* to be compared with figures obtained from tests on beech, birch and Scots pine, while those for *A. quadricornuta*, *C. halima*, and *P. prolifica* are first records.

REFERENCES

Bailey, I.W. & Vestal, M.R. (1937). *Journal of The Arnold Arboretum*, **18**, 196–205.

Barghoorn, E.S. & Linder, D.H. (1944). *Farlowia*, **1**, 395–467.

Bebout, B., Schatz, S., Kohlmeyer, J. & Haibach, M. (1987). *Journal of Experimental Marine Biology and Ecology*, **106**, 203–10.

Becker, G. & Kohlmeyer, J. (1958). *Journal of the Timber Dryer's and Preservers' Association of India*, **4**, 1–10.

Eaton, R.A. & Jones, E.B.G. (1971). *Material und Organismen*, **6**, 51–80.

Henningsson, M. (1976). *Material und Organismen Beih.*, **3**, 509–19.

Jones, E.B.G. (1971). In *Marine Borers, Fungi and Fouling Organisms of Wood*, ed. E.B.G. Jones & S.K. Eltringham. Organisation de Cooperation et de Development Economiques, Paris, pp. 237–58.

Kohlmeyer, J. (1984). *Marine Ecology (Pubbl. Stn. Zool. Napoli, I)*, **5**, 329–78.

Kohlmeyer, J. & Kohlmeyer, E. (1979). *Marine Mycology. The Higher Fungi*. Academic Press, New York, pp. 132, 315–18.

Lyman, J. & Fleming, R.H. (1940). *Journal of Marine Research*, **3**, 134–46.

Savory, J.G. (1954). *Annals of Applied Biology*, **41**, 336–47.

Vrijmoed, L.L.P., Hodgkiss, I.J. & Thrower, L.B. (1982). *Hydrobiologia*, **87**, 143–60.

An extended bibliography for this paper may be obtained from the senior author.

WASTES

Landfill Gas—a Global Review

K.M. Richards*

UK Energy Technology Support Unit Harwell, Didcot, Oxford, OX11 0RA, UK

ABSTRACT

A worldwide survey of landfill gas recovery and utilisation facilities shows there to be at least 146 commercial schemes currently in operation in a total of 15 countries, with a further 32 trials underway. At a very conservative estimate these projects are saving 825 000 tcepa by replacing sources of more conventional, non-renewable energy.

Some 77% of these projects (137 in total) are situated in the United States, West Germany or the United Kingdom. Of the remainder all but four schemes are located in either Western Europe or Canada.

The preferred routes to energy exploitation are through the generation of electricity using engines (56 schemes) a total of more than 170 MW capacity, burning in boilers (33 schemes), kilns/furnaces (15 schemes) or purification for main pipeline distribution or use as a vehicle fuel (13 schemes).

INTRODUCTION

Somewhere in the world in July 1987 the birth of a child brought the global human population up to 5 billion. Experts predict that the population will rise still further, perhaps doubling by the 21st century.

*The views expressed in this paper are those of the author and do not necessarily represent those of ETSU, UKAEA or the Department of Energy.

Such a vast and growing population makes an increasingly large demand upon the resources of our planet, in particular upon food, clean water, minerals and energy, all of which are finite. Five billion people also produce large amounts of waste which requires disposal or treatment if harm to the environment is to be minimised and resources protected.

Landfill gas

This paper addresses the progress being made in the production and exploitation of landfill gas, a technology which could assist in the better management of the world's resources. At present, the technology is limited in its global appeal, being most applicable to the richer, more profligate nations of the world. However, with further development it could also have an increasing impact for poorer countries.

The technology shows great promise in two globally significant respects. Firstly, the technology promotes the exploitation of a methane rich gas formed from a biomass which is naturally produced and renewable, hence contributing towards the world's energy needs and saving valuable non-renewable sources such as coal, oil, gas or nuclear power.

Secondly, the technology promotes 'good housekeeping' in terms of looking after the environment through better waste management.

The process of gas production

The natural bacterial decomposition of organic materials in the absence of oxygen results in the production of a methane rich biogas. This process of 'anaerobic digestion' has been known since historical times. Perhaps the most familiar example of this is 'marsh gas', a methane rich gas mixture formed in anaerobic river and pond sediments.

Anaerobic decomposition also takes place in modern day landfill sites which may, in many ways, be thought of as very large, batch, anaerobic bioreactors or digesters. Organic materials contained in a refuse are attacked by a consortium of bacteria under conditions of oxygen exclusion and generate a methane rich 'landfill gas'.

The bacterial decomposition process is extremely complicated and even today is only poorly understood. It is known that carbohydrates from paper, cardboard, etc., which form the major components of 'Western' refuse, decompose firstly to sugars, then to mainly acetic acid and finally to methane and carbon dioxide. A typical landfill gas is roughly a 50:50 mix of the two gases. Figure 1 shows, in a simplified

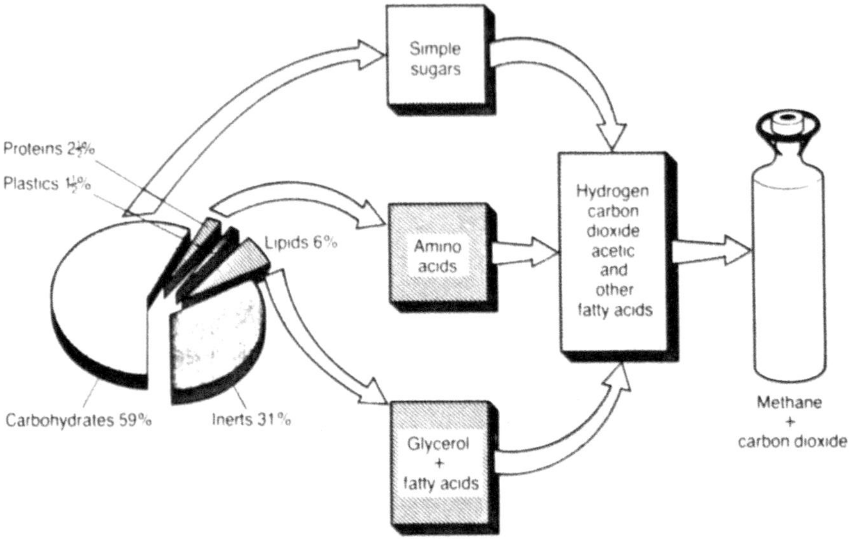

Fig. 1. Landfill gas production.

manner, the overall route taken during the bacterial decomposition process.

The 'discovery' of landfill gas in any great quantity is reported to have occurred in the 1960s in the United States and Germany. However the knowledge of the gas and the implications for its use as a fuel, is recent.

The technology

The United States have spearheaded much of the commercial endeavour in landfill gas abstraction and utilisation since the early 1970s and now leads the world in terms of numbers and scale of projects. West Germany and the United Kingdom are advancing rapidly and between them already have more than 60 schemes operational.

To understand why certain nations are rapidly exploiting this source of energy whilst others appear to be doing little requires more than just an historical perspective. An understanding of the whole strategy of waste arising and disposal is required.

WASTE DISPOSAL. THE LANDFILL GAS RESOURCE

In order to produce landfill gas in large enough volumes to merit commercial extraction for energy, at least a few thousand tonnes of predominantly organic wastes must be disposed of to a single site. This basic requirement necessitates there being:

- large settlements/towns/industries on a local level;
- a relatively affluent society.

First, large towns and industries are required to physically generate large amounts of waste, and as transporting waste in bulk is expensive, arisings must be local. (An exception to this will be discussed later.) Secondly, an affluent society is required for there to be biological materials in bulk which will be regarded as wastes and the richer the nation the greater the range of materials.

The US leads the list in our 'throw away society' with an estimated 200+ million tonnes of 'garbage' for disposal each and every year. Western European nations are equally wasteful, the UK, for example, producing an estimated 20+ million tonnes of organic wastes per annum. Such countries have massive waste disposal industries and feed between them thousands of large landfills.

In contrast, many poorer and 'Third World' countries generate much less waste and recover much of what they produce. They are also much less likely to have a centralised waste collection and disposal industry, making disposal to large sites unlikely. Such nations have little potential for utilising landfill gas, although digestion technology, using prefabricated tanks, may be relevant on a smaller scale.

Figure 2 illustrates the size of countries according to their Gross Domestic Product (GDP) which is a fair indicator of national wealth or affluence. North America is pre-eminent with some 40% of gross global product (1976 figures) and Western Europe with some 25%. Taken together with Japan, they produce at least two-thirds of the world's wealth.

Countries with large waste disposal industries and countries active in landfill gas exploitation are relatively large and wealthy. Landfill gas exploitation is then, at least for the time being, a technology more relevant to the rich.

Looking at the map there are notable exceptions, the USSR, China and Japan for example. No data are available from the USSR. However, it is known that China is interested in landfill gas technology, with at least one site, at Guangzho, currently under investigation. In addition, China already has much experience of conventional digester work, there being an estimated million plus digesters in the country.

Japan is a special case since it raises more general points about waste disposal and the perception of waste, and the further important factor of environmental acceptability.

Many countries consider landfilling as a relatively safe, cheap and

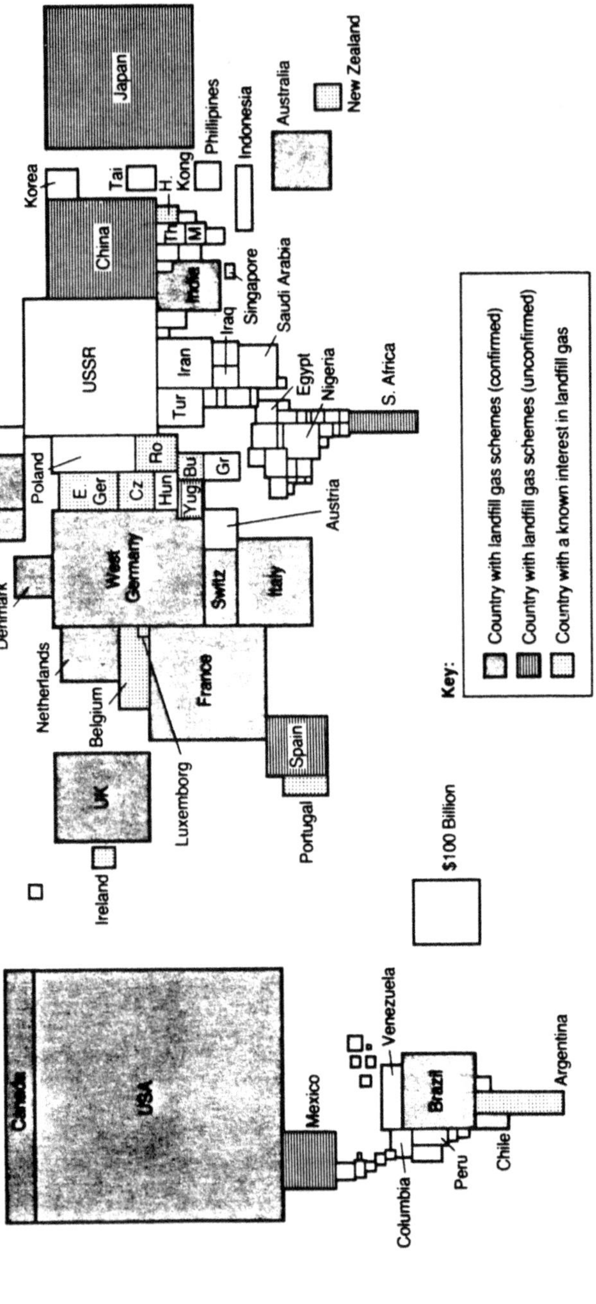

Fig. 2. Countries according to Gross Domestic Product (1976).

environmentally acceptable means of disposing of very large quantities of mixed household, commercial and industrial (non-toxic) waste, provided proper regulation and control are maintained. Further, landfilling is regarded positively as a means of reclaiming derelict, quarried or spoilt land. However, there are instances in some countries where pressures on land use make landfilling unattractive. Japan is a good example where land is in such short supply that disposing of wastes directly to landfill has become very expensive—more so than the alternatives, e.g. incineration. This has resulted in more than 67% of all household waste being incinerated with only 30% used for land 'reclamation' or landfilling. Compare this with the UK, where 90+ % of all waste is disposed of directly to landfill.

However, this is not the whole story. Japan also sees incineration with heat recovery as good 'housekeeping' and 'resource management' and a preferred technical option.

Environmental concerns coupled with expensive landfilling have also led to the development of the new industry of exporting wastes in bulk. At least two examples are known. Firstly, West Germany 'exports' large quantities of household refuse across the border to East Germany. Secondly, the USA is working on plans to export very large quantities of wastes to Central American and Caribbean countries.

It is not the purpose of this paper to discuss the merits or otherwise of this new departure, only to show that such practices are occurring. The impact on landfill gas is straightforward. As the resource (the waste) is moved in bulk, so the potential energy source is moved. Hence, small and often Third World nations may become large gas producers in the near future.

Disposal options

Generally, in the UK landfilling is popular with the Authorities, since it is simple and of low cost. Even so, like Japan, there are areas of the country where disposal to landfill is becoming more and more expensive as good sites become reclaimed. Here, incineration with heat recovery or refuse derived fuel (RDF) production may provide a better alternative for energy use. A number of commercial and development schemes are promoting these technologies. Figure 3 illustrates the various UK energy options for waste.

There is also scope for the development of schemes which 'marry' the best of digestion and RDF technologies. Such ideas are already being tried out in Norway and Sweden and it is hoped that the UK will soon be following suit when a good opportunity arises. Other possibilities being

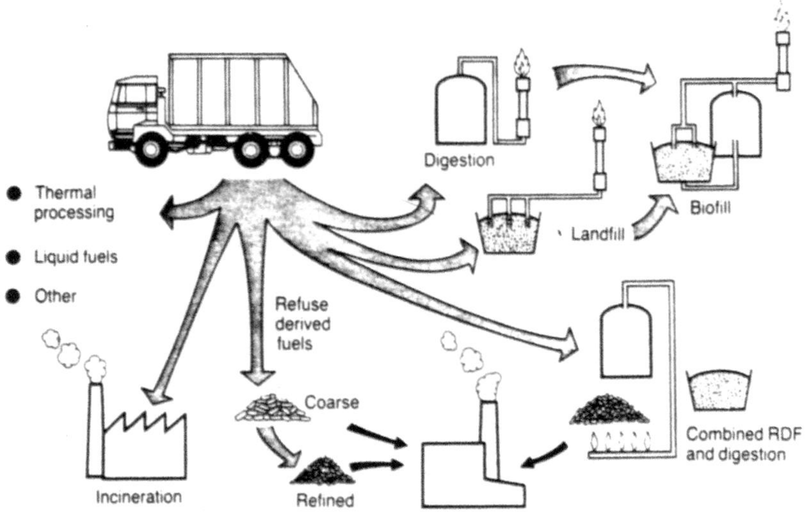

Fig. 3. UK energy options for waste.

seriously considered by the US and France include digestion of sorted putrescible waste in pre-fabricated tanks. Such technology could benefit poorer as well as richer nations. Thermal processing to produce a liquid fuel is a further option and there are still other possibilities being explored.

However, for all the major alternatives to landfill there still remain economic and technical questions. In particular:

- can the capital costs of necessary plant be reduced to make large schemes more economically attractive;
- can plant reliability be proven.

Also, where utilising waste as a fuel is concerned:

- can problems with emissions (e.g. dioxins) when waste is burnt be solved cheaply;
- can consumer resistance be overcome when selling products such as RDF or other reclaimed materials and can a good market be created for the fuel.

Environmental factors

Environmental issues surrounding the disposal of wastes to landfill (and other methods for that matter) and the resulting management problems continue to have a high profile in many countries. The United States is the best example of this, particularly California. Here, as a result of worries concerning the migration of gas laterally and to the atmosphere, with the associated odour problems, the Federal goverment has begun developing regulations for its control. Such moves are at first glance, likely to have a constraining effect on landfill gas operations.

However, the need for greater control of landfill operations is, in fact, proving beneficial to the landfill gas industry. Operators are required to adopt better management techniques, including an early provision for landfill gas abstraction and control. In this way environmental pressures promote and actually assist in the adoption of landfill gas exploitation for energy. Indeed so strong is the 'green' argument in the US that talk is of a move from an energy to a compliance industry.

Energy prices

One final global issue is that of falling energy prices. The impact in 1986 cannot be overstated. Energy prices in 1987 are approximately 65% of that 2 years previously. All energy industries have been hard hit, including the 'renewable-sources' which are at a delicate stage in their development in the Western World. In the UK a number of smaller landfill gas schemes have failed to proceed as a direct result of the fall. A similar situation holds true for the US and West German industries. However, those schemes with greater than a million therms output appear to be proceeding in Europe and, providing that further substantial falls in energy are not seen, should remain profitable.

Figure 4 shows the effect on the sale price of landfill gas to a typical UK exploiter during 1986. Since this time prices have continued to harden and most recently Brent crude has risen to over $20 a barrel, as a result of the continuing conflict in the Gulf. However, the future trend in oil prices remains highly uncertain and hence the price that 'landfill energy' can fetch, is difficult to predict.

LANDFILL GAS EXPLOITATION

Gas production

The mechanism for landfill gas production is biological and difficult to control, gas being produced 24 h a day, 365 days a year. This fact is universally true. Studies in the UK indicate that theoretical yields of up to 400 m^3 of biogas may be obtained from each tonne of landfilled waste.

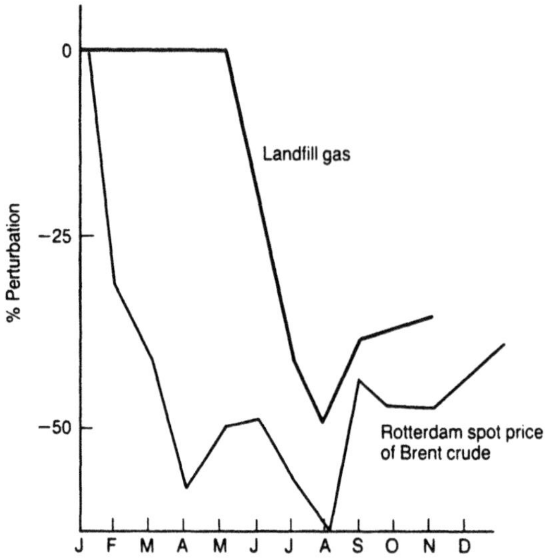

Fig. 4. Variation in the price of landfill gas in 1986—Direct Gas User Scheme.

However, in practice, only 25% of this may be recovered on a commercial basis over a period of 10–15 years.

The process of gas production is known to be affected by a variety of factors, including:

- site hydrogeology (i.e. moisture content);
- density of packing;
- waste composition;
- waste age;
- climate;
- cover material;
- site geometry.

Of these, moisture content and related climatic factors such as average ambient temperature, vary most from one part of the world to another. Comment has also been made on the variation of refuse between 'Western style' nations and poorer countries, the former having a higher carbohydrate (paper) content.

Moisture content

Gas production is roughly proportional to moisture content up to saturation. Hence, in the more arid parts of the world where rainfall is low, less than the optimal amount of moisture may gain access to the refuse and a relatively low gas yield may result. However, high amounts of rainfall are not always beneficial; excessive quantities of water may result in lower gas yields by stripping nutrients and by causing shock cooling.

Climate
Ambient temperature plays an important part in gas production, particularly for shallow, poorly insulated fills. Lower temperatures generally produce slower reactions and in extreme cases stop methanogenesis (methane production).

There are three recognised 'temperature ranges' which the necessary bacteria operate in, each having its own 'blend' of organisms. In temperate regions of the world mid-range, 'mesophilic' conditions are usually established within the bulk of the refuse with optimum temperatures of the order of 35°C being reached. In other hotter climates, and in one unusual case in the UK, warmer or 'thermophilic' (55°C) conditions prevail.

Waste composition
The higher the carbohydrate/polymeric content of waste and the lower the 'putrescible content', the slower will be the digestion process. Hence, 'Western Style' refuse may be expected to last longer than Third World waste for gas production purposes. Western waste will also be harder to digest in prefabricated tank systems.

Direct use of gas

In order to make the maximum use of the landfill gas resource, and to give maximum environmental control, exploitation needs to be on a continuous basis. Figure 5 shows a typical gas abstraction system which can be employed.

On most occasions a supply of gas on a dedicated basis via pipeline to an industrial consumer, in close proximity to the landfill, for direct use in kilns, boilers or furnaces will result in the simplest and most cost effective means of exploitation. In the UK this is by far the preferred technical and economic option (12 out of 20 schemes) and many notable successes have now been achieved. In West Germany it is also the most popular outlet for landfill gas (19 out of 43 schemes). In the United States, however, it is least popular (8 out of 54 schemes), possibly because of a lack of suitable consumers in close proximity to landfill sites. See Table 1 for a list of schemes by nation. A more detailed list of schemes by country is available from ETSU.

The US experience exposes the 'Achilles Heel' for direct use of landfill

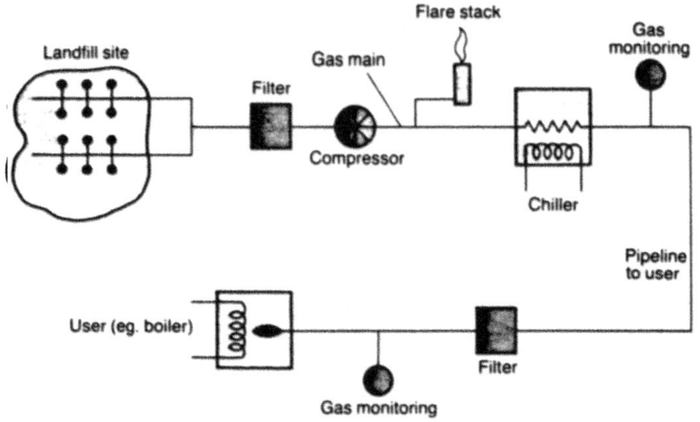

Fig. 5. Typical landfill gas abstraction plant layout. A number of gas 'wells' of simple design are constructed in the refuse using slotted or perforated piping. Each of these is interconnected via gas collection pipework to a pump or blower. Both well spacing and design are topics of current research activity around the world.

gas in kilns, furnaccs and boilcrs. Sincc whcrc local, good quality (continuous hightake) consumers are not available, pipeline costs to supply remote users, render such schemes non-viable.

Electricity generation

There will come a point when alternatives give a better economic prospect. Of these the globally preferred option is electricity generation with or without heat recovery. At the smaller scale of up to 1–2 MW output, converted internal combustion gas engines are preferred, whilst at multi-megawatt outputs gas turbines are also being used.

Technically, this option is much more demanding and riskier than direct consumption in boilers, etc. Landfill gas is not a pure methane, carbon dioxide mix, traces of many other gases, including potentially corrosive mercaptans and halocarbons are present. Problems can and have arisen, when raw landfill gas is burnt in engines, resulting in greatly enhanced chemical corrosion of apparatus leading to early plant failure. However, a more prudent approach is generally taken allowing for either some gas cleaning or more regular servicing and oil changing. With turbines, almost universally, gas cleaning to remove at least particulates and some corrosive gases is maintained.

Success in modifying engines and their running procedures can be gauged by their increasing use throughout the world. In the USA, 22 out of the existing 54 schemes generate electricity. Of those planned almost all will involve electricity generation.

TABLE 1
Use of Landfill Gas

Country	*Outlet*					*Total schemes*	*Trials/further schemes known to be planned*
	Boilers/ heating	*Kilns/ furnaces*	*Electricity generation/ (CHP)*	*Purification (pipeline/vehicle fuel)*	*Other/unknown application*		
United States	7	1	22[a]	10	14	54	13
West Germany	14[b]	5	17	—	7	43[b]	1
United Kingdom	5	7	6[c]	—	2	20	6
Sweden	2	—	4[d]	—	1	7	
Italy	—	—	2	1	3	6	
Holland	1	—	—	—	3	4	8
Denmark	—	—	3	—	—	3	
Canada	1	2	—	—	—	3	
France	1	—	—	—	—	1	
Norway	—	—	—	—	1	1	1
Switzerland	1	—	—	—	—	1	
Australia	—	—	1	—	—	1	1
Brazil	—	—	—	1	—	1	
India	—	—	1	—	—	1	1
Chile	—	—	—	1	—	1	1
	33	15	56	13	31	147	32

[a] One scheme generates electricity and sells gas.
[b] Includes one research project.
[c] Two schemes generate electricity and also use gas for heating.
[d] Four schemes are recorded as 'Boiler, CHP'.

In West Germany too, generation is popular, 17 out of 43 schemes choosing this option and in the UK 6 out of 20. West German schemes are generally smaller than either those of the UK or the US. An average of recorded US, UK and West German facilities shows the US to be 2·3 MW (largest planned facility at Puente Hills of 50 MW), the UK at 1·5 MW (largest 3·5 MW) and West Germany at only 0·5 MW (largest 1·08 MW).

In all the major Western nations interest in electricity generation is growing. In the UK and the USA operators are greatly attracted by more than straightforward technical reasons. Comments received put great store by:

- the increase in control for the operator;
- the greater security of the outlet;
- the reduction in overall 'hassle'.

First, the increase in control. When generating electricity sufficient capacity can be installed to utilise all the gas harnessed at the site. If exporting the electricity to a national utility, sufficient line capacity can also be installed. Unlike the situation where gas is piped to a consumer and where energy requirement may oscillate considerably, a utility offers to all intents and purposes, a total 'energy sink'. Hence, the landfill operator can determine how much gas he will use and when.

Secondly, the security of the outlet. Many direct users negotiate contracts stipulating gas take for a period of months or years. Re-negotiation is required at intervals to continue the supply and a positive outcome cannot always be guaranteed. In addition, a single dedicated consumer can go out of business or indeed change his line of business, in either case energy requirements may diminish leaving the operator with excess gas which he may be unable to sell.

Thirdly, the 'hassle' factor. Landfill operators are in the business of waste disposal and not energy management. Negotiation of contracts, keeping the customer satisfied, etc., causes headaches. With electricity generation, once discussions with the grid/utility are completed and a tariff agreed, running is relatively straightforward.

Total global electrical output from landfill gas fuelled installations will exceed 170 MW by the end of 1987, the USA contributing an expected 150 MW, the United Kingdom 8·9 MW, and West Germany 8·5 MW.

Gas purification

Of the three major options for landfill gas, purification and use is the least attractive. Gas clean up to provide a reasonably pure methane gas is complex and expensive. New systems employing semi-permeable membranes or molecular sieves are being developed, mostly in America.

In all but the United States the technology has taken a backseat, emphasis being laid on the more obvious routes to exploitation. However, in the USA, a number of purification plants have been constructed and supply considerable quantities of methane to local gas utility networks (10 out of 54 schemes).

The required specification for gas export is high and requires extensive pretreatment of the landfill gas. The specification includes:

—a minimum heating value of 970–980 BTU/SCF;
—0·7 lb of H_2O/MSCF;
—¼ grain of H_2S/100 SCF;
—total sulphur less than 9 grains/100 SCF;
—interchangeable with natural gas, i.e. acceptable heating value, relative density and oxygen level limits;
—no toxic, harmful or foreign materials.

To some degree the American picture can be put down to early enthusiasm in pursuing all options for landfill gas. However, this is not the whole story. Gas utilities in the past offered quite reasonable purchase tariffs to operators which encouraged the practice. This situation does not exist in other parts of the world.

A unique scheme exists in Santiago, Chile, where landfill gas is actually used to 'boost' an existing town's gas distribution facility (10 million standard cubic feet per day). Although the landfill gas is not scrubbed prior to injection no major problems have been encountered to date.

Treated gas may also be used in a compressed or liquified form, for use as a vehicle fuel, for example. This option has greater global appeal, although still marginal in terms of pure economics. Brazil and Italy are known to have at least one site each where this technology is being examined commercially. Even in the UK there are plans for a possible scheme, although again it must be emphasised that the technology is viewed as marginally acceptable.

Of other uses not considered separately, perhaps the most interesting is that of the production of a Refuse Derived Fuel (RDF). At least two countries, Norway and Sweden, are known to have commercial facilities. Landfill gas is used to assist in the drying of materials to be later pelletised for fuel. Both schemes are relatively small scale but offer much promise for the future.

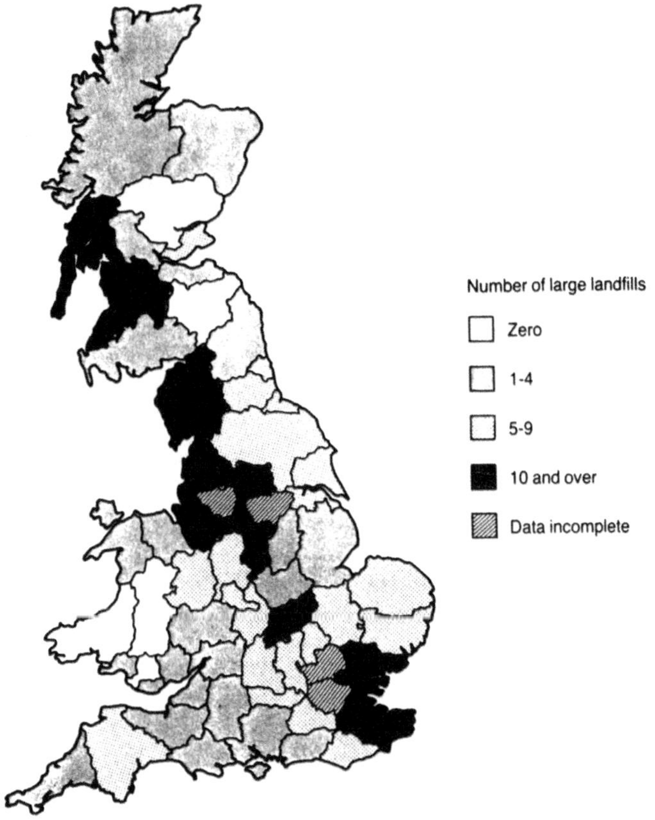

Fig. 6. Large landfill site distribution.

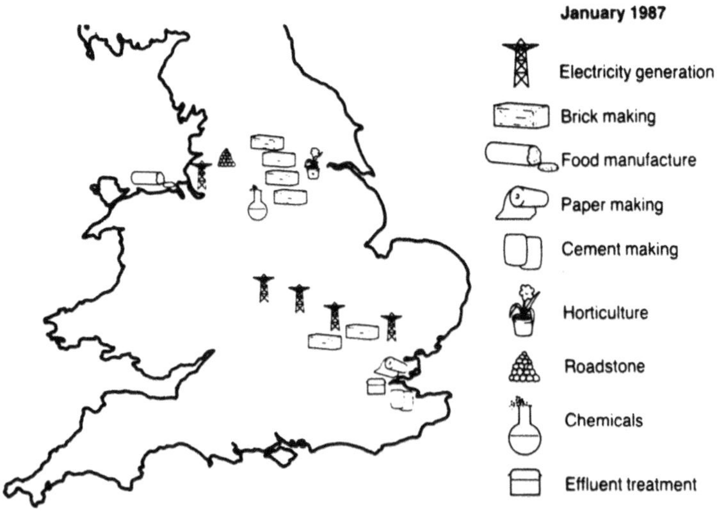

Fig. 7. Current use of landfill gas by industry.

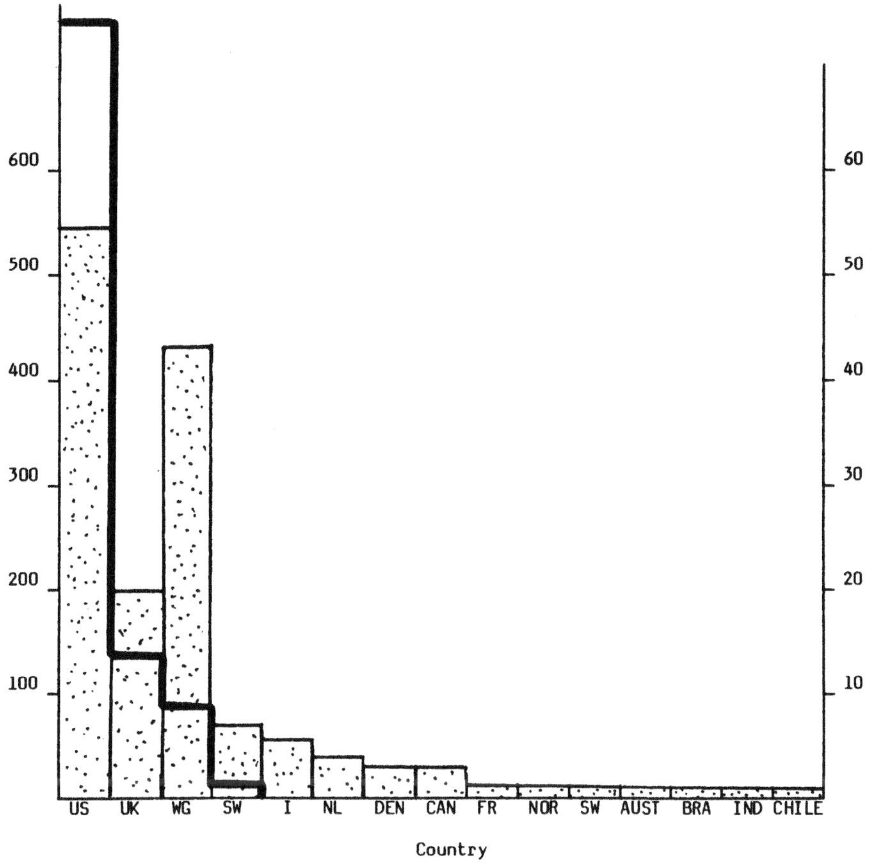

Fig. 8. The landfill gas exploitation league. (■), energy contribution (ktcepa); (▨), no. of commercial schemes.

United Kingdom

Figures 6 and 7 show the distribution of landfills in the United Kingdom and existing commercial schemes for gas exploitation. Landfill gas in the UK is already saving in excess of 140 000 tcepa which makes the UK second in the international 'league' of gas exploiters (see Fig. 8). In the UK there exist more than 300 sites with total energy producing potential of 2–3 million tonnes of coal equivalent.

The UK Department of Energy is funding a large programme of research, development and demonstration activities on landfill gas exploitation to enable the maximum benefit to be attained from this resource. The programme, currently managed by the author, from the Energy Technology Support Unit, consists of 15 schemes and is set to expand further. New areas for research include:

- studies of the fundamental processes of bacterial degradation in landfills;
- studies into the optimisation of conditions within landfills to promote the maximum gas production under controlled and environmentally acceptable states;
- the anaerobic digestion of leachates;
- biofilling. A complete waste management/gas production system. system.

COMMENT

The future progress of landfill gas as a fuel is obviously tied up with a number of issues of global significance. Not least of these is man's view of waste and his environment. Unless much more waste is recycled or materials reclaimed, then disposal to landfill will remain a popular option around the world. Exploiting landfill gas is common sense and good resource management.

ACKNOWLEDGEMENTS

I wish to thank a number of people for assisting my research into landfill gas schemes around the world. In particular to:

Jean Bogner of the Argonne National Laboratory, Chicago.
David Campbell of Harwell Laboratories.
Peter Ficke of Humphreys and Glasgow.

I also acknowledge Times Books Ltd who have kindly given permission to use their map 'concept' of the world (Fig. 2), taken from *The Times Concise Atlas of the World* (1979 Revision). © Times Books Ltd.

Methods for Sampling Airborne Micro-organisms at Solid Waste Disposal Sites

B. Crook,* S. Higgins & J. Lacey

Institute of Arable Crops Research, Rothamsted Experimental Station, Harpenden, Herts, UK

ABSTRACT

Andersen samplers, multistage liquid impingers (MSLI) and filters in aerosol monitors were compared for sampling airborne bacteria and fungi at domestic waste handling sites. MSLIs and polycarbonate filters gave similar results but Andersen samplers gave smaller counts, especially of Gram-negative bacteria. Of the collection fluids tested for use in the MSLI, ¼ strength Ringer's solution + 20/g litre inositol was found to be the best. Advantages and disadvantages of using the different samplers are summarised and recommendations made for future sampling in enclosed environments.

INTRODUCTION

Propagules of micro-organisms are easily liberated into the air either by specific dispersal mechanisms or by disturbance of the colonised substrate. Indoors, the airspora provides an inoculum for biodeterioration processes, e.g. damage to painted surfaces (Ross, 1963) and works of art (Strzelczyk, 1981). Also, large concentrations of airborne micro-organisms in enclosed environments may act as agents for occupational allergic diseases, e.g. farmer's lung. a form of occupational allergic alveolitis caused by the inhalation of dust from mouldy hay containing spores of thermophilic actinomycetes, especially *Faenia rectivirgula* (Lacey *et al.*, 1972).

*Present address: Health & Safety Executive, The Occupational, Medicine and Hygiene Laboratories, 403 Edgeware Rd, London NW2 6LN, UK.

Discarded household waste forms a ready nutrient source for the proliferation of micro-organisms and, especially in summer weather, decomposition may start before collection. Collected household waste is taken to transfer stations where it is stored in bulk, sometimes after sorting or pulverising, then either incinerated or loaded into large containers for transport by road, rail or river to sites where it is used for in-filling and land reclamation. Alternatively, paper and metal may be separated from the waste and recycled, while the remaining degradable fraction is composted. Tipping and separating the waste in transfer stations and bulldozers levelling waste at landfill sites generate much airborne dust which may contain micro-organisms. Isolating and identifying these airborne organisms enables the exposure of workers to large concentrations of infectious or allergenic agents to be determined.

Available air sampling methods collect airborne micro-organisms by either impacting them onto adhesive surfaces, impinging them into liquids or by filtration. The Andersen sampler, an impactor which deposits particles onto agar plates (Andersen, 1958), is widely used but is severely limited by the number of microbial colonies that can be cultivated on the exposed plates without overcrowding. This restricts sampling to a few seconds duration in highly contaminated environments and previous studies have indicated that fungi and bacteria may number as many as 10^5–10^6 colony-forming units (CFU)/m^3 in the air of refuse disposal facilities (Lembke & Kniseley, 1985). Consequently, alternative methods of sampling were sought. As part of a 3 year survey of airborne contaminants in waste disposal facilities in the UK, three sampling methods were evaluated; the Andersen sampler, collection of particles into liquid using a multistage liquid impinger and collecting particles onto the surface of polycarbonate membrane filters held in plastic aerosol monitors.

MATERIALS AND METHODS

Air sampling

Airborne micro-organisms were collected by the following methods:
Andersen samplers: Andersen impactors (Andersen 2000 Inc., Atlanta, GA), were operated at 25 litres/min for 20 s periods using a vacuum pump powered by mains electricity. Airborne particles were separated by aerodynamic size into six fractions and impacted successively onto the surface of a series of agar plates. After exposure, plates were incubated without further treatment. Numbers of colonies isolated on stages 3–6 were transformed to allow for the probability of more than one particle being deposited at any point on a plate but only producing one colony (Andersen, 1958).

Multistage liquid impinger (MSLI): MSLIs (Hants R and D Glassware, Southampton), were operated at 55 litres/min for 20 min periods, also using a vacuum pump powered by mains electricity. Airborne particles were separated by aerodynamic size into three fractions and impinged into a collection fluid (May, 1966).

Filtration: Disposable plastic three-piece aerosol monitors (Millipore UK, Harrow) housing polycarbonate membrane filters (Nuclepore; Sterilin, Feltham) were operated at 4 litres/min for 1 h periods using a battery operated vacuum pump. The filter holders were used 'open-face' with the top section of the monitor removed while sampling. After sampling, 5 ml of an aqueous wash fluid containing 10 g/litre peptone and 0·1 g/litre Tween 80 was pipetted into the chamber of the aerosol monitor, the top section replaced and particles washed from the surface of the filter by shaking on a Whirlimixer (Fisons Ltd) for 1 min. The resultant suspension of particles in the liquid was used to inoculate agar plates (Palmgren *et al.*, 1986).

Collection fluids

Recovery of micro-organisms by MSLIs was compared using different aqueous solutions to collect aerosols of *Escherichia coli* generated in a wind tunnel (Gregory, 1951). Collection fluids tested included: distilled water; lactose broth (Oxoid CM137); nutrient broth (CM1); nutrient broth No. 2 (CM67); peptone water (10 g/litre peptone, 5 g/litre NaCl); physiological saline (8·5 g/litre NaCl); ¼ strength Ringer's solution (¼R; 2·25 g/litre NaCl, 0·105 g/litre KCl, 0·12 g/litre $CaCl_2.6H_2O$ and 0·5 g/litre $NaHCO_3$) and ¼R + 20 g/litre inositol.

Culture methods

Cell suspensions collected from the impingers and aerosol monitors were diluted and 0·1 ml portions spread onto the surface of tryptone soya agar plates (TSA; Oxoid CM131) incubated at 37°C to isolate total bacteria, and 2% malt extract agar plates incubated at 25°C to isolate total fungi. The same agar media and incubation temperatures were also used for plates loaded into the Andersen samplers.

Comparison of samplers

Andersen samplers, MSLIs and filters in aerosol monitors were compared by taking simultaneous samples at two domestic waste transfer stations, two incineration plants and one recycling plant. Collection efficiencies

TABLE 1
Comparison of Media for Collection of Bacteria in MSLIs

Collection fluid	*Relative no. of bacteria collected* (%)[a]
¼R	100
Physiological saline	45
Distilled water	18
¼R + inositol	140
¼R	100
Peptone water	93
Nutrient broth No. 2	84
Nutrient broth	70
Lactose broth	51

[a] Mean values of yields expressed as a percentage of the number of CFU isolated when ¼R used as the collection fluid (n = 8).

for bacteria and fungi were assessed separately and paired comparisons made of log-transformed data (Bordner *et al.*, 1978).

RESULTS AND DISCUSSION

Comparison of collection fluids

The results of comparing the different collection fluids in two experiments are shown in Table 1. In the first, significantly more bacteria were recovered with ¼R than with physiological saline (P<0·001), which was in turn significantly better than distilled water (P<0·01). In the second experiment, ¼R + inositol gave a further significant increase over ¼R (P<0·01) while nutrient broth was significantly better than lactose broth (P<0·02). However, nutrient and lactose broths gave smaller catches than ¼R with or without inositol and also foamed, making removal of all the fluid from the sampler difficult. Therefore, ¼R + 20 g/litre inositol was used as the MSLI collection fluid in the subsequent comparison of air samplers.

Comparison of air samplers

The three methods of sampling air in several different waste handling facilities for viable micro-organisms were compared in atmospheres containing from 10^3–10^5 CFU/m^3 airborne bacteria and fungi. Aerosol monitors isolated significantly more bacteria ($P<0 \cdot 01$) and fungi ($P<0 \cdot 05$) than Andersen samplers. However, there were no significant differences between Andersen samplers and MSLIs, nor between MSLIs and aerosol monitors (Table 2). Collection in aerosol monitors and liquid impingers may have given slightly greater counts because clumps of cells, which formed one colony on Andersen sampler plates, were disrupted in the suspending liquids to give more than one colony when spread onto agar. Collection in aerosol monitors may have been further increased by gravitational deposition of particles onto the exposed filters. Such deposition would occur to a lesser extent in the other samplers, because of their restricted intake orifices.

Using Andersen samplers, MSLIs and aerosol monitors, respectively, Gram-negative bacteria averaged 5·8, 19·9 and 16·5% of the total bacteria isolated at 37°C. This suggests that Gram-negative bacteria were subjected to adverse conditions in the Andersen sampler and perhaps dehydration stress damaging surface structures and resulted in poorer survival (Cox, 1987). Also, direct impaction onto agar may not allow repair of damaged cells as may occur during suspension in liquid prior to plating. Such repair may be enhanced by adding simple sugars such as inositol to the liquid (Webb, 1960). If Gram-negative bacteria had been recovered by the Andersen sampler in similar proportions to those from MSLIs or aerosol monitors, the total yield of bacteria would have been between 35 037 and 36 113 CFU/m^3, close to the totals collected with the

TABLE 2
Paired Comparisons of three Methods for Sampling Airborne Micro-organisms in Waste Disposal Sites

	Mean CFU/m^3 air sampled	
Group of micro-organisms	*Total bacteria at 37°C*	*Total fungi at 25°C*
No. of samples	16	18
Andersen sampler		
mean	31 650	42 625
range	10 069–88 716	4 954–449 780
MSLI		
mean	33 161	47 128
range	5 129–82 604	3 365–805 378
Aerosol monitor		
mean	40 382	54 933
range	7 603–148 936	3 327–968 278

other samplers. *Penicillium* spp., the most commonly isolated fungi at 25°C, averaged 88·5, 82·0 and 87·0% of the total number and *Aspergillus* spp. 5·8, 6·3 and 5·2% respectively from Andersen samplers, MSLIs and aerosol monitors. Fungal propagules are more resistant than bacteria to desiccation stress and this is reflected in the similar proportions represented by each taxon in the three types of samplers.

Assessment of sampling methods

The Andersen sampler is well documented and frequently used, giving size separation of collected particles that can be related to the potential for deposition in the human lung. It provides direct collection onto agar media with the minimal handling of samples after collection, but may underestimate concentrations, especially of more delicate bacteria. The impactors are usually used with a mains powered vacuum pump, which restricts their portability. Sampling time is restricted in heavily contaminated environments if agar plates are not to become overloaded and uncountable.

The multistage liquid impinger also allows particle size separation into three fractions corresponding to deposition in the respiratory tract. Survival of micro-organisms can be increased by the choice of a suitable collection fluid. There is no limit to the sampling time although evaporation of the collection fluid may occur after long periods of sampling in a warm environment. The samples also require further handling after collection. Vacuum, supplied by a mains operated pump, limits portability and the glass construction makes it fragile.

The aerosol monitor is simple to use, robust and operated by a battery powered vacuum pump, allowing samples to be taken from one point or to be attached to workers for personal monitoring in the 'breathing zone'. There is no limit to sample collection time but, over a long period, delicate micro-organisms may be desiccated. No size separation of particles is possible and the samples require further handling after collection.

Conclusions and use

Because of their similar collecting efficiencies, MSLIs and aerosol monitors were selected to collect and measure airborne micro-organisms in an extensive survey of airborne microbial contamination at domestic waste disposal sites. Impingers provided information on particle sizes and aerosol monitors were used where mains power supply was unavailable, for example in cabs of cranes and bulldozers. During this survey, up to 9 × 10^6 CFU of airborne fungi/m^3, predominantly *Penicillium* spp. and up to 10^6 CFU of bacteria/m^3 of bacteria were found (Higgins *et al.*, 1987). On average, more than a quarter of the bacteria were Gram-negative and areas where workers were exposed to large concentrations were identified.

These air sampling techniques may have other applications where it is necessary to isolate, identify and enumerate airborne micro-organisms, e.g. in clean rooms to detect viable contaminants, or where perishable products are to be stored to determine levels of airborne micro-organisms capable of causing deterioration.

ACKNOWLEDGEMENTS

The authors wish to express their thanks to the various waste authorities for their co-operation in this study, which was supported by a Health and Safety Executive grant, No. 1/MS/126/643/82.

REFERENCES

Andersen, A.A. (1958). *Journal of Bacteriology*, **76**, 471–84.
Bordner, R.H., Winter, J. & Scarpino, P.V. (1978). *Microbiological Methods for Monitoring the Environment.* Springfield VA. EPA-600/8-78-017 National Technical Information Service, pp. 225–30.
Cox, C.S. (1987). *The Aerobiological Pathway of Micro-organisms.* Wiley-Interscience, Chichester, pp. 172–205, 238–43.
Gregory, P.H. (1951). *Annals of Applied Biology*, **38** (2), 357–76.
Higgins, S., Crook, B. & Lacey, J. (1987). In *Aerosols, their Generation, Behaviour and Applications, Proceedings 1st Conference Aerosol Society*, Loughborough, pp. 141–4.
Lacey, J., Pepys, J. & Cross, T. (1972). In *Safety in Microbiology*, ed. D.A. Shapton & R.G. Board. Society for Applied Bacteriology Series 6, pp. 151–84.
Lembke, L.L. & Kniseley, R.N. (1985). *Canadian Journal of Microbiology*, **31**, 198–205.
May, K.R. (1966). *Bacteriological Reviews*, **30** (3), 559–70.
Palmgren, U., Strom, G., Blomquist, G. & Malmberg, P. (1986). *Journal of Applied Bacteriology*, **61**, 401–6.
Ross, R.T. (1963). *Advances in Applied Microbiology*, **5**, 217–34.
Strzelczyk, A.B. (1981). In *Microbial Biodeterioration*, ed. A.H. Rose. Economic Microbiology Series 6, Academic Press, London, pp. 203–34.
Webb, S.J. (1960). *Canadian Journal of Microbiology*, **6**, 71–87.

The Effect of Incorporation of Refuse Derived Fuel Processing Fines into the Growing Media on Yield and other Physiological Characteristics of Crop Plants

I.C. Mathieson, R. Tinklin

School of Science, Humberside College of Higher Education, Cottingham Road, Hull HU6 7RT, UK

&

J. Mills

Sheffield City Polytechnic, Department of Biological Sciences, Pond Street, Sheffield S1 1WB, UK

ABSTRACT

The effect of waste fines on yield and availability of trace metals to crop plants was carried out using the plant as the best indicator of 'available' trace metals. Enzyme assays and measurement of other physiological characters were used to study the extent of trace metal pollution. There was no effect on dry weight yield until 24 days after emergence, after which yield was progressively reduced by increasing applications of waste fines. Yield reductions diminished with time until final yield was similar to that of controls, and in sugar beet exceeded them. Mean copper, lead and zinc contents were increased and the 95% confidence interval was widened in plants grown with added waste fines. Yield could not be related to trace metal content. In vitro addition of trace metals at 10^{-6} *mol/dm*^{3} *did not affect the activity of isolated enzymes. Phosphate and nitrate content were reduced in plants grown with added waste fines. It is suggested that the nutritional effects may be a result of the addition of this material to the rooting media.*

INTRODUCTION

In Grimsby, South Humberside, a refuse derived fuel (RDF) processing plant leaves a by-product of light industrial and domestic refuse. This fraction will pass through a 25 mm mesh, after pulverising and removal of ferrous metals. It is known as 'waste fines' and is currently sent for landfill. The local scarcity of landfill sites has prompted investigation of alternative disposal methods, including the possible spreading of waste fines on to suitable agricultural land. Its value as a fertiliser is minimal, there being less than 10 mg/g each of nitrogen, phosphate and potassium, but its trace metal content was shown to be considerable, up to 1200 μg/g lead, and 1600 μg/g zinc (MAFF, 1983; Tinklin & Langford, 1983). The aim of this work was to investigate the effects of waste fines on the physiology of selected crop plants when the waste fines was incorporated into soil as an alternative method of disposal.

MATERIALS AND METHODS

Various proportions of waste fines were thoroughly mixed with a commercially purchased clay-loam soil (pH 6·8) and the mixture placed into 225 mm diameter pots. Inorganic fertilisers were incorporated at rates equivalent to 125 kg nitrogen (as nitrate), 75 kg each phosphate and potassium per hectare and planted with barley seed (*Hordeum vulgare*, c.v. Triumph). Treatments were replicated tenfold with five plants per pot; the pots were arranged in randomised blocks, with guard pots. Shoots and roots were removed at appropriate time intervals, washed in running water, rinsed in distilled water, dried and weighed. Sugar beet (*Beta vulgaris*, c.v. Sharp's Klein monobeet) was grown in an identical trial with one plant per pot, and added fertiliser equivalent to 35 kg/ha each of nitrogen, phosphate and potassium. Weighed samples of dried roots and shoots from each replicate plant were extracted in 50% concentrated nitric acid, filtered and made up with distilled water to give a final ratio of 1:25. Copper, lead, nickel and zinc concentrations were determined using flame atomic absorption spectrophotometry. Appropriate acid standards were used to prepare calibration data and acid washed glassware was used throughout.

For the analysis of enzymes, phosphate, nitrate and chlorophyll content, barley seeds were planted in general purpose peat/sand compost (pH 6·7), to which waste fines and inorganic fertiliser had been added and incorporated as above. After 40 days growth in a heated glasshouse, roots and shoots were removed and washed in running water, followed by distilled water. Enzymes were extracted by grinding 1 g of plant material in the respective cold extraction media with 1 g of sand in a chilled mortar, filtering through muslin and centrifuging at 20 000 × g for 20 min (18 000 × g for 15 min for nitrate reductase) at 2°C. Roots were soaked in $2{\cdot}5 \times 10^{-3}$ mol/dm^3 EDTA for 2 h prior to extraction. Enzyme assays were carried out according to the methods of Lee *et al.* (1976), Wainwright & Woolhouse (1975) and Deane-Drummond *et al.* (1979). Total protein extracted was assayed colorimetrically, reading absorbance at 224 nm and 236 nm in quartz cells (Groves *et al.*, 1968). These procedures were repeated twice using five replicates from each treatment.

Weighed samples of dried roots and shoots were extracted in 50% concentrated nitric acid, filtered and made up with distilled water to give a final ratio of 1:25. Phosphate content was determined using the ammonium molybdate, copper acetate and *p*-methylaminophenol sulphate method (Plummer, 1978); appropriate blanks and calibration data were prepared. Identical samples were homogenised in a domestic food blender with distilled water. The filtered water extract was made up to give a final ratio of 1:5 and phosphate content was determined as described above. Identical weighed samples were ground in distilled water in a mortar, filtered and made up to give a final ratio of 1:25. Nitrate concentration was determined with a nitrate ion-selective electrode.

Dried, weighed samples were ground in a minimum volume of 80% v/v acetone with 0·5 g of sand in a mortar. The extract was filtered using a fibre-glass based filter paper and made up in a volumetric flask. Absorbance was measured in a scanning spectrophotometer between 200 nm and 750 nm and at 645 nm and 663 nm in quartz cells. Chlorophyll concentration was determined using a nomogram taken from Kirk (1968).

Analysis of variance was used to assign significant difference (at the 5% level) between treatments using the F ratio. Correlation was determined using the coefficient of product moment correlation (r). Regression analysis used the least squares method. Confidence intervals were calculated using the t-statistic and where appropriate for difference between means.

RESULTS AND DISCUSSION (Tables 1–4)

There was no yield difference in either crop until at least 24 days. The 5% treatment had least effect and the 25% treatment had the most deleterious effect. Root yield was reduced more than shoot yield. By 88 days all

TABLE 1

Enzyme Activity per Unit Protein of Enzymes Extracted from Barley (*Hordeum sativum*, c.v. Triumph) Grown in Compost with Added Waste Fines (Added by Volume to Compost) as a Percentage

% waste fines added	*Acid phosphatase*		*Malic dehydrogenase*		*Nitrate reductase*		*Glautamine synthetase*		*Peroxidase*		*α-Amylase*	
	Shoot	*Root*	*Shoot*	*Root*	*Shoot*	*Root*	*Shoot*	*Root*	*Shoot*	*Root*	*Shoot*	*Root*
0	100·0	100·0	100·0	100·0	100·0	100·0	100·0	100·0	100·0	100·0	100·0	100·0
5	204·0[a]	110·0[a]	101·6	109·7[a]	104·3	73·2[a]	96·0[a]	74·1[a]	100·0	236·9[a]	100·0	100·0
25	147·0[a]	109·0[a]	125·6[a]	121·9[a]	83·4[a]	56·5[a]	44·5[a]	51·7[a]	100·0	242·7[a]	100·0	100·0

[a] Denotes significance at the 5% level.
Standard error <2·1%.

TABLE 2

Nutrient Content (mg/g) of Barley (*Hordeum sativum*, c.v. Triumph) grown in different Proportions of Waste Fines and Sampled after 40 Days

% waste fines added by volume	*Phosphate content*				*Nitrate content*	
	(Acid extracted)		*(Water extracted)*			
	Shoot	*Root*	*Shoot*	*Root*	*Shoot*	*Root*
0	39·5	39·0	1·0	1·0	1·25	0·40
5	32·0[a]	34·0[a]	0·5[a]	0·5[a]	1·15	0·35
25	31·5[a]	32·0[a]	0·5[a]	0·3[a]	0·85[a]	0·26[a]

[a] Denotes significant difference at the 5% level.
Standard error <0·02.

treatments equalled control yield and regression analysis suggested that the effect of the treatment reduces with time. Sugar beet yield follows the same pattern as barley; however, after 113 and 134 days, the 25% treatment increased shoot yield.

The application of waste fines to the plants rooting media increased the means and widened the 95% confidence interval of trace metal

TABLE 3

Dry Weights (g) of Barley (*Hordeum sativum*, c.v. Triumph) and Sugar Beet (*Beta vulgaris*, c.v. Sharp's Klein monobeet) Grown with Waste Fines Added by Volume to Soil Outdoors

Days	*Percentage waste fines added by volume*							
	0		*5*		*15*		*25*	
	Shoot	*Root*	*Shoot*	*Root*	*Shoot*	*Root*	*Shoot*	*Root*
Barley (standard error >0·66)								
16	0·08	0·03	0·07	0·03	0·08	0·03	0·05	0·03
28	0·43	0·20	0·30[a]	0·19[a]	0·20[a]	0·05[a]	0·12[a]	0·03[a]
40	0·86	0·37	0·77	0·30	0·51[a]	0·11[a]	0·21[a]	0·07[a]
52	2·30	0·72	1·70	0·60	1·36[a]	0·38[a]	0·66[a]	0·22[a]
64	3·93	1·68	4·14	1·56	3·52	1·04[a]	1·35[a]	0·39[a]
76	6·83	2·56	5·64	1·67	6·21	1·46	4·50[a]	0·62[a]
88	7·41	3·10	8·96	3·20	6·88	2·57	5·42	1·67[a]
Sugar beet (standard error >0·74)								
113	11·75	15·65	12·39	13·38	10·79	11·27	14·59[a]	14·48
134	13·01	27·85	13·25	25·21	12·86	23·01	19·90[a]	30·98

[a] Denotes significance at the 5% level.

TABLE 4
Means and 95% Confidence Intervals of Copper, Lead and Zinc Content (μg/g) of Barley (*Hordeum sativum* c.v. Triumph) Grown with Waste Fines Added by Volume to Soil in Pots Outdoors

% waste fines added	*Copper Shoots*			*Lead Shoots*			*Zinc Shoots*		
	L	*X*	*U*	*L*	*X*	*U*	*L*	*X*	*U*
0	1·7	2·4	3·0	9·2	11·4	13·5	59·3	68·9	78·5
5	2·5	2·9	3·3	11·4	13·4	15·5	57·1	68·9	80·6
15	2·6	3·0	3·4	18·4	22·9	27·4	60·5	81·9	103·3
25	2·7	3·1	3·5	25·4	29·1	32·8	92·1	106·6	121·0
		Roots			*Roots*			*Roots*	
0	18·8	22·7	26·4	25·9	29·2	32·4	62·4	71·6	80·9
5	22·7	24·9	27·0	33·2	33·1	41·9	77·4	94·3	111·1
15	26·9	29·7	32·5	59·0	66·3	73·6	79·9	104·0	128·1
25	29·0	31·1	33·3	72·7	81·6	90·6	95·0	118·6	142·0

L, lower limit; X, mean; U, upper limit.

content in both crops. There was no relationship between yield and trace metal content at any time. The trace metal content of 40 day old barley did not exceed the upper critical limits (Cu 20, Ni 26, Pb 35, Zn 290, μg/g dry matter; Davis *et al.*, 1978) and this implies that the effect of trace metals is not additive and will not adversely effect yield.

Activity of shoot and root acid phosphatase, malic dehydrogenase and peroxidase in roots was increased in barley grown with waste fines. Shoot and root nitrate reductase and glutamine synthetase was reduced but α-amylase activity was unchanged. In vitro addition of 10^{-6} mol/dm^3 trace metal acetate solution did not affect enzyme activity. A zinc content of 140 μg/g in roots was the maximum trace metal content recorded. This implies that any effect of trace metals on enzyme activity was not direct. The reduction in water extractable phosphate was proportionally greater than reduction in acid extracted phosphate in treated plants. This reduction is the most likely possible cause of increased acid phosphatase activity. Similarly the reduced level of nitrate in treated plants could be via substrate-limiting inhibition of nitrate reductase (Hageman & Flesher, 1960), and in turn may limit glutamine synthetase activity. α-amylase was unaffected, while the activity of malic dehydrogenase was increased. Peroxidase activity was increased by the treatment in roots, while activity was unaffected by up to 10^{-4} mol/dm^3 application of trace metals. Peroxidase is a precursor for the production of lignin from phenolic alcohols (Salisbury & Ross, 1978). This result is consistent with the hypothesis that cell wall binding has a role in trace metal tolerance in plants (Thurman, 1981).

Chlorophyll content was unchanged qualitatively and quantitatively. These results imply that the physiological effects of soil application of this waste material are not those associated with trace metal pollution. The evidence suggests that these effects may be caused by changed nutritional status in plants grown with added waste fines.

REFERENCES

Davis, R.D., Beckett, P.H.T. & Woollon, E. (1978). *Plant and Soil*, **49**, 395–408.

Deane-Drummond, C.E., Clarckson, D.T. & Johnson, C.B. (1979). *Plant Physiology*, **64**, 660–2.

Groves, W.F., Davis, F.C. & Sells, B.H. (1968). *Analytical Biochemistry*, **22**, 195–210.

Hageman, R.H. & Flesher, D. (1960). *Plant Physiology*, **35**, 700–8.

Johnson, L.B. & Cunningham, B.A. (1972). *Phytochemistry*. **11**, 547–51.

Kirk, J. (1968). *Planta*, **78**, 200–7.

Lee, K.C., Cunningham, B.A., Chung, K.H., Paulsen, G.H. & Liang, G.H. (1976). *Journal of Environmental Quality*, **5**, 357–9.

MAFF (1983). Analysis of waste fines. Personal Communication.

Plummer, D.T. (1978). *Practical Biochemistry*, 2nd edn. McGraw-Hill, London.

Salisbury, F.B. & Ross, C.W. (1978). *Plant Physiology*, 2nd edn. Wadsworth, California.

Thurman, D.A. (1981). In *The Effect of Heavy Metal Pollution on Plants, Vol. 2*, ed. N.W. Lepp. Elsevier Applied Science, London.

Tinklin, R. & Langford, D.H. (1983). Consultancy project for Humberside County Council Technical Services Department. Report 3.

Wainwright, S.J. & Woolhouse, H.W. (1975). In *The Ecology of Resource Degradation and Renewal, 15th symposium of the British Ecological Society*, ed. M.J. Chadwick & G.T. Goodman. Blackwell, Oxford, pp. 231–59.

The Efficiency of the Insect *Musca domestica* L. in Recycling Organic Wastes as a Source of Protein

J. Ramos-Elorduy, R.J. Villegas & M.J.M. Pino

Instituto de Biolgía, Universidad Nacional Autonoma de México, D.F. Apdo, Postal 70–153, 04510 México, DF

ABSTRACT

The capability and efficiency of Musca domestica *L. for recycling different vegetable wastes, domestic organic trashes, ruminal manure with 4% ammonium hydroxide and compost has been studied during the optimum period of oviposition, for the purpose of obtaining a source of animal protein.* M. domestica *was able to recycle all materials but with differing efficiency. The quantities of pupae obtained after 7 days were higher at the end of the oviposition period (16 days), yielding the highest amount of fresh and dry weight and the highest crude protein. The best results were obtained with domestic organic trashes and ruminal manure and the lowest yield was provided by the compost.*

INTRODUCTION

Wastes cause a high degree of pollution and most are underutilised or not utilised at all. Biodegradation can help in solving the problem. The use of new and unusual sources of food and feeding stuffs is receiving increasing attention.

Insects have not been considered seriously as a source of human food, despite the fact that they are consumed all over the world and have a high nutritive value (Conconi *et al.*, 1981*a,b*, 1982*a,b*, 1983). They are eaten only where other sources of food are not abundant and studies of their use in animal feeds and in waste recycling are scarce (Defoliart *et al.*, 1982; Finke *et al.*, 1985).

Protein is the most expensive ingredient of both human and animal nutrition. Studies of insects as waste recyclers, mainly *Musca domestica* L., have only been carried out using animal wastes, e.g. poultry, pig, horse and cattle manure (Flores, 1981).

The study reported here has been carried out on vegetable wastes coming from domestic sources in order to establish the conversion efficiency and capability of *M. domestica* to recycle this material. The pupal stage of the fly has been used to feed rainbow trout alevins.

MATERIALS AND METHODS

Kitchen wastes have been used such as cooked meal residues, raw carrots, maize, potatoes, pea pods, pineapple, mango, tuna, banana and papaya peel. They were mixed and dried in an oven for 3 days at 60°C. Ruminal manure and compost were treated in the same way.

The dried wastes were ground to a mesh size of 20 and re-humidified to 60% before being inoculated with fly eggs.

A 100 g of substrate was used in each experiment and each had three replicates. The substrates were a milk medium, as control, domestic organic waste, ruminal manure with 4% ammonium hydroxide and compost.

Three hundred *M. domestica* were placed in each of the 60 × 60 × 60 cm cages and nourished according to Spiller (1966). The oviposition medium was a 1:1 mixture of milk and yeast. Eggs were collected from the media from the 6th to the 16th day after hatching, the optimum oviposition time. They were placed in the four re-humidified media and removed after 7 days.

Pupae were counted, weighed and dried at 60°C for 3 days and then stored.

All experiments were carried out in the spring and summer and cultures were maintained at 28 ± 2°C and 60–70% humidity.

Experiments with rainbow trout *Salmo gardnieri* were performed at The Pisciculture Station of the Fisheries Ministry at Huasca Village, Hidalgo State, Mexico. For each experiment two fishponds with 150 alevins were used. The dried pupae were ground and fed to the fish over a period of 8 weeks, until they reached adulthood. They were measured, weighed and compared with controls nourished on a special meal from Albamex Industries.

RESULTS AND DISCUSSION

Every waste was utilised by *Musca domestica* but with varying efficiency (Table 1).

TABLE 1

Values Obtained Recycling Different Wastes with *Musca domestica* L. (g/100 g)

Ovip. period (days)	*Pupae number*	*Mean of pupae obtained*	*Fresh weight (g)*	*Dry weight (g) 10% humidity*	*Crude protein dry basis (g)*
Domestic organic trashes					
6	900–1 000	1 000	21·3	10·7	6·6
7	1 001–1 500	1 142	24·3	12·1	7·5
8	1 501–2 000	1 925	40·9	20·5	12·7
9	2 001–2 500	2 115	44·9	22·5	13·9
10	2 501–3 000	2 930	62·3	31·1	19·3
11	3 001–3 500	3 500	74·4	37·2	23·1
12	3 501–4 000	3 900	82·9	41·4	25·7
13	4 001–4 500	4 500	95·6	47·8	29·6
14	4 501–5 000	5 000	106·3	53·1	32·9
15	5 001–5 500	5 200	110·5	55·3	34·3
16	5 501–6 000	6 000	127·5	67·8	39·6
Ruminal manure (+ 4% NH_4OH)					
6	1 000– 2 000	1 400	29·8	14·9	9·2
8	2 000– 3 000	2 775	58·3	29·1	18·0
10	3 001– 4 000	3 600	75·6	37·8	23·4
12	4 001– 5 000	4 200	67·3	33·6	20·8
14	6 001– 7 000	6 550	104·8	52·4	32·5
16	9 001–10 000	10 000	212·5	106·3	65·9
Composta (+4% NH_4OH)					
6	300– 400	368	7·0	3·5	2·2
8	400– 500	400	7·6	3·8	2·4
10	600– 700	632	12	6·0	3·7
12	800– 900	842	16	8·0	5·0
14	900–1 000	976	20	10	6·2
Milk medium					
10	3 001–4 000	3 333	70·8	35·4	22·0

TABLE 2
Conversion Efficiency of *Musca domestica* L. Recycling Different Wastes

Fresh weight	*Dry weight*	*Protein*
Domestic organic trashes (100 g)		
4·65 –1	9·41–1	15·19–1
4·12 –1	8·24–1	13·29–1
2·44 –1	4·88–1[a]	7·88–1
2·22 –1	4·44–1[a]	7·18–1
1·60 –1	3·21–1[a]	5·18–1
1·34 –1	2·68–1[a]	4·33–1
1·20 –1	2·41–1[a]	3·89–1
1·04 –1	2·09–1[a]	3·37–1
0·941–1	1·88–1[a]	3·03–1
0·905–1	1·80–1[a]	2·91–1
0·792–1	1·47–1[a]	2·53–1
Ruminal manure (100 g)		
3·36 –1	6·72 –1[a]	10·85–1
1·72 –1	3·43 –1[a]	5·53–1
1·32 –1	2·65 –1[a]	4·27–1
1·49 –1	2·98 –1[a]	4·80–1
0·95 –1	1·90 –1[a]	3·08–1
0·470–1	0·941–1[a]	1·52–1
Composta (100 g)		
14·28–1	28·57–1	37·03–1
13·15–1	26·31–1	42·37–1
8·33–1	16·66–1	26·88–1
6·25–1	12·5 –1	20·16–1
5·0 –1	10·0 –1	16·12–1
4·70–1	29·41–1	47·61–1
Milk (100 g)		
1·41–1	2·82–1[a]	4·55–1

[a] Conversion efficiency less than the chicken: 5·12–1 dry weight, 2·56–1 fresh weight.

Controls gave similar results at the 10th day of oviposition with a mean of 3333 pupae per 100 g of medium. Their weight was comparable to those on domestic waste at the 11th day of oviposition.

TABLE 3
Average Values of Weight and Size of the Rainbow Trout (*Salmo gardnieri*) Nourished with Fly Pupae (each test had 150 trout)

		$\overline{W}_i$ (g)	$\overline{W}_f$ (g)	A_W (%)	$\overline{L}_i$ (cm)	$\overline{L}_f$ (cm)	A_L (%)	Z (%)
Albamex	1	0·48	1·38	187·50	3·6	4·9	36·11	66
Fly pupae	1	0·52	2·34	350·00	3·6	5·8	61·11	43
Albamex	2	0·53	1·22	130·19	3·9	4·5	15·38	33
Fly pupae	2	0·60	1·74	190·00	3·8	5·2	36·84	42

W = weight (g); $\overline{W}_i$ = initial weight; $\overline{W}_f$ = final weight; A_W = increase; $\overline{L}_i$ = initial size; $\overline{L}_f$ = final size; A_L = increase; Z = mortality.

Ruminal manure gave poor results but when ammonium hydroxide was added it gave the best results of all.

There is a constant supply of domestic organic waste, it is abundant, cheap and of variable composition. Economically it could be used in the production of animal feed stuffs. It has been shown by converting the results to kg pupae/tonne of waste the production is significant.

Comparing the conversion efficiency with chicken (Table 2) the majority of the domestic organic wastes and all the ruminal manure results were better. Other insects providing similar values are the cockroach *Blatella germanica*, the larvae of the bee *Bombyx mori*, the beetle *Trilobium confusum* and another fly *Cochlyiomia hominivorax* (Taylor, 1975).

Comparing the results obtained with those of other authors (e.g. Metcalf & Flint, 1977) we find that those in this study are better.

Because fly pupae were palatable to chicks (Teotia & Miller, 1973) it was considered that they could be fed to fishes. The results show (Table 3) that the final weight of rainbow trout displayed an increase of 350% in the first experiment and 190% in the second both of which were superior to the control. The fish also were of greater length than the controls.

The percentage mortality was similar, 42·5% in those fed with pupae and 49·5% in the controls. The colour of the meat was superior; those fish fed on pupae being pink and more acceptable commercially.

CONCLUSION

The results demonstrate that insects reared on domestic organic waste can provide a low cost protein source for animal feedstuffs.

REFERENCES

Conconi, J.R.E., Pino, J.M. & González, O. (1981*a*). *Fol. Ent. Méx.*, **49**, 141-54.

Conconi, J.R.E., Bourges, H. & Pino, J.M. (1981*b*). *Fol. Ent. Méx.*, **48**, 101-2.

Conconi, J.R.E., Rincón, F., Bourges, H., Pino, J.M., Alvarado, M. & Escamilla, E. (1982*a*). *Proc. 9th Congr. Int. Union Study Soc.*, in Supplement 2.

Conconi, J.R.E., Bourges, H. & Pino, J.M. (1982*b*). *Fol. Ent. Méx.*, **53**, 111-18.

Conconi, J.R.E., Pino, J.M. & Bourges, H. (1983). Valor nutritivo y calidad de la proteina de algunos insectos comestibles de México. *XVIII Cong. Nac. de Ent. Resúmenes*, pp. 134-5.

Defoliart, G.R., Finke, M.D. & Sundae, M.L. (1982). *J. Econ. Ent.*, **75** (5), 848-52.

Finke, M.D., Sundae, M.L. & Defoliart, G.R. (1985). *Poultry Sci.*, **64**, 708-12.

Flores, F.J.M. (1981). Evaluacion de harina de mosca domestica (*Musca Domestica* L.) como substituto de harina de carne en una ración para pollos de engorda. *Tesis Prof.* I.T.E.S.M., 67 pp.

Metcalf, C.L. & Flint, W.L. (1977). *Insectos destructivos e insectos útiles sus costumbres y su control.* Ed CECSA Méx, 1208 pp.

Spiller, D. (1966). In *Insect Colonisation and Mass Rearing*, ed. C.N. Smith. Academic Press, New York, pp. 203-25.

Taylor, R.L. (1975). *Butterflies in my Stomach*. Woodbridge Press Publishing Co., California, 224 pp.

Teotia, J.S. & Miller, B.F. (1973). *Poultry Sci.*, **52**, 1830-5.

A fuller bibliography for this paper can be obtained from the senior author.

Exploitation of Refuse by Anaerobic Microbial Intervention

Eric Senior, Irene A. Watson-Craik & George B. Kasali*

Department of Bioscience and Biotechnology, University of Strathclyde, Glasgow G1 1XW, UK

ABSTRACT

Since total control of the solid-state refuse fermentation remains an elusive goal of waste biotechnology this study was initiated to examine the roles of two first-tier variables, moisture and chemical amendment, in the methanogenic fermentation. Although moisture amendment to field capacity affected methane content and methane generation rate-increases, accelerated acidogenesis also resulted thus representing an increased leachate pollution potential. Leachate recycle, even in the presence of phenolic waste-water, was, however, found to minimise this potential provided that account was taken of the critical factors of refuse age, fermentative activity and buffering capacity.

INTRODUCTION

For more than 5000 years domestic refuse landfilling has been a key component of solid waste management (White-Hunt, 1980). Unfortunately, perpetuated essentially by economic stringencies, the solid-state fermentation exploitation potential has remained largely in abeyance for two millennia. Recently, diminishing land resource, conflicing estimates of fossil fuel reserves, constant price oscillations and increasing waste disposal legislation have motivated a re-evaluation. Consequently, landfill site practices have changed to accommodate industrial liquid and

* Present address: National Council for Scientific Research, P.O. Box CH-158, Chelston, Lusaka, Zambia.

sludge effluent treatment (co-disposal) and promote the generation of energy, added-value chemicals, chemical feedstocks and carbon dioxide. Total control/exploitation of the landfill bioreactor does, however, necessitate a complete understanding of the fundamental biochemical and microbiological catabolic processes (Senior & Balba, 1987). Unfortunately, few definitive studies have, as yet, been reported and this recognised paucity (Filip & Küster, 1979) still necessitates an element of serendipity in fermentation control.

In this study two problems were addressed. First, the promotion of methanogenesis by manipulation of a single first-tier environmental variable, moisture, and, secondly, assessment of the efficacy of refuse to obviate the environmental impact potential of phenol. This compound was chosen as a representative industrial molecule since concentrations ranging from 0·2 to 20 mM characterise the high volume wastewaters of plastics, phenolic resin, coke and fibreboard production. The experimental approach adopted was laboratory investigation by use of models.

MATERIALS AND METHODS

Refuse

Pulverised domestic refuse was sampled from Kilgarth Landfill, Glasgow, 1 month after emplacement. After handsorting to remove inert components, the refuse was homogenised in a domestic blender (Krups) for 2 min.

Laboratory models

The laboratory models used in this study to examine refuse methanogenesis and co-disposal have previously been described (Kasali, 1986; Watson-Craik, 1987).

Batch cultures

The batch culture study method used was the same as that described by Kasali (1986).

Analytical methods

The volatile fatty acids and methane assay methods were the same as those described by Coutts *et al.* (1987) and Watson-Craik (1987).

RESULTS AND DISCUSSION

Optimisation of methanogenesis

Although equation- or kinetic model-derived estimates are often invalid, they do serve as goals for methanogenesis optimisation. Attempted optimisation through site variable manipulation does, however, necessitate an understanding of first-tier variable interactions in relation to the landfill ecosystem. Thus, for example, there is an interdependence between refuse density and moisture content. Despite this, it is generally accepted that refuse stabilisation may be accelerated by controlled moisture application (Leckie *et al.*, 1979) which enhances bacterial nutrient availability by mixing, dilution and flushing.

For aerobic species a refuse moisture content of 60% is optimum (Dobson, 1964). For methanogens, however, the equivalent figure required to maximise rate and concentration remains unresolved and is complicated by the recognition that moisture content and movement are separate variables affecting methanogenesis (Klink & Ham, 1982).

Laboratory model studies demonstrated highest cumulative methane evolution in the presence of an approximate field capacity amendment of 75% (w/w) (Fig. 1). At concentrations $> 60\%$, however, the resultant increase started to decline, possibly indicative of satisfaction of methanogen physiological requirements. Moisture promotes mass transfer/distribution of other methanogenic-enhancement precursors (Buivid *et al.*, 1981), thus once optimised methanogenesis should be independent of the moisture regime. From activation energy calculations (de Walle *et al.*, 1978) the conclusion was made that the overall reaction rate was chemically controlled and thus not subject to diffusional mass transfer limitations in the presence of high moisture contents.

In the presence of high moisture amendments oxygen ingression could result, thus both delaying initiation of methanogenesis and stimulating acidogenesis (Fig. 2) to the detriment of the slow-growing methanogens.

Caution must be exercised in extrapolating laboratory results to actual landfills although it may be speculated that early refuse emplacement waterlogging could be inhibitory such that flooded sites or fills extending below the water-table may be sub-optimal for methanogenesis.

Together with gas generation rate, moisture amendments mediate fermentation balance changes with low moisture fermentations characterised by carboxylic acids and gases (including hydrogen) (Rees & Grainger, 1982). Here, moisture influences acidogenic species growth,

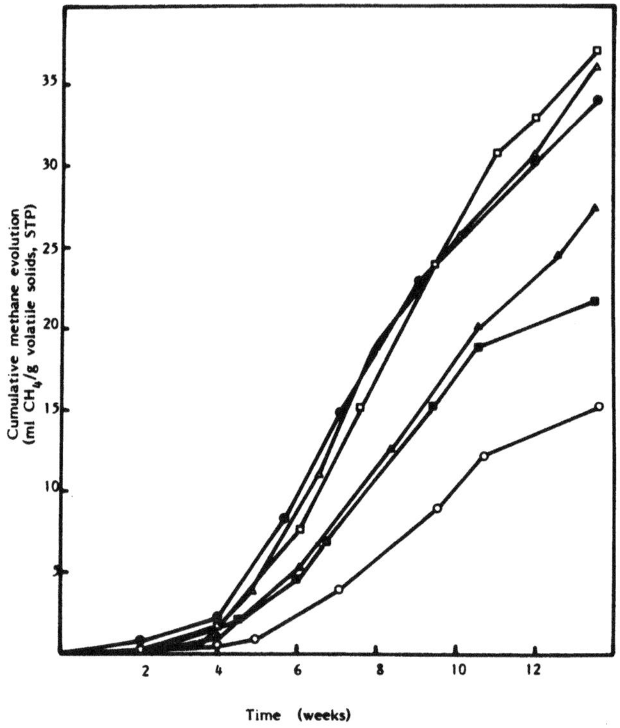

Fig. 1. Cumulative methane evolution from refuse columns poised at initial moisture contents of 55 (○), 60 (●), 65 (△), 70 (▲), 75 (□) and 80% (w/w) (■).

and thus acid production, or facilitates microenvironmental acid concentrations by dilution.

Laboratory model studies were initially characterised by a pattern of increased 'total' organic acid (acetate, propionate and butyrate) concentrations in response to elevated moisture regimes (Fig. 2). Subsequently, a step-wise sequential mode became apparent which, stoichiometrically, could not be accounted for by methanogenic rate changes, thus implicating product inhibition in the retardation of methanogenesis. Since the discrete concentrations recorded represent balances between genesis and trophy, interpretation is difficult, particularly when complicated by intermittent appearance of isobutyrate, isovalerate, valerate and caproate. The pH profiles did, however, mirror the 'total' organic acid profiles.

With elevated moisture, the methane:carbon dioxide ratio increased in favour of the former, possibly facilitated by the differential solubilities.

In view of the concomitant promotions of acidogenesis, methanogenesis and methane content in response to moisture amendment, provision, in the short term, may be necessitated to negate the increased leachate pollution potential.

Phenol co-disposal

Moisture amendment is an integral component of co-disposal, the admixing of industrial waste-waters and sludges with emplaced refuse; although, here, the environmental impact potential of the resulting leachate is compounded by the possible presence of additional chemical species which may also influence refuse catabolic processes. Despite this latter possibility, perfusion of 'fresh' (1 month) refuse columns, operated with leachate recycle, with phenol (2 mM) over a period of 80 days did not impair refuse catabolism compared with a distilled water control (Fig. 3). When, however, the 'fresh' refuse was replaced with 'mature' (1 year) refuse, within which a non-balanced anaerobic fermentation had

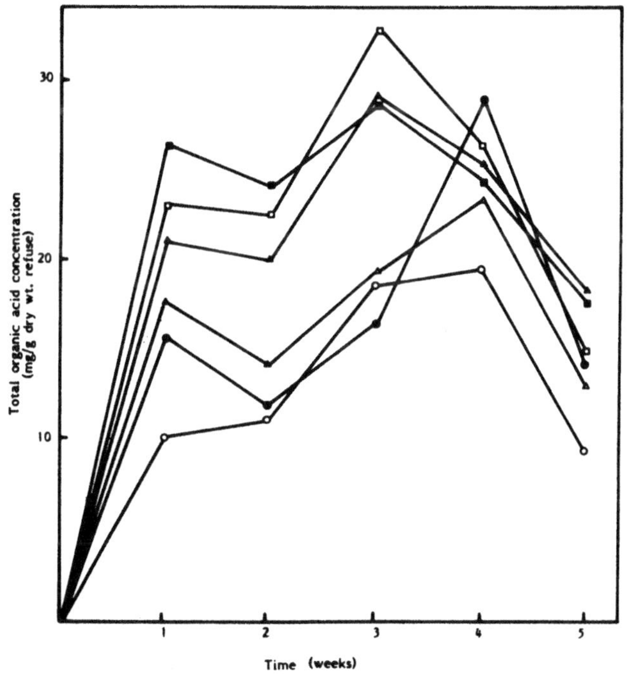

Fig. 2. Changes in discrete concentrations of 'total' organic acids (acetate, propionate, butyrate) in fermenting refuse poised at initial moisture contents of 55 (○), 60 (●), 65 (△), 70 (▲), 75 (□) and 80% (w/w) (■).

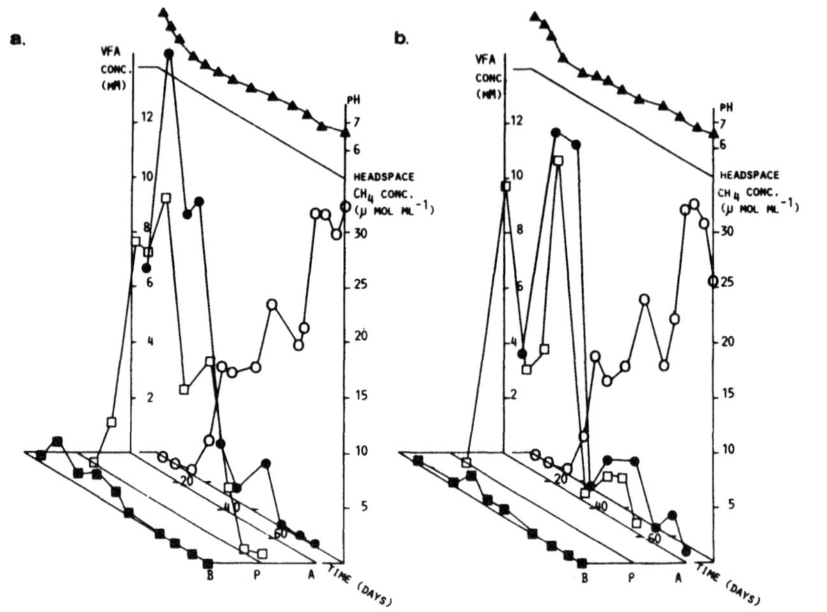

Fig. 3. Temporal concentration changes in headspace methane (○) and leachate acetate (●), propionate (□), butyrate (■) and pH (▲) in refuse columns operated with leachate recycle in the presence (a) and absence (b) of phenol supplementation.

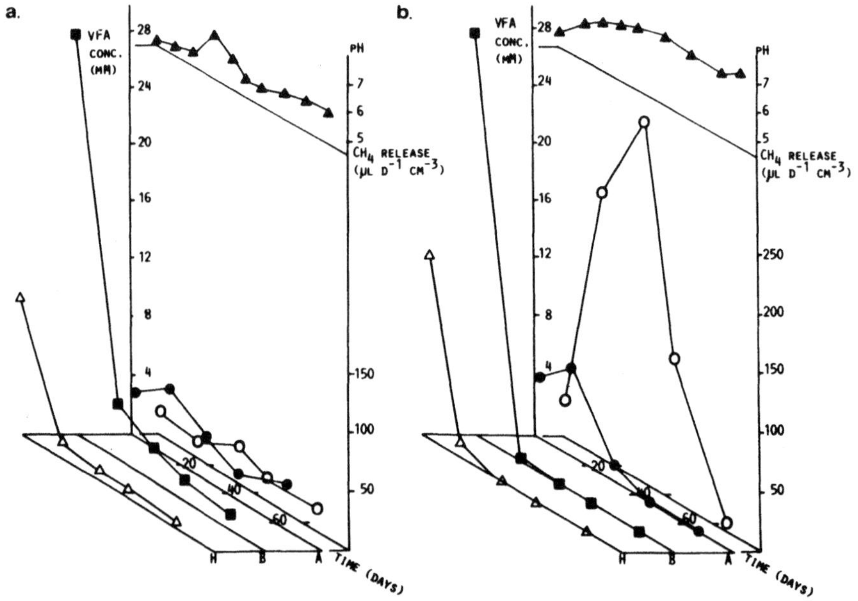

Fig. 4. Temporal changes in methane release rates (○) and in leachate acetate (●), butyrate (■), hexanoate (△) concentrations and pH (▲) in refuse columns operated with leachate recycle in the presence of (a) and absence (b) of phenol supplementation.

been induced, perfusion with distilled water resulted in the dissimilation of initially high concentrations of acetate, butyrate and hexanoate coincident with elevated leachate pH values and methane release (Fig. 4). Conversely, in the presence of phenol (2 mM) the leachate pH remained low and volatile fatty acid mineralisation decreased.

Despite the generation of high concentrations (> 12·5 mM) of individual volatile fatty acids from phenol-perfused 'fresh refuse', no toxic/inhibitory effects were observed although an inverse relationship exists with pH (Karabit *et al.*, 1985). Thus, it may be speculated that the buffering capacity (31 Meq H^+/kg dry weight) of the 'fresh' refuse facilitated the observed pH increase of one unit compared with the 'mature' (6 Meq H^+/kg dry weight) refuse. Thus, refuse age, fermentative activity and buffering capacity are critical variables of an effective co-disposal strategy. However, operation of landfills as anaerobic downflow filters, with leachate recycle, with final leachates posing minimum environmental impact potential, may necessitate accommodation of increased concentrations and elevated rates of methanogenesis.

ACKNOWLEDGEMENTS

This work was supported by scholarships, John Anderson and NCSR Zambia, to Irene A. Watson-Craik and George B. Kasali, respectively. Grateful thanks are also due to Cleanaway Limited and Cory Waste Management for additional funding.

REFERENCES

Buivid, M.G., Wise, D.L., Blanchet, M.J., Remedios, E.C., Jenkins, B.M., Boyd, W.F. & Pacey, J.G. (1981). *Resources Recovery and Conservation*, **6**, 3–20.

Coutts, D.A.P., Senior, E. & Balba, M.T.M. (1987). *Journal of Applied Bacteriology*, **62**, 251–60.

Dobson, A.L. (1964). Microbial decomposition investigations in sanitary landfills. Ph.D. Thesis, University of West Virginia.

Filip, Z. & Küster, E. (1979). *European Journal of Applied Microbiology and Biotechnology*, **7**, 371–9.

Karabit, M.S., Juneskans, O.T. & Lundgren, P. (1985). *Acta Pharmaceutica Suecica*, **22**, 281–90.

Kasali, G.B. (1986). Optimisation and Control of Methanogenesis in refuse fractions. Ph.D. Thesis, University of Strathclyde.

Klink, R.E. & Ham, R.K. (1982). *Resources Recovery and Conservation*, **8**, 29–41.

Leckie, O.J., Pacey, J.G. & Halvadakis, C. (1979). *Journal of the Environmental Engineering Division (ASCE)*, **105**, 337–54.
Rees, J.F. & Grainger, J.M. (1982). *Process Biochemistry*, **17**, 41–4.
Senior, E. & Balba, M.T.M. (1987). In *Biotechnology Applied to Environmental Problems*, ed. D.L. Wise. CRC Press, Boca Raton, Chap 9.
de Walle, F.B., Chian, E.S.K. & Hammerberg, E. (1978). *Journal of the Environmental Engineering Division (ASCE)*, **EE3**, 415–31.
Watson-Craik, I.A. (1987). Landfill as an anaerobic filter for the co-disposal of phenolic wastewaters. Ph.D. Thesis, University of Strathclyde.
White-Hunt, K. (1980). *Solid Wastes*, **70**, 609–15.

Biogas Production from Seaweed Waste Following Alginate Extraction

R.G.J. Edyvean, I.M. Stanley* & S.O. Stanley*

School of Materials, Department of Metallurgy, The University of Sheffield, Mappin St, Sheffield S1 3JD, UK

ABSTRACT

The results of an investigation into the use of seaweed waste, left after the extraction of alginate, as a substrate for methane production are described. Optimum conditions for gas production (376 litres/kg volatile solids (VS) biogas at 63% methane) were found from a two stage digester system operating at 2·5% solids loading and 40 days retention time. The temperature was 55°C in the first and 35°C in the second stage. In such a system acidogenesis and methanogenesis were completely separated into the first and second stages respectively. In a lower temperature system (40°C and 35°C) methane production was from both stages and biogas yields were only slightly lower (360 litres/kg VS at 55% methane), and the retention time could be reduced in this system to 20 days with little loss in gas production (320 litres/kg VS biogas at 62% methane). The results compare favourably with other substrates.

INTRODUCTION

Recently, the search for alternative energy sources has led to the study of methane generation from various substrates. This has led both to the processing of wastes, and the use of specifically grown organic substrates. Algae are highly productive organisms and several studies have

*Present address: Department of Sea Fisheries, Hobart, Tasmania, Australia.

investigated the feasibility of farming species such as *Macrocystis* spp. to produce methane, nitrogenous fertiliser, potassium chloride and trace metals by anaerobic fermentaion (Wilcox & Leese, 1976; Leese, 1977; Troiano *et al.*, 1976; Hart *et al.*, 1979; Eisenberg *et al.*, 1980). Such projects, while showing promise, suffer from the high costs of collecting or farming the algae and setting up production facilities. The present investigation utilises a substrate that has already been collected and chemically digested for alginate production, and many of the high cost constraints do not apply. However, as algin, one of the main volatile solids in algae, had been removed, and as only volatile solids are available to bacteria for methane production the suitability of this substrate for production of this gas was unknown.

MATERIALS AND METHODS

A range of macroalgal species, predominantly *Ascophyllum* spp. originally collected from a number of countries, is chemically and mechanically processed at the Kelco/AIL factory at Barcaldine, Scotland to remove alginate. The algal waste, a slurry of chopped residue in a neutral liquid, was washed in 50% seawater, drained and macerated to a uniform size. Solids loading was adjusted by the removal of supernatant or the addition of 50% seawater. Ammonium chloride was added at a rate of 0·2 g/litre.

Digester systems consisted of 20 litre glass vessels enclosed in constant temperature water baths. The digesters were airtight with ports allowing the addition and removal of substrate (using peristaltic pumps), stirring and the insertion of pH and E_h electrodes. The digesters were used either singly or in pairs as two-stage systems by passing the effluent from one digester into another (Edyvean *et al.*, in press). Two solids loadings were used, 4%, similar to the single stage digesters and 2·5%.

The original inoculum of methanogenic bacteria was a mud sample from a site in Loch Creran, Scotland where the discharge of waste had resulted in active methane generation. Subsequently, each new digester was inoculated with 1 litre of material from a previously stable digester.

All digesters were monitored daily for gas production, gas composition, sulphide, pH and E_h, and regular samples were taken to determine dry weight, volatile solids, percentage ash and analysis of acetic, propionic, butyric and lactic acids.

TABLE 1
Biogas Yields from Algal Waste in Two-Stage Digestion Systems

	Stage	*Retention time*	*Temp.*	*%Methane*	*Biogas (litres/kg VS)*	*Methane (litres/kg VS)*
First		20	55	0·3	36	0
Second		20	35	66	124	82
Total	4	40			160	82
First		20	40	48	55	26
Second		20	35	70	96	94
Total	4	40			151	94
First		20	55	5	33	2
Second		20	35	63	343	216
Total	2·5	40			376	218
First		20	40	46	148	68
Second		20	35	61	208	127
Total	2·5	40			356	195
First		10	55	10	25	2·5
Second		10	35	65	200	130
Total	2·5	20			225	132·5
First		10	40	59	188	111
Second		10	35	65	129	84
Total	2·5	20			317	195

RESULTS AND DISCUSSION

The anaerobic conversion of a substrate to methane can be divided into two stages of bacterial action, the acid producing stage and the methane producing stage. The immediate substrates for methanogenic bacteria are acetate, hydrogen and carbon dioxide (Hanaki *et al.*, 1981). The main advantage of two-stage systems is the buffering capacity of the first stage against variations in the substrate. Also differences in conditions and flora in the two stages reduce the chances of major setbacks.

The two-stage systems used in this work were found to give better gas yields and greater stability than single stage systems at the same overall retention time. This stability can be demonstrated with reference to pH. The levels in the first stage digesters varied between 6·52 and 8·47 at 55°C and 6·67 and 7·28 at 40°C and 35°C. The second stage digesters were very stable, only varying between pH 7·3 and 7·4.

Results averaged over 30-day periods from systems at different temperatures, retention times and solids loading are shown in Table 1. Physical breakdown of the substrate, considered essential to the success of anaerobic digestion (Wise *et al.*, 1978), had already taken place and an additional nitrogen source was found to be required for the efficient digestion. A considerable improvement in gas yields was achieved at a solids loading of 2·5% and while those of around 5% have been found optimal by several workers (Gosh *et al.*, 1977) physical factors such as clumping are known to have a marked effect (Wise *et al.*, 1978). Better mixing at the lower solids loading may also play an important part in gas production. In a two stage anaerobic digestion system, two approaches are possible. First, the complete separation of acidogenesis and methanogenesis by temperature and secondly, the production of methane from both stages, attempting to maximise the use of the substrate. Both approaches were adopted in this work and maximum yields were found in the system in which acidogenesis and methanogenesis were completely separated (55°C in the first stage and 35°C in the second) at a retention time of 40 days overall and solids loading of 2·5%. However, a lower temperature system (40° + 35°C) producing methane in both stages gave only slightly lower gas production. Reducing the retention time to 20 days overall resulted in a marked fall in gas and methane production in the higher temperature system; but methane, though not total gas production, was maintained in the lower temperature system. Under optimal conditions the composition of the gas averaged 62% methane, 20% carbon dioxide; the rest was nearly all nitrogen with traces of hydrogen (0·2%).

While the higher temperature system (55° + 35°C) 40 days overall retention time gives the best results, the costs of digester, size and heating are likely to make the lower temperatures and shorter retention time more efficient for full scale production (40° + 35°C) at 10 days overall).

Table 2 compares the optimum gas yield found in these experiments with yields reported for other substrates. The yield falls well within the general range.

The results show that this material is a viable substrate for methane generation. The removal of alginate appears to have no effect on the gas yield and the chemical and physical pretreatment may be of benefit to the digestion process. It is now realised that the production or harvesting of seaweed solely to produce biogas is not economic at present and that an integrated approach is needed to optimise resources on a commercial basis. The system described here uses a substrate that has been collected and used for another process so there are no collecting or production costs. The results demonstrate how a resource can be more efficiently

TABLE 2
A comparison of Gas Yields and Quality from Various Waste Products and Algae

Waste type	*Biogas (litres/kg VS)*	*%Methane*
Sewage sludge[a]	310–740	68
Cattle slurry[a]	94–310	65
Algal waste (this work)	318–376	62
Kelp (*Laminaria* spp.)[b]	310–340	65 (assumed)
Gracilaria tikvahiae[c]	350	65
Gracillaria ceae[d]	175	65 (assumed)
Ulva lactuca[d]	300	65 (assumed)

[a] Meynell (1986).
[b] Calculated from Troiano *et al.* (1976).
[c] Hanisak (1980).
[d] Calculated from Wise *et al.* (1979).

utilised and, with other established and experimental algal extraction methods, a whole series of processes could be combined into a total biotechnology package for exploiting this resource.

ACKNOWLEDGEMENTS

This work was funded by Alginate Industries Ltd (now Kelco/AIL Ltd.) and the authors would like to thank the company for permission to publish. Special thanks are due to John Johnson and Dougie McKinnes for their help. The work was carried out at the Dunstaffnage laboratories of the Scottish Marine Biological Association. RGJE was supported by SERC and is now the holder of the Sorby Research Fellowship of The Royal Society.

REFERENCES

Edyvean, R.G.J., Stanley, I.M. & Stanley, S.O. (in press). Methane production from seaweed waste left after alginate extraction. *Biorecovery*.

Eisenberg, D.M., Oswald, W.J., Benemann, J.R., Gobel, R.P. & Tiburzi, T.T. (1980). *1st International Symposium on Anaerobic digestion, Cardiff*, 1979, pp. 99–112.

Ghosh, S., Packer, M.L., Roess, A.C., Conrad, J.R. & Ryzin, E.M. (1977). *Journal of the Water Pollution Control Federation*, **49**, 1019–57.

Hanaki, K., Matsuo, T. & Nagase, M. (1981). *Biotechnology and Bioengineering*, **23**, 1591–610.
Hanisak, M.D. (1980). *Proceedings of the 10th International Seaweed Symposium, Goteborg*, August 1980. Walter de Gruyter, Berlin.
Hart, M.R., de Fremery, D., Lyon, C.K. & Kohler, G.O. (1979). *Proceedings 9th International Seaweed Symposium, Santa Barbara*. Science Press, Princeton, pp. 493–8.
Leese, T.M. (1977). *Symposium on Clean Fuels from Biomass, Sewage, Urban refuse and Agricultural wastes.* Orlando 1976.
Meynell, P.J. (1986). *Methane—Planning a Digester*. Prism Press.
Troiano, R.A., Wise, D.L., Augenstein, D.C. & Kispert,R.C. (1976). *Resource Recovery and Conservation*, **2**, 171–6.
Wilcox, H.A. & Leese, T.M. (1976). *Hydrocarbon Process*, April 1976, 86–9.
Wise, D.L., Wentworth, R.L. Augenstein, D.C. & Cooney, C.L. (1978). *Resource Recovery and Conservation,* **3**, 41–59.
Wise, D.L., Augenstein, D.C. & Ryther, J.H. (1979). *Resource Recovery and Conservation*, **4**, 217–37.

A Review of Aerobic Biodegradation of Solid Wastes

Anthony J. Biddlestone & Kenneth R. Gray

Department of Chemical Engineering, University of Birmingham, Birmingham B15 2TT, UK

ABSTRACT

The basic microbiology and biochemistry of aerobic degradation are briefly reviewed. Chemical and physical parameters which need consideration in controlled process systems are outlined and a summary given of the important design parameters. The major differences between processes arise in the biological degradation stage. Most recent installations incorporate either simple heaps, accelerated heaps with forced aeration or sophisticated mechanised digesters. In a controlled process the putrescible organic substrate undergoes chemical and physical transformation to give a reasonably stable product which can quickly equilibriate with the soil.

An alternative to the processing of the wastes before use is direct application of the untreated raw waste to the soil when degradation and stabilisation occur within the soil. This is possible, dependent upon the nature of the waste. Reported use of such techniques, from the Mazibuko trench system and 'alley cropping' in Africa to the addition of urban waste separates in the UK, are considered. Disadvantages of raw waste application can include initial disruption to the soil system by an imposed oxygen and nitrogen demand and the survival of pathogenic organisms.

INTRODUCTION

An enormous amount of organic matter, continually renewed by photosynthesis, undergoes microbial degradation and becomes available in the soil. During the process putrescible organic substrate undergoes chemical and physical transformation to give a stable humified end product. The product is of prime importance in maintaining soil fertility and productivity; especially in tropical regions where humus is rapidly oxidised. Wastes amenable to aerobic biodegradation vary from the heterogeneous organic/inorganic mixture in urban refuse to the more homogeneous farm manures, crop residues, sewage sludges and night soil

During the degradation process most of the oxygen demand of the wastes is met, the organic materials are converted to more stable products, carbon dioxide and water are released and heat is evolved. The process, when occurring in its natural environment on the soil surface at ambient temperature and mainly under aerobic conditions, is very slow. The natural process of breakdown can be accelerated by gathering the material into heaps to conserve part of the heat of fermentation so that the temperature of the mass rises and faster reaction rates are obtained. The accelerated process in controlled heaps is most commonly termed composting.

MICROBIOLOGICAL AND BIOCHEMICAL ASPECTS

The process of aerobic biodegradation of organic wastes is a complex interaction between the wastes, micro-organisms, moisture and oxygen. The waste material will normally have an indigenous mixed population of organisms present such that when the moisture content and oxygen concentration are brought to suitable levels microbial action takes place. In addition to oxygen and moisture, the micro-organisms require for their growth and reproduction a source of carbon, macronutrients such as nitrogen, phosphorus and potassium, and certain trace elements. These additional requirements are usually provided by the waste materials. A succession of microbial groups, each of which is appropriate to an environment of relatively limited duration, are responsible for the degradation. A list of the main classes of organisms involved is given in Table 1; these represent both the microbial and animal kingdoms. There are many different species, possibly 2000 bacteria and at least 50 fungi, within each genus. The species can be subdivided according to the temperature ranges of their activity. Psychrophiles prefer temperatures below 20°C, mesophiles 20–40°C and thermophiles above 40°C.

Studies on the populations of bacteria, actinomycetes and fungi during aerobic breakdown have been done by a number of workers. Yung Chang & Hudson (1967), Hayes & Lim (1979), Fermor & Wood

TABLE 1
Classes of Organisms in Aerobic Biodegradation of Organic Wastes

Microflora	Bacteria Actinomycetes Fungi Algae Protozoa
Macrofauna	Mites, Spring tails, Ants, Termites, Millepedes, Centipedes Spiders, Beetles Worms

(1979) and Atkey & Wood (1983) provide information on the microbial succession of a wide range of species of mesophilic and thermophilic bacteria, actinomycetes and fungi during the breakdown of straw in heaps. In the early stages thermophilic bacteria dominate, while in the later stages bacterial numbers decline and the actinomycete population increases. de Bertoldi *et al.* (1983) have reported the microbial changes during the degradation of urban solid waste (60%) with sewage sludge (40%) in static heaps with forced aeration. Over 50 days of breakdown they found an increase in cellulolytic fungi and actinomycetes with a corresponding decrease in the numbers of bacteria. Fungi isolated from the static heaps included 14 thermophilic species (at 50°C) and 43 mesophilic species (at 28°C). The mesophilic species were characterised according to their enzymatic activity; cellulolytic (35), pectinolytic (3), amylolytic (13) and ligninolytic (14).

Waste material that has been degraded in a heap passes through a temperature peak. When it has cooled it is accessible to a wide range of soil macrofauna as is material applied directly to the soil. These macrofauna feed upon other animals, animal excreta and the waste organic material. They normally require well aerated conditions, adequate moisture and prefer temperatures in the range of 7–13°C. Many of the soil animals make a major contribution to breakdown of the wastes due to physical maceration; breaking the material into smaller particles exposes new surface for subsequent attack by microflora. They also contribute to the mixing of the constituents. In temperate climates the earthworm plays a major role in the final stages of breakdown and in the subsequent incorporation of organic matter into the soil; in arid and

TABLE 2
Composition of Organic Wastes

Component	*% in dry weight*
Water solubles (sugars, starches, amino acids urea, ammonium salts)	2–30
Ether/alcohol solubles (fats, oils, waxes)	1–15
Proteins	5–40
Hemicelluloses	10–30
Cellulose	15–60
Lignin	5–30
Volatile matter	70–95

semi-arid climates this function is usually undertaken by the termite (Edwards, 1974). The macrofauna build up tissues rich in nitrogen which are easily decomposed, providing a reservoir of nitrogenous matter.

Organic waste materials, whether of industrial, urban or agricultural origin, are mixtures of sugars, proteins, fats, hemicelluloses, cellulose, lignin and minerals in a wide range of concentrations as shown in Table 2. Fresh green material contains much water, soluble matter, proteins and minerals but as plants age low molecular weight compounds are converted to higher molecular weight ones, especially hemicelluloses, cellulose and lignin. The composition of animal detritus will depend upon the type of animal and its feed. In the aerobic biodegradation of organic wastes the cell wall of the micro-organism attacking the waste is a significant factor. Soluble sugars and organic acids of low molecular weight can pass through the cell wall easily and be metabolised, being built up into larger polymers and providing energy. Longer chain components of the wastes cannot pass through the cell wall and thus cannot be used until broken down into simpler compounds by extra-cellular enzymes exuded from some of the micro-organisms.

Yung Chang (1967) studied the extent of biochemical changes during the degradation of wheat straw with added ammonium nitrate. The straw lost 50% of its dry weight in 60 days with the majority of the loss in the first 34 days. The loss could be accounted for almost entirely by the loss in hemicelluloses and cellulose. The ethanol-soluble fraction, which contains the simpler carbon compounds, remained fairly constant presumably being continually replaced by breakdown of the longer chain polymers whilst itself being consumed by the micro-organisms.

PROCESS FACTORS

When organic wastes are gathered together in a controlled heap for aerobic degradation, the insulating effect of the materials conserves the heat released by biological activity and causes a rise in temperature. In the first or mesophilic stage of breakdown, the indigenous micro-organisms multiply rapidly and the temperature rises from ambient to about 40 °C. Thermophilic strains take over as the temperature continues to rise but by about 60° C themophilic fungi have ceased activity and the reaction is continued by actinomycetes and spore-forming strains of bacteria. The rate of reaction slows and a temperature peak is reached; the rate of heat generation is then equal to the rate of heat loss from the mass. This marks the end of the thermophilic stage at which easily converted materials have been degraded and most of the high-rate oxygen demand met. During the cooling-down stage following the temperature peak thermophilic fungi reinvade the mass and, together with the actinomycetes, attack longer chain components, breaking them down into simpler sugars. The process finally enters the maturing stage in which mass loss and heat evolution are small.

The degradation of organic matter in a heap system is thus seen to be a dynamic and complex ecological process in which temperature and food availability are constantly changing. The numbers and species of organisms present change markedly during the process. The rate of progress towards a mature end product is dependent on several inter-related process factors. These include nutrient supply, particle size, moisture content, structural strength, aeration, agitation, pH and size of heap. A summary of the optimum values of the important parameters is given in Table 3. The complexity of the processing system, and the quality of the end product, will depend upon the nature of the organic waste to be degraded and the level of financial investment available.

Very frequently high temperatures have been considered necessary for high degradation rates. In fact degradation is suppressed at excessively high temperatures because the growth of most micro-organisms is inhibited; the threshold of suppression is approximately 60 °C. A peak temperature of the order of 60° C is useful in controlling thermosensitive pathogenic organisms but a slightly lower temperature is desirable to encourage the growth of those organisms which are the main decomposers of long chain polymers. An optimum of 55 °C is recommended. A forced aeration system removes heat by evaporative cooling and Finstein *et al.* (1983) recommend a temperature feed-back control system in which the air blow time is linked to a temperature sensor placed in the degrading mass. Such a system facilitates water removal in high moisture content

TABLE 3
Optimum Values of Process Parameters

Parameter	*Value*
C/N ratio of wastes	25–30/litre
Particle size	10–15 mm for agitated systems and forced aeration. 50 mm for windrow heaps and natural aeration
Moisture content	50–60% (higher value possible when using bulking agents such as straw or woodchips)
Air flow	0·6–1·8 m^3 air/day/kg volatile solids, or maintain oxygen level at 10–18%
Temperature	55°C
Agitation	No agitation to periodic turning in simple systems. Short bursts of vigorous agitation in mechanised systems
Heap size	Any length. 1·5 m high and 2·5 m wide for natural aeration heaps. Heap size can be increased for forced aeration

situations and has the advantage of controlling the important interrelation between temperature, moisture level and aeration requirements.

Water is essential to the degradation process and at a moisture content below 30% on a fresh weight basis the biological reactions slow down markedly. At too high a water content the voids within the matrix of the degrading mass become waterlogged, limiting access of oxygen to the micro-organisms. Some materials, such as paper and grass cuttings, collapse into an impervious mass when wet but others, such as straw-type materials, can tolerate high moisture contents without losing structural strength. The porosity of a heap of organic wastes can be defined as the ratio of the void volume to the total volume and the free air space within the matrix as the ratio of the gas volume to the total volume (Haug, 1980). Different waste materials can hold different moisture levels whilst still maintaining an adequate free air space. Experimental work by Jeris & Regan (1973) on the breakdown of mixed refuse suggests that a minimum free air space of 30% should be maintained.

Oxygen supplied by some form of aeration process is essential for the metabolism of aerobic species of micro-organisms. Aeration is possible by natural gaseous diffusion into the degrading mass, by turning the material regularly or by forced aeration from a fan. The theoretical stoichiometric chemical oxygen demand (COD) can be determined if the chemical composition of the organic waste and the extent of degradation are known. The theoretical requirement will vary from about 1·0 g oxygen/g of organic material for highly oxygenated wastes such as cellulose, to 4·0 g/g for saturated hydrocarbons. In practice most organic wastes will comprise a mixture of materials with differing theoretical oxygen demands and varying degradability such that typically only 40% of the organic matter may be oxidised. The various organic compounds present will also have different values for the heat of combustion. Three materials commonly found, proteins, carbohydrates and lipids have heats of combustion within the range of 9–40 kJ/g. This energy is released during biological oxidation and can be related to oxygen demand because most organic compounds have a heat of combustion of about 14·2 kJ/g COD of the organic material. In the absence of a forced aeration system oxygen supply can become a limiting factor to degradation as reported by Biddlestone *et al.* (1986) from experiments on the composting of dewatered sewage sludge and vegetable wastes with straw. Table 4 compares the temperatures at the centre of two 5 t heaps comprising vegetable waste (leeks) with 5% by weight of straw but with only one of the heaps being forced aerated. The heat release in the heap relying on natural aeration is significantly less than in the forced aeration heap.

The complexity of processing equipment used for aerobic biodegradation of organic wastes and the degree of approach to the optimum values of the process parameters vary considerably from the simple heap system to the highly sophisticated mechanised urban waste composting plant (Gray *et al.*, 1971; Biddlestone *et al.*, 1981).

TABLE 4
Centre Temperatures of 5 t Heaps of Leek Wastes with 5% by Weight of Straw

Days of Breakdown	*Centre temperature (°C)*	
	Natural aeration	*Forced aeration*
2	8	8
4	20	20
5	63	63
10	53	63
15	40	66
20	28	62
25	23	57
30	11	51

The forced aeration heap was blown for 7 min every hour by a fan delivering $2{\cdot}8^3$/min air at 160 mm water head.

PROCESS SYSTEMS

The aerobic biodegradation of waste organic matter in controlled heaps, termed composting, has been practised for many centuries. Historically the process was essentially a small-scale batch operation in heaps (FAO, 1977). Recent interest has been in response to the need to deal hygienically with large quantities of urban refuse and sewage sludge and the increasing need to recycle crop residues and animal manures in agriculture. For the degradation of agricultural, horticultural and garden wastes relatively simple processing schemes are still employed. Agitation is rare but forced aeration is sometimes used. Because of the fairly low monetary value of composts, sophisticated processes cannot be afforded in most agricultural situations and process conditions frequently fall short of the optimum values listed in Table 3. The need to process urban wastes in large quantities has resulted in some 30 different processing schemes being used, with varying success. Equipment for feed preparation and product finishing are similar in many of these processes. The biodegradation stage, however, has varied widely, being attempted in pits, cells, silos, digesters and drums. Very recently, large-scale mechanised systems with capacities in the range 200–500 t of refuse/day have been installed in various countries. Tenders and enquiries have been issued for 750 and 1000 t/day operations.

The composting of garden, small-holding or nursery wastes is widely practised, the system usually being a series of batch heaps handling 0·5–1·0 t of material/batch. Small batches have the problem of relatively high heat loss because of the large surface to volume ratio. It is preferable at this scale (Gray & Biddlestone, 1976) to have a walled enclosure with insulation, to cover the wastes with an air permeable top blanket, to arrange for air to permeate underneath the heap and percolate upwards by the 'chimney effect', and to collect sufficient waste material to build the heap in one operation.

In larger throughput systems the waste material is placed into long heaps known as windrows, either manually or using equipment such as front-end loaders. The windrows may be left standing for several months, until the temperature begins to fall, or they may be occasionally turned. The aims of turning the heaps are to provide aeration, reduce particle size, and to ensure that all material is subjected to the highest possible temperature. Turning is a time-consuming operation and some simple systems have used stakes or bamboo poles to create channels within the heaps to improve natural convective aeration. Several simple heap systems have been reported using a variety of waste materials. In Europe manure spreaders fitted with side ejectors have been used to build windrows (Pfirter *et al.*, 1981). In the USA Demmel (1980) reports the use of windrows about 2 m wide by 1·25 m high for farm manures. A study tour of organic waste recycling in China reported techniques for composting of human wastes, animal wastes and chopped plant stalks (FAO, 1977). A similar approach is used for the composting of a mixture of night soil, water hyacinths and leaves in Central Thailand (Polprasert *et al.*, 1980). Biddlestone *et al.* (1986) describe experiments on the composting of dewatered sewage sludge and celery/leek wastes when intimately mixed with straw, the straw being incorporated at a level of 5–10% of the total mass. A recently-published FAO Soils Bulletin (FAO, 1987) extensively reviews compost production and use in tropical and subtropical environments.

In order to increase the rate of decomposition and to remove the need for turning, some windrow systems aerate the heaps from channels or pipes beneath them. Several such aerated static pile systems have been installed in the USA over the last decade as shown in Table 5. Composting of dewatered sewage sludge with a bulking agent such as wood chips provides a means of stabilising the sludge and also effects a substantial weight reduction, resulting in significant savings in the cost of sludge haulage. Singley *et al.* (1982) have produced a detailed design and operating manual for aerated heap systems. Finstein *et al.* (1983) compare two 36 t heaps of sludge and woodchips, one with suction aeration, the other with blown aeration. Their recommendation is that a blown air system with feedback temperature control of the blower should be used. This view is supported by de Bertoldi *et al.* (1982) from experiments processing mixtures of dewatered sewage sludge and the organic fraction of urban refuse in 2 t heaps. Forced aeration heaps have been used by University experimenters in the UK (Stentiford *et al*, 1985; Biddlestone *et al.*, 1986) but there is no significant use of such processes

TABLE 5
Sewage Sludge Composting Sites in the USA

	Number of operations		
	Reactor	*Forced aeration heaps*	*Natural aeration heaps*
Operational	3	49	27
Under construction	12	9	2
Planned	16	17	5

Source of information: Goldstein (1985).

on a large scale in the UK. In contrast, there are presently over 100 composting plants operating in France (Clayton, 1985).

Modern large-scale urban waste biodegradation systems usually comprise refuse storage facilities, feed preparation equipment, the biological degradation stage and final product upgrading (Biddlestone *et al.*, 1981). The biological degradation stage is carried out in windrows, accelerated windrows or in more sophisticated mechanised units. Mechanised units vary from automated windrow systems with continuous automatic turning equipment, to totally enclosed silos which achieve extensive breakdown of the material within a few days. A single silo may contain up to 200 t of material which can release substantial amounts of heat, raising the temperature well beyond the optimum of 55–60°C. Excess air is often provided to effect cooling by evaporation of water into the air stream and, in order to prevent the material drying out, provision for water addition is usually made. Residence times are in the range 4–20 days, 8 days being most common. Facilities for subsequent windrowing may be made or a simple storage area of 2–3 weeks capacity provided. The majority of large scale systems erected in recent years have been installed in the Middle and Far East, processing city refuse and using the product for land reclamation and upgrading schemes.

DIRECT APPLICATION OF RAW WASTES TO SOIL

An alternative to the controlled processing of organic wastes before use is direct application of the untreated raw waste to the soil with degradation and stabilisation subsequently occurring within the soil. Several techniques have been reported.

The Mazibuko or 'fertility' trench system was developed to restore soil fertility in arid, hilly areas in the tropics and subtropics which had suffered so badly from soil erosion due to overgrazing that little food could be grown. The technique is to dig trenches 1·2 m deep and 2 m wide at regular intervals down the hillside. Each trench is then filled with alternate layers of soil and organic wastes, each about 300 mm deep. As commonly practised in village areas in less developed countries the organic wastes include cut grass, stems of millet and maize, animal manures and domestic wastes. In the filled trench is first sown Sunn hemp, a legume with deep roots which supply nitrogen to the below ground deposit of organic wastes thereby assisting the degradation process. When the rains fall the water soaks into the trench which acts like a sponge. Further soil erosion on the hillside is thus prevented. The second crop is madumbe, a variety of cocoyam or taro with edible leaves.

This is followed by maize, millet and several types of vegetables which can grow through the dry season drawing on the moisture held up in the organic 'sponge'. The stretches of hillside between trenches are sown with lucerne which can stabilise the soil provided the angle of the slope is not excessive. The Mazibuko trench system combines several functions, biodegrading the local organic wastes, preventing soil loss from exposed hillsides, providing a route for heavy rainfall to be channelled into the subsoil, and establishing high fertility areas with deep soils for crop growing.

A technique described as 'sheet composting' is being used by market gardeners in Europe. The raw materials, harvest residues remaining in the fields or green manures, are chopped *in situ* and mixed into the upper soil at a depth between 50 and 300 mm. The materials are left on the soil surface after chopping and remain there for a time before being worked in. Deep incorporation should be avoided as putrefaction rather than degradation would take place. A waiting time between incorporation and new planting of 1–2 weeks in summer and 2–3 weeks in autumn is recommended (Pfirter *et al.*, 1981). Sheet composting is distinguished from mulching in that the waste materials are incorporated into the soil using equipment such as rotovators whereas mulching consists of just leaving the materials on the soil surface. Considerable work on suitable equipment and operating techniques for straw incorporation into soil are currently in progress in the UK in order to reduce cereal straw burning at harvest.

Alley cropping is a recent agricultural practice with much potential in the tropics. Food crops are grown in alleys 4–6 mm wide formed by hedgerows of fast growing leguminous shrubs and trees planted in an orderly arrangement (Wilson & Kang, 1981). The hedgerows are cut back at the time of planting the alleys and kept pruned during cropping to prevent shading and to reduce competition with the food crops. The trees and shrubs of the hedgerow supply nitrogen to the soil via their root systems while their leaves provide waste organic matter which can be added to the soil in the alleys. In experiments in Africa on originally infertile sandy soil, alley cropping is giving maize yields higher than the average fertile soils. The International Institute of Tropical Agriculture in Nigeria have tested the leguminous tree/shrub species *Leucaena* and *Glyricidia* and have developed a leucaena/glyricidia-maize-cowpea and leucaena-maize-yam cropping system (Kang *et al.*, 1984). The technique maintains higher soil nutrient and moisture status and twice the amount of soil organic matter when the organic waste is added to the alleys compared to soil with no such addition.

Saull *et al.* (1985*a*) investigated the effect on selected soil characteristics of the addition of a putrescible organic fraction derived from a UK source of domestic refuse. The putrescible fraction used was a by-product from a refuse derived fuel plant located in Doncaster, Yorkshire, and was highly organic in nature. The organic fraction was added to impoverished cover soil at dosages up to 65 t/ha, incorporation being achieved using a mechanical rotavator. Biodegradation of the organic fraction was measured in terms of the release of ethanol/water solubles, decrease in organic carbon and nitrogen and the production and stabilisation of humic acid materials over a period. Soil pH, percentage moisture and water-holding capacity were also measured. It was concluded that the putrescible fraction could be successfully disposed of by direct soil incorporation. Biodegradation of the organics content was rapid, the level of humic materials built up and beneficial effects on soil physical properties were obtained. The same workers (Saull *et al.*, 1985*b*) also reported the effect of direct incorporation upon the growth of perennial ryegrass.

At a recent conference the results of adding municipal and industrial sludge wastes in forests was extensively reviewed (Cole *et al.*, 1986).

DISCUSSION

The importance of recycling organic wastes is being increasingly recognised. This review paper has shown that many techniques are available both for controlled biodegradation before soil application and for direct incorporation of raw wastes into the soil. The objective of the process varies from a requirement to dispose of organic wastes to the more long-term objective of providing a product essential to the improvement of soil structure and stability for increased food and biomass production. In tropical and subtropical areas organic waste materials in large quantities need to be recycled to the soil in order to increase productivity. The objective in these areas is to produce a product which is both a soil conditioner and to some extent a fertiliser. In contrast, there is a tendency in developed temperate areas to consider the process as a means of disposing of organic wastes; the product is of secondary importance, although with the rising popularity of organically grown food this situation may change.

There are advantages and disadvantages to be considered in relation to the application of stabilised or raw wastes. It can be argued that the soil should be used for growing crops, not as an environment for breaking down organic wastes, especially where increased food production is vital. If this view is accepted the wastes should be biodegraded in a controlled process. After a reasonable degree of maturity has been reached the product can be incorporated into the soil without causing a harmful effect on crop growth. A recent review (Gallardo-Lara & Nagales, 1987) extensively considers the effect of the application of town refuse compost on the soil-plant system. By contrast, the degradation of large quantities of raw organic wastes in the soil can interfere with the growth of many crops. During the initial stages of aerobic biodegradation there is a need for oxygen and possibly nitrogen; simple organic acids and ammonia are formed. A period of disruption to the soil processes can occur and seriously affect the germination of seeds and growth of young seedlings. The effect would be significantly less with established trees and shrubs. The disadvantages of using a controlled process before soil application are that an extra operation is introduced, involving both labour and equipment costs, and there is a requirement for organisational ability and practical skills. In considering the application of raw or stabilised

TABLE 6

Comparisons of Mineral Fertilisers, Raw Organic Wastes and Stabilised Organic Wastes

Advantage	*Disadvantage*
Mineral fertilisers	
1 Convenient	1 Easily leached
2 Transport and handling costs may be lower	2 Continuous use may lead to soil structure breakdown
3 Quick crop response	3 Supply major nutrients only
4 Concentrated nutrient source	
Raw organic wastes	
1 May be cheaper than mineral fertilisers or stabilised wastes	1 May be difficult to incorporate into soils
2 Improve soil fertility	2 If high C/N ratio slow decomposition and robbing of soil nitrogen
	3 If low C/N little effect on soil structure
	4 Dilute nutrient source
Stabilised organic wastes	
1 Improve soil structure	1 Dilute nutrient source
2 Control erosion	2 High labour/equipment usage
3 Supply range of nutrients	3 Transport costs more than mineral fertilisers
4 Hygienic disposal of pathogenic wastes	
5 Reduce transport needs of bulky wastes	

organic wastes in the context of improving food production, comparison with the use of mineral fertilisers should also be made. Table 6 presents some of the general comparisons.

An aspect of the processing of organic wastes that is of increasing concern is the survival rate of disease causing organisms. Organic wastes often carry some pathogens of plant, animal or human origin. It is important to note that above a certain temperature level thermal death of such organisms is time dependent. By exposing the wastes to the temperatures of 55–60°C normally reached in a controlled biodegradation process, there is an effective kill of most pathogens as well as weeds and seeds. There cannot be a guarantee of an entirely pathogen-free product. Some recent publications have extensively reviewed the literature on the presence and survival of such pathogens in aerobic biodegradation processing systems (Finstein *et al.*, 1982; Golueke, 1983; Higgins, 1983; Pereira-Neto *et al.*, 1986).

In conclusion it should be recognised that aerobic biodegradation of organic wastes is an extremely important process. It can be seen in the restricted context of a disposal process; it should be seen as part of the essential return of organic matter to the soil to combat soil erosion and to maintain and improve the productivity of soils used for food production.

REFERENCES

Atkey, P.T. & Wood, D.A. (1983). *J. Appl. Bact.*, **55**, 293–304.

Bertoldi, M.de., Vallini, G., Pera, A. & Zucconi, F. (1982). *Biocycle*, **23** (2), 45–50.

Bertoldi, M.de., Vallini, G. & Pera, A. (1983). *Waste Management and Res.*, **1**, 157–76.

Biddlestone, A.J., Ball, D. & Gray, K.R. (1981). In *Advances in Food Producing Systems for Arid and Semi-arid Lands*, ed. J.T. Manassah & E.J. Briskey. Academic Press, New York, pp. 191–224.

Biddlestone, A.J., Cooper, J.D. & Gray, K.R. (1986). In *Effluent Treatment and Disposal*, I. Chem. E. Symposium Series No. 96. The Institution of Chemical Engineers, pp. 355–64.

Clayton, R. (1985). *The Chem. Engr.*, October, 12.

Cole, A.W., Henry, C.L. & Nutter, W.L. (1986). *The Forest Alternative for Treatment and Utilisation of Municipal and Industrial Wastes.* University of Washington Press.

Demmel, D. (1980). *Compost Sci/Land Utilization*, **21** (1), 42–4.

Edwards, C.A. (1974). In *Biology of Plant Litter Decomposition*, ed. C.H. Dickinson and G.J.F. Pugh. Academic Press, New York, pp. 533–54.

FAO (1977). China: recycling of organic wastes in agriculture. *FAO Soils Bulletin*, 40. Food and Agricultural Organisation, United Nations.

FAO (1987). Soil management: compost production and use in tropical and subtropical environments, 177 pp. *FAO Soils Bulletin*, 56. Food and Agriculture Organisation, United Nations.
Fermor, T.R. & Wood, D.A. (1979). In *Straw Decay and its Effect on Disposal and Utilisation*, ed. E. Grossbard. John Wiley, New York, pp. 105–12.
Finstein, M.S., Wei Ru Lin, K. & Fischler, G.E. (1982). Sludge composting and utilization: review of the literature on temperature inactivation of pathogens, 29 pp. New Jersey Agricultural Experimental Station, Cook College, Rutgers University, New Brunswick, New Jersey.
Finstein, M.S., Miller, F.C., Strom, P.F., MacGregor, S.T. & Psarianos, K.M. (1983). *BioTechnology*, **1** (4), 347–53.
Gallardo-Lara, F. & Nagales, R. (1987). *Biological Wastes,* **19**, 35–62.
Goldstein, N. (1985). *Biocycle*, **26** (8), 19–24.
Golueke, C.G. (1983). *Biocycle,* **24** (4), 50–8.
Gray, K.R. & Biddlestone, A.J. (1976). *J. Roy. Hort. Soc.*, **101**, 540–4, 594–8.
Gray, K.R., Sherman, K. & Biddlestone, A.J. (1971). *Process Biochemistry*, **6** (10), 22–8.
Haug, R.T. (1980). *Compost Engineering: Principles and Practice*. Ann Arbor Science, Michigan.
Hayes, W.A. & Lim, W.C. (1979). In *Straw Decay and its Effect on Disposal and Utilisation*, ed. E. Grossbard. John Wiley, New York, pp. 85–94.
Higgins, A.J. (1983). *Biocycle*, **24** (1), 40–4.
Jeris, J.S. & Regan, R.W. (1973). *Compost Sci.*, **14** (2), 8–15.
Kang, B.T., Wilson, G.F. & Lawson, T.L. (1984). Alley cropping: a stable alternative to shifting cultivation, 22 pp. International Institute of Tropical Agriculture, Ibadan, Nigeria.
Pereira-Neto, J.T., Stentiford, E.I. & Mara, D.D. (1986). In *Effluent Treatment and Disposal*, I. Chem. E. Symposium Series No. 96, Institution of Chemical Engineers.
Pfirter, A., Von Hirschheydt, A., Ott, P. & Vogtmann, H. (1981). Composting: an introduction to the rational use of organic waste, 43 pp. Migros Co-operative Aargau/Soluthurn, Switzerland.
Polprasert, C., Wangsuphachart, S. & Muttamara, S. (1980). *Compost Sci/Land Utilization*, **21** (2), 25–7.
Saull, M.B., Mills, J. & Davis, B. (1985*a*). *International Biodeterioration*, **21** (3), 215–23.
Saull, M.B., Mills, J. & Davis, B. (1985*b*). *International Biodeterioration*, **21** (4), 295–300.
Singley, M.E., Higgins, A.J. & Rosengaus, M.F. (1982). Sludge composting and utilization: a design and operating manual, 295 pp. New Jersey Agricultural Experiment Station, Cook College, Rutgers University, New Brunswick, New Jersey.
Stentiford, E.I., Pereira-Neto, T.J., Taylor, P.L., Mara, D.D., Lopez-Real, J. & Witter, E. (1985). In *Advances in Water Engineering*, ed. T.H.Y. Tebbutt. Elsevier. Amsterdam.
Wilson, G.F. & Kang, B.T. (1981). In *Biological Husbandry: a Scientific Approach to Organic Farming*, ed. B. Stonehouse, Butterworths, London, pp. 193–203.
Yung Chang (1967). *Trans. Br. Mycol. Soc.*, **50** (4), 667–77.
Yung Chang & Hudson, H.J. (1967). *Trans. Br. mycol. Soc.*, **50** (4), 649–66.

Index of Contributors

The following papers, presented at the Seventh International Biodeterioration Symposium, will appear in Vol. 24, Nos 4 and 5 and Vol. 25, Nos 1 and 2 of the journal *International Biodeterioration*.

Volume 24, Nos 4 and 5

Volume 25, Nos 1 and 2

Industrial Case Histories of Microbiological Fuel Contamination—Cause, Effect and Treatment
D.G. PITCHER (UK)

Microbial Problems During Long-Term Storage of Petroleum Products Underground in Rock Caverns
R. ROFFEY (Sweden)

Microbial Influence on Cellulosic Textiles and Microbiological Testing
P. RASCHLE (Switzerland)

GPSR Compliance

The European Union's (EU) General Product Safety Regulation (GPSR) is a set of rules that requires consumer products to be safe and our obligations to ensure this.

If you have any concerns about our products, you can contact us on ProductSafety@springernature.com

In case Publisher is established outside the EU, the EU authorized representative is:

Springer Nature Customer Service Center GmbH
Europaplatz 3
69115 Heidelberg, Germany

Batch number: 10824463

Printed by Printforce, the Netherlands